NK Cell Receptors: Advances in Cell Biology and Immunology

NK Cell Receptors: Advances in Cell Biology and Immunology

Edited by Colton Williams

STATES
ACADEMIC PRESS
www.statesacademicpress.com

States Academic Press,
109 South 5th Street,
Brooklyn, NY 11249, USA

Visit us on the World Wide Web at:
www.statesacademicpress.com

ISBN: 978-1-63989-385-0

Cataloging-in-Publication Data

NK cell receptors : advances in cell biology and immunology / edited by Colton Williams.
p. cm.
Includes bibliographical references and index.
ISBN 978-1-63989-385-0
1. Killer cells. 2. Cell receptors. 3. Immunocompetent cells. 4. Cytology. 5. Immunology. I. Williams, Colton.
QR185.8.K54 N53 2022
616.079--dc23

Table of Contents

Preface

This book has been a concerted effort by a group of academicians, researchers and scientists, who have contributed their research works for the realization of the book. This book has materialized in the wake of emerging advancements and innovations in this field. Therefore, the need of the hour was to compile all the required researches and disseminate the knowledge to a broad spectrum of people comprising of students, researchers and specialists of the field.

NK cells, or natural killer cells belong to the group of innate lymphoid cells which are vital to the innate immune system. They are a type of cytotoxic lymphocytes. NK cells respond rapidly to virus infected cell and other intracellular pathogens. They also take action against the formation of tumors. There are different NK cell receptors which have different functions. They can directly cause cell death after binding themselves to Fas ligand which shows infection of a cell. The MHC-independent receptors use a different pathway to cause apoptosis in infected cells. The activation of NK cells depends on the balance of activating and inhibitory receptor stimulation. This book unfolds the innovative aspects of NK cell receptors, which will be crucial for the progress of this field in the future. It consists of contributions made by international experts. The book is appropriate for students seeking detailed information in this area as well as for experts.

At the end of the preface, I would like to thank the authors for their brilliant chapters and the publisher for guiding us all-through the making of the book till its final stage. Also, I would like to thank my family for providing the support and encouragement throughout my academic career and research projects.

Editor

1

Interferon Gamma Induces Changes in Natural Killer (NK) Cell Ligand Expression and Alters NK Cell-Mediated Lysis of Pediatric Cancer Cell Lines

*Arianexys Aquino-López[1,2], Vladimir V. Senyukov[1], Zlatko Vlasic[1], Eugenie S. Kleinerman[1] and Dean A. Lee[3]**

[1] Department of Pediatric Research, The University of Texas M.D. Anderson Cancer Center, Graduate School of Biomedical Sciences, Houston, TX, USA, [2] School of Medicine, University of Puerto Rico, San Juan, Puerto Rico, [3] Center for Childhood Cancer and Blood Disorders, The Research Institute, Nationwide Children's Hospital, Columbus, OH, USA

Edited by:
Antoine Toubert,
Paris Diderot University, France

Reviewed by:
Fabio Malavasi,
University of Turin, Italy
Norberto Walter Zwirner,
Institute of Biology and Experimental Medicine (IBYME-CONICET), Argentina

***Correspondence:**
Dean A. Lee
dean.lee@nationwidechildrens.org

Natural killer (NK) cells have therapeutic potential for cancer due to their capacity for targeting tumor cells without prior sensitization. Our laboratory has developed an NK cell expansion protocol that generates large quantities of NK cells for therapeutic infusion that secret 20 times the amount of interferon gamma (IFNγ) than resting NK cells. IFNγ can upregulate major histocompatibility complex (MHC)-class I, an inhibitory ligand for NK cells, but can also upregulate intercellular adhesion molecule 1 (ICAM-1) which promotes NK:target cell interaction for an efficient lysis. Due to the opposing effects reported for IFNγ on tumor sensitivity to NK cells, we evaluated a panel 22 tumor cell lines from the pediatric preclinical testing program corresponding to different tumor types. We determined the impact of IFNγ on their expression of NK cell activating and inhibitory ligands, death receptors, and adhesion molecules using mass cytometry. We also evaluated the effect of IFNγ on their sensitivity to NK cell-mediated lysis. Our results show upregulation of PD-L1, ICAM-1, MHC-class I, HLA-DR, CD95/FasR, and CD270/HVEM after IFNγ treatment, this upregulation is variable across different tumor types. We also observed a variable impact of IFNγ in NK cell-mediated lysis. For six of the cancer cell lines IFNγ resulted in increased resistance to NK cells, while for three of them it resulted in increased sensitivity. Modeling of the data suggests that the effect of IFNγ on NK cell-mediated tumor lysis is mostly dependent on changes in MHC-class I and ICAM-1 expression. For three of the cell lines with increased resistance, we observed higher upregulation of MHC-class I than ICAM-1. For the cell lines with increased sensitivity after IFNγ treatment, we observed upregulation of ICAM-1 exceeding MHC-class I upregulation. ICAM-1 upregulation resulted in increased conjugate formation between the NK cells and tumor cells, which can contribute to the increased sensitivity observed. However, the effects of MHC-class I and ICAM-1 are not readily predictable. Due to the high IFNγ secretion of NK cell infusion products, a better understanding of the NK ligands on tumor cells and how they are affected by IFNγ is essential to optimize NK cell immunotherapy.

Keywords: interferon gamma, natural killer cells, pediatric cancer, mass cytometry, natural killer ligands, immunotherapy, intercellular adhesion molecule 1, major histocompatibility complex

INTRODUCTION

Current chemotherapeutic approaches for pediatric cancer are associated with high morbidity and late effects in survivors. By the age of 50, many survivors experience cardiovascular, renal, and hepatic complications, in addition to the risk of subsequent malignancies (1). Therefore, there is a need to develop new therapies capable of eradicating disease with reduced late effects. Immunotherapy has emerged as a promising new approach for cancer treatment. Natural killer (NK) cells, an important component of our first-line innate immunity, have activity against tumor cells without prior sensitization and are increasingly recognized for their important role in preventing and eradicating cancer. Because NK cells have the ability to selectively target tumor cells without affecting healthy cells, they are an attractive approach for pediatric cancer therapy.

Natural killer cells are typically defined as CD56+/CD3− lymphocytes. Their activity is regulated by a series of inhibitory and activating receptors that recognize ligands on target cells. The balance between activating and inhibitory signals will determine whether an NK cell is activated. The three main receptor families present on NK cells include natural cytotoxicity receptors (NCRs), C-type lectin (CD94/NKG2), and killer cell immunoglobulin-like receptors (KIRs) (2). NCRs (NKp30, NKp44, NKp46) are activating receptors, as are CD244/2B4, CD226/DNAM-1, and CD314/NKG2D. KIRs and C-type lectin receptors may be activating or inhibitory. Among the most important NK cell, inhibitory receptors are the KIRs and NKG2A because they recognize major histocompatibility complex (MHC) on target cells as evidence of self, leading to inhibition of NK cell activity (3). Many tumor cells downregulate MHC to escape T cell immunity, and upregulate activating ligands, making them susceptible targets for NK cell attack (4–7). However, in tumor cells where both inhibitory and activating ligands are present, the balance of these signals determines whether the NK cell is activated.

We developed an expansion protocol that allows production of large quantities of NK cells for adoptive immunotherapy (8). We have observed that, as compared to primary NK cells or IL-15-expanded NK cells, IL-21-expanded NK cells secrete 20-fold or 100-fold more interferon gamma (IFNγ, median 2,493 vs. 24 or 111 pg/mL, respectively), in response to target recognition (8). IFNγ has been reported to upregulate the inhibitory MHC-class I in target cells, making them more resistant to NK cell-mediated lysis (9, 10). However, IFNγ has also been reported to promote NK:target cell interaction through upregulation of intercellular adhesion molecule 1 (ICAM-1), promoting increased target cell death (11, 12). These findings suggest that IFNγ can have opposing effects on tumor cell sensitivity to NK cell-mediated lysis. To optimize the use of expanded NK cells as an immunotherapy it is imperative that we better understand how IFNγ affects NK cell-mediated lysis of the tumor cells.

Opposing effects reported for IFNγ may be due to a focus on specific tumor types, therefore this study aimed to evaluate the effect of IFNγ on a broad selection of 22 tumor cell lines from the pediatric preclinical testing program (PPTP) *in vitro* panel. This panel was designed to evaluate new therapies against childhood leukemias and solid tumors and has already been used for *in vitro* testing of over 50 pediatric cancer therapies (13). Using these cancer cells corresponding to six different types of pediatric malignancies, we evaluated the effects of IFNγ treatment in tumor cell sensitivity to NK cell-mediated lysis. Also we evaluated the effects of IFNγ treatment on tumor expression of NK cell ligands, including activating and inhibitory ligands, death receptors, and adhesion molecules.

MATERIALS AND METHODS

Isolation and Expansion of Human NK Cells

Buffy coats from four anonymized donors were obtained from Gulf Coast Regional Blood Center (Houston, TX, USA). Exemption and waiver of consent for the research use of buffy coat fractions obtained from anonymized donors at Gulf Coast Regional Blood Center (Houston, TX, USA) was granted by the Institutional Review Board of the University of Texas MD Anderson Cancer Center under protocol PA13-0978. NK cells were isolated using the RossetteSep Human NK cell enrichment cocktail (Stem Cell Technologies) and expanded as described previously using K562 Clone9.mbIL21 as feeder cells for 21 days (8). Expanded NK cells were cryopreserved, and subsequently thawed and recovered for 1–2 days prior to their use. During recovery NK cells were cultured in NK cell media consisting of RPMI 1640 (Corning) supplemented with 50 IU/mL recombinant human IL-2 (Proleukin, Novartis Vaccines and Diagnostics, Inc.), 20% Fetal Bovine Serum (Thermofisher), L-glutamine (Gibco), and penicillin/streptomycin (Corning).

Tumor Cells

TC-71, NALM-6, and Ramos-RA1 were obtained as kind gifts from colleagues (Drs. Eugenie S. Kleinerman, L. J. N. Cooper, and J. Chandra, respectively). Karpas-299 was obtained from the German Collection of Microorganisms and Cell Cultures (DSMZ). RS4;11, MOLT-4, and CCRF-CEM were obtained from the America Type Culture Collection (ATCC). The remaining cell lines were obtained from the Children's Oncology Group (COG) Cell Line and Xenograft Repository. Brain tumor cell lines BT-12, SJ-GBM2, CHLA-266, Ewing sarcoma (EWS) cell lines CHLA-9, CHLA-10, CHLA-258, TC-71, neuroblastoma (NB) cell lines NB1643, NB-EBc1, CHLA-90, CHLA-136, rhabdomyosarcoma (RMS) cell line RD, and leukemia cell line COG-LL-317 were cultured in IMDM (Lonza) supplemented with 20% FBS (Thermofisher), 4 mM L-glutamine (Gibco), 1× ITS (Lonza), and penicillin/streptomycin (Corning). Lymphoma cell lines Karpas-299, Ramos-RA1, leukemia cell lines NALM-6, RS4;11, MOLT-4, CCRF-CEM, Kasumi-1, and RMS cell lines Rh41, Rh30, were cultured in RPMI 1640 (Corning) supplemented with 10% Fetal Bovine Serum (Thermofisher), L-glutamine (Gibco), and penicillin/streptomycin (Corning). Cultures were periodically tested to confirm absence of mycoplasma using MycoAlert Mycoplasma Detection Kit (Lonza). Identity was confirmed by STR DNA fingerprinting either using the AmpFlSTR Identifiler kit (Applied Biosystems) or the Power Plex 16HS Kit (Promega) according to manufacturer instructions. The STR profiles were compared to known fingerprints as published by ATCC or the COG cell STR Genotype Database (http://strdb.cogcell.org). STR profiles were last performed on March 2016 (SJ-GBM2, NB1643, MOLT-4), October 2015 (RD, Rh41, Rh30, BT-12, CHLA-10, NB-EBc1, NALM-6, and Ramos-RA1), or September 2012 (CHLA-266, CHLA-9, CHLA-258, TC-71, CHLA-90, CHLA-136, RS4;11, COG-LL-317, CCRF-CEM, Kasumi-1, and Karpas-299). Banks of STR validated, mycoplasma-free cell lines were cryopreserved. Cell lines were kept in culture no longer than eight passages or 4 weeks prior to use.

IFNγ Treatment of Tumor Cells

Cell lines that grow in suspension were seeded at 0.5×10^6 cells/mL and treated with 50 ng/mL of IFNγ (Peprotech) for 48 h. Adherent cells were cultured to a 60–70% confluence and treated with 50 ng/mL of IFNγ (Peprotech) for 48 h. Untreated tumor cells were seeded in parallel. After treatment cells were washed in IFNγ free media, and adherent cells were detached with non-enzymatic cell dissociation buffer (Gibco) to avoid degradation of cell surface proteins. Treated and untreated cells were evaluated for surface expression of NK cell ligands by mass cytometry and sensitivity to NK cell-mediated lysis by calcein release assay.

Cytotoxicity

The fluorescence based calcein release assay was used to assess cytotoxicity, as previously described (8, 14). Adherent cells were detached with non-enzymatic cell dissociation buffer (Gibco) and cells were filtered by using a 70 μm cell strainer (Corning) to obtain a single-cell suspension. Target cells were labeled with 5 μg/mL of calcein-AM (Sigma-Aldrich) for 1 h at 37°C. NK cells were cocultured with target cells at different effector to target (E:T) ratios (10:1, 5:1, 2.5:1, 1.25:1, 0.6:1, and 0.3:1) for 4 h at 37°C. Supernatant fluorescence was determined at 485 nm^{Exc}/530 nm^{Emm} using the SpectraMax $Plus^{384}$ spectrophotometer.

Mass Cytometry

Antibodies for mass cytometry were labeled with heavy metals using Maxpar-X8 labeling reagent kits (DVS Sciences) according to manufacturer's instructions and titrated for determination of optimal concentration. The antibodies and their respective heavy metal labeling can be found in Table S1 in Supplementary Material. Since NK cell receptors may have multiple ligands (e.g., NKG2D binds to MICA, MICB, and ULBP1-5), or unknown ligands, chimeric receptor:IgG-Fc fusion proteins were tagged with heavy metals and used for identification of ligands on tumor cells. Then, 1.5×10^6 cells were stained for viability with 2.5 μM cell ID cisplatin (Fluidigm, 201064) in serum free RPMI for 1 min and washed twice with complete media. Subsequently, surface staining was performed as previously described (15). Staining media were prepared by adding 5% FBS and 0.1% sodium-azide to PBS. During the intracellular staining step of tumor cells, two different isotopes of cisplatin Pt-194 (Fluidigm, 201194) and Pt-198 (Fluidigm, 201198) were used to barcode untreated and IFNγ-treated samples, respectively, allowing samples to be combined in a single tube, minimizing acquisition time, and variability between runs. Data were acquired on a CyTOF instrument (DVS Sciences). Files containing only live single cells were exported using FlowJo V10 Software and uploaded into Cytobank for further analysis (Figure S1 in Supplementary Material) (16).

Conjugation Assay

The determination of effector conjugation to target cells was performed as described by Burshtyn et al., with some minor modifications (17). Briefly, NK cells and tumor cells were stained with green dye PKH67-GL (Sigma, MINI67) and red dye PKH26-GL (Sigma, MINI26), respectively, in 5 μM dye at 5×10^6 cells/mL for 5 min at room temperature. Dye staining was stopped by adding two volumes of FBS and two volumes of complete media. Cells were washed twice with complete media and let rest for at least 1 h at 37°C. Then, 10^5 NK cells were combined with 2×10^5 tumor cells in 200 μL, centrifuged at 20 *g* for 1 min (to initiate contact), and incubated at 37°C for 30 min. Cells were resuspended by gentle vortexing, fixed with 200 μL of 4% formaldehyde, and analyzed by flow cytometry. For antibody blocking experiments, NK cells were pre-incubated with 5 μL of Fc blocker (Biolegend, 422302) for 10 min to avoid antibody-dependent cell cytotoxicity. ICAM-1 was blocked on tumor cells by adding 10 μg/mL of anti-CD54/ICAM-1 clone HCD54 (Biolegend, 322703) for 20 min at room temperature.

Statistical Analysis

Statistics were performed in GraphPad Prism Software. For determination of Δ in % Lysis after IFNγ treatment (**Figure 1**), we used data from six different E:T ratios. For each cell line, we calculated the average difference in lysis after IFNγ treatment (%lysis IFNγ treated − %lysis untreated) using four NK cell donors. Significance was determined by using the *t*-test with a hypothetical value of 0 for comparison. Heatmaps were generated using Cytobank (16). Conjugation data significance was determined by using the *t*-test, statistical significance was determined by a $p < 0.05$.

RESULTS

IFNγ Has a Variable Impact on Tumor Cell Sensitivity to NK Cell-Mediated Lysis

We determined whether tumor IFNγ treatment affected NK cell-mediated lysis for 22 pediatric cancer cell lines from the PPTP *in vitro* panel. Tumor cell lines evaluated were derived from brain tumors, EWS, NB, leukemia, lymphoma, and RMS. The Rh18 cell line was excluded because of repeated problems maintaining the cell line in culture. We observed that 6 of the 22 cell lines evaluated show a significant decrease in NK cell-mediated lysis after IFNγ treatment (**Figure 1**). The group of cell lines with decreased lysis after IFNγ treatment includes leukemia cell lines Molt-4 ($p = 0.0030$) and Kasumi-1 ($p = 0.0231$), EWS cell lines CHLA-9

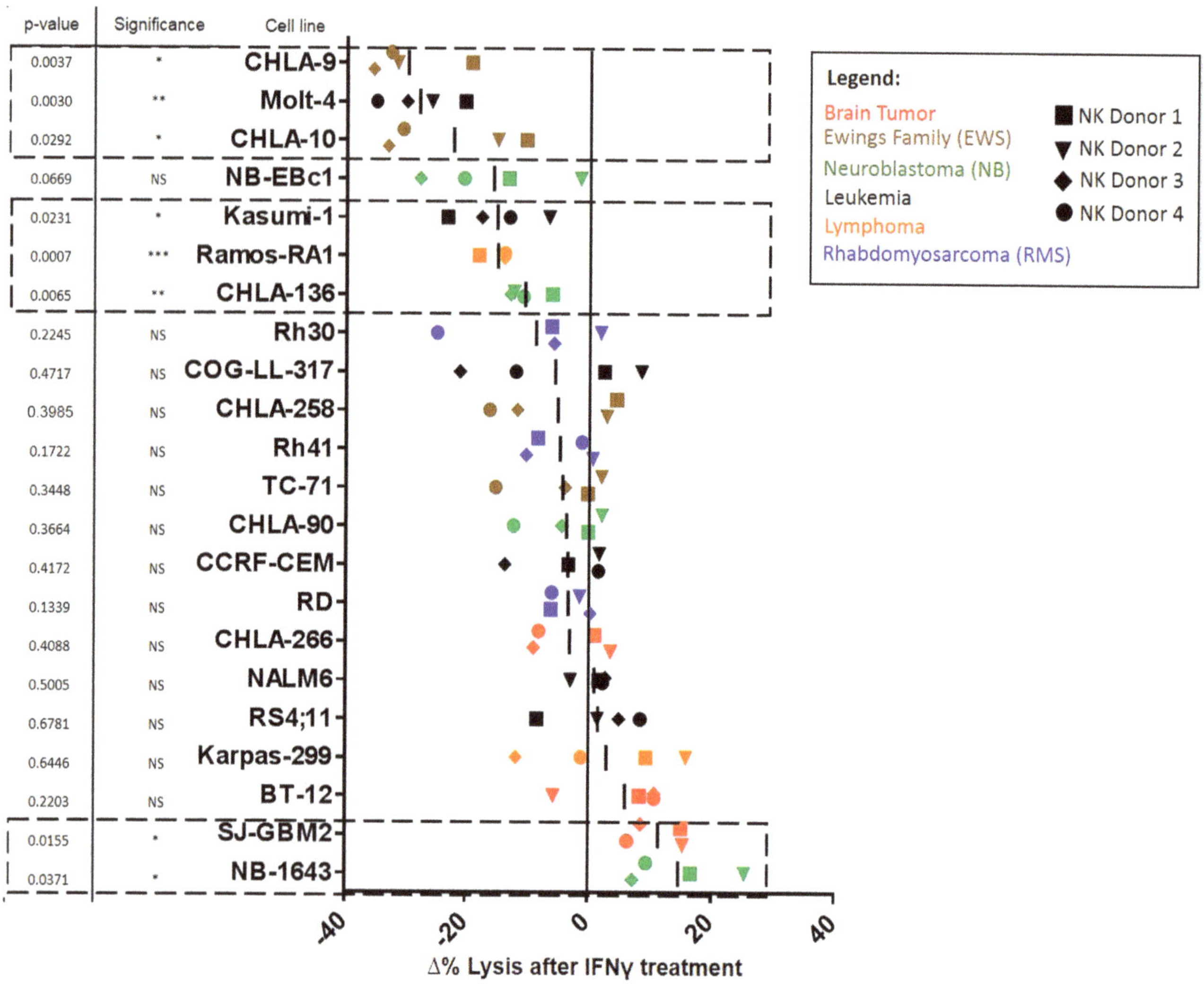

FIGURE 1 | Waterfall plot of change in lysis by natural killer (NK) cells after treatment of pediatric cancer cell lines with IFNγ. NK cell cytotoxic activity toward IFNγ-treated and -untreated cancer cells was evaluated using calcein release assays. Changes in lysis (Δ% Lysis) of treated compared to untreated cancer cells were quantified for each NK cell donor. Each shape represents a different NK cell donor (the same four donors were used for all cell lines). Color coding corresponds to cell line cancer type. *p*-Values are for probability that $\Delta\% \neq 0$ (*t*-test). Squares with dashed lines indicate cancer cell lines with significant ($p \leq 0.05$) changes in NK cell-mediated lysis after IFNγ treatment.

($p = 0.0037$) and CHLA-10 ($p = 0.0292$), lymphoma cell line Ramos-RA1 ($p = 0.0007$), and NB cell line CHLA-136 ($p = 0.0065$). The NB cell line NB-EBc1 appears to have a decreased lysis after treatment, however, it was not statistically significant. When decreased lysis after treatment was observed, it was consistent for all four donors tested and across E:T ratios (**Figure 2A**).

By contrast, two of the cell lines showed an increase in NK cell-mediated lysis after IFNγ treatment. These cell lines include the glioblastoma cell line SJ-GBM2 ($p = 0.0155$) and the NB cell line NB1643 ($p = 0.0371$) (**Figure 1**). In addition, although the brain tumor cell line BT-12 shows a non-significant increase in sensitivity, we observed that for three of the four NK cell donors there was an increased sensitivity after IFNγ treatment, this increase is significant if these three donors are evaluated ($p = 0.0061$). Increased lysis after IFNγ treatment was a consistent finding with most donors mostly at high NK cell doses (**Figure 2B**). The remaining cell lines evaluated showed no differences in NK cell-mediated lysis after IFNγ treatment, and this group includes cell lines corresponding to all the tumor types evaluated (**Figure 1**). After stratifying our data by tumor type, we can observe that the only tumor type for which IFNγ had no effect on lysis was RMS. In EWS, leukemia, and lymphoma IFNγ had no effect or resulted in decreased lysis, whereas IFNγ had no effect or resulted in increased lysis for brain tumors. Interestingly, for NB cell lines, the effect of IFNγ on NK cell-mediated lysis was variable. Lysis of some NB cell lines was unaffected by IFNγ treatment, some became more resistant, and some more sensitive.

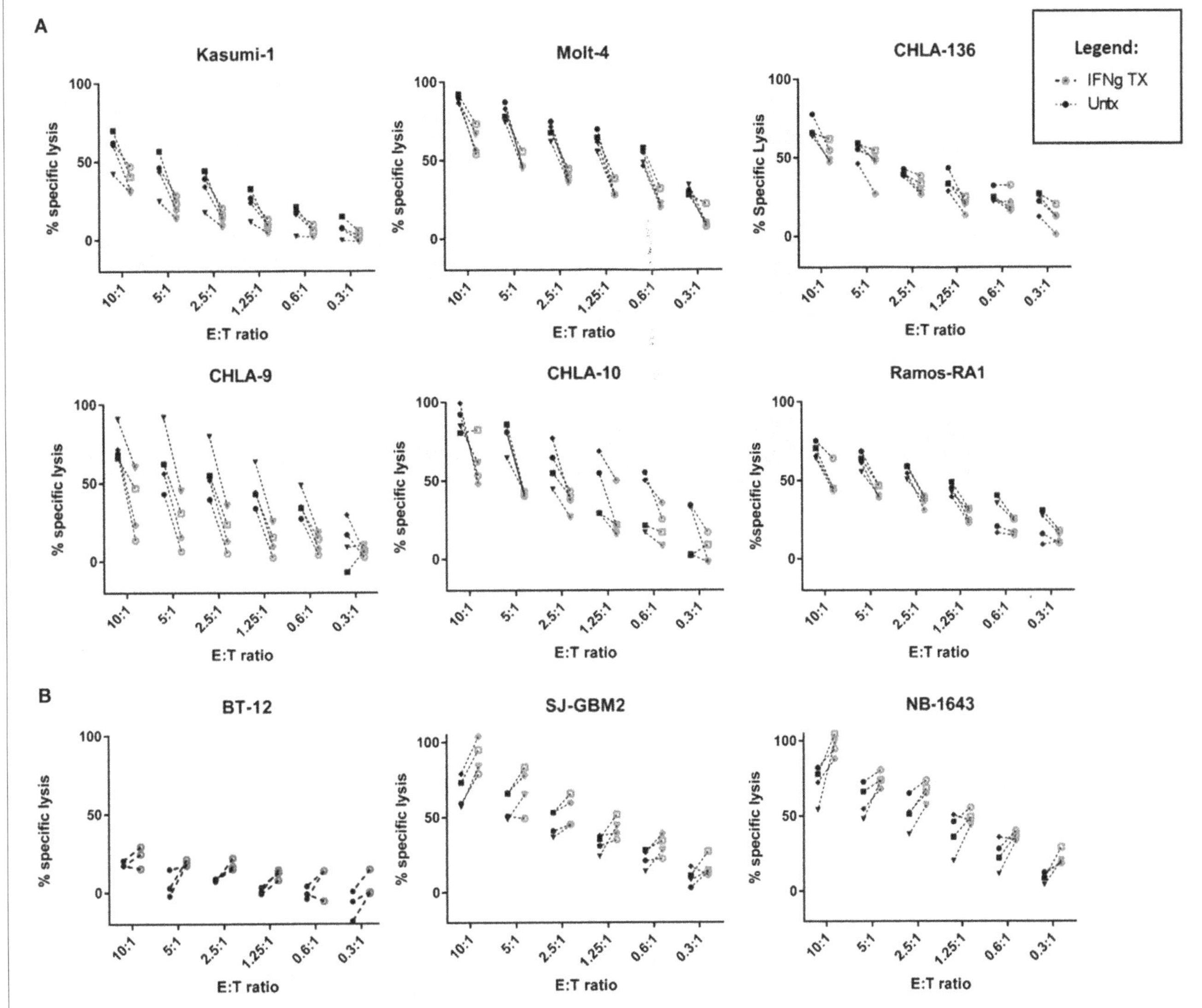

FIGURE 2 | IFNγ has a variable impact on natural killer (NK) cell-mediated cancer lysis. NK cell cytotoxic activity toward IFNγ-treated (gray) and -untreated (black) cancer cells was evaluated at six different effector to target (E:T) ratios using calcein release assays. **(A)** Cell lines showing decreased lysis [leukemia (Molt-4, Kasumi-1), Ewing sarcoma (CHLA-9, CHLA-10), lymphoma (Ramos-RA1), and neuroblastoma (NB) (CHLA-136)] and **(B)** cell lines showing increased lysis [brain tumors (BT-12, SJ-GBM2) and NB (NB1643)] after IFNγ treatment. The four NK cell donors are represented for all cell lines, except for BT-12 with three donors represented.

IFNγ Alters Surface Expression of NK Cell Ligands in Pediatric Cancer Cell Lines

The 22 pediatric cancer cell lines obtained from the PPTP *in vitro* panel were evaluated by mass cytometry for the expression of 16 NK cell ligands. Baseline expression of NK cell ligands in terms of median expression and percentage positive cells is provided in **Figure 3**. The inhibitory ligands MHC-class I and HLA-E are homogeneously expressed (>70%) for all PPTP cell lines with the exception of the RMS cell line Rh41 where MHC-class I was expressed in only 35% of the cells (**Figure 3**, right panel). Interestingly, we observe that solid tumor cell lines (RMS, brain tumor, EWS, and NB) have higher median expression levels of TRAIL receptor CD262/DR5, and ligands for the activating receptors NKG2D, DNAM-1, and NCRs, when compared to leukemia cell lines (**Figure 3**, left panel).

Baseline expression levels of NK cell ligands were compared to the levels after IFNγ treatment and changes were quantified. The percentage change in mean mass intensity (MMI) after IFNγ treatment was determined for each of the 22 cell lines (**Figure 4A**). CD274/PD-L1, CD54/ICAM-1, HLA-DR, MHC-class I, CD95/FasR, and CD270/HVEM were the most affected by IFNγ (**Figure 4A**), with at least three cell lines showing fivefold increase or more in MMI. Data were also evaluated in terms of median expression. IFNγ induced changes in median expression were quantified, and a heatmap corresponding to the fold increase in median expression (obtained from the arcsinh ratio

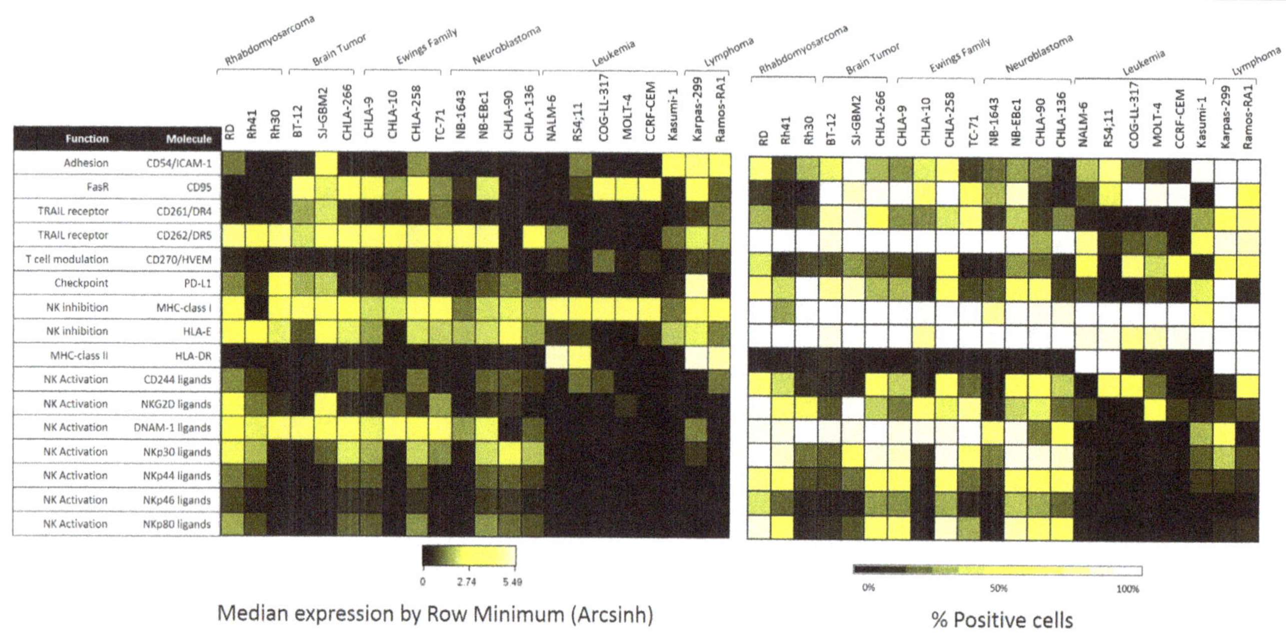

FIGURE 3 | Expression of natural killer (NK) cell ligands in pediatric cancer cell lines. Pediatric cancer cell lines were evaluated by mass cytometry for the baseline expression level of multiple NK cell ligands. The evaluated markers and their function can be found at the table above. Left panel: median expression level normalized by row minimum (obtained by evaluating the arcsinh ratio of the median expression compared to the row minimum). Right panel: percent positive cells for each given marker.

of median) was generated (**Figure 4B**). The changes in median expression were similar to changes in mean expression. CD274/PD-L1, CD54/ICAM-1, HLA-DR, MHC-class I, CD95/Fas, and CD270/HVEM were the markers mostly upregulated by IFNγ (**Figure 4B**). When stratified by tumor type (**Figure 5**; Figure S2 in Supplementary Material), we observe that PD-L1 was most upregulated by IFNγ in solid tumor cell lines (RMS, brain, EWS, and NB), but was relatively unaffected on leukemia and lymphoma cell lines. ICAM-1 upregulation was higher for brain tumors, EWS, NB and leukemia, and lower for RMS and lymphomas. HLA-DR was upregulated by IFNγ on some brain tumor, EWS, and NB cell lines. MHC-class I upregulation was variable even within the same tumor type. However, IFNγ-induced MHC-class I upregulation was consistently observed in all NB cell lines. Finally, CD270/HVEM upregulation was induced by IFNγ in some brain tumor, EWS, and NB cell lines. Our data show that lymphoma cell lines were the least affected by IFNγ in terms of NK cell ligand expression (**Figure 5**; Figure S2 in Supplementary Material).

IFNγ-Induced Upregulation of MHC-Class I and ICAM-1 Correlates with Changes NK Cell-Mediated Lysis

Next, we wanted to determine whether there was a correlation between changes in ligand expression after IFNγ treatment and changes in NK cell-mediated tumor lysis. We first assessed our expanded NK cells for expression of the receptors corresponding to the most upregulated ligands—PD-L1, ICAM-1, and MHC-class I. Expression of PD-1, the receptor for PD-L1, was observed on only 7% of the expanded NK cells (**Figure 6**). This low percentage of PD-1+ NK cells in our expanded product suggests that PD-L1 upregulation is unlikely to play a role in the IFNγ-induced changes in lysis we observed.

By contrast, LFA-1, the integrin that binds ICAM-1, was expressed in 99.6% of our expanded NK cells (**Figure 6**), and inhibitory KIR receptors were also highly expressed in our expanded NK cells. KIR2DL2/3 expression was 83.9%, KIR2DL1/2DS5 was 97.18%, and KIR3DL1 was 89.15% (**Figure 6**). Also we observed expression of the inhibitory receptor NKG2A in 97% of expanded NK cells. Knowing that the majority of our expanded NK cells express LFA-1, KIR, and NKG2A receptors, we focused on changes in ICAM-1 and MHC-class I expression (**Figure 7**). The ratio of change in MHC-class I over change in ICAM-1 after IFNγ treatment was evaluated and plotted for cell lines with altered sensitivity (**Figure 7C**). We observed that all the cell lines that became more resistant after IFNγ treatment had an increase in MHC-class I expression (**Figures 2A** and **7A**). However, MHC-class I upregulation was also observed in some of the cell lines with increased sensitivity after IFNγ treatment (**Figures 2B** and **7B**). Interestingly, MHC-class I/ICAM-1 change ratio was <1 and, indicating that upregulation of ICAM-1 exceeded MHC-class I upregulation, for all the cell lines with increased sensitivity after IFNγ treatment (**Figures 7B,C**). The opposite pattern, with an MHC-class I/ICAM-1 change ratio >1 and MHC-class I upregulation exceeding ICAM-1 upregulation, was observed in three of the six cell lines where IFNγ induced resistance (Kasumi-1, MOLT-4, and CHLA-136) (**Figures 7A,C**). However, ICAM-1 upregulation exceeded MHC-class I upregulation (ratio <1) in

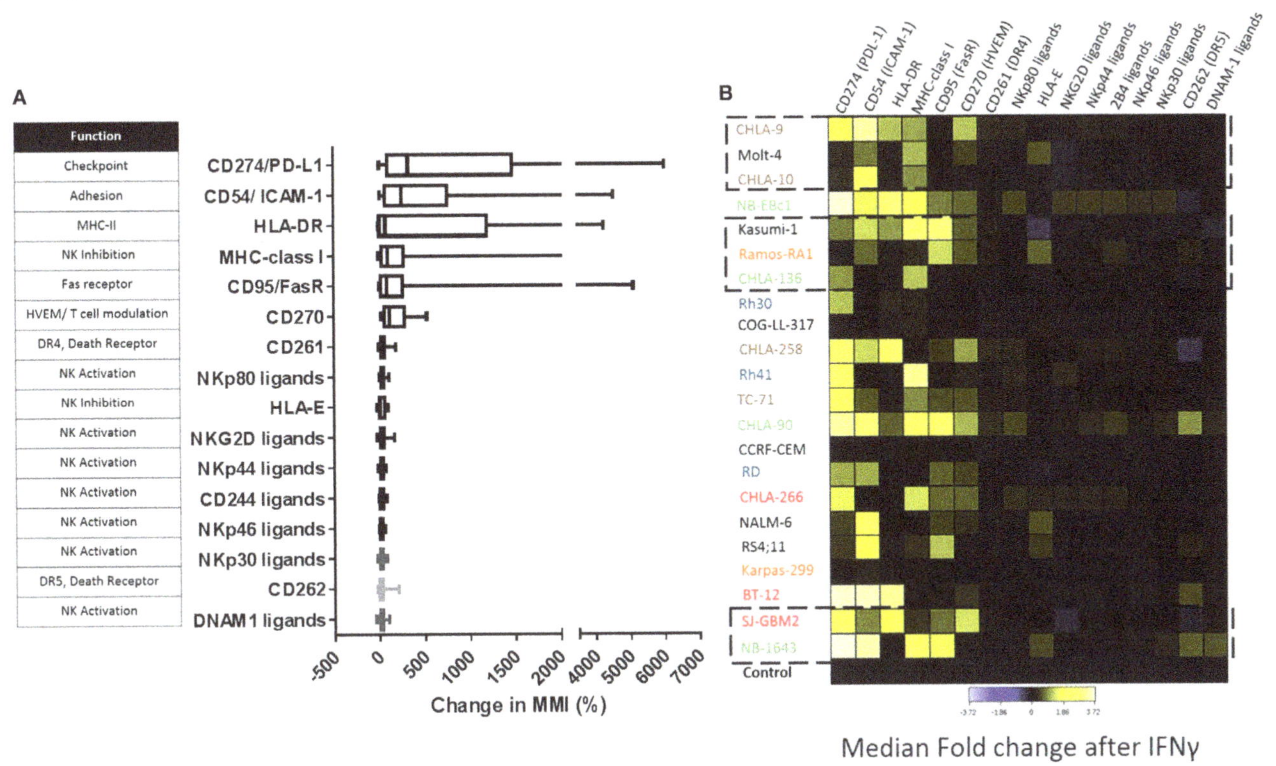

FIGURE 4 | Impact of IFNγ treatment on the surface expression of natural killer (NK) cell ligands for pediatric cancer cell lines. (A) Change in expression level [mean mass intensity (MMI)] was determined by quantifying the mean of each parameter at baseline and after IFNγ treatment for each cell line. Percent change in MMI after IFNγ treatment was calculated for each cell line and box plots were generated for each parameter using data obtained from all 22 cell lines. **(B)** Fold change in median expression for each parameter in all cell lines. Heat corresponds to the fold change in median expression after IFNγ treatment (obtained as the arcsinh ratio of median expression for the given markers compared to untreated cell line control). Cancer type color coding is the same as in **Figure 1**. Squares with dashed lines indicate cancer cell lines with significant changes in NK cell-mediated lysis after IFNγ treatment.

the EWS cell lines CHLA-9 and CHLA-10 and the lymphoma cell line Ramos-RA1, for which IFNγ treatment resulted in decreased NK cell-mediated lysis.

IFNγ-Induced ICAM-1 Upregulation Increases Conjugate Formation for Cell Lines with Increased Sensitivity

Since IFNγ is known to enhance MHC-class I expression, resulting in resistance of target cells to NK cell-mediated lysis (9, 10), we investigated possible mechanisms for the enhanced sensitivity observed after treatment for the cell lines SJ-GBM2, NB1643, and BT-12. These showed increased sensitivity to NK cell-mediated lysis after IFNγ treatment despite presence of high levels of MHC-class I. Since upregulation of the adhesion molecule ICAM-1 exceeded MHC-class I upregulation for these tumors, we determined whether IFNγ-mediated ICAM-1 upregulation was capable of overcoming MHC-class I inhibition through increased conjugate formation between the NK cells and the target cells (**Figure 8A**). For comparison, we also evaluated IFNγ-mediated changes in conjugate formation between NK cells and Ramos-RA1, a cell line having high ICAM-1 expression but for which IFNγ resulted in more resistance. We observed that treatment of BT-12, SJ-GBM2, and NB1643 with IFNγ resulted in increased conjugate formation (**Figure 8B**) and this was statistically significant for BT-12 ($p = 0.026$) and SJ-GBM2 ($p = 0.011$). By contrast, IFNγ treatment did not increase conjugate formation for Ramos-RA1. Next, to confirm whether the increased conjugate formation was specifically mediated by ICAM-1 upregulation, we used monoclonal antibodies to block ICAM-1 on IFNγ-treated cells (BT-12, SJ-GBM2, and NB1643). Blocking of ICAM-1 resulted in a significant decrease in conjugate formation for BT-12 ($p = 0.042$), SJ-GBM2 ($p = 0.008$), and NB1643 ($p = 0.041$) (compared to isotype control, **Figure 8C**).

DISCUSSION

Natural killer cells are an attractive approach for cancer immunotherapy. Our laboratory has developed an *ex vivo* NK cell expansion platform that allows us to generate large quantities of NK cells for patient infusion. We have shown that our expanded NK cells, currently used in several clinical trials for myeloid malignancies and posterior fossa tumors (NCT01787474, NCT01904136, NCT01823198, NCT02271711), secrete large amounts of IFNγ compared to primary NK cells (8). Similarly, memory-like NK cells used for adoptive transfer to AML patients exhibited enhanced IFNγ production when compared to control NK cells

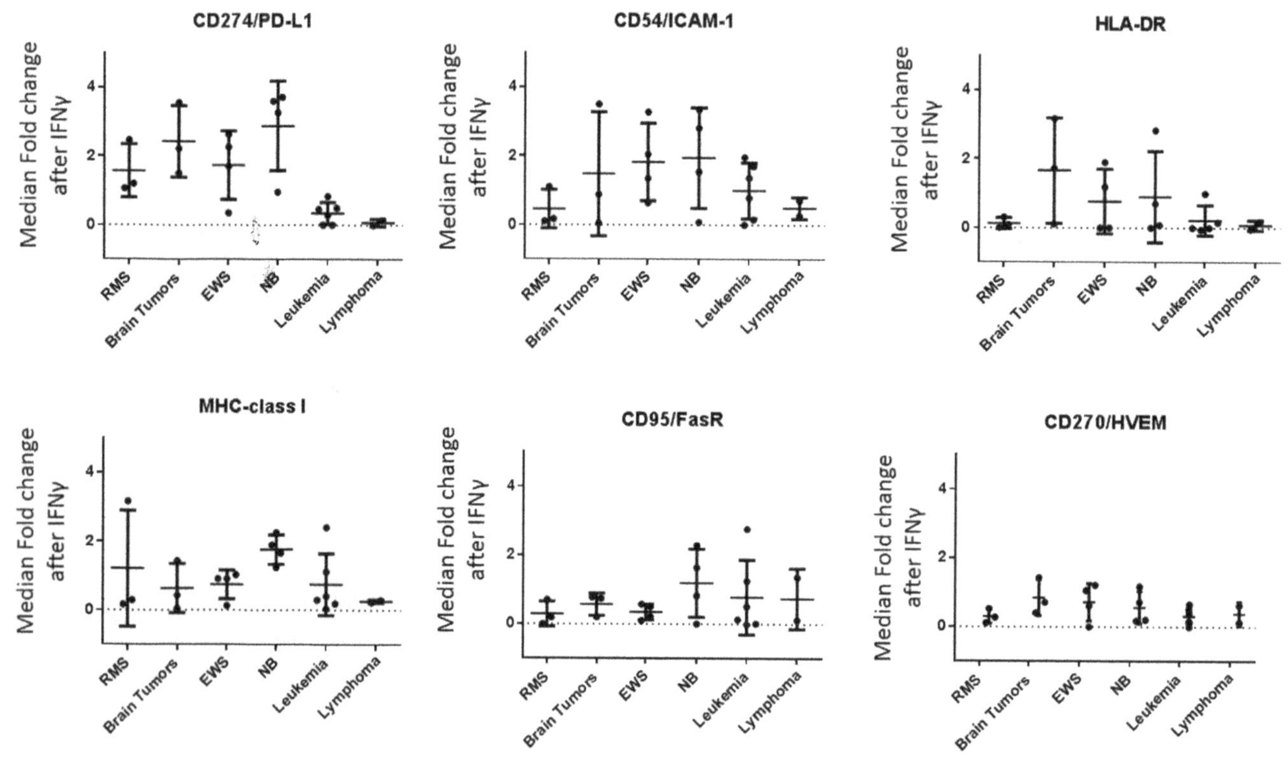

FIGURE 5 | Major ligands affected by IFNγ classified by tumor type. IFNγ induced changes in NK cell ligand expression were evaluated by cancer type. *Y*-axis corresponds to the fold increase in marker median expression after IFNγ treatment, this was obtained by evaluating the arcsinh ratio of median expression for the given marker compared to untreated cell line.

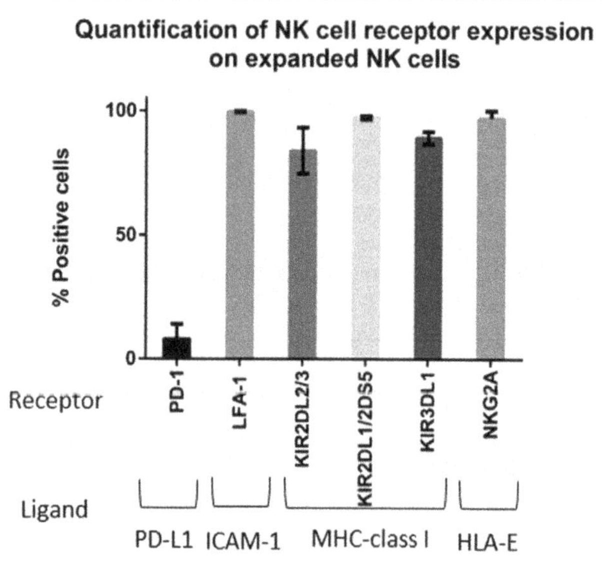

FIGURE 6 | Phenotyping of expanded natural killer (NK) cells. Expression of PD-1, LFA-1, killer cell immunoglobulin-like receptors (KIRs), and NKG2A on the expanded NK cells was obtained by mass cytometry (*n* = 4 donors).

(18). Due to the high levels of IFNγ secreted by NK cell infusion products, we sought to determine the effects of IFNγ on NK cell interactions with the tumor cells.

Through this study we evaluated a broad selection of pediatric tumor cells, representing at least six different types of malignancies, for the effect of IFNγ on their sensitivity to NK cell-mediated lysis using expanded NK cells from four independent donors. This enabled a broad study across tumor types demonstrating opposing effects of IFNγ, a finding that may not have been evident in other studies that focus on a single tumor type or modulation of a single surface ligand. Of the 22 cell lines evaluated, six showed a significant decrease in NK cell-mediated lysis after IFNγ treatment, including leukemia, EWS, lymphoma, and NB cells. By contrast, treatment with IFNγ resulted in enhanced sensitivity to NK cells for three cell lines, BT-12, SJ-GBM2, and NB1643 (two brain tumors and a NB, respectively). For the remaining cell lines, IFNγ treatment did not significantly affect NK cell-mediated lysis, though some showed trends that may be significant with more donor replicates. The effect of IFNγ treatment was variable within the same tumor type, with the exception of RMS cell lines for which IFNγ treatment had no effect. These results suggest that the effect of IFNγ on NK cell-mediated lysis of tumor cells is variable and cell line dependent. Our findings warrant more focused investigation and validation for specific tumor types using primary tumor samples or patient-derived xenografts where feasible.

To better understand this variability, we used mass cytometry to evaluate the effect of IFNγ on expression of NK cell ligands by cancer cells. First, we observed broad heterogeneity in baseline expression levels of ligands between and within tumor types (**Figure 3B**). Solid tumor cell lines had higher median expression of TRAIL receptor

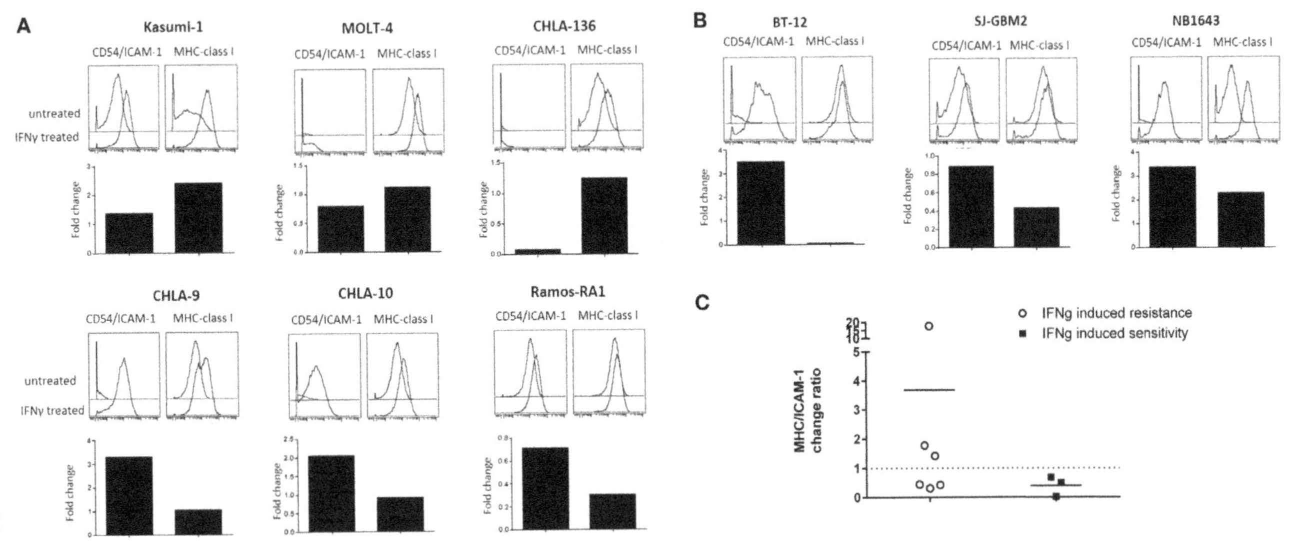

FIGURE 7 | Effect of IFNγ treatment on intercellular adhesion molecule 1 (ICAM-1) and major histocompatibility complex (MHC)-class I expression for cancer cell lines with altered sensitivity. The effect of IFNγ on ICAM-1 and MHC-class I expression for cell lines in which IFNγ treatment conferred **(A)** resistance or **(B)** sensitivity to NK cell-mediated lysis. Histograms compare ICAM-1 (left) and MHC-class I (right) expression for untreated cancer cells (top) and IFNγ-treated cells (bottom). Bar graphs represent ICAM-1 and MHC-class I fold change in median expression after IFNγ treatment. **(C)** Ratio of MHC-class I/ICAM-1 change after IFNγ treatment for cell lines with altered sensitivity. Fold change in median expression after IFNγ treatment for each parameter was quantified and the ratio was calculated. Ratio <1 indicates higher ICAM-1 upregulation and ratio >1 indicates higher MHC-class I upregulation after IFNγ treatment.

CD262/DR5 and ligands for the activating receptors NKG2D, DNAM-1, and NCRs, compared to leukemia cell lines (**Figure 3A**). This suggests NK cells may be a promising therapy for solid tumors. We also observed that none of these ligands were downregulated by IFNγ; however, CD274/PD-L1, CD54/ICAM-1, HLA-DR, MHC-class I, CD95/FasR, and CD270/HVEM were upregulated in a variety of tumor types. Other than CD270/HVEM, these findings are consistent with previously published studies (11, 12, 19–24). HVEM is involved in T cell regulation (25), but no studies have yet reported regulation of this ligand by IFNγ and its role in NK cell biology has not been well described.

Although IFNγ-mediated upregulation of MHC-class I, PD-L1, and ICAM-1 has been previously described, this study uncovers the variability of IFNγ responses across different pediatric tumor types. In terms of PD-L1, our results show upregulation by IFNγ for most pediatric solid cancers (RMS, brain tumors, EWS, and NB), but no effect on pediatric leukemia and lymphoma cells. On the other hand, the adhesion molecule ICAM-1 was upregulated by IFNγ on brain tumors, EWS, NB, and leukemia, but not on RMS and lymphoma cells. Although IFNγ-induced ICAM-1 upregulation has been previously described in NB and leukemia cells (11, 12), to our knowledge no studies have shown its upregulation on EWS.

According to our data, the ligands most upregulated by IFNγ were PD-L1, ICAM-1, and MHC-class I; therefore, we determined whether changes in their expression correlated to changes in tumor lysis after IFNγ treatment. PD-1/PD-L1 is associated with immune cell suppression; however, we observed that <7% of expanded NK cells express PD-1. In addition, changes in expression of PD-L1 did not correlate with changes in cytotoxicity, suggesting that PD-L1 did not play a role in our model. By contrast, changes in ICAM-1 and MHC-class I cooperate in affecting NK cell responsiveness. For the cell lines in which IFNγ increased sensitivity to NK cells, we observed ICAM-1 upregulation exceeding MHC-class I upregulation (MHC-class I/ICAM-1 < 1). This suggests that the LFA-1/ICAM-1 interaction augments NK cell-mediated lysis even in the presence of high levels of inhibitory MHC-class I molecules. MHC-class I molecules mediate NK cell inhibition through binding with KIRs (HLA-A, HLA-B, HLA-C) and NKG2A (HLA-E). The anti-MHC-class I antibody clone used in this study, W6/32, recognizes both classical (HLA-A, HLA-B, HLA-C) and non-classical (HLA-E) HLA (26). We observed that for three of the six cell lines with increased resistance after IFNγ treatment, MHC-class I upregulation exceeded ICAM-1 upregulation (MHC-class I/ICAM-1 >1). This suggests that the NK cell balance was shifted toward inhibition due to the increased expression of MHC-class I, which binds NK cell inhibitory receptors. However, in the other three cell lines with increased resistance after IFNγ treatment we observed that ICAM-1 upregulation exceeded MHC-class I upregulation (MHC-class I/ICAM-1 <1) for which this model would have predicted increased sensitivity after IFNγ treatment. Our results suggest that the differential effects on MHC-class I and ICAM-1 expression can explain some but not all of the effects of IFNγ on sensitivity to NK cell lysis.

We investigated possible mechanisms for the enhanced sensitivity observed after IFNγ treatment for the cell lines SJ-GBM2, NB1643, and also BT-12. After IFNγ treatment these cell lines showed increased sensitivity to NK cell-mediated lysis, even in the presence of high levels of MHC-class I (**Figures 2B** and **7B**). Since IFNγ treatment caused ICAM-1 upregulation exceeding MHC-class I upregulation in these cells lines (MHC-class I/

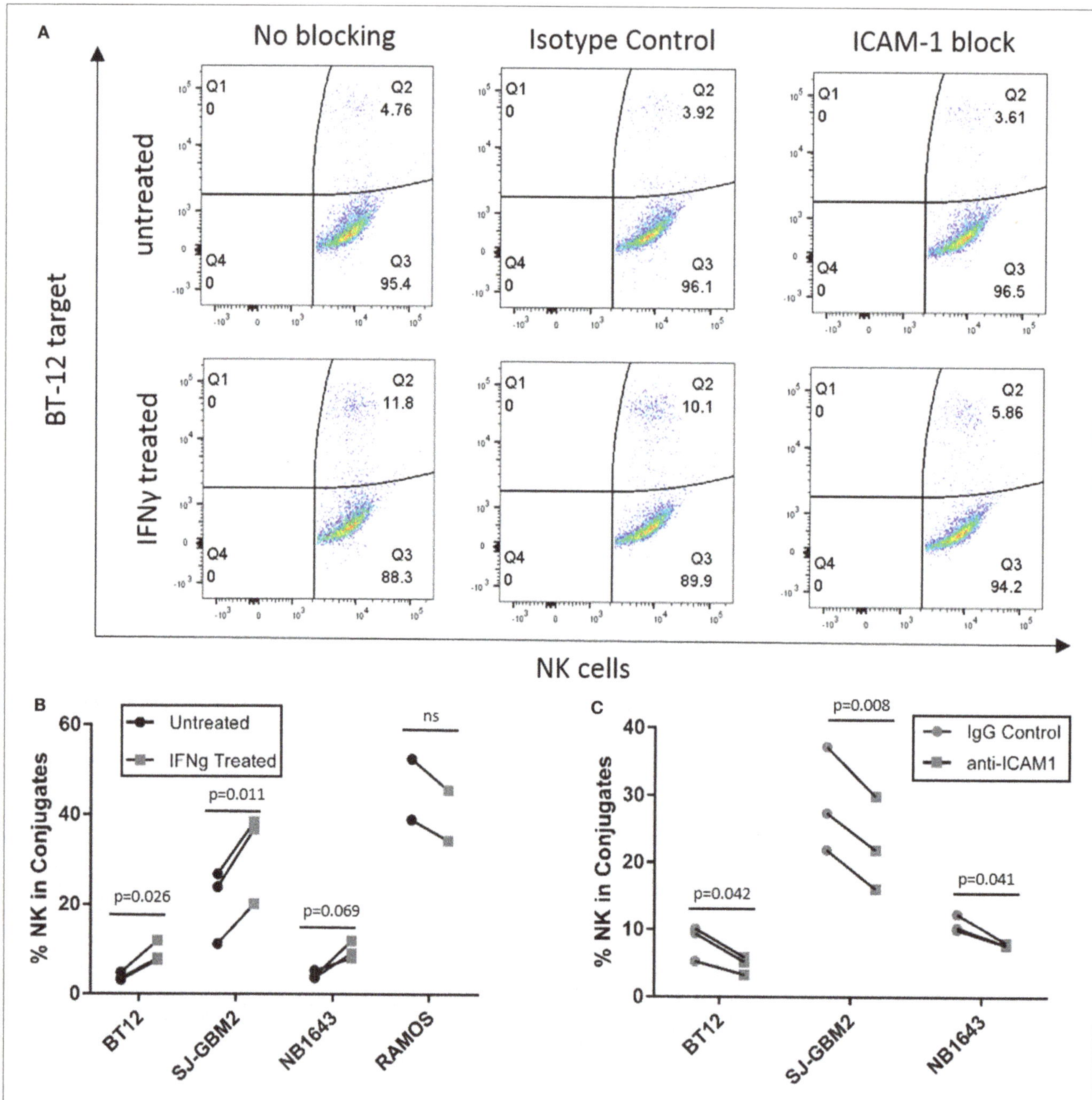

FIGURE 8 | **Conjugation assay for cell lines with increased sensitivity after IFNγ treatment. (A)** *y*-Axis represents fluorescence of labeled target cells and *x*-axis represents fluorescence of labeled natural killer (NK) cells, and Q2 indicates dual fluorescence of NK cells conjugated with target cells. **(B)** Quantification of %NK cells in conjugate with BT-12, SJ-GBM2, NB1643, and Ramos-RA1. **(C)** Quantification of %NK cells in conjugate after ICAM-1 block for IFNγ-treated BT-12, SJ-GBM2, and NB1643 (n = 3 donors, Student's t-test).

ICAM-1 <1), we evaluated conjugate formation after IFNγ treatment. Our results show an increase in the formation of conjugates between the NK cells and the target cells SJ-GBM2, BT-12, and NB1643 after IFNγ treatment, and conjugate formation was decreased for IFNγ-treated cells after blocking ICAM-1 on the target cells. Ramos-RA1 did not show increased conjugate formation or killing in response to IFNγ, possibly because it has such high baseline ICAM-1 expression and conjugate formation. This suggests that the mechanism of increased sensitivity from IFNγ treatment is, at least in part, mediated by increased ICAM-1 upregulation leading to enhanced effector-target conjugation.

Our study uncovers the complexity behind cancer cell responses to IFNγ. Published literature has shown increased resistance to NK cell-mediated lysis due to MHC-class I upregulation in some cancers (9, 10), but increased sensitivity to NK cell-mediated lysis due to ICAM-1 upregulation in others (11, 12). These studies reported IFNγ effects on MHC-class I or ICAM-1 expression individually, and the contradictory results stem from analysis of a small number of different targets or focus on particular tumor types. This is, to our knowledge, the first study to evaluate the effect of IFNγ on a broad range of NK cell ligands and across nearly two dozen cell lines representing six broad tumor types. We observed variable effects of IFNγ on tumor sensitivity to NK cells that are cell line dependent and attributable to individual MHC-class I and ICAM-1 responses, suggesting that IFNγ effects on cancer cell sensitivity to NK cells should not be broadly generalized, although further experiments with cell lines and patient samples may determine whether specific tumor types have generalizable IFNγ responses. Our data suggest that increased IFNγ secretion from expanded NK cells can mediate ICAM-1 upregulation and enhanced NK cell conjugate formation in brain tumors and NB, enhancing NK cell activity in adoptive immunotherapy. Thus, a better understanding of the effects of NK cell-mediated inflammation on the tumor microenvironment is essential to optimize cellular immunotherapy of cancer.

AUTHOR CONTRIBUTIONS

Conceived and designed the experiments: AA-L, VS, EK, and DL. Performed the experiments and acquired data: AA-L, VS, and ZV. Analyzed the data: AA-L, EK, and DL. Wrote the paper: AA-L and DL. All authors read and accepted the final version of the manuscript.

ACKNOWLEDGMENTS

Mass cytometry was performed by the MD Anderson Flow Cytometry Core, with assistance from Duncan H. Mak. The flow core is funded by NCI Cancer Center Support Grant P30CA16672. STR DNA fingerprinting was performed by the CCSG-funded Characterized Cell Line Core, NCI # CA016672. The authors would like to acknowledge the American Legion Auxiliary Fellowship in Cancer Research and the George M. Stancel, Ph.D. Fellowship in Biomedical Sciences for supporting AA-L.

REFERENCES

1. Armstrong GT, Kawashima T, Leisenring W, Stratton K, Stovall M, Hudson MM, et al. Aging and risk of severe, disabling, life-threatening, and fatal events in the childhood cancer survivor study. *J Clin Oncol* (2014) 32(12):1218–27. doi:10.1200/JCO.2013.51.1055
2. Cheent K, Khakoo SI. Natural killer cells: integrating diversity with function. *Immunology* (2009) 126(4):449–57. doi:10.1111/j.1365-2567.2009.03045.x
3. Ljunggren HG, Karre K. In search of the 'missing self': MHC molecules and NK cell recognition. *Immunol Today* (1990) 11(7):237–44. doi:10.1016/0167-5699(90)90097-S
4. Garcia-Lora A, Algarra I, Garrido F. MHC class I antigens, immune surveillance, and tumor immune escape. *J Cell Physiol* (2003) 195(3):346–55. doi:10.1002/jcp.10290
5. Haworth KB, Leddon JL, Chen CY, Horwitz EM, Mackall CL, Cripe TP. Going back to class I: MHC and immunotherapies for childhood cancer. *Pediatr Blood Cancer* (2015) 62(4):571–6. doi:10.1002/pbc.25359
6. Raulet DH, Guerra N. Oncogenic stress sensed by the immune system: role of natural killer cell receptors. *Nat Rev Immunol* (2009) 9(8):568–80. doi:10.1038/nri2604
7. Vivier E, Tomasello E, Baratin M, Walzer T, Ugolini S. Functions of natural killer cells. *Nat Immunol* (2008) 9(5):503–10. doi:10.1038/ni1582
8. Denman CJ, Senyukov VV, Somanchi SS, Phatarpekar PV, Kopp LM, Johnson JL, et al. Membrane-bound IL-21 promotes sustained ex vivo proliferation of human natural killer cells. *PLoS One* (2012) 7(1):e30264. doi:10.1371/journal.pone.0030264
9. de Fries RU, Golub SH. Characteristics and mechanism of IFN-gamma-induced protection of human tumor cells from lysis by lymphokine-activated killer cells. *J Immunol* (1988) 140(10):3686–93.
10. Moore M, White WJ, Potter MR. Modulation of target cell susceptibility to human natural killer cells by interferon. *Int J Cancer* (1980) 25(5):565–72. doi:10.1002/ijc.2910250504
11. Naganuma H, Kiessling R, Patarroyo M, Hansson M, Handgretinger R, Gronberg A. Increased susceptibility of IFN-gamma-treated neuroblastoma cells to lysis by lymphokine-activated killer cells: participation of ICAM-1 induction on target cells. *Int J Cancer* (1991) 47(4):527–32. doi:10.1002/ijc.2910470410
12. Wang R, Jaw JJ, Stutzman NC, Zou Z, Sun PD. Natural killer cell-produced IFN-gamma and TNF-alpha induce target cell cytolysis through up-regulation of ICAM-1. *J Leukoc Biol* (2012) 91(2):299–309. doi:10.1189/jlb.0611308
13. Houghton PJ, Morton CL, Tucker C, Payne D, Favours E, Cole C, et al. The pediatric preclinical testing program: description of models and early testing results. *Pediatr Blood Cancer* (2007) 49(7):928–40. doi:10.1002/pbc.21078
14. Somanchi SS, Senyukov VV, Denman CJ, Lee DA. Expansion, purification, and functional assessment of human peripheral blood NK cells. *J Vis Exp* (2011) (48):e2540. doi:10.3791/2540
15. Bendall SC, Simonds EF, Qiu P, Amir el AD, Krutzik PO, Finck R, et al. Single-cell mass cytometry of differential immune and drug responses across a human hematopoietic continuum. *Science* (2011) 332(6030):687–96. doi:10.1126/science.1198704
16. Kotecha N, Krutzik PO, Irish JM. Web-based analysis and publication of flow cytometry experiments. *Curr Protoc Cytom* (2010) Chapter 10:Unit10.17. doi:10.1002/0471142956.cy1017s53
17. Burshtyn DN, Shin J, Stebbins C, Long EO. Adhesion to target cells is disrupted by the killer cell inhibitory receptor. *Curr Biol* (2000) 10(13):777–80. doi:10.1016/S0960-9822(00)00568-6
18. Romee R, Rosario M, Berrien-Elliott MM, Wagner JA, Jewell BA, Schappe T, et al. Cytokine-induced memory-like natural killer cells exhibit enhanced responses against myeloid leukemia. *Sci Transl Med* (2016) 8(357):357ra123. doi:10.1126/scitranslmed.aaf2341
19. Lee SJ, Jang BC, Lee SW, Yang YI, Suh SI, Park YM, et al. Interferon regulatory factor-1 is prerequisite to the constitutive expression and IFN-gamma-induced upregulation of B7-H1 (CD274). *FEBS Lett* (2006) 580(3):755–62. doi:10.1016/j.febslet.2005.12.093
20. Dong H, Strome SE, Salomao DR, Tamura H, Hirano F, Flies DB, et al. Tumor-associated B7-H1 promotes T-cell apoptosis: a potential mechanism of immune evasion. *Nat Med* (2002) 8(8):793–800. doi:10.1038/nm0902-1039c
21. Steimle V, Siegrist CA, Mottet A, Lisowska-Grospierre B, Mach B. Regulation of MHC class II expression by interferon-gamma mediated by the transactivator gene CIITA. *Science* (1994) 265(5168):106–9. doi:10.1126/science.8016643

22. Propper DJ, Chao D, Braybrooke JP, Bahl P, Thavasu P, Balkwill F, et al. Low-dose IFN-gamma induces tumor MHC expression in metastatic malignant melanoma. *Clin Cancer Res* (2003) 9(1):84–92.
23. Munker R, Andreeff M. Induction of death (CD95/FAS), activation and adhesion (CD54) molecules on blast cells of acute myelogenous leukemias by TNF-alpha and IFN-gamma. *Cytokines Mol Ther* (1996) 2(3):147–59.
24. Li JH, Kluger MS, Madge LA, Zheng L, Bothwell AL, Pober JS. Interferon-gamma augments CD95(APO-1/Fas) and pro-caspase-8 expression and sensitizes human vascular endothelial cells to CD95-mediated apoptosis. *Am J Pathol* (2002) 161(4):1485–95. doi:10.1016/S0002-9440(10)64424-0
25. del Rio ML, Lucas CL, Buhler L, Rayat G, Rodriguez-Barbosa JI. HVEM/LIGHT/BTLA/CD160 cosignaling pathways as targets for immune regulation. *J Leukoc Biol* (2010) 87(2):223–35. doi:10.1189/jlb.0809590
26. Nattermann J, Nischalke HD, Hofmeister V, Ahlenstiel G, Zimmermann H, Leifeld L, et al. The HLA-A2 restricted T cell epitope HCV core 35-44 stabilizes HLA-E expression and inhibits cytolysis mediated by natural killer cells. *Am J Pathol* (2005) 166(2):443–53. doi:10.1016/S0002-9440(10)62267-5

2

The Smac Mimetic BV6 Improves NK Cell-Mediated Killing of Rhabdomyosarcoma Cells by Simultaneously Targeting Tumor and Effector Cells

Kyra Fischer[1,2], Sara Tognarelli[1,2], Stefanie Roesler[3,4,5], Cathinka Boedicker[3,4,5], Ralf Schubert[6,7], Alexander Steinle[8], Thomas Klingebiel[2,6], Peter Bader[1,2], Simone Fulda[3,4,5†] and Evelyn Ullrich[1,2†]*

[1] *University Hospital Frankfurt, Department for Children and Adolescents Medicine, Division of Stem Cell Transplantation and Immunology, Goethe University, Frankfurt, Germany,* [2] *LOEWE Center for Cell and Gene Therapy, Goethe University, Frankfurt, Germany,* [3] *Institute for Experimental Cancer Research in Pediatrics, Goethe University, Frankfurt, Germany,* [4] *German Cancer Consortium (DKTK), Heidelberg, Germany,* [5] *German Cancer Research Center (DKFZ), Heidelberg, Germany,* [6] *University Hospital Frankfurt, Department for Children and Adolescents Medicine, Goethe University, Frankfurt, Germany,* [7] *University Hospital Frankfurt/Main, Department for Children and Adolescents Medicine, Division of Pulmonology, Allergy and Cystic Fibrosis, Goethe University, Frankfurt, Germany,* [8] *University Hospital Frankfurt, Department for Molecular Medicine, Goethe University, Frankfurt, Germany*

Edited by:
Gianfranco Pittari, Hamad Medical Corporation, Qatar

Reviewed by:
Ralf Dressel, Universitätsmedizin Göttingen, Germany
Nicolas Dulphy, Paris Diderot University, France
David Hildeman, Cincinnati Children's Hospital, USA
Robert G. Korneluk, Children's Hospital of Eastern Ontario, Canada

***Correspondence:**
Evelyn Ullrich
evelyn.ullrich@kgu.de

†Shared senior authorship.

Rhabdomyosarcoma (RMS), the most common cancer of connective tissues in pediatrics, is often resistant to conventional therapies. One underlying mechanism of this resistance is the overexpression of Inhibitor of Apoptosis (IAP) proteins, leading to a dysfunctional cell death program within tumor cells. Smac mimetics (SM) are small molecules that can reactivate the cell death program by antagonizing IAP proteins and thereby compensating their overexpression. Here, we report that SM sensitize two RMS cell lines (RD and RH30) toward natural killer (NK) cell-mediated killing on the one hand, and increase the cytotoxic potential of NK cells on the other. The SM-induced sensitization of RH30 cells toward NK cell-mediated killing is significantly reduced through blocking tumor necrosis factor-related apoptosis-inducing ligand (TRAIL) on NK cells prior to coculture. In addition, the presence of zVAD.fmk, a pancaspase inhibitor, rescues tumor cells from the increase in killing, indicating an apoptosis-dependent cell death. On the NK cell side, the presence of SM in addition to IL-2 during the *ex vivo* expansion leads to an increase in their cytotoxic activity against RH30 cells. This effect is mainly TNFα-dependent and partially mediated by NK cell activation, which is associated

Abbreviations: (A/E)RMS, (alveolar/embryonal) rhabdomyosarcoma; CIK cells, cytokine-induced killer cells; CFSE, carboxyfluorescein succinimidyl ester; DAPI, 4′,6-diamidino-2-phenylindole; DC, dendritic cell; DLI, donor lymphocyte infusions; DNAM, DNAX accessory molecule; FasL, Fas ligand; FBS, fetal bovine serum; ICAM, intercellular adhesion molecule; IFN, interferon; KIR, killer immunoglobulin like receptor; LFA, lymphocyte function-associated antigen; MIC, MHC class I polypeptide-related sequence; MHC, major compatibility complex; NF-κB, nuclear factor "kappa-light-chain-enhancer" of activated B-cells; NK cells, natural killer cells; NKG2D, natural killer group 2, member D; PenStrep, penicillin/streptomycin; PVR, poliovirus receptor; RING, really interesting new gene; RPMI, Roswell Park Memorial Institute 1640; rt-PCR, real-time polymerase chain reaction; Smac, second mitochondria-derived activator of caspases; SM, Smac mimetic; TNFα, tumor necrosis factor α; TRAIL, tumor necrosis factor-related apoptosis-inducing ligand; ULBP, UL16-binding protein; (X/c) IAPs, (X-linked/cellular) inhibitor of apoptosis proteins.

with transcriptional upregulation of NF-κB target genes such as IκBα and RelB. Taken together, our findings implicate that SM represent a novel double-hit strategy, sensitizing tumor and activating NK cells with one single drug.

Keywords: natural killer cells, second mitochondria-derived activator of caspases mimetic, rhabdomyosarcoma, RH30 cells, RD cells, BV6

INTRODUCTION

Cancer is the second leading cause of death in children after accidents (1). The most common type of cancer is leukemia, followed by cancer of the brain and nervous system, and soft tissue sarcomas including rhabdomyosarcomas (RMS) (2). The survival of pediatric cancer has improved through progress in treatment (1, 3). However, the outcome greatly depends on the type of cancer. RMS ranks second to last in the 5-year survival rate, four out of five high-risk patients are defeated by their disease (4, 5). Especially the histological subgroup referred to as alveolar RMS (ARMS) is known for its aggressive growth, while embryonal sarcomas (ERMS) are correlated to a slightly better outcome (3). The low overall survival rate is due to resistance against established therapy regimens comprising chemo- and radiotherapy in first place or relapse after initial therapy, underlining the necessity of new treatment approaches. Here, we address two possibilities of improving anticancer therapy for ARMS with one single drug: sensitizing tumor cells on the one hand, and activating immune effector cells on the other.

The essential idea of the first approach is to overcome one of the key features of malignant cells: the resistance to apoptosis (6). This resistance is frequently caused by an imbalance between pro- and antiapoptotic proteins, resulting in a dysfunctional intrinsic cell death program. Since anticancer therapies such as chemo- and radiotherapy function by triggering cell death pathways, the intrinsic defect within the cell leads to treatment failure, associated with tumor progression and poor survival (7). Proteins known to play a role in the regulation of apoptosis are the IAP proteins and their physiological antagonists: second mitochondria-derived activator of caspases (Smac). Overexpression of IAP proteins (8–11), as well as the reduced expression of Smac (12, 13) in human malignancies have been correlated with treatment resistance and tumor growth. This has led to the development of small-molecule compounds that antagonize IAP proteins, i.e., Smac mimetics (SM). SM assume the function of the endogenous Smac by neutralizing IAP proteins, thereby rendering cancer cells more susceptible to the induction of apoptosis (14). While X-linked IAP (XIAP) proteins directly bind and inhibit caspases (15–17), cellular IAP (cIAP) proteins play an important role in regulating signaling pathways within a cell through an intrinsic ubiquitin ligase activity in their C-terminal RING domain (18). One signaling pathway known to be regulated by cIAP proteins is the nuclear factor "kappa-light-chain-enhancer" of activated B-cells (NF-κB) pathway. The NF-κB family comprises several transcription factors that regulate genes that are involved in cell survival, as well as innate and adaptive immune response (19) such as the tumor necrosis factor α (TNFα), TRAIL, interferon γ (IFN-γ), and MHC class I (20).

Research over the last years revealed the broad field of immune therapy as a promising anticancer strategy (21, 22). Especially NK cells, immune effector cells that physiologically recognize and eliminate cancer cells, are of great interest for adoptive cell therapy. NK cells are equipped with a broad range of receptors on their surface, allowing the discrimination between healthy and malignantly transformed or virus-infected cells. Once activated, NK cells have different strategies to kill their targets such as exocytosis of secretory lysosomes containing cytolytic proteins (perforin, granzymes, Fas ligand) (23, 24), induction of apoptosis in the death receptor-bearing target cell through ligation with FasL or TRAIL expressed on the NK cell surface (25, 26), or further immune activation through the secretion of cytokines such as IFN-γ and TNFα (27). Considering the current state of knowledge and understanding, NK cell infusions can be referred to as safe in autologous and allogeneic settings (28, 29). Some antitumor effects have been shown for hematological as well as several solid tumors (30–32). Current limitations of NK cell therapy are the exhaustion of cellular cytotoxicity (33) and the resistance of tumor cells to the induction of apoptosis *per se*. In this work, we investigate the potential of the small-molecule SM BV6 to influence both effector and target cells simultaneously, enhancing the susceptibility of tumor cells to NK cell-mediated killing on the one hand and increasing the cytotoxic activity of NK cells on the other.

MATERIALS AND METHODS

Cell Culture

All work with cell lines was performed under sterile conditions, using sterile media, buffer, and material. Both RMS cell lines used, RH30 as an alveolar and RD as an embryonal RMS cell line, were cultured in Roswell Park Memorial Institute 1640 GlutaMAX medium (RPMI 1640) with 10% fetal bovine serum and 1% penicillin/streptomycin (PenStrep). The cells were split approximately twice a week depending on the growth rate and stored in an incubator at 37°C, 5% CO_2, and 90% relative humidity.

This study was approved by the Ethics Committee of the Goethe University Frankfurt, Germany and carried out in accordance with the Declaration of Helsinki. All subjects gave written informed consent in accordance with the Declaration of Helsinki. NK cells were isolated out of freshly generated male donor buffy coats provided by the GRC-Blood Donor Service (DRK-Blutspendedienst) in Frankfurt, using immunomagnetic negative selection (EasySep Human NK Cell Enrichment Kit,

StemCell Technologies, Canada) according to manufacturer's instructions. The purity of the resulting cell suspension was analyzed through flow cytometry, using fluorochrome-conjugated antibodies against CD56 (clone: HCD56), CD19 (clone: HIB19), CD14 (clone: HCD14), CD16 (clone:3G8), CD3 (clone: UCHT1) (Biolegend, USA), and CD45 (clone: HI30) (Invitrogen, UK). Only cell suspensions that showed more than 85% NK cells were taken into culture and used for further experiments. The freshly isolated cells were cultured at a concentration of 2×10^6 cells/ml of the hematopoietic medium xVivo (Lonzagroup, CH), which was enriched with 5% heat-inactivated human plasma, 1% PenStrep, 100 U/ml IL-2, and the Smac mimetic BV6 if necessary. The Smac mimetic BV6, which neutralizes XIAP, cIAP1, and cIAP2 (34), was kindly provided by Genentech, Inc. (South San Francisco, CA, USA). NK cells were fed approximately every 3 days through discarding half of the medium and adding the equivalent amount of fresh medium, containing the doubled amount of additives.

Determination of Cell Viability and Proliferation

To determine at which concentrations SM become toxic to RH30 and NK cells, the cells were challenged with increasing doses of SM, harvested, and stained with 4′,6-diamidino-2-phenylindole (DAPI; BioLegend, USA) after 24 and 48 h. The cell suspensions were measured by flow cytometry using a BD FACSCanto10c™ instrument (BD Bioscience, San Diego, CA, USA) and data were analyzed using FlowJo (FlowJo LLC, Ashland, Oregon, USA), first gating on single cells, then defining the DAPI negative population as viable cells. To analyze the effect of SM on the proliferation of NK cells, the cells were isolated and taken into culture as described above with increasing doses of SM in addition to IL-2. On days 0, 3, 6, and 10, NK cells were harvested and counted using the COULTER® Ac·T diff™ Analyzer (Beckman Coulter, Germany), an automatic cell counter.

Surface Marker Profile Analysis

In search of SM-induced alterations in the expression of surface markers, unstimulated and SM-stimulated RH30, RD, and NK cells were stained with a panel of fluorochrome-conjugated antibodies and analyzed using flow cytometry. NK cells cultured with IL-2 alone, as well as cells that were additionally stimulated with 5 and 10 µM SM were harvested on day 7 of culture, which complies with the timepoint of harvest for cytotoxicity assays. First, the harvested cells were incubated with 50 µg/ml human IgG (Kiovig, Baxter, Germany) in order to saturate the Fc receptors of NK cells prior to the staining with fluorochrome-conjugated antibodies. This procedure ensured that the antibodies added afterwards marked their specific antigen, instead of being captured by their Fc fragment through the Fc receptors, which would cause a false-positive rate of the particular antigen. Then, a complex staining pattern was established including following antibodies: CD3/19/14-V450, CD45-BV510 (clone: HI30), CXCR4-PE-Cy7 (clone: 12G5), DNAX accessory molecule (DNAM)-1-FITC (clone: DX11), NKp44-PE (clone: p44-8.1), CD25-PE (clone: 2A3), CD62L-APC (clone: DREG-56), CD69-BV605 (clone: FN50), and CD107a-APC-H7 (clone: H4A3) from BD Biosciences, San Diego, CA, USA; CD16-PE-Cy7, CD16-PE, and CD253(TRAIL)-PE (clone: RIK-2) from Biolegend, USA; NKp30-AlexaFluor488 (clone: 210845), and CCR7-FITC (clone: 150505) from R&D Systems; KIR2D-FITC (clone: NKVFS1), CD158e/k-PE (clone: 5.133), and NKp46-APC (clone: 9E2) from Miltenyi, Germany; NKG2A-APC (clone: 1D11) from Beckman Coulter, Germany; CX3CR1-PerCPeFluor710 (clone: 2A91) from eBiosciences, USA, and DAPI (Biolegend, USA), serving as a life–dead stain.

The data were analyzed using FlowJo (FlowJo LLC, Ashland, Oregon, USA), analyzing the antigen of interest in the population defined as NK cells (DAPI$^-$/CD3$^-$/CD19$^-$/CD45$^+$/CD56$^+$).

For the analysis of the surface marker profile of RH30 and RD cells, cells were pretreated according to the conditions used as targets in cytotoxicity assays: 0, 5, or 10 µM SM for 24 h. Afterwards, these cells were harvested and stained with a staining panel comprising the following antibodies: MICA (clone: AMO1), MICB (clone: BMO2), UL16-binding protein (ULBP)-1 (clone: AUMO3), -2 (clone: BUMO1), -3 (clone: CUMO3), and -4 (clone: 3B6) according to Ref. (35), B7-H6 (kind gift of Prof. Adelheid Cerwenka DKFZ, Germany (36)), followed by the secondary antibody GAM-APC from Jackson Immunoresearch, USA; intercellular adhesion molecule (ICAM)-1-PE (clone: HA58), ICAM-2-FITC (clone: CBR-IC2/2), ICAM-3-APC (clone: CBR-IC3/1), CD262 (TRAIL-R)-PE, and Nectin-2-PE (clone: TX31) from Biolegend, USA; poliovirus receptor (PVR)-FITC (clone: 300907) from R&D Systems, as well as DAPI (Biolegend, USA). The established data were also analyzed using FlowJo (FlowJo LLC, Ashland, Oregon, USA).

Cytotoxicity Assays

All cytotoxicity assays were performed using a flow cytometry-based method and an effector to target (E: T) ratio of 10:1. Herein, the target cells were stained with carboxyfluorescein succinimidyl ester (CFSE; Life Technologies, USA) for 5 minutes at room temperature, then washed three times prior to coculture with unstained effector cells. Coculture was completed in a round-bottom 96-well plate for four (pretreated tumor cells) or 16 h (pretreated NK cells). The RMS target cells were pretreated with 5 or 10 µM SM for 3 or 24 h, harvested, and the cell suspension was adjusted to a concentration of 0.25×10^6/ml. The NK cells used as effectors were cultured with 100 U/ml IL-2 alone or with 5 or 10 µM SM in addition to IL-2, harvested on day 6 or 7 of culture and the cell suspension was adjusted to a concentration of 2.5×10^6/ml. For the analysis of NK cell mediated cytotoxicity, 100 µl of the target and 100 µl of the effector cell suspension were pipetted into one well. For each combination, three wells were filled, representing technical replicates. In addition, control wells for all tumor conditions used in the experiment were added, containing target cells only in order to determine the spontaneous lysis. In the final evaluation of the experiment, the specific lysis was calculated as the percentage of dead cells in the wells containing target and effector cells minus the specific lysis of the respective tumor cell condition. Through this calculation of the specific lysis, the percentage of killed tumor cells was attributed completely to the NK cells. In addition, three wells were added containing tumor cells and ethanol, serving as a positive control for the live-dead stain with DAPI. Once the coculture time was over, 100 µl supernatant was taken from each well for further analyses (cytometric bead assay) and immediately frozen at −80°C. Then, each well was harvested using 50 µl Trypsin (Gibco Invitrogen, Germany), resuspended in 400 µl of a 1:6,000 DAPI dilution, measured at the flow cytometer, and analyzed with FlowJo.

The cytotoxicity assays as described above were repeated with different additives, with the aim to elucidate mechanisms behind the SM-induced increase in NK cell-mediated killing. The role of TNFα was analyzed through the addition of 0.25 mg/ml Enbrel (Pfizer, Germany); the role of caspases was analyzed through the addition of zVAD.fmk (Sigma, Germany) at a concentration of 40 µM; and the role of TRAIL was analyzed through blocking TRAIL on NK cells with an antibody from BD Biosciences, USA (cat. 550515) at a concentration of 50 µg/ml for 2 hours prior to coculture. The efficiency of the block was tested by staining the blocked cells with an antibody of the same clone.

Determination of mRNA Levels of TNFα, TRAIL, RelB, IκBα, TRAIL-R1 (DR4), and TRAIL-R2 (DR5)

TRAIL mRNA levels were determined by quantitative real-time PCR analysis. Total RNA extraction and cDNA synthesis were performed as previously described (37). TRAIL and TNFα mRNA levels were assessed by Taqman Gene Expression Assay purchased from Life Technologies (TRAIL: Hs00921974_m1; TNFα: Hs01113624_g1) and the levels of RelB, IκBα, TRAIL-R1 (DR4), TRAIL-R2 (DR5), and 28S rRNA by SYBR®Green qPCR assay from Applied Biosystems (Darmstadt, Germany) according to the manufacturer's instructions using the 7900HT fast real-time PCR system from Applied Biosystems (Darmstadt, Germany); RelB forward primer: GCTCTACTTGCTCTGCAGACA; reverse primer: GGCCTGGGAGAAGTCAGC; IκBα forward primer: GTCAAGGAGCTGCAGGAGAT; reverse primer: ATGGCCAAGTGCAGGAAC; DR4 forward primer: GGGTCCACAAGACCTTCAAGT; reverse primer: TGCAGCTGAGCTAGGTACGA; DR5 forward primer: AGACCCTTGTGCTCGTTGTC; reverse primer: TTGTTGGGTGATCAGAGCAG; 28S rRNA forward primer: TTGAAAATCCGGGGGAGAG; reverse primer: ACATTGTTCCAACATGCCAG.

The relative expression of the target gene transcript and reference gene transcript was calculated as $\Delta\Delta C_t$. 28S rRNA was used as reference gene.

Cytometric Bead Array

In order to screen supernatants of NK and RMS cells in culture as well as after coculture in cytotoxicity assays for changes in secreted cytokines through treatment with SM, supernatants were frozen at −80°C. Cytokine concentrations in these culture supernatants were determined by flow cytometry using the BD™ CBA Flex Set System (BD Bioscience, San Diego, CA, USA). Tests were performed according to the manufacturer's instructions using a mixture of PE-conjugated antibodies against the cytokines listed above. Data were acquired with the BD FACSVerse™ Bioanalyzer and analysis was carried out by using the FCAP Array™ software (v3.0.1).

Statistical Analysis

Prior to analyses, the Shapiro–Wilk normality test was performed in order to prove normal distribution of the analyzed data. The statistical analyses of cytotoxicity assays as well as SM-induced changes in NK cell surface markers and viability experiments were performed using the repeated measures one-way ANOVA with Dunnett's multiple comparison in GraphPad Prism 6 (Inc. LA JOLLA, CALIFORNIA, USA). In case of the blocking experiments, which require the comparison of each group with every other group rather than the comparison to a control column, Turkey's multiple comparisons test was added to the repeated measures one-way ANOVA. Changes in mRNA levels and amounts of secreted cytokines upon stimulation with SM were analyzed through paired *t*-tests. All significant differences with a *p*-value less than 0.1 are marked in the results.

RESULTS

NK Cells Tolerate Higher SM Concentrations than RMS Cells

When testing a substance for its therapeutic potential, such as SM for the sensitization of tumor and the stimulation of NK cells, a first important piece of information is the optimal dose and the concentration of SM that becomes toxic to the respective cells. In order to investigate this, NK and RMS cells were incubated with increasing concentrations of SM and cell viability was analyzed after 24 and 48 h using flow cytometry. RH30 cells already reacted to 10 µM SM with a significantly decreased viability after 24 h (IC50 24 h: 15 µM), RD cells showed similar results (data not shown). NK cells, however, tolerated doses of up to 30 µM (IC50 24 h: 68 µM) (**Figures 1A,B**). Considering the objective of using SM to optimize NK cell therapy, the impact of SM on the proliferative behavior of NK cells is also of interest. Up to a concentration of 2.5 µM, SM did not affect the proliferation of NK cells (**Figure 1C**). Increasing the dose to 5 µM SM led to a balance between cell death and proliferation, while the presence of 10 µM resulted in a significantly decreased cell number (**Figure 1C**).

Pretreatment of either RMS or NK Cells with SM Increases NK Cell-Mediated Killing of RMS Cells

Following the hypothesis, that SM can restore the defective apoptotic machinery in tumor cells and therefore sensitize them toward NK cell-mediated killing, cytotoxicity assays were performed using targets that were pretreated with SM prior to coculture. Indeed, pretreating the ARMS cell line RH30 cells for 24 h with subtoxic concentrations of SM that had no or little effects on cell viability significantly increased the percentage of lysed cells from 24 to 54% regarding pretreatment with 10 µM SM (**Figure 2A**). Also, embryonal RD cells were rendered more susceptible to NK cell-mediated killing by SM as

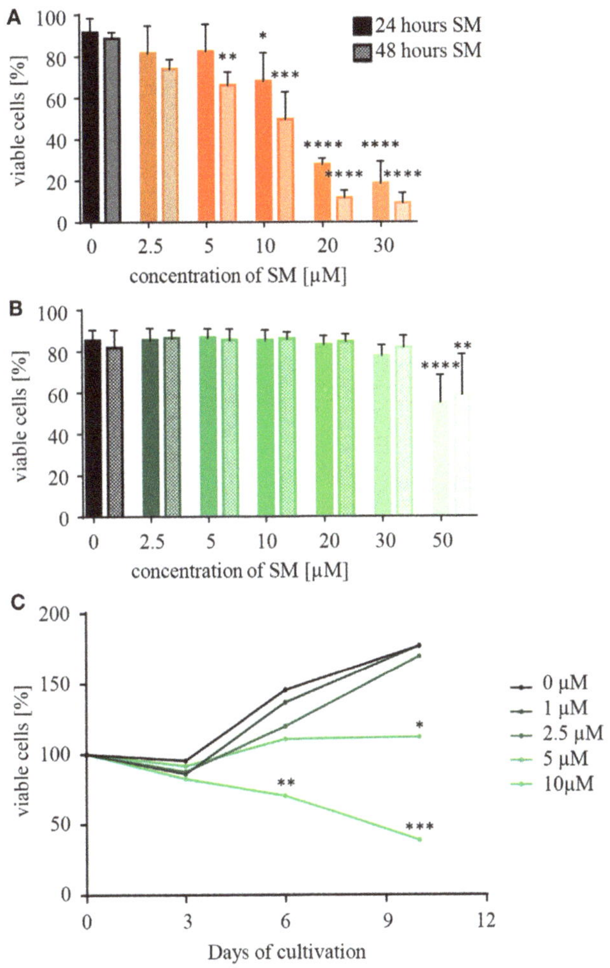

FIGURE 1 | Natural killer (NK) cells tolerate higher concentrations of Smac mimetics (SM) than RH30 cells. (A) RH30 and **(B)** NK cells were challenged with increasing doses of SM (0, 2.5, 5, 10, 20, 30, and 50 μM). The percentage of viable cells was determined with a life–dead stain (DAPI) in flow cytometry after 24 and 48 h. **(C)** NK cells were isolated and cultured in medium containing increasing doses of SM in addition to IL-2. On day 3, 6, and 9, they were fed with fresh medium including both additives and their proliferative behavior was tracked through cell counts. The determined cell numbers were considered in relation to the absolute number of cells on day 0. Data concerning the impact of SM on RMS cells are always depicted in orange (RH30) or red (RD), and NK cells in green color. Three individual experiments were performed for each setting ($n = 3$). Statistical analysis with repeated measures one-way ANOVA + Dunnett's multiple comparison, $^{*}p < 0.05$, $^{**}p < 0.01$, $^{***}p < 0.001$, $^{****}p < 0.0001$, each referred to 0 μM.

pretreatment with 10 μM SM increased the specific lysis from 24 to 45% (Figure S1A in Supplementary Data). To ensure that the effectors, IL-2-stimulated NK cells, were responsible for the killing, specific lysis was calculated as the difference between the percentage of all dead cells (absolute lysis) and the percentage of RMS cells that simply died due to pretreatment in absence of effectors during the coculture period (spontaneous lysis). Furthermore, we observed that a 3-h pretreatment is insufficient to sensitize RH30 cells toward NK cell-mediated killing (Figure S1B in Supplementary Data), similar observations were made regarding RD cells (data not shown).

In a next step, we investigated whether SM alter the cytotoxic potential of NK cells. For this purpose, cytotoxicity assays were performed using NK cells which were stimulated with SM in addition to IL-2 during their expansion period as effectors, and untreated RH30 cells as targets. The presence of SM during the 7 days of culture significantly increased the NK cell-mediated killing of targets compared to the killing through NK cells stimulated with IL-2 alone (**Figure 2B**). On average, NK cells that were stimulated with 10 μM SM in addition to IL-2 killed 67% of the untreated targets, while IL-2-stimulated NK cells only killed 43% (**Figure 2B**). This SM-stimulated increase in the cytotoxic potential of NK cells was seen in every tested donor, although the extent of the effect differed. The most impressive increase was observed during the experiment with donor four, where pretreated cells killed more than twice as much tumor cells as the untreated counterpart. We further observed that the increase in the cytotoxic potential is limited to the late phase killing. Evaluating the same cytotoxicity assays after a coculture time of only 4 instead of 16 h did not show comparable results (Figure S1C in Supplementary Data). Interestingly, the positive effect of pretreating NK cells was restricted to the more aggressive histological subtype of RMS tested: the ARMS cell line RH30. Testing the pretreated NK cells against the untreated ERMS cell line RD did not show a comparable result (Figure S1D in Supplementary Data).

TRAIL Signaling Contributes to SM-Induced Sensitization of RMS Cells to NK Cell Killing

Next, we aimed at unveiling the mechanisms that are responsible for the SM-conferred sensitization of RMS cells toward NK cell-mediated killing. Motivated by the hypothesis, that SM induce changes within the surface marker profile of RMS cells, untreated as well as pretreated cells were stained with fluorochrome-conjugated antibodies against a broad range of ligands for NK cell receptors: ULBP-1, -2, -3, -4, MHC class I polypeptide-related sequence (MIC)A, and -B as ligands for natural killer group2, member D (NKG2D); ICAM-1, -2, and -3 as ligands for lymphocyte function-associated antigen-1; Nectin-2 and PVR as ligands for DNAM-1, as well as B7-H6 as a ligand for NKp30. The markers were screened for their surface expression and changes in their mean fluorescence intensity normalized to the untreated control. However, none of the tested surface markers showed a significant change in their expression level upon treatment with SM (Figure S2 in Supplementary Data). The analyses were performed with RH30 (Figure S2A in Supplementary Data) as well as RD cells (Figure S2B in Supplementary Data), leading to similar results.

Since SM have been described to stimulate TNFα production, which contributes to SM-induced cell death in an autocrine/paracrine manner (34), we tested whether the increased susceptibility of RH30 cells is dependent on TNFα by adding Enbrel, the inhibiting soluble TNF receptor (TNFR2) fused with an IgG1 Fc part, to the coculture. However, the addition of Enbrel did not protect RH30 cells from NK cell-induced killing (**Figure 3A**),

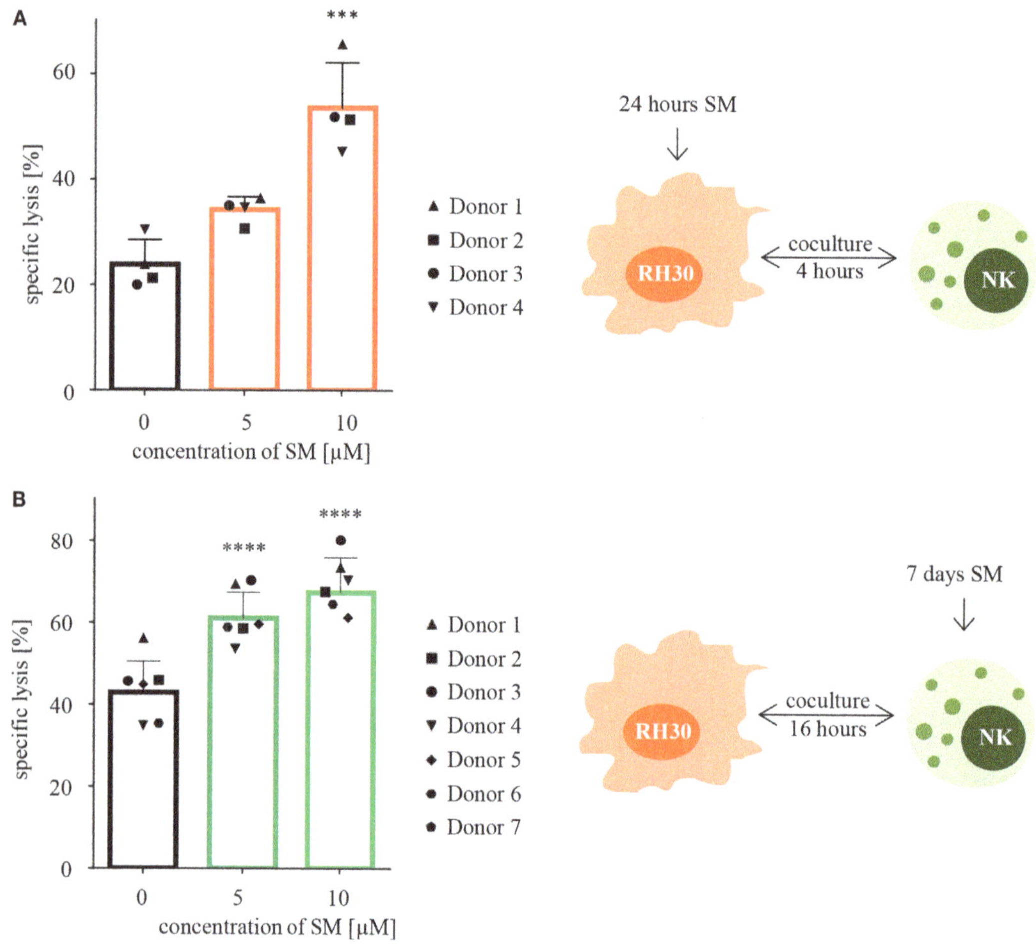

FIGURE 2 | Pretreatment with SM effects both RH30 and NK cells. (A) SM sensitize RH30 cells toward NK cell mediated killing. RH30 cells were pretreated with 0, 5, or 10 μM SM for 24 h prior to being used as targets for IL-2-stimulated NK cells on day six of culture (as depicted in the scheme on the right-hand side). The cytotoxicity assay was repeated with NK cells of four different donors ($n = 4$). E:T ratio = 10:1, coculture time 4 h. **(B)** SM increase the cytotoxic potential of NK cells. NK cells were cultured with different doses of SM (0, 5, and 10 μM) in addition to IL-2 for 7 days (as depicted in the scheme on the right-hand side). On day seven, cytotoxicity assays were performed using untreated RH30 cells as targets. The experiment was repeated with NK cells from seven different donors ($n = 7$). E:T ratio = 10:1, coculture time 16 h. Statistical analysis through repeated measures one-way ANOVA + Dunnett's multiple comparison, $^{***}p < 0.001$, $^{****}p < 0.0001$, each referred to 0 μM.

while Enbrel prevented SM-induced cell death in MDA-MB231 cells that were used as a positive control, since they have been reported to die in a TNFα-dependent manner upon treatment with SM (38) (Figure S6A in Supplementary Data). Also, the amount of secreted cytokines during the described cytotoxicity assays was not altered by pretreatment of RMS with SM (Figure S3 in Supplementary Data).

Further, we recently identified TRAIL receptor ligand signaling as another critical mediator of SM-induced cell death (38). We therefore asked whether TRAIL is required for the SM-mediated sensitization of RH30 cells to NK cell-mediated killing. To address this question, we used a TRAIL-blocking antibody to neutralize TRAIL on NK cells prior to coculture (Figure S6B in Supplementary Data). The presence of the TRAIL-blocking antibody significantly reduced the SM-conferred sensitization of RH30 cells to NK cell-mediated killing (**Figure 3B**). While neither surface expression of TRAIL-R1 or TRAIL-R2 on RH30 cells, nor TRAIL expression on NK cells was altered by treatment with SM (Figure S4 in Supplementary Data), SM significantly increased mRNA levels of the NF-κB target genes TRAIL-R1, TRAIL-R2 and TRAIL in RH30 cells (**Figure 4**). SM-stimulated upregulation of TRAIL-R1, TRAIL-R2, and TRAIL mRNA expression was confirmed in RD cells (Figure S5 in Supplementary Data).

Furthermore, SM have been reported to promote activation of caspases by neutralizing XIAP proteins (34), wherefore we tested whether caspases are required for the SM-mediated sensitization of RH30 cells. To this extent, we added the pan-caspase inhibitor zVAD.fmk to the medium during cytotoxicity assays after functionality of this caspase-blocking inhibitor on RH30 cells has been proven in standardized cell death assays (Figure S6C in Supplementary Data). Importantly, the presence of zVAD.fmk significantly rescued RH30 cells from NK cell-induced killing (**Figure 3C**).

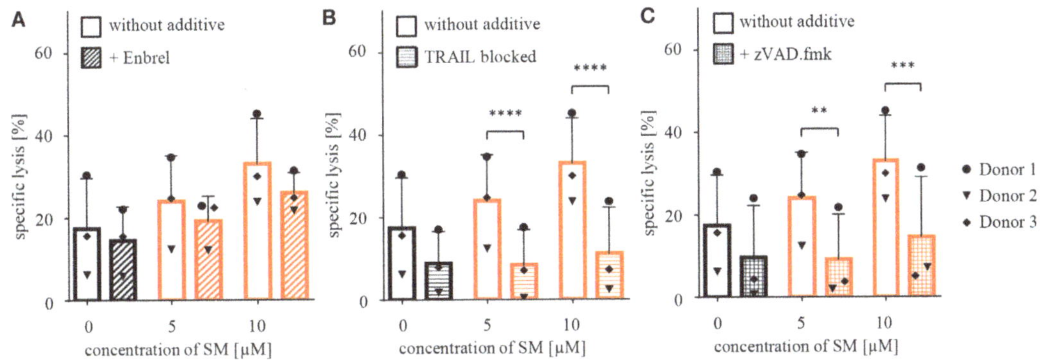

FIGURE 3 | Defining underlying mechanisms behind the SM-induced sensitization of RH30 cells. Cytotoxicity assays with pretreated RH30 cells as targets for IL-2-stimulated natural killer (NK) cells as shown in **Figure 2A** were performed with the additional presence of **(A)** Enbrel, a TNFα blocker (concentration 250 µg/ml) or **(C)** zVAD.fmk (concentration 40 µM) during coculture. In panel **(B)**, TRAIL on NK cells was blocked through incubation with an antibody (concentration 50 µg/ml) prior to coculture. Each experiment was repeated with NK cells from three different donors ($n = 3$), E:T ratio = 10:1, coculture 4 h. Statistical analysis with repeated measures one-way ANOVA with Turkey's multiple comparison, $^{**}p < 0.01$, $^{***}p < 0.001$, $^{****}p < 0.0001$.

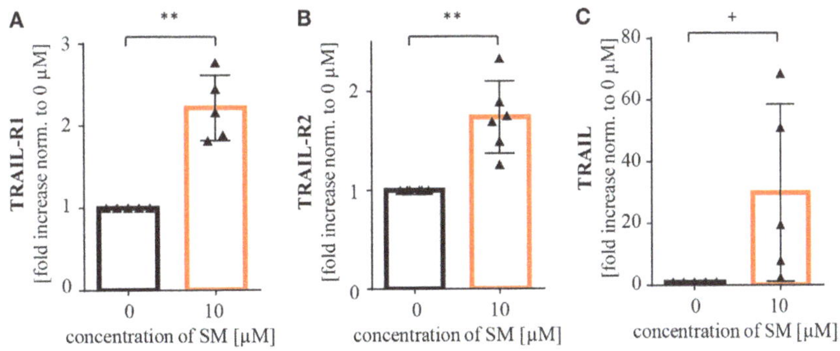

FIGURE 4 | SM induces upregulation of NF-κB target genes in RH30 cells. Tumor cells pretreated with 10 µM SM for 24 h as well as untreated cells were analyzed for mRNA levels of the NF-κB target genes **(A)** TRAIL-R1, **(B)** TRAIL-R2, and **(C)** TRAIL through real time (rt)-PCR. The experiment was repeated with RH30 cells from four to six different cell passages, depicted is the fold increase normalized to 0 µM. Statistical analysis through paired t-test, $^{**}p < 0.01$, $^{+}p < 0.1$.

TNFα Signaling Contributes to SM-Induced Increase in the Cytotoxic Potential of NK Cells against RH30 Cells

To elucidate which mechanisms are responsible for the SM-induced increase in the cytotoxic potential of NK cells, we first screened a broad range of NK cell surface molecules for changes upon treatment with SM: activating NK cell receptors (NKp30, NKp44, NKp46, NKG2D, DNAM-1, and CD16) (Figure S7A in Supplementary Data), chemokine receptors (CCR7, CX3CR1, and CXCR4) (Figure S7B in Supplementary Data), inhibitory NK cell receptors (NKG2A and KIR2D) (Figure S7C in Supplementary Data), and activation markers (CD25, CD107a, CD69, CD62L) (Figure S7D in Supplementary Data). Generally speaking, we observed a donor-dependent influence of SM on the expression of NK cell surface proteins. Only two of the tested molecules showed a consistent, donor-independent change: CX3CR1, a chemokine receptor, was upregulated, while NKp46 was downregulated upon treatment with SM.

In further search of underlying mechanisms, we investigated the role of TNFα during cytotoxicity assays with SM-pretreated NK cells against RH30 cells by adding Enbrel to the medium during coculture. In fact, the SM-induced increase in NK cell-mediated killing was reduced by the presence of Enbrel (**Figure 5A**). Although this protection by Enbrel was partial, it was consistently found in each of the four tested donors (**Figure 5A**). Correspondingly, we detected a significant increase in secreted TNFα in supernatants of SM-pretreated compared to untreated NK cells at the end of the coculture period of the cytotoxicity assays. Besides TNFα, the levels of IFN-γ were significantly increased in these supernatants (**Figure 5B**), while IFN-α and FasL levels remained unchanged (data not shown). Interestingly, we could only observe this increase in TNFα and IFN-γ levels when the untreated and SM-pretreated NK cells were challenged with their target cells and not upon SM-induction alone (Figure S8 in Supplementary Data).

There are probably further mechanisms contributing to the increased level of activation of NK cells. In line with our findings

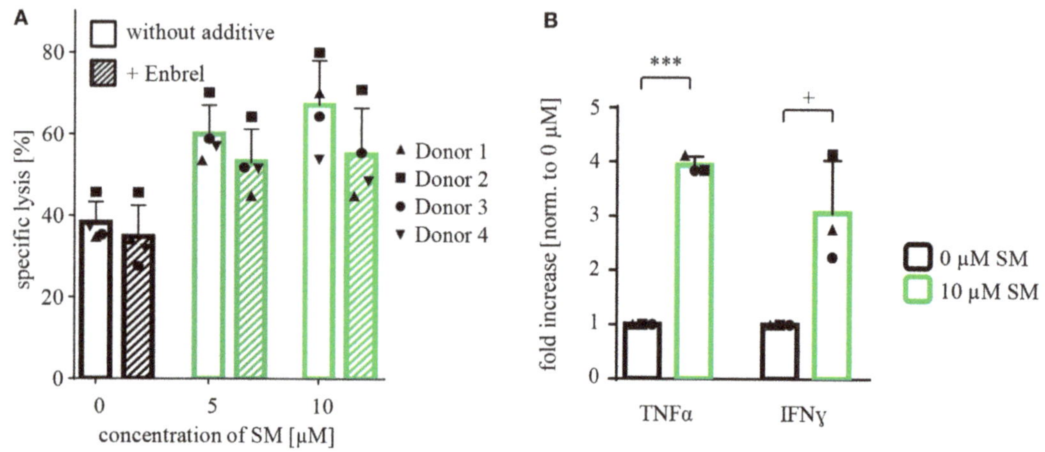

FIGURE 5 | Defining mechanisms behind the SM-mediated stimulation of NK cells. (A) Cytotoxicity assays with pretreated NK cells (SM in addition to IL-2 for 7 days) against untreated RH30 cells (as shown in **Figure 2B**) were repeated under the presence of Enbrel, a TNFα blocker (concentration 250 μg/ml). The experiment was performed with NK cells of four different donors (n = 4), E:T ratio = 10:1, coculture time 16 h. Statistical analysis through repeated measures one-way ANOVA were not significant. **(B)** At the end of coculture of the experiments described under **(A)**, supernatants were taken away and analyzed using cytometric bead array. The graph shows the fold increase in the amount of TNFα and IFN-γ normalized to 0 μM. The experiment was repeated from three different cytotoxicity assays ($n = 3$). Statistical analysis through paired t-test, $^{+}p < 0.1$, $^{***}p < 0.001$.

in tumor cells, we detected an increase in RelB and IκBα mRNA levels, also indicating an activation of NF-κB signaling in SM-treated NK cells. Also, the NF-κB target gene TRAIL was found to be upregulated, while mRNA levels of TNFα were lower than in the untreated control (**Figure 6**).

DISCUSSION

Since immune effector cells including NK cells mainly function by inducing apoptosis in their targets, the efficacy of immunotherapy critically depends on intact apoptosis signaling pathways within the targeted cancer cells. Here, we report that SM, which antagonize IAP proteins, can prime RMS cells toward NK cell-mediated cytotoxicity (as shown for RD and RH30 cells), and increase the cytotoxic potential of NK cells toward RH30 cells. Of note, the simultaneous targeting of tumor and immune cells with one single drug was restricted to the more aggressive alveolar RH30 cell line.

Importantly, we found that TRAIL signaling contributes to SM-induced sensitization of RMS cells toward NK cell-mediated cytotoxicity, as the addition of a neutralizing TRAIL antibody on NK cells prior to coculture with tumor cells significantly reduced tumor lysis. While we confirmed the overall expression of TRAIL on IL-2-stimulated NK cells, as well as the upregulation of the NF-κB target gene TRAIL in RMS cells upon treatment with SM, we did not detect any changes in neither TRAIL-R1 or TRAIL-R2 expression on RMS cells, nor TRAIL expression on NK cells upon exposure to SM. Overall, our findings are consistent with our previous reports showing that SM or Smac peptides can prime cancer cells toward TRAIL *in vitro* and *in vivo* (39, 40). In addition, we recently identified TRAIL receptor ligand signaling as one critical mediator of SM-induced cell death (38). Also, cooperative TRAIL production has been shown to mediate SM/IFNα-induced cell death in TNFα-resistant solid cancer cells (41). By comparison, TRAIL signaling turned out to be dispensable for SM/glucocorticoid-induced cell death in leukemia cells (42) or in SM/temozolomide-triggered cell death in glioblastoma cells (43). This indicates that the TRAIL system contributes to SM-induced cell death in a context-dependent manner.

Interestingly, we found a differential role of TNFα in SM-imposed sensitization of RMS cells to NK cell killing, depending on whether RMS or NK cells were pretreated with SM. TNFα contributes, at least to some extent, to the enhanced cytotoxicity when NK cells were pretreated with SM, since the addition of TNFα-blocking Enbrel to the medium during the killing assay significantly, although partially, decreased the NK cell-mediated killing of RH30 cells. In addition, SM-pretreated NK cells produce significantly higher amounts of TNFα and IFN-γ than their untreated counterpart, when cocultured with their tumor target cells. However, beside the possible relevance of TNFα, there are likely additional mechanisms contributing to the SM-induced activation of NK cells, for example, activation of NF-κB signaling through SM.

On the contrary, TNFα was found to be dispensable for the enhanced cytotoxicity of NK cells when RMS cells were pretreated with SM, since the addition of Enbrel to cytotoxicity assays failed to rescue RMS cells from NK cell-mediated killing. This is underlined by the fact that the supernatants of pretreated RMS cells did not contain more TNFα than the untreated cells. These findings are consistent with previous studies on a context-dependent impact of TNFα as a mediator of SM-induced cytotoxicity. On the one hand, there are several studies showing that an autocrine/paracrine TNFα loop plays a critical role in SM-induced cell death (34, 44–48). On the other hand, blockage of TNFα signaling has also been reported to fail in providing protection against SM in other settings (38, 42, 43). We previously demonstrated that

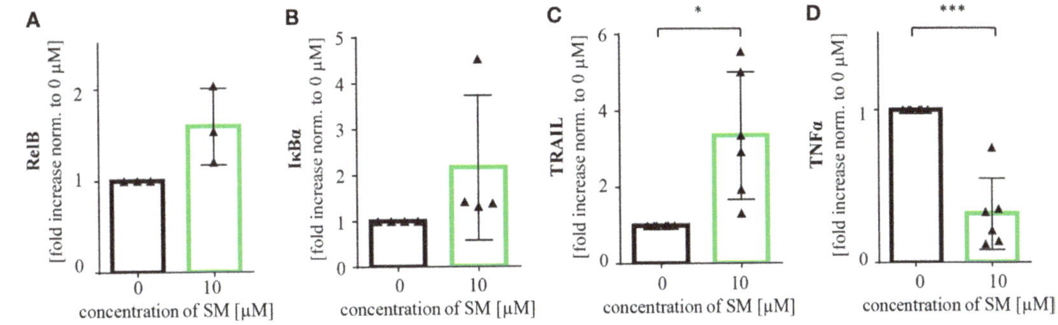

FIGURE 6 | SM induces upregulation of NF-κB target genes in natural killer (NK) cells. NK cells cultured for 7 days with IL-2 alone and in addition with SM were analyzed for mRNA levels of the NF-κB target genes **(A)** RelB, **(B)** IκBα, **(C)** TRAIL, and **(D)** TNFα through rt-PCR. The experiment was repeated with NK cells from three to six different donors, depicted is the fold increase normalized to 0 µM. Statistical analysis through paired *t*-test, $^{*}p < 0.05$, $^{***}p < 0.001$.

cell type-dependent sensitivity to TNFα can determine whether a cell line depends on TNFα signaling to mediate BV6-induced cell death (41). In TNFα-resistant types of cancer, we showed that TRAIL as another death receptor ligand can mediate SM-induced cell death instead of TNFα (41). Also, differential upregulation of TNFα upon SM might explain TNFα dependency in some but not other instances.

While the monovalent SM LCL161 has previously been reported to upregulate ligands for the activating NK cell receptor NKG2D such as MICA and MICB (49), we did not detect changes in NK cell receptor ligands on RMS cells upon treatment with the SM BV6, which might be due to different tumor types or different SM.

Moreover, caspase-dependent as well as caspase-independent effector pathways may be involved in the SM-conferred increased sensitivity of RH30 cells toward NK cell killing. Our finding that the presence of zVAD.fmk, a pancaspase inhibitor, rescues tumor cells from the increase in killing indicates an apoptosis-dependent cell death in line with previous reports that zVAD. fmk rescues tumor cells from increased cytotoxicity of NK or cytokine-induced killer cells (50, 51).

It is also interesting to note that a 24-h pretreatment with BV6 was necessary to adequately sensitize RMS cells to NK cell cytotoxicity, while a pretreatment of only three hours turned out to be insufficient. By comparison, we previously reported that a 4-h pretreatment with BV6 primed RMS cells for CIK cell-mediated killing (50). One possible explanation for the different requirement of preincubation time is a difference in the cytotoxicity of NK versus CIK cells. Alternatively, these findings may indicate that neither the depletion of cIAPs, nor the direct release of caspases from XIAP proteins, which has been reported to occur within minutes or hours upon exposure to BV6 (34), are responsible for the BV6-conferred sensitization of RMS cells to NK cell-mediated killing. Rather, induction and subsequent expression of proteins or cytokines, for example, as the result of SM-stimulated engagement of alternative NF-κB signaling upon cIAP1/2 depletion, might be necessary which requires some time. Consistently, we found upregulation of several NF-κB target genes, including proapoptotic genes such as TRAIL-R1, TRAIL-R2, and TRAIL, in SM-treated RMS cells.

On the NK cell side, the expression of killer immunoglobulin like receptors (KIRs) was slightly decreased as detected by an antibody recognizing the common extracellular Ig-like domains of the KIR2D receptor family—a family whose greater part belongs to the inhibitory receptors. Therefore, the decrease observed might minimize the inhibitory signals within the NK cell and lower the threshold for NK cell activation. This in addition to the increase of CX3CR1 might contribute to the activating effect, even though it is difficult to envisage one of these molecules operating as a key player. In contrast, NKp46, another activating receptor, shows a clear tendency to be downregulated upon exposure to SM, possibly lowering the state of activation in NK cells. All in all, as the increase of the cytotoxic potential through stimulation with SM observed during the cytotoxicity assays was clearly donor-independent, we conclude that the activating effect of SM on NK cell function is not absolutely attributable to changes in the expression of surface molecules.

Consistent with our findings demonstrating that SM at non-toxic concentrations enhance NK cell cytotoxicity, there is increasing evidence showing that SM can potentiate cancer immunotherapy not only by promoting apoptosis of cancer cells but also by modulating immune cell functions without inducing cell death in the majority of immune cells. For example, SM have been shown to augment human and mouse T-cell responses to physiologically relevant stimuli via activation of alternative NF-κB signaling (52). In addition, SM were described to increase *in vitro* expansion of antigen-specific naive and memory T cells to enhance T-lymphocyte function (53), to trigger phenotypic maturation of monocyte-derived dendritic cells (DCs) (53), and to stimulate maturation of immature DCs (54).

Taken together, SM represent an interesting strategy in optimizing NK cell therapy for the treatment of RMS by sensitizing the tumor cells to NK cell-mediated cell death on the one hand, and by directly activating NK cells on the other (**Figure 7**).

AUTHOR CONTRIBUTIONS

KF, ST, CB, SR, and RS performed experiments and analyzed data. KF, ST, SR, AS, PB, and TK discussed data. SF and EU designed

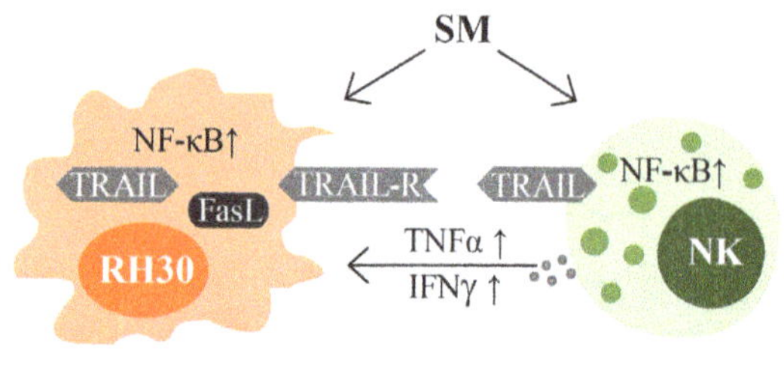

FIGURE 7 | Impacts of SM on RMS and NK cells. Schematic picture illustrating the mechanisms of interaction between NK cells and their targets that are influenced by SM.

the project. KF, SF, and EU wrote the manuscript with support of all other co-authors. All the authors agreed to be accountable for the content of the work.

ACKNOWLEDGMENTS

The authors thank Domagoj Vucic (Genentech Inc., South San Francisco, CA, USA) for kindly providing BV6, Halvard Bönig from the BSD Frankfurt, Germany, for providing healthy donor blood samples, Adelheid Cerwenka from the DKFZ, Germany, for providing antibodies, Katja Stein and Christina Born for technical support, as well as Christina Hugenberg and Sandra Kunz for excellent secretarial assistance.

REFERENCES

1. Jemal A, Siegel R, Ward E, Hao Y, Xu J, Murray T, et al. Cancer statistics, 2008. *CA Cancer J Clin* (2008) 58(2):71–96. doi:10.3322/CA.2007.0010
2. Sun X, Guo W, Shen JK, Mankin HJ, Hornicek FJ, Duan Z. Rhabdomyosarcoma: advances in molecular and cellular biology. *Sarcoma* (2015) 2015:232010. doi:10.1155/2015/232010
3. Fletcher CDM, Bridge JA, Hogendoorn P, Mertens F. *WHO Classification of Tumours of Soft Tissue and Bone*. 4th ed. (Vol. 5). Paris, France: IARC Press (2013).
4. Hawkins DS, Spunt SL, Skapek SX, COG Soft Tissue Sarcoma Committee. Children's Oncology Group's 2013 blueprint for research: soft tissue sarcomas. *Pediatr Blood Cancer* (2013) 60(6):1001–8. doi:10.1002/pbc.24435
5. Hettmer S, Li Z, Billin AN, Barr FG, Cornelison DD, Ehrlich AR, et al. Rhabdomyosarcoma: current challenges and their implications for developing therapies. *Cold Spring Harb Perspect Med* (2014) 4(11):a025650. doi:10.1101/cshperspect.a025650
6. Hanahan D, Weinberg RA. Hallmarks of cancer: the next generation. *Cell* (2011) 144(5):646–74. doi:10.1016/j.cell.2011.02.013
7. Fulda S. Tumor resistance to apoptosis. *Int J Cancer* (2009) 124(3):511–5. doi:10.1002/ijc.24064
8. Che X, Yang D, Zong H, Wang J, Li X, Chen F, et al. Nuclear cIAP1 overexpression is a tumor stage- and grade-independent predictor of poor prognosis in human bladder cancer patients. *Urol Oncol* (2012) 30(4):450–6. doi:10.1016/j.urolonc.2010.12.016
9. Ponnelle T, Chapusot C, Martin L, Bonithon-Kopp C, Bouvier AM, Plenchette S, et al. Subcellular expression of c-IAP1 and c-IAP2 in colorectal cancers: relationships with clinicopathological features and prognosis. *Pathol Res Pract* (2003) 199(11):723–31. doi:10.1078/0344-0338-00488
10. Qi S, Mogi S, Tsuda H, Tanaka Y, Kozaki K, Imoto I, et al. Expression of cIAP-1 correlates with nodal metastasis in squamous cell carcinoma of the tongue. *Int J Oral Maxillofac Surg* (2008) 37(11):1047–53. doi:10.1016/j.ijom.2008.06.004
11. Tanimoto T, Tsuda H, Imazeki N, Ohno Y, Imoto I, Inazawa J, et al. Nuclear expression of cIAP-1, an apoptosis inhibiting protein, predicts lymph node metastasis and poor patient prognosis in head and neck squamous cell carcinomas. *Cancer Lett* (2005) 224(1):141–51. doi:10.1016/j.canlet.2004.11.049
12. Endo K, Kohnoe S, Watanabe A, Tashiro H, Sakata H, Morita M, et al. Clinical significance of Smac/DIABLO expression in colorectal cancer. *Oncol Rep* (2009) 21(2):351–5.
13. Pluta P, Cebula-Obrzut B, Ehemann V, Pluta A, Wierzbowska A, Piekarski J, et al. Correlation of Smac/DIABLO protein expression with the clinico-pathological features of breast cancer patients. *Neoplasma* (2011) 58(5):430–5. doi:10.4149/neo_2011_05_430
14. Fulda S, Galluzzi L, Kroemer G. Targeting mitochondria for cancer therapy. *Nat Rev Drug Discov* (2010) 9(6):447–64. doi:10.1038/nrd3137
15. Deveraux QL, Takahashi R, Salvesen GS, Reed JC. X-linked IAP is a direct inhibitor of cell-death proteases. *Nature* (1997) 388(6639):300–4. doi:10.1038/40901
16. Scott FL, Denault JB, Riedl SJ, Shin H, Renatus M, Salvesen GS. XIAP inhibits caspase-3 and -7 using two binding sites: evolutionarily conserved mechanism of IAPs. *EMBO J* (2005) 24(3):645–55. doi:10.1038/sj.emboj.7600544
17. Shiozaki EN, Chai J, Rigotti DJ, Riedl SJ, Li P, Srinivasula SM, et al. Mechanism of XIAP-mediated inhibition of caspase-9. *Mol Cell* (2003) 11(2):519–27. doi:10.1016/S1097-2765(03)00054-6
18. Estornes Y, Bertrand MJ. IAPs, regulators of innate immunity and inflammation. *Semin Cell Dev Biol* (2015) 39:106–14. doi:10.1016/j.semcdb.2014.03.035
19. Vallabhapurapu S, Karin M. Regulation and function of NF-kappaB transcription factors in the immune system. *Annu Rev Immunol* (2009) 27:693–733. doi:10.1146/annurev.immunol.021908.132641
20. Pahl HL. Activators and target genes of Rel/NF-kappaB transcription factors. *Oncogene* (1999) 18(49):6853–66. doi:10.1038/sj.onc.1203239
21. Dougan M, Dranoff G. Immune therapy for cancer. *Annu Rev Immunol* (2009) 27:83–117. doi:10.1146/annurev.immunol.021908.132544
22. Kirkwood JM, Butterfield LH, Tarhini AA, Zarour H, Kalinski P, Ferrone S. Immunotherapy of cancer in 2012. *CA Cancer J Clin* (2012) 62(5):309–35. doi:10.3322/caac.20132
23. Orange JS. Formation and function of the lytic NK-cell immunological synapse. *Nat Rev Immunol* (2008) 8(9):713–25. doi:10.1038/nri2381
24. Trapani JA, Smyth MJ. Functional significance of the perforin/granzyme cell death pathway. *Nat Rev Immunol* (2002) 2(10):735–47. doi:10.1038/nri911
25. Ashkenazi A. Targeting death and decoy receptors of the tumour-necrosis factor superfamily. *Nat Rev Cancer* (2002) 2(6):420–30. doi:10.1038/nrc821
26. Vujanovic NL. Role of TNF superfamily ligands in innate immunity. *Immunol Res* (2011) 50(2–3):159–74. doi:10.1007/s12026-011-8228-8
27. Papamichail M, Perez SA, Gritzapis AD, Baxevanis CN. Natural killer lymphocytes: biology, development, and function. *Cancer Immunol Immunother* (2004) 53(3):176–86. doi:10.1007/s00262-003-0478-4
28. deMagalhaes-Silverman M, Donnenberg A, Lembersky B, Elder E, Lister J, Rybka W, et al. Posttransplant adoptive immunotherapy with activated natural killer cells in patients with metastatic breast cancer. *J Immunother* (2000) 23(1):154–60. doi:10.1097/00002371-200001000-00018
29. Ishikawa E, Tsuboi K, Saijo K, Harada H, Takano S, Nose T, et al. Autologous natural killer cell therapy for human recurrent malignant glioma. *Anticancer Res* (2004) 24(3b):1861–71.

30. Miller JS, Soignier Y, Panoskaltsis-Mortari A, McNearney SA, Yun GH, Fautsch SK, et al. Successful adoptive transfer and in vivo expansion of human haploidentical NK cells in patients with cancer. *Blood* (2005) 105(8):3051–7. doi:10.1182/blood-2004-07-2974
31. Stern M, Passweg JR, Meyer-Monard S, Esser R, Tonn T, Soerensen J, et al. Pre-emptive immunotherapy with purified natural killer cells after haploidentical SCT: a prospective phase II study in two centers. *Bone Marrow Transplant* (2013) 48(3):433–8. doi:10.1038/bmt.2012.162
32. Terme M, Ullrich E, Delahaye NF, Chaput N, Zitvogel L. Natural killer cell-directed therapies: moving from unexpected results to successful strategies. *Nat Immunol* (2008) 9(5):486–94. doi:10.1038/ni1580
33. Sutlu T, Alici E. Natural killer cell-based immunotherapy in cancer: current insights and future prospects. *J Intern Med* (2009) 266(2):154–81. doi:10.1111/j.1365-2796.2009.02121.x
34. Varfolomeev E, Blankenship JW, Wayson SM, Fedorova AV, Kayagaki N, Garg P, et al. IAP antagonists induce autoubiquitination of c-IAPs, NF-kappaB activation, and TNFalpha-dependent apoptosis. *Cell* (2007) 131(4):669–81. doi:10.1016/j.cell.2007.10.030
35. Salih HR, Antropius H, Gieseke F, Lutz SZ, Kanz L, Rammensee HG, et al. Functional expression and release of ligands for the activating immunoreceptor NKG2D in leukemia. *Blood* (2003) 102(4):1389–96. doi:10.1182/blood-2003-01-0019
36. Fiegler N, Textor S, Arnold A, Rölle A, Kanz L, Oehme I, et al. Downregulation of the activating NKp30 ligand B7-H6 by HDAC inhibitors impairs tumor cell recognition by NK cells. *Blood* (2013) 122(5):684–693. doi:10.1182/blood-2013-02-482513
37. Tchoghandjian A, Jennewein C, Eckhardt I, Rajalingam K, Fulda S. Identification of non-canonical NF-kappaB signaling as a critical mediator of Smac mimetic-stimulated migration and invasion of glioblastoma cells. *Cell Death Dis* (2013) 4:e564. doi:10.1038/cddis.2013.70
38. Eckhardt I, Roesler S, Fulda S. Identification of DR5 as a critical, NF-kappaB-regulated mediator of Smac-induced apoptosis. *Cell Death Dis* (2013) 4:e936. doi:10.1038/cddis.2013.457
39. Fulda S, Wick W, Weller M, Debatin KM. Smac agonists sensitize for Apo2L/TRAIL- or anticancer drug-induced apoptosis and induce regression of malignant glioma in vivo. *Nat Med* (2002) 8(8):808–15. doi:10.1038/nm735nm735
40. Fakler M, Loeder S, Vogler M, Schneider K, Jeremias I, Debatin KM, et al. Small molecule XIAP inhibitors cooperate with TRAIL to induce apoptosis in childhood acute leukemia cells and overcome Bcl-2-mediated resistance. *Blood* (2009) 113(8):1710–22. doi:10.1182/blood-2007-09-114314
41. Roesler S, Eckhardt I, Wolf S, Fulda S. Cooperative TRAIL production mediates IFNalpha/Smac mimetic-induced cell death in TNFalpha-resistant solid cancer cells. *Oncotarget* (2016) 7(4):3709–25. doi:10.18632/oncotarget.6915
42. Belz K, Schoeneberger H, Wehner S, Weigert A, Bonig H, Klingebiel T, et al. Smac mimetic and glucocorticoids synergize to induce apoptosis in childhood ALL by promoting ripoptosome assembly. *Blood* (2014) 124(2):240–50. doi:10.1182/blood-2013-05-500918
43. Wagner L, Marschall V, Karl S, Cristofanon S, Zobel K, Deshayes K, et al. Smac mimetic sensitizes glioblastoma cells to Temozolomide-induced apoptosis in a RIP1- and NF-kappaB-dependent manner. *Oncogene* (2013) 32(8):988–97. doi:10.1038/onc.2012.108
44. Fulda S. Smac mimetics as IAP antagonists. *Semin Cell Dev Biol* (2015) 39:132–8. doi:10.1016/j.semcdb.2014.12.005
45. Wang L, Du F, Wang X. TNF-alpha induces two distinct caspase-8 activation pathways. *Cell* (2008) 133(4):693–703. doi:10.1016/j.cell.2008.03.036
46. Loeder S, Fakler M, Schoeneberger H, Cristofanon S, Leibacher J, Vanlangenakker N, et al. RIP1 is required for IAP inhibitor-mediated sensitization of childhood acute leukemia cells to chemotherapy-induced apoptosis. *Leukemia* (2012) 26(5):1020–9. doi:10.1038/leu.2011.353
47. Bake V, Roesler S, Eckhardt I, Belz K, Fulda S. Synergistic interaction of Smac mimetic and IFNalpha to trigger apoptosis in acute myeloid leukemia cells. *Cancer Lett* (2014) 355(2):224–31. doi:10.1016/j.canlet.2014.08.040
48. Chromik J, Safferthal C, Serve H, Fulda S. Smac mimetic primes apoptosis-resistant acute myeloid leukaemia cells for cytarabine-induced cell death by triggering necroptosis. *Cancer Lett* (2014) 344(1):101–9. doi:10.1016/j.canlet.2013.10.018
49. Sauer M, Reiners KS, Hansen HP, Engert A, Gasser S, von Strandmann EP. Induction of the DNA damage response by IAP inhibition triggers natural immunity via upregulation of NKG2D ligands in Hodgkin lymphoma in vitro. *Biol Chem* (2013) 394(10):1325–31. doi:10.1515/hsz-2013-0161
50. Rettinger E, Glatthaar A, Abhari BA, Oelsner S, Pfirrmann V, Huenecke S, et al. SMAC mimetic BV6 enables sensitization of resistant tumor cells but also affects cytokine-induced killer (CIK) cells: a potential challenge for combination therapy. *Front Pediatr* (2014) 2:75. doi:10.3389/fped.2014.00075
51. Brinkmann K, Hombach A, Seeger JM, Wagner-Stippich D, Klubertz D, Kronke M, et al. Second mitochondria-derived activator of caspase (SMAC) mimetic potentiates tumor susceptibility toward natural killer cell-mediated killing. *Leuk Lymphoma* (2014) 55(3):645–51. doi:10.3109/10428194.2013.807925
52. Dougan M, Dougan S, Slisz J, Firestone B, Vanneman M, Draganov D, et al. IAP inhibitors enhance co-stimulation to promote tumor immunity. *J Exp Med* (2010) 207(10):2195–206. doi:10.1084/jem.20101123
53. Knights AJ, Fucikova J, Pasam A, Koernig S, Cebon J. Inhibitor of apoptosis protein (IAP) antagonists demonstrate divergent immunomodulatory properties in human immune subsets with implications for combination therapy. *Cancer Immunol Immunother* (2013) 62(2):321–35. doi:10.1007/s00262-012-1342-1
54. Muller-Sienerth N, Dietz L, Holtz P, Kapp M, Grigoleit GU, Schmuck C, et al. SMAC mimetic BV6 induces cell death in monocytes and maturation of monocyte-derived dendritic cells. *PLoS One* (2011) 6(6):e21556. doi:10.1371/journal.pone.0021556

3

The Rise of Allogeneic Natural Killer Cells as a Platform for Cancer Immunotherapy: Recent Innovations and Future Developments

John P. Veluchamy[1,2†], *Nina Kok*[2†], *Hans J. van der Vliet*[1], *Henk M. W. Verheul*[1], *Tanja D. de Gruijl*[1] and *Jan Spanholtz*[2*]

[1] Department of Medical Oncology, VU University Medical Center, Cancer Center Amsterdam, Amsterdam, Netherlands,
[2] Glycostem Therapeutics, Oss, Netherlands

Edited by:
Gianfranco Pittari,
Hamad Medical Corporation, Qatar

Reviewed by:
Evelyn Ullrich,
Goethe University Frankfurt,
Germany
Bjarne Kuno Møller,
Aarhus University Hospital,
Denmark

***Correspondence:**
Jan Spanholtz
jan@glycostem.com

†These authors have contributed equally to this work.

Natural killer (NK) cells are critical immune effector cells in the fight against cancer. As NK cells in cancer patients are highly dysfunctional and reduced in number, adoptive transfer of large numbers of cytolytic NK cells and their potential to induce relevant antitumor responses are widely explored in cancer immunotherapy. Early studies from autologous NK cells have failed to demonstrate significant clinical benefit. In this review, the clinical benefits of adoptively transferred allogeneic NK cells in a transplant and non-transplant setting are compared and discussed in the context of relevant NK cell platforms that are being developed and optimized by various biotech industries with a special focus on augmenting NK cell functions.

Keywords: hematopoietic stem cell transplantation, autologous natural killer cells, allogeneic natural killer cells, adoptive natural killer cell therapy, natural killer cell biotech companies, natural killer cell combinatorial studies

NATURAL KILLER (NK) CELLS IN ONCOLOGY

So far T cells have been the mainstay of cancer immunotherapy; however, it is generally recognized that NK cells also play an essential role in antitumor immunity. Certainly in the prevention of metastases through the elimination of circulating cancer stem cells with a high metastatic potential, NK cells are recognized as main immune effector cells (1). Moreover, as solid tumors have a propensity to particularly down-regulate MHC-I, NK cells provide a failsafe mechanism in these circumstances where cytotoxic T cells, which depend on MHC-I for tumor recognition and elimination, are debilitated. NK cells have recently been more intensely explored as a viable therapeutic platform next to T cell-based approaches. This review aims to summarize the latest developments in the clinical translation of adoptive transfer of NK cells in the oncology field.

NK CELLS AND THEIR ACTIVATING AND INHIBITORY RECEPTORS

Human NK cells are generally categorized by their level of CD56 and CD16 expression into two subsets: $CD56^{bright}CD16^{dim}$ and $CD56^{dim}CD16^{bright}$ NK cells. Most NK cells in the peripheral blood and spleen are $CD56^{dim}CD16^{bright}$ and are cytotoxic against a variety of tumor cells, whereas $CD56^{bright}CD16^{dim}$ NK cells are immune regulatory in function and constitute the majority in secondary lymphoid tissues, producing abundant cytokines but exerting weak cytotoxicity compared to $CD56^{dim}CD16^{bright}$

NK cells (2). The ability of NK cells to discriminate between a cancer cell and a healthy cell is regulated by a balance between its activating and inhibitory receptors. NK-activating receptors such as DNAM-1 and NKG2D; natural cytotoxicity receptors (NCRs) such as NKp30, NKp44, NKp46, CD94/NKG2C, CD94/NKG2E, and CD16a; and activating killer cell-immunoglobulin like receptors (KIRs) contribute to NK cell activation, triggering the release of cytotoxic granules and proinflammatory cytokines such as interferon gamma (IFNγ) from NK cells to lyse cancer cells (3). The NK cell-activating receptor NKG2D (CD314) recognizes MHC class-I-chain related proteins A and B (MICA and MICB) and ULBPs (1–6), while DNAM-1 binds to CD112 (Nectin-2) and CD155 (poliovirus receptor) (5) on stressed, infected, and cancer cells. The ligands for NCRs are widely expressed on cells infected by viruses or by intracellular bacteria and on tumor cells, but their exact modes of action are yet to be characterized to define their role in NK cytotoxicity (6). In addition to this, the heterodimers of the NKG2 family; CD94/NKG2C and CD94/NKG2E recognize the non-classical MHC class I molecule HLA-E and associate with DAP-12 molecule to trigger an NK activation signal (7, 8). Another very important activation mechanism of NK cells is through the interaction of CD16a (FcγRIIIa, a low affinity Fc receptor) with the Fc portion of IgG_1 antibodies, forming an immunological synapse to engage antibody opsonized targets for NK cell-mediated antibody-dependent cell mediated cytotoxicity (ADCC) (9). Besides engaging activating receptors, NK cells also induce target cell death using tumor necrosis factor α (TNF-α), Fas ligand, and TNF-related apoptosis-inducing ligand (TRAIL) (10). The most prominent NK cell inhibitory receptors include inhibitory KIRs that recognize MHC class I (HLA-ABC) molecules, which are universally expressed on healthy tissues. Similarly, CD94/NKG2A, an inhibitory receptor from the NKG2 family, binds to HLA-E and induces NK cell tolerance through the activation of an intracellular immunoreceptor tyrosine-based inhibitory motif (ITIM) (8). Hence, knowing that NK cell functions are determined by an array of receptors, which can either potentiate an activating or inhibitory signal, depending on different ligand interactions with tumor cells, it is critical to shift the balance in a therapeutic setting toward an activating NK phenotype to expedite enhanced NK tumor killing mechanisms.

NK CELL DYSFUNCTIONALITY IN CANCER

Natural killer cells can control circulating tumor cells and prevent formation of tumor metastases (11). However, tumors employ different strategies to evade killing by NK cells. Upregulation of inhibitory ligands such as MHC class I molecules (HLA-ABC, HLA-G and HLA-E) has been associated with a stronger inhibitory signal to NK cells (12–15). Furthermore, increased expression of the inhibitory NKG2A receptor reported in renal cell carcinoma resulted in decreased functionality of tumor infiltrating NK cells (16). On the other hand, downregulation of NK-activating ligands for NKG2D such as MICA and MICB and increased shedding of tumor-derived soluble MIC also impair NKG2D-mediated NK cell tumor recognition (17). Another important necessity for optimal NK cell function is the ability to home and migrate to tumor sites. Several studies have correlated increased homing of NK cells to tumor tissues with improved treatment outcomes in solid tumors (18–22). However, the immunosuppressive tumor stroma comprising regulatory T cells (T-regs) (23), myeloid-derived suppressor cells (MDSCs) (24), M2 macrophages (25), and immature dendritic cells severely restricts NK cell functionality and their entry into solid tumors. In chronic diseases, such as those associated with human immunodeficiency virus and cytomegalovirus infections, mainly exhausted NK cells with decreased cytokine production and reduced cytolytic activity are observed (26, 27). In a study with breast cancer patients, the NK cell expression levels of activating receptors (NKG2D, DNAM, CD16, and NKp30) were decreased, whereas inhibitory receptor (NKG2A) expression levels were increased and this apparent dysfunctionality of NK cells was found to directly affect NK cell cytotoxicity (28). Similarly, the effector subset of NK cells ($CD56^{dim}CD16^{+}$) from head and neck and breast cancer patients, when tested *in vitro*, was highly prone to apoptosis, thus pointing to low NK cell activity in these patients (29). Impaired NK cell functionality may result from tumor-imposed suppressive mechanisms and presents a major hurdle for NK cell-targeted immunotherapies. Therefore, approaches to restore or replace impaired NK cell cytotoxicity may prove essential for an effective host defense against cancers.

NK CELLS IN THE CLINIC

Novel NK cell-based immunotherapeutic strategies are being developed to overcome the functional limitations of the use of cancer patients' autologous NK cells. To increase the number of functional NK cells even in case of a high tumor load, adoptive transfer of autologous NK cells served as a very feasible approach, as this ruled out the need for immunosuppression, HLA-matching, and prevented the risk of graft versus host disease (GvHD). These advantages sparked the initiation of large-scale expansion protocols and clinical trials using autologous NK cells as a treatment modality for cancer. Though adoptive transfer of autologous NK cells resulted in an increased number of circulating NK cells in peripheral blood, it failed to produce significant therapeutic effects in hematological malignancies, metastatic melanoma, and renal cell carcinoma patients due to the inhibition by self-HLA molecules (30–32). Moreover, the expansion efficiency and functional status of autologous NK cells were still limited when compared to allogeneic NK cells, as autologous cells were often obtained from heavily pretreated patients (33). In addition to this, it was difficult to track infused autologous NK cells in patients and to study their antitumor effects from peripheral blood analyses due to the inability to differentiate *ex vivo* manipulated and transferred autologous NK cells from the non-manipulated circulating NK cells. These limitations motivated researchers shifting their focus to allogeneic NK cells to treat cancer.

In patients with leukemia undergoing allogeneic hematopoietic stem cell transplantation (HSCT), NK cells, being the first lymphoid subset to appear after allogeneic HSCT (34), play a crucial role in controlling host defense against infections and residual cancer cells before T cells are reconstituted (35). These donor T cells are prime mediators of GvHD (36), and the life-threatening complications that arise due to GvHD have completely overshadowed the beneficial effects of alloreactive NK and T cells, fueling efforts to use T cell depleted grafts (37). Further, this led to the development of NK cell-based therapies coupled with T cell depleted HSCs to enhance the graft versus tumor effect (GvT) without causing GvHD. Unlike autologous NK cells, allogeneic NK cells are not restricted by the patient's tumor's HLA expression, which is an added advantage to mount an improved anti-tumor effect (38, 39). Current translational efforts that are explored as anticancer therapies include adoptive transfer of *ex vivo* activated and/or expanded allogeneic NK cells, either alone or in combination with HSCT.

SOURCES OF ALLOGENEIC NK CELLS USED IN THE CLINIC

Commonly used allogeneic NK cells are apheresis products collected from haploidentical and unrelated donor PBMC (40). Another source is umbilical cord blood (UCB), where NK cells are generated from CD34+ progenitor cells that undergo expansion and differentiation using cytokines and growth factors and thereby mature into cytolytic NK cells (41). Apart from PBMC and UCB, NK cells have also been obtained from the clonal cell line NK-92, derived from immortalized lymphoma NK cells (42, 43).

ALLOGENEIC NK CELL THERAPY IN A TRANSPLANT SETTING

Autologous or allogeneic HSCT serves as a curative regimen by reconstituting the immune system in hematological malignancies. At an earlier stage post HSCT, NK and T cells developing from the graft are immature and less in number with reduced functionality. Under those circumstances, the infusion of purified allogeneic NK cells was explored as a viable option to target minimal residual disease (MRD), prevent graft failure, and relapse. Grafts for allogeneic HSCT and allogeneic NK cell treatments were obtained from HLA matched/mismatched and related/unrelated donors (38, 39). Earlier clinical trials performed by Passweg et al. (44), Koehl et al. (45), Shi et al. (46), Yoon et al. (47), Rizzieri et al. (48), and Brehm et al. (49) have shown that NK cells can be safely administered prior to or post HSCT in patients with different types of hematological diseases. Immune suppression is a prerequisite prior to most of the allogeneic HSCT and NK-cell infusions. A non-myeloablative conditioning regimen usually consisting of cyclophosphamide (Cy) and fludarabine (Flu) was found to facilitate NK cell persistence and expansion *in vivo* (50). High doses of Cy/Flu caused pancytopenia and resulted in high plasma IL-15 levels, which also correlated with the detection of adoptively transferred NK cells up to 14 days after infusion, thus suggesting that excess IL-15 was probably utilized by the NK cells to proliferate and persist longer *in vivo* (51). A summary of clinical trials with allogeneic NK-cell infusions in a HSCT setting with published data is summarized in **Table 1**, and selected clinical trials from recent years are reviewed below.

In 2013, Stern et al. treated acute myeloid leukemia (AML), Acute Lymphocytic Leukemia, Hodgkin's lymphoma (HL), and sarcoma patients with allogeneic NK cells (CD3 depleted and CD56 selected) after a haploidentical HSCT, using the same donor as NK cell source. An overall survival (OS) of 25% was achieved during a median follow-up of 5.8 years. And 4/16 patients developed acute GvHD (aGvHD) due to high T-cell impurities present in two NK cell products and two from stem cell grafts, both containing $\geq 0.5 \times 10^5$ cells/kg T cells (52). Although this prospective Phase II study reported the safety and feasibility of NK-cell infusion following allo-HSCT, it failed to yield results in support of anti-leukemia effects, raising questions as to whether the NK cell dosage of $0.3–3.8 \times 10^7$ cells/kg used was sufficient to induce a clinical effect. In the same year, Klingemann et al. published data highlighting the safety and alloreactivity of HLA-mismatched (CD3 depleted) NK cells, transfused after autologous HSCT in multiple myeloma (MM), non-Hodgkin's lymphoma (NHL), and HL patients. In this study, 13 patients were enrolled; 6/13 relapsed and 7/13 were in remission during a follow-up between 144 and 1,158 days following autologous stem cell transplantation. The allogeneic NK cells were well tolerated without GvHD. In addition, this study also demonstrated that NK cells generated and processed at distant centers can be shipped and transfused without significantly affecting the viability and cytotoxicity of the NK cell product (53).

Choi et al. (54) summarized their observations from a study, in which allogeneic *ex vivo* expanded and activated NK cells derived from the same donor were administered 14 and 21 days post HLA haploidentical HSCT to patients with hematological malignancies ($n = 41$). The data set from this study was compared with a group of 31 patients, who underwent only HLA haploidentical HSCT. A significantly higher progression-free survival (PFS) was seen in the HSCT + NK group compared to HSCT only group (74% versus 46%). In addition to this, the occurrence of chronic GvHD (cGvHD) (15% versus 10%) and transplant-related mortality (27% versus 19%) was reduced in the HSCT + NK group compared to the HSCT only group (54). In another study by Killig et al. (55), AML patients were treated with haploidentical HSCT followed by NK-cell infusions (CD3 depleted and CD56 selected) from the same donor, on days +1 and +2 post HSCT. aGvHD was highly prevalent in 20/24 patients in this study and histological analysis of skin revealed that GvHD was associated with infiltration by perforin$^+$CD8$^+$ T cells. Allogeneic NK cells contributed to an increased OS in the HSCT + NK group compared to the HSCT only group (37% versus 14%) over a median follow-up of 2 years (55).

Subsequently, Shah et al. published data from a Phase I study treating patients with Ewing sarcoma, rhabdomyosarcoma, and desmoplastic small round cell tumors ($n = 9$), using donor-derived IL-15/4-1BBL-activated allogeneic NK cells (CD3 depleted and CD56 selected) following allogeneic HSCT from the same donor. aGvHD was highly prevalent in the patient group that received stem cells from matched unrelated donors and was directly linked with a faster T cell recovery and higher T cell chimerism from reconstituted HSCT grafts. About 4/9 patients were alive in this study with a median follow-up of 23.1 months (56). Lee et al. reported results from a Phase I study

TABLE 1 | Summary of allogeneic NK cell clinical trials in a transplantation setting.

Study	Malignancy	Clinical trial design	Culture method[a]	Infused dose NK cells	Final product characteristics	Outcome
Phase I (NCT01729091) Shah et al. (59)	MM (*n* = 12)	Conditioning with Mel on day 7 and Lnd from days 8 to 2 prior to UCB-NK-cell infusion (day 5), followed by autologous-HSCT on day 0	*Ex vivo* expanded MNCs from unrelated UCB donors. Culture duration: 14 days with irradiated K562 clone 9.mbIL-21 aAPCs and IL-2 [a]CD3 depleted (on day 7)	Four escalating doses: 5×10^6, 1×10^7, 5×10^7, and 1×10^8 cells/kg	Mean purity: 98.9% CD56+/CD3− cells	Well tolerated. No GvHD. 4/12 progressed or relapsed (median of 21 months follow-up)
Phase I (NCT01795378) Choi et al. (58)	AML (*n* = 45) and ALL (*n* = 6)	Haplo-HSCT followed by DNKI from the same donor on days 6, 9, 13, and 20 post HSCT	*Ex vivo* expanded and activated PBNK cells from haploidentical donors. Culture duration: 2–3 weeks with IL-15 and IL-21	Four escalating doses: median DNKIs are 5×10^7, 5×10^7, 1×10^8, and 2×10^8 cells/kg	Median viability: 80%. Purity: 48–98% CD56+ CD122+ cells. 0–22% CD3+ CD56+ cells. 0–10.4% CD3+ CD56− cells	Toxicity observed in 73% of patients, 9/45 aGvHD. 29/51 CR (9.3–34.7 months follow-up), 35/51 PD
Phase I (NCT00402558) Phase II (NCT01390402) Lee et al. (57)	AML (*n* = 8), MDS (*n* = 6), and CML (*n* = 7)	Conditioning with Flu/Bu prior to haplo-allo NK-cell infusion, followed by IL-2 therapy (5×, daily); conditioning with Thy/Tac prior to HLA-matched related unrelated allo-HSCT	*Ex vivo* expanded and activated PBNK cells from haploidentical donors. Culture duration: o/n with IL-2. [a]CD3 depleted and CD56 selected (in three infusions)	Four escalating doses: 1×10^6, 5×10^6, 3×10^7, and 3×10^7 cells/kg in Phase I study. Four escalating doses of 5×10^6 cells/kg in Phase II study	Median purity: 0.02% CD3+ cells. 11.41% CD14+ cells. 21.84% CD19+ cells. 14.1% CD56+ CD3− cells	Well tolerated, no GvHD. 5/21 CR, 5/21 died of transplantation related issues and 11/21 died of relapse
Phase I (NCT01287104) Shah et al. (56)	EWS (*n* = 5), DSRCT (*n* = 3), RMS (*n* = 1)	HLA matched haplo- or unrelated allo-HSCT followed by aNK-DLI from the same donor on day 7 and 35 post HSCT	*Ex vivo* expanded and activated PBNK cells from haploidentical donors. Culture duration: 9–11 days with KT64.4-BBL artificial antigen presenting cells. [a]CD3 depleted and CD56 selected	Repeated doses (2× doses 1, 2, and 3): 1×10^5 cells/kg (dose 1), 1×10^6 cells/kg (dose 2), and 1×10^7 cells/kg (dose 3)	Median purity: CD3+ cells 1.4×10^4 cells/kg. CD56+ cells ≥90%. Viability: ≥70%	5/9 aGvHD. 2/9 SD, 7/9 CR. 4/9 are still alive (12.5–27.4 months after treatment)
Phase I/II (NCT01220544) Killig et al. (55)	AML (*n* = 24)	Haplo-HSCT followed by NK-cell infusion from same donor and OKT3 treatment from days −5 to +3	PBNK cells from haploidentical donors. [a]CD3 depleted and CD56 selected	Single dose: 1.61–32.2 × 10^6 CD56+/CD3− cells/kg	Purity: CD56+ CD3− cells 99.97%. CD3+ cells 0.95–7.4 × 10^4 cells/kg	Toxicity correlated with haplo-HSCT. Deaths: 2/24 GvHD, 6/24 infections and 7/24 died of relapse. 9/24 CR (0.1–8.6-year follow-up)
Phase I/II (NCT00823524) Choi et al. (54)	AML (*n* = 32), ALL (*n* = 7), MDS (*n* = 1), DLBCL (*n* = 1)	HLA haplo-HSCT followed by DNKI from the same donor, 14 days and 21 days after HSCT	*Ex vivo* expanded and activated PBNK cells from haploidentical donor. Culture duration: 13–20 days with IL-15, IL-21, and hydrocortisone	Escalating doses (2×): 0.2×10^8 cells/kg (3 pts), 0.5×10^8 cells/kg (3 pts), 1.0×10^8 cells/kg (8 pts), and ≥1.0×10^8 cells/kg (27 pts)	Viability: 71–85%. Median purity: CD56+ CD122+ cells >90%. CD3+ CD56+ cells <3%. Fold expansion: 0.8–70 (after 13–20 days of culture)	Well tolerated. 9/41 aGvHD, 10/41 cGVHD. In total, 11 patients died of TRM. In AML (21/29) (4/8) ALL/lymphoma are in CR
Phase I (IND # 12971) Klingemann et al. (53)	NHL (*n* = 6), MM (*n* = 5), and HL (*n* = 2)	MHC-mismatched haploidentical NK-MC infusion, 49–191 days post auto-HSCT	*Ex vivo* expanded and activated PBNK cells from haploidentical donors. Culture duration: o/n with IL-2	4 Escalating doses: 1×10^5, 1×10^6, 1×10^7, and 2×10^7 MC/kg	Median purity: 26% CD56+ CD3− cells. 0.15% CD3+ cells. Median viability: 95% post wash	Well tolerated. No GvHD. 6/13 relapsed and 7/13 in remission
Phase II (NCT01386619) Stern et al. (52)	AML (n = 8), ALL (*n* = 5), HL (*n* = 2) sarcoma (*n* = 1)	Haplo-HSCT followed by NK-DLI from the same donor, +day 3, +day 40, and +day 100 post HSCT	PBNK cells from haploidentical donors. [a]CD3 depleted and CD56 selected	Repeated doses (2–3): 0.3–3.8 × 10^7 cells/kg	Median purity: CD3+ cells 0.03×10^5 cells/kg. Median viability: 84%	Safe and feasible. 4/16 aGvHD. Median follow-up of 5.8 years 4/16 are alive. 3/16 died from graft failure

(Continued)

TABLE 1 | Continued

Study	Malignancy	Clinical trial design	Culture method[a]	Infused dose NK cells	Final product characteristics	Outcome
Phase I/II (NCT01386619) Brehm et al. (49)	AML (n = 6), ALL (n = 5), NB (n = 5), RMS (n = 1) HL (n = 1)	Haplo-HSCT followed by IL-2 stimulated NK-cell infusion (cryo) or unstimulated NK-cell infusion (fresh) from the same donor, +da 3, +day 40, and +day 100 post HSCT	*Ex vivo* expanded and activated PBNK cells from haploidentical donors. Culture duration: 9–14 days with (group II) or without (group I) IL-2 (fresh or cryo). [a]CD3 depleted and CD56 selected	Repeated doses (1–3 doses): Group I: 3.2–38.3 × 10⁶ cells/kg Group II: 6.0–45.1 × 10⁶ cells/kg	Purity: CD56+ CD3− cells 84.4–98.6%. CD3+ cells group I: 0.4–53.4 × 10³ cells/kg. CD3+ cells group II: 7.7–98.3 × 10³ cells/kg. Viability: freshly NK-cell unstimulated median 93%. Cryo NK-cell IL-2 stim 30–70%	Well tolerated without GvHD >grade II. Group I: 5/9 died (126–498 days post SCT), 3/9 CR (742–2,218 days). Group II: 5/9 died (27–373), 2/9 CR, and 2/9 in remission
Phase I (NCT00586690) Rizzieri et al. (48)	Lymphoma (n = 30)	3–6/6 HLA-matched haploidentical NK-cell infusion, ~8 weeks post haplo-HSCT from the same donor	PBNK cells from haploidentical donors. [a]Only CD56 selected	Repeated dose (1–3): median dose in 3–5/6 HLA match: 9.21 × 10⁶ CD3+/CD56− cells/kg, median dose 6/6 HLA match: 10.6 × 10⁶ CD3+/CD56− cells/kg	6/6 HLA-matched: Purity: 87–100% CD56+ cells. 0.53 ± 1.1 × 10⁶ cells/kg CD3+ CD56−. 3–5/6 HLA-matched: Purity: 86–100% CD56+ cells. 0.27 ± 0.78 × 10⁶ cells/kg CD3+ CD56−	Safe. Low toxicity. 6/6 HLA-matched: 6/14 aGvHD (1 severe) and median OS 12 months. 3–5/6 HLA-matched: 8/16 aGvHD and median OS 27 months
Phase I (NCT00569283) Yoon et al. (47)	AML (n = 12) MDS (n = 2)	HLA-mismatched HSCT followed by allo NK-cell infusion from the same donor	*Ex vivo* expanded, differentiated and activated CD34+ progenitor cells (PB-derived) from haploidentical donors. Culture duration: 21 days with FLt3, IL-7 and hydrocortisone followed by 21 days with IL-15, IL-21 and hydrocortisone	Single dose: 0.33–24.5 × 10⁶ cells/kg	Mean purity: CD56+ CD122+ cells 64%. CD3+ cells 1.0%. Mean viability: 88%	1/14 aGvHD and 4/14 cGvHD. 9/14 died (between 1.7 and 15.5 months), 4/14 CR (between 16.2 and 21.6 months) 1/14 PD (25.9 months)
(BB-IND-11347) Shi et al. (46)	MM (n = 10)	Conditioning with Flu/Dex/Mel followed by haplo-KIR-ligand-mismatched NK-cell infusion on day 0 and day +2; IL-2 therapy daily (11×) starting on day +1 after NK-cell infusion; auto-HSCT on day +14	*Ex vivo* expanded and activated PBNK cells from haploidentical donors. Culture duration: o/n with IL-2 (pts 1–5) and brief incubation with IL-2 and anti-CD3 beads (pts 5–10)	Combined dose (day 0 and day +2): 2.7–92 × 10⁶ cells/kg	Purity: median CD3+ cells 5.5 × 10⁴ cells/kg. Viability: 95%	Safe and no GvHD. 5/10 CR, 1/10 PR, 1/10 MR, 1/10 SD, and 2/10 PD. 4/10 are alive at 1.4, 1.5, 2.3, and 3 years post NK-cell therapy
Pilot study Koehl et al. (45)	AML (n = 1) ALL (n = 2)	Haplo-HSCT followed by KIR-mismatched NK-cell infusion on day +1 or +2 post HSCT and additional infusions every 4–6 weeks; IL-2 therapy +2 days post HSCT, every second day for 2–4 weeks	*Ex vivo* expanded and activated PBNK cells from haploidentical donors. Culture duration: >12 days with IL-2. [a]CD3 depleted and CD56 selected (in three infusions)	Repeated doses (1–3): 8.9–29.5 × 10⁶ cells/kg (first infusion), 3.3 and 11.1 × 10⁶ cells/kg (second infusion), 33.8 × 10⁶ cells/kg (third infusion)	Purity: CD56+ CD3− cells 95%. Median CD3+ cells 0.04%, 45–1,100 × 10³ cells. Viability: 95%. Fold expansion: median 5 (after 14 days of culture)	Well tolerated, no GvHD. 1/3 CR (152 days), 2/3 died (80 days and 45 days after NK-cell infusion)
Passweg et al. (44)	AML (n = 4), CML (n = 1)	Haplo-HSCT followed by NK-DLI from the same donor 3–12 months post HSCT	PBNK cells from haploidentical donors. [a]CD3 depleted and CD56 selected	Single dose: 0.21–1.41 × 10⁷ cells/kg	Median purity: CD56+ CD3− cells 97.3%. T-cell 0.22 × 10⁵ cells/kg	Well tolerated and feasible. 4/5 continuous remission (8–18 months), 1/5 PD

CNS, central nervous system; MPDs, myeloproliferative disorders; LPD, lymphoproliferative disorder; MM, multiple myeloma; MDS, myelodysplastic syndromes; MDN, myelodysplastic neoplasms; MPN, myeloproliferative neoplasms; AML, acute myeloid leukemia; LBLL, lymphoblastic leukemia-lymphoma; ALL, acute lymphoblastic leukemia; NB, neuroblastoma; RMS, rhabdomyosarcoma; CML, chronic myelogenous leukemia; NHL, non-Hodgkin's lymphoma; MCL, mantle cell lymphoma; DLBCL, diffuse large B cell lymphoma; FL, follicular lymphoma; ALCL, anaplastic large cell lymphoma; HL, Hodgkin's lymphoma; RCC, renal cell cancer; SCLC, small cell lung cancer; CLL, chronic lymphocytic leukemia; HCC, hepatocellular carcinoma; PNET, primitive neuroectodermal tumor; ACC, adrenal cortical carcinoma; EWS, Ewing sarcoma; DSRCT. desmoplastic small round cell tumor; Cy, cyclophosphamide; Flu, fludarabine; Bor, bortezomib; Dex, dexamethasone; Clo, clofarabine; Eto, etoposide; Cis, cisplatin; Pac, paclitaxel; Doc, docetaxel; Vin, vinorelbine; Gem, gemcitabine; Car, carboplatin; Pem, pemetrexed; TBI, total body irradiation; Tac, tacrolimus; Pred, prednisolone; mPred, methylprednisolone; Thy, thymoglobulin; Vin, vincristine; Adr, adriamycin; Predn, prednisone; Mel, melphalan; OKT3, muromonab-CD3; HSPC, human stem and progenitor cells; aGvHD, acute graft versus host disease; DNKI, donor NK-cell infusion; Stim, stimulated; Unstim, unstimulated; DFS, disease-free survival, CR, complete remission; PR, partial response; MR, minimal response; SD, stable disease; PD, progressive disease; aNK-DLI, donor-derived IL-15/4-1BBL activated NK-cell infusion; TRM, transplant-related mortality; MC, mononuclear cell; TLS, tumor lysis syndrome; PLS, passenger lymphocyte syndrome; NE, not evaluable.

[a]Culture method displays CD3 depleted PBMC's, otherwise deviated selection method is mentioned.

in patients with AML, myelodysplastic syndromes (MDS), and chronic myelogenous leukemia (CML), in which alloreactive haploidentical NK cells (CD3 depleted and CD56 selected) were administered along with IL-2 injections, followed by thymoglobulin conditioning and allogeneic HSCT. Thymoglobulin was administered to prevent NK cells from hampering engraftment of allogeneic HSCT. Out of 20 evaluable patients, 16 had GvHD (10/16 aGvHD and 6/16 cGvHD) after transplantation. In this study, GvHD was not directly associated with donor T cell or NK cell contents. From this study, also it was concluded that the lack of anti-leukemic effect was mainly due to the low dose of infused NK cells and it was further suggested that thymoglobulin conditioning could also have potentially affected NK cells survival *in vivo* (57).

Later, Choi et al. presented results from a modified treatment protocol of four consecutive infusions of *ex vivo* activated and expanded haploidentical NK cells after HLA-matched HSCT and compared the outcomes to their previous study, in which they administered two infusions. In the subsequent study, additional donor NK-cell infusions were given on days 6 and 9 (i.e., at days 6, 9, 13, and 20). Out of 51 patients with ALL (n = 6) and AML (n = 45), 24/51 (47%) had four NK infusions. Out of 45 evaluable patients, the 3-year OS rate was 9% in AML and 21% for ALL and 9/45 had aGvHD. Early administration of NK cells after HSCT caused significant toxicities with no improvements in anti-leukemic effects, compared to the previous study. In this study group, a higher CR rate correlated with higher expression levels of NK activating receptors NKG2D and NCRs (NKp44, NKp46, and NKp30) on donor NK cells. In addition, NKp30 expression was significantly higher than that of NKG2D and other NCRs, thus suggesting a role for NKp30 as a predictive biomarker for anti-leukemic effects of NK cells (58).

In 2017, Shah et al. published data from a Phase I study, treating MM patients with UCB-derived NK cells (day 5) along with autologous-HSCT (day 0), following high-dose chemotherapy and low-dose lenalidomide. Mononuclear cells (MNCs) isolated from UCB units (CD3 depleted) were cultured with K562-based artificial antigen presenting cells (aAPCs) expressing membrane bound IL-21. No treatment related toxicities or GVHD was reported in this study. During a median follow-up of 21 months in 12 patients, 4/12 patients had progressive disease (PD) or relapsed. Stable expression of NKG2D and increased expression of CD16 and NKp30 of UCB-NK cells were observed in six patients. This study further reiterates the safety of NK-cell infusions in high doses; however, due to combinatorial set up with HSCT and lenalidomide, it is difficult to interpret the clinical efficacy of UCB-NK cells alone from this study (59).

Taken together, it is evident from these studies, as well as from many others, that GvHD, which is mainly caused by T cells from transplanted grafts, is a major concern in the field of allogeneic HSCT. Under these circumstances, it is difficult to reliably study the safety of allogeneic NK-cell infusions. The timing of NK-cell infusion, NK cell dosage and NK cell promoting conditioning regimens are critical factors that need to be more extensively studied to assess the safety and efficacy of allogeneic NK-cell infusions.

ADOPTIVE NK CELL THERAPY IN A NON-TRANSPLANT SETTING

To gain a better understanding of the safety and efficacy of allogeneic NK cell transfer, investigators started to study NK cells in a non-transplant setting. Landmark clinical trials were performed by Miller et al. (50), Iliopoulou et al. (60), Rubnitz et al. (61), Bachanova et al. (62) Curti et al. (63), and Geller et al. (33) predominantly in hematological malignancies, but also in various solid tumors. These studies demonstrated the safety and in part the efficacy of allogeneic NK-cell infusions in the absence of GvHD. A summary of allogeneic NK cell clinical trials in a non-transplant setting with published results is presented in **Table 2**.

Here, we focus on the latest reports from clinical trials using allogeneic NK cells in a non-transplant setting. Bachanova et al. developed a recombinant cytotoxic protein, i.e., an IL-2/diphteria toxin fusion protein (IL2DT), which functions by selectively depleting the IL-2 receptor CD25 expressing cells, including regulatory T cells (T-regs). In total, 57 AML patients were treated with KIR and HLA-mismatched haploidentical NK cells and 15 of them in cohort 3 received IL2DT, 1 or 2 days prior to NK-cell infusion, to deplete T-regs. In addition to IL2DT treatment, three different processing methods were used, i.e., a CD3-depleted cohort (cohort 1, n = 32), a cohort using CD3 depletion followed by CD56 selection (cohort 2, n = 10), and a cohort using CD3 and CD19 depletion (cohort 3, n = 15). Higher NK cell doses were obtained from cohort 3, and it could possibly be the reason for the observed longer disease-free survival (DFS) (33% versus 5% at 6 months) in cohort 3 compared to cohorts 1 and 2. In this study, endogenous IL-15 serum levels correlated with reduced T-reg levels in patients treated with IL2DT (64).

Szmania et al. investigated the effect of infusing cryopreserved and freshly prepared NK cells (CD3 depleted), either allogeneic or autologous, given after bortezomib with or without lymphodepletion in high risk relapsed MM patients. NK cells were cocultured with K562-mb15-41BBL cells for 8–9 days. Initially, 6 out of 8 NK products were cultured with low dose IL-2 (10 U/ml), however, increasing the dosage of IL-2 to 500 U/ml resulted in enhanced expression of NKp30 and NK cytolytic activity without affecting the T cell content of the NK cell product and, therefore the last 2 patients in this study were treated with high dose IL-2 cultured NK cells. The highest post-transfer number of circulating NK cells was observed in the high dose IL-2 group. Patients treated with fresh NK cells showed a median 21-fold increase in peripheral NK cell rates by day 7, while no *in vivo* expansion of NK cells was seen in patients treated with cryopreserved NK cells. Overall, the NK-cell infusions were well tolerated and no GvHD was observed. From 7 evaluable patients, 6 had (PD), and 1 had a partial response (PR) for up to 6 months post infusion (65).

In the same year, Kottaridis et al. presented data from a clinical trial in AML, for the first time using tumor-primed NK cells from related haploidentical donors. During a 6-month follow-up period (n = 7), three patients in CR remained in CR, one patient in PR achieved CR, two patients relapsed, and one patient died. Median OS was 468 days post NK-cell infusion (66).

Rubnitz and his team reported on the safety and feasibility of haploidentical NK cell therapy in children with relapsed or

TABLE 2 | Summary of allogeneic NK cell clinical trials in a non-transplantation setting.

Study	Malignancy	Clinical Trial design	Culture method[a]	Infused dose NK cells	Final product characteristics	Outcome
Phase I (EudraCT number: 2010-018988-41) Dolstra et al. (73)	AML (n = 10)	Conditioning with Cy/Flu followed by KIR-mismatched UCB-NK-cell infusion	*Ex vivo* expanded, differentiated and activated UCB-NK cells from unrelated donors. Culture duration: 42 days with GM-CSF, G-SCF, IL-6, SCF, Flt3L, TPO, IL-7, IL-2, and IL-15 [a]CD34+ selected HSPC's	Escalating doses: 3 × 10^6 cells/kg (cohort 1), 10 × 10^6 cells/kg (cohort 2), and 30 × 10^6 cells/kg (cohort 3)	Mean purity: 74 ± 13% CD34+ cell product. 75 ± 12% generated CD56+ CD3− NK cells. 0.03 ± 0.04% CD3+ cells. 0.16 ± 0.21% CD19+ cells. Mean viability: 94%	Well tolerated, no GvHD nor toxicity. 4/10 DFS for 55, 47, 17, and 12 months after infusion
Phase I (NCT01898793) Romee et al. (72)	AML (n = 13)	Conditioning with Cy/Flu followed by cytokine-induced memory-like NK-cell infusion and subsequent IL-2 therapy (every other day, 6×)	*Ex vivo* expanded and activated PBNK cells from haploidentical donors. Culture duration: 12–16 h with IL-15, IL-12, and IL-18. [a]CD3 depleted and CD56 selected	Repeated dose: level 1: 0.5 × 10^6 NK cells/kg level 2: 1 × 10^6 NK cells/kg level 3: 10 × 10^6 NK cells/kg	Purity: >90% CD56+ CD3− cells	Well tolerated, no GvHD. 4/13 NE, 4/13 TF-PD, 3/13 CR, 1/13 Cri, and 1/13 MLFS
Phase I (NCT00799799) Curti et al. (71)	AML (n = 16)	Conditioning with Cy/Flu followed by KIR ligand-mismatched NK-cell infusion; IL-2 therapy (3× weekly for 2 weeks)	PBNK cells from haploidentical donors. [a]CD3 depleted and CD56 selected	Single dose: 1.29–5.53 × 10^6 cells/kg	Median purity: infused CD3+ cells: 0.65 × 10^5 cells/kg. Mean viability: 95%	Feasible study, moderate toxicity. 9/16 DFS, 7/16 in relapse (3–51 months), 1/16 died of bacterial pneumonia
Phase II (NCT00526292) Shaffer et al. (70)	AML (n = 6) and MDS (n = 2)	Conditioning with Cy/Flu followed by HLA-mismatched NK-cell infusion; IL-2 therapy (6×) starting 1 day before and after NK-cell infusion	PBNK cells from haploidentical donors. [a]CD3 depleted and CD56 selected	Single dose: 4.3–22.4 × 10^6 cells/kg	Purity: ≥90% CD3− CD56+ cells. CD3+ cells <0.1%. Viability: 82–100%	No GvHD. 3/8 PR, 5/8 no response. Median survival is 12.9 months
Phase I (NCT01212341) Yang et al. (69)	Lymphoma (n = 2) and solid tumor (n = 18)	KIR ligand-mismatched NK-cell infusion	*Ex vivo* expanded and activated PBNK cells from unrelated donors. Culture duration: 14 days with irradiated auto-PBMCs, OKT3 and IL-2	Single dose: 1 × 10^6 cells/kg (cohort 1) 1 × 10^7 cells/kg (cohort 2) Repeated dose: 1 × 10^6 cells/kg (cohort 3) 3 × 10^6 cells/kg (cohort 4) 1 × 10^7 cells/kg (cohort 5), and 3 × 10^7 cells/kg (cohort 6)	Purity: CD16 +/CD56+ cells: 98.13 ± 1.98%; CD3+ cells: 0.41 ± 0.43%; CD14+ cells: 0.40 ± 0.37%; CD19+ cells: 0.15 ± 0.25%. Fold expansion: 757.5 ± 232.2. Viability: 92.9 ± 2.1%	No GvHD nor severe toxicities. 8/20 SD, 9/20 PD, 3/20 NE. Median PFS in SD patients: 4 months (2–18 months)
Phase I (NKAML: NCT00697671) Pilot study (NKHEM: NCT00187096) Rubnitz et al. (67)	Relapsed leukemia post HSCT (n = 15) Refractory/ relapsed leukemia (no prior HSCT) (n = 14)	Conditioning with Clo/Eto/Cy followed by KIR-matched or -mismatched NK-cell infusion; IL-2 therapy (6×) starting 1 day before and after NK-cell infusion	*Ex vivo* expanded PBNK cells from haploidentical donors. Culture duration: >12 h. [a]CD3 depleted and CD56 selected	Single dose: 3.5–103 × 10^6 cells/kg	Median purity: 98.4% CD56+ cells. 0% CD3+ CD56− T cells. 0.31% CD19+ B-cells	Well tolerated, no GvHD. 6/29 PR, 14/29 CR, 8/29 no response, and 1/29 NE. 4/29 are alive and DFS
Phase I (EudracT number: 2005-006087-62) Kottaridis et al. (66)	AML (n = 7)	Conditioning with Flu and TBI followed by haploidentical tumor primed NK-cell infusion	*Ex vivo* expanded and activated PBNK cells from haploidentical donors. Culture duration: o/n with CTV-1 lysate and cryopreserved for infusion. [a]Only CD56 selected	Single dose: 1 × 10^6 cells/kg	Purity: CD56+ cells 97.17% of which 80% CD56+ CD3− cells	Serious adverse reactions, no GvHD. 3/7 in CR remained in remission, 1/7 in PR achieved CR, 2/7 relapsed and 1/7 died (6 months follow-up). Median OS: 141–910 days

(Continued)

TABLE 2 | Continued

Study	Malignancy	Clinical Trial design	Culture method[a]	Infused dose NK cells	Final product characteristics	Outcome
Phase I (BB-IND-14560) Szmania et al. (65)	MM (*n* = 8)	Conditioning with Bor (+/−Cy/Flu/Dex) followed by fresh haplo-(*n* = 6) or cryopreserved auto (*n* = 2) NK cells	*Ex vivo* expanded and activated PBNK cells from haploidentical (fresh) and autologous (cryopreserved) donors. Culture: 8–9 days with K562-mb15-41BBL stimulator cells and IL-2	Single dose: 2×10^7–1×10^8 cells/kg	Median purity: 78% CD3− CD56+ cells. CD3+/CD56− 0.1%. Viability cryopreserved: 94%. Viability fresh: 93%. Recovery cryopreserved: 16%. Recovery fresh: 119%	Feasible and safe. 1/8 PR, 6/8 PD, 1/8 NE, and 3/8 died between days 11 and 98 after NK-cell infusion
Phase II (NCT00274846) Bachanova et al. (64)	AML (*n* = 57)	Conditioning with Cy/Flu; IL2DT in cohort 3 followed by haploidentical NK-cell infusion 1 day later; IL-2 therapy (14×, daily)	*Ex vivo* expanded and activated PBNK cells from haploidentical donors. Culture duration: o/n with IL-2. [a]CD3 depleted (cohort 1) or CD3 depleted/CD56 selected (cohort 2) or CD3/CD19 depleted (cohort 3)	Single dose: $0.96 \pm 0.3 \times 10^7$ cells/kg (cohort 1) $0.34 \pm 0.05 \times 10^7$ cells/kg (cohort 2) $2.6 \pm 1.5 \times 10^7$ cells/kg (cohort 3)	Purity: NK cells 39 ± 9%, T cells: 0.7% (cohort 1) NK cells 75 ± 6%, T cells: 1.3% (cohort 2) NK cells 54 ± 16%, T cells: 0.3% (cohort 3)	Well tolerated, no GvHD and mild toxicities. 9/42 in remission (1.8–15 months) (cohorts 1 and 2, *n* = 42). 8/15 in remission (1–32 months) (cohort 3, *n* = 15). DFS: 5% (cohorts 1 and 2) and 33% in cohort 3
Tonn et al. (43)	Solid tumors/ sarcoma (*n* = 12) Leukemia/ lymphoma (*n* = 2)	Pretreatment with mPred following NK-92 cell infusion	*Ex vivo* expanded and activated allogeneic NK-92 cells. Culture duration: 100–300 h with IL-2. [a]No selection	Repeated doses (2 × 48 h apart): 1×10^9 (cohort 1), 3×10^9 (cohort 2) and 1×10^6 (cohort 3) cells/m^2 and additional dose level of 10^{10} cells/m^2 in some patients	Viability: >80%. Fold expansion: 32	Infusion of 10^{10} NK-92 cells/m^2 were well tolerated. 12/15 PD, 2/15 MR, 1/15 SD for 2 years, OS: 13–801 days
Pilot study (NCT00799799) Curti et al. (63)	AML (*n* = 13)	Conditioning with Cy/Flu followed by KIR ligand-mismatched NK-cell infusion; IL-2 therapy (3× weekly for 2 weeks)	PBNK cells from haploidentical donors. [a]CD3 depleted and CD56 selected	Single dose: 1.11–5 × 10^6 CD3− CD56+ cells/kg	Mean viability: 95%. Median purity: 93.5% NK cells. Maximum T-cell dose 10^5 cells/kg	Feasible and safe, no GvHD. 5/13 active disease: 1/5 CR (6 months), 4/5 died of PD. 3/6 treated in CR are DFS (34, 32, and 18 months), 2/13 in MR in CR (4 and 9 months)
Phase II (BB-IND 8847) Geller et al. (33)	Refractory metastatic breast cancer (*n* = 14) Ovarian cancer (*n* = 6)	Conditioning with Cy/Flu with or without TBI followed by allogeneic NK-cell infusion; IL-2 therapy (3× weekly for 2 weeks)	*Ex vivo* expanded and activated PBNK cells from haploidentical donors. Culture duration: o/n with IL-2	Single dose: 8.33 × 10^6–3.94×10^7 cells/kg	Viability: >70%. Median T cells: 0.11% CD3+ cells	TLS and PLS and limited infusion or IL-2 related toxicities. 1/20 died due to grade 5 toxicity. 4/20 PR, 12/20 SD, and 3/20 PD (between 31 and 109 days)
Pilot study Bachanova et al. (62)	B-cell NHL (*n* = 6)	Conditioning with Cy/Flu and mAb (rituximab, 4×) before and after haplo NK-cell infusion followed by IL-2 therapy (6×, every other day)	*Ex vivo* expanded and activated PBNK cells from haploidentical donors. Culture duration: 8–16 h with IL-2	Single dose: $21 \pm 19 \times 10^6$ NK cells/kg	Purity: 43 ± 11% NK cells. 0.16 ± 0.12% T cells	Feasible and safe. 2/6 CR, 2/6 relapsed at 6 months, 2/6 died
Pilot study NKAML Rubnitz et al. (61)	AML (*n* = 10)	Conditioning with Cy/Flu followed by KIR-mismatched NK-cell infusion; IL-2 therapy (6×) starting 1 day before and after NK-cell infusion	PBNK cells from haploidentical donors. [a]CD3 depleted and CD56 selected	Single dose: 5–81 × 10^6 cells/kg	Median purity: B-cells 0.097×10^6 cells/kg. T cells 1×10^3 cells/kg	Feasible and safe. 10/10 in remission (569–1,162 days)

(Continued)

TABLE 2 | Continued

Study	Malignancy	Clinical Trial design	Culture method[a]	Infused dose NK cells	Final product characteristics	Outcome
Phase I (EudraCT number: 2005-005125-58) Iliopoulou et al. (60)	Non-SCLC (n = 16)	Haploidentical NK-cell infusion after chemotherapy	*Ex vivo* expanded and activated PBNK cells from haploidentical donors. Culture duration: 21–23 days with IL-15 followed by 1 h with IL-15 and hydrocortisone. [a]Only CD56 selected	Repeated doses (2–4): 0.2–29 × 10^6 cells/kg per dose	Median purity: (T cells) CD3+ CD56+ CD28– 0.12 × 10^6 cells/kg. CD56+ CD3 cells 97.9% (after 20 days culture). Fold expansion: 23	Safe, no GvHD. 2/16 PR, 6/16 SD, 7/16 PD, 1/16 not treated. 1-year OS 56% (9/16), 2-year OS 19% (4/16)
Phase I Arai et al. (42)	Metastatic RCC (n = 11) or Malignant Melanoma (n = 1)	NK-92-cell infusion	*Ex vivo* expanded and activated allogeneic NK 92 cells. Culture duration: 15–17 days with or without IL-2. [a]No selection	Repeated doses (3× in cohort): 1 × 10^8 (cohort 1), 3 × 10^8 (cohort 2), 1 × 10^9 (cohort 3), and 3 × 10^9 (cohort 4) cells/m^2	Fold expansion: 200 over 15–17 days. Viability: ≥80%	Safe and feasible, mild toxicities (1 grade 4, hypoglycemia). 10/12 PD (died between day 101 and 1,059), 1/12 alive (1,450 days) and 1/12 died of bronchopneumonia (day 832)
Phase I (BB-IND 8847) Miller et al. (50)	Metastatic Melanoma (n = 10), Metastatic RCC (n = 13), Refractory HL (n = 1), and AML (n = 19)	Conditioning with low Cy/mPred or Flu or high-Cy/Flu followed by NK-cell infusion; IL-2 therapy (14×, daily)	*Ex vivo* expanded and activated PBNK cells from haploidentical donors. Culture duration: o/n with IL-2	Escalating doses: low cy/mPred: 1 × 10^5, 1 × 10^6, 1 × 10^7, or 2 × 10^7 cells/kg (at least three per cohort). Flu or high-Cy/Flu: 2 × 10^7 cells/kg	Viability: >70%. Purity: NK cells 40 ± 2%. T cells 1.75 ± 0.3 × 10^5 cells/kg is 0.9 ± 0.1%. Monocytes 25 ± 1.6% and B-cells 19 ± 2%	Feasible and tolerated without toxicities. Low-Cy/mPred: 2/17 with MRCC SD for 20 and 21 months. 4/17 with MM SD for 4–9 months (n = 17) High-Cy/Flu: 5/19 AML pts in CR (n = 19)

CNS, central nervous system; MPDs, myeloproliferative disorders; LPD, lymphoproliferative disorder; MM, multiple myeloma; MDS, myelodysplastic syndromes; MDN, myelodysplastic neoplasms; MPN, myeloproliferative neoplasms; AML, acute myeloid leukemia; LBLL, lymphoblastic leukemia-lymphoma; ALL, acute lymphoblastic leukemia; NB, neuroblastoma; RMS, rhabdomyosarcoma; CML, chronic myelogenous leukemia; NHL, non-Hodgkin's lymphoma; MCL, mantle cell lymphoma; DLBCL, diffuse large B cell lymphoma; FL, follicular lymphoma; ALCL, anaplastic large cell lymphoma; HL, Hodgkin's lymphoma; RCC, renal cell cancer; SCLC, small cell lung cancer; CLL, chronic lymphocytic leukemia; HCC, hepatocellular carcinoma; PNET, primitive neuroectodermal tumor; ACC, adrenal cortical carcinoma; EWS, Ewing sarcoma; DSRCT, desmoplastic small round cell tumor; Cy, cyclophosphamide; Flu, fludarabine; Bor, bortezomib; Dex, dexamethasone; Clo, clofarabine; Eto, etoposide; Cis, cisplatin; Pac, paclitaxel; Doc, docetaxel; Vin, vinorelbine; Gem, gemcitabine; Car, carboplatin; Pem, pemetrexed; TBI, total body irradiation; Tac, tacrolimus; Pred, prednisolone; mPred, methylprednisolone; Thy, thymoglobulin; Vin, vincristine; Adr, adriamycin; Predn, prednisone; Mel, melphalan; OKT3, muromonab-CD3; HSPC, human stem and progenitor cells; aGvHD, acute graft versus host disease; DNKI, donor NK-cell infusion; Stim, stimulated; Unstim, unstimulated; DFS, disease-free survival, CR, complete remission; PR, partial response; MR, minimal response; SD, stable disease; PD, progressive disease; aNK-DLI, donor-derived IL-15/4-1BBL activated NK-cell infusion; TRM, transplant-related mortality; MC, mononuclear cell; TLS, tumor lysis syndrome; PLS, passenger lymphocyte syndrome; NE, not evaluable.

[a]*Culture method displays CD3 depleted PBMC's, otherwise deviated selection method is mentioned in product characteristics.*

refractory leukemia. NK cells were administered along with IL-2 injections in 29 patients, out of which 14 had not undergone HSCT (cohort I) and the other 15 have relapsed after HSCT (cohort II). In total, 90% of the NK cell donors were KIR mismatched and when the outcomes from both cohorts were combined, 14/29 were in CR, 6/29 showed PR, and 8/29 patients showed no response to the treatment (67).

The first clinical trial to adoptively transfer allogeneic NK cells derived from peripheral blood of unrelated donors without immune suppression was performed by Yang et al. In this study, allogeneic IL-2-activated NK cells (MG4101) were expanded at Green Cross Lab Cell (68) and administered to patients with advanced lymphoma and recurrent solid tumors. Following NK cell-adoptive transfer, an increase in NKG2D expression levels on $CD8^+$ T cells, a reduction in the number of T-regs, and MDSCs followed by a decrease in serum levels of transforming growth factor-beta were noted. An enhanced PFS was noted in KIR-mismatched NK cell recipients. In addition, a KIR B haplotype was associated with a higher incidence of stable disease (SD). This study demonstrated that KIR-ligand mismatched donor NK cells can be safely administered without any sign of GvHD and with a GvT effect. Though an antitumor effect of the adoptively transferred NK cells could be observed, their persistence *in vivo* was shorter (between 1 and 4 days) in comparison to other clinical trials. This stresses the potential need for an effective NK cell-promoting conditioning regimen, to increase the life span and migration of NK cells in patients (69).

Shaffer et al. published results from a Phase II study in patients with relapsed or progressive AML ($n = 6$) or MDS ($n = 2$), treated with allogeneic NK-cell infusions and supported by IL-2 injections *in vivo*. NK cell donor chimerism was not detected post infusion, and no signs of GvHD were reported from this study. About 3/8 patients achieved PR of which 1/6 patients with AML and 1/2 patients with MDS achieved a CR after treatment but relapsed within 2 months. Of note, both these patients survived for 20.2 months post NK infusion, while the remaining 5 patients without response had a median survival of 5.4 months (70).

Around the same time, Curti et al. published results from a Phase I trial using KIR ligand-mismatched haploidentical NK cells to treat AML patients in CR ($n = 16$). About 7/16 patients relapsed, while 9/16 remained disease free at a median follow-up of 22.5 months. Overall, 69% (11/16) responded to therapy, with no signs of GvHD. Prolonged DFS was higher in patients with an absolute increase in the number of circulating alloreactive NK cells in this study (71).

Romee et al. investigated the antitumor effects of cytokine induced memory-like NK cells, which were adoptively transferred in relapsed or refractory AML patients after overnight activation with IL-12, IL-15, and IL-18. They were defined as memory-like NK cells based on their enhanced responsiveness upon restimulation with cytokines. NK-cell infusions were safe and well tolerated and no GvHD was reported in this study. Out of nine evaluable patients, four had CR and five showed disease responses (72).

Most recently, a first clinical trial using UCB CD34+ progenitor cell-derived NK cells was published by Dolstra et al. Allogeneic NK cells were generated from UCB CD34+ cells using an *ex vivo* expansion and differentiation method developed by Glycostem Therapeutics (41). In this study, 10 AML patients in morphologic CR, who were ineligible for HSCT transplantation, received partially HLA-matched (5/10 KIR ligand-ligand mismatched and 7/10 KIR receptor-ligand mismatched) UCB-NK cells (oNKord®). Following Cy/Flu conditioning, lymphocytopenia was induced and found to correlate with elevated IL-15 levels, which peaked at day 6 after NK-cell infusion. Out of 10 treated patients, 5 were alive and 4 had a DFS of 60, 52, 22, and 16 months after infusion. About 2/4 patients with very poor prognosis (i.e., with detectable MRD in bone marrow before NK infusion) became MRD negative for 6 months post NK infusion, indicating a potential GvT effect. UCB-NK-cell infusions were safe and well tolerated without signs of GvHD. Interestingly, UCB-NK cells expressing low levels of KIRs and CD16a at the end of the *ex vivo* culture, underwent further maturation post-transfer *in vivo*, resulting in the upregulation of KIRs and CD16a, but continued to preserve the activated phenotype denoted by high expression of NKp30, NKp44, NKp46, NKG2D, and DNAM (73).

In addition to NK cells from PBNK and UCB-NK, two clinical studies reported on the use of the NK-92 cell line. A Phase I trial conducted by Arai et al. investigated the safety and feasibility of allogeneic NK-92 cells in advanced renal cancer and melanoma. A total of 12 patients were evaluated and 6/12 had PD, 4/12 had SD, while 1/12 had minor response and 1/12 had mixed response, 4 weeks post infusion (42). Similarly, in another study with NK-92 conducted by Tonn et al., 15 patients were included with advanced solid ($n = 13$) and hematological malignancies ($n = 2$). About 1/7 tested patients produced antibodies against the HLA antigens expressed by NK-92 cells, 1/15 showed SD, 1/15 a mixed response, and the rest of the patient group had disease progression, being treated with a maximum tolerated dose of 10^{10} cells/m^2. NK-92 cells had a very short persistence (48 h) *in vivo* (43). As NK-92 cells are derived from cancer (lymphoma) cells and require irradiation before infusion, which could hamper their ability to proliferate and home *in vivo*, potentially limiting their efficacy.

Overall, analyzing the data from adoptive allogeneic NK cell therapy trials in a non-transplant setting, we conclude that such treatments are very safe and well tolerated and efficacious in hematological malignancies, especially in AML, but as yet relatively ineffective in solid tumors. Trials using allogeneic NK cells alone yielded valuable information on the *in vivo* persistence, donor chimerism, and antitumor potential in different indications. Furthermore, unlike combined approaches with HSCT, the absence of life-threatening GvHD and major treatment-related toxicities makes this method advantageous and provides an opportunity to further enhance the cytotoxic effects of allogeneic NK cells.

APPROACHES TO AUGMENT NK CELL FUNCTIONS: A VIEW ON BIOTECH INDUSTRIES

As reviewed above, various clinical trials have been published, mainly initiated by academia, proposing allogeneic NK cells as an effective therapeutic option. As a result of these studies, interest in NK cell-based immunotherapy strategies has been engendered in an increasing number of biotech companies. Clinical trials conducted in academia are often restricted to Phase I or II, as progression of experimental therapies to Phase III clinical trials and further on to commercialization and marketing requires a level of funding that surpasses the capacity of academic institutions. The financing of market enabling studies is coming mainly from industry. Although NK cells can be effective in some types of cancer as a monotherapy, considering their heterogeneity, complex networking, and the inherent adaptability of several tumors to evade killing by immune cells, one believe is that it is necessary to improve on the efficacy of currently available NK cell products. In this respect, it is worthwhile to consider combinatorial approaches of different treatment strategies involving NK cell functions. A summary of biotech companies involved in NK cell research is listed in **Table 3** and **Figure 1**. Here, we review for a selected group of NK cell companies, which develop NK cell-specific treatments, the underlying scientific principles and findings of their product pipelines, revealing highly innovative concepts that herald future clinical applications.

Fc OPTIMIZED MONOCLONAL ANTIBODIES (mAbs)

The potential of NK cells to mediate ADCC with therapeutic mAbs has been well described over the years (74). However, concerns have been voiced based on results from certain clinical trials, showing that polymorphisms in NK CD16 (V158V, V158F, and F158F) could influence the efficacy of mAb treatment and ADCC (75). To address this issue and limit the variations between different CD16 sequences, Fc glyco-engineered (defucosylated) mAbs with enhanced binding affinities to NK CD16a were developed. The Fc optimized anti-CCR4 mAb mogamulizumab (76) (Kyowa Hakko Kirin) has entered Phase III clinical testing in patients with adult T cell leukemia, emerging as the lead NK cell ADCC product to reach the market soon. Fc-optimized anti-CD20 mAbs Obinutuzumab (Genentech) (77) and Ocaratuzumab (Mentrik Biotech, LLC) (78) are currently tested in patients with chronic lymphocytic leukemia and follicular lymphoma. Similarly, the Fc-optimized anti-EGFR mAb imgatuzumab (Roche Glycart) is tested in Phase I/II clinical trials for head and neck cancer and in KRAS mutant colorectal cancer (79, 80). Although Fc-engineered mAbs address NK-mAb-binding issues, reports of serious side effects, like from the imgatuzumab study (81), have made the scientists rethink this strategy and call for the careful study of the advantages and disadvantages of this approach.

BISPECIFIC ANTIBODIES

In the last decade, several bispecific and trispecific Ab platforms, simultaneously targeting immune cells and tumor cells, have been developed in the field of cancer immunotherapy (82). To date, the majority of bispecific Abs that has been developed targets T cells, while only a limited number of bispecific approaches targets NK cells (83). Affimed is a clinical stage pharmaceutical company developing bifunctional antibodies that recruit immune cells such as T and NK cells to tumor sites. These bispecifics (TandAbs) are tetravalent in nature, thus offering four binding sites, two aimed at tumor antigens and two aimed at immune cells. Currently, Affimed's AFM13 that targets CD30 on cancer cells and CD16a on NK cells is in clinical Phase II testing in patients with HL. In Phase I studies AFM13 was found to be safe and well tolerated and resulted in an overall response rate of 23%. Furthermore, AFM13 treatment resulted in an increase in NK cell activation and a decrease in soluble CD30 levels in peripheral blood (NCT01221571) (84). Further, two other bispecific CD16a-based tumor targeting antibodies are in preclinical phase development, i.e., AFM22 and AFM24 that bind to EGFRvIII expressed by several solid tumors, including glioblastoma (GBM), and wild-type EGFR, respectively. Another promising NK cell-focused bispecific platform is developed by AvidBiotics to target tumors that evade NK killing *via* downregulation or shedding of the NKG2D ligand MICA, which is a major limiting step in NK-mediated tumor targeting. To overcome this, AvidBiotics designed MicAbody proteins that bind to the NK cell NKG2D receptor with high affinity. Further, this MicAbody was engineered with an additional binding site to target tumor antigens of interest, thus enabling recruitment of NK cells to tumors (85).

NK CELL CHECKPOINT INHIBITORS

Another strategy to increase NK cell functionality is the disruption or blocking of NK inhibitory signals. Innate Pharma is a clinical stage pharmaceutical company focused on developing NK cell checkpoint inhibitors. Lirilumab (IPH2102/BMS 986015) is a fully humanized IgG_4 anti-KIR mAb against the inhibitory KIRs KIR2DL1, L2, and L3, which are expressed predominantly on NK cells and on some T cells. Lirilumab induced significant anti-tumor activity of NK cells against HLA-C-expressing tumor cells, contributing to increased survival in lirilumab-treated mice (86). Similar to KIRs, the NK cell inhibitory receptor NKG2A binds to its ligand HLA-E on tumor cells resulting in an inhibition of NK cell function. HLA-E is overexpressed in colon, cervical, and ovarian cancers, thus serving as an escape mechanism for NK killing in these tumors (87, 88). The anti-NKG2A mAb monalizumab was developed to block the interaction between NKG2A and HLA-E and is currently under clinical investigation. IPH4102, which targets KIR3DL2, is under Phase I clinical investigation in cutaneous T cell lymphoma (CTCL). Clinical trials testing lirilumab, monalizumab, and IPH4102 are listed in **Table 3**.

GENETIC MODIFICATION OF NK CELLS

In addition to successful expansion, differentiation, and demonstrable anti-tumor effects of NK cells, NK cell tumor targeting can be made more specific by employing chimeric antigen receptors (CARs) as demonstrated for T cell adoptive transfer strategies (89). CARs are recombinant Ab-based molecules that upon expression in immune effector cells bind antigens of interest on target cells, resulting in immune activation and enhanced immune effector cell survival through specific

TABLE 3 | List of biotech NK cellular therapies and NK cell function enhancing compounds.

Company	NK cell product	Product characteristics	Disease target	Product stage
Fortress Biotech Inc.	CNDO-109	Tumor primed NK cells	AML	Phase I/II
Multimmune GmbH	ENKASTIM-ev	A synthetic peptide which mimics Hsp70 and activates NK cells *ex vivo*	Metastatic colon and non-small cell lung cancer	Phase II
Glycostem Therapeutics	oNKord®	NK cells derived from umbilical cord blood (UCB) progenitor cells	AML and solid tumors	Phase I (AML)
NantKwest Inc.	Activated NK-92 cells (aNK cell)	IL-2-dependent tumor cell-derived NK cell line	Solid tumors and hematological malignancies	Phase I
	High affinity NK cells (haNK)	aNK cells genetically modified to express CD16 for ADCC with therapeutic mAbs	Ideally in combination with IgG_1 therapeutic mAbs in solid tumors (e.g., cetuximab) and hematological malignancies (e.g., rituximab)	Preclinical
	Target-activated NK cells (taNK)	aNK cells genetically modified to express CARs	NK-92 CARs are developed targeting tumor antigens in neuroblastoma, melanoma, breast cancer, MM and leukemias	Preclinical
Green Cross Lab Cell	MG4101	*Ex vivo* expanded NK cells derived from CD3 depleted unrelated donors	Solid tumors and lymphoma (NCT01212341)	Phase I
Gamida Cell	NAM-NK cells	Nicotinamide-based PBNK cell culture system	Solid tumors and hematological malignancies	Preclinical
Celgene Cellular Therapeutics	NK cells	NK cells derived from UCB and placenta.	Solid tumors and hematological malignancies	Preclinical
Fate Therapeutics Inc.	iNK cells	NK cells derived from induced pluripotent stem cells	Solid tumors and hematological malignancies	Preclinical
Sorrento Therapeutics Inc	CARs to enhance tumor homing of NK-92 cells	NK-92 cells CAR targeting programming death ligand-1 and NK-92 CAR targeting receptor tyrosine kinase like orphan receptor to increase NK-92 tumor homing	Solid tumors and hematological malignancies	Preclinical
Nkarta Therapeutics	NKG2D CARs	NKG2D CARs developed with NK-92 and PBNK to enhance the functions of NKG2D receptor in NK cells	Osteosarcoma and hepatocellular carcinoma	Preclinical
Ziopharm Oncology Inc.	HLA gene editing	Zinc finger nuclease technology to delete HLA-A sequences from allogeneic NK cells, allowing them to evade recipient T cell killing	Solid tumors and hematological malignancies	Preclinical
Company	**NK cell enhancing products**	**Product characteristics**	**Disease target**	**Product stage**
Kyowa Hakko Kirin	Mogamulizumab	Fc optimized anti CCR4 CD20 mAb	CTCL	Phase III
Genentech	Obinutuzumab	Fc optimized anti CD20 mAb	CLL	Phase II
Mentrik Biotech, LLC	Ocaratuzumab	Fc optimized anti CD20 mAb	CLL	Phase II
Roche Glycart	Imgatuzumab	Fc optimized anti EGFR mAb	Head and neck and KRAS mutant colorectal cancer	Phase I/II
Affimed N.V.	AFM13	Bispecific antibody binding to CD16a on NK cells and CD30 on tumor cells	Hodgkin's lymphoma and lymphomas	Phase II
	AFM22	Bispecific antibody binding to CD16a on NK cells and EGFR vIII on tumor cells	Head and neck and solid tumors	Preclinical
	AFM24	Bispecific antibody binding to CD16a on NK cells and wild type EGFR on tumor cells	EGFR-expressing solid tumors	Preclinical
Innate Pharma S. A.	Lirilumab	mAb to block NK cell inhibitory signaling from KIRs (KIR2DL1–3)	As monotherapy (Phase II, NCT02399917), with nivolumab (Phase I, NCT01592370), with ipilliumab (Phase I, NCT01750580), 5-azacytidine (Phase I, NCT02399917), with nivolumab + 5-azacytidine (Phase II, NCT02599649), with elotuzumab (NCT02252263) and with rituximab (Phase I, NCT02481297)	Phase I/II

(Continued)

TABLE 3 | **Continued**

Company	NK cell product	Product characteristics	Disease target	Product stage
Innate Pharma S. A.	Monalizumab	mAb to block NK cell inhibitory receptor NKG2A	As monotherapy (Phase I/II, NCT02459301, NCT02331875) with cetuximab (NCT02643550), with ibrutinib (NCT02557516) and with durvalumab (NCT02671435)	Phase I/II
	IPH4102	mAb to block NK cell inhibitory receptor KIR3DL2	As monotherapy in CTCL (NCT02593045)	Phase I
	IPH4301	mAb to target NKG2D ligands MICA/MICB and it also mediates ADCC with NK cells	Solid tumors and hematological malignancies	Preclinical
Altor Biosciences corporation	ALT-803	IL-15 super agonist reported to stably express IL-15. Increases NK cell proliferation *in vivo*, also enhances expansion of migratory NK subsets	Advanced solid tumors (NCT01946789), MM (NCT02099539), HIV patients (NCT02191098), with nivolumab in NSCLC (NCT02523469), with rituximab (NCT02384954) in B cell Non-Hodgkin Lymphoma (NHL) (NCT02384954), with (BCG) in Non-Muscle Invasive Bladder Cancer (NCT02138734), with chemotherapy drugs gemcitabine and Nab-paclitaxel in advanced pancreatic cancer (NCT02559674)	Phase I/II
NOXXON Pharma	NOX-A12	Functions as chemokine receptor CXCL12 inhibitor, enables the release of CXCL12 from the surface of tumor stromal cells, thus facilitating migration of tumor cells toward NK cells	Solid tumors and MM	Preclinical
AvidBiotics	MicAbody proteins	Dual role: binds to NKG2D receptor in NK cells and to target antigens of interest simultaneously	Solid tumors and hematological malignancies	Preclinical

KIRs, killer cell immunoglobulin-like receptors; ADCC, antibody-dependent cell mediated cytotoxicity; aNK, activated NK cells; haNK, high affinity NK cells; taNK, target activated NK cells; mAbs, monoclonal antibodies; NAM, nicotinamide; CARs, chimeric antigen receptors; EpCAM, epithelial cell adhesion molecule; AML, acute myeloid leukemia; KIR2DL1–3, killer cell immunoglobulin like receptor two domains long cytoplasmic tail 1–3; KIR3DL2, killer cell immunoglobulin like receptor three domains long cytoplasmic tail 2; MICA/MICB, MHC class-I-chain related protein A and B; HIV, human immunodeficiency virus; NSCLC, non-small-cell lung cancer; BCG, Bacillus Calmette Guerin; MM, multiple myeloma; CLL, chronic lymphocytic leukemia; CTCL, cutaneous T cell lymphoma; EGFR, epidermal growth factor receptor.

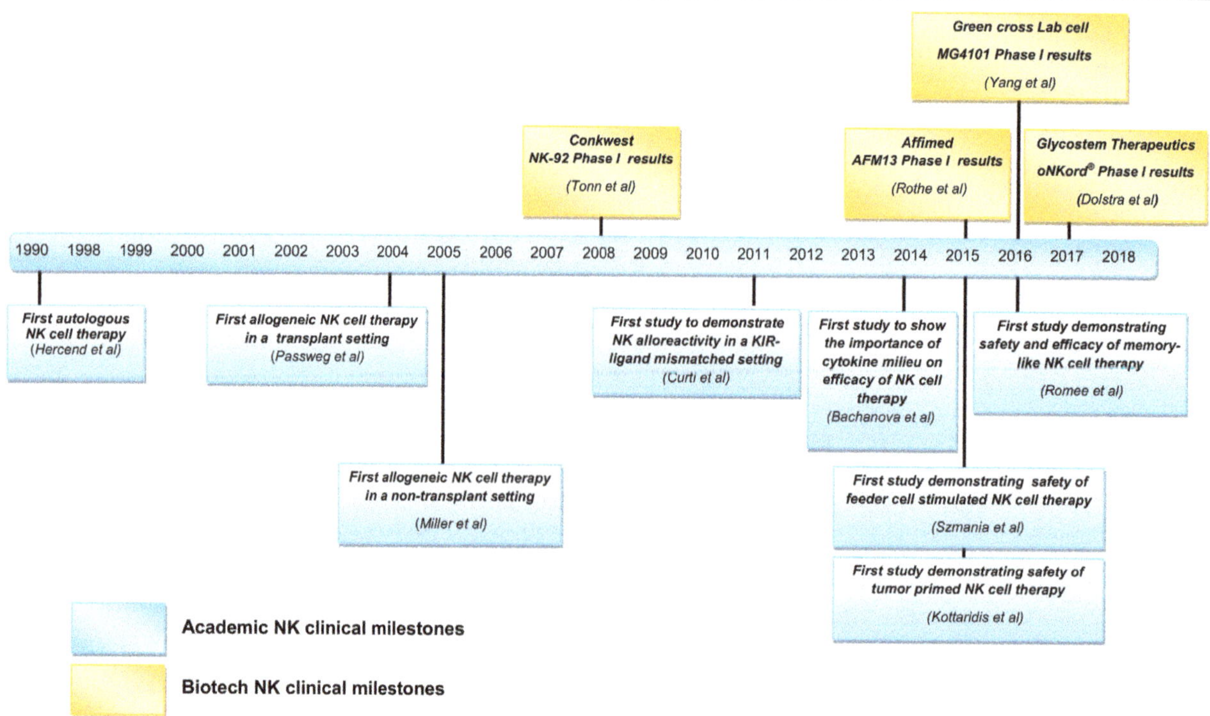

FIGURE 1 | **Summary of natural killer (NK) cell clinical milestones from academia and biotech industries.**

intracellular signaling motifs fused to the antigen binding domain [usual a single-chain Fv fragment (scFv)]. PBNK-CARs against breast cancer (HER-2), NB (CD244), and CD19 + B-cell precursor cell ALL (CD19) (90) have demonstrated efficacy in preclinical studies, while two clinical trials are ongoing using modified haplo-identical PBNK cells with anti-CD19 CARs in B cell malignancies (NCT00995137 and NCT01974479) (89). NantKwest, is actively involved in enhancing the functions of its lead product, parental NK-92 cells (activated NK cells, aNK), through gene modifications employing CARs to make them target specific. NK-92 CARs (taNK) are developed against tumor markers in NB (GD2), melanoma (GPA7) (91), breast cancer (EpCAM, HER-2, EGFR) (92, 93), MM [CS1 (94), CD138 (95)], and leukemias (CD19, CD20) (96) and have shown efficacy in preclinical studies. In an alternative approach, NK-92 cells have also been modified to express CD16a (high affinity NK cells, haNK) to promote ADCC (97). NantKwest has also partnered with Sorrento Therapeutics to develop NK-92 CARs targeting programmed death-ligand1 (PD-L1) (98) and receptor tyrosine kinase-like orphan receptor 1 (ROR-1) (99).

Besides specific targeting of tumor antigens and strategies to promote ADCC, Nkarta therapeutics developed NKG2D CARs (NKG2D-CD3ζ-DAP10) using NK-92 cells and PBNK cells, which exhibited enhanced cytotoxicity against osteosarcoma and hepatocellular carcinoma when compared to activated and expanded PBNK cells (100, 101). mRNA-based genetic engineering has been used to enhance migration of NK cells to tumors.

Apart from gene modification, gene editing is also widely used to overexpress or knock out genes of interest to augment NK cell function. Expression of HLA-A on allogeneic NK cells leads to rejection of allogeneic NK cells by the recipient's T and NK cells. Cooper and colleagues from Ziopharm Oncology used zing finger nuclease (ZFN) technology to remove HLA-A sequences from allogeneic NK cells, thus enabling these immune effector cells to escape rejection from recipient T cells. However, in that case, there is yet a high probability of being attacked by endogenous NK cells targeting HLA-A negative allogeneic cells. This was further addressed by retaining HLA-B and HLA-C genes in donor NK cells (102–104). To increase NK cell persistence *in vivo*, scientists at oNKo-innate identified a group of proteins called suppressor of cytokine signaling (CIS, SOCS 1–7), which negatively regulate CIS pathways. SOCS1 and SOCS3 bind to JAK1, JAK2, and TYK2 molecules and inhibit JAK activity. Similarly, CIS protein binds to JAK1 and suppresses IL-15 signaling in NK cells. It became evident from *in vivo* studies in mice with *Cish*$^{-}$/$^{-}$ knockout NK cells that loss of CIS led to prolonged IL-15 signaling, resulting in an increased proliferation, survival, and functionality of NK cells (105).

NK CELLS FROM iPSCs

In recent years, NK cells generated from induced pluripotent stem cells (iPSC-NK) and human embryonic stem cells (hESC-NK) have been gaining more interest as an NK cell therapeutic product. Fate Therapeutics developed a platform technology to generate NK cells from iPSC. hESC/iPSC were made into aggregates by centrifugation to form so-called embryoid bodies (spin EBs) (106), giving rise to hematopoietic progenitor cells expressing CD34 and CD45, which were then differentiated into mature NK cells using a specific cytokine cocktail. iPSC/hESC-derived NK cells were shown to express common NK cell markers, such as KIRs, CD16, NKp44, NKp46, NKG2D, and TRAIL, and were cytotoxic against several hematological and solid tumor cells *in vitro* (107, 108). In the next stage, iPSC/hESC-derived NK cells were successfully expanded using IL-2 and K562-based aAPCs with membrane-bound IL-21 to generate sufficiently high numbers for clinical applications (109).

NK CELLS FROM HUMAN UCB CELLS

Stem cell progenitors from cord blood offer a unique platform to be expanded and differentiated into cytotoxic NK cells. The low immunogenicity of cord blood cells strongly reduces the risk of relapse and GvHD after transplantation (110). Considering the advantages of using cord blood, Glycostem Therapeutics, a clinical stage biotech company, which in the last decade has developed a flexible platform technology to expand and differentiate NK cells from CD34+ cells (111), upgraded this into a large scale GMP UCB-NK platform for clinical implementation (oNKord®) (41). UCB-NK cells were infused at up to 30 × 10^6 cells/kg/bodyweight in elderly AML patients, resulting in excellent safety and initial efficacy in a Phase I trial. Infused oNKord® cells showed active migration to the marrow and further matured in the absence of any exogenous cytokine injections. This confirms previous findings from a preclinical model, showing migration to the bone marrow and upregulation of KIRs and CD16a *in vivo* as well as antileukemic activity (112). oNKord® is well characterized and was found to have a similar functionality and gene expression profile as PBNK cells (113). Furthermore, oNKord® is highly cytotoxic against solid tumor targets such as cervical cancer cells, in which killing was independent of HLA expression levels, tumor histology and HPV types (114), or colorectal cancer cells, in which killing was independent of tumor EGFR levels, and RAS and RAF mutations (115), thus paving the way for oNKord® as immunotherapy for advanced solid tumors.

CYTOKINES TO ENHANCE NK CELL FUNCTIONS

To improve the antitumor activity of autologous NK cells, systemic administration of clinical grade recombinant IL-2 (rIL-2) and single chain IL-15 (scIL-15) has been used in high doses and this has resulted in severe grade 3/4 toxicities (116–118). Since then, their safety and efficacy have been tested in low doses following NK cell-adoptive transfer in cancer patients (50, 63, 119). However, IL-2 resulted in expansion and mobilization of inhibitory T-regs, severely limiting NK cell cytotoxicity (120). This shifted the focus toward the use of IL-15 for clinical trials involving NK cells. Currently more potent and advanced heterodimeric IL-15, which has a longer shelf life than scIL-15, is being tested in several studies (121). IL-15 is known to be more effective in membrane-bound form (i.e., bound to its receptor), engaging target immune cells in a cell contact dependent manner. Campana and his team (from Nkarta Therapeutics) addressed this by stably transducing the membrane bound IL-15 (mbIL-15) gene into proliferating PBNK cells, which were stimulated with K562-mb15-41BBL. mbIL-15 resulted in increased survival, proliferation, and enhanced cytotoxic functions of NK cells (122). Further, Cyto-Sen Therapeutics compared mbIL-15 to K562-based aAPCs with mbIL-21. From their findings, it was evident that mbIL-21 NK cells have a significantly higher expansion and proliferation ability compared to mbIL-15 NK cells (123). Cyto-Sen also developed plasma membrane particles (PM21) engineered from K562-mb21-41BBL cells and found that these PM21 particles stimulated efficient NK cell expansion in AML patient's PBMC samples (124).

Compared to IL-2, the use of IL-15 minimizes capillary leak syndromes and has less side effects overall, thus providing a strong rationale to use IL-15 instead of IL-2. However, the use of mammalian recombinant IL-15 in the clinic has been limited due to its short half-life and decreased functional activity *in vivo*. Altor BioSciences corporation came up with a unique design to overcome these limitations. It developed an IL-15 super agonist known as ALT-803. It consists of a human IL-15 mutant N72D variant, which is stably complexed with a soluble human IL-15Rα sushi-Fc dimer protein. Enhanced biological activity of ALT-803 was reported in several preclinical studies showing durable anti-tumor activity in various solid and hematological malignancies (125–128). Furthermore, ALT-803 facilitated expansion of effector and migratory NK cell subsets and significantly decreased the metastatic activity of tumor cells in a murine colon cancer pulmonary metastasis model (129). ALT-803 stimulated primary human NK cells to exhibit increased degranulation, IFNγ production, and ADCC when exposed to B cell lymphoma cell lines coated with IgG_1 therapeutic anti-CD20 mAbs (130). Several clinical trials are currently ongoing with ALT-803 as monotherapy in patients with advanced solid tumors, hematological malignancies, and AIDS as summarized in **Table 3**.

PRIMING NK CELLS TO ENHANCE TUMOR KILLING

Mark Lowdell and his team proposed that for a NK cell to be able to kill tumor cells, it requires a priming and triggering signal. NK cells failing to kill tumor cells, though they are exposed to the triggering signal, remain inactive due to the absence of a priming ligand. To address this, Fortress Biotech (previously known

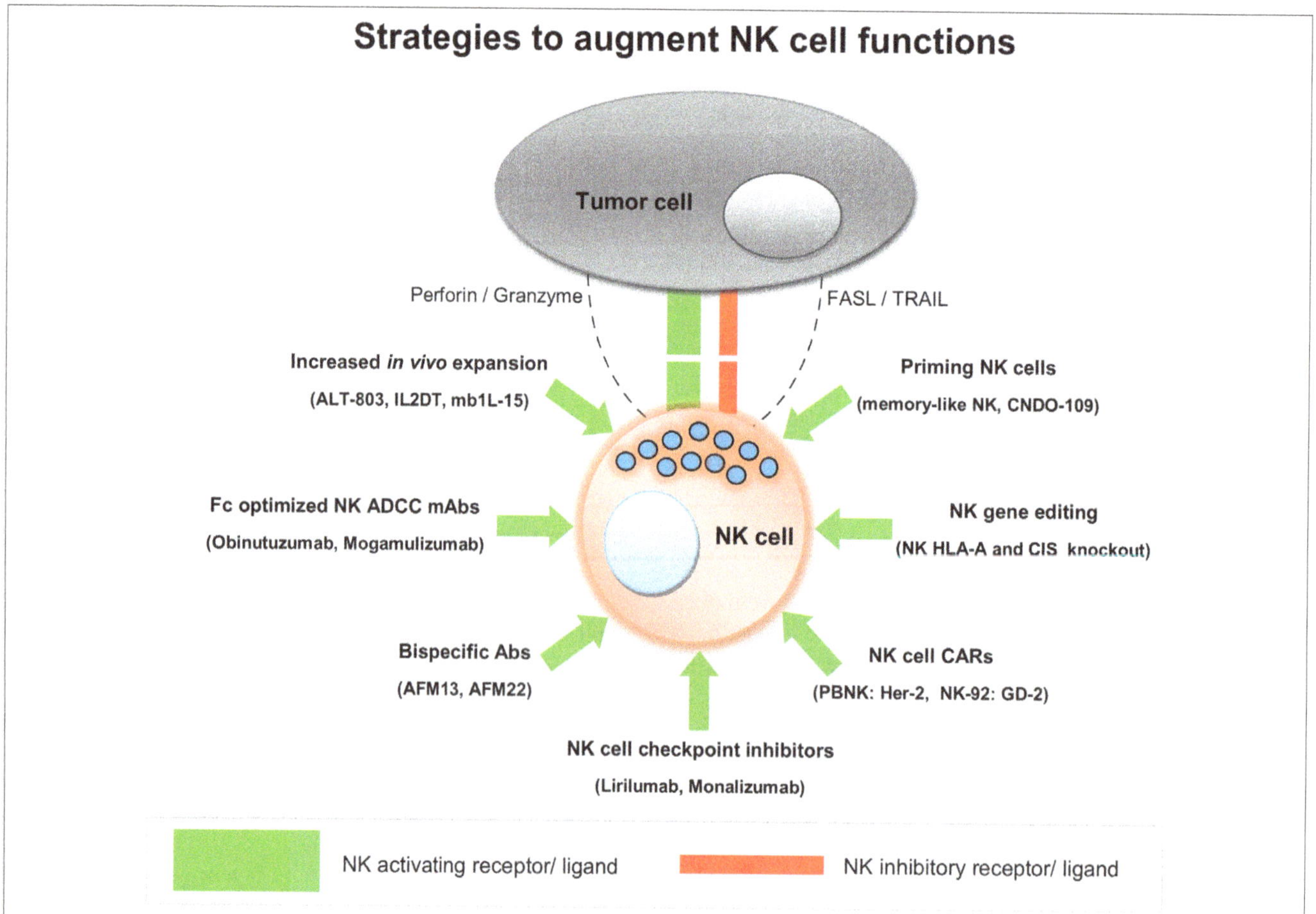

FIGURE 2 | **Strategies to augment natural killer (NK) cell functions.**

as Coronado Biosciences) developed a technology to increase NK cell tumor killing using cell lysates from the leukemia cell line CTV-1, known as CNDO-109, to prime NK cells. A Phase I/II clinical trial of activated PBNK cells from haploidentical donors coincubated with CNDO-109, infused at doses of up to 3×10^6 kg/recipient/body weight was tolerable without any adverse reactions. Out of seven evaluable patients, four remained disease relapse free for more than 1 year (131).

Another NK cell-activating product is ENKASTIM-ev, developed by Multimmune GmbH, which mimics the functions of heat shock protein 70 (Hsp70). ENKASTIM-ev resulted in NK specific activation and actively targeted Hsp70 expressing tumors. Safety of Hsp70-activated autologous NK cells has been documented in a Phase I study in patients with metastatic colorectal and non-small cell lung cancer (132).

ENHANCING NK CELL HOMING FUNCTIONS

Gamida-cell developed a feeder cell-free NK cell culture and expansion system containing nicotinamide (NAM) to generate NK cells from PBMC apheresis products. Nicotinamide, a derivative of vitamin B3, serves as a potent inhibitor of NAD dependent enzymes. Results from *in vivo* studies in mice showed that PBNK cells expanded with NAM in feeder free cultures exhibited increased homing potential toward lymphoid organs, with a significant increase in the expression of CD62L (L-selectin) compared to cultures without NAM (133).

TUMOR DISRUPTIVE TECHNOLOGY AIDING NK TUMOR RECOGNITION

NOXXON Pharma target chemokine receptor CXCL12, with the aim of increasing the sensitivity of tumor cells to drugs and immune cells. Their product NOX-A12 functions as a CXCL12 inhibitor and enabled the release of CXCL12 from the surface of tumor stromal cells and blocked its interaction with cell surface receptors CXCR4 and CXCR7. This mechanism facilitated the mobilization of CXCR4-expressing tumor cells from their tissue niches to areas, where they become easily accessible by NK cells or T cells (134, 135). Using tumor spheroids, increased mobilization of T and NK cells toward tumor cells in the tumor microenvironment was demonstrated. NOX-A12 also enhanced NK killing of obinutuzumab-coated Raji cells *in vitro*, mediated by ADCC (136).

CONCLUSION

From this literature review, we conclude that adoptive transfer of allogeneic NK cells in a non-transplant setting is safe and shows early signs of clinical efficacy against hematological and certain solid tumors. Current data are mostly based on Phase I clinical trials, and hence it is still too early to get an overall picture of NK cell alloreactivity in different kinds of cancer. Most of the clinical studies conducted so far have used primary NK cells but with limited efficacy, pointing to the need to improve the functionality of these NK cells after their transfer to patients. The growing opportunities to augment NK cell functions have attracted several biotech companies to invest in NK cell research, spearheading NK therapy development with different innovative approaches. This review also stresses the need for combining adoptive transfer of allogeneic NK cells with NK function-augmenting products to achieve a maximum anti-tumor effect. As NK cells are safe to infuse, the use of CAR-NK cells may be instrumental in providing a much safer but still very effective platform, to bring CAR-based therapies to broader clinical applications. It may also facilitate effective tumor targeting of NK cells. oNKord® and iPSC-derived NK cells could serve as alternative allogeneic platforms to develop CAR-NK products, besides NK cell lines. In a solid tumor setting, NK cells are challenged by several factors that affect their homing and penetration into the tumor tissues. Moreover, they should achieve and maintain an activated effector state, even in the face of immune suppressive conditions, that are prevalent in patients with cancer. To overcome these bottlenecks in NK therapy of solid tumors, a plethora of creative solutions are being pursued by numerous research labs as well as by biotech companies in clinical or close to clinical phase. Strategies to enhance NK cell functions from leading NK cell products are summarized in **Figure 2**. With all these exciting developments, NK cells are set to make a considerable impact on the future treatment of patients with hematological as well as with solid tumors.

AUTHOR CONTRIBUTIONS

JV, NK, HVV, TG, and JS wrote the paper. HVV, TG, HV, and JS reviewed the paper.

REFERENCES

1. Luna JI, Grossenbacher SK, Murphy WJ, Canter RJ. Targeting cancer stem cells with natural killer cell immunotherapy. *Expert Opin Biol Ther* (2017) 17(3):313–24. doi:10.1080/14712598.2017.1271874
2. Cooper MA, Fehniger TA, Caligiuri MA. The biology of human natural killer-cell subsets. *Trends Immunol* (2001) 22(11):633–40. doi:10.1016/S1471-4906(01)02060-9
3. Campbell KS, Hasegawa J. NK cell biology: an update and future directions. *J Allergy Clin Immunol* (2013) 132(3):536–44. doi:10.1016/j.jaci.2013.07.006
4. Zhang J, Basher F, Wu JD. NKG2D ligands in tumor immunity: two sides of a coin. *Front Immunol* (2015) 6:97. doi:10.3389/fimmu.2015.00097
5. de Andrade LF, Smyth MJ, Martinet L. DNAM-1 control of natural killer cells functions through nectin and nectin-like proteins. *Immunol Cell Biol* (2014) 92(3):237–44. doi:10.1038/icb.2013.95
6. Kruse PH, Matta J, Ugolini S, Vivier E. Natural cytotoxicity receptors and their ligands. *Immunol Cell Biol* (2014) 92(3):221–9. doi:10.1038/icb.2013.98
7. Borrego F, Ulbrecht M, Weiss EH, Coligan JE, Brooks AG. Recognition of human histocompatibility leukocyte antigen (HLA)-E complexed with HLA class I signal sequence-derived peptides by CD94/NKG2 confers protection

from natural killer cell-mediated lysis. *J Exp Med* (1998) 187(5):813–8. doi:10.1084/jem.187.5.813
8. Pegram HJ, Andrews DM, Smyth MJ, Darcy PK, Kershaw MH. Activating and inhibitory receptors of natural killer cells. *Immunol Cell Biol* (2011) 89(2):216–24. doi:10.1038/icb.2010.78
9. Leibson PJ. Signal transduction during natural killer cell activation: inside the mind of a killer. *Immunity* (1997) 6(6):655–61. doi:10.1016/S1074-7613(00)80441-0
10. Zamai L, Ahmad M, Bennett IM, Azzoni L, Alnemri ES, Perussia B. Natural killer (NK) cell-mediated cytotoxicity: differential use of TRAIL and Fas ligand by immature and mature primary human NK cells. *J Exp Med* (1998) 188(12):2375–80. doi:10.1084/jem.188.12.2375
11. Krasnova Y, Putz EM, Smyth MJ, Souza-Fonseca-Guimaraes F. Bench to bedside: NK cells and control of metastasis. *Clin Immunol* (2015). doi:10.1016/j.clim.2015.10.001
12. Ljunggren HG, Karre K. In search of the 'missing self': MHC molecules and NK cell recognition. *Immunol Today* (1990) 11(7):237–44. doi:10.1016/0167-5699(90)90097-S
13. Parham P, Moffett A. Variable NK cell receptors and their MHC class I ligands in immunity, reproduction and human evolution. *Nat Rev Immunol* (2013) 13(2):133–44. doi:10.1038/nri3370
14. Rouas-Freiss N, Moreau P, Menier C, LeMaoult J, Carosella ED. Expression of tolerogenic HLA-G molecules in cancer prevents antitumor responses. *Semin Cancer Biol* (2007) 17(6):413–21. doi:10.1016/j.semcancer.2007.07.003
15. Lee N, Llano M, Carretero M, Ishitani A, Navarro F, Lopez-Botet M, et al. HLA-E is a major ligand for the natural killer inhibitory receptor CD94/NKG2A. *Proc Natl Acad Sci U S A* (1998) 95(9):5199–204. doi:10.1073/pnas.95.9.5199
16. Schleypen JS, Von Geldern M, Weiss EH, Kotzias N, Rohrmann K, Schendel DJ, et al. Renal cell carcinoma-infiltrating natural killer cells express differential repertoires of activating and inhibitory receptors and are inhibited by specific HLA class I allotypes. *Int J Cancer* (2003) 106(6):905–12. doi:10.1002/ijc.11321
17. Jinushi M, Takehara T, Tatsumi T, Hiramatsu N, Sakamori R, Yamaguchi S, et al. Impairment of natural killer cell and dendritic cell functions by the soluble form of MHC class I-related chain A in advanced human hepatocellular carcinomas. *J Hepatol* (2005) 43(6):1013–20. doi:10.1016/j.jhep.2005.05.026
18. Hsia JY, Chen JT, Chen CY, Hsu CP, Miaw J, Huang YS, et al. Prognostic significance of intratumoral natural killer cells in primary resected esophageal squamous cell carcinoma. *Chang Gung Med J* (2005) 28(5):335–40.
19. Kondo E, Koda K, Takiguchi N, Oda K, Seike K, Ishizuka M, et al. Preoperative natural killer cell activity as a prognostic factor for distant metastasis following surgery for colon cancer. *Dig Surg* (2003) 20(5):445–51. doi:10.1159/000072714
20. Ishigami S, Natsugoe S, Tokuda K, Nakajo A, Che X, Iwashige H, et al. Prognostic value of intratumoral natural killer cells in gastric carcinoma. *Cancer* (2000) 88(3):577–83. doi:10.1002/(SICI)1097-0142(20000201)88:3<577::AID-CNCR13>3.3.CO;2-M
21. Villegas FR, Coca S, Villarrubia VG, Jimenez R, Chillon MJ, Jareno J, et al. Prognostic significance of tumor infiltrating natural killer cells subset CD57 in patients with squamous cell lung cancer. *Lung Cancer* (2002) 35(1):23–8. doi:10.1016/S0169-5002(01)00292-6
22. Burke S, Lakshmikanth T, Colucci F, Carbone E. New views on natural killer cell-based immunotherapy for melanoma treatment. *Trends Immunol* (2010) 31(9):339–45. doi:10.1016/j.it.2010.06.003
23. Ghiringhelli F, Menard C, Martin F, Zitvogel L. The role of regulatory T cells in the control of natural killer cells: relevance during tumor progression. *Immunol Rev* (2006) 214:229–38. doi:10.1111/j.1600-065X.2006.00445.x
24. Hoechst B, Voigtlaender T, Ormandy L, Gamrekelashvili J, Zhao F, Wedemeyer H, et al. Myeloid derived suppressor cells inhibit natural killer cells in patients with hepatocellular carcinoma via the NKp30 receptor. *Hepatology* (2009) 50(3):799–807. doi:10.1002/hep.23054
25. Gabrilovich DI, Ostrand-Rosenberg S, Bronte V. Coordinated regulation of myeloid cells by tumours. *Nat Rev Immunol* (2012) 12(4):253–68. doi:10.1038/nri3175
26. Gill S, Vasey AE, De Souza A, Baker J, Smith AT, Kohrt HE, et al. Rapid development of exhaustion and down-regulation of eomesodermin limit the antitumor activity of adoptively transferred murine natural killer cells. *Blood* (2012) 119(24):5758–68. doi:10.1182/blood-2012-03-415364
27. Schafer JL, Müller-Trutwin MC, Reeves RK. NK cell exhaustion: bad news for chronic disease? *Oncotarget* (2015) 6(26):21797–8. doi:10.18632/oncotarget.5490
28. Mamessier E, Sylvain A, Thibult ML, Houvenaeghel G, Jacquemier J, Castellano R, et al. Human breast cancer cells enhance self tolerance by promoting evasion from NK cell antitumor immunity. *J Clin Invest* (2011) 121(9):3609–22. doi:10.1172/jci45816
29. Bauernhofer T, Kuss I, Henderson B, Baum AS, Whiteside TL. Preferential apoptosis of CD56dim natural killer cell subset in patients with cancer. *Eur J Immunol* (2003) 33(1):119–24. doi:10.1002/immu.200390014
30. Parkhurst MR, Riley JP, Dudley ME, Rosenberg SA. Adoptive transfer of autologous natural killer cells leads to high levels of circulating natural killer cells but does not mediate tumor regression. *Clin Cancer Res* (2011) 17(19):6287–97. doi:10.1158/1078-0432.CCR-11-1347
31. Rosenberg SA, Lotze MT, Muul LM, Leitman S, Chang AE, Ettinghausen SE, et al. Observations on the systemic administration of autologous lymphokine-activated killer cells and recombinant interleukin-2 to patients with metastatic cancer. *N Engl J Med* (1985) 313(23):1485–92. doi:10.1056/NEJM198512053132327
32. Rosenberg SA, Restifo NP, Yang JC, Morgan RA, Dudley ME. Adoptive cell transfer: a clinical path to effective cancer immunotherapy. *Nat Rev Cancer* (2008) 8(4):299–308. doi:10.1038/nrc2355
33. Geller MA, Cooley S, Judson PL, Ghebre R, Carson LF, Argenta PA, et al. A phase II study of allogeneic natural killer cell therapy to treat patients with recurrent ovarian and breast cancer. *Cytotherapy* (2011) 13(1):98–107. doi:10.3109/14653249.2010.515582
34. Minculescu L, Marquart HV, Friis LS, Petersen SL, Schiødt I, Ryder LP, et al. Early natural killer cell reconstitution predicts overall survival in T cell-replete allogeneic hematopoietic stem cell transplantation. *Biol Blood Marrow Transplant* (2016) 22(12):2187–93. doi:10.1016/j.bbmt.2016.09.006
35. Wu CJ, Ritz J. Induction of tumor immunity following allogeneic stem cell transplantation. *Adv Immunol* (2006) 90:133–73. doi:10.1016/S0065-2776(06)90004-2
36. Ho VT, Soiffer RJ. The history and future of T-cell depletion as graft-versus-host disease prophylaxis for allogeneic hematopoietic stem cell transplantation. *Blood* (2001) 98(12):3192–204. doi:10.1182/blood.V98.12.3192
37. Antin JH. T-cell depletion in GVHD: less is more? *Blood* (2011) 117(23):6061–2. doi:10.1182/blood-2011-04-348409
38. Ruggeri L, Capanni M, Urbani E, Perruccio K, Shlomchik WD, Tosti A, et al. Effectiveness of donor natural killer cell alloreactivity in mismatched hematopoietic transplants. *Science* (2002) 295(5562):2097–100. doi:10.1126/science.1068440
39. Ruggeri L, Capanni M, Casucci M, Volpi I, Tosti A, Perruccio K, et al. Role of natural killer cell alloreactivity in HLA-mismatched hematopoietic stem cell transplantation. *Blood* (1999) 94(1):333–9.
40. Koepsell SA, Miller JS, McKenna DH Jr. Natural killer cells: a review of manufacturing and clinical utility. *Transfusion* (2013) 53(2):404–10. doi:10.1111/j.1537-2995.2012.03724.x
41. Spanholtz J, Preijers F, Tordoir M, Trilsbeek C, Paardekooper J, de Witte T, et al. Clinical-grade generation of active NK cells from cord blood hematopoietic progenitor cells for immunotherapy using a closed-system culture process. *PLoS One* (2011) 6(6):e20740. doi:10.1371/journal.pone.0020740
42. Arai S, Meagher R, Swearingen M, Myint H, Rich E, Martinson J, et al. Infusion of the allogeneic cell line NK-92 in patients with advanced renal cell cancer or melanoma: a phase I trial. *Cytotherapy* (2008) 10(6):625–32. doi:10.1080/14653240802301872
43. Tonn T, Schwabe D, Klingemann HG, Becker S, Esser R, Koehl U, et al. Treatment of patients with advanced cancer with the natural killer cell line NK-92. *Cytotherapy* (2013) 15(12):1563–70. doi:10.1016/j.jcyt.2013.06.017
44. Passweg JR, Tichelli A, Meyer-Monard S, Heim D, Stern M, Kuhne T, et al. Purified donor NK-lymphocyte infusion to consolidate engraftment after haploidentical stem cell transplantation. *Leukemia* (2004) 18(11):1835–8. doi:10.1038/sj.leu.2403524
45. Koehl U, Esser R, Zimmermann S, Tonn T, Kotchetkov R, Bartling T, et al. Ex vivo expansion of highly purified NK cells for immunotherapy after haploidentical stem cell transplantation in children. *Klin Padiatr* (2005) 217(6):345–50. doi:10.1055/s-2005-872520

46. Shi J, Tricot G, Szmania S, Rosen N, Garg TK, Malaviarachchi PA, et al. Infusion of haplo-identical killer immunoglobulin-like receptor ligand mismatched NK cells for relapsed myeloma in the setting of autologous stem cell transplantation. *Br J Haematol* (2008) 143(5):641–53. doi:10.1111/j.1365-2141.2008.07340.x
47. Yoon SR, Lee YS, Yang SH, Ahn KH, Lee J-H, Lee J-H, et al. Generation of donor natural killer cells from CD34+ progenitor cells and subsequent infusion after HLA-mismatched allogeneic hematopoietic cell transplantation: a feasibility study. *Bone Marrow Transplant* (2010) 45(6):1038–46. doi:10.1038/bmt.2009.304
48. Rizzieri DA, Storms R, Chen D-F, Long G, Yang Y, Nikcevich DA, et al. Natural killer cell enriched donor lymphocyte infusions from a 3-6/6 HLA matched family member following non-myeloablative allogeneic stem cell transplantation. *Biol Blood Marrow Transplant* (2010) 16(8):1107–14. doi:10.1016/j.bbmt.2010.02.018
49. Brehm C, Huenecke S, Quaiser A, Esser R, Bremm M, Kloess S, et al. IL-2 stimulated but not unstimulated NK cells induce selective disappearance of peripheral blood cells: concomitant results to a phase I/II study. *PLoS One* (2011) 6(11):e27351. doi:10.1371/journal.pone.0027351
50. Miller JS, Soignier Y, Panoskaltsis-Mortari A, McNearney SA, Yun GH, Fautsch SK, et al. Successful adoptive transfer and in vivo expansion of human haploidentical NK cells in patients with cancer. *Blood* (2005) 105(8):3051–7. doi:10.1182/blood-2004-07-2974
51. Copelan EA. Hematopoietic Stem-Cell Transplantation. *New Engl J Med* (2006) 354(17):1813–26. doi:10.1056/NEJMra052638
52. Stern M, Passweg JR, Meyer-Monard S, Esser R, Tonn T, Soerensen J, et al. Pre-emptive immunotherapy with purified natural killer cells after haploidentical SCT: a prospective phase II study in two centers. *Bone Marrow Transplant* (2013) 48(3):433–8. doi:10.1038/bmt.2012.162
53. Klingemann H, Grodman C, Cutler E, Duque M, Kadidlo D, Klein AK, et al. Autologous stem cell transplant recipients tolerate haploidentical related-donor natural killer cell enriched infusions. *Transfusion* (2013) 53(2):412–8. doi:10.1111/j.1537-2995.2012.03764.x
54. Choi I, Yoon SR, Park SY, Kim H, Jung SJ, Jang YJ, et al. Donor-derived natural killer cells infused after human leukocyte antigen-haploidentical hematopoietic cell transplantation: a dose-escalation study. *Biol Blood Marrow Transplant* (2014) 20(5):696–704. doi:10.1016/j.bbmt.2014.01.031
55. Killig M, Friedrichs B, Meisig J, Gentilini C, Bluthgen N, Loddenkemper C, et al. Tracking in vivo dynamics of NK cells transferred in patients undergoing stem cell transplantation. *Eur J Immunol* (2014) 44(9):2822–34. doi:10.1002/eji.201444586
56. Shah NN, Baird K, Delbrook CP, Fleisher TA, Kohler ME, Rampertaap S, et al. Acute GVHD in patients receiving IL-15/4-1BBL activated NK cells following T-cell–depleted stem cell transplantation. *Blood* (2015) 125(5):784–92. doi:10.1182/blood-2014-07-592881
57. Lee DA, Denman CJ, Rondon G, Woodworth G, Chen J, Fisher T, et al. Haploidentical natural killer cells infused before allogeneic stem cell transplantation for myeloid malignancies: a phase I trial. *Biol Blood Marrow Transplant* (2016) 22(7):1290–8. doi:10.1016/j.bbmt.2016.04.009
58. Choi I, Yoon SR, Park SY, Kim H, Jung SJ, Kang YL, et al. Donor-derived natural killer cell infusion after human leukocyte antigen-haploidentical hematopoietic cell transplantation in patients with refractory acute leukemia. *Biol Blood Marrow Transplant* (2016) 22(11):2065–76. doi:10.1016/j.bbmt.2016.08.008
59. Shah N, Li L, McCarty J, Kaur I, Yvon E, Shaim H, et al. Phase I study of cord blood-derived natural killer cells combined with autologous stem cell transplantation in multiple myeloma. *Br J Haematol* (2017) 177(3):457–66. doi:10.1111/bjh.14570
60. Iliopoulou EG, Kountourakis P, Karamouzis MV, Doufexis D, Ardavanis A, Baxevanis CN, et al. A phase I trial of adoptive transfer of allogeneic natural killer cells in patients with advanced non-small cell lung cancer. *Cancer Immunol Immunother* (2010) 59(12):1781–9. doi:10.1007/s00262-010-0904-3
61. Rubnitz JE, Inaba H, Ribeiro RC, Pounds S, Rooney B, Bell T, et al. NKAML: a pilot study to determine the safety and feasibility of haploidentical natural killer cell transplantation in childhood acute myeloid leukemia. *J Clin Oncol* (2010) 28(6):955–9. doi:10.1200/JCO.2009.24.4590
62. Bachanova V, Burns LJ, McKenna DH, Curtsinger J, Panoskaltsis-Mortari A, Lindgren BR, et al. Allogeneic natural killer cells for refractory lymphoma. *Cancer Immunol Immunother* (2010) 59(11):1739–44. doi:10.1007/s00262-010-0896-z
63. Curti A, Ruggeri L, D'Addio A, Bontadini A, Dan E, Motta MR, et al. Successful transfer of alloreactive haploidentical KIR ligand-mismatched natural killer cells after infusion in elderly high risk acute myeloid leukemia patients. *Blood* (2011) 118(12):3273–9. doi:10.1182/blood-2011-01-329508
64. Bachanova V, Cooley S, Defor TE, Verneris MR, Zhang B, McKenna DH, et al. Clearance of acute myeloid leukemia by haploidentical natural killer cells is improved using IL-2 diphtheria toxin fusion protein. *Blood* (2014) 123(25):3855–63. doi:10.1182/blood-2013-10-532531
65. Szmania S, Lapteva N, Garg T, Greenway A, Lingo J, Nair B, et al. Ex vivo expanded natural killer cells demonstrate robust proliferation in vivo in high-risk relapsed multiple myeloma patients. *J Immunother* (2015) 38(1):24–36. doi:10.1097/CJI.0000000000000059
66. Kottaridis PD, North J, Tsirogianni M, Marden C, Samuel ER, Jide-Banwo S, et al. Two-stage priming of allogeneic natural killer cells for the treatment of patients with acute myeloid leukemia: a phase I trial. *PLoS One* (2015) 10(6):e0123416. doi:10.1371/journal.pone.0123416
67. Rubnitz JE, Inaba H, Kang G, Gan K, Hartford C, Triplett BM, et al. Natural killer cell therapy in children with relapsed leukemia. *Pediatr Blood Cancer* (2015) 62(8):1468–72. doi:10.1002/pbc.25555
68. Lim O, Lee Y, Chung H, Her JH, Kang SM, Jung MY, et al. GMP-compliant, large-scale expanded allogeneic natural killer cells have potent cytolytic activity against cancer cells in vitro and in vivo. *PLoS One* (2013) 8(1):e53611. doi:10.1371/journal.pone.0053611
69. Yang Y, Lim O, Kim TM, Ahn YO, Choi H, Chung H, et al. Phase I study of random healthy donor-derived allogeneic natural killer cell therapy in patients with malignant lymphoma or advanced solid tumors. *Cancer Immunol Res* (2016) 4(3):215–24. doi:10.1158/2326-6066.CIR-15-0118
70. Shaffer BC, Le Luduec J-B, Forlenza C, Jakubowski AA, Perales M-A, Young JW, et al. Phase II study of haploidentical natural killer cell infusion for treatment of relapsed or persistent myeloid malignancies following allogeneic hematopoietic cell transplantation. *Biol Blood Marrow Transplant* (2016) 22(4):705–9. doi:10.1016/j.bbmt.2015.12.028
71. Curti A, Ruggeri L, Parisi S, Bontadini A, Dan E, Motta MR, et al. Larger size of donor alloreactive NK cell repertoire correlates with better response to NK cell immunotherapy in elderly acute myeloid leukemia patients. *Clin Cancer Res* (2016) 22(8):1914–21. doi:10.1158/1078-0432.CCR-15-1604
72. Romee R, Rosario M, Berrien-Elliott MM, Wagner JA, Jewell BA, Schappe T, et al. Cytokine-induced memory-like natural killer cells exhibit enhanced responses against myeloid leukemia. *Sci Transl Med* (2016) 8(357):ra123–357. doi:10.1126/scitranslmed.aaf2341
73. Dolstra H, Roeven MW, Spanholtz J, Hangalapura BN, Tordoir M, Maas F, et al. Successful transfer of umbilical cord blood CD34+ hematopoietic stem and progenitor-derived NK cells in older acute myeloid leukemia patients. *Clin Cancer Res* (2017). doi:10.1158/1078-0432.ccr-16-2981
74. Wang W, Erbe AK, Hank JA, Morris ZS, Sondel PM. NK cell-mediated antibody-dependent cellular cytotoxicity in cancer immunotherapy. *Front Immunol* (2015) 6:368. doi:10.3389/fimmu.2015.00368
75. Mellor J, Brown M, Irving H, Zalcberg J, Dobrovic A. A critical review of the role of Fc gamma receptor polymorphisms in the response to monoclonal antibodies in cancer. *J Hematol Oncol* (2013) 6(1):1. doi:10.1186/1756-8722-6-1
76. Ogura M, Ishida T, Hatake K, Taniwaki M, Ando K, Tobinai K, et al. Multicenter phase II study of mogamulizumab (KW-0761), a defucosylated anti-CC chemokine receptor 4 antibody, in patients with relapsed peripheral T-cell lymphoma and cutaneous T-cell lymphoma. *J Clin Oncol* (2014) 32(11):1157–63. doi:10.1200/JCO.2013.52.0924
77. Goede V, Klein C, Stilgenbauer S. Obinutuzumab (GA101) for the treatment of chronic lymphocytic leukemia and other B-cell non-hodgkin's lymphomas: a glycoengineered type II CD20 antibody. *Oncol Res Treat* (2015) 38(4):185–92. doi:10.1159/000381524
78. Cheney CM, Stephens DM, Mo X, Rafiq S, Butchar J, Flynn JM, et al. Ocaratuzumab, an Fc-engineered antibody demonstrates enhanced antibody-dependent cell-mediated cytotoxicity in chronic lymphocytic leukemia. *mAbs* (2014) 6(3):749–55. doi:10.4161/mabs.28282
79. Oppenheim DE, Spreafico R, Etuk A, Malone D, Amofah E, Peña-Murillo C, et al. Glyco-engineered anti-EGFR mAb elicits ADCC by NK cells from

colorectal cancer patients irrespective of chemotherapy. *Br J Cancer* (2014) 110(5):1221–7. doi:10.1038/bjc.2014.35

80. Delord J-P, Tabernero J, García-Carbonero R, Cervantes A, Gomez-Roca C, Bergé Y, et al. Open-label, multicentre expansion cohort to evaluate imgatuzumab in pre-treated patients with KRAS-mutant advanced colorectal carcinoma. *Eur J Cancer* (2014) 50(3):496–505. doi:10.1016/j.ejca.2013.10.015
81. Ochoa MC, Minute L, Rodriguez I, Garasa S, Perez-Ruiz E, Inoges S, et al. Antibody-dependent cell cytotoxicity (ADCC): immunotherapy strategies enhancing effector NK cells. *Immunol Cell Biol* (2017). doi:10.1038/icb.2017.6
82. Kontermann RE, Brinkmann U. Bispecific antibodies. *Drug Discov Today* (2015) 20(7):838–47. doi:10.1016/j.drudis.2015.02.008
83. Fan G, Wang Z, Hao M, Li J. Bispecific antibodies and their applications. *J Hematol Oncol* (2015) 8(1):130. doi:10.1186/s13045-015-0227-0
84. Rothe A, Sasse S, Topp MS, Eichenauer DA, Hummel H, Reiners KS, et al. A phase 1 study of the bispecific anti-CD30/CD16A antibody construct AFM13 in patients with relapsed or refractory Hodgkin lymphoma. *Blood* (2015) 125(26):4024–31. doi:10.1182/blood-2014-12-614636
85. Martin DW, Williams SR. *Non-Natural MIC Proteins*. Google Patents (2014).
86. Sola C, Chanuc F, Thielens A, Fuseri N, Morel Y, Bléry M, et al. Anti-tumoral efficacy of therapeutic human anti-KIR antibody (Lirilumab/BMS-986015/IPH2102) in a preclinical xenograft tumor model. *J Immunother Cancer* (2013) 1(Suppl 1):P40. doi:10.1186/2051-1426-1-S1-P40
87. Levy EM, Bianchini M, Von Euw EM, Barrio MM, Bravo AI, Furman D, et al. Human leukocyte antigen-E protein is overexpressed in primary human colorectal cancer. *Int J Oncol* (2008) 32(3):633–41.
88. Gooden M, Lampen M, Jordanova ES, Leffers N, Trimbos JB, van der Burg SH, et al. HLA-E expression by gynecological cancers restrains tumor-infiltrating CD8(+) T lymphocytes. *Proc Natl Acad Sci U S A* (2011) 108(26):10656–61. doi:10.1073/pnas.1100354108
89. Glienke W, Esser R, Priesner C, Suerth JD, Schambach A, Wels WS, et al. Advantages and applications of CAR-expressing natural killer cells. *Front Pharmacol* (2015) 6:21. doi:10.3389/fphar.2015.00021
90. Altvater B, Landmeier S, Pscherer S, Temme J, Schweer K, Kailayangiri S, et al. 2B4 (CD244) signaling by recombinant antigen-specific chimeric receptors costimulates natural killer cell activation to leukemia and neuroblastoma cells. *Clin Cancer Res* (2009) 15(15):4857–66. doi:10.1158/1078-0432.CCR-08-2810
91. Zhang G, Liu R, Zhu X, Wang L, Ma J, Han H, et al. Retargeting NK-92 for anti-melanoma activity by a TCR-like single-domain antibody. *Immunol Cell Biol* (2013) 91(10):615–24. doi:10.1038/icb.2013.45
92. Sahm C, Schonfeld K, Wels WS. Expression of IL-15 in NK cells results in rapid enrichment and selective cytotoxicity of gene-modified effectors that carry a tumor-specific antigen receptor. *Cancer Immunol Immunother* (2012) 61(9):1451–61. doi:10.1007/s00262-012-1212-x
93. Chen X, Han J, Chu J, Zhang L, Zhang J, Chen C, et al. A combinational therapy of EGFR-CAR NK cells and oncolytic herpes simplex virus 1 for breast cancer brain metastases. *Oncotarget* (2016) 7(19):27764–77. doi:10.18632/oncotarget.8526
94. Chu J, Deng Y, Benson DM, He S, Hughes T, Zhang J, et al. CS1-specific chimeric antigen receptor (CAR)-engineered natural killer cells enhance in vitro and in vivo antitumor activity against human multiple myeloma. *Leukemia* (2014) 28(4):917–27. doi:10.1038/leu.2013.279
95. Jiang H, Zhang W, Shang P, Zhang H, Fu W, Ye F, et al. Transfection of chimeric anti-CD138 gene enhances natural killer cell activation and killing of multiple myeloma cells. *Mol Oncol* (2014) 8(2):297–310. doi:10.1016/j.molonc.2013.12.001
96. Boissel L, Betancur-Boissel M, Lu W, Krause DS, Van Etten RA, Wels WS, et al. Retargeting NK-92 cells by means of CD19- and CD20-specific chimeric antigen receptors compares favorably with antibody-dependent cellular cytotoxicity. *Oncoimmunology* (2013) 2(10):e26527. doi:10.4161/onci.26527
97. Klingemann H, Boissel L, Toneguzzo F. Natural killer cells for immunotherapy–advantages of the NK-92 cell line over blood NK cells. *Front Immunol* (2016) 7:91. doi:10.3389/fimmu.2016.00091
98. Adam J, Planchard D, Marabelle A, Soria JC, Scoazec JY, Lantuejoul S. [PD-L1 expression: an emerging biomarker in non-small cell lung cancer]. *Ann Pathol* (2016) 36(1):94–102. doi:10.1016/j.annpat.2015.11.004
99. Borcherding N, Kusner D, Liu G-H, Zhang W. ROR1, an embryonic protein with an emerging role in cancer biology. *Protein Cell* (2014) 5(7):496–502. doi:10.1007/s13238-014-0059-7
100. Chang Y-H, Connolly J, Shimasaki N, Mimura K, Kono K, Campana D. A chimeric receptor with NKG2D specificity enhances natural killer cell activation and killing of tumor cells. *Cancer Res* (2013) 73(6):1777–86. doi:10.1158/0008-5472.can-12-3558
101. Kamiya T, Chang YH, Campana D. Expanded and activated natural killer cells for immunotherapy of hepatocellular carcinoma. *Cancer Immunol Res* (2016) 4(7):574–81. doi:10.1158/2326-6066.CIR-15-0229
102. Torikai H, Mi T, Gragert L, Maiers M, Najjar A, Ang S, et al. Genetic editing of HLA expression in hematopoietic stem cells to broaden their human application. *Sci Rep* (2016) 6:21757. doi:10.1038/srep21757
103. Carlsten M, Childs RW. Genetic manipulation of NK cells for cancer immunotherapy: techniques and clinical implications. *Front Immunol* (2015) 6:266. doi:10.3389/fimmu.2015.00266
104. Torikai H, Reik A, Soldner F, Warren EH, Yuen C, Zhou Y, et al. Toward eliminating HLA class I expression to generate universal cells from allogeneic donors. *Blood* (2013) 122(8):1341–9. doi:10.1182/blood-2013-03-478255
105. Delconte RB, Kolesnik TB, Dagley LF, Rautela J, Shi W, Putz EM, et al. CIS is a potent checkpoint in NK cell-mediated tumor immunity. *Nat Immunol* (2016) 17(7):816–24. doi:10.1038/ni.3470
106. Ng ES, Davis RP, Azzola L, Stanley EG, Elefanty AG. Forced aggregation of defined numbers of human embryonic stem cells into embryoid bodies fosters robust, reproducible hematopoietic differentiation. *Blood* (2005) 106(5):1601–3. doi:10.1182/blood-2005-03-0987
107. Woll PS, Martin CH, Miller JS, Kaufman DS. Human embryonic stem cell-derived NK cells acquire functional receptors and cytolytic activity. *J Immunol* (2005) 175(8):5095–103. doi:10.4049/jimmunol.175.8.5095
108. Woll PS, Grzywacz B, Tian X, Marcus RK, Knorr DA, Verneris MR, et al. Human embryonic stem cells differentiate into a homogeneous population of natural killer cells with potent in vivo antitumor activity. *Blood* (2009) 113(24):6094–101. doi:10.1182/blood-2008-06-165225
109. Knorr DA, Ni Z, Hermanson D, Hexum MK, Bendzick L, Cooper LJ, et al. Clinical-scale derivation of natural killer cells from human pluripotent stem cells for cancer therapy. *Stem Cells Transl Med* (2013) 2(4):274–83. doi:10.5966/sctm.2012-0084
110. Shaim H, Yvon E. Cord blood: a promising source of allogeneic natural killer cells for immunotherapy. *Cytotherapy* (2015) 17(1):1–2. doi:10.1016/j.jcyt.2014.12.001
111. Spanholtz J, Tordoir M, Eissens D, Preijers F, Meer A, Joosten I, et al. High log-scale expansion of functional human natural killer cells from umbilical cord blood CD34-positive cells for adoptive cancer immunotherapy. *PLoS One* (2010) 5(2):e9221. doi:10.1371/journal.pone.0009221
112. Cany J, van der Waart AB, Tordoir M, Franssen GM, Hangalapura BN, de Vries J, et al. Natural killer cells generated from cord blood hematopoietic progenitor cells efficiently target bone marrow-residing human leukemia cells in NOD/SCID/IL2Rgnull mice. *PLoS One* (2013) 8(6):e64384. doi:10.1371/journal.pone.0064384
113. Lehmann D, Spanholtz J, Osl M, Tordoir M, Lipnik K, Bilban M, et al. Ex vivo generated natural killer cells acquire typical natural killer receptors and display a cytotoxic gene expression profile similar to peripheral blood natural killer cells. *Stem Cells Dev* (2012) 21(16):2926–38. doi:10.1089/scd.2011.0659
114. Veluchamy JP, Heeren AM, Spanholtz J, van Eendenburg JD, Heideman DA, Kenter GG, et al. High-efficiency lysis of cervical cancer by allogeneic NK cells derived from umbilical cord progenitors is independent of HLA status. *Cancer Immunol Immunother* (2017) 66(1):51–61. doi:10.1007/s00262-016-1919-1
115. Veluchamy JP, Lopez-Lastra S, Spanholtz J, Bohme F, Kok N, Heideman DAM, et al. In vivo efficacy of umbilical cord blood stem cell-derived NK cells in the treatment of metastatic colorectal cancer. *Front Immunol* (2017) 8:87. doi:10.3389/fimmu.2017.00087
116. Rosenberg SA. Interleukin-2 and the development of immunotherapy for the treatment of patients with cancer. *Cancer J Sci Am* (2000) 6(Suppl 1):S2–7.
117. West WH, Tauer KW, Yannelli JR, Marshall GD, Orr DW, Thurman GB, et al. Constant-infusion recombinant interleukin-2 in adoptive immunotherapy of advanced cancer. *N Engl J Med* (1987) 316(15):898–905. doi:10.1056/NEJM198704093161502
118. Conlon KC, Lugli E, Welles HC, Rosenberg SA, Fojo AT, Morris JC, et al. Redistribution, hyperproliferation, activation of natural killer cells and

CD8 T cells, and cytokine production during first-in-human clinical trial of recombinant human interleukin-15 in patients with cancer. *J Clin Oncol* (2015) 33(1):74–82. doi:10.1200/JCO.2014.57.3329
119. Miller JS. Therapeutic applications: natural killer cells in the clinic. *Hematology Am Soc Hematol Educ Program* (2013) 2013:247–53. doi:10.1182/asheducation-2013.1.247
120. Ito S, Bollard CM, Carlsten M, Melenhorst JJ, Biancotto A, Wang E, et al. Ultra-low dose interleukin-2 promotes immune-modulating function of regulatory T cells and natural killer cells in healthy volunteers. *Mol Ther* (2014) 22(7):1388–95. doi:10.1038/mt.2014.50
121. Steel JC, Waldmann TA, Morris JC. Interleukin-15 biology and its therapeutic implications in cancer. *Trends Pharmacol Sci* (2012) 33(1):35–41. doi:10.1016/j.tips.2011.09.004
122. Imamura M, Shook D, Kamiya T, Shimasaki N, Chai SM, Coustan-Smith E, et al. Autonomous growth and increased cytotoxicity of natural killer cells expressing membrane-bound interleukin-15. *Blood* (2014) 124(7):1081–8. doi:10.1182/blood-2014-02-556837
123. Denman CJ, Senyukov VV, Somanchi SS, Phatarpekar PV, Kopp LM, Johnson JL, et al. Membrane-bound IL-21 promotes sustained ex vivo proliferation of human natural killer cells. *PLoS One* (2012) 7(1):e30264. doi:10.1371/journal.pone.0030264
124. Oyer JL, Pandey V, Igarashi RY, Somanchi SS, Zakari A, Solh M, et al. Natural killer cells stimulated with PM21 particles expand and biodistribute in vivo: clinical implications for cancer treatment. *Cytotherapy* (2016) 18(5):653–63. doi:10.1016/j.jcyt.2016.02.006
125. Xu W, Jones M, Liu B, Zhu X, Johnson CB, Edwards AC, et al. Efficacy and mechanism-of-action of a novel superagonist interleukin-15: interleukin-15 receptor α/Fc fusion complex in syngeneic murine models of multiple myeloma. *Cancer Res* (2013) 73(10):3075–86. doi:10.1158/0008-5472.CAN-12-2357
126. Rhode PR, Egan JO, Xu W, Hong H, Webb GM, Chen X, et al. Comparison of the superagonist complex, ALT-803, to IL15 as cancer immunotherapeutics in animal models. *Cancer Immunol Res* (2016) 4(1):49–60. doi:10.1158/2326-6066.CIR-15-0093-T
127. Mathios D, Park C-K, Marcus WD, Alter S, Rhode PR, Jeng EK, et al. Therapeutic administration of IL-15 superagonist complex ALT-803 leads to long-term survival and durable antitumor immune response in a murine glioblastoma model. *Int J Cancer* (2016) 138(1):187–94. doi:10.1002/ijc.29686
128. Gomes-Giacoia E, Miyake M, Goodison S, Sriharan A, Zhang G, You L, et al. Intravesical ALT-803 and BCG treatment reduces tumor burden in a carcinogen induced bladder cancer rat model; a role for cytokine production and NK cell expansion. *PLoS One* (2014) 9(6):e96705. doi:10.1371/journal.pone.0096705
129. Kim PS, Kwilas AR, Xu W, Alter S, Jeng EK, Wong HC, et al. IL-15 superagonist/IL-15RalphaSushi-Fc fusion complex (IL-15SA/IL-15RalphaSu-Fc; ALT-803) markedly enhances specific subpopulations of NK and memory CD8+ T cells, and mediates potent anti-tumor activity against murine breast and colon carcinomas. *Oncotarget* (2016) 7(13):16130–45. doi:10.18632/oncotarget.7470
130. Rosario M, Liu B, Kong L, Collins LI, Schneider SE, Chen X, et al. The IL-15-based ALT-803 complex enhances FcgammaRIIIa-triggered NK cell responses and in vivo clearance of B cell lymphomas. *Clin Cancer Res* (2016) 22(3):596–608. doi:10.1158/1078-0432.CCR-15-1419
131. Fehniger TA, Stuart RK, Cooley SA, Miller JS, Curtsinger J, Hillman TM, et al. Preliminary results of a phase 1/2 clinical trial of Cndo-109-activated allogeneic natural killer cells in high risk acute myelogenous leukemia patients in first complete remission. *Blood* (2014) 124(21):2320.
132. Krause SW, Gastpar R, Andreesen R, Gross C, Ullrich H, Thonigs G, et al. Treatment of colon and lung cancer patients with ex vivo heat shock protein 70-peptide-activated, autologous natural killer cells: a clinical phase i trial. *Clin Cancer Res* (2004) 10(11):3699–707. doi:10.1158/1078-0432.CCR-03-0683
133. Frei GM, Persi N, Lador C, Peled A, Cohen YC, Nagler A, et al. Nicotinamide, a form of vitamin B3, promotes expansion of natural killer cells that display increased in vivo survival and cytotoxic activity. *Blood* (2011) 118(21):4035.
134. Roccaro AM, Mishima Y, Sacco A, Moschetta M, Tai Y-T, Shi J, et al. CXCR4 regulates extra-medullary myeloma through epithelial-mesenchymal transition-like transcriptional activation. *Cell Rep* (2015) 12(4):622–35. doi:10.1016/j.celrep.2015.06.059
135. Marasca R, Maffei R. NOX-A12: mobilizing CLL away from home. *Blood* (2014) 123(7):952–3. doi:10.1182/blood-2013-12-542480
136. Chou T-C. Drug combination studies and their synergy quantification using the Chou-Talalay method. *Cancer Res* (2010) 70(2):440–6. doi:10.1158/0008-5472.can-09-1947

4

Umbilical Cord Blood Natural Killer Cells, their Characteristics, and Potential Clinical Applications

*Anushruti Sarvaria[1,2], Dunia Jawdat[3], J. Alejandro Madrigal[1,2] and Aurore Saudemont[1,2]**

[1] Anthony Nolan Research Institute, London, UK, [2] Cancer Institute, University College London, London, UK,
[3] King Abdullah International Medical Research Center, Riyadh, Saudi Arabia

Natural killer (NK) cells are lymphocytes of the innate immune system able to kill different targets such as cancer cells and virally infected cells without prior activation making then attractive candidates for cancer immunotherapy. Umbilical cord blood (UCB) has become a source of hematopoietic stem cells for transplantation but as we gain a better understanding of the characteristics of each immune cell that UCB contains, we will also be able to develop new cell therapies for cancer. In this review, we present what is currently known of the phenotype and functions of UCB NK cells and how these cells could be used in the future for cancer immunotherapy.

Keywords: natural killer cells, umbilical cord blood, immunotherapy, cancer, hematopoietic stem cells

Edited by:
Antoine Toubert,
Paris Diderot University, France

Reviewed by:
Subramaniam Malarkannan,
Medical College of Wisconsin, USA
Amir Ahmed Toor,
Virginia Commonwealth University, USA

Correspondence:
Aurore Saudemont
aurore.saudemont@anthonynolan.org

INTRODUCTION

Natural killer (NK) cells are lymphocytes of the innate immune system that exhibit cytotoxicity toward cancer cells and virus-infected cells and have the capacity to produce cytokines such as interferon-γ (IFN-γ) and tumor necrosis factor-α (TNF-α) in response to stimuli. NK cells are defined as $CD56^+CD3^-$ cells and can be divided into two main subsets according to their expression of CD56 and CD16. $CD56^{dim}CD16^+$ NK cells ($CD56^{dim}$ NK cells) are cytotoxic NK cells capable to mediate direct killing of target cells *via* exocytosis of granules containing granzyme B and perforin, activation of cell death pathways such as TRAIL or FAS/FAS-L or *via* antibody-dependent cellular cytotoxicity. $CD56^{bright}CD16^{-/low}$ NK cells ($CD56^{bright}$ NK cells) are the main cytokine-producing NK cells (1). In peripheral blood (PB), up to 90% of NK cells are $CD56^{dim}$ NK cells while most NK cells are $CD56^{bright}$ NK cells in lymph nodes.

Natural killer cell functions are regulated by signals delivered through activating and inhibitory receptors. As opposite to T cells, NK cells are "ready to go" and can eliminate target cells without prior stimulation. However, stimulation of NK cells by cytokines leads to NK cell activation and enhanced functions, in particular enhanced cytolytic activity and proliferation. NK cells have long been considered potential candidates for cancer immunotherapy and their versatility makes them attractive cells to explore. Phase I clinical trials showed autologous NK cell therapies to be feasible and safe without adverse effects in patients with breast cancer or non-Hodgkin's lymphoma; however, these therapies had no or little impact on relapse rates (2). The potential impact of NK cell alloreactivity in hematopoietic stem cell transplantation (HSCT) was suggested by Valiante and Parham (3). The first evidence that allogeneic NK cells could exert strong anti-leukemic activity and impact on the outcome of haploidentical transplantation stems from the study of Ruggeri et al. (4) who reported NK cell alloreactivity against leukemic cells while reducing the risk of graft-versus-host disease (GvHD) in the context of human leukocyte antigen (HLA) mismatch settings. Other trials have showed that allogeneic NK cells alone can target different types of cancers such as acute myeloid leukemia (AML), melanoma, renal cell carcinoma, Hodgkin lymphoma (5), breast and ovarian cancer (6), or refractory lymphoma (7). The same group has shown the importance of NK cell expansion *in vivo*, which can be accomplished by infusion of interleukin (IL)-2. However, regulatory T cells were also found to compete for this cytokine and beneficial effects on NK cell expansion were observed when regulatory T cells could be depleted (8). Interestingly, other studies also have indicated that NK cell therapy could also be of interest to treat glioma (9) or neuroblastoma (10).

Umbilical cord blood (UCB) has become an established source of hematopoietic stem cells (11) for transplantation. Advantages for the use of UCB include low risk of viral transmission from donor to recipient, rapid availability of UCB units serving as an immediate "off-the-shelf" product, less stringent requirements for HLA matching, and lower risk of GvHD. However, UCB contains between 10- and 100-fold fewer nucleated cells than other sources of HSC, limiting how many cells of interest can be retrieved from one UCB unit. Interestingly, NK cells are the first lymphocytes to recover after HSCT including after umbilical cord blood transplantation (UCBT) (12). In addition, NK cells are key effectors of the graft-versus-leukemia (GvL) effect. Especially after UCBT, as T cell immune reconstitution is delayed and there is no increased incidence of relapse, it is likely that NK cells are actually the main effectors of the GvL effect in the first year post-UCBT. However, UCB also contains different types of immune cells including NK cells and as we learn more about their specific characteristics, we will identify the conditions which might benefit of an UCB NK cell therapy. This review focuses on providing an overview of the characteristics of UCB NK cells compared to NK cells from PB and explain how they could be used as a cell therapy to cancer.

CHARACTERISTICS OF UCB NK CELLS

Natural killer cells constitute up to 10% of lymphocytes in PB and up to 30% in UCB (13, 14), and both CD56^{dim} NK cells and CD56^{bright} NK cells can be found in PB and UCB with some groups reporting similar proportions of both subsets or higher frequency of CD56^{bright} NK cells in UCB (14–16). Regarding the phenotype and functions of UCB NK cells, some groups have identified differences when compared to PB NK cells while others found them to be similar to PB NK cells (17) (**Figure 1**).

Advantages of UCB-Derived NK Cells

Aside from the higher percentage of NK cells present in UCB, the ability to cryopreserve UCB together with the ease of collecting UCB units offers a unique clinical advantage of making UCB an off-the-shelf source for NK cell immunotherapy. Moreover, a more rapid recovery of NK cells was reported after UCBT than PB HSCT (18, 19). This faster recovery could be explained by the fact that UCB contains different NK cell progenitor populations that have the capacity to differentiate into NK cells and are typically absent in PB (20–22). Further, PB and UCB NK cells produced similar amounts of IFN-γ and TNF-α in response to different stimuli (14, 23) and could proliferate in response to cytokines such as IL-2 or IL-15 (14, 16, 24) despite UCB NK cells exhibiting lower expression of the IL-2 receptor subunits and lower phosphorylation of STAT5 (25). Additionally, UCB NK cells have also been described to have a higher expression of the bone marrow homing receptor, CXCR4, compared to PB NK cells indicating that UCB NK cells may contain a greater potential to home to the bone marrow (14). Finally, IL-15 activated UCB NK cells have been reported to impact positively on UCB HSC engraftment by enhancing their migration and clonogenic capacity, and their engraftment in humanized animal model (26).

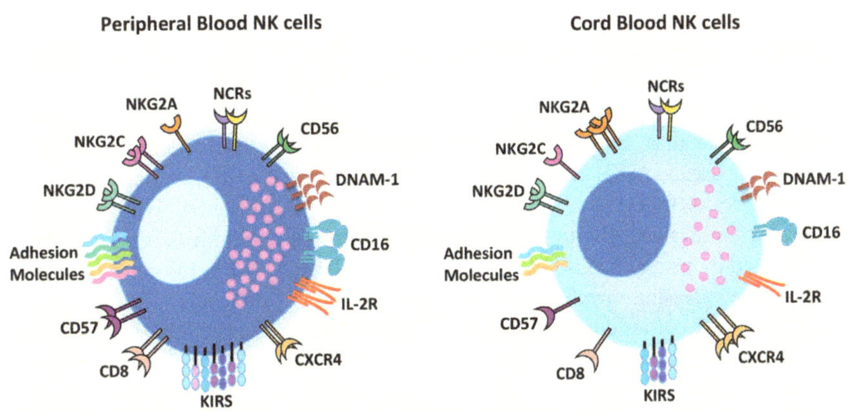

FIGURE 1 | Comparison of phenotypic characteristics between umbilical cord blood (UCB) natural killer (NK) cells and peripheral blood (PB) NK cells. In comparison to PB NK cells, UCB NK cells exhibit similar levels of CD56, NCRs (NKp46 and NKp30), and NKG2D but a lower expression of CD16, adhesion molecules (e.g., CD2, CD11a, CD18, CD62L), KIRs, DNAM-1, NKG2C, IL-2R, and CD57 and CD8 (receptors associated with terminal NK cell maturation) together with a higher expression of inhibitory receptor NKG2A indicating that UCB NK cells possess an immature phenotype and reduced cytotoxicity compared to PB NK cells. Further UCB NK cells have a higher expression of the bone marrow homing receptor, CXCR4, compared to PB NK cells proposing that cord blood NK cells may contain a greater potential to home to the bone marrow. Abbreviations: KIRs, killer-cell immunoglobulin-like receptors; NCRs, natural cytotoxicity receptors.

Drawbacks of UCB-Derived NK Cells

The use of cord blood (CB) as a source of NK cells for immunotherapy, however, is also limited as a result of the low numbers and immaturity of CB NK cells. Although, UCB NK cells have been reported to be fully mature and functional (16, 27), some groups found them to have an immature phenotype (14, 28), exhibiting normal levels of degranulation but lower cytotoxicity against K562 cells as compared to PB NK cells (14, 23). This lower activity could be explained by the fact that UCB NK cells have decreased expression of certain adhesion molecules on their surface such as CD2, CD11a, CD18, and CD62L (15, 16), decreased expression of CD16 (15), decreased expression of perforin and granzyme B (14, 23), and lower killer-cell immunoglobulin-like receptors (KIRs) expression together with a higher expression of inhibitory molecules such as NKG2A when compared to PB NK cells indicating an immature phenotype (14, 23). However, activation with cytokines such as IL-2 or IL-15 or the combination of IL-15 with IL-2 or IL-18 was able to restore or enhance their cytotoxicity to the levels observed for PB NK cells (14, 16, 23, 25, 29). Moreover, although the frequencies of NK cells present in UCB is greater than PB (14), low numbers of UCB NK cells are obtained as a result of the limited volume of an UCB unit, which is a major obstacle in obtaining sufficient numbers of NK cells for clinical application. However, different strategies to increase NK cell doses have been developed.

EXPANSION OF UCB NK CELLS

A number of studies have recently explored different platforms to expand UCB NK cells. Increased NK cell numbers can be achieved either by large-scale expansion techniques using artificial antigen-presenting cell (aAPC) or cytokines including IL-2, IL-15, and/or FLT-3 ligand. One such strategy employed to expand purified UCB-derived NK cells on a large scale has been reported using good manufacturing practice (GMP)-grade K562-based aAPCs expressing membrane-bound IL-21 (30). Shah and colleagues have shown that following 14 days of culture in a gas permeable culture system, a 2,389-mean fold expansion of NK cells derived from frozen UCB was achieved. The expanded NK cells presented >95% purity of $CD56^{+}CD3^{-}$ NK cells and displayed efficient killing capacity against multiple myeloma *in vitro* and *in vivo*, highlighting the use of aAPCs as an attractive approach to generate large numbers of functionally competent UCB NK cells. A further strategy to evaluate the potential use of expanded NK cells was reported by using aAPCs in the form of genetically modified K562 cells expressing membrane-bound IL-15 and 41BBL (31). The aAPCs were cultured with CB mononuclear cells for 7 days, which led to the generation of expanded UCB NK cells that displayed increased expression of NK cell activating receptors, increased perforin and granzyme expression, and increased cytotoxicity against B-cell non-Hodgkin lymphoma *in vitro* and *in vivo*. The study merits the use of expanded NK cells for adoptive cellular therapy specifically to target relapse or refractory disease after UCBT. Finally, the use of irradiated Epstein–Barr virus-transformed lymphoblastoid cell lines and IL-2 was also recently reported to generate large numbers of $CD56^{+}$ NK cells derived frozen UCB (32). The generated NK cells exhibited higher levels of cytotoxicity against K562 leukemic cells than expanded PB-derived NK cells (32). The unique advantage of this platform is that only 1 ml of the UCB unit is selectively used to generate expanded NK cells for adoptive therapy and the remaining UCB from the same unit can be cryopreserved and used for future transplantation. It would be interesting to assess whether the use of the same UCB for early NK cell adoptive therapy and transplantation can help to prevent relapse and augment GvL post-UCBT.

DIFFERENTIATION OF NK CELLS FROM UCB CD34+ CELLS

Natural killer cells can be directly isolated from PB and UCB but an alternative to these cell sources is the differentiation of NK cells from HSC as a way to generate high numbers of cells (33). NK cells can be differentiated from $CD34^{+}$ cells from the bone marrow, from embryonic stem cells, mobilized PB, or UCB $CD34^{+}$ cells. The expansion of NK cells derived from both fresh and frozen UCB $CD34^{+}$ cells using a cocktail of cytokines in a culture system has also been described as an efficient system to generate large numbers of NK cells. We and others have reported the characteristics of NK cells produced *in vitro* from UCB $CD34^{+}$ cells (34–36). These cells are mostly similar to PB NK cells with the exception that they express low levels of inhibitory receptors. However, NK cells produced in such a way have been shown to be functional, able to kill leukemic cell lines and patient cells *in vitro* and *in vivo* and produce cytokines in response to diverse stimuli (34, 36–38). Interestingly, NK cells produced *in vitro* have been shown to expand to high numbers while preserving their phenotype and functions after cryopreservation (39). Thus, frozen UCB $CD34^{+}$ cells were found to be the best source of NK cells when compared to fresh UCB-derived $CD34^{+}$ cells and frozen PB $CD34^{+}$ cells and could therefore be a readily available off-the-shelf product for NK cell immunotherapy.

NK Cells Alloreactivity in UCBT Setting

Umbilical cord blood NK cells express both inhibitory and activating receptors, which are highly important in mediating self-tolerance or NK cell activity (40). Inhibitory receptors are part of the immunoglobulin superfamily including the KIRs, the immunoglobulin-like transcripts, and C-type lectin receptors CD94/NKG2A. Inhibitory receptors recognize the classical MHC class I molecules on target cells and inhibit NK cell lysis (41). Most KIRs are inhibitory receptors but a limited number of KIRs also function as activating receptors; however, the function and ligands of the later are less well understood. Since KIR genes are not on the same chromosome as HLA, these genes are inherited independently. This allows for donor and recipient HLA-matched UCBT and mismatching between KIRs and their ligands, maintaining the appropriate matching required for HSCT but providing NK cell alloreactivity, which triggers NK cell activation leading to tumor cell lysis (42). This phenomenon of NK cell alloreactivity was proposed as beneficial in reducing relapse after HSCT; however, variable results have been reported from different studies (4, 43–47). In UCBT setting, only few studies have evaluated the outcome of UCBT using mismatched KIR and its ligands (48–51) with only some of them reporting beneficial results (52, 53). KIR haplotype has also been shown to influence the outcome of HSCT. In this context, the higher the number of activating KIR a donor has the higher NK cell alloreactivity might be. Some studies have reported the beneficial effect of the donor B haplotype that contains more activating gene than a A haplotype on HSCT outcome in particular showing a lower incidence of relapse for patients with AML or lower GvHD incidence depending on the study considered (47, 54–56). Whether KIR haplotype can also influence UCBT outcome needs to be investigated.

Finally, NK cell licensing (57), arming/disarming (58), or education (59) is another factor to be considered. NK cells can express one or more inhibitory receptors recognizing HLA molecules. The process by which NK cells become functional and tolerant to self-HLA can be referred to as NK cell licensing and is defined by the fact that to be functional NK cells must express inhibitory receptors recognizing self-HLA. This concept has been well studied in mice and there are now also evidence in humans (59, 60). However, it has been reported that unlicensed NK cells are able to mount an immune response against cytomegalovirus in mice (61) and can kill neuroblastoma cells in humans (10). Therefore, moving forward it will be essential to gain a better understanding of the impact of NK cell licensing on their functions especially in the context of HSCT including UCBT.

CURRENT CLINICAL STUDIES INVOLVING UCB NK CELLS

Natural killer cells can be isolated from UCB based on CD56 purification methods. One step isolation method can be used in UCB as opposite to PB where two steps are needed in order to eliminate NKT cells. This is not necessary when considering UCB as it contains a very low percentage of that cell subset. In addition, UCB has the advantage of being readily available as UCB is cryopreserved and can be obtained from accredited UCB banks. Therefore, a NK cell product derived from UCB has the potential to be off-the-shelf. Another advantage of UCB is that HLA is less stringent, although it is not clear what level of matching will be necessary to develop a third party NK cell product from UCB. However, because of the limited volume of blood collected from the umbilical cord there are only a limited number of NK cells that can be isolated from UCB. In addition, as they are immature and have lower functionality as compared to PB NK cells; taking UCB NK cells to the clinics will require a prior activation/expansion step. Several clinical trials are currently ongoing to evaluate the safety and feasibility of UCB NK cells as an "off the shelf product" in transplant and non-transplant settings (**Table 1**). GMP grade expansion methods for UCB NK cells are currently available as previously described. Notably, only a handful of clinical trials are currently ongoing and recruiting patients using the latest method to expand UCB NK cells to reach the cell dose required. Two clinical phase I studies aim to use expanded UCB NK cells for the treatment of patients with chronic lymphocytic leukemia (NCT01619761, NCT02280525), while another aims to evaluate NK cell therapy in the context of autologous HSCT for patients with myeloma (NCT01729091).

Only a few groups have focused on developing cell therapy approaches based on the differentiation of NK cells from HSC *in vitro*. However, NK cells produced *in vitro* have been shown to be safe and their use feasible when considered in the context of allogeneic HSCT (62). In addition, another trial, oNKord®, is currently ongoing testing the use of NK cells produced *in vitro* from UCB $CD34^+$ cells in patients with AML (EudraCT number 2010-018988-41).

CONCLUDING REMARKS

Immunotherapy is a promising treatment for different types of cancer allowing the possibility of personalized medicine for each cancer patient. UCB provides distinct advantages and is an increasingly attractive source for HSCT and cellular therapy. Despite low NK cell numbers within a single UCB unit and their immature phenotype, strategies to expand UCB NK cells using aAPCs or cytokines and feeder cells are

TABLE 1 | UCB NK cells currently in the clinic.

Clinical trial identifier	Diseases	Trial phase	Type of transplant	Conditioning	Method of expansion	Sponsor
NCT01619761	ALL, AML, CLL, CML, HL, MDS, MM, NHL, SLL	I	Double umbilical cord blood transplantation	Fludarabin, melphalan, lenalidomide ± rituximab	*Ex vivo* expansion of NK cells from 20% UCB unit fraction	MD Anderson Cancer Center
NCT02280525	CLL, ALL, AML, CML, NHL, HL	I	Non-HSCT	Fludarabin, cyclophosphamide, lenalidomide, and rituximab	*Ex vivo* expansion of NK cells from thawed from UCB unit	MD Anderson Cancer Center
NCT01729091	MM	I/II	Autologous	Melphalan, lenalidomide	*Ex vivo* expansion of NK cells from thawed from UCB unit	MD Anderson Cancer Center
EudraCT number 2010-018988-41	AML	I	Non-HSCT	Fludarabin, cyclophosphamide	NK cells generated *in vitro* from UCB progenitor cells	Radboud Medical Centre, Nijmegen, Netherlands

ALL, acute lymphoblastic leukemia; AML, acute myeloid leukemia; CB, cord blood; CLL, chronic lymphoblastic leukemia; CML, chronic myeloid leukemia; HL, Hodgkin lymphoma; HSCT, hematopoietic stem cell transplantation; MDS, myelodysplastic syndromes; MM, multiple myeloma; NHL, non-Hodgkin lymphoma; NK, natural killer; SLL, small lymphocytic lymphoma; UCB, umbilical cord blood.

paving the way for NK cell adoptive immunotherapy. NK cells have shown great potential in eliminating different types of cancer cells *in vitro* and in animal models. A few clinical trials are currently underway to evaluate the safety and feasibility of using UCB NK cells as an "off the shelf" product for the prevention of relapse. The results from these studies will help in understanding how to maximize the beneficial potential of UCB NK cells for the treatment of hematological malignancies and solid tumors.

AUTHOR CONTRIBUTIONS

All the authors contributed to writing and reviewing the manuscript.

REFERENCES

1. Cooper MA, Fehniger TA, Caligiuri MA. The biology of human natural killer-cell subsets. *Trends Immunol* (2001) 22:633–40. doi:10.1016/S1471-4906(01)02060-9
2. Burns LJ, Weisdorf DJ, Defor TE, Vesole DH, Repka TL, Blazar BR, et al. IL-2-based immunotherapy after autologous transplantation for lymphoma and breast cancer induces immune activation and cytokine release: a phase I/II trial. *Bone Marrow Transplant* (2003) 32:177–86. doi:10.1038/sj.bmt.1704086
3. Valiante NM, Parham P. Natural killer cells, HLA class I molecules, and marrow transplantation. *Biol Blood Marrow Transplant* (1997) 3:229–35.
4. Ruggeri L, Capanni M, Urbani E, Perruccio K, Shlomchik WD, Tosti A, et al. Effectiveness of donor natural killer cell alloreactivity in mismatched hematopoietic transplants. *Science* (2002) 295:2097–100. doi:10.1126/science.1068440
5. Miller JS, Soignier Y, Panoskaltsis-Mortari A, Mcnearney SA, Yun GH, Fautsch SK, et al. Successful adoptive transfer and in vivo expansion of human haploidentical NK cells in patients with cancer. *Blood* (2005) 105:3051–7. doi:10.1182/blood-2004-07-2974
6. Geller MA, Cooley S, Judson PL, Ghebre R, Carson LF, Argenta PA, et al. A phase II study of allogeneic natural killer cell therapy to treat patients with recurrent ovarian and breast cancer. *Cytotherapy* (2011) 13:98–107. doi:10.3109/14653249.2010.515582
7. Bachanova V, Burns LJ, Mckenna DH, Curtsinger J, Panoskaltsis-Mortari A, Lindgren BR, et al. Allogeneic natural killer cells for refractory lymphoma. *Cancer Immunol Immunother* (2010) 59:1739–44. doi:10.1007/s00262-010-0896-z
8. Bachanova V, Cooley S, Defor TE, Verneris MR, Zhang B, Mckenna DH, et al. Clearance of acute myeloid leukemia by haploidentical natural killer cells is improved using IL-2 diphtheria toxin fusion protein. *Blood* (2014) 123:3855–63. doi:10.1182/blood-2013-10-532531
9. Ishikawa E, Tsuboi K, Saijo K, Harada H, Takano S, Nose T, et al. Autologous natural killer cell therapy for human recurrent malignant glioma. *Anticancer Res* (2004) 24:1861–71.
10. Tarek N, Le Luduec JB, Gallagher MM, Zheng J, Venstrom JM, Chamberlain E, et al. Unlicensed NK cells target neuroblastoma following anti-GD2 antibody treatment. *J Clin Invest* (2012) 122:3260–70. doi:10.1172/JCI62749
11. Sun J, Wang J, Pefanis E, Chao J, Rothschild G, Tachibana I, et al. Transcriptomics identify CD9 as a marker of murine IL-10-competent regulatory B cells. *Cell Rep* (2015) 13:1110–7. doi:10.1016/j.celrep.2015.09.070
12. Komanduri KV, St John LS, De Lima M, Mcmannis J, Rosinski S, Mcniece I, et al. Delayed immune reconstitution after cord blood transplantation is characterized by impaired thymopoiesis and late memory T-cell skewing. *Blood* (2007) 110:4543–51. doi:10.1182/blood-2007-05-092130
13. Kotylo PK, Baenzinger JC, Yoder MC, Engle WA, Bolinger CD. Rapid analysis of lymphocyte subsets in cord blood. *Am J Clin Pathol* (1990) 93:263–6. doi:10.1093/ajcp/93.2.263
14. Luevano M, Daryouzeh M, Alnabhan R, Querol S, Khakoo S, Madrigal A, et al. The unique profile of cord blood natural killer cells balances incomplete maturation and effective killing function upon activation. *Hum Immunol* (2012) 73:248–57. doi:10.1016/j.humimm.2011.12.015
15. Tanaka H, Kai S, Yamaguchi M, Misawa M, Fujimori Y, Yamamoto M, et al. Analysis of natural killer (NK) cell activity and adhesion molecules on NK cells from umbilical cord blood. *Eur J Haematol* (2003) 71:29–38. doi:10.1034/j.1600-0609.2003.00081.x
16. Dalle JH, Menezes J, Wagner E, Blagdon M, Champagne J, Champagne MA, et al. Characterization of cord blood natural killer cells: implications for transplantation and neonatal infections. *Pediatr Res* (2005) 57:649–55. doi:10.1203/01.PDR.0000156501.55431.20
17. Verneris MR, Miller JS. The phenotypic and functional characteristics of umbilical cord blood and peripheral blood natural killer cells. *Br J Haematol* (2009) 147:185–91. doi:10.1111/j.1365-2141.2009.07768.x
18. Cooley S, Mccullar V, Wangen R, Bergemann TL, Spellman S, Weisdorf DJ, et al. KIR reconstitution is altered by T cells in the graft and correlates with clinical outcomes after unrelated donor transplantation. *Blood* (2005) 106:4370–6. doi:10.1182/blood-2005-04-1644
19. Nguyen S, Kuentz M, Vernant JP, Dhedin N, Bories D, Debre P, et al. Involvement of mature donor T cells in the NK cell reconstitution after haploidentical hematopoietic stem-cell transplantation. *Leukemia* (2008) 22:344–52. doi:10.1038/sj.leu.2405041
20. Gaddy J, Broxmeyer HE. Cord blood CD16+56- cells with low lytic activity are possible precursors of mature natural killer cells. *Cell Immunol* (1997) 180:132–42. doi:10.1006/cimm.1997.1175
21. Perez SA, Sotiropoulou PA, Gkika DG, Mahaira LG, Niarchos DK, Gritzapis AD, et al. A novel myeloid-like NK cell progenitor in human umbilical cord blood. *Blood* (2003) 101:3444–50. doi:10.1182/blood-2002-05-1501
22. Rutella S, Bonanno G, Marone M, De Ritis D, Mariotti A, Voso MT, et al. Identification of a novel subpopulation of human cord blood CD34-CD133-CD7-CD45+lineage- cells capable of lymphoid/NK cell differentiation after in vitro exposure to IL-15. *J Immunol* (2003) 171:2977–88. doi:10.4049/jimmunol.171.6.2977
23. Wang Y, Xu H, Zheng X, Wei H, Sun R, Tian Z. High expression of NKG2A/CD94 and low expression of granzyme B are associated with reduced cord blood NK cell activity. *Cell Mol Immunol* (2007) 4:377–82.
24. Satwani P, Van De Ven C, Ayello J, Cairo D, Simpson LL, Baxi L, et al. Interleukin (IL)-15 in combination with IL-2, fms-like tyrosine kinase-3 ligand and anti-CD3 significantly enhances umbilical cord blood natural killer (NK) cell and NK-cell subset expansion and NK function. *Cytotherapy* (2011) 13:730–8. doi:10.3109/14653249.2011.563292
25. Alnabhan R, Madrigal A, Saudemont A. Differential activation of cord blood and peripheral blood natural killer cells by cytokines. *Cytotherapy* (2015) 17:73–85. doi:10.1016/j.jcyt.2014.08.003
26. Escobedo-Cousin M, Jackson N, Laza-Briviesca R, Ariza-Mcnaughton L, Luevano M, Derniame S, et al. Natural killer cells improve hematopoietic stem cell engraftment by increasing stem cell clonogenicity in vitro and in a humanized mouse model. *PLoS One* (2015) 10:e0138623. doi:10.1371/journal.pone.0138623
27. Schonberg K, Fischer JC, Kogler G, Uhrberg M. Neonatal NK-cell repertoires are functionally, but not structurally, biased toward recognition of self HLA class I. *Blood* (2011) 117:5152–6. doi:10.1182/blood-2011-02-334441
28. Gaddy J, Risdon G, Broxmeyer HE. Cord blood natural killer cells are functionally and phenotypically immature but readily respond to interleukin-2 and interleukin-12. *J Interferon Cytokine Res* (1995) 15:527–36. doi:10.1089/jir.1995.15.527
29. Lin SJ, Kuo ML. Cytotoxic function of umbilical cord blood natural killer cells: relevance to adoptive immunotherapy. *Pediatr Hematol Oncol* (2011) 28:640–6. doi:10.3109/08880018.2011.613092
30. Shah N, Martin-Antonio B, Yang H, Ku S, Lee DA, Cooper LJ, et al. Antigen presenting cell-mediated expansion of human umbilical cord blood yields log-scale expansion of natural killer cells with anti-myeloma activity. *PLoS One* (2013) 8(10):e76781. doi:10.1371/journal.pone.0076781

31. Ayello J, Hochberg J, Flower A, Chu Y, Baxi LV, Quish W, et al. Genetically re-engineered K562 cells significantly expand and functionally activate cord blood natural killer cells: potential for adoptive cellular immunotherapy. *Exp Hematol* (2017) 46:38–47. doi:10.1016/j.exphem.2016.10.003
32. Vasu S, Berg M, Davidson-Moncada J, Tian X, Cullis H, Childs RW. A novel method to expand large numbers of CD56(+) natural killer cells from a minute fraction of selectively accessed cryopreserved cord blood for immunotherapy after transplantation. *Cytotherapy* (2015) 17(11):1582–93. doi:10.1016/j.jcyt.2015.07.020
33. Luevano M, Madrigal A, Saudemont A. Generation of natural killer cells from hematopoietic stem cells in vitro for immunotherapy. *Cell Mol Immunol* (2012) 9:310–20. doi:10.1038/cmi.2012.17
34. Spanholtz J, Tordoir M, Eissens D, Preijers F, Van Der Meer A, Joosten I, et al. High log-scale expansion of functional human natural killer cells from umbilical cord blood CD34-positive cells for adoptive cancer immunotherapy. *PLoS One* (2010) 5:e9221. doi:10.1371/journal.pone.0009221
35. Spanholtz J, Preijers F, Tordoir M, Trilsbeek C, Paardekooper J, De Witte T, et al. Clinical-grade generation of active NK cells from cord blood hematopoietic progenitor cells for immunotherapy using a closed-system culture process. *PLoS One* (2011) 6:e20740. doi:10.1371/journal.pone.0020740
36. Luevano M, Domogala A, Blundell M, Jackson N, Pedroza-Pacheco I, Derniame S, et al. Frozen cord blood hematopoietic stem cells differentiate into higher numbers of functional natural killer cells in vitro than mobilized hematopoietic stem cells or freshly isolated cord blood hematopoietic stem cells. *PLoS One* (2014) 9:e87086. doi:10.1371/journal.pone.0087086
37. Cany J, Van Der Waart AB, Tordoir M, Franssen GM, Hangalapura BN, De Vries J, et al. Natural killer cells generated from cord blood hematopoietic progenitor cells efficiently target bone marrow-residing human leukemia cells in NOD/SCID/IL2Rg(null) mice. *PLoS One* (2013) 8:e64384. doi:10.1371/journal.pone.0064384
38. Cany J, Van Der Waart AB, Spanholtz J, Tordoir M, Jansen JH, Van Der Voort R, et al. Combined IL-15 and IL-12 drives the generation of CD34+-derived natural killer cells with superior maturation and alloreactivity potential following adoptive transfer. *Oncoimmunology* (2015) 4:e1017701. doi:10.1080/2162402X.2015.1017701
39. Domogala A, Madrigal JA, Saudemont A. Cryopreservation has no effect on function of natural killer cells differentiated in vitro from umbilical cord blood CD34(+) cells. *Cytotherapy* (2016) 18:754–9. doi:10.1016/j.jcyt.2016.02.008
40. Vivier E, Nunes JA, Vely F. Natural killer cell signaling pathways. *Science* (2004) 306:1517–9. doi:10.1126/science.1103478
41. Vilches C, Parham P. KIR: diverse, rapidly evolving receptors of innate and adaptive immunity. *Annu Rev Immunol* (2002) 20:217–51. doi:10.1146/annurev.immunol.20.092501.134942
42. Leung W. Use of NK cell activity in cure by transplant. *Br J Haematol* (2011) 155:14–29. doi:10.1111/j.1365-2141.2011.08823.x
43. Bishara A, De Santis D, Witt CC, Brautbar C, Christiansen FT, Or R, et al. The beneficial role of inhibitory KIR genes of HLA class I NK epitopes in haploidentically mismatched stem cell allografts may be masked by residual donor-alloreactive T cells causing GVHD. *Tissue Antigens* (2004) 63:204–11. doi:10.1111/j.0001-2815.2004.00182.x
44. Hsu KC, Gooley T, Malkki M, Pinto-Agnello C, Dupont B, Bignon JD, et al. KIR ligands and prediction of relapse after unrelated donor hematopoietic cell transplantation for hematologic malignancy. *Biol Blood Marrow Transplant* (2006) 12:828–36. doi:10.1016/j.bbmt.2006.04.008
45. Ruggeri L, Mancusi A, Capanni M, Urbani E, Carotti A, Aloisi T, et al. Donor natural killer cell allorecognition of missing self in haploidentical hematopoietic transplantation for acute myeloid leukemia: challenging its predictive value. *Blood* (2007) 110:433–40. doi:10.1182/blood-2006-07-038687
46. Vago L, Forno B, Sormani MP, Crocchiolo R, Zino E, Di Terlizzi S, et al. Temporal, quantitative, and functional characteristics of single-KIR-positive alloreactive natural killer cell recovery account for impaired graft-versus-leukemia activity after haploidentical hematopoietic stem cell transplantation. *Blood* (2008) 112:3488–99. doi:10.1182/blood-2007-07-103325
47. Venstrom JM, Gooley TA, Spellman S, Pring J, Malkki M, Dupont B, et al. Donor activating KIR3DS1 is associated with decreased acute GVHD in unrelated allogeneic hematopoietic stem cell transplantation. *Blood* (2010) 115:3162–5. doi:10.1182/blood-2009-08-236943
48. Brunstein CG, Wagner JE, Weisdorf DJ, Cooley S, Noreen H, Barker JN, et al. Negative effect of KIR alloreactivity in recipients of umbilical cord blood transplant depends on transplantation conditioning intensity. *Blood* (2009) 113:5628–34. doi:10.1182/blood-2008-12-197467
49. Willemze R, Ruggeri A, Purtill D, Rodrigues CA, Gluckman E, Rocha V, et al. Is there an impact of killer cell immunoglobulin-like receptors and KIR-ligand incompatibilities on outcomes after unrelated cord blood stem cell transplantation? *Best Pract Res Clin Haematol* (2010) 23:283–90. doi:10.1016/j.beha.2010.05.005
50. Garfall A, Kim HT, Sun L, Ho VT, Armand P, Koreth J, et al. KIR ligand incompatibility is not associated with relapse reduction after double umbilical cord blood transplantation. *Bone Marrow Transplant* (2013) 48:1000–2. doi:10.1038/bmt.2012.272
51. Tanaka J, Morishima Y, Takahashi Y, Yabe T, Oba K, Takahashi S, et al. Effects of KIR ligand incompatibility on clinical outcomes of umbilical cord blood transplantation without ATG for acute leukemia in complete remission. *Blood Cancer J* (2013) 3:e164. doi:10.1038/bcj.2013.62
52. Willemze R, Rodrigues CA, Labopin M, Sanz G, Michel G, Socie G, et al. KIR-ligand incompatibility in the graft-versus-host direction improves outcomes after umbilical cord blood transplantation for acute leukemia. *Leukemia* (2009) 23:492–500. doi:10.1038/leu.2008.365
53. Sekine T, Marin D, Cao K, Li L, Mehta P, Shaim H, et al. Specific combinations of donor and recipient KIR-HLA genotypes predict for large differences in outcome after cord blood transplantation. *Blood* (2016) 128:297–312. doi:10.1182/blood-2016-03-706317
54. Cooley S, Trachtenberg E, Bergemann TL, Saeteurn K, Klein J, Le CT, et al. Donors with group B KIR haplotypes improve relapse-free survival after unrelated hematopoietic cell transplantation for acute myelogenous leukemia. *Blood* (2009) 113:726–32. doi:10.1182/blood-2008-07-171926
55. Cooley S, Weisdorf DJ, Guethlein LA, Klein JP, Wang T, Le CT, et al. Donor selection for natural killer cell receptor genes leads to superior survival after unrelated transplantation for acute myelogenous leukemia. *Blood* (2010) 116:2411–9. doi:10.1182/blood-2010-05-283051
56. Cooley S, Weisdorf DJ, Guethlein LA, Klein JP, Wang T, Marsh SG, et al. Donor killer cell Ig-like receptor B haplotypes, recipient HLA-C1, and HLA-C mismatch enhance the clinical benefit of unrelated transplantation for acute myelogenous leukemia. *J Immunol* (2014) 192:4592–600. doi:10.4049/jimmunol.1302517
57. Kim S, Poursine-Laurent J, Truscott SM, Lybarger L, Song YJ, Yang L, et al. Licensing of natural killer cells by host major histocompatibility complex class I molecules. *Nature* (2005) 436:709–13. doi:10.1038/nature03847
58. Shifrin N, Raulet DH, Ardolino M. NK cell self tolerance, responsiveness and missing self recognition. *Semin Immunol* (2014) 26:138–44. doi:10.1016/j.smim.2014.02.007
59. Anfossi N, Andre P, Guia S, Falk CS, Roetynck S, Stewart CA, et al. Human NK cell education by inhibitory receptors for MHC class I. *Immunity* (2006) 25:331–42. doi:10.1016/j.immuni.2006.06.013
60. Cooley S, Xiao F, Pitt M, Gleason M, Mccullar V, Bergemann TL, et al. A subpopulation of human peripheral blood NK cells that lacks inhibitory receptors for self-MHC is developmentally immature. *Blood* (2007) 110:578–86. doi:10.1182/blood-2006-07-036228
61. Orr MT, Murphy WJ, Lanier LL. 'Unlicensed' natural killer cells dominate the response to cytomegalovirus infection. *Nat Immunol* (2010) 11:321–7. doi:10.1038/ni.1849
62. Yoon SR, Lee YS, Yang SH, Ahn KH, Lee JH, Lee JH, et al. Generation of donor natural killer cells from CD34(+) progenitor cells and subsequent infusion after HLA-mismatched allogeneic hematopoietic cell transplantation: a feasibility study. *Bone Marrow Transplant* (2010) 45:1038–46. doi:10.1038/bmt.2009.304

5

Chimeric Antigen Receptor-Engineered NK-92 Cells: An Off-the-Shelf Cellular Therapeutic for Targeted Elimination of Cancer Cells and Induction of Protective Antitumor Immunity

Congcong Zhang[1,2,3], *Pranav Oberoi*[1], *Sarah Oelsner*[1], *Anja Waldmann*[1], *Aline Lindner*[1], *Torsten Tonn*[3,4,5,6] *and Winfried S. Wels*[1,2,3]*

[1] *Georg-Speyer-Haus, Institute for Tumor Biology and Experimental Therapy, Frankfurt am Main, Germany,* [2] *German Cancer Consortium (DKTK), Partner Site Frankfurt/Mainz, Frankfurt am Main, Germany,* [3] *German Cancer Research Center (DKFZ), Heidelberg, Germany,* [4] *German Red Cross Blood Donation Service North-East, Institute for Transfusion Medicine, Dresden, Germany,* [5] *Medical Faculty Carl Gustav Carus, TU Dresden, Dresden, Germany,* [6] *German Cancer Consortium (DKTK), Partner Site Dresden, Dresden, Germany*

Edited by:
Gianfranco Pittari,
Hamad Medical Corporation, Qatar

Reviewed by:
Adelheid Cerwenka,
German Cancer Research Center, Germany
Michael Bachmann,
Helmholz Zentrum Dresden Rossendorf (HZDR), Germany

Correspondence:
Winfried S. Wels
wels@gsh.uni-frankfurt.de

Significant progress has been made in recent years toward realizing the potential of natural killer (NK) cells for cancer immunotherapy. NK cells can respond rapidly to transformed and stressed cells and have the intrinsic potential to extravasate and reach their targets in almost all body tissues. In addition to donor-derived primary NK cells, also the established NK cell line NK-92 is being developed for adoptive immunotherapy, and general safety of infusion of irradiated NK-92 cells has been established in phase I clinical trials with clinical responses observed in some of the cancer patients treated. To enhance their therapeutic utility, NK-92 cells have been modified to express chimeric antigen receptors (CARs) composed of a tumor-specific single chain fragment variable antibody fragment fused *via* hinge and transmembrane regions to intracellular signaling moieties such as CD3ζ or composite signaling domains containing a costimulatory protein together with CD3ζ. CAR-mediated activation of NK cells then bypasses inhibitory signals and overcomes NK resistance of tumor cells. In contrast to primary NK cells, CAR-engineered NK-92 cell lines suitable for clinical development can be established from molecularly and functionally well-characterized single cell clones following good manufacturing practice-compliant procedures. In preclinical *in vitro* and *in vivo* models, potent antitumor activity of NK-92 variants targeted to differentiation antigens expressed by hematologic malignancies, and overexpressed or mutated self-antigens associated with solid tumors has been found, encouraging further development of CAR-engineered NK-92 cells. Importantly, in syngeneic mouse tumor models, induction of endogenous antitumor immunity after treatment with CAR-expressing NK-92 cells has been demonstrated, resulting in cures and long-lasting immunological memory protecting against tumor rechallenge at distant sites. Here, we summarize the current status and future prospects of CAR-engineered NK-92 cells as off-the-shelf cellular therapeutics, with special emphasis on ErbB2 (HER2)-specific NK-92 cells that are approaching clinical application.

Keywords: natural killer cells, NK-92, chimeric antigen receptor, adoptive cancer immunotherapy, leukemia, lymphoma, solid tumors

INTRODUCTION

Natural killer (NK) cells are specialized effectors of the innate immune system and central players in the defense against viral infections and cancer. Natural cytotoxicity of NK cells can be triggered rapidly upon appropriate stimulation and is regulated by a complex balance of signals from germ-line encoded activating and inhibitory cell surface receptors (1, 2). The antitumoral activity of NK cells has been well documented in mouse models (3, 4). In humans, a correlation between low peripheral blood NK-cell activity and an increased cancer risk was demonstrated (5), and numbers and phenotype of tumor-infiltrating NK cells likely influence the course of the disease (6–8). Mechanisms involved in tumor immune evasion can be diverse and include upregulation of the non-classical MHC molecules HLA-E and HLA-G that trigger inhibitory NK-cell receptors (9), selective loss of ligands for activating NK-cell receptors (10, 11), as well as shedding of soluble forms of MHC class I polypeptide-related sequence A/B (MICA/B) and B7-H6 (12–14). Furthermore, the tumor microenvironment plays a crucial role in preventing infiltration by NK and other immune cells and interfering with the activity of NK cells already present in the tumor (15, 16). Hypoxia as well as immunosuppressive factors such as transforming growth factor (TGF)-β, indoleamine 2,3-deoxygenase (IDO), prostaglandin E2, nitric oxide (NO), and reactive oxygen species (ROS), which are produced by regulatory immune cells like regulatory T (T_{reg}) cells and myeloid-derived suppressor cells, by stromal cells like cancer-associated fibroblasts, and by tumor cells themselves can inhibit expression of activating NK-cell receptors, disrupt the interactions between NK and other immune cells, and avert the contact of NK cells with tumor cells (17).

To bypass deficiencies in endogenous NK-cell activity, current NK-cell therapies are typically based on adoptive transfer of *ex vivo*-expanded allogeneic NK cells derived from a suitable donor (18–20). While displaying graft-versus-leukemia (GvL) or graft-versus-tumor (GvT) activity, such donor-derived NK cells do not carry a high risk of inducing graft-versus-host-disease (GvHD) frequently associated with donor lymphocyte infusion (DLI) of allogeneic T cells (20). In addition, antibodies that block inhibitory NK-cell receptors such as killer cell immunoglobulin-like receptors (KIRs) and NKG2A/CD94, or link activating NK-cell receptors to tumor cell surface antigens are being investigated as activity enhancers for endogenous or adoptively transferred NK cells (21, 22). Sparked by the clinical success of chimeric antigen receptor (CAR)-engineered T cells in the treatment of B-cell malignancies, genetic modification of NK cells with CAR constructs is receiving increasing attention. CAR engagement in NK cells can override inhibitory signals deployed by tumor cells and directly trigger the effector cells' intrinsic cytolytic effector functions as well as the release of pro-inflammatory cytokines (23, 24). Nevertheless, despite the close similarity of NK cells to T cells with respect to their cytotoxic mechanisms, the development of CAR-engineered NK cells for adoptive cancer immunotherapy is still in its early stages, owing mainly to the complexity of isolating, activating, expanding, and manufacturing large numbers of peripheral blood-derived NK cells, the lower efficiency of gene transfer when compared to T cells, and the limited *in vivo* proliferation and persistence in recipients. While efforts are being made to overcome these hurdles by improving *ex vivo* expansion of NK cells to allow multiple infusions (25), results from clinical trials with CAR NK cells are not yet available.

Continuously expanding NK cell lines provide an unlimited source of effector cells to investigate and improve concepts for genetic engineering of NK cells (23, 26–29) but also hold potential for development as standardized off-the-shelf therapeutics for adoptive cancer immunotherapy. Different human NK cell lines have been established, including NK-92, HANK-1, KHYG-1, NK-YS, NKG, YT, YTS, NKL, and NK3.3 (30). Among them, NK-92 cells (also termed "aNK" for activated NK) have been investigated most thoroughly and already been applied in a clinical setting (31, 32). NK-92 express many activating NK-cell receptors such as NKp30, NKp46, and NKG2D but lack most of the inhibitory KIRs, except for low levels of KIR2DL4 (33, 34). Other inhibitory receptors expressed by NK-92 are Ig-like transcript 2 (ILT-2) and NKG2A/CD94. This unique profile renders NK-92 cells highly cytotoxic against a broad spectrum of malignant cells of hematologic origin and other cancers (32). General safety of infusion of irradiated NK-92 cells has been established in phase I clinical trials in patients with advanced cancers (35, 36), and results from other phase I and phase II studies may soon become available (NCT00990717, NCT00900809, NCT02465957; https://clinicaltrials.gov).

As outlined in the following sections, the robust *ex vivo* expansion of NK-92 cells to high cell numbers, their exquisite safety profile, as well as the ease of genetic modification make this cell line an ideal platform for the development of CAR-engineered variants. Here, we provide an overview of the diverse approaches that have been taken to date to target NK-92 cells to various hematological malignancies and solid tumors, summarize preclinical *in vitro* and *in vivo* studies with special emphasis on ErbB2 (HER2)-specific CAR NK-92 cells (NK-92/5.28.z) that are ready to enter clinical trials, and discuss general advantages and challenges associated with the use of CAR NK-92 cells as an off-the-shelf cellular therapeutic.

ADVANCES FROM THE CAR T CELL FIELD ENABLING THE GENERATION OF TUMOR-SPECIFIC NK CELLS

Since introduction of the basic CAR design with a single chain fragment variable (scFv) antibody for target recognition fused to CD3ζ or FcεRIγ chains for signaling (first-generation CARs) by Eshhar and colleagues (37), many groups have contributed to further improve and develop this concept, facilitating the clinical success of CAR T cell therapy seen today (38, 39). The most significant CAR modification was thereby the inclusion of costimulatory protein domains derived from CD27, CD28, CD134 (OX40), CD137 (4-1BB), CD244 (2B4) or CD278 (ICOS) (second-generation CARs), or their combinations (third-generation CARs) in addition to CD3ζ to improve T-cell activation, proliferation, and persistence (40). Other advances enhancing CAR functionality and providing additional benefits with respect to stimulating innate immunity, improving safety, or alleviating tumor immune escape have been reviewed extensively by Fesnak et al. (41). They include for instance interleukin (IL)-12-armed T cells redirected for universal cytokine-mediated killing (TRUCKs) (42), universal CARs activated by modular antibody-based targeting molecules (43), and dual-targeting tandem CARs (TanCARs) (44, 45). In patients with lymphomas and leukemias of B-cell origin, remarkable efficacy was demonstrated and durable responses were achieved with both, T-cell products harboring CD19-specific second-generation CD28- or CD137-containing CARs. While in experimental models CD28-CD3ζ CARs led to stronger T-cell activation, CD137-CD3ζ CARs prolonged *in vivo* T-cell persistence and reduced exhaustion (46, 47).

Already early on, it was postulated for first-generation CARs that they would be functional in NK cells (37), which was formally demonstrated for a CAR-like CD4-CD3ζ fusion receptor in human NK3.3 cells (26). In the first report proposing CAR-engineered NK-92 cells as a continuously expanding off-the-shelf cell therapeutic, we also applied a first-generation CAR consisting of an ErbB2-specific scFv antibody fused to CD3ζ through a CD8α hinge region, which resulted in high and specific cytotoxicity of the genetically modified cells toward ErbB2-expressing breast cancer cells and other targets of solid tumor origins (23). Similar first-generation CAR designs were successfully used in subsequent studies with NK-92 cells targeting the B-cell differentiation antigens CD19 and CD20 (48–53), CD138 for recognition of multiple myeloma (54), and various surface antigens expressed by solid tumors including the disialoganglioside GD2, epithelial cell adhesion molecule (EpCAM), and a peptide epitope of the melanoma antigen gp100 in complex with HLA-A2 (55–59) (**Table 1**). In studies with CD19- and GD2-targeted primary human NK cells, inclusion of costimulatory CD137 or CD244 domains in the CAR in addition to CD3ζ enhanced both specific cytotoxicity and production of interferon (IFN)-γ and granulocyte-macrophage colony stimulating factor (GM-CSF) when compared to first-generation CARs (24, 60). This clearly demonstrates that at least primary NK cells benefit from CAR-induced costimulatory signals. In preclinical studies, also an ErbB2-specific CD28-CD3ζ CAR and a CD20-specific CD137-CD3ζ CAR were shown to be functional in donor-derived human NK cells, but no comparison with respective CD3ζ-only CARs was performed (61, 62). Clinical trials with CAR-engineered primary NK cells for the treatment of B-cell acute lymphoblastic leukemia (B-ALL) employ CD19-specific CD137-CD3ζ receptors (NCT00995137, NCT01974479; https://clinicaltrials.gov), but results from these trials are not yet available.

INFLUENCE OF THE CAR DESIGN ON FUNCTIONALITY OF RETARGETED NK-92 CELLS

In the presence of IL-2, NK-92 cells persistently exhibit a phenotype similar to activated NK cells (33). Hence, CAR-engineered NK-92 variants may be less dependent on costimulation than T cells and primary NK cells (80). Nevertheless, second-generation CARs employing a composite CD28-CD3ζ signaling domain have been shown to be functional in NK-92 cells targeting EpCAM and ErbB2 on breast cancer cells (28, 67, 68), epidermal growth factor receptor (EGFR) on glioblastoma cells and breast cancer brain metastases (70–72), EGFRvIII, a glioblastoma-specific mutant form of EGFR arising from an in-frame deletion of exons 2-7 of the receptor (71, 72), CD19 on B-cell malignancies (74), CS1 on multiple myeloma cells (77), and CD33 on acute myeloid leukemia cells (81). Likewise, second-generation CARs harboring CD137-CD3ζ domains and targeting ErbB2 (67), CD19 (74), or peptide epitopes of Epstein–Barr virus (EBV) latent protein EBNA3C, and Wilms tumor protein in complex with HLA-A2 (78, 79) have been used successfully with NK-92 cells as well as third generation CD28-CD137-CD3ζ CARs that recognize CD3 or CD5 for elimination of malignant T cells (75, 76) (**Table 1**).

Only two reports compared the functionality of NK-92 cells harboring CD3ζ-based first-generation or CD28-CD3ζ- and CD137-CD3ζ-based second-generation CARs directly (67, 74), using a general CAR design as depicted in **Figure 1A**. NK-92 cells express high levels of CD3ζ and moderate levels of CD28 and CD137 (23, 34, 67) (**Figure 1B**), suggesting that the CARs could readily link to respective endogenous signaling pathways. Indeed, while differences were relatively small, ErbB2-targeted NK-92 cells expressing CD28-CD3ζ and CD137-CD3ζ CARs displayed more pronounced cytotoxicity in short-term assays when compared to a corresponding CD3ζ-only CAR (67). Conversely, CD19-targeted NK-92 cells harboring a CD137-CD3ζ CAR were much less effective in cell killing than cells expressing a CD3ζ-only or a CD28-CD3ζ CAR containing the same cell targeting domain (74). With respect to cytokine production, highest amounts of IFN-γ were found in cultures of CD19-specific NK-92 expressing a CD28-CD3ζ CAR, while less pronounced levels were secreted upon CAR activation by cells harboring a CD3ζ-only CAR, and only marginally enhanced levels by cells carrying the CD137-CD3ζ CAR.

The first-generation CARs included in these studies for comparison utilized the endogenous transmembrane domain of CD3ζ. This allowed formation of both, disulfide-linked CAR homodimers, and heterodimers of the CAR with endogenous CD3ζ of NK-92 cells (67, 74). Such preformed receptor complexes may get activated more rapidly and by lower target antigen densities than the CARs with CD28-CD3ζ and CD137-CD3ζ

TABLE 1 | Preclinical studies with CAR NK-92 cells.

Target	Antibody	Hinge	TM	Signaling	Gene transfer	Cancer type	*In vivo* model	Treatment	Reference
ErbB2 (HER2)	FRP5	mCD8α	mCD3ζ	mCD3ζ	Retrovirus	Breast ca. Ovarian ca. SCC	CD-1 nude	Local co-injection	Uherek et al. (23)
ErbB2 (HER2)	FRP5	mCD8α	mCD3ζ	mCD3ζ	Retrovirus	Breast ca.	BALB/c nude	Systemic	Daldrup-Link et al. (63), Meier et al. (64)
ErbB2 (HER2)	FRP5	mCD8α	mCD3ζ	mCD3ζ	Retrovirus	Brain metastasis	Athymic nude rats	Systemic (with FUS)	Alkins et al. (65), Alkins et al. (66)
ErbB2 (HER2)	4D5-8	hIgG2	hFcεRIγ	hFcεRIγ	Retrovirus	Breast ca.	NSG	Systemic	Clemenceau et al. (58)
ErbB2 (HER2)	FRP5	hCD8α	hCD3ζ hCD28 hCD137	hCD3ζ hCD28-CD3ζ hCD137-CD3ζ	Lentivirus	Breast ca. Ovarian ca. Melanoma RCC	NSG	Systemic	Schönfeld et al. (67)
ErbB2 (HER2)	n.s.	hCD8α	hCD28	hCD28-CD3ζ	Electroporation	Breast ca.	BALB/c nude	Systemic	Liu et al. (68)
ErbB2 (HER2)	FRP5	hCD8α	hCD28	hCD28-CD3ζ	Lentivirus	GBM	NSG C57BL/6	Local	Zhang et al. (69)
Epidermal growth factor receptor (EGFR)	528	n.s.	n.s.	hCD28-CD3ζ	Lentivirus	Brain metastasis	NSG	Local (combined with HSV-1)	Chen et al. (70)
EGFR	R-1	hCD8α	hCD28	hCD28-CD3ζ	Lentivirus	GBM	NSG	Local	Genßler et al. (71)
EGFR/EGFRvIII	528	n.s.	hCD28	hCD28-CD3ζ	Lentivirus	GBM	NSG	Local	Han et al. (72)
EGFR/EGFRvIII	225	hCD8α	hCD28	hCD28-CD3ζ	Lentivirus	GBM	NSG	Local	Genßler et al. (71)
EGFRvIII	MR1-1	hCD8α	hCD28	hCD28-CD3ζ	Lentivirus	GBM	NSG	Local	Genßler et al. (71)
GD2	ch14.18	mCD8α	mCD3ζ	mCD3ζ	Retrovirus	NB Breast ca. Melanoma	NSG	Local	Esser et al. (56), Seidel et al. (59)
Epithelial cell adhesion molecule (EpCAM)	MOC31	mCD8α	mCD3ζ	mCD3ζ	Retrovirus	Prostate ca.	Athymic nude rats	Systemic	Tavri et al. (55), Meier et al. (73)
EpCAM	MOC31	hCD8α	hCD28	hCD28-CD3ζ	Lentivirus	Breast ca.	–	–	Sahm et al. (28)
CD19	FMC63	mCD8α	mCD3ζ	mCD3ζ	Retrovirus	B-ALL	–	–	Romanski et al. (48), Romanski et al. (53)
CD19	FMC63	mCD8α	mCD3ζ	mCD3ζ	mRNA transfection	B-ALL CLL	–	–	Boissel et al. (50)
CD19	FMC63	mCD8α	mCD3ζ	mCD3ζ	mRNA transfection Lentivirus	B-ALL CLL Burkitt's lymphoma	–	–	Boissel et al. (51)
CD19	FMC63	mCD8α	mCD3ζ	mCD3ζ	Lentivirus	B-ALL CLL	NOD/SCID NSG	Local Systemic	Boissel et al. (52)
CD19	FMC63	hCD8α	hCD3ζ hCD28 hCD137	hCD3ζ hCD28-CD3ζ hCD137-CD3ζ	Lentivirus	B-ALL Burkitt's lymphoma	NSG	Systemic	Oelsner et al. (74)
CD20	Leu-16	mCD8α	mCD3ζ	mCD3ζ	Retrovirus	B-ALL CLL Burkitt's lymphoma	NSG	Local co-injection	Müller et al. (49)
CD20	Leu-16	mCD8α	mCD3ζ	mCD3ζ	mRNA transfection Lentivirus	B-ALL CLL Burkitt's lymphoma	–	–	Boissel et al. (51)
CD20	Leu-16	mCD8α	mCD3ζ	mCD3ζ	Lentivirus	B-ALL CLL	NOD/SCID NSG	Local Systemic	Boissel et al. (52)

(Continued)

TABLE 1 | Continued

Target	Antibody	Hinge	TM	Signaling	Gene transfer	Cancer type	*In vivo* model	Treatment	Reference
CD3	n.s.	hCD8α	hCD8α	hCD28-CD137-CD3ζ	Lentivirus	PTCL T-ALL	NSG	Systemic	Chen et al. (75)
CD5	n.s.	hCD8α	hCD8α	hCD28-CD137-CD3ζ	Lentivirus	PTCL T-ALL Sézary syndrome	NSG	Systemic	Chen et al. (76)
CD138	4B3	hCD8α	hCD3ζ	hCD3ζ	Lentivirus	MM	NOD/SCID	Systemic	Jiang et al. (54)
CS1	Luc90	n.s.	n.s.	hCD28-CD3ζ	Lentivirus	MM	NSG	Systemic	Chu et al. (77)
EBNA3C peptide	EBNA Clone 315	hCD8α	hCD8α	hCD137-CD3ζ	Retrovirus	BLCL	–	–	Tassev et al. (78)
$gp100_{209-217}$ peptide	GPA7	n.s.	HLA-A2	hCD3ζ	Electroporation	Melanoma	NOD/SCID	Systemic	Zhang et al. (57)
$WT1_{126}$ peptide	Q2L	hCD8α	hCD8α	hCD137-CD3ζ	Retrovirus	B-ALL AMoL NB	–	–	Zhao et al. (79)

TM, transmembrane domain; n.s., not specified; m, murine; h, human; SCC, squamous cell carcinoma; RCC, renal cell carcinoma; GBM, glioblastoma; NB, neuroblastoma; B-ALL, B-cell acute lymphoblastic leukemia; CLL, chronic lymphocytic leukemia; PTCL, peripheral T-cell lymphoma; T-ALL, T-cell acute lymphoblastic leukemia; MM, multiple myeloma; BLCL, Epstein–Barr virus (EBV)-transformed lymphoblastoid B cell line; AMoL, acute monocytic leukemia; FUS, MRI-guided focused ultrasound.

domains, which contained the transmembrane domains of CD28 and CD137 and did not form covalent dimers as assessed by SDS-PAGE and immunoblot analysis (74) (**Figure 1C**). Of note, while this was not the case for the CD19-specific CD137-CD3ζ CAR tested in NK-92 cells, a different CD19-specific CD137-CD3ζ CAR that contained the transmembrane domain of CD8α and formed covalent CAR homodimers showed enhanced activity in comparison to a respective CD3ζ-only CAR in primary NK cells (24). Also sterical effects such as distance of the target epitope to the cell surface and CAR accessibility can play a role in determining the activation threshold of individual CARs (82, 83). This may explain why in NK-92 cells otherwise identical ErbB2- and CD19-targeted CD137-CD3ζ CARs in one case led to higher and in the other case to lower-specific cell killing when compared to the respective CD3ζ-only CAR (67, 74). Hence, while the data available so far suggest that inclusion of a costimulatory protein domain in the CAR can be beneficial at least for particular functions of NK-92 cells, continuing research efforts are needed to clarify whether cytotoxicity, cytokine production, and resistance to immunosuppressive mechanisms can be improved with a single, generalized CAR design. Possibly, the most optimal CAR composition has to be determined experimentally in each case, taking into consideration CAR-binding affinity, location of the binding epitope within the target antigen, length of hinge region, and nature of the transmembrane domain (84).

CONTINUOUS EXPANSION OF CAR NK-92 CELLS

Isolation and *ex vivo* expansion of peripheral blood-derived NK cells for therapeutic applications can be demanding, time-consuming, and costly (85). Since KIR-mismatched allogeneic NK cells are superior to autologous cells, a suitable donor needs to be identified to allow for efficient GvL or GvT activity (18–20). Moreover, owing to the intricate heterogeneity of human NK cells with respect to cytotoxic and regulatory activity, NK-cell licensing, unlicensing, and memory, selecting the most appropriate NK subpopulations for cancer therapy is difficult (86, 87). Sufficient numbers of NK cells are critical for a better clinical outcome, which is complicated by the limited *ex vivo* expansion potential of NK cells that remains a challenge despite the development of genetically engineered feeder cells supporting NK-cell growth and improved protocols for cytokine stimulation (25, 88, 89). These issues are also relevant for the development of CAR-engineered primary NK cells, which may explain the slow progress in this field with respect to CAR T cells.

Chimeric antigen receptor-engineered NK-92 could offer a valid and cost-effective alternative to primary CAR NK or T cells, in particular, in cases, where a suitable donor is not available or the sophisticated infrastructure needed for cell isolation, expansion, and genetic modification is missing. Methodology for continuous good manufacturing practice (GMP)-compliant expansion from an established master cell bank has been validated in the framework of early phase clinical trials with unmodified NK-92 cells and can easily be adapted for large-scale production in centralized facilities (32, 90). This advantage may readily be extended to CAR-engineered NK-92 variants. In contrast to CAR approaches based on autologous or donor-derived primary cells, genetic modification of NK-92 cells is thereby not performed in a patient-individual setting under tight time constraints. Instead, a molecularly and functionally well-characterized cell product can be established for a particular target specificity independent from the time point of therapeutic application. The resulting cells are stable with respect to CAR expression and functionality during extended expansion, as recently demonstrated for ErbB2-specific NK-92/5.28.z cells (also termed "HER2.taNK" for HER2-specific target-activated NK), a single-cell clone derived under GMP-compliant conditions that is intended for clinical use (67).

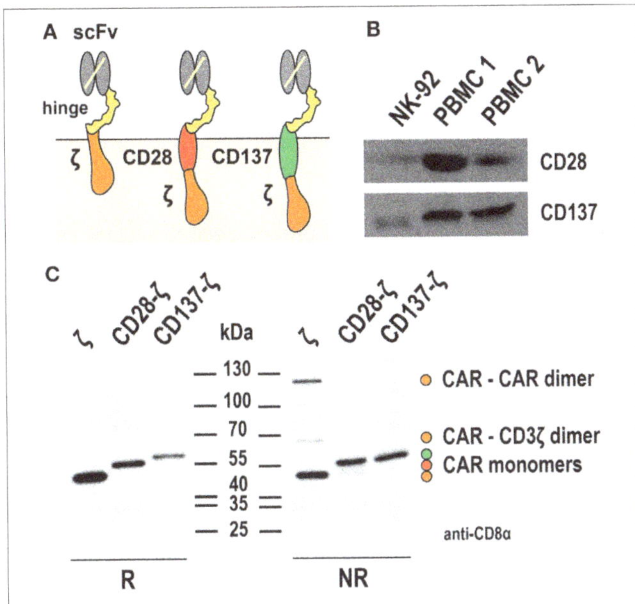

FIGURE 1 | Expression of first- and second-generation chimeric antigen receptors (CARs) in NK-92 cells. (A) Schematic representation of first- and second-generation CARs for expression in NK-92 cells that consist of an extracellular single chain fragment variable (scFv) antibody domain for target recognition fused *via* a hinge region derived from CD8α (hinge) to transmembrane and intracellular domains of CD3ζ (left), transmembrane and intracellular domains of CD28, and the intracellular domain of CD3ζ (middle), or transmembrane and intracellular domains of CD137 (4-1BB) and the intracellular domain of CD3ζ (right). **(B)** To assess endogenous expression of CD28 and CD137, lysates of NK-92 cells were subjected to SDS-PAGE and subsequent immunoblotting with CD28- and CD137-specific antibodies as indicated. Lysates of peripheral blood mononuclear cells (PBMCs) from healthy donors were included for comparison. **(C)** For analysis of CAR expression, lysates of NK-92 cells transduced with lentiviral vectors that encode CD19-specific CARs containing CD3ζ, composite CD28-CD3ζ, or CD137-CD3ζ signaling domains as represented in **(A)** were subjected to SDS-PAGE under reducing (R, left panel) or non-reducing conditions (NR, right panel) and subsequent immunoblotting with CD8α-specific antibody, which detects the hinge domain. The positions of CAR monomers, homodimers of the CD3ζ CAR, and heterodimers of the CD3ζ CAR with endogenous CD3ζ are indicated. Data in panel **(C)** are from Oelsner et al. (74).

As NK cells, NK-92 as well as their CAR-expressing derivatives are dependent on exogenous IL-2 for growth and maintenance of their activated phenotype (23, 33). To ease cell expansion, different groups have engineered NK-92 by retroviral transduction or particle-mediated non-viral gene transfer to ectopically produce IL-2, leading to IL-2 secretion and growth of the cells in the absence of IL-2 supplementation (91, 92). Similarly, IL-15 and stem cell factor (SCF) have been ectopically expressed in NK-92 using plasmid DNA transfection (93, 94). While the resulting cells proliferated in medium with lower IL-2 concentrations than parental NK-92, in contrast to the IL-2-engineered variants, they were not completely independent from exogenous cytokines. In humans, high concentrations of IL-2 are associated with severe toxicity. Furthermore, in contrast to IL-15, IL-2 preferentially enhances the activity of T_{reg} cells, which is not desired in the context of cancer immunotherapy (95). Hence, a modified version of IL-2 was developed for expression in NK-92, which carries a C-terminal KDEL endoplasmic reticulum retention signal. This still allowed activation of IL-2 receptor complexes in the secretory pathway of the producer cells but limited release of IL-2 and availability to bystander cells (96, 97). A similar effect was achieved by expression of unmodified IL-15 in NK-92 using a lentiviral vector, which supported growth in the absence of exogenous IL-2 but also restricted cytokine activity to the producer cells. When IL-15 was coexpressed with an EpCAM-specific CAR from a bicistronic lentiviral vector, transduced cells could be enriched merely by IL-2 withdrawal, with the selected CAR NK-92 cells displaying high and specific cytotoxicity in the absence of exogenous cytokines (28).

CAR-ENGINEERED NK-92 CELLS EXHIBIT ANTIBODY-DEPENDENT CELL-MEDIATED CYTOTOXICITY (ADCC)-LIKE ACTIVITY AND SERIAL KILLING

Antibody-dependent cell-mediated cytotoxicity of NK cells is triggered by FcγRIIIa (CD16), which associates with CD3ζ and FcεRIγ that are linked to overlapping as well as distinct intracellular signaling pathways (98, 99). NK-92, which is phenotypically CD16-negative, readily mediates ADCC in the presence of a suitable IgG antibody when engineered to express FcγRIIIa (27, 97, 100). This has sparked efforts to clinically develop genetically modified NK-92 cells that harbor the high affinity V158 variant of CD16 (termed haNK) in combination with antibodies of IgG1 isotype (32, 97). Initial safety assessment of such cells in cancer patients is expected to begin soon (NCT03027128; https://clinicaltrials.gov). Interestingly, side by side comparison of NK-92 cells carrying a CD20-specific first-generation CAR with a CD3ζ domain showed more pronounced killing of otherwise NK-resistant primary CLL cells than CD16-engineered NK-92 applied together with rituximab (52). Similarly, NK-92 cells harboring an EBV EBNA3C-specific CAR lysed peptide-pulsed B-cell lymphoblastic cells more efficiently than CD16-engineered NK-92 in the presence of an anti-EBNA3C-Fc fusion protein (78) and NK-92 cells expressing a trastuzumab-based ErbB2-specific CAR with an FcεRIγ signaling domain displayed more enhanced cytotoxicity against breast carcinoma cells than NK-92 harboring a CD16-FcεRIγ hybrid receptor in combination with trastuzumab antibody (58).

Successful triggering of ADCC through CD16 requires its non-covalent interaction with the Fc portion of an antibody that is simultaneously bound to its antigen on the surface of a neighboring target cell, as well as association with intracellular CD3ζ and FcεRIγ. Direct linkage of extracellular target recognition and intracellular signaling functions in one molecule as implemented in a CAR can bypass such complex stoichiometry and intermolecular interactions, likely accelerating kinetics of NK-cell activation. CAR signal strength is further enhanced by integrating CD3ζ, which in monomeric form contributes three immunoreceptor tyrosine-based activation motifs (ITAMs) that are crucial for downstream signaling, while an FcεRIγ monomer only provides one ITAM sequence (101). Accordingly, specific target-cell recognition by CAR NK-92 results in immediate and effective ADCC-like activity, characterized by orientation of cytotoxic granules toward the immunological synapse, release of high levels of perforin and granzyme B, and rapid induction of target-cell apoptosis as demonstrated for various tumor-associated antigens (23, 49, 67, 71, 74). Live cell imaging and cytotoxicity experiments at effector to target ratios below 1:1 showed that one CAR-engineered NK-92 cell can thereby kill multiple targets within a few hours (49, 67, 74). This includes tumor cells exhibiting only moderately enhanced expression of the chosen target antigen, as demonstrated for established and tumor-initiating primary glioblastoma cells exposed to ErbB2-specific NK-92/5.28.z cells (69). NK-92/5.28.z cells also killed trastuzumab-sensitive and trastuzumab-resistant ErbB2-positive breast carcinoma cells to a similar extent (**Figure 2**), attesting to the different mode of action of the retargeted NK cells and suggesting their application in a disease setting with existing resistance to other targeted therapies.

IN VIVO ANTITUMOR ACTIVITY OF CAR NK-92 CELLS

Initial studies performed with ErbB2-, CD20-, and GD2-targeted CAR NK-92 cells showed that these cells retain specific cytotoxicity in simplified *in vivo* models in immunocompromised nude and NOD-SCID IL2R γ^{null} (NSG) mice, where effector cells were either subcutaneously coinjected together with tumor cells, or established subcutaneous tumors treated by peritumoral NK-cell injection. This resulted in delayed tumor onset and extended survival when compared to animals receiving parental NK-92 cells (23, 49, 59). Similar intratumoral treatment may be an option for cancer indications such as glioblastoma and brain metastasis, where disease is locally restricted. This has been investigated with NK-92 cells-expressing second-generation CARs targeting ErbB2, EGFR, or mutant EGFRvIII, which are expressed by a large proportion of human glioblastomas. In orthotopic xenograft models in NSG mice, repeated stereotactic injection of ErbB2-specific NK-92/5.28.z cells into the tumor area effectively inhibited tumor progression and resulted in a marked extension of survival, while parental NK-92 cells were ineffective (69). Similar effects were seen upon local application of NK-92 cells equipped with CARs that recognize EGFR, mutant EGFRvIII, or both antigens against orthotopic EGFR- and/or EGFRvIII-positive glioblastoma xenografts or breast cancer brain metastases growing in NSG mice (70–72). In contrast to EGFR- or EGFRvIII-targeted monospecific NK-92 variants, dual targeting of EGFR and EGFRvIII with a cetuximab-based CAR recognizing a common epitope of the receptors, thereby circumvented immune escape in mixed tumors that similar to the clinical situation, consisted of EGFR-positive and EGFR/EGFRvIII-double positive glioblastoma cells (71).

For broad applicability in metastatic and disseminated disease, CAR effector cells must cross tissue barriers and reach distant tumor sites to be effective. Magnetic resonance imaging, bioluminescence imaging, and positron emission tomography experiments as well as direct analysis of tumor infiltration revealed rapid and specific accumulation of intravenously injected NK-92 carrying first- and second-generation ErbB2-specific or EpCAM-specific CARs in orthotopic breast and subcutaneous prostate carcinoma xenografts in rodents (55, 63, 64, 67, 73), while parental NK-92 cells showed no tumor homing and were mainly localized to spleen and liver (55). Focused ultrasound has been demonstrated to allow systemically applied CAR NK-92 cells to cross the blood–brain barrier and reach breast cancer brain metastases in a xenograft model in immunocompromised rats (65, 66). Specific antitumor activity of intravenously applied CAR NK-92 cells has also been found in a model of locally growing breast carcinoma (68), in an experimental renal cell carcinoma metastasis model (67), and models of disseminated leukemia, lymphoma, and

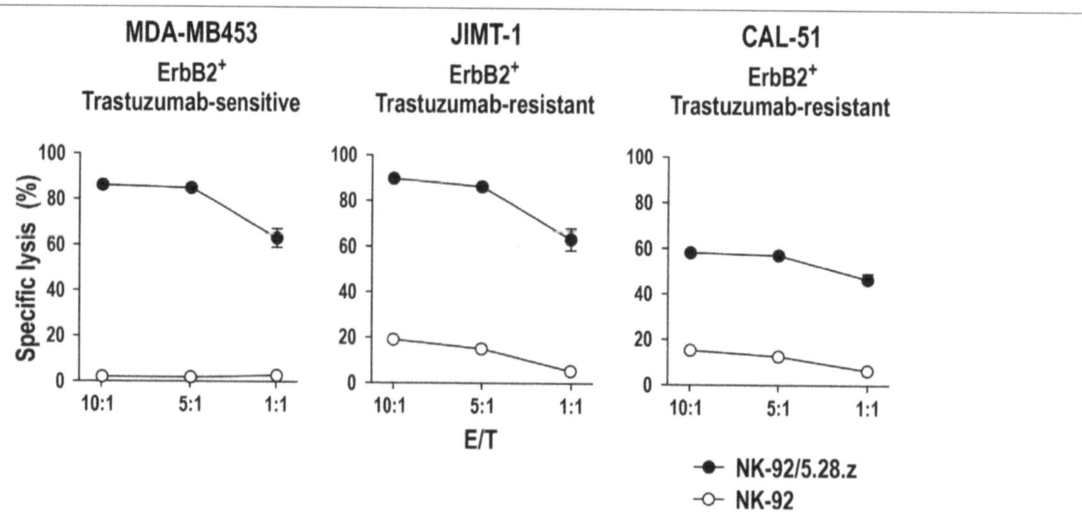

FIGURE 2 | Activity of NK-92/5.28.z against ErbB2-expressing breast carcinoma cells. Cytotoxicity of CAR-engineered ErbB2-specific NK-92/5.28.z cells (filled circles) against ErbB2-overexpressing and trastuzumab-sensitive MDA-MB453 (left), or ErbB2-overexpressing and trastuzumab-resistant JIMT-1 (middle) and CAL-51 (right) breast carcinoma cells was investigated in flow cytometry-based cytotoxicity assays after coincubation of NK cells and tumor cells at different effector to target ratios (E/T) for 2 h. Parental NK-92 cells were included for comparison (open circles). Mean values ± SEM are shown; $n = 3$.

multiple myeloma (52, 74–77), underscoring the potential of CAR-engineered NK-92 cells for the treatment of a large variety of different cancers.

NK CELLS: A BRIDGE BETWEEN INNATE AND ADAPTIVE ANTITUMOR IMMUNITY

Natural killer cells do not only play a critical role in antitumor immunity by directly eliminating malignant cells, but also by regulating tumor-specific adaptive immune responses through cross talk with other immune cells. In particular, the interaction between NK cells and dendritic cells (DCs) is important in this context (**Figure 3**). On the one hand, DCs enhance the direct antitumor activity of NK cells (102). On the other hand, NK cells regulate DC maturation, thereby determining the effectiveness of subsequent DC-mediated T-cell activation (103, 104). Once activated by target cells or soluble factors, NK cells secrete high amounts of IFN-γ and tumor necrosis factor (TNF)-α, which synergistically contribute to the maturation of immature DCs (iDCs). This leads to enhanced expression of costimulatory molecules such as CD80, CD83, and CD86 by the DCs and favors Th1 polarization during subsequent DC-mediated T-cell activation (105–107). Mature DCs (mDCs) release IL-12, IL-15, and IL-18, which in turn enhance IFN-γ expression by NK cells and NK-cell cytotoxicity against virus-infected and tumor cells (103, 108). Likewise, cytotoxicity of NK cells can be boosted by type I interferons such as IFN-α secreted by plasmacytoid DCs (pDCs) (109).

Dendritic cell maturation and reciprocal NK-cell activation are also strongly dependent on the engagement of activating receptors like NKp30, NKG2D, and NKp46 on NK cells (107, 110–113). Concurrent with inducing DC maturation, NK cells control the quality of the mDC population by killing iDCs (DC editing), which can otherwise induce immune tolerance through T-cell depletion or T_{reg} expansion. Discrimination and lysis of iDCs by NK cells is mainly regulated by activating signals through NKp30 and inhibitory signals through KIRs and the NKG2A/CD94 complex (107, 114). Accordingly, inhibition of NKp30 signaling or reduced NKp30 expression results in impaired NK-mediated killing of iDCs (110, 115, 116). Both mDCs and iDCs express NKp30 ligands. However, higher amounts of HLA class I and HLA-E molecules expressed by mDCs protect them from lysis

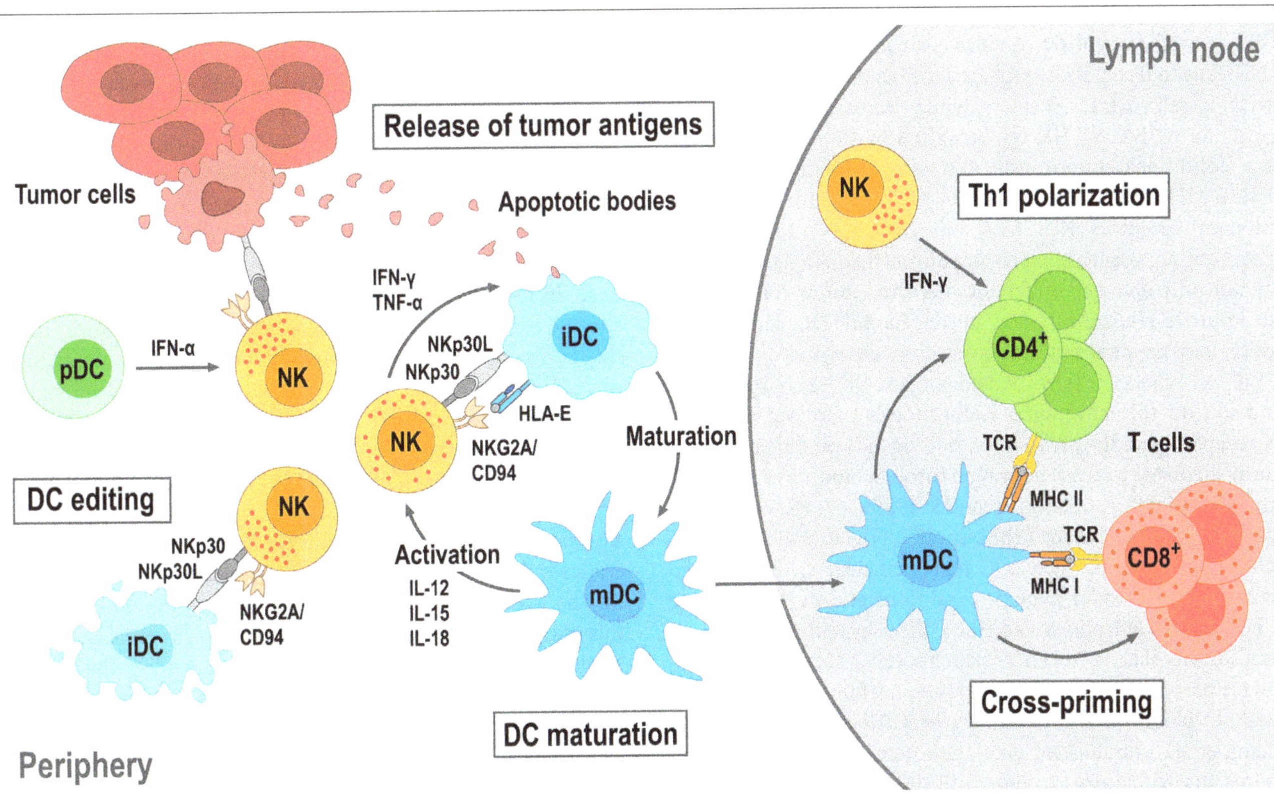

FIGURE 3 | Reciprocal natural killer (NK)—dendritic cell (DC) cross talk. Upon activation by target tumor cells or cytokines, NK cells produce IFN-γ and tumor necrosis factor (TNF)-α that can promote DC maturation. DC maturation is also strongly dependent on the engagement of activating receptors on NK cells such as NKp30 and NKG2D. Mature DCs (mDCs) will in turn produce interleukin (IL)-12, IL-15, and IL-18, which enhance cytotoxicity and IFN-γ secretion of NK cells. NK cells can also distinguish immature (iDC) and mDCs through activating NKp30 and inhibitory killer cell immunoglobulin-like receptors and NKG2A/CD94 and eliminate immature DCs (iDCs), thereby maintaining the quality of the mDC population (DC editing). NK-cell cytotoxicity can be further augmented by IFN-α secreted by plasmacytoid DCs (pDCs). NK-induced tumor cell lysis provides antigens, which can be taken up by DCs for antigen presentation. Once maturated, antigen-loaded mDCs will migrate into tumor-draining lymph nodes, cross-present tumor antigens to naïve T cells, and induce their differentiation toward tumor-specific CD8+ cytotoxic T cells and CD4+ T helper 1 (Th1) cells.

by NK cells, whereas lower HLA class I and HLA-E expression makes iDCs vulnerable (114, 117). Importantly, by direct lysis of malignant cells, NK cells also provide tumor antigens for uptake and processing by DCs, which upon maturation and migration into a tumor-draining lymph node can cross-present such antigens to T cells, thereby inducing Th1 polarization of CD4+ T cells and differentiation of CD8+ T cells into tumor-specific cytotoxic T-lymphocytes (CTLs). NK cells can also migrate into tumor-draining lymph nodes and provide an early source of IFN-γ for Th1 polarization (118).

CAR-ENGINEERED NK-92 CELLS OVERCOME IMMUNOSUPPRESSIVE MECHANISMS AND ENHANCE ADAPTIVE ANTITUMOR IMMUNITY

As discussed above, efficient NK-cell activation is a prerequisite for productive NK–DC interaction. However, in cancer patients, NK-cell abnormalities are frequently found, including reduced NK-cell numbers, impaired cytotoxicity, and inefficient tumor infiltration (119). Especially in solid tumors, NK-cell activity is negatively affected by immunosuppressive factors in the tumor microenvironment (17, 120). High levels of TGF-β, IDO, and PGE2, as well as hypoxic conditions strongly inhibit the ability of NK cells to upregulate cytokine production and expression of activating cell surface receptors, while decreasing expression of ligands for activating NK-cell receptors by tumor cells. Under these conditions, tumor cells can also upregulate the non-classical MHC class I molecule HLA-G, a ligand for the NK-cell inhibitory receptors KIR2DL4 and ILT-2 (17, 121). Hence, therapeutic approaches that restore diminished NK-cell function may not only enhance direct NK-mediated tumor cell lysis but also improve clinical outcome by reinforcing DC activity and induction of adaptive antitumor immune responses.

Killing of cancer cells by CAR-engineered NK-92 is largely independent from the activation of endogenously expressed activating NK receptors and the presence of their ligands on target cells but mainly mediated by CAR-activation through binding to a cognate tumor-associated surface antigen (23, 49, 56, 67, 69, 71, 74). As recently demonstrated for ErbB2-specific NK-92/5.28.z carrying a CD28 CD3ζ CAR, such cells retain efficient CAR-mediated cell killing even under hypoxic conditions and in the presence of TGF-β concentrations exceeding the elevated TGF-β levels found in the plasma of cancer patients (69, 122). Furthermore, target tumor cells ectopically overexpressing human HLA-G were unable to block specific cell killing by CAR-engineered NK-92 (Zhang et al., unpublished data), although NK-92 cells express the immunoregulatory receptors KIR2DL4 and ILT-2, which are activated by HLA-G (34, 123). These findings show that activated CAR NK-92 cells can maintain their cytotoxic potential in an immunosuppressive environment similar to the one found within a solid tumor. In addition, NK-92 readily express activating NK receptors such as NKp30 and NKG2D while most of the inhibitory KIRs are absent (34), which may make CAR NK-92 cells particularly effective in aiding DC maturation and editing, and enhancing DC-mediated cross-priming of tumor-specific T cells and induction of adaptive antitumor immunity.

We recently investigated this possibility in an immunocompetent mouse model for glioblastoma and could indeed demonstrate the induction of endogenous antitumor immunity following therapy with CAR-engineered NK-92 cells (69). In this model, the majority of mice carrying syngeneic intracranial GL261/ErbB2 glioblastomas were cured upon repeated intratumoral injection of ErbB2-specific NK-92/5.28.z cells, while unmodified parental NK-92 cells were unable to inhibit tumor progression (**Figure 4A**). Human NK-92 and CAR NK-92 cells do not permanently engraft in immunodeficient mice and are quickly rejected by immunocompetent animals (67, 69, 124, 125). Nevertheless, without any further treatment, all mice that were cured from their initial tumors also rejected a rechallenge with GL261/ErbB2 cells injected into the other brain hemisphere 4 months after initial therapy, mediated by an endogenous memory immune response induced in the animals by initial treatment with NK-92/5.28.z (69). Sera from these mice contained IgG antibodies reactive with both GL261/ErbB2 and ErbB2-negative, but otherwise isogenic GL261 cells (**Figure 4B**), indicating that the induced protective antitumor immune response was broadly directed against the glioblastoma cells and not limited to the CAR target antigen. Accordingly, mice that were cured of GL261/ErbB2 tumors also rejected a subsequent challenge with GL261 cells (Zhang et al., unpublished data). When animals that had rejected the initial tumor and the first rechallenge with GL261/ErbB2 were injected once again with GL261/ErbB2 cells but this time after depletion of CD4+ and CD8+ T cells, tumors formed in a large proportion of the mice. This demonstrates that protective immunity induced by initial treatment with NK-92/5.28.z cells was also dependent on T-cell memory (**Figure 4C**). Similarly, in a later study by Boissel et al., intratumoral injection with NK-92 cells expressing a CAR specific for murine CD19 induced protective antitumor immunity in a syngeneic A20 lymphoma model in immunocompetent mice (126).

These data suggest that the release of tumor antigens through the cytotoxic activity of CAR NK-92 cells, most likely augmented by the demonstrated CAR-induced production of high levels of pro-inflammatory cytokines (67, 69, 71, 74), can not only induce a humoral immune response directed against the tumor but also enhance cross-presentation of tumor antigens by murine DCs for the activation of tumor-specific CTLs. In a clinical setting, this may be further enhanced by IFN-γ, which is released at high levels by activated CAR NK-92. In murine models, the effects of human IFN-γ are limited due to the species-specificity of IFN-γ/IFNGR1 interactions (127, 128). Of note, apoptotic tumor cells have been shown to be superior to cell lysates or tumor cell RNA in inducing a tumor-specific T-cell response (129, 130). Hence, tumor-cell apoptosis induced by the release of cytotoxic granules from activated CAR NK-92 cells provides tumor antigens in a most effective form for uptake, processing, and presentation by DCs. Whether this vaccine or adjuvant effect of CAR NK-92 cells is a particular consequence of direct intratumoral administration as performed in our study and the study by Boissel et al. (69, 126) is subject of ongoing investigations.

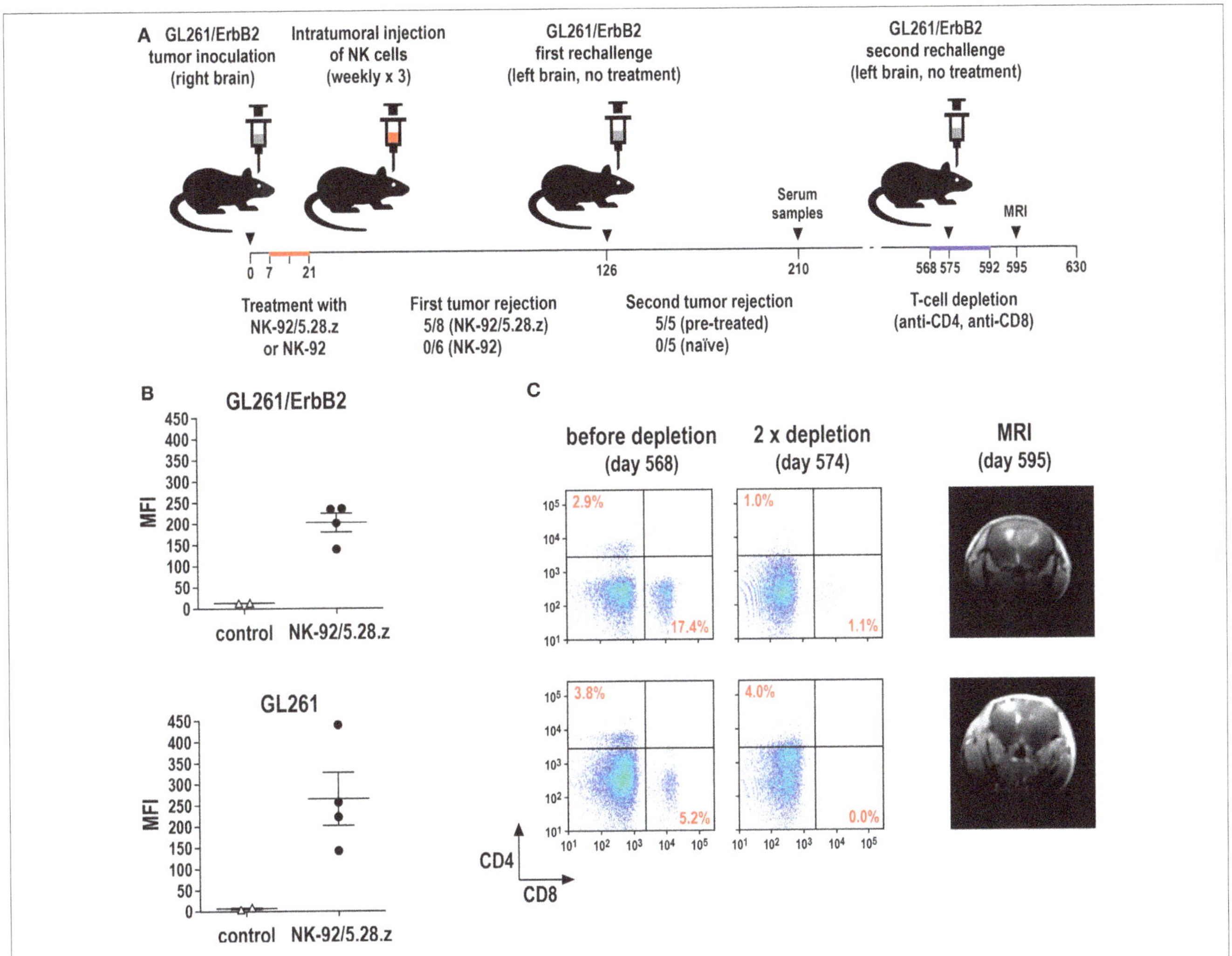

FIGURE 4 | Induction of protective antitumor immunity by CAR-engineered NK-92 cells. (A) Murine GL261/ErbB2 glioblastoma cells (5×10^3) stably expressing human ErbB2 were stereotactically injected into the right striatum of syngeneic C57BL/6 mice. Seven days later, animals were treated once per week for 3 weeks by intratumoral injection of 2×10^6 parental NK-92 ($n = 6$) or NK-92/5.28.z cells ($n = 8$), which express an ErbB2-specific CAR with CD28 and CD3ζ signaling domains. Animals that were cured upon NK-92/5.28.z treatment ($n = 5$) were rechallenged at day 126 by injection of 5×10^3 GL261/ErbB2 cells into the left brain hemisphere without further NK-cell therapy and symptom-free survival was followed. Naïve C57BL/6 mice injected into the brain with GL261/ErbB2 cells at day 126 served as a control ($n = 5$) (69). **(B)** Induction of IgG serum antibodies against glioblastoma cells in NK-92/5.28.z-treated animals ($n = 4$) from the experiment summarized in **(A)** was investigated by flow cytometry with GL261/ErbB2 (upper panel) and ErbB2-negative parental GL261 cells (lower panel) using sera collected at day 210. Sera from naïve C57BL/6 mice ($n = 2$) served as controls. MFI, mean fluorescence intensity (geometric mean). Data for GL261/ErbB2 cells are from Zhang et al. (69). **(C)** T cells in the NK-92/5.28.z-pretreated and rechallenged mice were depleted by intravenous injection of CD4- and CD8-specific antibodies at days 568, 573, 578, 582, 587, and 592 (left and middle panels). At day 575, these animals were rechallenged a second time with 5×10^3 GL261/ErbB2 cells injected into the left brain hemisphere without further NK-cell therapy. Tumor development was assessed by MRI at day 595 (right panels).

GENERAL SAFETY ASPECTS OF CAR NK-92 CELLS

In addition to the well-defined chimeric receptor, polyclonal CAR T cells carry endogenous, MHC-restricted T-cell receptors (TCRs) of unknown specificity. In an autologous setting, it can be expected that due to thymic selection, only very few autoreactive T cells are present in the periphery. Nevertheless, in donor-derived CAR T cells, CAR-induced activation and expansion may accidentally result in increased TCR-mediated reactivity with the recipient's tissues, leading to severe GvHD (131). Unlike T cells, NK cells do not carry genetically rearranged clonogenic receptors. Ligands for germline-encoded activating NK-cell receptors are typically upregulated only by stressed cells after virus infection or malignant transformation, which is the basis of the NK cells' intrinsic antitumor activity (2). Hence, the activating receptors NKp30, NKp46, and NKG2D expressed by NK-92 can be expected to contribute to the antitumor activity of CAR-engineered variants rather than causing adverse effects (34). The natural cytotoxicity of NK-92 is largely retained by CAR-expressing variants as demonstrated in different studies using K562 cells as targets (23, 49).

Donor lymphocyte infusion with allogeneic NK cells is mostly performed in the context of hematopoietic stem cell transplantation and generally considered safe, without a high risk of inducing GvHD (20). Nevertheless, depending on donor selection and *ex vivo* activation, development of acute GvHD after NK DLI has been described, attributed to NK-dependent augmentation of T-cell alloreactivity (132). In two clinical trials with NK-92 cells, only mild infusion-related side effects were noted, while no severe treatment-related toxicities were observed even at a cell dose as high as $1 \times 10^{10}/m^2$ body surface (31, 35, 36). While HLA-specific antibodies reactive with the allogeneic cells were found in some patients after NK-92 therapy, this was not linked to adverse effects. To prevent potential engraftment of NK-92 that was initially derived from a non-Hodgkin lymphoma patient (33), the cells were irradiated with 10 Gy prior to infusion, prohibiting further proliferation but only resulting in a gradual decline of cytotoxicity over several days (31). Since it is presently unknown whether non-irradiated NK-92 cells have the potential to form secondary lymphoma in a human host, irradiation of the cells before infusion will also be included as a safety measure in a planned phase I clinical trial investigating intratumoral injection of ErbB2-specific NK-92/5.28.z cells in patients with recurrent ErbB2-positive glioblastoma (69, 133). Like parental NK-92, CAR-engineered NK-92 variants transiently retain specific cytotoxicity after γ-irradiation with 10 Gy, with unchanged *in vitro* and *in vivo* antitumor activity (23, 56, 67, 69). This extends to the immunostimulatory activity of CAR NK-92 cells, which was also found for irradiated ErbB2-specific NK-92/5.28.z cells in an immunocompetent glioblastoma mouse model similar to the one described in **Figure 4** (Zhang et al., unpublished data). Hence, in contrast to CAR T cells, which are capable of uncontrolled *in vivo* expansion, the effective dose of CAR NK-92 cells can be tightly managed to establish a therapeutic window, albeit at the price of potentially higher cell numbers and more frequent treatment intervals needed.

A major concern with CAR T-cell therapy is cytokine release syndrome (CRS) frequently observed in clinical trials with CD19-specific effector cells, which can be severe and even cause fatalities. IL-6 production and IL-6 trans-signaling after massive activation of infused CAR T cells were found to play a critical role in CRS (134). CRS can be managed with the IL-6 receptor (IL-6R) blocking antibody tocilizumab or steroid treatment (134–136). While the latter inhibits CAR T-cell expansion and activity, it constitutes an important option for patients who do not respond to the IL-6R antagonist. In contrast to CAR T cells, activated CAR NK-92 cells do not produce measurable amounts of IL-6 and IL-4 as demonstrated for NK-92 variants targeted to EGFR, EGFRvIII, ErbB2, or CD19 (67, 69, 71, 74). Instead, upon CAR activation, these cells secrete high levels of IFN-γ, macrophage inflammatory protein (MIP)-1α (CCL3), GM-CSF, and moderate levels of TNF-α. This cytokine/chemokine profile appears more favorable and less likely to induce CRS, while supporting the CAR NK-92-induced activation of endogenous antitumor immunity described above.

POTENTIAL ON-TARGET/OFF-TUMOR TOXICITY

Chimeric antigen receptor effector cells specifically targeting mutated tumor antigens and viral antigens not expressed in normal tissues do not carry the risk of inducing on-target/off-tumor toxicity. This safety feature is given for CAR NK-92 cells selective for the tumor-specific EGFR mutant EGFRvIII frequently expressed in glioblastoma (71) and genetically modified NK-92 cells recognizing an epitope of the EBV latent protein EBNA3C in complex with HLA-A2 (78). However, most CARs currently available are directed to non-mutated self-antigens differentially expressed by the cancer cells. Consequently, there is a possibility for on-target/off-tumor activity against antigen-positive healthy tissues, which can result in severe toxicities. B-cell aplasia is typically observed after CD19 CAR T-cell therapy but can easily be managed by infusion of immunoglobulins. This may be different if a tumor-associated target antigen is also present in vital tissues. In a clinical trial conducted at the National Cancer Institute, a fatal adverse event occurred after infusion of autologous T cells modified to express an ErbB2-specific third generation CAR based on trastuzumab (137). Although antigen-independent CAR activation due to the combination of three signaling domains (CD137, CD28, CD3ζ) cannot be excluded (46), massive T-cell activation and respiratory failure immediately after CAR T-cell infusion may have been triggered at least in part by ErbB2 expressed at low levels on normal lung epithelium.

In addition to CAR affinity, the location of the CAR-binding epitope within the target antigen can play a decisive role in effector cell activation and influence on-target/off-tumor effects (82, 138). In the case of CAR T cells, CARs directed to membrane-distal epitopes were shown to be superior in binding but less potent in mediating activation than CARs directed to membrane-proximal epitopes of the same antigen (82, 83). We recently showed a similar effect for CAR NK-92 cells targeting EGFR, where a second-generation CAR based on antibody cetuximab, which interacts with domain III of EGFR mediated more potent cytotoxicity than an otherwise identical CAR based on antibody R1 that recognizes an epitope within the N-terminal EGFR domain I (71). The trastuzumab-derived ErbB2-specific scFv antibody fragment used by Morgan et al. binds to the juxtamembrane region (domain IV) of the target receptor (137, 139). In contrast, antibody FRP5 used for the generation of ErbB2-specific NK-92/5.28.z cells recognizes a discontinuous epitope within domain I of ErbB2 facing away from the cell surface (44, 140). Consequently, FRP5-based CARs are less likely than trastuzumab-based CARs to get activated by ErbB2 expressed at moderate levels, which is supported by data from a clinical trial in sarcoma patients with ErbB2-specific T cells carrying an FRP5-based second-generation CAR, where no on-target/off-tumor toxicities were observed (141).

Irradiated CAR NK-92 cells do not expand and persist *in vivo*. Hence, they may even be applicable in a clinical setting to target more abundantly expressed self-antigens such as non-mutated EGFR (70–72), since side effects due to reactivity with normal tissues would be expected to be transient. In the future, more sophisticated safety measures like expression of inducible caspase-9 (iCasp9) as a suicide gene may thereby replace γ-irradiation of NK-92 and CAR NK-92 cells, allowing to rapidly eliminate the cells in case of toxicities, but also to extend *in vivo* activity with a reduction in the cell dose and treatment frequency needed. iCasp9 represents a fusion of the human FK506-binding protein FKBP12 harboring an F36V mutation, and truncated human caspase-9 lacking the caspase activation and recruitment domain (142). In the presence of otherwise inert FK506 analogs such as

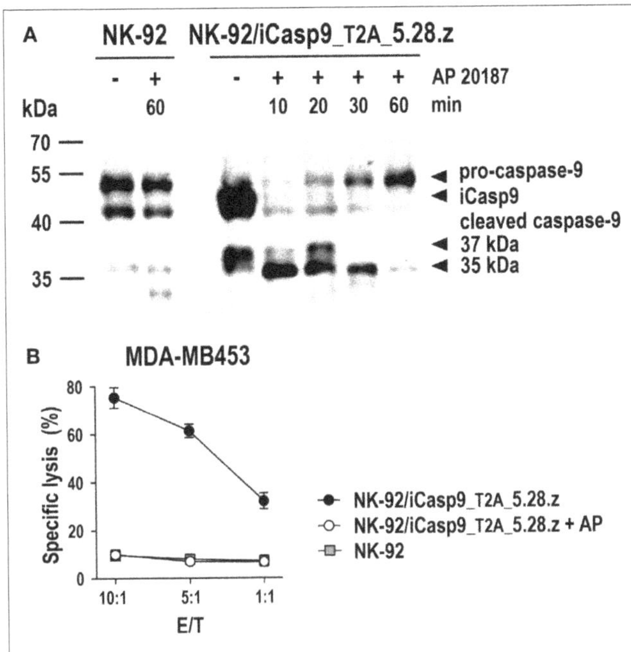

FIGURE 5 | Inducible caspase-9 (iCasp9) as a safety switch for CAR-engineered NK-92 cells. (A) NK-92 cells transduced with a lentiviral vector that encodes iCasp9 and ErbB2-specific CAR 5.28.z separated by a *Thosea asigna* virus self-cleaving peptide (T2A) (NK-92/iCasp9_T2A_5.28.z) were incubated in the presence of 10 nM of the homodimerizer AP20187 for iCasp9 activation. Lysates of cells collected after 10, 20, 30, or 60 min of exposure to AP20187 were subjected to SDS-PAGE and subsequent immunoblotting with a caspase-9-specific antibody. Lysates of NK-92/iCasp9_T2A_5.28.z cells kept without dimerizer and parental NK-92 cells incubated in the absence or presence of AP20187 served as controls. **(B)** Cytotoxicity of NK-92/iCasp9_T2A_5.28.z cells against ErbB2-overexpressing MDA-MB453 breast carcinoma cells was investigated in flow cytometry-based cytotoxicity assays after co-incubation of NK cells and tumor cells at different effector to target ratios (E/T) for 2 h in the absence (filled circles) or presence of AP20187 (open circles). Parental NK-92 cells were included for comparison (gray boxes). Mean values ± SEM are shown; $n = 2$.

AP1903 or AP20187, iCasp9 dimerizes, inducing caspase activation and apoptotic cell death. NK-92 cells coexpressing iCasp9 and a CAR are viable in the absence of dimerizer and retain high and specific CAR-mediated cytotoxicity. In contrast, addition of AP20187 rapidly induces activation of iCasp9 and cleavage of endogenous caspases, precluding any further cell killing by the CAR-engineered cells (**Figure 5**).

CONCLUSION AND FUTURE PERSPECTIVES

Over the past 25 years, NK-92 cells have transformed from a readily available model for studies on human NK cell biology to a promising cell therapeutic for applications in adoptive cancer immunotherapy. As outlined above, genetic modification of NK-92 with CARs has emerged as a successful strategy to enhance the cells' intrinsic antitumor activity and provide them with the capacity for selective target recognition. The ability of CAR NK-92 cells to bypass the immunosuppressive effects of TGF-β and hypoxia in preclinical studies and to enhance or initiate adaptive antitumor immunity is encouraging (69). Nevertheless, the potential impact of immunosuppressive factors like IDO, PGE2, IL-4, NO, and ROS abundant in the tumor microenvironment has not yet been investigated. Concurrent interference with these mechanisms may offer an opportunity to further improve the direct antitumor activity of CAR NK-92 and enhance their immunostimulatory potential. Also, clarifying the relevance of checkpoint regulators such as PD-1 for CAR NK-92 functionality (74) and investigating combination therapies with checkpoint inhibitors and other immunomodulatory regimens appears warranted. Specific cytotoxicity of the NK cells may be enhanced by ectopic expression of components of the cytolytic machinery (143). Chemoattractants like CXCR3 ligands and chemerin can increase accumulation of NK cells at tumor sites (144, 145) and modulation of chemokine receptor expression would likely augment the tumor homing capability of CAR NK-92 cells (146).

Clinical responses seen in individual patients treated with parental NK-92 and ease of improvement by genetic modification with Fc receptors and CARs not only increased efforts of academic researchers to design tailor-made variants for specific disease entities and target antigens but also sparked commercial interest, which is essential to address the challenges associated with standardization of such cell products, large-scale expansion, logistics for distribution, and advanced clinical development (32, 90). Current efforts in the CAR T-cell field are aimed at generating similar universal cell products by eliminating endogenous TCR and MHC with the help of sophisticated gene editing procedures (147–149). This underscores the relevance of truly off-the-shelf CAR cell products like CAR NK-92 for broader applicability of this therapeutic strategy. Early phase clinical trials with CAR NK-92 cells are expected to commence in the near future. Insights from these studies will be essential to judge the therapeutic potential of CAR NK-92 in comparison to *ex vivo* expanded and CAR-engineered primary NK cells and determine the direction of further development.

ETHICS STATEMENT

Animal experiments were approved by the responsible government committee (Regierungspräsidium Darmstadt, Darmstadt, Germany) and were conducted according to the applicable guidelines and regulations.

AUTHOR CONTRIBUTIONS

All authors made substantial contributions to the conception and design of this review article and critically evaluated the cited literature. CZ and SO performed experiments and analyzed data. CZ, PO, AW, AL, and WW drafted the initial version of the manuscript with support from all other co-authors. All authors revised the manuscript and approved the final version of this study. All authors agree to be accountable for the content of this work.

ACKNOWLEDGMENTS

The authors thank Michael C. Burger and Joachim P. Steinbach for their essential contributions to the evaluation of CAR NK-92 cells in glioblastoma models, NantKwest Inc. for kindly providing NK-92 cells, Malcolm K. Brenner and Gianpietro Dotti for kindly providing iCasp9 cDNA, Stefan Stein and Tefik Merovci for flow cytometric cell sorting, Barbara Uherek and Thorsten Geyer for technical assistance, and the staff at the animal facility of the Georg-Speyer-Haus for their support.

Frankfurt (CGT) (HMWK III L 5-518/17.004 2013, HMWK II L 5-518/17.004 2016), German Federal Ministry of Education and Research (BMBF) (Ci3; FKZ 131A009A, 131A009C), Deutsche Forschungsgemeinschaft (DFG) (GRK1172), German Cancer Consortium (DKTK), Alfons und Gertrud Kassel-Stiftung, institutional funds of the Georg-Speyer-Haus, and a fellowship from Stiftung Polytechnische Gesellschaft, Frankfurt am Main to Sarah Oelsner. The Georg-Speyer-Haus is funded jointly by the German Federal Ministry of Health and the Ministry of Higher Education, Research and the Arts of the State of Hessen (HMWK).

REFERENCES

1. Lanier LL. Up on the tightrope: natural killer cell activation and inhibition. *Nat Immunol* (2008) 9(5):495–502. doi:10.1038/ni1581
2. Koch J, Steinle A, Watzl C, Mandelboim O. Activating natural cytotoxicity receptors of natural killer cells in cancer and infection. *Trends Immunol* (2013) 34(4):182–91. doi:10.1016/j.it.2013.01.003
3. Smyth MJ, Hayakawa Y, Takeda K, Yagita H. New aspects of natural-killer-cell surveillance and therapy of cancer. *Nat Rev Cancer* (2002) 2(11):850–61. doi:10.1038/nrc928
4. Vesely MD, Kershaw MH, Schreiber RD, Smyth MJ. Natural innate and adaptive immunity to cancer. *Annu Rev Immunol* (2011) 29:235–71. doi:10.1146/annurev-immunol-031210-101324
5. Imai K, Matsuyama S, Miyake S, Suga K, Nakachi K. Natural cytotoxic activity of peripheral-blood lymphocytes and cancer incidence: an 11-year follow-up study of a general population. *Lancet* (2000) 356(9244):1795–9. doi:10.1016/S0140-6736(00)03231-1
6. Carrega P, Morandi B, Costa R, Frumento G, Forte G, Altavilla G, et al. Natural killer cells infiltrating human nonsmall-cell lung cancer are enriched in CD56 bright CD16(-) cells and display an impaired capability to kill tumor cells. *Cancer* (2008) 112(4):863–75. doi:10.1002/cncr.23239
7. Halama N, Braun M, Kahlert C, Spille A, Quack C, Rahbari N, et al. Natural killer cells are scarce in colorectal carcinoma tissue despite high levels of chemokines and cytokines. *Clin Cancer Res* (2011) 17(4):678–89. doi:10.1158/1078-0432.CCR-10-2173
8. Eckl J, Buchner A, Prinz PU, Riesenberg R, Siegert SI, Kammerer R, et al. Transcript signature predicts tissue NK cell content and defines renal cell carcinoma subgroups independent of TNM staging. *J Mol Med (Berl)* (2012) 90(1):55–66. doi:10.1007/s00109-011-0806-7
9. Kochan G, Escors D, Breckpot K, Guerrero-Setas D. Role of non-classical MHC class I molecules in cancer immunosuppression. *Oncoimmunology* (2013) 2(11):e26491. doi:10.4161/onci.26491
10. Guerra N, Tan YX, Joncker NT, Choy A, Gallardo F, Xiong N, et al. NKG2D-deficient mice are defective in tumor surveillance in models of spontaneous malignancy. *Immunity* (2008) 28(4):571–80. doi:10.1016/j.immuni.2008.02.016
11. Elboim M, Gazit R, Gur C, Ghadially H, Betser-Cohen G, Mandelboim O. Tumor immunoediting by NKp46. *J Immunol* (2010) 184(10):5637–44. doi:10.4049/jimmunol.0901644
12. Reiners KS, Topolar D, Henke A, Simhadri VR, Kessler J, Sauer M, et al. Soluble ligands for NK cell receptors promote evasion of chronic lymphocytic leukemia cells from NK cell anti-tumor activity. *Blood* (2013) 121(18): 3658–65. doi:10.1182/blood-2013-01-476606
13. Schlecker E, Fiegler N, Arnold A, Altevogt P, Rose-John S, Moldenhauer G, et al. Metalloprotease-mediated tumor cell shedding of B7-H6, the ligand of the natural killer cell-activating receptor NKp30. *Cancer Res* (2014) 74(13):3429–40. doi:10.1158/0008-5472.CAN-13-3017
14. Klöss S, Chambron N, Gardlowski T, Arseniev L, Koch J, Esser R, et al. Increased sMICA and TGFbeta1 levels in HNSCC patients impair NKG2D-dependent functionality of activated NK cells. *Oncoimmunology* (2015) 4(11):e1055993. doi:10.1080/2162402x.2015.1055993
15. Bruno A, Ferlazzo G, Albini A, Noonan DM. A think tank of TINK/TANKs: tumor-infiltrating/tumor-associated natural killer cells in tumor progression and angiogenesis. *J Natl Cancer Inst* (2014) 106(8):dju200. doi:10.1093/jnci/dju200
16. Melero I, Rouzaut A, Motz GT, Coukos G. T-cell and NK-cell infiltration into solid tumors: a key limiting factor for efficacious cancer immunotherapy. *Cancer Discov* (2014) 4(5):522–6. doi:10.1158/2159-8290.CD-13-0985
17. Vitale M, Cantoni C, Pietra G, Mingari MC, Moretta L. Effect of tumor cells and tumor microenvironment on NK-cell function. *Eur J Immunol* (2014) 44(6):1582–92. doi:10.1002/eji.201344272
18. Velardi A, Ruggeri L, Mancusi A, Aversa F, Christiansen FT. Natural killer cell allorecognition of missing self in allogeneic hematopoietic transplantation: a tool for immunotherapy of leukemia. *Curr Opin Immunol* (2009) 21(5):525–30. doi:10.1016/j.coi.2009.07.015
19. Davis ZB, Felices M, Verneris MR, Miller JS. Natural killer cell adoptive transfer therapy: exploiting the first line of defense against cancer. *Cancer J* (2015) 21(6):486–91. doi:10.1097/ppo.0000000000000156
20. Rezvani K, Rouce RH. The application of natural killer cell immunotherapy for the treatment of cancer. *Front Immunol* (2015) 6:578. doi:10.3389/fimmu.2015.00578
21. Vivier E, Ugolini S, Blaise D, Chabannon C, Brossay L. Targeting natural killer cells and natural killer T cells in cancer. *Nat Rev Immunol* (2012) 12(4):239–52. doi:10.1038/nri3174
22. Vallera DA, Felices M, McElmurry R, McCullar V, Zhou X, Schmohl JU, et al. IL15 trispecific killer engagers (TriKE) make natural killer cells specific to CD33+ targets while also inducing persistence, in vivo expansion, and enhanced function. *Clin Cancer Res* (2016) 22(14):3440–50. doi:10.1158/1078-0432.CCR-15-2710
23. Uherek C, Tonn T, Uherek B, Becker S, Schnierle B, Klingemann HG, et al. Retargeting of natural killer-cell cytolytic activity to ErbB2-expressing cancer cells results in efficient and selective tumor cell destruction. *Blood* (2002) 100(4):1265–73.
24. Imai C, Iwamoto S, Campana D. Genetic modification of primary natural killer cells overcomes inhibitory signals and induces specific killing of leukemic cells. *Blood* (2005) 106(1):376–83. doi:10.1182/blood-2004-12-4797
25. Shimasaki N, Coustan-Smith E, Kamiya T, Campana D. Expanded and armed natural killer cells for cancer treatment. *Cytotherapy* (2016) 18(11):1422–34. doi:10.1016/j.jcyt.2016.06.013
26. Tran AC, Zhang D, Byrn R, Roberts MR. Chimeric zeta-receptors direct human natural killer (NK) effector function to permit killing of NK-resistant tumor cells and HIV-infected T lymphocytes. *J Immunol* (1995) 155(2):1000–9.
27. Binyamin L, Alpaugh RK, Hughes TL, Lutz CT, Campbell KS, Weiner LM. Blocking NK cell inhibitory self-recognition promotes antibody-dependent cellular cytotoxicity in a model of anti-lymphoma therapy. *J Immunol* (2008) 180(9):6392–401. doi:10.4049/jimmunol.180.9.6392
28. Sahm C, Schönfeld K, Wels WS. Expression of IL-15 in NK cells results in rapid enrichment and selective cytotoxicity of gene-modified effectors that carry a tumor-specific antigen receptor. *Cancer Immunol Immunother* (2012) 61(9):1451–61. doi:10.1007/s00262-012-1212-x
29. Oberoi P, Jabulowsky RA, Bähr-Mahmud H, Wels WS. EGFR-targeted granzyme B expressed in NK cells enhances natural cytotoxicity and mediates specific killing of tumor cells. *PLoS One* (2013) 8(4):e61267. doi:10.1371/journal.pone.0061267

30. Glienke W, Esser R, Priesner C, Suerth JD, Schambach A, Wels WS, et al. Advantages and applications of CAR-expressing natural killer cells. *Front Pharmacol* (2015) 6:21. doi:10.3389/fphar.2015.00021
31. Tonn T, Becker S, Esser R, Schwabe D, Seifried E. Cellular immunotherapy of malignancies using the clonal natural killer cell line NK-92. *J Hematother Stem Cell Res* (2001) 10(4):535–44. doi:10.1089/15258160152509145
32. Klingemann H, Boissel L, Toneguzzo F. Natural killer cells for immunotherapy – advantages of the NK-92 cell line over blood NK cells. *Front Immunol* (2016) 7:91. doi:10.3389/fimmu.2016.00091
33. Gong JH, Maki G, Klingemann HG. Characterization of a human cell line (NK-92) with phenotypical and functional characteristics of activated natural killer cells. *Leukemia* (1994) 8(4):652–8.
34. Maki G, Klingemann HG, Martinson JA, Tam YK. Factors regulating the cytotoxic activity of the human natural killer cell line, NK-92. *J Hematother Stem Cell Res* (2001) 10(3):369–83. doi:10.1089/152581601750288975
35. Arai S, Meagher R, Swearingen M, Myint H, Rich E, Martinson J, et al. Infusion of the allogeneic cell line NK-92 in patients with advanced renal cell cancer or melanoma: a phase I trial. *Cytotherapy* (2008) 10(6):625–32. doi:10.1080/14653240802301872
36. Tonn T, Schwabe D, Klingemann HG, Becker S, Esser R, Koehl U, et al. Treatment of patients with advanced cancer with the natural killer cell line NK-92. *Cytotherapy* (2013) 15(12):1563–70. doi:10.1016/j.jcyt.2013.06.017
37. Eshhar Z, Waks T, Gross G, Schindler DG. Specific activation and targeting of cytotoxic lymphocytes through chimeric single chains consisting of antibody-binding domains and the gamma or zeta subunits of the immunoglobulin and T-cell receptors. *Proc Natl Acad Sci U S A* (1993) 90(2):720–4. doi:10.1073/pnas.90.2.720
38. Gill S, June CH. Going viral: chimeric antigen receptor T-cell therapy for hematological malignancies. *Immunol Rev* (2015) 263(1):68–89. doi:10.1111/imr.12243
39. Davila ML, Sadelain M. Biology and clinical application of CAR T cells for B cell malignancies. *Int J Hematol* (2016) 104(1):6–17. doi:10.1007/s12185-016-2039-6
40. Maus MV, Grupp SA, Porter DL, June CH. Antibody-modified T cells: CARs take the front seat for hematologic malignancies. *Blood* (2014) 123(17):2625–35. doi:10.1182/blood-2013-11-492231
41. Fesnak AD, June CH, Levine BL. Engineered T cells: the promise and challenges of cancer immunotherapy. *Nat Rev Cancer* (2016) 16(9):566–81. doi:10.1038/nrc.2016.97
42. Chmielewski M, Hombach AA, Abken H. Of CARs and TRUCKs: chimeric antigen receptor (CAR) T cells engineered with an inducible cytokine to modulate the tumor stroma. *Immunol Rev* (2014) 257(1):83–90. doi:10.1111/imr.12125
43. Cartellieri M, Feldmann A, Koristka S, Arndt C, Loff S, Ehninger A, et al. Switching CAR T cells on and off: a novel modular platform for retargeting of T cells to AML blasts. *Blood Cancer J* (2016) 6(8):e458. doi:10.1038/bcj.2016.61
44. Grada Z, Hegde M, Byrd T, Shaffer DR, Ghazi A, Brawley VS, et al. TanCAR: a novel bispecific chimeric antigen receptor for cancer immunotherapy. *Mol Ther Nucleic Acids* (2013) 2:e105. doi:10.1038/mtna.2013.32
45. Hegde M, Mukherjee M, Grada Z, Pignata A, Landi D, Navai SA, et al. Tandem CAR T cells targeting HER2 and IL13Ralpha2 mitigate tumor antigen escape. *J Clin Invest* (2016) 126(8):3036–52. doi:10.1172/JCI83416
46. Milone MC, Fish JD, Carpenito C, Carroll RG, Binder GK, Teachey D, et al. Chimeric receptors containing CD137 signal transduction domains mediate enhanced survival of T cells and increased antileukemic efficacy in vivo. *Mol Ther* (2009) 17(8):1453–64. doi:10.1038/mt.2009.83
47. Kawalekar OU, O'Connor RS, Fraietta JA, Guo L, McGettigan SE, Posey AD Jr, et al. Distinct signaling of coreceptors regulates specific metabolism pathways and impacts memory development in CAR T cells. *Immunity* (2016) 44(2):380–90. doi:10.1016/j.immuni.2016.01.021
48. Romanski A, Uherek C, Bug G, Müller T, Rössig C, Kampfmann M, et al. Re-targeting of an NK cell line (NK92) with specificity for CD19 efficiently kills human B-precursor leukemia cells. *Blood* (2004) 104(11):751A–751A.
49. Müller T, Uherek C, Maki G, Chow KU, Schimpf A, Klingemann HG, et al. Expression of a CD20-specific chimeric antigen receptor enhances cytotoxic activity of NK cells and overcomes NK-resistance of lymphoma and leukemia cells. *Cancer Immunol Immunother* (2008) 57(3):411–23. doi:10.1007/s00262-007-0383-3
50. Boissel L, Betancur M, Wels WS, Tuncer H, Klingemann H. Transfection with mRNA for CD19 specific chimeric antigen receptor restores NK cell mediated killing of CLL cells. *Leuk Res* (2009) 33(9):1255–9. doi:10.1016/j.leukres.2008.11.024
51. Boissel L, Betancur M, Lu W, Wels WS, Marino T, Van Etten RA, et al. Comparison of mRNA and lentiviral based transfection of natural killer cells with chimeric antigen receptors recognizing lymphoid antigens. *Leuk Lymphoma* (2012) 53(5):958–65. doi:10.3109/10428194.2011.634048
52. Boissel L, Betancur-Boissel M, Lu W, Krause DS, Van Etten RA, Wels WS, et al. Retargeting NK-92 cells by means of CD19- and CD20-specific chimeric antigen receptors compares favorably with antibody-dependent cellular cytotoxicity. *Oncoimmunology* (2013) 2(10):e26527. doi:10.4161/onci.26527
53. Romanski A, Uherek C, Bug G, Seifried E, Klingemann H, Wels WS, et al. CD19-CAR engineered NK-92 cells are sufficient to overcome NK cell resistance in B-cell malignancies. *J Cell Mol Med* (2016) 20(7):1287–94. doi:10.1111/jcmm.12810
54. Jiang H, Zhang W, Shang P, Zhang H, Fu W, Ye F, et al. Transfection of chimeric anti-CD138 gene enhances natural killer cell activation and killing of multiple myeloma cells. *Mol Oncol* (2014) 8(2):297–310. doi:10.1016/j.molonc.2013.12.001
55. Tavri S, Jha P, Meier R, Henning TD, Müller T, Hostetter D, et al. Optical imaging of cellular immunotherapy against prostate cancer. *Mol Imaging* (2009) 8(1):15–26. doi:10.2310/7290.2009.00002
56. Esser R, Müller T, Stefes D, Kloess S, Seidel D, Gillies SD, et al. NK cells engineered to express a GD2-specific antigen receptor display built-in ADCC-like activity against tumour cells of neuroectodermal origin. *J Cell Mol Med* (2012) 16(3):569–81. doi:10.1111/j.1582-4934.2011.01343.x
57. Zhang G, Liu RZ, Zhu XK, Wang L, Ma J, Han HM, et al. Retargeting NK-92 for anti-melanoma activity by a TCR-like single-domain antibody. *Immunol Cell Biol* (2013) 91(10):615–24. doi:10.1038/icb.2013.45
58. Clemenceau B, Valsesia-Wittmann S, Jallas AC, Vivien R, Rousseau R, Marabelle A, et al. In vitro and in vivo comparison of lymphocytes transduced with a human CD16 or with a chimeric antigen receptor reveals potential off-target interactions due to the IgG2 CH2-CH3 CAR-spacer. *J Immunol Res* (2015) 2015:482089. doi:10.1155/2015/482089
59. Seidel D, Shibina A, Siebert N, Wels WS, Reynolds CP, Huebener N, et al. Disialoganglioside-specific human natural killer cells are effective against drug-resistant neuroblastoma. *Cancer Immunol Immunother* (2015) 64(5):621–34. doi:10.1007/s00262-015-1669-5
60. Altvater B, Landmeier S, Pscherer S, Temme J, Schweer K, Kailayangiri S, et al. 2B4 (CD244) signaling by recombinant antigen-specific chimeric receptors costimulates natural killer cell activation to leukemia and neuroblastoma cells. *Clin Cancer Res* (2009) 15(15):4857–66. doi:10.1158/1078-0432.CCR-08-2810
61. Kruschinski A, Moosmann A, Poschke I, Norell H, Chmielewski M, Seliger B, et al. Engineering antigen-specific primary human NK cells against HER-2 positive carcinomas. *Proc Natl Acad Sci U S A* (2008) 105(45):17481–6. doi:10.1073/pnas.0804788105
62. Chu Y, Hochberg J, Yahr A, Ayello J, van de Ven C, Barth M, et al. Targeting CD20+ aggressive B-cell non-Hodgkin lymphoma by anti-CD20 CAR mRNA-modified expanded natural killer cells in vitro and in NSG mice. *Cancer Immunol Res* (2015) 3(4):333–44. doi:10.1158/2326-6066.CIR-14-0114
63. Daldrup-Link HE, Meier R, Rudelius M, Piontek G, Piert M, Metz S, et al. In vivo tracking of genetically engineered, anti-HER2/neu directed natural killer cells to HER2/neu positive mammary tumors with magnetic resonance imaging. *Eur Radiol* (2005) 15(1):4–13. doi:10.1007/s00330-004-2526-7
64. Meier R, Piert M, Piontek G, Rudelius M, Oostendorp RA, Senekowitsch-Schmidtke R, et al. Tracking of [18F]FDG-labeled natural killer cells to HER2/neu-positive tumors. *Nucl Med Biol* (2008) 35(5):579–88. doi:10.1016/j.nucmedbio.2008.02.006
65. Alkins R, Burgess A, Ganguly M, Francia G, Kerbel R, Wels WS, et al. Focused ultrasound delivers targeted immune cells to metastatic brain tumors. *Cancer Res* (2013) 73(6):1892–9. doi:10.1158/0008-5472.CAN-12-2609
66. Alkins R, Burgess A, Kerbel R, Wels WS, Hynynen K. Early treatment of HER2-amplified brain tumors with targeted NK-92 cells and focused ultrasound improves survival. *Neuro Oncol* (2016) 18(7):974–81. doi:10.1093/neuonc/nov318

67. Schönfeld K, Sahm C, Zhang C, Naundorf S, Brendel C, Odendahl M, et al. Selective inhibition of tumor growth by clonal NK cells expressing an ErbB2/HER2-specific chimeric antigen receptor. *Mol Ther* (2015) 23(2):330–8. doi:10.1038/mt.2014.219
68. Liu H, Yang B, Sun T, Lin L, Hu Y, Deng M, et al. Specific growth inhibition of ErbB2-expressing human breast cancer cells by genetically modified NK92 cells. *Oncol Rep* (2015) 33(1):95–102. doi:10.3892/or.2014.3548
69. Zhang C, Burger MC, Jennewein L, Genssler S, Schönfeld K, Zeiner P, et al. ErbB2/HER2-specific NK cells for targeted therapy of glioblastoma. *J Natl Cancer Inst* (2016) 108(5):djv375. doi:10.1093/jnci/djv375
70. Chen X, Han J, Chu J, Zhang L, Zhang J, Chen C, et al. A combinational therapy of EGFR-CAR NK cells and oncolytic herpes simplex virus 1 for breast cancer brain metastases. *Oncotarget* (2016) 7(19):27764–77. doi:10.18632/oncotarget.8526
71. Genßler S, Burger MC, Zhang C, Oelsner S, Mildenberger I, Wagner M, et al. Dual targeting of glioblastoma with chimeric antigen receptor-engineered natural killer cells overcomes heterogeneity of target antigen expression and enhances antitumor activity and survival. *Oncoimmunology* (2016) 5(4):e1119354. doi:10.1080/2162402X.2015.1119354
72. Han J, Chu J, Keung Chan W, Zhang J, Wang Y, Cohen JB, et al. CAR-engineered NK cells targeting wild-type EGFR and EGFRvIII enhance killing of glioblastoma and patient-derived glioblastoma stem cells. *Sci Rep* (2015) 5:11483. doi:10.1038/srep11483
73. Meier R, Golovko D, Tavri S, Henning TD, Knopp C, Piontek G, et al. Depicting adoptive immunotherapy for prostate cancer in an animal model with magnetic resonance imaging. *Magn Reson Med* (2011) 65(3):756–63. doi:10.1002/mrm.22652
74. Oelsner S, Friede ME, Zhang C, Wagner J, Badura S, Bader P, et al. Continuously expanding CAR NK-92 cells display selective cytotoxicity against B-cell leukemia and lymphoma. *Cytotherapy* (2017) 19(2):235–49. doi:10.1016/j.jcyt.2016.10.009
75. Chen KH, Wada M, Firor AE, Pinz KG, Jares A, Liu H, et al. Novel anti-CD3 chimeric antigen receptor targeting of aggressive T cell malignancies. *Oncotarget* (2016) 7(35):56219–32. doi:10.18632/oncotarget.11019
76. Chen KH, Wada M, Pinz KG, Liu H, Lin KW, Jares A, et al. Preclinical targeting of aggressive T-cell malignancies using anti-CD5 chimeric antigen receptor. *Leukemia* (2017). doi:10.1038/leu.2017.8
77. Chu J, Deng Y, Benson DM, He S, Hughes T, Zhang J, et al. CS1-specific chimeric antigen receptor (CAR)-engineered natural killer cells enhance in vitro and in vivo antitumor activity against human multiple myeloma. *Leukemia* (2014) 28(4):917–27. doi:10.1038/leu.2013.279
78. Tassev DV, Cheng M, Cheung NKV. Retargeting NK92 cells using an HLA-A2-restricted, EBNA3C-specific chimeric antigen receptor. *Cancer Gene Ther* (2012) 19(2):84–100. doi:10.1038/cgt.2011.66
79. Zhao Q, Ahmed M, Tassev DV, Hasan A, Kuo TY, Guo HF, et al. Affinity maturation of T-cell receptor-like antibodies for Wilms tumor 1 peptide greatly enhances therapeutic potential. *Leukemia* (2015) 29(11):2238–47. doi:10.1038/leu.2015.125
80. Klingemann H. Are natural killer cells superior CAR drivers? *Oncoimmunology* (2014) 3:e28147. doi:10.4161/onci.28147
81. Rafiq S, Purdon TJ, Schultz L, Klingemann H, Brentjens RJ. NK-92 cells engineered with anti-CD33 chimeric antigen receptors (CAR) for the treatment of acute myeloid leukemia (AML). *Cytotherapy* (2015) 17(6):S23–23. doi:10.1016/j.jcyt.2015.03.384
82. Hombach AA, Schildgen V, Heuser C, Finnern R, Gilham DE, Abken H. T cell activation by antibody-like immunoreceptors: the position of the binding epitope within the target molecule determines the efficiency of activation of redirected T cells. *J Immunol* (2007) 178(7):4650–7. doi:10.4049/jimmunol.178.7.4650
83. James SE, Greenberg PD, Jensen MC, Lin Y, Wang J, Till BG, et al. Antigen sensitivity of CD22-specific chimeric TCR is modulated by target epitope distance from the cell membrane. *J Immunol* (2008) 180(10):7028–38. doi:10.4049/jimmunol.180.10.7028
84. Sadelain M, Brentjens R, Riviere I. The basic principles of chimeric antigen receptor design. *Cancer Discov* (2013) 3(4):388–98. doi:10.1158/2159-8290.CD-12-0548
85. Klingemann H. Challenges of cancer therapy with natural killer cells. *Cytotherapy* (2015) 17(3):245–9. doi:10.1016/j.jcyt.2014.09.007
86. Vivier E. What is natural in natural killer cells? *Immunol Lett* (2006) 107(1):1–7. doi:10.1016/j.imlet.2006.07.004
87. Sun JC, Lanier LL. NK cell development, homeostasis and function: parallels with CD8(+) T cells. *Nat Rev Immunol* (2011) 11(10):645–57. doi:10.1038/nri3044
88. Childs RW, Berg M. Bringing natural killer cells to the clinic: ex vivo manipulation. *Hematology Am Soc Hematol Educ Program* (2013) 2013:234–46. doi:10.1182/asheducation-2013.1.234
89. Koehl U, Brehm C, Huenecke S, Zimmermann SY, Kloess S, Bremm M, et al. Clinical grade purification and expansion of NK cell products for an optimized manufacturing protocol. *Front Oncol* (2013) 3:118. doi:10.3389/fonc.2013.00118
90. Suck G, Odendahl M, Nowakowska P, Seidl C, Wels WS, Klingemann HG, et al. NK-92: an 'off-the-shelf therapeutic' for adoptive natural killer cell-based cancer immunotherapy. *Cancer Immunol Immunother* (2016) 65(4):485–92. doi:10.1007/s00262-015-1761-x
91. Nagashima S, Mailliard R, Kashii Y, Reichert TE, Herberman RB, Robbins P, et al. Stable transduction of the interleukin-2 gene into human natural killer cell lines and their phenotypic and functional characterization in vitro and in vivo. *Blood* (1998) 91(10):3850–61.
92. Tam YK, Maki G, Miyagawa B, Hennemann B, Tonn T, Klingemann HG. Characterization of genetically altered, interleukin 2-independent natural killer cell lines suitable for adoptive cellular immunotherapy. *Hum Gene Ther* (1999) 10(8):1359–73. doi:10.1089/10430349950018030
93. Zhang J, Sun R, Wei H, Zhang J, Tian Z. Characterization of interleukin-15 gene-modified human natural killer cells: implications for adoptive cellular immunotherapy. *Haematologica* (2004) 89(3):338–47.
94. Zhang J, Sun R, Wei H, Zhang J, Tian Z. Characterization of stem cell factor gene-modified human natural killer cell line, NK-92 cells: implication in NK cell-based adoptive cellular immunotherapy. *Oncol Rep* (2004) 11(5):1097–106. doi:10.3892/or.11.5.1097
95. Waldmann TA, Dubois S, Tagaya Y. Contrasting roles of IL-2 and IL-15 in the life and death of lymphocytes: implications for immunotherapy. *Immunity* (2001) 14(2):105–10. doi:10.1016/S1074-7613(01)00093-0
96. Konstantinidis KV, Alici E, Aints A, Christensson B, Ljunggren HG, Dilber MS. Targeting IL-2 to the endoplasmic reticulum confines autocrine growth stimulation to NK-92 cells. *Exp Hematol* (2005) 33(2):159–64. doi:10.1016/j.exphem.2004.11.003
97. Jochems C, Hodge JW, Fantini M, Fujii R, Morillon YM II, Greiner JW, et al. An NK cell line (haNK) expressing high levels of granzyme and engineered to express the high affinity CD16 allele. *Oncotarget* (2016) 7(52):86359–73. doi:10.18632/oncotarget.13411
98. Lanier LL, Yu G, Phillips JH. Analysis of Fc gamma RIII (CD16) membrane expression and association with CD3 zeta and Fc epsilon RI-gamma by site-directed mutation. *J Immunol* (1991) 146(5):1571–6.
99. Vivier E, Rochet N, Ackerly M, Petrini J, Levine H, Daley J, et al. Signaling function of reconstituted CD16: zeta: gamma receptor complex isoforms. *Int Immunol* (1992) 4(11):1313–23. doi:10.1093/intimm/4.11.1313
100. Clemenceau B, Vivien R, Pellat C, Foss M, Thibault G, Vie H. The human natural killer cytotoxic cell line NK-92, once armed with a murine CD16 receptor, represents a convenient cellular tool for the screening of mouse mAbs according to their ADCC potential. *MAbs* (2013) 5(4):587–94. doi:10.4161/mabs.25077
101. Isakov N. Immunoreceptor tyrosine-based activation motif (ITAM), a unique module linking antigen and Fc receptors to their signaling cascades. *J Leukoc Biol* (1997) 61(1):6–16.
102. Fernandez NC, Lozier A, Flament C, Ricciardi-Castagnoli P, Bellet D, Suter M, et al. Dendritic cells directly trigger NK cell functions: cross-talk relevant in innate anti-tumor immune responses in vivo. *Nat Med* (1999) 5(4):405–11. doi:10.1038/7403
103. Walzer T, Dalod M, Robbins SH, Zitvogel L, Vivier E. Natural-killer cells and dendritic cells: "l'union fait la force". *Blood* (2005) 106(7):2252–8. doi:10.1182/blood-2005-03-1154
104. Van Elssen CH, Oth T, Germeraad WT, Bos GM, Vanderlocht J. Natural killer cells: the secret weapon in dendritic cell vaccination strategies. *Clin Cancer Res* (2014) 20(5):1095–103. doi:10.1158/1078-0432.CCR-13-2302
105. Gerosa F, Baldani-Guerra B, Nisii C, Marchesini V, Carra G, Trinchieri G. Reciprocal activating interaction between natural killer cells and dendritic cells. *J Exp Med* (2002) 195(3):327–33. doi:10.1084/jem.20010938

106. Piccioli D, Sbrana S, Melandri E, Valiante NM. Contact-dependent stimulation and inhibition of dendritic cells by natural killer cells. *J Exp Med* (2002) 195(3):335–41. doi:10.1084/jem.20010934
107. Vitale M, Della Chiesa M, Carlomagno S, Pende D, Arico M, Moretta L, et al. NK-dependent DC maturation is mediated by TNFalpha and IFNgamma released upon engagement of the NKp30 triggering receptor. *Blood* (2005) 106(2):566–71. doi:10.1182/blood-2004-10-4035
108. Ni J, Miller M, Stojanovic A, Garbi N, Cerwenka A. Sustained effector function of IL-12/15/18-preactivated NK cells against established tumors. *J Exp Med* (2012) 209(13):2351–65. doi:10.1084/jem.20120944
109. Colonna M, Trinchieri G, Liu YJ. Plasmacytoid dendritic cells in immunity. *Nat Immunol* (2004) 5(12):1219–26. doi:10.1038/ni1141
110. Ferlazzo G, Tsang ML, Moretta L, Melioli G, Steinman RM, Munz C. Human dendritic cells activate resting natural killer (NK) cells and are recognized via the NKp30 receptor by activated NK cells. *J Exp Med* (2002) 195(3):343–51. doi:10.1084/jem.20011149
111. Jinushi M, Takehara T, Kanto T, Tatsumi T, Groh V, Spies T, et al. Critical role of MHC class I-related chain A and B expression on IFN-alpha-stimulated dendritic cells in NK cell activation: impairment in chronic hepatitis C virus infection. *J Immunol* (2003) 170(3):1249–56. doi:10.4049/jimmunol.170.3.1249
112. Draghi M, Pashine A, Sanjanwala B, Gendzekhadze K, Cantoni C, Cosman D, et al. NKp46 and NKG2D recognition of infected dendritic cells is necessary for NK cell activation in the human response to influenza infection. *J Immunol* (2007) 178(5):2688–98. doi:10.4049/jimmunol.178.5.2688
113. Wai LE, Garcia JA, Martinez OM, Krams SM. Distinct roles for the NK cell-activating receptors in mediating interactions with dendritic cells and tumor cells. *J Immunol* (2011) 186(1):222–9. doi:10.4049/jimmunol.1002597
114. Della Chiesa M, Vitale M, Carlomagno S, Ferlazzo G, Moretta L, Moretta A. The natural killer cell-mediated killing of autologous dendritic cells is confined to a cell subset expressing CD94/NKG2A, but lacking inhibitory killer Ig-like receptors. *Eur J Immunol* (2003) 33(6):1657–66. doi:10.1002/eji.200323986
115. Castriconi R, Cantoni C, Della Chiesa M, Vitale M, Marcenaro E, Conte R, et al. Transforming growth factor beta 1 inhibits expression of NKp30 and NKG2D receptors: consequences for the NK-mediated killing of dendritic cells. *Proc Natl Acad Sci U S A* (2003) 100(7):4120–5. doi:10.1073/pnas.0730640100
116. Fauriat C, Moretta A, Olive D, Costello RT. Defective killing of dendritic cells by autologous natural killer cells from acute myeloid leukemia patients. *Blood* (2005) 106(6):2186–8. doi:10.1182/blood-2005-03-1270
117. Ferlazzo G, Morandi B, D'Agostino A, Meazza R, Melioli G, Moretta A, et al. The interaction between NK cells and dendritic cells in bacterial infections results in rapid induction of NK cell activation and in the lysis of uninfected dendritic cells. *Eur J Immunol* (2003) 33(2):306–13. doi:10.1002/immu.200310004
118. Martin-Fontecha A, Thomsen LL, Brett S, Gerard C, Lipp M, Lanzavecchia A, et al. Induced recruitment of NK cells to lymph nodes provides IFN-gamma for T(H)1 priming. *Nat Immunol* (2004) 5(12):1260–5. doi:10.1038/ni1138
119. Sutlu T, Alici E. Natural killer cell-based immunotherapy in cancer: current insights and future prospects. *J Intern Med* (2009) 266(2):154–81. doi:10.1111/j.1365-2796.2009.02121.x
120. Baginska J, Viry E, Paggetti J, Medves S, Berchem G, Moussay E, et al. The critical role of the tumor microenvironment in shaping natural killer cell-mediated anti-tumor immunity. *Front Immunol* (2013) 4:490. doi:10.3389/fimmu.2013.00490
121. Favier B, Lemaoult J, Lesport E, Carosella ED. ILT2/HLA-G interaction impairs NK-cell functions through the inhibition of the late but not the early events of the NK-cell activating synapse. *FASEB J* (2010) 24(3):689–99. doi:10.1096/fj.09-135194
122. Narai S, Watanabe M, Hasegawa H, Nishibori H, Endo T, Kubota T, et al. Significance of transforming growth factor beta1 as a new tumor marker for colorectal cancer. *Int J Cancer* (2002) 97(4):508–11. doi:10.1002/ijc.1631
123. Kirwan SE, Burshtyn DN. Killer cell Ig-like receptor-dependent signaling by Ig-like transcript 2 (ILT2/CD85j/LILRB1/LIR-1). *J Immunol* (2005) 175(8):5006–15. doi:10.4049/jimmunol.175.8.5006
124. Yan Y, Steinherz P, Klingemann HG, Dennig D, Childs BH, McGuirk J, et al. Antileukemia activity of a natural killer cell line against human leukemias. *Clin Cancer Res* (1998) 4(11):2859–68.
125. Tam YK, Miyagawa B, Ho VC, Klingemann HG. Immunotherapy of malignant melanoma in a SCID mouse model using the highly cytotoxic natural killer cell line NK-92. *J Hematother* (1999) 8(3):281–90. doi:10.1089/106161299320316
126. Boissel L, Klingemann H, Khan J, Soon-Shiong P. Intra-tumor injection of CAR-engineered NK cells induces tumor regression and protection against tumor re-challenge. *Blood* (2016) 128(22):466.
127. Green JD, Terrell TG. Utilization of homologous proteins to evaluate the safety of recombinant human proteins – case study: recombinant human interferon-gamma (rhIFN-γ). *Toxicol Lett* (1992) 6(4–65):321–7. doi:10.1016/0378-4274(92)90204-W
128. Schroder K, Hertzog PJ, Ravasi T, Hume DA. Interferon-gamma: an overview of signals, mechanisms and functions. *J Leukoc Biol* (2004) 75(2):163–89. doi:10.1189/jlb.0603252
129. Schnurr M, Scholz C, Rothenfusser S, Galambos P, Dauer M, Robe J, et al. Apoptotic pancreatic tumor cells are superior to cell lysates in promoting cross-priming of cytotoxic T cells and activate NK and gammadelta T cells. *Cancer Res* (2002) 62(8):2347–52.
130. Kokhaei P, Choudhury A, Mahdian R, Lundin J, Moshfegh A, Osterborg A, et al. Apoptotic tumor cells are superior to tumor cell lysate, and tumor cell RNA in induction of autologous T cell response in B-CLL. *Leukemia* (2004) 18(11):1810–5. doi:10.1038/sj.leu.2403517
131. Jacoby E, Yang Y, Qin H, Chien CD, Kochenderfer JN, Fry TJ. Murine allogeneic CD19 CAR T cells harbor potent antileukemic activity but have the potential to mediate lethal GVHD. *Blood* (2016) 127(10):1361–70. doi:10.1182/blood-2015-08-664250
132. Shah NN, Baird K, Delbrook CP, Fleisher TA, Kohler ME, Rampertaap S, et al. Acute GVHD in patients receiving IL-15/4-1BBL activated NK cells following T-cell-depleted stem cell transplantation. *Blood* (2015) 125(5):784–92. doi:10.1182/blood-2014-07-592881
133. Burger MC, Mildenberger I, Zhang C, Ihrig K, Wagner M, Mittelbronn M, et al. The CAR2BRAIN study: a monocentric phase I trial with ErbB2-specific NK-92/5.28.z cells in recurrent glioblastoma. *Neuro Oncol* (2016) 18:24–5. doi:10.1093/neuonc/now188.083
134. Teachey DT, Lacey SF, Shaw PA, Melenhorst JJ, Maude SL, Frey N, et al. Identification of predictive biomarkers for cytokine release syndrome after chimeric antigen receptor T-cell therapy for acute lymphoblastic leukemia. *Cancer Discov* (2016) 6(6):664–79. doi:10.1158/2159-8290.CD-16-0040
135. Lee DW, Gardner R, Porter DL, Louis CU, Ahmed N, Jensen M, et al. Current concepts in the diagnosis and management of cytokine release syndrome. *Blood* (2014) 124(2):188–95. doi:10.1182/blood-2014-05-552729
136. Maude SL, Barrett D, Teachey DT, Grupp SA. Managing cytokine release syndrome associated with novel T cell-engaging therapies. *Cancer J* (2014) 20(2):119–22. doi:10.1097/ppo.0000000000000035
137. Morgan RA, Yang JC, Kitano M, Dudley ME, Laurencot CM, Rosenberg SA. Case report of a serious adverse event following the administration of T cells transduced with a chimeric antigen receptor recognizing ERBB2. *Mol Ther* (2010) 18(4):843–51. doi:10.1038/mt.2010.24
138. Liu X, Jiang S, Fang C, Yang S, Olalere D, Pequignot EC, et al. Affinity-tuned ErbB2 or EGFR chimeric antigen receptor T cells exhibit an increased therapeutic index against tumors in mice. *Cancer Res* (2015) 75(17):3596–607. doi:10.1158/0008-5472.can-15-0159
139. Cho HS, Mason K, Ramyar KX, Stanley AM, Gabelli SB, Denney DW Jr, et al. Structure of the extracellular region of HER2 alone and in complex with the Herceptin Fab. *Nature* (2003) 421(6924):756–60. doi:10.1038/nature01392
140. Gerstmayer B, Altenschmidt U, Hoffmann M, Wels W. Costimulation of T cell proliferation by a chimeric B7-2 antibody fusion protein specifically targeted to cells expressing the erbB2 proto-oncogene. *J Immunol* (1997) 158(10):4584–90.
141. Ahmed N, Brawley VS, Hegde M, Robertson C, Ghazi A, Gerken C, et al. Human epidermal growth factor receptor 2 (HER2)-specific chimeric antigen receptor-modified T cells for the immunotherapy of HER2-positive sarcoma. *J Clin Oncol* (2015) 33(15):1688–96. doi:10.1200/jco.2014.58.0225
142. Di Stasi A, Tey SK, Dotti G, Fujita Y, Kennedy-Nasser A, Martinez C, et al. Inducible apoptosis as a safety switch for adoptive cell therapy. *N Engl J Med* (2011) 365(18):1673–83. doi:10.1056/NEJMoa1106152

143. Oberoi P, Wels WS. Arming NK cells with enhanced antitumor activity: CARs and beyond. *Oncoimmunology* (2013) 2(8):e25220. doi:10.4161/onci.25220
144. Wendel M, Galani IE, Suri-Payer E, Cerwenka A. Natural killer cell accumulation in tumors is dependent on IFN-gamma and CXCR3 ligands. *Cancer Res* (2008) 68(20):8437–45. doi:10.1158/0008-5472.CAN-08-1440
145. Pachynski RK, Zabel BA, Kohrt HE, Tejeda NM, Monnier J, Swanson CD, et al. The chemoattractant chemerin suppresses melanoma by recruiting natural killer cell antitumor defenses. *J Exp Med* (2012) 209(8):1427–35. doi:10.1084/jem.20112124
146. Müller N, Michen S, Tietze S, Topfer K, Schulte A, Lamszus K, et al. Engineering NK cells modified with an EGFRvIII-specific chimeric antigen receptor to overexpress CXCR4 improves immunotherapy of CXCL12/SDF-1alpha-secreting glioblastoma. *J Immunother* (2015) 38(5):197–210. doi:10.1097/cji.0000000000000082
147. Torikai H, Reik A, Liu PQ, Zhou Y, Zhang L, Maiti S, et al. A foundation for universal T-cell based immunotherapy: T cells engineered to express a CD19-specific chimeric-antigen-receptor and eliminate expression of endogenous TCR. *Blood* (2012) 119(24):5697–705. doi:10.1182/blood-2012-01-405365
148. Ren J, Liu X, Fang C, Jiang S, June CH, Zhao Y. Multiplex genome editing to generate universal CAR T cells resistant to PD1 inhibition. *Clin Cancer Res* (2016). doi:10.1158/1078-0432.ccr-16-1300
149. Qasim W, Zhan H, Samarasinghe S, Adams S, Amrolia P, Stafford S, et al. Molecular remission of infant B-ALL after infusion of universal TALEN gene-edited CAR T cells. *Sci Transl Med* (2017) 9(374):eaaj2013. doi:10.1126/scitranslmed.aaj2013

6

Redirected Primary Human Chimeric Antigen Receptor Natural Killer Cells as an "Off-the-Shelf Immunotherapy" for Improvement in Cancer Treatment

*Olaf Oberschmidt, Stephan Kloess and Ulrike Koehl**

Institute of Cellular Therapeutics, Hannover Medical School, Hannover, Germany

Edited by:
Alok Srivastava,
Christian Medical College and Hospital, India

Reviewed by:
Jianhua Yu,
The Ohio State University Columbus, United States
Raquel Tarazona,
University of Extremadura, Spain

***Correspondence:**
Ulrike Koehl
koehl.ulrike@mh-hannover.de

Primary human natural killer (NK) cells recognize and subsequently eliminate virus infected cells, tumor cells, or other aberrant cells. However, cancer cells are able to develop tumor immune escape mechanisms to undermine this immune control. To overcome this obstacle, NK cells can be genetically modified to express chimeric antigen receptors (CARs) in order to improve specific recognition of cancer surface markers (e.g., CD19, CD20, and ErbB2). After target recognition, intracellular CAR domain signaling (CD3ζ, CD28, 4-1BB, and 2B4) leads to activation of PI3K or DNAX proteins (DAP10, DAP12) and finally to enhanced cytotoxicity, proliferation, and/or interferon γ release. This mini-review summarizes both the first preclinical trials with CAR-engineered primary human NK cells and the translational implications for "off-the-shelf immunotherapy" in cancer treatment. Signal transduction in NK cells as well as optimization of CAR signaling will be described, becoming more and more a focal point of interest in addition to redirected T cells. Finally, strategies to overcome off-target effects will be discussed in order to improve future clinical trials and to avoid attacking healthy tissues.

Keywords: natural killer cells, chimeric antigen receptor, chimeric antigen receptor-associated signaling domain, intracellular chimeric antigen receptor-dependent signaling, cancer immunotherapy

INTRODUCTION

Natural killer (NK) cells are peripheral blood lymphocytes that mediate immune surveillance in regard to virus infected and malignant cells (1–3). For early disease detection and killing NK cells rely on several mechanisms such as inflammatory cytokine secretion [e.g., interferon gamma (IFNγ), tumor necrosis factor alpha (TNF-α), interleukin-10 (IL-10)], receptor ligand binding (e.g., tumor necrosis factor-related apoptosis inducing ligand, Fas ligand) (4), or release of cytoplasmic granule toxins (e.g., perforin, granzyme A, granzyme B, and granulysin) (5, 6) as a result of antibody-dependent cellular cytotoxicity (ADCC) (7).

Recognition of aberrant and stressed cells occurs by means of activating cell surface receptors including natural killer group 2 member D (NKG2D) (CD314), NKp30 (CD337), NKp46 (CD335), and NKp44 (CD336) (8), receptor complex CD94/NKG2C (9), or FCγRIII (CD16) for ADCC (10–12). The counterpart of these activating complexes comprises various inhibitory receptors that usually bind to a variety of different major histocompatibility complex I (MHC I) molecules. Examples for these receptors are several receptors of the killer cell Ig-like receptors (KIRs) family (CD158), NKG2A that pairs with CD94 to a heterodimer (binding the non-classical MHC molecule HLA-E), leukocyte immunoglobulin-like receptor (LILR), natural killer cell receptor protein 1 (CD161), sialic acid-binding immunoglobulin-like lectin-7 (CD328), leukocyte-associated Ig-like receptor 1 (LAIR-1; CD305), killer cell lectin-like receptor G1, carcinoembryonic antigen-related cell adhesion molecule (CD66a), paired immunoglobulin-like receptor α, and CD300a. Each NK cell expresses individually a composition of inhibitory and activating receptors (9). In the resting state, NK cells are in balance receiving signals from activating and inhibitory ligands and no signaling pathway dominates. After adaption to self-MHC I environment, NK cells respond to ligands for activating receptors resulting in killing of malignant cells. Presence of self-MHC I demonstrates inhibitory response. Contrarily, lack of constitutive self-MHC I enhances elimination of aberrant cells (13). At least, NK cell activation by ligand receptor interaction sum up signals received from inhibitory and activation receptors, which cumulates to release perforin and granzymes (cytotoxicity) as well as cytokine production (e.g., IFNγ and TFN-α) mediated by adaptor proteins (DNAX activation proteins DAP10 and DAP12, CD3ζ). These peptides contain immunoreceptor tyrosine-based activation motifs (ITAMs) that become phosphorylated by Src kinase family members and result in at least cytotoxicity and cytokine production.

However, tumors can develop tumor immune escape mechanisms to protect themselves from NK cell attack, e.g., by matrix metalloproteinase-dependent proteolytic cleavage of MHC class I polypeptide-related sequence A and B (MICA and MICB) (14). These soluble immunosuppressive molecules decrease NK cell cytotoxicity by reduction of NKG2D expression that leads to attenuated recognition of target cells. Strategies has been developed to overcome this inhibition using cell modifications such as vector transduction (15) or antibodies bound to the NK cell surface. These bi- and trispecific killer engagers recognize, e.g., CD33 *in vivo* on myelodysplastic syndrome target cells, and induce cell lysis (16). Also, a promising approach is the use of chimeric antigen receptors (CARs) to improve NK cell cytotoxicity. CARs consist of an external recognition domain [single-chain variable fragment (scFv)] combined with a transmembrane domain followed by one or more signaling domains. It has been shown that CARs using CD3ζ and CD28 domains and/or additional 4-1BB (CD137) or 2B4 domains demonstrate an enhanced killing activity (see **Table 1**).

Most published preclinical and clinical studies with CAR-modified immune cells comprise T cells. On the NK cell side, publications are mainly restricted to NK cell lines as reviewed in Ref. (31, 32). Less is known about CAR-engineered primary human NK cells as alternative effector cells since the advantages of NK cells are the limited lifespan of several weeks or months (2, 33) and the absent formation of memory cells that persist in patients as observed in CAR T cells. That means multiple dose of CAR NK cells might be safely administered to patients. The present review will discuss the use of primary NK cells isolated from peripheral blood for CAR engineering.

SIGNAL TRANSDUCTION IN NK CELLS

There is a competitive equilibrium between different opposing pathways (13, 34) that culminate at least in activation or inhibition of NK cells depending on the cell surface complexes that are formed by non-covalent associations between distinct transmembrane ligand-binding and signaling adaptor proteins. The Src (sarcoma) family kinases seem to be essential in these interactions because the enzymes are involved in receptor clustering in these microdomains that may facilitate receptor phosphorylation (35, 36).

Starting with NK-cell–target-cell interactions on the surface, this leads to induction of signaling pathways and at least to release of cytotoxic granules (e.g., perforin, granzyme A/B, and granulysine) and/or secretion of cytokines (e.g., IFNγ and TNF-α).

Activation Receptors

Natural cytotoxicity receptors (NCRs) as NKp30 and NKp46 can couple to CD3ζ that contains several ITAMs (37). NKp44 instead associates with the ITAM-bearing adaptor DAP12. In the next step, tyrosine residues of the ITAM sequences are phosphorylated by protein tyrosine kinases of the Src family. This leads to recruitment of protein tyrosine kinases of the Syk family (e.g., Syk or ZAP70; spleen-associated tyrosine kinase or zeta-chain-associated protein kinase 70) and transmembrane adaptor molecules (e.g., linker for activation of T cells and non-T cell activation linker) that provide multiple docking sites for Syk family kinases. These associations of different signaling partners initialize activation and phosphorylation of multiple partners of signaling pathways such as PI3K (phosphatidylinositol-4,5-bisphosphate 3-kinase) or members of Vav family resulting in release of lytic granules and leading to cytotoxicity (9). The activation of a single NCR seems to start an activation cascade in which different NCRs cross talk to each another for amplifying activating signals (e.g., cross talk between NKp30, NKp44, and NKp46) (38).

Natural killer group 2 member D (CD314) is non-covalent associated with transmembrane adaptor protein DAP10. This pathway is independent of Syk family tyrosine protein kinases (39) and involved PI3K in its signaling cascade. After ligand binding (MICA, MICB, or divers UL16-binding proteins), phosphorylation of a tyrosine-based DAP10 motif by Src family kinases creates binding sites for p85 subunit of PI3K or for the adaptor protein complex Grb2–Vav1 (growth factor receptor-bound protein 2-vav guanine nucleotide exchange factor 1). The result is exocytosis of lytic granules (e.g., perforin, granzyme A/B, and CD107a) in response to PLC (phospholipase C)-γ2-induction (40–43).

For NK cell activation, Vav proteins are essential. Depending of the NCR and of the DNAX proteins, different Vav proteins are involved, e.g., Vav1 is part of the signaling with NKG2D/DAP10 (39–41), whereas Vav2 and Vav3 take part of the DAP12 signaling cascade (44). Vav proteins are involved in a GTPase-dependent reorganization of the cytoskeleton to mediate the directed release of the granules (9).

The NK cell-activating receptor CD226 (DNAX accessory molecule 1) lacks any ITAM. Instead, intracellular signaling starts with phosphorylation of a serine and a tyrosine residue by

TABLE 1 | Preclinical and clinical investigations of CAR-modified primary human natural killer cells.

	Antigen	Signaling domain	Target cells	Efficacy	Reference or ClinicalTrials.gov identifier
Preclinical studies with cell lines as targets	CD19	4-1BB/CD3ζ	Acute lymphatic leukemia cell lines	+++	(15)
	HER-2	CD28/CD3ζ	Ovarian cancer cell line and breast cancer cell line	+	(17)
	Disialoganglioside 2 (GD2)	2B4/CD3ζ	Neuroblastoma cell line	+++	(18)
	CD19	2B4/CD3ζ	ALL cell lines	+++	(18)
	CD19	4-1BB/CD3ζ	B-ALL cell line	+++	(19)
	CD19	4-1BB/CD3ζ	B-ALL cell lines and B cell lymphoma cell lines	++ to +++	(20)
	Natural killer group 2 member D ligands	DAP10/CD3ζ	ALL cell lines and several solid tumor cell lines	+ to +++	(21)
	HER-2	CD28/CD3ζ	HER-2-expressing cell lines	n.a.	(22)
	CD19	4-1BB/CD3ζ	B-ALL cell lines	+	(23)
	CS1	CD28/CD3ζ	Myeloma cell lines	Data not shown	(24)
	CD20	4-1BB/CD3ζ	CD20+ B-cell non-Hodgkin lymphoma cell lines	++ to +++	(25)
	Epidermal growth factor receptor (EGFR)	CD28/CD3ζ	Glioblastoma cell lines	+	(26)
	Prostate stem cell antigen (PSCA)	DAP12	several PSCA+ tumor cells	(+) to +++	(27)
	CD19	CD28/4-1BB/CD3ζ	CD19+ leukemia cell line	+ to +++	(28)
	EGFR	CD28/CD3ζ	Breast cancer cell lines	+	(29)
	GD2	CD28/4-1BB/CD3ζ	Ewing sarcoma cell lines	+ to ++	(30)
Preclinical studies with patient malignant cells as targets	CD19	4-1BB/CD3ζ	Acute lymphatic leukemia	+++	(15)
	CD19	2B4/CD3ζ	Acute lymphatic leukemia	+++	(18)
	CD19	4-1BB/CD3ζ	B-CLL cells	+++	(19)
	CD19	4-1BB/CD3ζ	B-ALL cells	+	(23)
	EGFR	CD28/CD3ζ	Glioblastoma stem cells	(+)	(26)
Clinical trials	CD19	4-1BB/CD3ζ	B-lineage acute lymphoblastic leukemia	n.a.	NCT 00995137
	CD19	4-1BB/CD3ζ	B-lineage acute lymphoblastic leukemia	n.a.	NCT 01974479
	CD19	CD28/CD3ζ	B-lymphoid malignancies	n.a.	NCT 03056339

(+), cytotoxicity <25%; +, cytotoxicity 25–49%, ++, cytotoxicity 50–75%, +++, cytotoxicity >75%.

protein kinase C. This step is critical for association of CD226 to lymphocyte function-associated antigen 1 at the cell surface and facilitates simultaneously cytoplasmic signaling involving Src kinase, Vav1, and PLC-γ2 leading to NK cell activation (45).

The receptor 2B4 has been characterized as costimulatory for activation receptors (e.g., for CD226). A complex formed of CD226 and 2B4 triggers NK cell degranulation, activates PLC-γ2, and increases Ca^{2+} intracellular flux (46). On one hand, 2B4 is sufficient to induce IFNγ release alone (47), on the other hand, 2B4 demonstrated enhanced cytokine secretion after cross linking with NKG2D (48).

Inhibitory Receptors

After binding of MHC class I molecules to inhibitory receptors, the inhibitory signaling cascade starts with phosphorylation of one or more ITIM sequences (immunoreceptor tyrosine-based inhibitory motif). Detailed mechanism of phosphorylation is unknown but tyrosine kinases that are involved in activation pathways have been expected. After tyrosine phosphorylation, the phosphatases Src homology 2 domain-containing protein tyrosine phosphatase 1 (SHP1) and SHP2 bind to ITIMs (49) and subsequently recruit additional molecules such as inhibitory C-terminal Src kinase Crk (for LILR and LAIR-1) or β-arrestin 2 (for KIRs). ITIM-bound SHP starts to dephosphorylize specifically Vav1 or other pivotal proteins to inhibit clustering of receptors and cytoskeleton rearrangements (50).

Killer cell Ig-like receptor and CD94/NKG2A initiate an alternative signaling pathway that also results in inhibition. The binding of the tyrosine kinase c-Abl and the subsequent phosphorylation of adaptor protein Crk (CT10 regulator of kinase) by c-Abl cause dissociation of Crk from protein complexes that are involved in NK cell activation (51, 52). Inhibition of NK cells is achieved and lysis of target cells decreased.

A second MHC class I independent pathway is composed of the inhibitory receptor T cell immunoglobulin and ITIM domain (TIGIT) and the putative weak-activating receptor CD96. For inhibitory signaling of TIGIT, the intracellular motifs immunoglobulin tail tyrosine and/or ITIM are phosphorylated after ligand binding following recruitment of SHP1 and Grb2 that result in blocking the pathways of PI3K and mitogen-activated protein kinase. CD96 contains a cytoplasmic ITIM as well as a YXXM motif that is a putative binding sites for the p85 subunit of PI3K that may lead to NK cell activation. Both receptors, TIGIT and CD96, were found to counterbalance the costimulatory receptor CD226 and limit NK cell-mediated cytotoxicity and IFNγ release (46, 53).

OPTIMIZATION OF CAR SIGNALING IN NK CELLS

Chimeric antigen receptors contain an extracellular region of a scFv that was fused to transmembrane domain and cytoplasmic signaling components. The antibody-derived scFv domain is involved in antigen recognition and immune synapse formation, whereas the endodomains are responsible for cell activation. Because CAR constructs are premised on a modular system, it is feasible to combine any scFv with any signaling or cosignaling domain. First-generation CARs that included only signaling motifs derived from CD3 (ζ or γ chain) (13, 54) were fully capable to activate murine CTL hybridoma cells (55), although no additional intracellular signaling region was added. But some tumors were able to inactivate CAR-engineered cells and leading them to anergy (54). To prevent this effect and to improve CAR functionality, subsequent CAR designs incorporated additional costimulatory domains (CD28, 4-1BB, OX40, and 2B4) and evolved to CARs of the second- (addition of one costimulatory domain) or third-generation (addition of more than one costimulatory domain) (**Figure 1A**).

The motif CD28 is most common in these CAR constructs but is not naturally expressed in human NK cells (56). In T cells, the mode of action of this costimulatory molecule starts with phosphorylation of its intracellular tyrosine residues by PI3K following recruitment of Grb2 and results in activation of protein kinase B (PKB/Akt) and in IL-2 production (57). The advantage for NK cells is still in discussion although for T cells CD28 demonstrates high effectivity (58).

4-1BB is a surface protein discovered on activated T cells (59) that is often used in CAR constructs for NK cells (28). For domain 4-1BB (CD137), costimulation could be clearly detected in T cells (60), but there are conflicting data for NK cells. Navabi et al. demonstrated neither improved NK cell cytotoxicity nor enhanced IFNγ production (61) after NK cell stimulation by 4-1BB ligands in contrast to augmented NK cell-killing capacity as reported in Ref. (62, 63).

The transmembrane adaptor polypeptide DAP10 is originally associated with NKG2D. Comparing the CAR constructs anti-CD19-DAP10 and anti-CD19-CD3ζ, both CARs evoke NK cell cytotoxicity but anti-CD19-CD3ζ exhibited higher antitumor activity than anti-CD19-DAP10 molecules (15). The combination of both signaling domains DAP10 and CD3ζ resulted in secretion of several cytokines (e.g., IFNγ and TNF-α) as well as in a vast release of cytotoxic granules that both increased NK cell cytotoxicity (21).

DAP12 is involved in signal transduction of activated NK cells and is associated with activating receptors such as NKG2C or NKp44. Transmission of intracellular signaling occurs *via* a single ITAM compared to CD3ζ containing three ITAMs (9). Therefore, DAP12 provides an alternative signaling pathway resulting in antitumor activity of NK cells. First investigations assessed DAP12-based CARs in NK cell line YTS (64) as well as in primary human NK cells (27). Combinations of scFv against prostate stem cell antigen (PSCA) with DAP12 exhibit an improved cytotoxicity and increased IFNγ release in primary NK cells compared to CAR NK cells expressing the first-generation CD3ζ-based construct anti-PSCA-CD3ζ (53). This concept without CD3ζ-signaling domain may promise new opportunities to redirect NK cells to resistant target cells.

2B4 (CD244) is a member of the signaling lymphocytic activation molecule family and contains four immunoreceptor tyrosine-based switch motifs (ITSMs) of which the first and second is associated with activation of stimulatory pathways in NK cells (65). Altvater et al. (18) investigated the signaling component 2B4 combined with CD3ζ in primary human NK cells and compared this CAR construct with CAR molecules incorporated either 2B4 or CD3ζ signaling element. As a result, induction of cytokine secretion failed when 2B4 is the sole signaling compound in CAR molecules.

Instead, combination of the domains 2B4 and CD3ζ demonstrated enhanced cytokine secretion (IFNγ and TNF-α) and release of cytolytic granules. In addition, comparable results were observed for a 4-1BB-CD3ζ CAR construct demonstrating equality of 2B4 and 4-1BB signaling domains in combination with CD3ζ.

Similar to 4-1BB, OX40 (CD134) is a TNF receptor on the surface of lymphatic cells (e.g., T cells, NK cells, and NK-like T cells) (66). This costimulatory molecule is involved in recruitment of TNF receptor-associated factor adaptor proteins and leads to cell survival and cytokine release (67, 68). OX40 is often part of third-generation CARs in T cells that show improved signaling capacities based on putative upregulation of PI3K pathway and lead to enhanced cytokine production and cytotoxicity (69), but was not integrated yet in CAR constructs neither for NK cell lines nor for primary NK cells. Because of its costimulatory potential, OX40 may present a promising candidate for improved endogenous CAR signaling in NK cells.

PRECLINICAL INVESTIGATIONS WITH PRIMARY HUMAN CAR NK CELLS

To date, several preclinical studies have been investigated primary human CAR-modified NK cells directed against various antigens (**Table 1**). However, compared to CAR T cells that already entered clinical studies, there is only a small number of clinical investigations using CAR NK cells (**Table 1**).

Most preclinical data describe primary human CAR NK cells directed against CD19 and few against CD20, human epidermal growth factor receptor 2, disialoganglioside 2, epidermal growth factor receptor, and PSCA (references see **Table 1**).

Mostly, second-generation CARs use CD3ζ in combination with 4-1BB, DAP10, or 2B4, respectively, and result in strong efficacy based on upregulation of the PI3K/AKT pathway. By contrast, CD3ζ constructs with CD28 led to less cytotoxicity. High efficacy could also be revealed by third-generation CARs (CD28/4-1BB/CD3ζ) (28, 30) and a DAP12-based first-generation CAR (27). There is a long-standing discussion that costimulatory domain combines best to CD3ζ. For CAR T cells, investigations suggest that constructs containing 4-1BB may be superior (70), but this has not been yet evaluated for CAR NK cells. In addition, so far safety aspects have not been addressed extensively in CAR NK cells and are under discussion.

Although feasibility and efficacy could be shown for all mentioned constructs in **Table 1**, safety aspects have to be clarified in detail in an ongoing discussion.

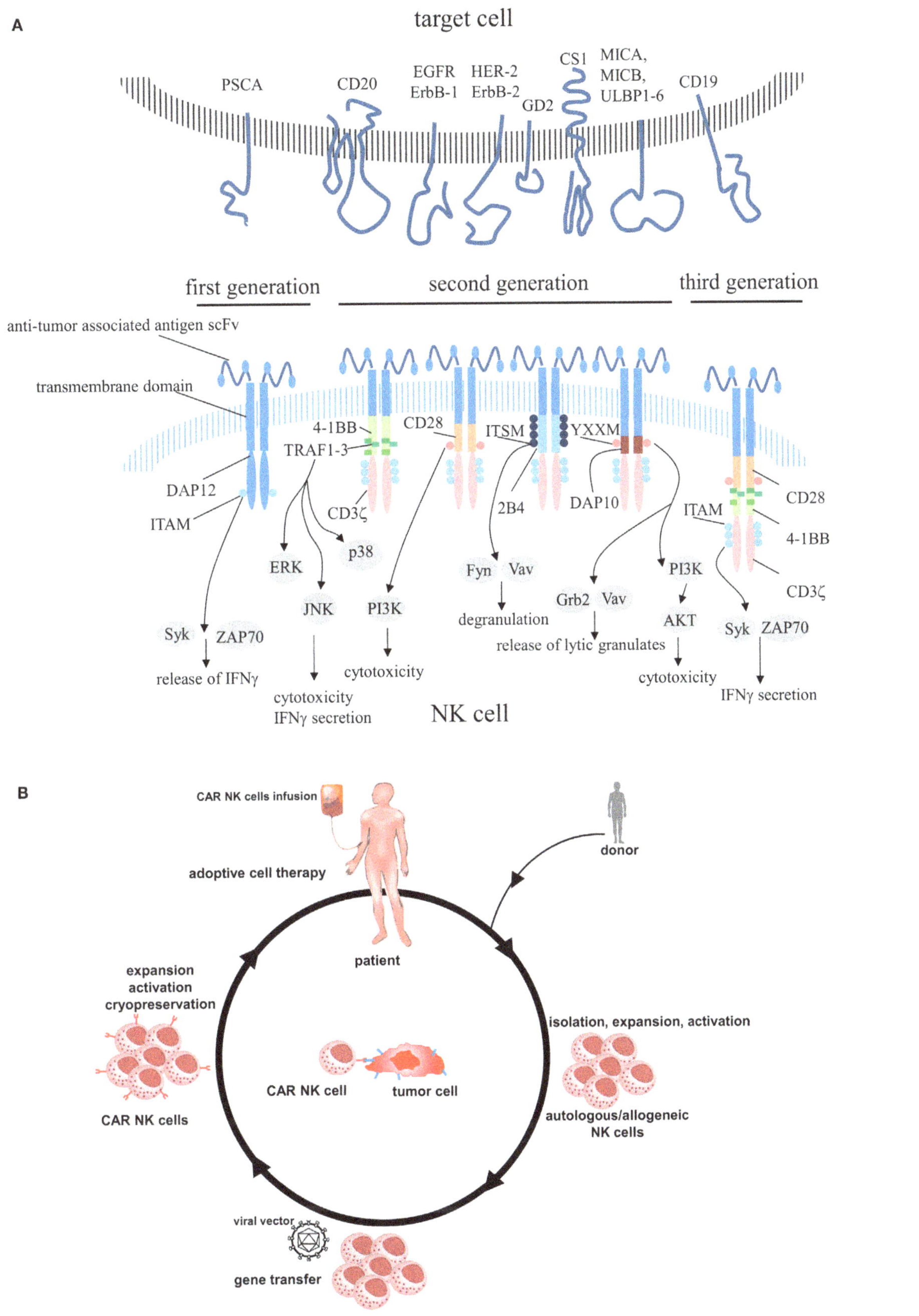

FIGURE 1 | Continued

FIGURE 1 | Continued
(A) Schematic structures of various chimeric antigen receptors applied in engineered primary human NK cells including its intracellular signaling domains. **(B)** CAR NK cell therapy. Autologous NK cells or donor NK cells (allogeneic) are isolated, expanded, and activated by cytokines. After modification of NK cells to express CAR, NK cells are expanded, activated, and administered to the patient or frozen for long-term preservation. PSCA, prostate stem cell antigen; EGFR, epidermal growth factor receptor; HER-2, human epidermal growth factor receptor 2; GD2, disialoganglioside 2; CS1, CD2 subset 1; MICA/B, MHC class I polypeptide-related sequence A/B; ULBP1-6, UL16-binding proteins 1–6; DAP, DNAX-activation protein; ITAM, immunoreceptor tyrosine-based activation motif; Syk, spleen-associated tyrosine kinase; ZAP70, zeta-chain-associated protein kinase 70; TRAF, tumor necrosis factor receptor-associated factor; ERK, extracellular signal-regulated kinase; JNK, c-Jun N-terminal kinase; I3K, phosphatidylinositol-4,5-bisphosphate 3-kinase; ITSM, immunoreceptor tyrosine-based switch motif; Fyn, Src family tyrosine kinase; Vav, vav guanine nucleotide exchange factor; YXXM, phosphorylation motif; Grb2, growth factor receptor-bound protein 2; AKT, protein kinase B.

First clinical studies followed the success of CAR T cell trials redirecting NK cells against CD19. These antiCD19-4-1BB-CD3ζ CAR NK cells were administered to patients with B-ALL (NCT 00995137; NCT 01974479) but results have not been published to date. The first study comprises expansion of donor-derived NK cells cocultured with irradiated and gene-modified K562 cells that expressed surface bound IL-15 and 4-1BB 1. The second trial expands IL-2-activated haploidentical NK cells before administering to pediatric and adult patients. Recently, a third study (NCT 03056339) started for patients suffering from relapsed and/or refractory B-cell lymphoma or leukemia. Genetically engineered NK cells derive from umbilical cord blood (CB) and express antiCD19-CD28-CD3ζ CAR, the iCasp9 safety switch as well as IL-15.

OFF-THE-SHELF (OTS) IMPLICATIONS FOR CANCER TREATMENT

Antigen specificity of CAR NK cells is independent of the recipient's human leukocyte antigen (HLA) type. This feature is the prerequisite for targeting the same antigen on several tumor types even if recipients demonstrate a high variability of HLA. There may be no need any more to customize individual therapies for each patient. Implementation of a cell bank with cryopreserved immune cells that are allogeneic and genetically modified may solve availability and reduce cost of treatment. Developing OTS therapies means that portions of immune cells will be manufactured (and modified) in advance, stored in cryopreservation, and infused on demand as required by attending physicians (**Figure 1B**). It has been shown in several studies that administration of haploidentical NK cells to patients with relapsed acute myelogenous leukemia cause good clinical effects without graft versus host disease (GvHD) as reviewed in Ref. (71). For this reason, it seems to be a successful strategy to set a strong focus on CAR NK cell-based immunotherapies (see **Table 1**), although Shah et al. recently observed GvHD after infusion of *ex vivo* expanded activated allogeneic NK cells (72).

The ideal source for CAR NK cells as OTS products is still in discussion. The cell line NK92 has been described as an option that can be easily transduced and irradiated before administration (73, 74). On the other hand, umbilical CB is well known to be a good source for primary NK cells (75, 76). But limitations as immature phenotype or restriction of NK cell amount should be kept in mind (77, 78), which might be circumvent by refined protocols for primary NK cell *ex vivo* expansion and activation (79) especially in regard to GMP compliance.

A subset of NK cells has been described in mouse and human that demonstrated long-lived capacity for several months (80). These "memory-like" NK cells respond to antigens in second confrontation and show enhanced effector function and expansion. They even may prevent leukemia relapse by a robust cytokine production (81, 82) and may therefore be beneficial in general in long-term antitumor responses. For safety reason, a CAR suicide system should be integrated in CAR NK cells to limit circulation of CAR effector cells in patients (83) and to restrict putative toxic side effects as demonstrated for CAR T cells (84, 85).

STRATEGIES TO OVERCOME OFF-TARGET TOXICITIES

The choice of tumor antigens that can be recognized by CARs depends on the unique and selective character of the antigen for target cancer cells. These regular antigens mainly show increased expression on tumor tissues but are also detectable on normal tissues, often in a minute amount. For this reason, on-target toxicities may appear in clinical studies that have been described for CAR T cells (86–88). On the other hand, off-target toxicity attacks tissues and organs that do not express the antigen but CAR constructs can bind unspecifically. For primary human CAR NK cells, toxicity reports have not been published yet but recognition of specific tumor targets are the base for safe and effective CAR constructs.

To increase selectivity of CAR molecules and reduce putative off-target effects, different strategies have been developed, e.g., combination of two extracellular domains in a tandem structure (89, 90) or of two independent constructs to form bispecific CAR molecules (91). A second concept describes CAR constructs that triggers the release of pro-inflammatory IL-12. The composition of CAR resulted in expression of IL-12 after antigen binding to the extracellular CAR domain (92). Recently, Wu et al. developed a split CAR construct that needs a dimerizing small molecule to form a functional unit. This new strategy promises control of timing, location, and dosage of CAR activity and thereby a possible mitigation of toxicities (93). A similar concept demonstrates the use of an inducible molecular switch off (94). When exposed to a dimerizing drug, the fusion protein iCasp9 is activated and triggers apoptosis in all gene-modified cells. In general, all strategies have been shown for engineered CAR T cells, except the last one that has also been evaluated in the murine model using modified

NK cells (95) and even for primary CB-derived NK cells expressing antiCD19 CAR molecules [unpublished data mentioned in Ref. (96)].

CONCLUSION

In the next years, the possibility of unlimited access to cryopreserved NK cells from CB or third party donors may revolutionize therapy options for cancer patients. Although discussions about best source of NK cells and the question of long-living NK cells have not been finished yet, generation of redirected NK cells against new targets is in rapid progress. Demonstrated results using CAR technologies are auspiciously and may improve cancer therapy also by implemented novel safety strategies. Furthermore, combined immunotherapies using checkpoint blockade monoclonal antibodies to overcome inhibitory signals (e.g., anti-KIR or anti-TIGIT) may enhance CAR NK cell activity.

AUTHOR CONTRIBUTIONS

OO performed the review of the literature and wrote the manuscript. SK and UK edited the manuscript.

REFERENCES

1. Moretta L, Pietra G, Vacca P, Pende D, Moretta F, Bertaina A, et al. Human NK cells: from surface receptors to clinical applications. *Immunol Lett* (2016) 178:15–9. doi:10.1016/j.imlet.2016.05.007
2. Miller JS, Soignier Y, Panoskaltsis-Mortari A, McNearney SA, Yun GH, Fautsch SK, et al. Successful adoptive transfer and in vivo expansion of human haploidentical NK cells in patients with cancer. *Blood* (2005) 105(8):3051–7. doi:10.1182/blood-2004-07-2974
3. Cerwenka A, Lanier LL. Natural killer cells, viruses and cancer. *Nat Rev Immunol* (2001) 1(1):41–9. doi:10.1038/35095564
4. Srivastava S, Lundqvist A, Childs RW. Natural killer cell immunotherapy for cancer: a new hope. *Cytotherapy* (2008) 10(8):775–83. doi:10.1080/14653240802648181
5. Smyth MJ, Cretney E, Kelly JM, Westwood JA, Street SE, Yagita H, et al. Activation of NK cell cytotoxicity. *Mol Immunol* (2005) 42(4):501–10. doi:10.1016/j.molimm.2004.07.034
6. Lanier LL. Up on the tightrope: natural killer cell activation and inhibition. *Nat Immunol* (2008) 9(5):495–502. doi:10.1038/ni1581
7. Becknell B, Caligiuri MA. Natural killer cells in innate immunity and cancer. *J Immunother* (2008) 31(8):685–92. doi:10.1097/CJI.0b013e318182de23
8. Farag SS, Caligiuri MA. Human natural killer cell development and biology. *Blood Rev* (2006) 20(3):123–37. doi:10.1016/j.blre.2005.10.001
9. Watzl C, Long EO. Signal transduction during activation and inhibition of natural killer cells. *Curr Protoc Immunol* (2010) Chapter 11:Unit 11 9B. doi:10.1002/0471142735.im1109bs90
10. Lanier LL. NK cell recognition. *Annu Rev Immunol* (2005) 23:225–74. doi:10.1146/annurev.immunol.23.021704.115526
11. Moretta A, Bottino C, Vitale M, Pende D, Cantoni C, Mingari MC, et al. Activating receptors and coreceptors involved in human natural killer cell-mediated cytolysis. *Annu Rev Immunol* (2001) 19:197–223. doi:10.1146/annurev.immunol.19.1.197
12. Cooper MA, Fehniger TA, Caligiuri MA. The biology of human natural killer-cell subsets. *Trends Immunol* (2001) 22(11):633–40. doi:10.1016/S1471-4906(01)02060-9
13. Long EO, Kim HS, Liu D, Peterson ME, Rajagopalan S. Controlling natural killer cell responses: integration of signals for activation and inhibition. *Annu Rev Immunol* (2013) 31:227–58. doi:10.1146/annurev-immunol-020711-075005
14. Kloess S, Huenecke S, Piechulek D, Esser R, Koch J, Brehm C, et al. IL-2-activated haploidentical NK cells restore NKG2D-mediated NK-cell cytotoxicity in neuroblastoma patients by scavenging of plasma MICA. *Eur J Immunol* (2010) 40(11):3255–67. doi:10.1002/eji.201040568
15. Imai C, Iwamoto S, Campana D. Genetic modification of primary natural killer cells overcomes inhibitory signals and induces specific killing of leukemic cells. *Blood* (2005) 106(1):376–83. doi:10.1182/blood-2004-12-4797
16. Gleason MK, Verneris MR, Todhunter DA, Zhang B, McCullar V, Zhou SX, et al. Bispecific and trispecific killer cell engagers directly activate human NK cells through CD16 signaling and induce cytotoxicity and cytokine production. *Mol Cancer Ther* (2012) 11(12):2674–84. doi:10.1158/1535-7163.MCT-12-0692
17. Kruschinski A, Moosmann A, Poschke I, Norell H, Chmielewski M, Seliger B, et al. Engineering antigen-specific primary human NK cells against HER-2 positive carcinomas. *Proc Natl Acad Sci U S A* (2008) 105(45):17481–6. doi:10.1073/pnas.0804788105
18. Altvater B, Landmeier S, Pscherer S, Temme J, Schweer K, Kailayangiri S, et al. 2B4 (CD244) signaling by recombinant antigen-specific chimeric receptors costimulates natural killer cell activation to leukemia and neuroblastoma cells. *Clin Cancer Res* (2009) 15(15):4857–66. doi:10.1158/1078-0432.CCR-08-2810
19. Li L, Liu LN, Feller S, Allen C, Shivakumar R, Fratantoni J, et al. Expression of chimeric antigen receptors in natural killer cells with a regulatory-compliant non-viral method. *Cancer Gene Ther* (2010) 17(3):147–54. doi:10.1038/cgt.2009.61
20. Shimasaki N, Fujisaki H, Cho D, Masselli M, Lockey T, Eldridge P, et al. A clinically adaptable method to enhance the cytotoxicity of natural killer cells against B-cell malignancies. *Cytotherapy* (2012) 14(7):830–40. doi:10.3109/14653249.2012.671519
21. Chang YH, Connolly J, Shimasaki N, Mimura K, Kono K, Campana D. A chimeric receptor with NKG2D specificity enhances natural killer cell activation and killing of tumor cells. *Cancer Res* (2013) 73(6):1777–86. doi:10.1158/0008-5472.CAN-12-3558
22. Alsamah W, Romia Y. Modification of natural killer cells to target tumors. *Int J Pharm Clin Res* (2014) 6(1):97–100.
23. Cho FN, Chang TH, Shu CW, Ko MC, Liao SK, Wu KH, et al. Enhanced cytotoxicity of natural killer cells following the acquisition of chimeric antigen receptors through trogocytosis. *PLoS One* (2014) 9(10):e109352. doi:10.1371/journal.pone.0109352
24. Chu J, Deng Y, Benson DM, He S, Hughes T, Zhang J, et al. CS1-specific chimeric antigen receptor (CAR)-engineered natural killer cells enhance in vitro and in vivo antitumor activity against human multiple myeloma. *Leukemia* (2014) 28(4):917–27. doi:10.1038/leu.2013.279
25. Chu Y, Hochberg J, Yahr A, Ayello J, van de Ven C, Barth M, et al. Targeting CD20+ aggressive B-cell non-Hodgkin lymphoma by anti-CD20 CAR mRNA-modified expanded natural killer cells in vitro and in NSG mice. *Cancer Immunol Res* (2015) 3(4):333–44. doi:10.1158/2326-6066.CIR-14-0114
26. Han J, Chu J, Keung Chan W, Zhang J, Wang Y, Cohen JB, et al. CAR-engineered NK cells targeting wild-type EGFR and EGFRvIII enhance killing of glioblastoma and patient-derived glioblastoma stem cells. *Sci Rep* (2015) 5:11483. doi:10.1038/srep11483
27. Topfer K, Cartellieri M, Michen S, Wiedemuth R, Muller N, Lindemann D, et al. DAP12-based activating chimeric antigen receptor for NK cell tumor immunotherapy. *J Immunol* (2015) 194(7):3201–12. doi:10.4049/jimmunol.1400330

28. Suerth JD, Morgan MA, Kloess S, Heckl D, Neudorfl C, Falk CS, et al. Efficient generation of gene-modified human natural killer cells via alpharetroviral vectors. *J Mol Med (Berl)* (2016) 94(1):83–93. doi:10.1007/s00109-015-1327-6
29. Chen X, Han J, Chu J, Zhang L, Zhang J, Chen C, et al. A combinational therapy of EGFR-CAR NK cells and oncolytic herpes simplex virus 1 for breast cancer brain metastases. *Oncotarget* (2016) 7(19):27764–77. doi:10.18632/oncotarget.8526
30. Kailayangiri S, Altvater B, Spurny C, Jamitzky S, Schelhaas S, Jacobs AH, et al. Targeting Ewing sarcoma with activated and GD2-specific chimeric antigen receptor-engineered human NK cells induces upregulation of immune-inhibitory HLA-G. *Oncoimmunology* (2017) 6(1):e1250050. doi:10.1080/2162402X.2016.1250050
31. Glienke W, Esser R, Priesner C, Suerth JD, Schambach A, Wels WS, et al. Advantages and applications of CAR-expressing natural killer cells. *Front Pharmacol* (2015) 6:21. doi:10.3389/fphar.2015.00021
32. Klingemann H. Are natural killer cells superior CAR drivers? *Oncoimmunology* (2014) 3:e28147. doi:10.4161/onci.28147
33. Vivier E, Tomasello E, Baratin M, Walzer T, Ugolini S. Functions of natural killer cells. *Nat Immunol* (2008) 9(5):503–10. doi:10.1038/ni1582
34. Rezvani K, Rouce RH. The application of natural killer cell immunotherapy for the treatment of cancer. *Front Immunol* (2015) 6:578. doi:10.3389/fimmu.2015.00578
35. Endt J, McCann FE, Almeida CR, Urlaub D, Leung R, Pende D, et al. Inhibitory receptor signals suppress ligation-induced recruitment of NKG2D to GM1-rich membrane domains at the human NK cell immune synapse. *J Immunol* (2007) 178(9):5606–11. doi:10.4049/jimmunol.178.9.5606
36. Watzl C, Long EO. Natural killer cell inhibitory receptors block actin cytoskeleton-dependent recruitment of 2B4 (CD244) to lipid rafts. *J Exp Med* (2003) 197(1):77–85. doi:10.1084/jem.20020427
37. Moretta L, Bottino C, Pende D, Mingari MC, Biassoni R, Moretta A. Human natural killer cells: their origin, receptors and function. *Eur J Immunol* (2002) 32(5):1205–11. doi:10.1002/1521-4141(200205)32:5<1205::AID-IMMU1205>3.0.CO;2-Y
38. Augugliaro R, Parolini S, Castriconi R, Marcenaro E, Cantoni C, Nanni M, et al. Selective cross-talk among natural cytotoxicity receptors in human natural killer cells. *Eur J Immunol* (2003) 33(5):1235–41. doi:10.1002/eji.200323896
39. Billadeau DD, Upshaw JL, Schoon RA, Dick CJ, Leibson PJ. NKG2D-DAP10 triggers human NK cell-mediated killing via a Syk-independent regulatory pathway. *Nat Immunol* (2003) 4(6):557–64. doi:10.1038/ni929
40. Upshaw JL, Arneson LN, Schoon RA, Dick CJ, Billadeau DD, Leibson PJ. NKG2D-mediated signaling requires a DAP10-bound Grb2-Vav1 intermediate and phosphatidylinositol-3-kinase in human natural killer cells. *Nat Immunol* (2006) 7(5):524–32. doi:10.1038/ni1325
41. Graham DB, Cella M, Giurisato E, Fujikawa K, Miletic AV, Kloeppel T, et al. Vav1 controls DAP10-mediated natural cytotoxicity by regulating actin and microtubule dynamics. *J Immunol* (2006) 177(4):2349–55. doi:10.4049/jimmunol.177.4.2349
42. Caraux A, Kim N, Bell SE, Zompi S, Ranson T, Lesjean-Pottier S, et al. Phospholipase C-gamma2 is essential for NK cell cytotoxicity and innate immunity to malignant and virally infected cells. *Blood* (2006) 107(3):994–1002. doi:10.1182/blood-2005-06-2428
43. Tassi I, Klesney-Tait J, Colonna M. Dissecting natural killer cell activation pathways through analysis of genetic mutations in human and mouse. *Immunol Rev* (2006) 214:92–105. doi:10.1111/j.1600-065X.2006.00463.x
44. Cella M, Fujikawa K, Tassi I, Kim S, Latinis K, Nishi S, et al. Differential requirements for Vav proteins in DAP10- and ITAM-mediated NK cell cytotoxicity. *J Exp Med* (2004) 200(6):817–23. doi:10.1084/jem.20031847
45. de Andrade LF, Smyth MJ, Martinet L. DNAM-1 control of natural killer cells functions through nectin and nectin-like proteins. *Immunol Cell Biol* (2014) 92(3):237–44. doi:10.1038/icb.2013.95
46. Martinet L, Smyth MJ. Balancing natural killer cell activation through paired receptors. *Nat Rev Immunol* (2015) 15(4):243–54. doi:10.1038/nri3799
47. Bryceson YT, March ME, Ljunggren HG, Long EO. Synergy among receptors on resting NK cells for the activation of natural cytotoxicity and cytokine secretion. *Blood* (2006) 107(1):159–66. doi:10.1182/blood-2005-04-1351
48. Fauriat C, Long EO, Ljunggren HG, Bryceson YT. Regulation of human NK-cell cytokine and chemokine production by target cell recognition. *Blood* (2010) 115(11):2167–76. doi:10.1182/blood-2009-08-238469
49. Long EO. Negative signaling by inhibitory receptors: the NK cell paradigm. *Immunol Rev* (2008) 224:70–84. doi:10.1111/j.1600-065X.2008.00660.x
50. Abeyweera TP, Merino E, Huse M. Inhibitory signaling blocks activating receptor clustering and induces cytoskeletal retraction in natural killer cells. *J Cell Biol* (2011) 192(4):675–90. doi:10.1083/jcb.201009135
51. Peterson ME, Long EO. Inhibitory receptor signaling via tyrosine phosphorylation of the adaptor Crk. *Immunity* (2008) 29(4):578–88. doi:10.1016/j.immuni.2008.07.014
52. Liu D, Peterson ME, Long EO. The adaptor protein Crk controls activation and inhibition of natural killer cells. *Immunity* (2012) 36(4):600–11. doi:10.1016/j.immuni.2012.03.007
53. Blake SJ, Dougall WC, Miles JJ, Teng MW, Smyth MJ. Molecular pathways: targeting CD96 and TIGIT for cancer immunotherapy. *Clin Cancer Res* (2016) 22(21):5183–8. doi:10.1158/1078-0432.CCR-16-0933
54. Marcus A, Eshhar Z. Allogeneic chimeric antigen receptor-modified cells for adoptive cell therapy of cancer. *Expert Opin Biol Ther* (2014) 14(7):947–54. doi:10.1517/14712598.2014.900540
55. Eshhar Z, Waks T, Gross G, Schindler DG. Specific activation and targeting of cytotoxic lymphocytes through chimeric single chains consisting of antibody-binding domains and the gamma or zeta subunits of the immunoglobulin and T-cell receptors. *Proc Natl Acad Sci U S A* (1993) 90(2):720–4. doi:10.1073/pnas.90.2.720
56. Lang S, Vujanovic NL, Wollenberg B, Whiteside TL. Absence of B7.1-CD28/CTLA-4-mediated co-stimulation in human NK cells. *Eur J Immunol* (1998) 28(3):780–6. doi:10.1002/(SICI)1521-4141(199803)28:03<780::AID-IMMU780>3.3.CO;2-#
57. Alegre ML, Frauwirth KA, Thompson CB. T-cell regulation by CD28 and CTLA-4. *Nat Rev Immunol* (2001) 1(3):220–8. doi:10.1038/35105024
58. Kowolik CM, Topp MS, Gonzalez S, Pfeiffer T, Olivares S, Gonzalez N, et al. CD28 costimulation provided through a CD19-specific chimeric antigen receptor enhances in vivo persistence and antitumor efficacy of adoptively transferred T cells. *Cancer Res* (2006) 66(22):10995–1004. doi:10.1158/0008-5472.CAN-06-0160
59. Kwon BS, Weissman SM. cDNA sequences of two inducible T-cell genes. *Proc Natl Acad Sci U S A* (1989) 86(6):1963–7. doi:10.1073/pnas.86.6.1963
60. Habib-Agahi M, Phan TT, Searle PF. Co-stimulation with 4-1BB ligand allows extended T-cell proliferation, synergizes with CD80/CD86 and can reactivate anergic T cells. *Int Immunol* (2007) 19(12):1383–94. doi:10.1093/intimm/dxm106
61. Navabi S, Doroudchi M, Tashnizi AH, Habibagahi M. Natural killer cell functional activity after 4-1BB costimulation. *Inflammation* (2015) 38(3):1181–90. doi:10.1007/s10753-014-0082-0
62. Fujisaki H, Kakuda H, Shimasaki N, Imai C, Ma J, Lockey T, et al. Expansion of highly cytotoxic human natural killer cells for cancer cell therapy. *Cancer Res* (2009) 69(9):4010–7. doi:10.1158/0008-5472.CAN-08-3712
63. Zhang H, Cui Y, Voong N, Sabatino M, Stroncek DF, Morisot S, et al. Activating signals dominate inhibitory signals in CD137L/IL-15 activated natural killer cells. *J Immunother* (2011) 34(2):187–95. doi:10.1097/CJI.0b013e31820d2a21
64. Muller N, Michen S, Tietze S, Topfer K, Schulte A, Lamszus K, et al. Engineering NK cells modified with an EGFRvIII-specific chimeric antigen receptor to overexpress CXCR4 improves immunotherapy of CXCL12/SDF-1alpha-secreting glioblastoma. *J Immunother* (2015) 38(5):197–210. doi:10.1097/CJI.0000000000000082
65. Garni-Wagner BA, Purohit A, Mathew PA, Bennett M, Kumar V. A novel function-associated molecule related to non-MHC-restricted cytotoxicity mediated by activated natural killer cells and T cells. *J Immunol* (1993) 151(1):60–70.
66. Croft M. Control of immunity by the TNFR-related molecule OX40 (CD134). *Annu Rev Immunol* (2010) 28:57–78. doi:10.1146/annurev-immunol-030409-101243
67. Croft M. The role of TNF superfamily members in T-cell function and diseases. *Nat Rev Immunol* (2009) 9(4):271–85. doi:10.1038/nri2526
68. Gramaglia I, Jember A, Pippig SD, Weinberg AD, Killeen N, Croft M. The OX40 costimulatory receptor determines the development of CD4 memory by regulating primary clonal expansion. *J Immunol* (2000) 165(6):3043–50. doi:10.4049/jimmunol.165.6.3043
69. Pule MA, Straathof KC, Dotti G, Heslop HE, Rooney CM, Brenner MK. A chimeric T cell antigen receptor that augments cytokine release and supports

clonal expansion of primary human T cells. *Mol Ther* (2005) 12(5):933–41. doi:10.1016/j.ymthe.2005.04.016

70. Milone MC, Fish JD, Carpenito C, Carroll RG, Binder GK, Teachey D, et al. Chimeric receptors containing CD137 signal transduction domains mediate enhanced survival of T cells and increased antileukemic efficacy in vivo. *Mol Ther* (2009) 17(8):1453–64. doi:10.1038/mt.2009.83
71. Koehl U, Kalberer C, Spanholtz J, Lee DA, Miller JS, Cooley S, et al. Advances in clinical NK cell studies: donor selection, manufacturing and quality control. *Oncoimmunology* (2016) 5(4):e1115178. doi:10.1080/2162402X.2015.1115178
72. Shah NN, Baird K, Delbrook CP, Fleisher TA, Kohler ME, Rampertaap S, et al. Acute GVHD in patients receiving IL-15/4-1BBL activated NK cells following T-cell-depleted stem cell transplantation. *Blood* (2015) 125(5):784–92. doi:10.1182/blood-2014-07-592881
73. Suck G, Odendahl M, Nowakowska P, Seidl C, Wels WS, Klingemann HG, et al. NK-92: an 'off-the-shelf therapeutic' for adoptive natural killer cell-based cancer immunotherapy. *Cancer Immunol Immunother* (2016) 65(4):485–92. doi:10.1007/s00262-015-1761-x
74. Suck G, Linn YC, Tonn T. Natural killer cells for therapy of leukemia. *Transfus Med Hemother* (2016) 43(2):89–95. doi:10.1159/000445325
75. Shah N, Martin-Antonio B, Yang H, Ku S, Lee DA, Cooper LJ, et al. Antigen presenting cell-mediated expansion of human umbilical cord blood yields log-scale expansion of natural killer cells with anti-myeloma activity. *PLoS One* (2013) 8(10):e76781. doi:10.1371/journal.pone.0076781
76. Shaim H, Yvon E. Cord blood: a promising source of allogeneic natural killer cells for immunotherapy. *Cytotherapy* (2015) 17(1):1–2. doi:10.1016/j.jcyt.2014.12.001
77. Dalle JH, Menezes J, Wagner E, Blagdon M, Champagne J, Champagne MA, et al. Characterization of cord blood natural killer cells: implications for transplantation and neonatal infections. *Pediatr Res* (2005) 57(5 Pt 1):649–55. doi:10.1203/01.PDR.0000156501.55431.20
78. Fan YY, Yang BY, Wu CY. Phenotypic and functional heterogeneity of natural killer cells from umbilical cord blood mononuclear cells. *Immunol Invest* (2008) 37(1):79–96. doi:10.1080/10826080701690714
79. Alnabhan R, Madrigal A, Saudemont A. Differential activation of cord blood and peripheral blood natural killer cells by cytokines. *Cytotherapy* (2015) 17(1):73–85. doi:10.1016/j.jcyt.2014.08.003
80. Sun JC, Beilke JN, Bezman NA, Lanier LL. Homeostatic proliferation generates long-lived natural killer cells that respond against viral infection. *J Exp Med* (2011) 208(2):357–68. doi:10.1084/jem.20100479
81. Romee R, Rosario M, Berrien-Elliott MM, Wagner JA, Jewell BA, Schappe T, et al. Cytokine-induced memory-like natural killer cells exhibit enhanced responses against myeloid leukemia. *Sci Transl Med* (2016) 8(357):357ra123. doi:10.1126/scitranslmed.aaf2341
82. Jin F, Lin H, Gao S, Wang H, Yan H, Guo J, et al. Characterization of IFNgamma-producing natural killer cells induced by cytomegalovirus reactivation after haploidentical hematopoietic stem cell transplantation. *Oncotarget* (2017) 8(1):51–63. doi:10.18632/oncotarget.13916
83. Jones BS, Lamb LS, Goldman F, Di Stasi A. Improving the safety of cell therapy products by suicide gene transfer. *Front Pharmacol* (2014) 5:254. doi:10.3389/fphar.2014.00254
84. Stauss HJ, Morris EC, Abken H. Cancer gene therapy with T cell receptors and chimeric antigen receptors. *Curr Opin Pharmacol* (2015) 24:113–8. doi:10.1016/j.coph.2015.08.006
85. Maus MV, Grupp SA, Porter DL, June CH. Antibody-modified T cells: CARs take the front seat for hematologic malignancies. *Blood* (2014) 123(17):2625–35. doi:10.1182/blood-2013-11-492231
86. Morgan RA, Yang JC, Kitano M, Dudley ME, Laurencot CM, Rosenberg SA. Case report of a serious adverse event following the administration of T cells transduced with a chimeric antigen receptor recognizing ERBB2. *Mol Ther* (2010) 18(4):843–51. doi:10.1038/mt.2010.24
87. Brentjens R, Yeh R, Bernal Y, Riviere I, Sadelain M. Treatment of chronic lymphocytic leukemia with genetically targeted autologous T cells: case report of an unforeseen adverse event in a phase I clinical trial. *Mol Ther* (2010) 18(4):666–8. doi:10.1038/mt.2010.31
88. Lamers CH, Sleijfer S, van Steenbergen S, van Elzakker P, van Krimpen B, Groot C, et al. Treatment of metastatic renal cell carcinoma with CAIX CAR-engineered T cells: clinical evaluation and management of on-target toxicity. *Mol Ther* (2013) 21(4):904–12. doi:10.1038/mt.2013.17
89. Genssler S, Burger MC, Zhang C, Oelsner S, Mildenberger I, Wagner M, et al. Dual targeting of glioblastoma with chimeric antigen receptor-engineered natural killer cells overcomes heterogeneity of target antigen expression and enhances antitumor activity and survival. *Oncoimmunology* (2016) 5(4):e1119354. doi:10.1080/2162402X.2015.1119354
90. Hegde M, Mukherjee M, Grada Z, Pignata A, Landi D, Navai SA, et al. Tandem CAR T cells targeting HER2 and IL13Ralpha2 mitigate tumor antigen escape. *J Clin Invest* (2016) 126(8):3036–52. doi:10.1172/JCI83416
91. Kloss CC, Condomines M, Cartellieri M, Bachmann M, Sadelain M. Combinatorial antigen recognition with balanced signaling promotes selective tumor eradication by engineered T cells. *Nat Biotechnol* (2013) 31(1):71–5. doi:10.1038/nbt.2459
92. Zhang L, Kerkar SP, Yu Z, Zheng Z, Yang S, Restifo NP, et al. Improving adoptive T cell therapy by targeting and controlling IL-12 expression to the tumor environment. *Mol Ther* (2011) 19(4):751–9. doi:10.1038/mt.2010.313
93. Wu CY, Roybal KT, Puchner EM, Onuffer J, Lim WA. Remote control of therapeutic T cells through a small molecule-gated chimeric receptor. *Science* (2015) 350(6258):aab4077. doi:10.1126/science.aab4077
94. Di Stasi A, Tey SK, Dotti G, Fujita Y, Kennedy-Nasser A, Martinez C, et al. Inducible apoptosis as a safety switch for adoptive cell therapy. *N Engl J Med* (2011) 365(18):1673–83. doi:10.1056/NEJMoa1106152
95. Leboeuf C, Mailly L, Wu T, Bour G, Durand S, Brignon N, et al. In vivo proof of concept of adoptive immunotherapy for hepatocellular carcinoma using allogeneic suicide gene-modified killer cells. *Mol Ther* (2014) 22(3):634–44. doi:10.1038/mt.2013.277
96. Burga RA, Nguyen T, Zulovich J, Madonna S, Ylisastigui L, Fernandes R, et al. Improving efficacy of cancer immunotherapy by genetic modification of natural killer cells. *Cytotherapy* (2016) 18(11):1410–21. doi:10.1016/j.jcyt.2016.05.018

7

In Vivo Efficacy of Umbilical Cord Blood Stem Cell-Derived NK Cells in the Treatment of Metastatic Colorectal Cancer

John P. Veluchamy[1,2†], Silvia Lopez-Lastra[3,4,5†], Jan Spanholtz[2], Fenna Bohme[2], Nina Kok[2], Daniëlle A. M. Heideman[6], Henk M. W. Verheul[1], James P. Di Santo[3,4], Tanja D. de Gruijl[1] and Hans J. van der Vliet[1]*

[1] Department of Medical Oncology, VU University Medical Center, Cancer Center Amsterdam, Amsterdam, Netherlands, [2] Glycostem Therapeutics, Oss, Netherlands, [3] Innate Immunity Unit, Institut Pasteur, Paris, France, [4] Institut National de la Santé et de la Recherche Médicale (INSERM) U1223, Paris, France, [5] Université Paris-Sud (Paris-Saclay), Paris, France, [6] Department of Pathology, VU University Medical Center, Amsterdam, Netherlands

Edited by:
Ulrike Koehl,
Hannover Medical School, Germany

Reviewed by:
Olaf Penack,
Charité – Universitätsmedizin Berlin, Germany
Winfried S. Wels,
Georg-Speyer-Haus Institute for Tumor Biology and Experimental Therapy, Germany

***Correspondence:**
Hans J. van der Vliet
jj.vandervliet@vumc.nl

†These authors have contributed equally to this work.

Therapeutic monoclonal antibodies against the epidermal growth factor receptor (EGFR) act by inhibiting EGFR downstream signaling and by eliciting a natural killer (NK) cell-mediated antitumor response. The IgG_1 mAb cetuximab has been used for treatment of RAS^{wt} metastatic colorectal cancer (mCRC) patients, showing limited efficacy. In the present study, we address the potential of adoptive NK cell therapy to overcome these limitations investigating two allogeneic NK cell products, i.e., allogeneic activated peripheral blood NK cells (A-PBNK) and umbilical cord blood stem cell-derived NK cells (UCB-NK). While cetuximab monotherapy was not effective against $EGFR^-$ RAS^{wt}, $EGFR^+$ RAS^{mut}, and $EGFR^+$ $BRAF^{mut}$ cells, A-PBNK were able to initiate lysis of $EGFR^+$ colon cancer cells irrespective of RAS or BRAF status. Cytotoxic effects of A-PBNK (but not UCB-NK) were further potentiated significantly by coating $EGFR^+$ colon cancer cells with cetuximab. Of note, a significantly higher cytotoxicity was induced by UCB-NK in $EGFR^-RAS^{wt}$ (42 ± 8 versus 67 ± 7%), $EGFR^+$ RAS^{mut} (20 ± 2 versus 37 ± 6%), and $EGFR^+$ $BRAF^{mut}$ (23 ± 3 versus 43 ± 7%) colon cancer cells compared to A-PBNK and equaled the cytotoxic efficacy of the combination of A-PBNK and cetuximab. The antitumor efficacy of UCB-NK cells against cetuximab-resistant human $EGFR^+$ RAS^{mut} colon cancer cells was further confirmed in an *in vivo* preclinical mouse model where UCB-NK showed enhanced antitumor cytotoxicity against colon cancer independent of EGFR and RAS status. As UCB-NK have been proven safe in a recently conducted phase I clinical trial in acute myeloid leukemia, a fast translation into clinical proof of concept for mCRC could be considered.

Keywords: EGFR, RAS mutation, cetuximab, metastatic colorectal cancer, A-PBNK, UCB-NK, allogeneic NK cell immunotherapy

INTRODUCTION

Colorectal cancer (CRC) is the fourth leading cause of cancer-related deaths in the world (1). Despite substantial advances in the treatment of metastatic CRC (mCRC) over the last decades that have contributed to better survival rates (2, 3), the disease is still frequently fatal. Monoclonal antibodies (mAbs) targeting the epidermal growth factor receptor (EGFR) pathway, such as panitumumab and cetuximab, are approved for the treatment of patients with advanced CRC either in combination with chemotherapy or, as monotherapy, in chemorefractory conditions (4). Cetuximab (CET) and panitumumab block the interaction between EGFR and its ligands, thus inhibiting the downstream RAS-signaling cascade and tyrosine kinase activation (5). However, mutations in tumor suppressor genes and proto-oncogenes in EGFR signaling pathways, such as in RAS, BRAF, and PIK3CA, are common in patients with CRC. These mutations represent a poor prognostic marker and render anti-EGFR mAbs ineffective, leaving 42% of the chemorefractory mCRC population without this standard treatment option (6, 7).

Besides the blockade of the EGFR–ligand interaction on tumor cells, therapeutic mAbs can also interact with natural killer (NK) cells triggering antibody-dependent cell-mediated cytotoxicity (ADCC) (8–10), and this can translate into superior antitumor effects (11). Two NK cell subsets can be identified based on the expression of CD16, the low affinity FcγRIIIa receptor. The majority of NK cells are $CD56^{dim}CD16^{+}$ and play an active role in NK cell cytotoxicity and are capable of performing ADCC upon IgG_1 engagement via CD16, whereas $CD56^{bright}CD16^{-}$ NK cells are mainly immune regulatory in function, secreting cytokines, and are less cytotoxic than $CD56^{dim}$ cells (12). NK cell functions are tightly regulated by a delicate balance between activating receptors (like the natural cytotoxicity receptors NKp46, NKp30, and NKp44, or C-type lectin-like receptor NKG2D) (13) and major histocompatibility complex (MHC) class I binding inhibitory receptors, including killer cell immunoglobulin-like receptors (KIRs), LIR1/ILT2, and NKG2A/CD94 (14). The importance of NK cells in controlling tumors has been extensively demonstrated since their identification 40 years ago (15–17).

Several studies have shown a dysfunctional phenotype and poor infiltration of NK cells in the CRC tissue from early stages on, together with an immunosuppressive tumor microenvironment (18, 19). Hence, various strategies, e.g., using cytokines or therapeutic ADCC enhancing mAbs, have been explored to increase NK cell numbers and function and to enhance their trafficking to tumor sites (20). Another approach entails the adoptive transfer of *in vitro* manipulated and expanded autologous or allogeneic NK cells. Autologous NK cells so far have failed to demonstrate significant therapeutic benefits in solid tumors (21–23). Therefore, the focus has shifted to the development of allogeneic NK cells as a potential adoptive cell therapy for treatment in solid tumors. Previously, we demonstrated that the combination of allogeneic activated peripheral blood NK cells (A-PBNK) and CET can effectively target RAS mutant (RAS^{mut}) CRC tumors (24).

Here, we compared two feeder cell-free allogeneic NK cell products, i.e., A-PBNK and umbilical cord blood stem-cell derived NK cells (UCB-NK), alone or in combination with cetuximab for antitumor effects against RAS^{mut} CRC.

MATERIALS AND METHODS

Cell Lines

Cell lines A431 (epidermoid carcinoma), COLO320, SW480, and HT-29 (colon carcinoma) were obtained from American Type Culture Collection and cultured in Dulbecco's modified medium (DMEM; Invitrogen, Carlsbad, CA, USA) containing 100 U/ml penicillin, 100 µg/ml streptomycin, and 10% fetal calf serum (FCS; Integro, Zaandam, The Netherlands). Cell cultures were passaged every 5 days and maintained in a 37°C, 95% humidity, 5% CO_2 incubator.

PBNK Isolation and Activation

Peripheral blood mononuclear cells (PBMCs) were isolated from the heparinized blood of healthy donors (six males, four females, age range = 56–64 years and CRC patients (eight males, two females, age range = 66–74 years) after written informed consent and according to protocols approved by the institutional review board of VU University Medical Center, Amsterdam (NCT01792934). Blood samples were collected at baseline and after the first cycle of first-line palliative chemotherapy consisting of oral capecitabine (1,000 mg/m^2, bid, days 1–14), i.v. oxaliplatin (130 mg/m^2, day 1), and i.v. bevacizumab (7.5 mg/kg, day 1, in 4/10 mCRC patients). PBMCs were isolated using Lymphoprep™ (STEMCELL Technologies, Cologne, Germany) density gradient centrifugation. $CD56^{+}$ NK cells were isolated from PBMC using a MACS Human NK cell isolation kit (Miltenyi Biotec, Bergisch Gladbach, Germany) according to the manufacturer's instructions. PBNK cells purity and viability were checked using CD3 VioBlue, CD56 APC Vio 770, and CD16 APC (Miltenyi Biotech) and 7-AAD (Sigma Aldrich, Zwijndrecht, The Netherlands). Isolated PBNK cells were activated overnight with 1,000 U/ml IL-2 (Proleukin®; Chiron, München, Germany) and 10 ng/ml IL-15 (CellGenix) for use in cytotoxicity assays. The parameters compared before and after stimulation with cytokines were NK purity (87 ± 5 versus $84 \pm 2\%$), NK $CD16^{+}$, (92 ± 12 versus $88 \pm 8\%$) and NK viability (89 ± 5 versus $84 \pm 8\%$), respectively.

Flow Cytometry

The antibody staining mix for the assessment of NK cell functionality consisted of CD45 VioGreen, CD14 VioBlue, CD19 VioBlue, and SYTOX® Blue, together with CD3 PerCP-Vio 700 and TCRγδ PerCP-Vio700 to exclude dead cells, debris, and non-NK populations from PBMCs. NK cells were identified by the expression of $CD45^{+}CD3^{-}CD56^{+}$ cells, and further characterized for NK functionality by plotting against CD16 APC, CD25 VioBrightFITC, CD107a PE, and NKp44 PE-Vio770 and for NK cell phenotype by plotting against NKG2A PE-Vio770, NKG2C PE, NKG2D PerCP-Cy5.5, and PanKIR2D FITC. All antibodies were supplied by Miltenyi Biotec except SYTOX® Blue (Thermo Fisher Scientific, Berlin, Germany).

UCB-NK Cultures

Allogeneic NK cells (UCB-NK) were generated from cryopreserved umbilical cord blood (UCB) hematopoietic stem cells as previously described (25). $CD34^{+}$ UCB cells from six UCB-donors were plated (4×10^5/ml) into 12-well tissue culture plates (Corning Incorporated, Corning, New York, NY, USA) in Glycostem Basal Growth Medium (GBGM®) (Clear Cell Technologies, Beernem, Belgium) supplemented with 10% human serum (HS; Sanquin Bloodbank, Amsterdam, The Netherlands), 25 ng/ml of SCF, Flt-3L, TPO, and IL-7 (CellGenix, Freiburg, Germany). In the expansion phase II, from day 9 to day 14, TPO was replaced with 20 ng/ml IL-15 (CellGenix). During the first 14 days of culture, low molecular weight heparin (Clivarin®; Abbott, Wiesbaden, Germany) in a final concentration of 20 µg/ml and a low-dose cytokine cocktail consisting of 10 pg/ml GM-CSF (Neupogen), 250 pg/ml G-CSF and 50 pg/ml IL-6 (CellGenix) were added to the expansion cultures. Cells were refreshed with new medium twice a week and maintained at 37°C, 5% CO_2. On day 14, the NK cell differentiation process was initiated by addition of NK cell differentiation medium consisting of the same basal medium with 2% HS but with high-dose cytokine cocktail consisting of 20 ng/ml of IL-7, SCF, IL-15 (CellGenix), and 1,000 U/ml IL-2 (Proleukin®; Chiron, München, Germany). Cultures were refreshed every 2–3 days and maintained till day 42. Five UCB-NK cultures were used for cytotoxicity assays and one UCB-NK culture for *in vivo* studies (both with a $CD56^+$ cell purity of >95%). UCB-NK CD16 levels in matured UCB-NK cells were monitored using an antibody mix of human CD45VioGreen (1:11), CD56 APC-Vio770 (1:11), and CD16 APC (1:11). Similarly, UCB-NK CD16 expression in BRGS mice was monitored using an antibody mix of BV650 anti-mouse CD45 (clone 30-F11), Alexa Fluor® 700 anti-human CD45 (clone HI30), PE-CF594 anti-human CD56 (clone B159), all from BD, and APC-Vio770 anti-human CD56 (clone REA196) and APC CD16 (clone REA423) both from Miltenyi Biotec.

NK Cell Cytotoxicity Assays

Flow cytometry was used for the readout of cytotoxicity assays. Target cells (COLO320, SW480, and HT-29) were labeled with 5 µM pacific blue succinimidyl ester (PBSE; Molecular Probes Europe, Leiden, The Netherlands) at a concentration of 1×10^7 cells/ml for 10 min at 37°C. The reaction was terminated by adding an equal volume of FCS, followed by incubation at room temperature for 5 min after which stained cells were washed twice and suspended in DMEM + 10% FCS to a final concentration of 5×10^5/ml. Overnight activated PBNK cells and UCB-NK cells were washed with PBS and suspended in GBGM + 2% FCS to a final concentration of 5×10^5/ml. Target cells were cocultured with effector cells at an E:T ratio of 1:1 in a total volume of 250 µl in 96-well flat-bottom plates (5×10^4 targets in 100 µl of DMEM + 10% FCS incubated with 5×10^4 effectors in 100 µl of GBGM + 2% FCS, further supplemented with 25 µl of GBGM + 2% FCS and DMEM + 10% FCS medium). NK cells and target cells alone were plated out in triplicate as negative controls. Target cells were coated with 5 µg/ml cetuximab (Merck, Darmstadt, Germany) for 1 h at 4°C. To measure degranulation of NK cells, anti-CD107a PE (Miltenyi Biotech) was added in 1:20 dilution at the beginning of the assay. After incubation for 4 h at 37°C, cells were harvested and stained with CD56 APC Vio 770 (1:25) and CD16 APC (1:25) (Miltenyi Biotech) and 7-AAD (1:500) (Sigma Aldrich). Degranulation of NK cells was measured by detecting cell surface expression of CD107a.

In Vivo Studies

The $EGFR^+RAS^{mut}$ SW480 cell line and $EGFR^{+++}RAS^{wt}$ A431 cell line were stably transduced with Gaussia Luciferase (Gluc) for *in vivo* studies. Lentiviral (LV) supernatant of Cerulean Fluorescent Protein (CFP)-positive Gluc virus (LV-CFP-Gluc) was kindly provided by Dr. Tom Würdinger (26). SW480 and A431 cells with Gluc expression of 95% were used for mouse studies.

Immunodeficient BRGS mice (BALB/c *Rag2*^{tm1Fwa} *Il2rg*^{tm1Cgn} *Sirpa*^{NOD}) were used in this study. Twenty-four adult mice (male, 8 weeks old) received an intravenous (i.v.) tail vein injection with 0.5×10^6 SW480 Gluc cells at day 0 and were randomized into four groups. Group A only received SW480 cells, group B received SW480 in combination with cetuximab intraperitoneally (i.p., 0.5 mg, days 1, 4, and 7), group C received SW480 in combination with UCB-NK i.v. (1×10^7, days 1, 4, and 7), and group D received SW480 cells in combination with UCB-NK i.v. (1×10^7, days 1, 4, and 7) and cetuximab i.p. (0.5 mg, days 1, 4, and 7). Groups C and D received i.p. 0.5 µg IL-15 + 7.5 µg IL-15Rα every 2–3 days from day 0 till day 14. Further, three adult mice received i.v. tail vein injection of 0.5×10^6 A431 Gluc cells at day 0 and were treated with 0.5 mg cetuximab (i.p., 0.5 mg, days 1, 4, and 7), which were used as a cetuximab efficacy control. Treatment effects were monitored using blood Gluc levels and bioluminescence imaging (BLI). All manipulations of BRGS mice were performed under laminar flow conditions.

Blood Gluc Quantification *In Vitro*

Secreted Gluc was measured according to a protocol described previously (27). A total of 10 µl of blood was collected by capillarity into EDTA containing Microvette® CB tubes. Blood samples were distributed in 96-well black plates then mixed with 100 µl of 100 mM Gluc substrate native coelenterazine in PBS (P.J.K. GmbH, Kleinblittersdorf, Germany), and 5 min later, light emission was quantified. Blood that was withdrawn before tumor inoculation served to determine a baseline value. Measurements were done twice a week until day 35. Gluc activity was measured using IVIS spectrum luminescence detector (PerkinElmer, Villebon-sur-Yvette, France). Data obtained were quantified using Living Image 4.0 software (PerkinElmer, Villebon-sur-Yvette, France).

BLI *In Vivo*

Mice were anesthetized using isofluorane gas in an induction chamber at a gas flow of 2.5 pm. Retro-orbital injection of coelenterazine (4 mg/kg body weight) was administered and mice were placed in the anesthesia manifold inside the imaging chamber and imaged within 5 min following substrate injection. Mice were placed into the light chamber and overlay images were collected for a period of 15 min using IVIS spectrum *in vivo* imaging system (PerkinElmer, Villebon-sur-Yvette, France). Images were then analyzed using Living Image 4.0 software (PerkinElmer, Villebon-sur-Yvette, France).

Ethics Statement

Animals were housed in isolators under pathogen-free conditions with humane care and anesthesia was performed using inhalational isoflurane anesthesia to minimize suffering. Experiments were approved by the Institut Pasteur's ethical committee for animal use in research, Comité d'étique en expérimentation animale (CETEA) #89, protocol reference # 2007–006 and validated by the French Ministry of Education and Research (Reference # 02162.01).

Statistical Analysis

Data were analyzed using GraphPad Prism version 6 (GraphPad Software, San Diego, CA). Differences between conditions were determined using one-way ANOVA or two-way ANOVA with multiple comparisons between column means, unpaired *t* test and log-rank (Mantel–Cox) test as deemed appropriate. A *P*-value of <0.05 was considered statistically significant.

RESULTS

Highly Dysfunctional NK Cells in CRC Patients

Flow cytometry was used to determine the frequency, phenotype, and functionality of NK cells in PBMC of healthy volunteers ($n = 10$, age range 56–64 years, 6 males/4 females) and patients with metastatic CRC ($n = 10$, age range 66–74 years, 8 males/2 females) before and after the first cycle of first-line palliative chemotherapy consisting of oral capecitabine (1,000 mg/m^2, bid, days 1–14), i.v. oxaliplatin (130 mg/m^2, day 1), and i.v. bevacizumab (7.5 mg/kg, day 1, in 4/10 mCRC patients). As illustrated in **Figure 1A**, mCRC patients harbored on average a 20% lower percentage of CD3$^-$CD56$^+$NK cells in the total CD45$^+$ lymphocyte population as compared to healthy controls ($P < 0.05$). These lower NK rates, which are in line with a previous report in CRC (28), further declined after the first cycle of chemotherapy ($P < 0.01$).

We next evaluated whether this quantitative NK cell defect was also accompanied by functional defects in the NK cell population. For this purpose, the ability of NK cells from healthy volunteers and mCRC patients to induce both natural cytotoxicity and mediate ADCC of the epidermoid carcinoma cell line A431 (MHC-I^{low}, EGFR^{high}, KRAS^{wt}) was assessed. For ADCC, tumor target cells were coated with cetuximab before the addition of NK cells. It was evident that the cytotoxic potential of NK cells from mCRC patients, as reflected by degranulation (i.e., CD107a surface expression), was highly impaired both before chemotherapy and after the first cycle of chemotherapy. Though NK cells of mCRC patients were capable of ADCC, as evidenced by significant increases in degranulation when target cells were coated with cetuximab ($P < 0.05$), levels were still low compared to those observed in healthy volunteers (**Figure 1B**). Of note, although the NK cells of healthy volunteers and mCRC patients expressed similar levels of CD16 (**Figure 1C**), this did not translate into comparable levels of ADCC. NKp44 expression, known to reflect the activation status of NK cells, was similar between the HD and mCRC groups used in NK cytotoxicity experiments (**Figure 1D**). Furthermore, no significant differences were observed in expression levels of NK activating (NKG2D, NKG2C) and NK inhibiting

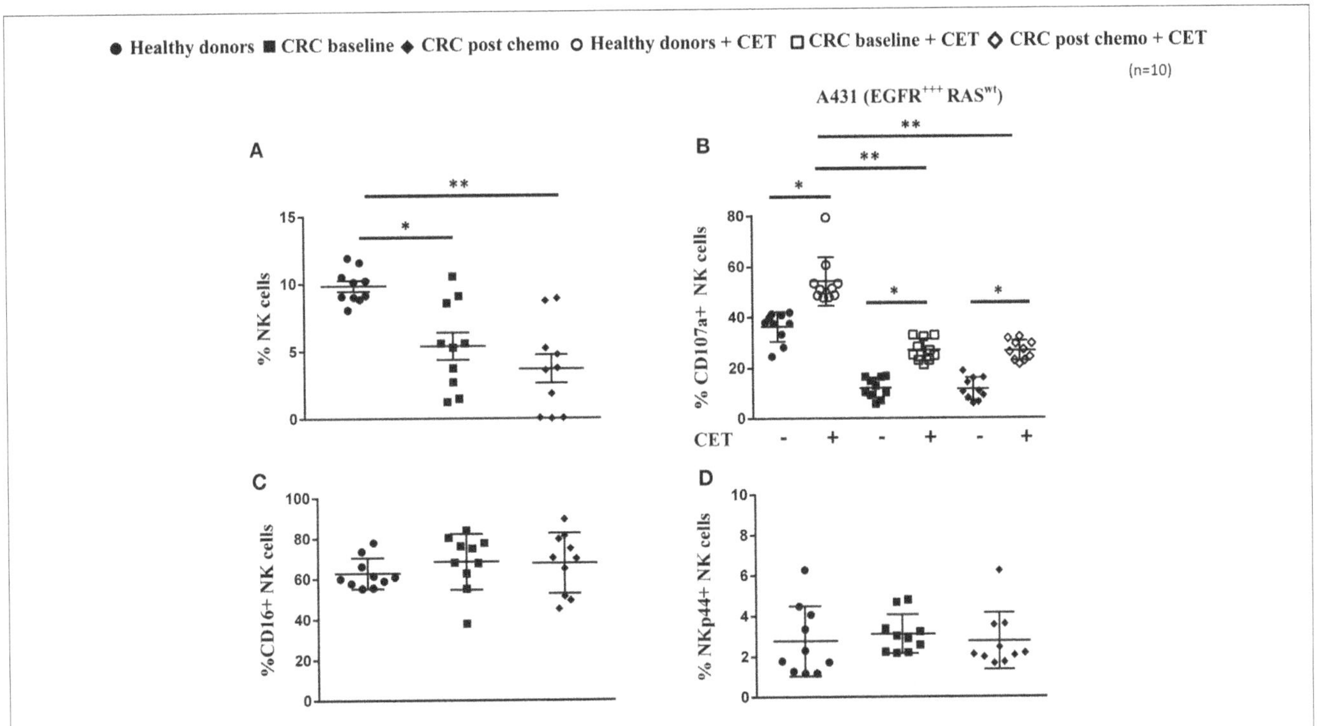

FIGURE 1 | Low prevalence and functionally impaired natural killer (NK) cells in colorectal cancer (CRC) patients. **(A)** Frequency of NK cells within peripheral blood mononuclear cells from healthy controls and from metastatic CRC (mCRC) patients at baseline and after the first cycle of chemotherapy. **(B)** NK cell degranulation in healthy controls and mCRC patients after a 4-h coculture of resting NK cells with A431 cells in the presence (open symbols) or absence (closed symbols) of cetuximab at an E:T ratio of 1:1. **(C)** Expression levels of resting NK cell CD16 and **(D)** NKp44 in healthy controls and in mCRC patients before and after one cycle of chemotherapy. Data represent mean ± SEM from 10 mCRC patients and 10 age- and sex-matched healthy controls. **P* < 0.05, ***P* < 0.01, ****P* < 0.005, calculated with one-way ANOVA, multiple comparison between column means.

(NKG2A, KIR2D) receptors between healthy controls and CRC patients (Figure S1 in Supplementary Material).

Enhanced *In Vitro* Cytotoxicity of Colon Cancer Cells Mediated by UCB-NK Cells

In order to explore novel therapies to replace dysfunctional NK cells in patients with advanced CRC, we tested two different sources of allogeneic NK cell products (A-PBNK and UCB-NK) that could eventually be used for adoptive transfer strategies. We next compared the activity of A-PBNK cells (age range 22–37 years) and UCB-NK cells using a flow based NK cell cytotoxicity assay based on detection of 7-AAD accumulation in tumor cells. Three different cell lines of colon cancer origin were compared, i.e., COLO320 ($EGFR^- RAS^{wt}$), SW480 ($EGFR^+ RAS^{mut}$), and HT-29 ($EGFR^+ RAS^{wt} BRAF^{mut}$). As expected, addition of cetuximab to $EGFR^- RAS^{wt}$ COLO320 cells did not result in increased killing. Of interest, lysis was consistently and significantly higher ($P < 0.01$) using UCB-NK compared to A-PBNK. As reported previously, the combination of cetuximab and A-PBNK resulted in increased killing of $EGFR^+RAS^{mut}$ SW480 and $EGFR^+ BRAF^{mut}$ HT-29 via ADCC (24). CD16 was expressed by 88 ± 8% ($n = 5$) of A-PBNK after overnight stimulation with cytokines and by 7 ± 2% ($n = 5$) of UCB-NK cells at the end of the 35-day culture period. No added effect of cetuximab was observed when using UCB-NK cells, which is possibly related to their lower *in vitro* CD16 levels (29). Of note, tumor cell lysis induced by UCB-NK cells was comparable to that observed with the combination of A-PBNK and cetuximab (**Figures 2A,B,C**). Measurements of NK cell degranulation reflected equivalent trends observed for tumor cell lysis (**Figures 2D,E,F**). These results show that UCB-NK cells have superior cytotoxic efficacy over A-PBNK cells against cetuximab-resistant colon cancer cells *in vitro*.

UCB-NK Cells Inhibit *In Vivo* Tumor Growth and Increase Survival

To address whether UCB-NK cells exhibit similar antitumor efficacy *in vivo*, we transferred Gluc transduced SW480 cells to immunodeficient mice (BRGS; see Materials and Methods). SW480 cells are $EGFR^+RAS^{mut}$ and cetuximab monotherapy resistant. Mice were divided into four groups of six mice per group: SW480 only (group A), SW480 + cetuximab (group B), SW480 + UCB-NK (group C), and SW480 + UCB-NK + cetuximab (group D). Gaussia luciferase activity in whole blood was measured every 3 days to monitor the tumor burden (Figure S2 in Supplementary Material). These data confirmed our *in vitro* observations that SW480 cells were resistant to cetuximab-mediated growth inhibition (blue line). Of note, while treatment with UCB-NK cells alone significantly decreased the tumor load (green line), this effect was not increased by combining UCB-NK cells with cetuximab and thereby further confirmed both the

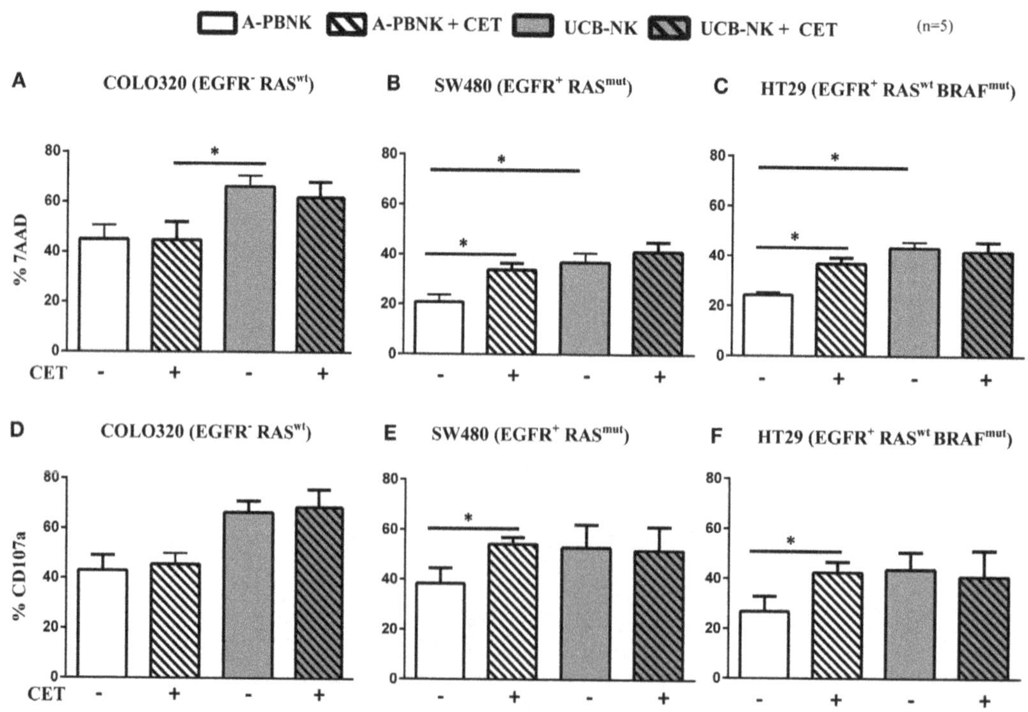

FIGURE 2 | ***Ex vivo*** **cytotoxic efficacy of A-PBNK and UCB-NK cells against colorectal cancer (CRC) cells**. CRC cell lines COLO320 ($EGFR^-$, RAS^{wt}), SW480 ($EGFR^+$, RAS^{mut}), and HT-29 ($EGFR^+$, RAS^{wt}, $BRAF^{mut}$) were subjected to natural killer (NK) killing using two allogeneic NK cell products, i.e., A-PBNK and UCB-NK cells. 7-AAD **(A,B,C)** and CD107a **(D,E,F)** were measured after a 4-h coculture of A-PBNK and UCB-NK cells with CRC targets in the presence or absence of cetuximab at an E:T ratio of 1:1. Experiments were carried out in triplicate. Bars represent mean ± SEM, $n = 5$. *$P < 0.05$ and **$P < 0.01$, calculated with two-way ANOVA, multiple comparison between column means.

inefficacy of cetuximab in treating RAS mutated tumors as well as the inability of cetuximab to induce ADCC of UCB-NK cells *in vivo* (orange line) (**Figure 3**). CD16 expression levels on UCB-NK cells were monitored in two mice upon adoptive transfer and increased from 6.0% before transfer to 14.0% (mouse 1) and 19.1% (mouse 2) at day 5 post UCB-NK cell infusion (data not shown).

While the blood Gluc assay measurements provided evidence of a reduction in the total tumor burden after UCB-NK treatment, we wanted to explore the impact of the therapy on the localization and size of the metastases. For that purpose, BLI was performed at day 35 after tumor inoculation. **Figure 4A** depicts four representative BLI images from each group at day 35 posttumor injection and average radiance from range of interest measurements are shown in **Figure 4B**. It is clear that mice from groups A and B showed a higher and more diffuse tumor load compared to mice treated with UCB-NK alone or in combination with cetuximab. In order to demonstrate the possibility of antitumor efficacy of cetuximab in the BRGS mouse model, we performed a similar tumor challenge using the cetuximab-sensitive A431 cell line, which bears wild-type RAS and overexpresses EGFR. A significant decrease in tumor load was observed when A431 tumors were treated with the same concentration of cetuximab as in the SW480 study (**Figure 4C**), confirming the *in vivo* functionality of cetuximab. We next assessed whether treatment of SW480 bearing mice with UCB-NK cells alone or in combination with cetuximab translated into a survival advantage (**Figure 5**). Indeed, treatment of mice with UCB-NK cells alone resulted in a significant prolongation in their life span ($P = 0.01$), whereas combinatorial therapy did not add significantly to this. Treatment with cetuximab alone did not translate into a significant survival advantage, consistent with the observed effects on tumor growth.

DISCUSSION

In order to test the cytotoxic potential of NK cells for treating advanced CRC patients, we compared their functional status before and after chemotherapy. We observed that peripheral blood NK cell numbers were reduced in mCRC patients and that residual NK cells were dysfunctional and unable to mount a strong effector response when stimulated with an NK cell sensitive tumor target. Though an increase in NK cell cytotoxicity was observed when tumor target cells were coated with the anti-EGFR mAb cetuximab, reflecting a capacity for ADCC, cytotoxicity was still significantly lower (both before and after chemotherapy) than that observed in healthy controls. These data indicate a decreased functional state of NK cells in patients with mCRC, which is in line with studies in mice where the cytokine production and antitumor activity of adoptively transferred NK cells were highly affected following long-term exposure to tumors (30). Through recognition of MHC class I molecules, KIRs prevent NK cells from targeting healthy cells while allowing them to detect tumor or

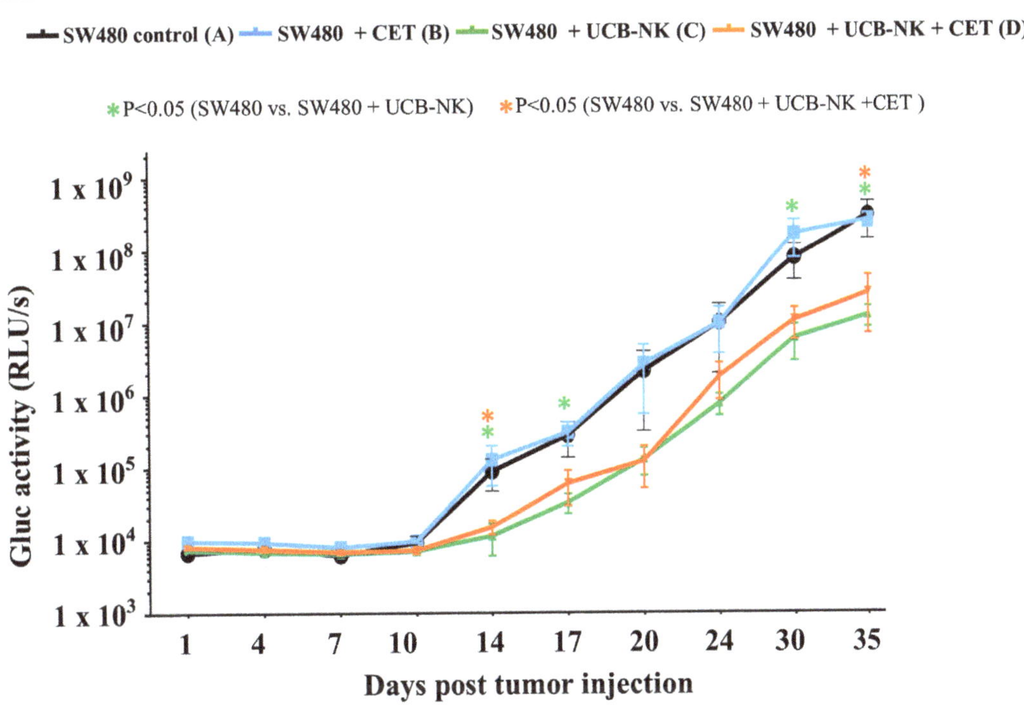

FIGURE 3 | **Significant antitumor effects of UCB-NK cells *in vivo*.** Real-time monitoring of tumor progression and treatment response was performed measuring Gluc levels from mice blood twice a week. Baseline Gluc values were obtained from all mice a day before tumor injection (day 1), and further monitoring continued until day 35. Blood Gluc levels were compared between control SW480 only (A) group and treatment groups SW480 + cetuximab (B), SW480 + UCB-NK (C), and SW480 + UCB-NK + cetuximab (D) for statistical significance. Data presented is from six mice per group ($n = 6$). Scatter plots represent mean ± SEM. *$P < 0.05$, calculated with unpaired t test.

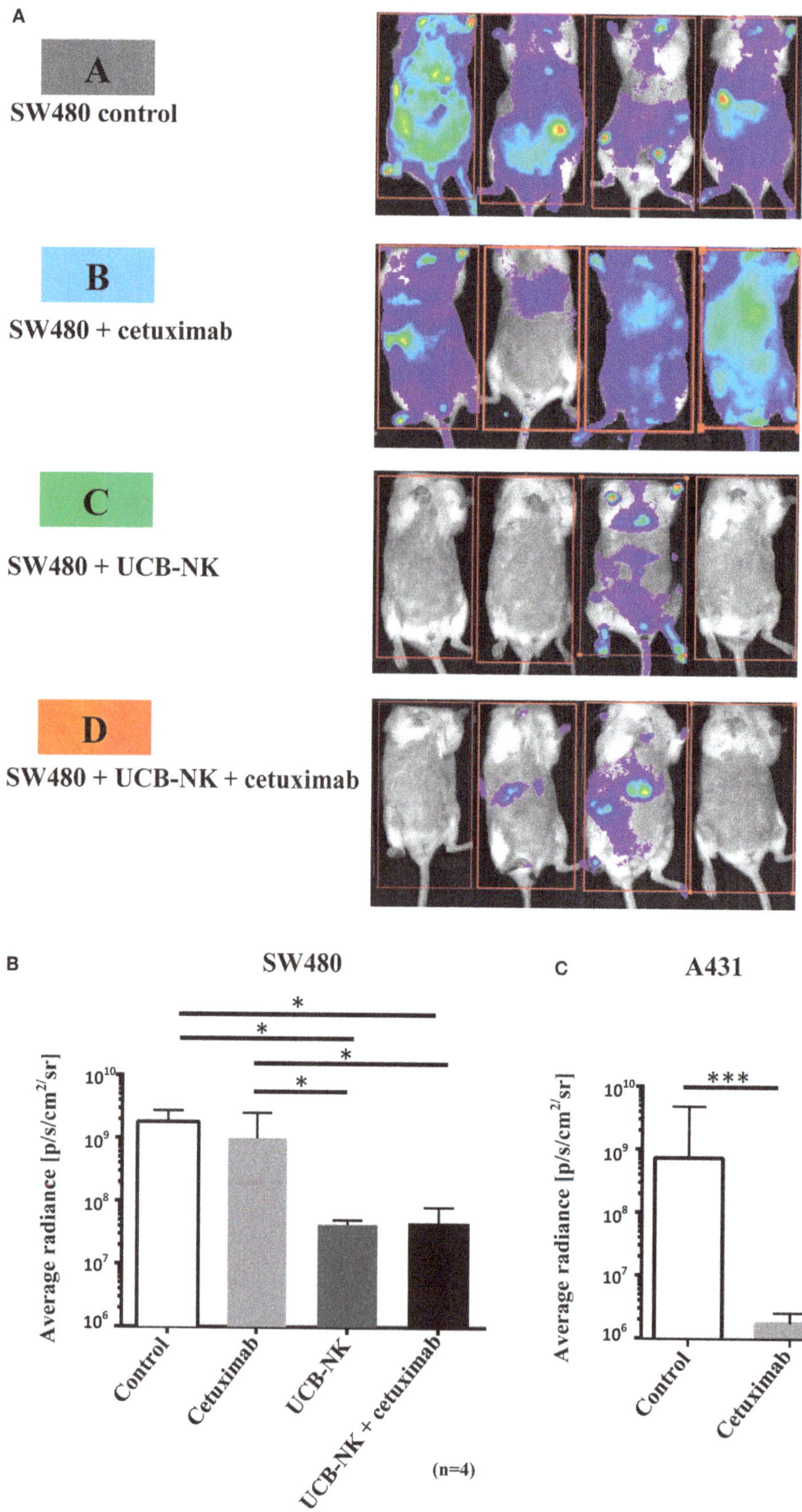

FIGURE 4 | Continued

FIGURE 4 | Continued
Successful tumor elimination by UCB-NK cells as revealed by bioluminescence imaging *in vivo*. (A) Four mice from control and treatment groups were imaged at day 35 for tumor load and distribution. Mice were injected retro-orbitally with Gluc substrate coelenterazine and images were acquired for 5 min. In SW480 control and SW480 + cetuximab groups, tumor growth was extensive and highly disseminated, spreading to most parts of the body. However, in UCB-NK and UCB-NK + cetuximab groups, there was a significantly lower tumor load, which was further verified by calculating the average radiance between groups as shown in panel **(B)** ($n = 4$ mice per group). **(C)** Cetuximab functionality against $EGFR^{+++}$ RAS^{wt} A431 cells was tested in parallel to SW480 studies in BRGS mice ($n = 3$ mice per group). For panels **(B,C)**, bars represent mean ± SEM. *$P < 0.05$ for panels **(B,C)** was calculated with unpaired *t* test.

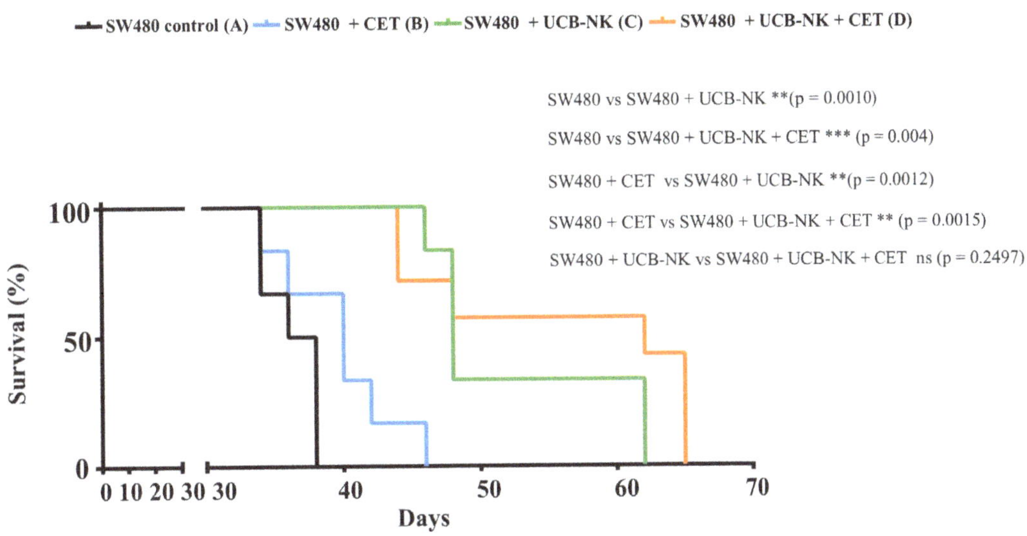

Median survival (days)			
SW480 control	SW480 + CET	SW480 + UCB-NK	SW480 + UCB-NK + CET
37	40	48	62

FIGURE 5 | Significant survival benefit in cetuximab-resistant RAS mutant tumor bearing mice treated with UCB-NK cells. Kaplan–Meier survival curves were plotted for the total experimental study period from day 0 until day 65. Survival rates of SW480 ($EGFR^{+}$, RAS^{mut}) tumor-bearing mice ($n = 6$ per group) following treatment with PBS only (black line), cetuximab only (blue line), UCB-NK only (green line), and UCB-NK + cetuximab (orange line) were plotted over time to monitor treatment outcome. Statistical differences between groups were calculated using log-rank (Mantel–Cox) test and indicated in the figure.

infected cells with low or downregulated expression of MHC class I in a process known as "missing self" (31). Severely diminished or aberrant expression of MHC class I has been reported in the majority of colorectal adenocarcinomas (32, 33), which makes them an ideal target for NK cell-mediated killing. Although NK cells are infrequent in colorectal tissues (18), several independent studies investigated the clinical impact of NK cell infiltration on the prognosis of CRC, as well as in other types of carcinoma. These clinical studies, including a recent tissue microarray of 1,414 CRC biopsies, led to the conclusion that NK cell infiltration in tumors correlated with better overall response rates and progression-free survival in CRC patients (34–37), suggesting that therapies aimed at boosting NK cell functions could be beneficial in mCRC and possibly also in other types of cancer.

We evaluated and compared the cytotoxic efficacy of two different sources of feeder cell free allogeneic NK cells, i.e., A-PBNK cells and *in vitro* expanded and differentiated UCB-NK cells. *In vitro* NK cell cytotoxicity experiments revealed that the cytotoxic activity of UCB-NK cells against CRC cells was significantly higher than that of A-PBNK cells and in addition demonstrated that, while an increase in cytotoxicity through ADCC was not evident with UCB-NK cells, their cytotoxic potential was still comparable to that observed with A-PBNK potentiated by cetuximab-mediated ADCC. It is possible that the stronger cytotoxic effects of UCB-NK cells result from a more intense stimulation with cytokines in comparison to A-PBNK cells. The failure to observe ADCC-enhanced cytotoxicity with UCB-NK cells *in vitro* can be explained by their low expression levels of CD16 (29). As we previously observed *in vivo* upregulation of CD16 on UCB-NK cells upon their transfer to NOD/SCID/IL2Rgnull (NSG) mice (38), we decided to also test the efficacy of cetuximab treatment in combination with UCB-NK cells in an *in vivo* model. Treatment of SW480 RAS^{mut} tumors in BRGS mice with UCB-NK cells resulted in control of disease progression and translated into a significantly longer survival. As expected, cetuximab monotherapy did not result in a decreased SW480 tumor load or improvement in survival, recapitulating the clinical data from patients bearing RAS^{mut} CRC tumors. Unexpectedly, we failed to demonstrate superior *in vivo* antitumor effects or survival when we combined the transfer of UCB-NK cells with cetuximab infusions. The underlying causes for this latter finding remain obscure but may be related to suboptimal *in vivo* upregulation of CD16 in the used mouse model or CD16 polymorphisms in the employed batch of UCB-NK cells, both of which could have hampered efficient ADCC.

Taken together, UCB-NK cells displayed significant antitumor efficacy, suggesting a potential beneficial role for UCB-NK cells in the treatment of RAS and BRAF mutant CRC. As an important present limitation in treating mCRC patients is related to resistance to anti-EGFR mAbs, adoptive transfer of cytolytic UCB-NK cells could thus constitute a viable treatment option. Our *in vitro* and *in vivo* data demonstrating that adoptive transfer of UCB-NK cells alone was as effective as the combination of A-PBNK and cetuximab raises the possibility that UCB-NK administration could obviate the use of cetuximab in RAS^{wt} mCRC. Furthermore, UCB-NK can also lyse RAS^{mut} CRC cells at levels higher than those observed with A-PBNK. Importantly, allogeneic NK cells have demonstrated their safety in clinical trials in several solid tumors (39, 40), and more specifically, the UCB-NK cell product used in our experiments was found to be safe in a clinical trial in acute myeloid leukemia (AML) patients (Dolstra et al., 2016 manuscript submitted).

Several features make UCB-NK attractive for further clinical development. For example, our GMP-based expansion and differentiation protocol reproducibly resulted in a more than 10,000-fold expansion of cytotoxic UCB-NK cells from single donors. Furthermore, UCB-NK cells can be supplied as an "off the shelf" product, stored in large aliquots facilitating multiple infusions. Also, the low immunogenicity by UCB grafts prevents adverse reactions that are prevalent after repeated PBNK transfusions (41). In this respect, it is relevant to mention that while NK cells in general are often inhibited by recognition of MHC class I molecules on the surface of tumor cells, UCB-NK display relatively low levels of KIRs supporting their ability to effectively lyse MHC class I-expressing tumor cells (29). Finally, the ability of UCB-NK cells to proliferate and home to liver, lungs, spleen, and bone marrow after adoptive transfer has been previously demonstrated in NSG mice (38), though additional studies are required to determine whether UCB-NK cells have a similar migratory pattern upon adoptive transfer in solid tumor patients. Together, these features and observations provide UCB-NK cells with several unique advantages for further development as a universal NK cell platform.

Considering the size and heterogeneity of the tumor mass in advanced stages of CRC and other types of cancer, UCB-NK may not provide a sufficient therapeutic effect as a single agent. However, rational combinations of UCB-NK cells with existing drugs or drugs that are in clinical development can be envisioned to further increase their efficacy. Previous studies have pointed out that the proteasome inhibitor (bortezomib) (42) and the immunomodulatory drug (lenalidomide) (43) sensitize tumor cells to NK-mediated killing. In addition, UCB-NK cell application together with bispecific or trispecific antibodies that bind to tumor and UCB-NK cell-activating receptors can also increase NK cell tumor specificity (44). Though we did not specifically assess ADCC induced by other mAbs, it is very likely that the failure of UCB-NK to mediate ADCC is a more general phenomenon as this depends on binding to CD16/FcγRIII, which was found to be expressed at only low levels in the UCB-NK cell product. However, recent data from a clinical phase 1 study with the same UCB-NK cell product in patients with AML revealed significant upregulation of CD16 on UCB-NK cells post transfusion suggesting that the UCB-NK cell product may acquire the capacity to mediate ADCC in patients following adoptive transfer (Dolstra et al., manuscript submitted). Further, this phenomenon may also provide a strong rationale for combining UCB-NK cells with bispecific or trispecific killer cell engagers (45). Taken together, these approaches can substantially increase UCB-NK cell responses to advanced solid tumors, including mCRC.

In conclusion, in this study, we have demonstrated the *in vitro* efficacy of UCB-NK cells against multiple CRC cell lines independent of EGFR expression and EGFR downstream signaling mutations, and in addition have demonstrated the *in vivo* antitumor efficacy of adoptively transferred UCB-NK cells against $EGFR^{+}RAS^{mut}$ tumors. As the adoptive transfer of UCB-NK cells (oNKord®) has been shown to be safe in patients with AML (CCMO no. NL31699 and Dutch trial register no 2818), our data provide a rationale for the clinical exploration of UCB-NK cells in the treatment of mCRC.

ETHICS STATEMENT

Approval for human subjects use: this study was carried out in accordance with the recommendation of the institutional review board of VU University Medical Center, Amsterdam (NCT01792934) with written informed consent from all subjects. All subjects gave written informed consent in accordance with the Declaration of Helsinki. The protocol was approved by the Committee for Scientific research of the VU University Medical Center, Cancer Center Amsterdam. Approval for animal subjects use: this study was carried out in accordance with the recommendations of the ethical committee at the Institut Pasteur (Reference # 2007-006). The protocol was approved by the French Ministry of Education and Research (Reference # 02162.01).

AUTHOR CONTRIBUTIONS

Conceived and designed the experiments: JV, SL-L, JPD, TG, HVV, JS, and HV. Performed the experiments: JV, SL-L, NK, and FB. Analyzed the data: JV and SL-L. Contributed reagents/materials/analysis tools: DH. Wrote the paper: JV and SL-L.

ACKNOWLEDGMENTS

The authors are grateful to M. A. Nicola (Imagopole, Institut Pasteur) for her support with the mouse imaging studies. The authors like to thank all the healthy volunteers and CRC patients who participated in this study.

REFERENCES

1. Siegel RL, Miller KD, Jemal A. Cancer statistics, 2016. *CA Cancer J Clin* (2016) 66(1):7–30. doi:10.3322/caac.21332
2. Fakih MG. Metastatic colorectal cancer: current state and future directions. *J Clin Oncol* (2015) 33(16):1809–24. doi:10.1200/jco.2014.59.7633
3. Dozois EJ, Boardman LA, Suwanthanma W, Limburg PJ, Cima RR, Bakken JL, et al. Young-onset colorectal cancer in patients with no known genetic predisposition: can we increase early recognition and improve outcome? *Medicine* (2008) 87(5):259–63. doi:10.1097/MD.0b013e3181881354
4. Petrelli F, Borgonovo K, Cabiddu M, Ghilardi M, Barni S. Cetuximab and panitumumab in KRAS wild-type colorectal cancer: a meta-analysis. *Int J Colorectal Dis* (2011) 26(7):823–33. doi:10.1007/s00384-011-1149-0
5. Graham J, Muhsin M, Kirkpatrick P. Cetuximab. *Nat Rev Drug Discov* (2004) 3(7):549–50. doi:10.1038/nrd1445
6. Karapetis C, Khambata-Ford S, Jonker D, O'Callaghan C, Tu D, Tebbutt N, et al. K-ras mutations and benefit from cetuximab in advanced colorectal cancer. *N Engl J Med* (2008) 359(17):1757–65. doi:10.1056/NEJMoa0804385
7. Sorich MJ, Wiese MD, Rowland A, Kichenadasse G, McKinnon RA, Karapetis CS. Extended RAS mutations and anti-EGFR monoclonal antibody survival benefit in metastatic colorectal cancer: a meta-analysis of randomized, controlled trials. *Ann Oncol* (2015) 26(1):13–21. doi:10.1093/annonc/mdu378
8. Garnock-Jones KP, Keating GM, Scott LJ. Trastuzumab: a review of its use as adjuvant treatment in human epidermal growth factor receptor 2 (HER2)-positive early breast cancer. *Drugs* (2010) 70(2):215–39. doi:10.2165/11203700-000000000-00000
9. Garcia-Foncillas J, Diaz-Rubio E. Progress in metastatic colorectal cancer: growing role of cetuximab to optimize clinical outcome. *Clin Transl Oncol* (2010) 12(8):533–42. doi:10.1007/s12094-010-0551-3
10. Winter MC, Hancock BW. Ten years of rituximab in NHL. *Expert Opin Drug Saf* (2009) 8(2):223–35. doi:10.1517/14740330902750114
11. Wang W, Erbe AK, Hank JA, Morris ZS, Sondel PM. NK cell-mediated antibody-dependent cellular cytotoxicity in cancer immunotherapy. *Front Immunol* (2015) 6:368. doi:10.3389/fimmu.2015.00368
12. Vivier E, Tomasello E, Baratin M, Walzer T, Ugolini S. Functions of natural killer cells. *Nat Immunol* (2008) 9(5):503–10. doi:10.1038/ni1582
13. Moretta A, Bottino C, Vitale M, Pende D, Cantoni C, Mingari MC, et al. Activating receptors and coreceptors involved in human natural killer cell-mediated cytolysis. *Annu Rev Immunol* (2001) 19:197–223. doi:10.1146/annurev.immunol.19.1.197
14. Lanier LL. NK cell receptors. *Annu Rev Immunol* (1998) 16:359–93. doi:10.1146/annurev.immunol.16.1.359
15. Herberman RB, Nunn ME, Lavrin DH. Natural cytotoxic reactivity of mouse lymphoid cells against syngeneic acid allogeneic tumors. I. Distribution of reactivity and specificity. *Int J Cancer* (1975) 16(2):216–29. doi:10.1002/ijc.2910160205
16. Cheng M, Chen Y, Xiao W, Sun R, Tian Z. NK cell-based immunotherapy for malignant diseases. *Cell Mol Immunol* (2013) 10(3):230–52. doi:10.1038/cmi.2013.10
17. Hayakawa Y, Smyth MJ. Innate immune recognition and suppression of tumors. *Adv Cancer Res* (2006) 95:293–322. doi:10.1016/S0065-230X(06)95008-8
18. Halama N, Braun M, Kahlert C, Spille A, Quack C, Rahbari N, et al. Natural killer cells are scarce in colorectal carcinoma tissue despite high levels of chemokines and cytokines. *Clin Cancer Res* (2011) 17(4):678–89. doi:10.1158/1078-0432.CCR-10-2173
19. Larsen SK, Gao Y, Basse PH. NK cells in the tumor microenvironment. *Crit Rev Oncog* (2014) 19(0):91–105. doi:10.1615/CritRevOncog.2014011142
20. Terme M, Ullrich E, Delahaye NF, Chaput N, Zitvogel L. Natural killer cell-directed therapies: moving from unexpected results to successful strategies. *Nat Immunol* (2008) 9(5):486–94. doi:10.1038/ni1580
21. Parkhurst MR, Riley JP, Dudley ME, Rosenberg SA. Adoptive transfer of autologous natural killer cells leads to high levels of circulating natural killer cells but does not mediate tumor regression. *Clin Cancer Res* (2011) 17(19):6287–97. doi:10.1158/1078-0432.CCR-11-1347
22. deMagalhaes-Silverman M, Donnenberg A, Lembersky B, Elder E, Lister J, Rybka W, et al. Posttransplant adoptive immunotherapy with activated natural killer cells in patients with metastatic breast cancer. *J Immunother* (2000) 23(1):154–60. doi:10.1097/00002371-200001000-00018
23. Ishikawa E, Tsuboi K, Saijo K, Harada H, Takano S, Nose T, et al. Autologous natural killer cell therapy for human recurrent malignant glioma. *Anticancer Res* (2004) 24(3b):1861–71.
24. Veluchamy JP, Spanholtz J, Tordoir M, Thijssen VL, Heideman DAM, Verheul HMW, et al. Combination of NK cells and cetuximab to enhance anti-tumor responses in RAS mutant metastatic colorectal cancer. *PLoS One* (2016) 11(6):e0157830. doi:10.1371/journal.pone.0157830
25. Spanholtz J, Tordoir M, Eissens D, Preijers F, van der Meer A, Joosten I, et al. High log-scale expansion of functional human natural killer cells from umbilical cord blood CD34-positive cells for adoptive cancer immunotherapy. *PLoS One* (2010) 5(2):e9221. doi:10.1371/journal.pone.0009221
26. Wurdinger T, Badr C, Tannous B. Gaussia Luciferase Blood Level as an Index of Cell Growth and Proliferation. *Nat Protoc Exchange* (2008). doi:10.1038/nprot.2008.26
27. Tannous BA. Gaussia luciferase reporter assay for monitoring biological processes in culture and in vivo. *Nat Protoc* (2009) 4(4):582–91. doi:10.1038/nprot.2009.28
28. Rafique M, Adachi W, Koike S, Kajikawa S, Yazawa K, Sugenoya A, et al. Adverse effects of intraportal chemotherapy on natural killer cell activity in colorectal cancer patients. *J Surg Oncol* (1997) 64(4):324–30. doi:10.1002/(SICI)1096-9098(199704)64:4<324::AID-JSO14>3.0.CO;2-2
29. Lehmann D, Spanholtz J, Osl M, Tordoir M, Lipnik K, Bilban M, et al. Ex vivo generated natural killer cells acquire typical natural killer receptors and display a cytotoxic gene expression profile similar to peripheral blood natural killer cells. *Stem Cells Dev* (2012) 21(16):2926–38. doi:10.1089/scd.2011.0659
30. Gill S, Vasey AE, De Souza A, Baker J, Smith AT, Kohrt HE, et al. Rapid development of exhaustion and down-regulation of eomesodermin limit the antitumor activity of adoptively transferred murine natural killer cells. *Blood* (2012) 119(24):5758–68. doi:10.1182/blood-2012-03-415364
31. Ljunggren HG, Karre K. In search of the 'missing self': MHC molecules and NK cell recognition. *Immunol Today* (1990) 11(7):237–44. doi:10.1016/0167-5699(90)90097-S
32. Sandel MH, Speetjens FM, Menon AG, Albertsson PA, Basse PH, Hokland M, et al. Natural killer cells infiltrating colorectal cancer and MHC class I expression. *Mol Immunol* (2005) 42(4):541–6. doi:10.1016/j.molimm.2004.07.039
33. Maleno I, Aptsiauri N, Cabrera T, Gallego A, Paschen A, Lopez-Nevot MA, et al. Frequent loss of heterozygosity in the beta2-microglobulin region of chromosome 15 in primary human tumors. *Immunogenetics* (2011) 63(2):65–71. doi:10.1007/s00251-010-0494-4
34. Coca S, Perez-Piqueras J, Martinez D, Colmenarejo A, Saez MA, Vallejo C, et al. The prognostic significance of intratumoral natural killer cells in patients with colorectal carcinoma. *Cancer* (1997) 79(12):2320–8. doi:10.1002/(SICI)1097-0142(19970615)79:12<2320::AID-CNCR5>3.0.CO;2-P
35. Peng Y-P, Zhu Y, Zhang J-J, Xu Z-K, Qian Z-Y, Dai C-C, et al. Comprehensive analysis of the percentage of surface receptors and cytotoxic granules positive natural killer cells in patients with pancreatic cancer, gastric cancer, and colorectal cancer. *J Transl Med* (2013) 11(1):1–10. doi:10.1186/1479-5876-11-262
36. Villegas FR, Coca S, Villarrubia VG, Jimenez R, Chillon MJ, Jareno J, et al. Prognostic significance of tumor infiltrating natural killer cells subset CD57 in patients with squamous cell lung cancer. *Lung cancer* (2002) 35(1):23–8. doi:10.1016/S0169-5002(01)00292-6
37. Ishigami S, Natsugoe S, Tokuda K, Nakajo A, Che X, Iwashige H, et al. Prognostic value of intratumoral natural killer cells in gastric carcinoma. *Cancer* (2000) 88(3):577–83. doi:10.1002/(SICI)1097-0142(20000201)88:3<577::AID-CNCR13>3.3.CO;2-M
38. Cany J, van der Waart AB, Tordoir M, Franssen GM, Hangalapura BN, de Vries J, et al. Natural killer cells generated from cord blood hematopoietic progenitor cells efficiently target bone marrow-residing human leukemia cells in NOD/SCID/IL2Rgnull mice. *PLoS One* (2013) 8(6):e64384. doi:10.1371/journal.pone.0064384
39. Miller JS, Soignier Y, Panoskaltsis-Mortari A, McNearney SA, Yun GH, Fautsch SK, et al. Successful adoptive transfer and in vivo expansion of human haploidentical NK cells in patients with cancer. *Blood* (2005) 105(8):3051–7. doi:10.1182/blood-2004-07-2974

40. Geller MA, Cooley S, Judson PL, Ghebre R, Carson LF, Argenta PA, et al. A phase II study of allogeneic natural killer cell therapy to treat patients with recurrent ovarian and breast cancer. *Cytotherapy* (2011) 13(1):98–107. doi:10.3109/14653249.2010.515582
41. Gluckman E, Rocha V, Boyer-Chammard A, Locatelli F, Arcese W, Pasquini R, et al. Outcome of cord-blood transplantation from related and unrelated donors. *New Engl J Med* (1997) 337(6):373–81. doi:10.1056/NEJM199708073370602
42. Armeanu S, Krusch M, Baltz KM, Weiss TS, Smirnow I, Steinle A, et al. Direct and natural killer cell-mediated antitumor effects of low-dose bortezomib in hepatocellular carcinoma. *Clin Cancer Res* (2008) 14(11):3520–8. doi:10.1158/1078-0432.CCR-07-4744
43. Wu L, Adams M, Carter T, Chen R, Muller G, Stirling D, et al. Lenalidomide enhances natural killer cell and monocyte-mediated antibody-dependent cellular cytotoxicity of rituximab-treated CD20+ tumor cells. *Clin Cancer Res* (2008) 14(14):4650–7. doi:10.1158/1078-0432.ccr-07-4405
44. Wu J, Fu J, Zhang M, Liu D. AFM13: a first-in-class tetravalent bispecific anti-CD30/CD16A antibody for NK cell-mediated immunotherapy. *J Hematol Oncol* (2015) 8:96. doi:10.1186/s13045-015-0188-3
45. Gleason MK, Verneris MR, Todhunter DA, Zhang B, McCullar V, Zhou SX, et al. Bispecific and trispecific killer cell engagers directly activate human NK cells through CD16 signaling and induce cytotoxicity and cytokine production. *Mol Cancer Ther* (2012) 11(12):2674–84. doi:10.1158/1535-7163.mct-12-0692

8

Triplebody Mediates Increased Anti-Leukemic Reactivity of IL-2 Activated Donor Natural Killer (NK) Cells and Impairs Viability of their CD33-Expressing NK Subset

Stephan Kloess[1], Alessa Ede Valverde da Silva[1], Olaf Oberschmidt[1], Tanja Gardlowski[1], Nadine Matthies[1], Maulik Vyas[2], Lubomir Arseniev[1], Michael Heuser[3], Elke Pogge von Strandmann[4] and Ulrike Köhl[1]*

[1] Institute for Cellular Therapeutics, IFB-Tx, Hannover Medical School (MHH), Hannover, Germany, [2] Department I of Internal Medicine, University Hospital of Cologne, Cologne, Germany, [3] Department of Hematology, Hemostasis, Oncology, and Stem Cell Transplantation, Hannover Medical School (MHH), Hannover, Germany, [4] Experimental Tumor Research, Center for Tumor Biology and Immunology, Philipps University Marburg, Marburg, Germany

Edited by:
Hermann Einsele, University of Würzburg, Germany

Reviewed by:
Alejandro Madrigal, Anthony Nolan, United Kingdom
Marion Subklewe, Ludwig-Maximilians-Universität München, Germany

***Correspondence:**
Stephan Kloess
kloess.stephan@mh-hannover.de

Natural killer cells (NK) are essential for the elimination of resistant acute myeloid and acute lymphoblastic leukemia (AML and ALL) cells. NK cell-based immunotherapies have already successfully entered for clinical trials, but limitations due to immune escape mechanisms were identified. Therefore, we extended our established NK cell protocol by integration of the previously investigated powerful trispecific immunoligand ULBP2-aCD19-aCD33 [the so-called triplebodies (TBs)] to improve the anti-leukemic specificity of activated NK cells. IL-2-driven expansion led to strongly elevated natural killer group 2 member D (NKG2D) expressions on donor NK cells which promote the binding to ULBP2$^+$ TBs. Similarly, CD33 expression on these NK cells could be detected. Dual-specific targeting and elimination were investigated against the B-cell precursor leukemia cell line BV-173 and patient blasts, which were positive for myeloid marker CD33 and B lymphoid marker CD19 exclusively presented on biphenotypic B/myeloid leukemia's. Cytotoxicity assays demonstrated improved killing properties of NK cells pre-coated with TBs compared to untreated controls. Specific NKG2D blocking on those NK cells in response to TBs diminished this killing activity. On the contrary, the observed upregulation of surface CD33 on about 28.0% of the NK cells decreased their viability in response to TBs during cytotoxic interaction of effector and target cells. Similar side effects were also detected against CD33$^+$ T- and CD19$^+$ B-cells. Very preliminary proof of principle results showed promising effects using NK cells and TBs against primary leukemic cells. In summary, we demonstrated a promising strategy for redirecting primary human NK cells in response to TBs against leukemia, which may lead to a future progress in NK cell-based immunotherapies.

Keywords: natural killer cells, natural killer group 2 member D, triplebodies, ULBP2, CD19, CD33, immunoligands, acute myeloid leukemia

INTRODUCTION

Natural killer (NK) cells are a subset of lymphoid effector cells within the innate immune response and have been shown to be a suitable tool for adoptive immunotherapy because of their ability of anti-tumor surveillance (1–6). In contrast to T cells, NK cells identify and eliminate malignant and virus-infected target cells in a major histocompatibility complex (MHC)-unrestricted way by engaging natural cytotoxicity receptors (NCRs), such as NKp30, NKp44, NKp46, and the activating receptor NKG2D (natural killer group 2 member D) which recognizes a variety of well-defined ligands expressed by transformed cells (7–9).

Major histocompatibility complex class I-related chain A and B (MICA/B) and the UL-16 binding protein family are cancer cell surface ligands that interact with NKG2D on NK cells. These specific bindings between NKG2D and their corresponding ligands (NKG2DL) on cancer cells are responsible for improved cytotoxic properties of NK cells against tumor and leukemia cells (10–12). Based on these receptor–ligand bindings between effector and target cells, an increased secretion of granzyme A (GraA) and B (GraB), granulysin, and perforin induced in NK cells could be demonstrated (13–15).

However, several types of cancer have developed a broad spectrum of immune escape mechanisms that down-modulate the NKG2D-mediated immune surveillance by metalloproteinases-driven proteolytic shedding and release of soluble NKG2DLs (16–20). In addition, elevated DNA-“hyper”-methylations for NKG2DLs could be detected in some malignant cells, mainly in acute myeloid leukemia (AML) cells, resulted in a clearance of NKG2DL surface cell expression, also detected for MICA, ULBP1/2 in AML patients (21–23). Enhanced tumor-shedding and DNA-methylation could contribute to an unhampered proliferation and evasion of immune control in AML patients (21).

Previous reports indicated that human leukocyte antigen (HLA) class I diversities could be responsible for induction of NK cell alloreactions by KIR (killer-cell immunoglobulin-like receptors)-ligand mismatch as shown in acute leukemia patients. The efficacy of this donor NK cell alloreactivity in mismatched hematopoietic transplants resulted in strong graft-versus-leukemia effects, prevented graft rejection and graft versus host disease and protected against AML relapse (24–26). Increased eliminations of AML blasts could be also shown by adoptive transfer of haploidentical NK cells and IL-2 infusions to stimulate *ex vivo* donor NK cell expansion. However, limitations have been observed by lacking of antigen specificity and long-lasting increase of immunosuppressive regulatory T cells that resulted in a reduction of NK cell proliferations and/or cytotoxic properties (27–30).

Some of the current anti-leukemia therapy studies focus on developing antibody constructs that target activated NK cells to specific leukemia antigens to overcome those limitations listed here on the functionality, expansion, and persistence of NK cells. Recent advances were made, including manipulation of receptor-mediated activation, augmentation of antibody-dependent cellular cytotoxicity reactions, gene-modified NK cells engineered by chimeric antigen receptors or, finally, mono-, bi-, and tri-specific engagers for antigen retargeting on cancer cells (31).

In the past, therapeutic monoclonal antibodies (mAbs) [e.g., rituximab (anti-CD20), cetuximab (anti-EGFR), lintuzumab (anti-CD33), and alemtuzumab (anti-CD52)] against the corresponding surface antigens on leukemia cells have positively contributed to the treatment but still lead to the development of resistance and an unsatisfactory response rate. Moreover, several high expressed antigens appear on non-transformed cells and, thus, therapeutic antibodies that recognize those target molecules may be scavenged and turned ineffective (32–37). Recently, with the advance in recombinant DNA technology, bispecific (CD16 × CD19 or CD16 × CD33) and trispecific killer engager (CD16 × CD19 × CD22) were developed to redirect NK cell cytotoxicity toward malignant cells, demonstrating significant increase of NK cell cytotoxicity and cytokine release against several CD19 expressing B cell lines. Miller et al. have shown that efficacy with CD16 × CD33 bispecific (BiKE) or IL-15-trispecific killer cell engagers (TriKE) successfully reversed CD33-positive myeloid-derived suppressor cells and stimulated NK cell-induced target cell lysis (38, 39).

Vyas et al. showed clearly that trispecific immunoligands (ULBP2-aCD19-aCD33 and ULBP2-aCD19-aCD19), designated as triplebodies (TBs), successfully retargeted short-time-activated (24 h) NK cells demonstrating increased NK cell-dependent killing activities of several target cells (MEC1, HL60, BV-173, and SEM) by using ULBP2 as a natural ligand to induce high expression levels of NKG2D receptors on activated NK cells. Moreover, activated NK cells in response to control TBs without ULPB2 domains showed a reduced IFNγ release and killing properties compared to full-constructed TBs (ULBP2-aCD19-aCD33) (40).

Based on our review from a clinical phase I/II study using IL-2 activated haploidentical NK cell for adaptive immunotherapy (Clin-Gov-No-NCT01386619) showing not only benefits but also limitations due to tumor immune escape mechanisms (TIEMs), we focused on those TBs in response to NK cells to overcome TIEMs (6, 41, 42). All experiments were performed to investigate specifically the efficacy of the employed ULBP2-aCD19-aCD33 against only CD19/CD33-expressing leukemia cells, which are mainly found in resistant antigen loss variants especially described as mixed lineage leukemia (MLL). In combination with primary donor NK cells, activated up to 14 days, we analyzed the TB-dependent improvement of retargeted recognition and cytotoxicity of those effector cells. In addition, possible side effects due to activated NK cells in the crosslink to these TBs should be evaluated.

MATERIALS AND METHODS

Construction, Expression, and Purification of the Trispecific Immunoligand ULBP2-aCD19-aCD33

The ULBP2-aCD19-aCD33 TBs, kindly provided by Prof. Elke Pogge von Strandmann and Dr. Maulik Vyas, was constructed from immunoligands with high specificity for NKG2D receptors on NK cells and for CD19 and CD33 on AML cells that were efficiently expressed and secreted by HEK293T cells as previously reported (40).

BV-173 Cell Line

The B cell precursor leukemia cell line BV-173 was purchased from Leibnitz Institute DSMZ (German Collection of Microorganisms and Cell Cultures) and maintained in RPMI-1640 Medium supplemented with 10–20% fetal calf serum (FCS). Cells were split every 3 days under cell culture conditions (37°C, 5% CO_2). For functional assays, the cells were washed once with phosphate-buffered saline (PBS), centrifuged and adjusted to a final concentration of 2.5 × 10^5/ml in TexMACS (Miltenyi Biotec) containing 5% human serum albumin (HSA).

Toxicity Studies Containing T and B Cells

For toxicity experiments in response to TBs (ULBP2-aCD19-aCD33), T and B cells were isolated from fresh whole blood of healthy donors. The EasySep™ HLA Whole Blood B Cell and CD3 Positive Selection Kit (STEMCELL™ TECHNOLOGIES, Germany) was used to separate $CD19^+$/$CD20^+$ B or $CD3^+$ T cells, respectively, by positive selection according to the manufacturer's recommendations. The isolated cells were expanded in RPMI-1640 medium containing 10% FCS and in presence of a cytokine composition [final concentration: 50 IU/ml (IL-2), 100 IU/ml (IL-4), and 20 IU/ml (IL-10)]. The cells were split every 2–3 days under culture conditions (37°C, 5% CO_2). These cells were washed once with PBS and adjusted to a final concentration of 2.5 × 10^5/ml in TexMACS (Miltenyi Biotec) supplemented with 5% HSA. For toxicity assays, CD33 on T cells and CD19 surface expression on B cells were characterized by 10-color flow cytometry (FCM) analysis and then co-incubated with IL-2 activated NK cells (E/T ratio: 1:1) pretreated with 1 µg/ml TBs.

Thawed Primary Human AML Cells

Thawed primary AML samples from three different patients [French-American-British classification system: M0 or M5, respectively, kindly provided from Prof. M. Heuser, Hannover Medical School (MHH)] disclosing myeloid CD33 and B lymphoid CD19 surface marker expression were used as examples for antigen loss variants such as MLLs. These primary blasts were washed twice with PBS containing 10% FCS and treated with DNase I to avoid cell clumping. Cells were cultured up to 2 weeks in IMDM supplemented with 10% FCS, Penicillin/Streptavidin, L-Glutamine, and 20 ng/ml each of IL-3, IL-6, SCF, G-CSF, and GM-CSF. As an essential control for differentiation, marker expression, and stability, CD33, CD19 surface levels and cell viability of thawed AML samples were monitored every 2–3 days over a time period about 2 weeks (**Figure 1A**). Afterward, primary leukemic cells were washed with PBS and adjusted to 2.5 × 10^5/ml in TexMACS (Miltenyi Biotec) containing 5% HSA and used for cytotoxicity assays in different E/T ratios in response to NK cells and TBs. Additional mono-cultured primary target cell controls were monitored during cytotoxicity assays under normal culture conditions (37°C, 5% CO_2) and analyzed by FCM to estimate the viability, CD33 and CD19 surface expressions.

Untouched Isolation and Expansion of Primary $CD56^+CD3^-$ NK Cells

Up to 30 ml anticoagulated whole blood from different healthy donors was used to separate "non-touched" primary human NK cells without density gradient centrifugation using MACSxpress® NK Cell Isolation Kit (Miltenyi Biotec, Germany) according to the manufacturer's recommendations. Based on the expansion protocol from the previous clinical phase I/II NK cell study (42), we improved the protocol and expanded these freshly isolated NK cells (purity: 97.8 ± 1.4%) in NK MACS® basal medium (Miltenyi Biotec, Germany) containing 5% AB serum (human) and 1,000 IU/ml IL-2 up to 14 days (d) as described previously (43).

Cytotoxicity Assay

To assess the NK cell-mediated killing activity in the presence and absence of TBs (ULBP2-aCD19-aCD33), we optimized a no-wash, single platform cytotoxic assay based on FCM (Navios, Beckman Coulter, Germany). This functional assay is based on the recovery of the viable effector and target cells after cytotoxic interaction within a predefined period of time (4 h). Initially, the surface expression of relevant antigens on cultured effector and target cells were determined by FCM as an essential control prior to each approach of this cytotoxicity assay. These phenotypic determinations included specific markers, such as CD45, CD56, CD16, CD33, and NKG2D (CD314) for NK cells and CD9, CD19, CD33, and HLA-DR for target leukemia cells. Further on, freshly isolated and cultured (0–14 days of expansion) NK cells were pre-incubated with various TB doses (TBs: 0.1–30 µg/ml). To determine TB-mediated cytotoxic effects against $CD33^+$/$CD19^+$ leukemia target cells, pre-coated or non-treated (control) NK cells were co-incubated in different ratios (E:T ratio: 1:1, 5:1) with the leukemic cell line BV-173 and/or primary leukemia blasts. To prevent insufficient stirring of incubated samples or cell sedimentations during cytotoxic cell contacts the co-cultured suspensions were shaken in an CO_2-incubator (CO2 cell, 170-400 Plus, RS Biotech, Scotland) for up to 4 h (37°C, 5% CO_2, 250 rpm). Afterward, effector cells were stained with mAbs by using CD45 KO (Krome Orange), CD56 PC-7 (Phycoerythrin-Cyanine-7) and CD16 APC (Allophycocyanin) in order to exclude the effector cells from leukemia cells stained with CD9 FITC (Fluorescein Isothiocyanate), CD34 PE (Phycoerythrin) and HLA-DR PB (Pacific Blue). Toxicity against effector and/or target cells with and without TBs was calculated as the increased loss of viable cells (43–46):

$$\text{Cytotoxicity} = \left(\begin{array}{l} 1 - \text{concentration}_{[\text{co-cultured target cells}/\mu\text{l}]} \\ \text{over concentration}_{[\text{target control cells}/\mu\text{l}]} \end{array} \right) \times 100\%$$

To prove TB specificities, the primary NK cells were pre-incubated (20 min, 37°C, 5% CO_2, 250 rpm) with 1 µg/ml anti-NKG2D to block the redirected cytotoxicity in response to TBs against leukemia blasts.

CD107a-Degranulation Assay

Concurrent to our cytotoxicity assays, we assessed the NK cell degranulation by monitoring the cell surface expression of the lysosomal protein CD107a *via* FCM. NK cells were also co-incubated with leukemia cells at the same E:T ratios in response to TBs (ULBP2-aCD19-aCD33). Cells were stained with PE-conjugated anti-CD107a mAbs and incubated for 1 h at 37°C, 5% CO_2. Phorbol 12-myristate 13-acetate and Ionomycin (I) (Cell stimulation cocktail from eBioscience) were used as a positive control whereas NK cells alone served as unstimulated baseline parameter. After stimulation, Monensin (1:1,000; eBiosciences) and GolgiPlug (1:1,000; BD Biosciences) were added to the samples. These batches were incubated for additional 3 h. Subsequently, cells were washed, stained and analyzed by FCM (see chapter: "Cytotoxicity Assay").

Cytokine Analysis

The multi-analyte flow assay kit (LEGENDPLEX™, BioLegend®, USA) was used for detection of soluble cytokines and pro-apoptotic markers, especially IFNγ, TNFα, perforin, GraA and GraB, and granulysin. Two sets of beads with known size and fluorescence allowed detections of those soluble molecules in supernatants that previously contained co-cultured effector and target cells. All analysis and evaluations were carried out according to manufacturer's recommendations.

Time-Lapse Microscopy

Redirected cell contacts and interactions between effector and target (E/T) or effector and effector (E/E) cells in presence of TBs could be monitored and followed by fluorescence scanning microscope (IX81, Olympus, USA). As a control before starting for those imaging experiments, surface expression levels of IL-2-expanded NK cells and cultured leukemia cells (BV-173) were characterized for CD3 (PB), CD9 (FITC), NKG2D (PE), CD33 (PE), CD56 (PC-7), CD16 (APC), CD19 (ECD), 7-AAD (PC-5.5), HLA-DR (PB), and CD45 (KO). Afterward, NK cells and BV-173 cells were intracellular stained with cell proliferation dyes (CFSE/eFluor® 450, affymetrix eBiosience, USA). In the following, NK and BV-173 cells were co-incubated (E/T ratios: 5:1) on chamber slides over a time period of 8 h in response to 10 μg/ml TBs under culture conditions (37°C, 5% CO_2). Beside time-lapse movie experiments to follow specific cell migrations and interactions by designed tracking protocols (time-lapse movie: see Figure S1 in Supplementary Material), it was also possible to evaluate all recorded images containing specific E/T- and E/E-cell contacts and cluster formations by quantitative analyses using the Olympus scanR automated image and data analysis (quantitative evaluations/gating strategy: see Figure S2 in Supplementary Material).

Statistical Analyses

Statistical analysis has been performed using GraphPad Prism v6.02 (GraphPad Software, San Diego, CA, USA). Results of different cytotoxic experiments were compared by the paired Student's *t*-test in order to assess the significance of the NK cell-mediated cytotoxicity incubated in absence and presence of TBs (ULBP2-aCD19-aCD33). Statistical evaluations of surface expression levels are indicated as median with range in the individual text parts. Differences were stated significant for a $p \leq 0.05$ and $p \leq 0.01$ (indicated as * and **, respectively). Minor differences were defined as statistically non-significant (n.s.). Unless otherwise declared, results of statistical evaluations from functional assays are indicated as mean ± SD and represent 4–6 independent experiments and measured in duplicates by FCM.

RESULTS

TBs Increase Killing Activities of IL-2-Expanded NK Cells against Leukemia Cells

The capability of ULBP2-aCD19-aCD33 TBs to induce specific NK cell cytotoxicity against human leukemic cells was determined using the CD19- and CD33-double-positive BV-173 cell line with pre-B phenotype. Purified NK cells (97.8 ± 1.4% $CD56^+CD3^-$) were IL-2 activated and expanded for 14 days. The moderate-to-low expression levels of the NCRs on freshly isolated NK cells were markedly increased approximately 5.2-, 4.9-, and 1.4-fold for NKp30, NKp44, and NKp46, respectively (data not shown). Concomitantly, NKG2D revealed higher median expression levels on these expanded NK cells with 86.4% (range: 64.5–99.2%; 10–14 days) and 57.2% (range: 15.1–97.3%; 6–9 days) compared to unstimulated and early cultured NK cells (**Figure 1A**, right graph). Activated NK cell cytotoxicity rises with increasing duration of expansion time which correlates also with elevated NKG2D levels on these NK cells (**Figure 1A**). However, the NK cell-mediated cytotoxicity against BV-173 cells could be further enhanced at increased expansion periods by pre-incubation of NK cells with 1 μg/ml TBs (ULBP2-aCD19-aCD33) (**Figure 1B**). Accordingly, the NK cell killing activity in presence of TBs reached a maximum of cytotoxic average value of 33.2 ± 5.1% (E/T: 1:1) and 55.2 ± 8.8% (E/T: 5:1) with IL-2-cultured NK cells expanded for 10–14 days compared to significant lower cytotoxic levels of untreated NK cells [24.8 ± 9.4% (E/T: 1:1) and 26.8 ± 5.0% (E/T: 5:1)] (**Figure 1B**).

In opposite to corresponding control TBs containing depleted ligand (ULPB2) or receptors (anti-CD33/anti-CD19), respectively, specific blocking antibodies were used to inhibit TB-induced cytotoxicity. For these NKG2D blocking experiments, different concentrations of TBs were used to show specific competitive inhibition by saturation of the target epitope with defined concentrations of blocking antibodies (anti-NKG2D) adjusted in several pre-experiments by titration of anti-NKG2D. Thus, inhibition of TB-dependent cytotoxicity by specific blocking of the receptor-ligand-(NKG2D-ULPB2)-binding sites could be achieved partially by pre-incubation of IL-2-expanded NK cells (10–14 days) with anti-NKG2D mAbs (1 μg/ml, 20 min) following treatment with 1 and 10 μg/ml TBs. This resulted in a reduction of cytotoxicity against BV-173 of 20.2-fold (1 μg/ml TB) or 25.8-fold (10 μg/ml TB) (**Figure 1D**). These results of cytotoxicity assays against BV-173 were largely consistent with the data from the degranulation assays. NK cells in different expansion periods were pre-incubated in absence or presence of 1 μg/ml TBs and co-cultured for 4 h (E/T ratios: 1:1, 5:1) to detect the lysosomal-associated membrane protein-1 (LAMP-1/CD107a) on NK cells as an mobilized cell surface marker following stimulation-induced granule exocytosis. TBs were able to elevate the degranulating subpopulation of NK cells in response to BV-173 cells at the indicated ratios with a maximum degranulation average of 11.1 ± 6.3% (E/T: 1:1) and 9.9 ± 4.6% (E/T: 5:1) on NK cells expanded for 10–14 days (**Figure 2A**). Interestingly, in some cases, non-significant lower

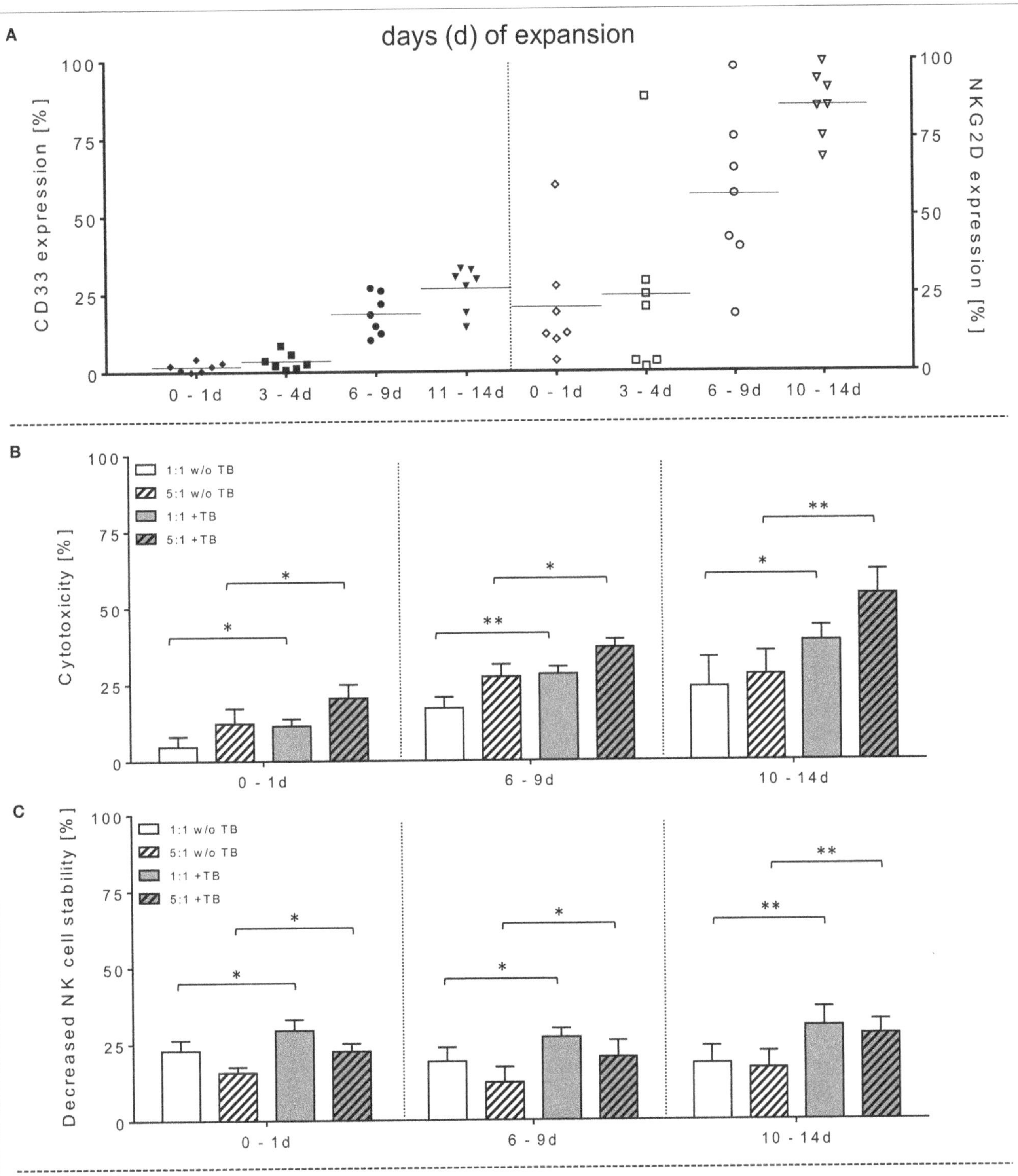

FIGURE 1 | Continued

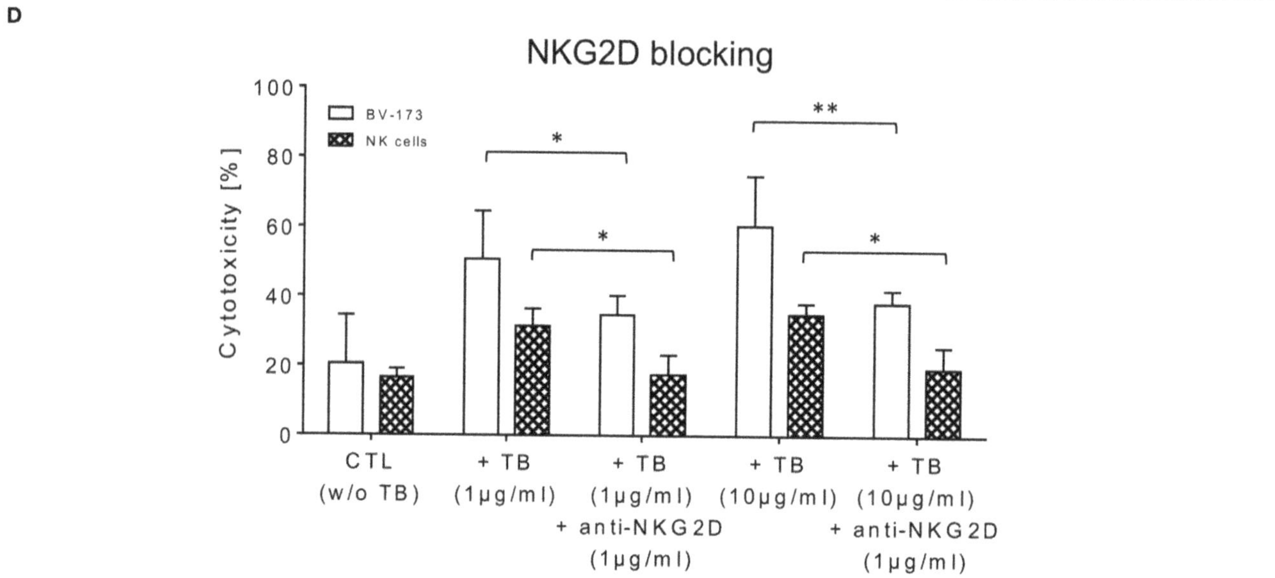

FIGURE 1 | Phenotyping, effector cell stability and cytotoxic properties of activated natural killer (NK) cells alone or in combination with triplebodies (TBs) (ULBP2-aCD19-aCD33)-inducing effects against BV-173. **(A)** *In vitro* rearrangement of the NK cell phenotype was assayed immediately after effector cell separation as well as every 2–3 days within IL-2 expansion (1,000 IU/ml) for the following 14 days. In addition to monitoring of natural cytotoxicity receptors, co-expression of CD33, and natural killer group 2 member D (NKG2D) was also quantified for these unstimulated and expanded effector cells. **(B)** Effector cells were co-incubated for 4 h (37°C, 5% CO_2, 250 rpm) with BV-173 cells at 1:1 or 5:1 E/T ratios, respectively, and killing activities (%) were determined in presence (+TB) and absence (w/o TB) of 1 µg/ml TBs (ULBP2-aCD19-aCD33) by flow cytometry. **(C)** Impact of 1 µg/ml TBs on the NK cell stability during effector-target-interactions corresponding to E/T samples. **(D)** To inhibit TB effects on NK cell killing and stability, receptor–ligand-(NKG2D-ULPB2)-binding to activated NK cells (expansion period: 10–14 days) were blocked by pre-incubation with anti-NKG2D monoclonal antibodies (1 µg/ml for 20 min) in presence of 1 or 10 µg/ml TBs, respectively. NKG2D-dependent killing rates (white columns) of those incubated NK cells in response to BV-173 cells and NK cells stability (columns with squared pattern) were analyzed after 4 h ("NKG2D blocking," ratio: 5:1, 37°C, 5% CO_2). Data of cytotoxic results are shown as mean ± SD from 4 to 6 experiments for each sample in duplicates. Statistically significant difference: $^*p \leq 0.05$ and $^{**}p \leq 0.01$.

degranulation of the effector cell at higher (5:1) compared to lower E/T ratios (1:1) in a TB-independent manner was observed (**Figure 2A**). Correspondingly, the analysis of cytokines and apoptotic markers released from TB-incubated NK cells during cytotoxic interaction against BV-173 cells showed increased concentrations for perforin, GraA and GraB, but no significant alterations of TNFα, IFNγ, or granulysin could be detected (**Figure 2C**).

TBs Improve Killing Activity against Native AML Blasts by IL-2-Activated NK Cells

In order to demonstrate that TBs-(ULBP2-aCD19-aCD33)-treated NK cells also promote a cytotoxic effect against primary AML blasts, three different patient samples were thawed and NK cell cytotoxicity was assessed by FCM analyses. In the presence of TBs, the NK cell-mediated cytotoxicity against primary blasts from three AML patients was significantly enhanced compared to NK cell killing activities in absence of TBs. Moreover, the TB-mediated cytotoxic response was more pronounced at higher E/T ratios and TB concentrations. This resulted in improved killing activities of 1.4-fold (1.3-fold) with 1 µg/ml TBs and 1.8-fold (1.6-fold) with 10 µg/ml at E/T ratios of 1:1 or 5:1, respectively (**Figure 3A**). In accordance with our previous degranulation assays in response to BV-173 cells, increased CD107a-positive NK cell subsets could be identified during cytotoxic interaction at the indicated ratios with a maximum degranulation mean of 9.7 ± 5.2% (E/T: 1:1, 10 µg/ml TBs) and 8.9 ± 4.3% (E/T: 5:1, 10 µg/ml TBs) (**Figure 2B**). Cytokine and apoptotic marker detections in response to AML blasts revealed that pre-incubation of NK cells from expansion period 10–14 days with 1 µg/ml TBs resulted in elevated levels of perforin and GraA and GraB without changes in the amount of TNFα, IFNγ, and granulysin (**Figure 2D**). Corresponding to previous experiments in response to BV-173 cells, TNFα, and IFNγ showed a non-significant tendency to lower cytokine release at higher (5:1) compared to lower E/T ratios (1:1) which was also independent of the impact from TBs (**Figure 2D**).

TBs Decrease NK Cell Viability during Cytotoxic Interactions against Leukemia Cells

Surface expression levels of CD33 were monitored within NK cell expansion over 14 days (**Figure 1A**, left graph). Beside increased NCRs and NKG2D, CD33 levels were also elevated on IL-2-activated NK cells with median expression levels of 27.9% (range: 15.8–33.8%; 10–14 days) and 21.6% (range: 11.0–28.2%; 6–9 days) compared to very low amounts in early expansion stages (**Figure 1A**, left graph). This led to the upcoming question whether the cytotoxic potential of the applied TBs analogous to the shown directed killing effects against AML blasts also has an unfavorable impact on the expanded NK cells themselves. Therefore, in addition to cytotoxic determinations against AML cells, the stability of the NK cells during cytotoxic interactions in the presence and absence of TBs was investigated in all cytotoxic experiments. Interestingly, in all E/T ratios pre-incubated with TBs, a marked decrease of these effector cells could be demonstrated both in response to BV-173 cells or AML blasts. Accordingly, TB-induced effector cell decrease reached

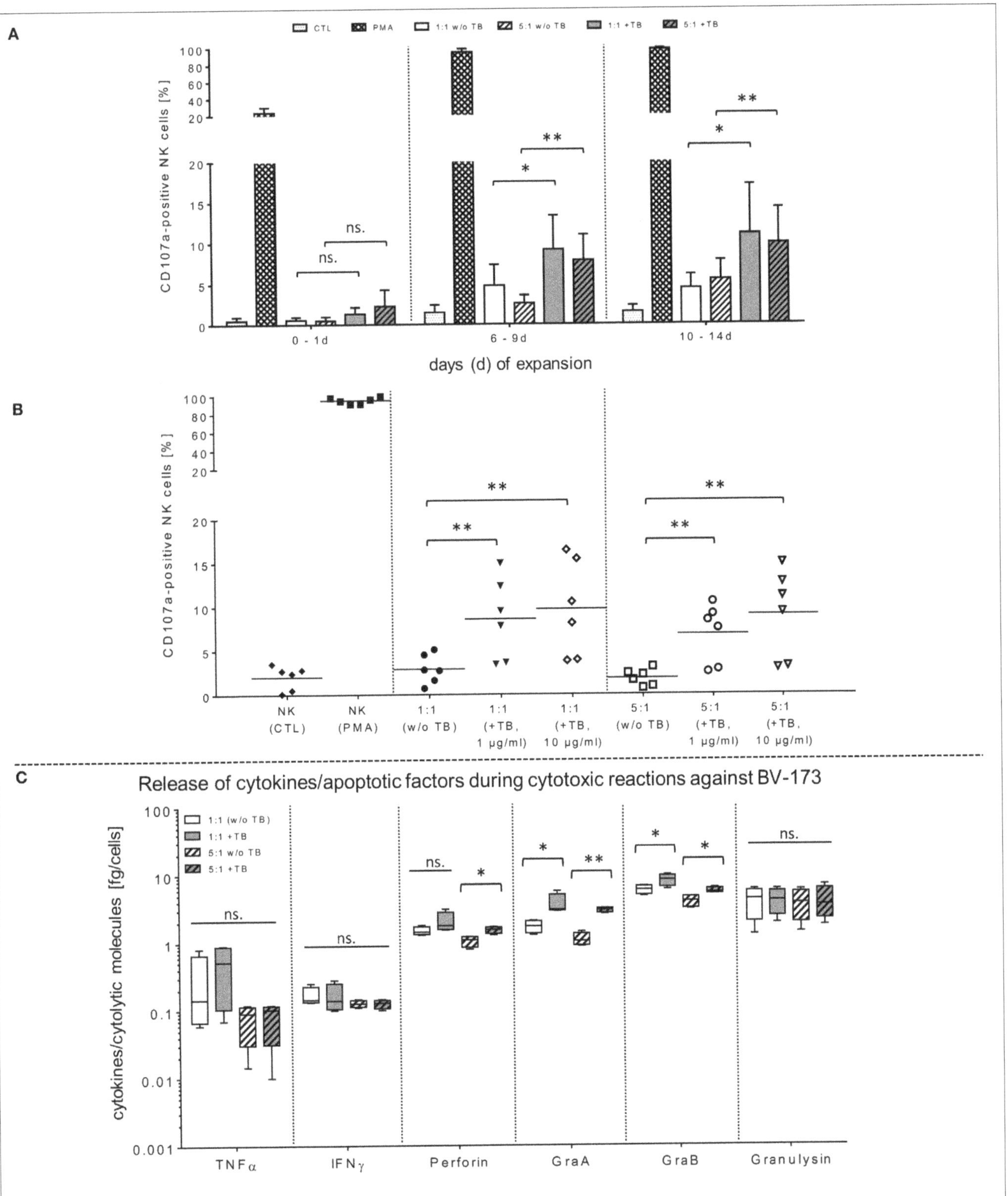

FIGURE 2 | Continued

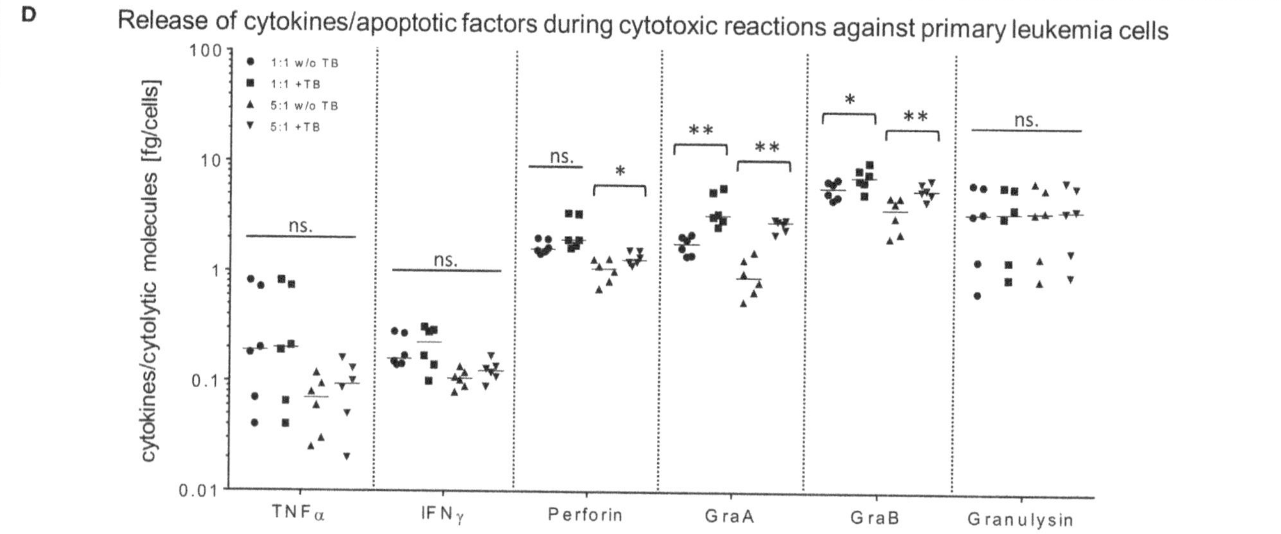

FIGURE 2 | Functional activities of natural killer (NK) cells against BV-173 and primary leukemia cells in response to triplebodies (TBs). **(A,B)** Activated NK cells from latest expansion periods (10–14 days) were co-cultured with BV-173 cell line or with primary blasts from all acute myeloid leukemia patients, respectively. Afterward, the NK cell degranulation were detected by flow cytometry (FCM) analysis using the lysosomal protein CD107a in presence of 1 or 10 μg/ml TBs, respectively, under same experimental conditions as described in **Figure 1**. **(C,D)** Supernatants of different effector–target cell ratios against BV-173 **(C)** or primary patient blasts **(D)**, respectively, were collected after co-incubations over 4 h. Afterward, NK cell-mediated secretion of cytokines and pro-apoptotic markers were quantified by FCM at the indicated ratios in presence or absence of 1 μg/ml TBs. Data show mean ± SD from six experiments measured in duplicates. Statistically significant difference: $^{*}p \leq 0.05$ and $^{**}p \leq 0.01$.

a maximum average of 31.0 ± 6.4% (E/T: 1:1; 1 μg/ml TBs) co-cultivated with BV-173 cells (**Figure 1C**) or 24.9 ± 5.8% (E/T: 1:1, 10 μg/ml TBs) in response to primary blasts (**Figure 3B**), respectively. Interestingly, the NK cell viability seemed to be more reduced in 1:1 than in 5:1 E/T ratios only in presence of 1 μg/ml TBs (**Figures 1C** and **3B**). By blocking of the receptor–ligand-(NKG2D-ULPB2)-binding (anti-NKG2D mAbs, 1 μg/ml, 20 min) in presence of TBs and BV-173 cells, this unfavorable effect on the stability of the NK cells could be almost completely abolished (**Figure 1D**).

Dose Escalations of TBs (ULBP2-aCD19-aCD33) in Regard to Effector and Target Cell Stability

In order to estimate in which concentrations those TBs (ULBP2-aCD19-aCD33) affect efficiently CD33^{+}/CD19^{+} target cells and the stability of CD33^{+} effector cells, several dose-escalation experiments with highly activated NK (expansion period: 10–14 days) containing elevated CD33 levels in response to BV-173 cells were performed. Dose-escalation experiments of at least 0.1–30 μg/ml resulted in a maximum cytotoxic against target cells average of 60.2 ± 2.6% (52.4 ± 0.7%) from NK cells in response to BV-173 cells starting from a TBs concentration of at least ≥1 μg/ml and an E/T ratio of 5:1 or ≥7.5 μg/ml and an E/T ratio of 1:1 (**Figure 3C**). Moreover, a maximum toxicity mean response of 41.2 ± 1.5% (33.1 ± 0.8%) against co-cultured NK cells could be achieved from TBs concentrations of at least ≥7.5 μg/ml and an E/T ratio of 5:1 or ≥1 μg/ml at an E/T ratio of 1:1 (**Figure 3D**). Further toxicity experiments with activated NK cells and without any other target cells in presence of 1 or 10 μg/ml TBs (4 h), respectively, also demonstrated a pronounced decrease in the viability of the effector cells. Accordingly, a maximum reduction of 31.8% (24.0%) for effector cell stability after 4 h could be determined by 10 μg/ml (1 μg/ml) TBs (**Figure 4A**). However, pretreatment of NK cells with anti-NKG2D mAbs (1 μg/ml, 20 min) could partially neutralize the TB-mediated destabilization effects with a maximal blocking efficiency of 62.7% (64.0%) after 4 h in presence of 10 μg/ml (1 μg/ml) TBs (**Figure 4A**). These toxicity data raised the question whether these TBs, in addition to the demonstrated toxicity against activated NK cells, also revealed side effects against other lymphocytes, especially CD33^{+} T cells or CD19^{+} B cells. Therefore, we co-incubated (E/T ratios: 1:1) activated NK cells pretreated with 1 μg/ml TBs in response to T or B cells. This resulted in a moderate decrease of T cells, exhibiting only a weak CD33 surface expression, by approximately 12.1% after 4 h in contrast to higher reductions of 25.1% for B cells containing high CD19 expression levels (**Figure 4B**). However, pre-coating (20 min) of both lymphocyte subsets with the respective AK constructs (anti-CD33 or anti-CD19, respectively; each with 1 μg/ml) allowed blocking of the TB-induced toxicity (1 μg/ml TBs) down to 0.3% for T cells and 10.2% for B cells after 4 h co-incubations (**Figure 4B**).

TBs (ULBP2-aCD19-aCD33) Promote Cell Cluster Formations between NK and Leukemia Cells

Previous toxicity studies showed that our TBs also bind high-activated CD33^{+} NK cells resulted in reduced effector cell stability during the cytotoxic attack on leukemia cells. This led to the hypothesis that, in addition to increased cytotoxic contacts between NK and leukemia cells, elevated effector-to-effector cell contacts are responsible for higher effector cluster formations in the presence of TBs. This could partially neutralize the improved effect of TB-mediated NK cell killing activity. Therefore, $CD33^+$ NK cells from late expansion periods (10–14 days) and BV-173 cells were intracellular stained with cell proliferation dyes (see Materials and Methods). Before fluorescent microscopy

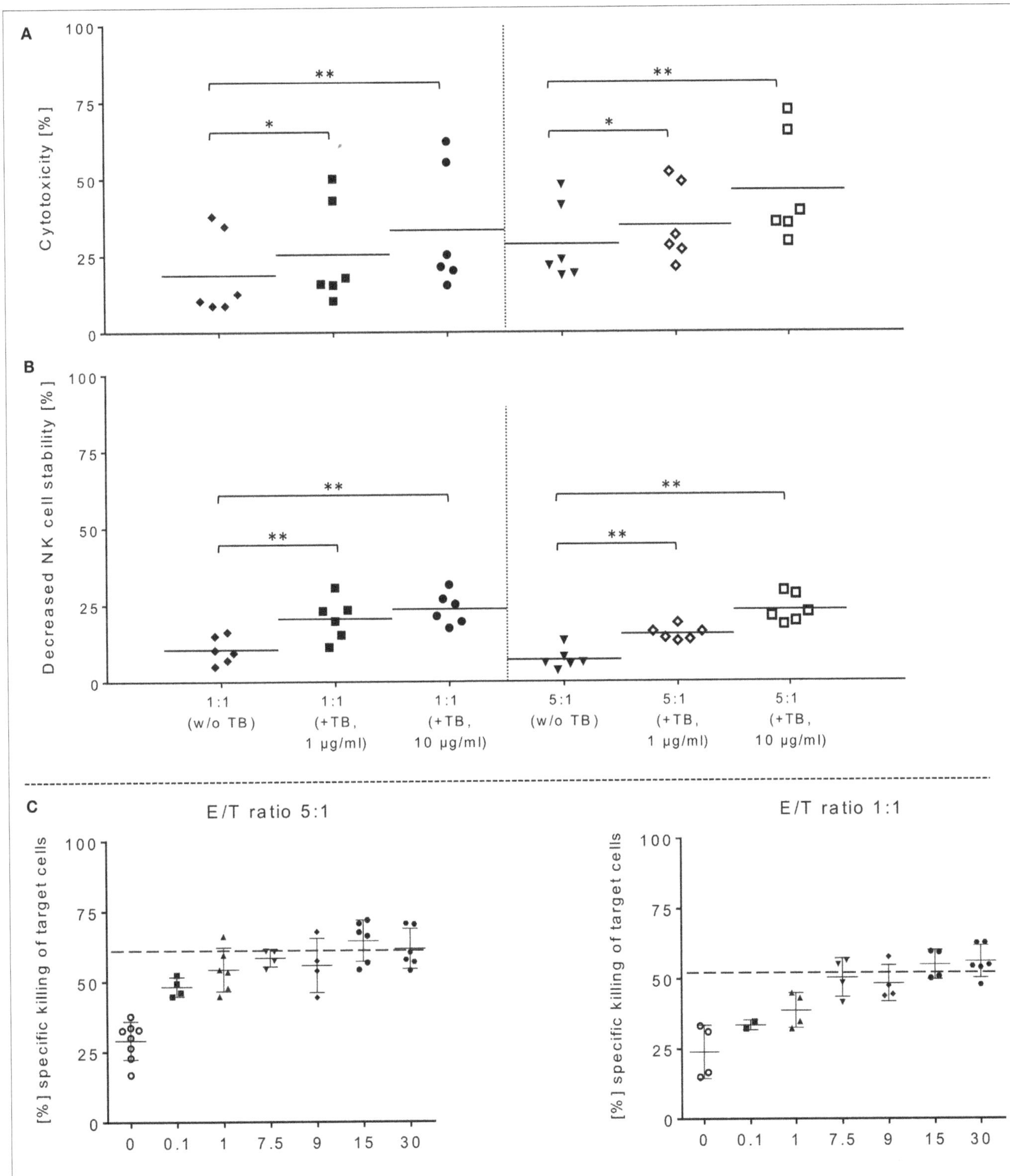

FIGURE 3 | Continued

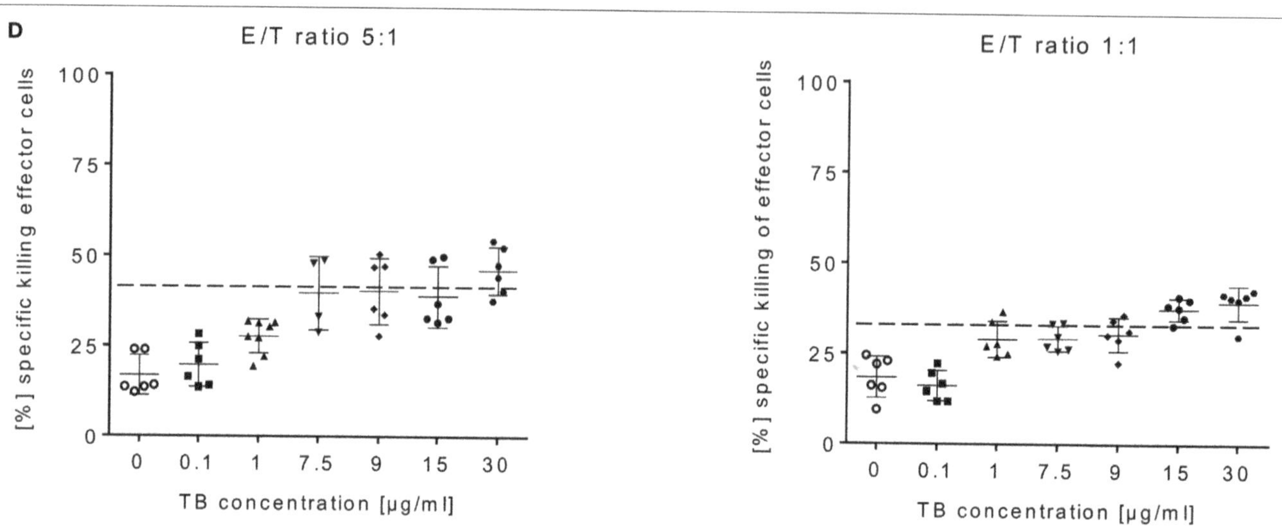

FIGURE 3 | ULBP2-aCD19-aCD33-mediated natural killer (NK) cell killing of primary acute myeloid leukemia (AML) cells. **(A)** Activated NK cells were co-incubated with primary blasts from three different AML patients in presence of 1 or 10 µg/ml triplebodies (TBs), respectively, under same experimental conditions as described in **Figure 1**. **(B)** TBs-induced effects on the stability of NK cells were also determined corresponding to E/T samples (1:1, 5:1). **(C,D)** Estimate the maximum efficiency of ULBP2-aCD19-aCD33 against effector and target cells, different concentrations (0.1–30 µg/ml) of TBs were applied to activated NK cells (time period: 10–14 days) and BV-173 cells at the indicated ratios after 4 h (37°C, 5% CO_2). Data present mean ± SD from six independent experiments measured in duplicates for each patient's sample. Statistically significant difference: $^*p \leq 0.05$ and $^{**}p \leq 0.01$.

experiments were started, activated NK cells were also analyzed for NKG2D or CD33 (PE), respectively, CD56 (PC-7), CD16 (APC), CD3 (PB), and CD45 (KO) surface expression levels. Similarly, target cells were examined for following markers: CD9 (FITC), CD33 (PE), CD19 (ECD), 7-AAD (PC-5.5), HLA-DR (PB), and CD45 (KO). Subsequently, NK and BV-173 cells (E/T ratios: 5:1) were co-cultured over 8 h in presence of TBs (10 µg/ml) monitored by designed tracking protocols. Generated transmission and fluorescent images were quantitatively evaluated by described Olympus scanR acquisition analysis (quantitative evaluations/gating strategy: see Figure S2 in Supplementary Material). It was shown that specific E/T and also E/E cell contacts had increased significantly in presence to TBs (**Figure 5A**) compared to time-limited and unspecific/confused cell contacts in untreated controls (**Figure 5B**), exemplarily shown for two separated tracking runs (time-lapse movie: see Figure S1 in Supplementary Material). Accordingly, subsequent quantitative analyses confirmed the results of time-lapse monitoring by elevated numbers of E/T and/or E/E cell contacts in presence of TBs and resulted in higher cell cluster formations containing up to eight different effector and/or target cells shown in **Tables 1** and **3**. By contrast, only unspecific cell clumping and lower cell clustering containing smaller E/T or E/E cell numbers could be detected in absence of TBs (**Tables 2** and **4**) exemplarily presented for three independent experiments.

DISCUSSION

In our experiments, we could confirm the effectiveness of the (ULBP2-aCD19-aCD33) TBs in the crosslink with activated NK cells showing increased specific killing against a leukemia cell line (BV-173) and primary AML samples from three different patients compared to single use of NK cells only. Successful targeting was directed against both, the CD19 and CD33 antigen. The transmembrane glycoprotein CD19 (95 kDa) and the early myelopoietic antigen CD33 (approximately 67–75 kDa), seemed to be suitable and prominent surface markers to distinguish myelogenous leukemia cells from lymphoid or erythroid leukemia and were also clinically validated antigens for development of antibody-based immunotherapeutic (bi- or tri-specific) construct's (47–49). However, it should be noted that human-activated NK cells also show a diversity of CD33 surface expression levels within different developmental stages (50–52).

Similar to our cytotoxic assays with the (ULBP2-aCD19-aCD33) TBs in response to leukemia cell lines and primary AML blasts, designed NKG2D-stimulating TBs that contained targeting against CD19 antigens (ULBP2-aCD19-aCD19) only displayed strong affinity to CD19 surface molecules on CLL cells. Vyas et al. (40) showed a significantly higher NK cell-mediated cytolytic activity in response to TBs (ULBP2-aCD19-aCD33 and ULBP2-aCD19-aCD19) against both, target cell lines (MEC1, BV-173, and SEM) and primary CLL blasts. The effects were independent from different E/T ratios (40). In our study as a proof-of-principle-experiment, we were able to inhibit the TB-induced cytotoxic specificity of IL-2-activated NK cells against BV-173 cells by blocking of NKG2D using anti-NKG2D mAbs. However, specific inhibition of these cytotoxic reactions could be achieved only partially and not fully by 1 µg/ml anti-NKG2D. This shows that in addition to the TB-induced killing activity, other cytotoxic mechanisms of activated NK cells are also present, which are NKG2D independent and could not be blocked in those analyses. In analogous experiments, Vyas et al. (40) achieved a decreasing specificity of TBs by pre-blocking the target antigens CD19 and CD33 on the surface of BV-173 cells. It was also shown that the NK cell-mediated cytotoxicity was strictly NKG2D dependent because control constructs lacking the ULBP2 domain could not induce IFNγ secretion and killing activity of co-incubated NK cells in response to leukemia cells (40). Further functional experiments revealed correlations of TBs-stimulated NK cell degranulations detected by increased CD107a+ effector cell populations. IFNγ secretion were also enhanced in presence of CD33-/CD19-expressing target cells that were inhibited by control constructs lacking the natural ligand ULBP2 for retargeting the NK cells *via* NKG2D receptors (40). By contrast, in our study, the functional assays showed no significant alterations for IFNγ or TNFα in collected supernatant samples after cytotoxic reactions. However, we detected TBs-dependent elevated NK secretion levels of apoptotic markers (perforin, GraA, and GraB) in response to

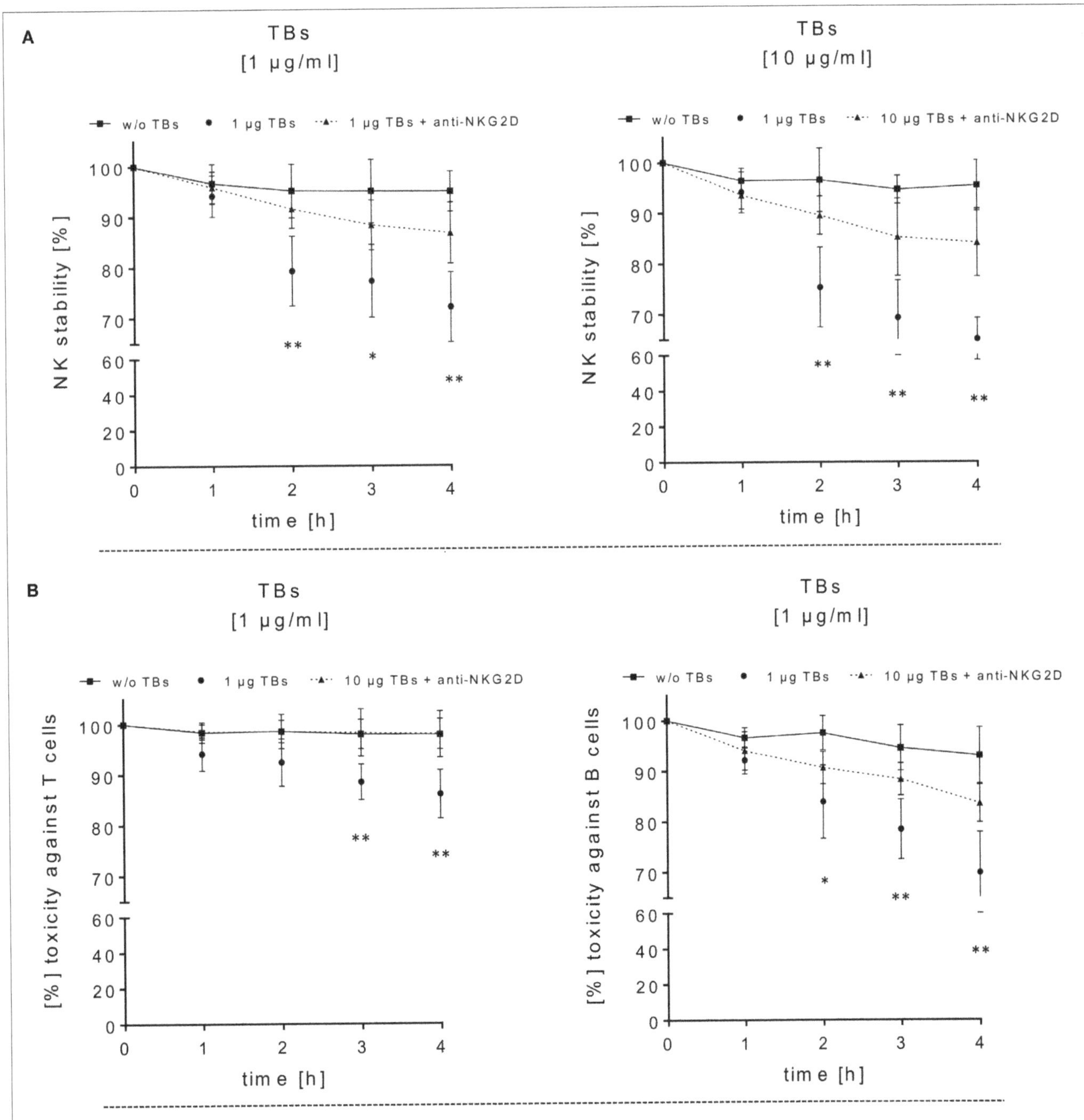

FIGURE 4 | Continued

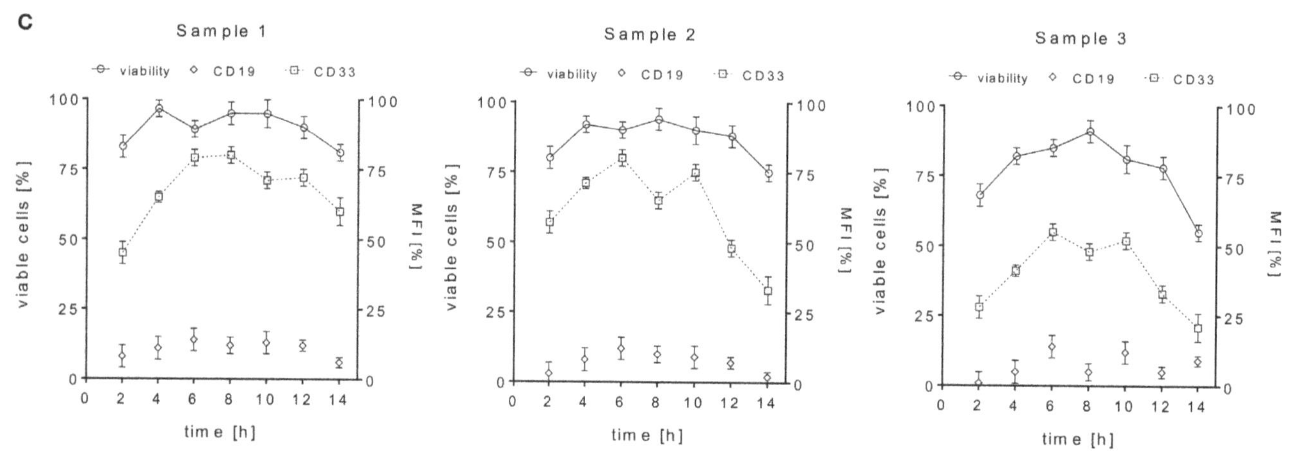

FIGURE 4 | Adverse effects of ULBP2-aCD19-aCD33 against CD19- or CD33-positive lymphocytes. **(A)** Cell viability/stability experiments with activated natural killer (NK) cells alone in presence (dotted lines) or absence (black line) of 1 or 10 μg/ml triplebodies (TBs) (4 h), respectively, were performed for 4 h under culture conditions. To inhibit TB-induced toxicity effects, NK cells were pretreated with anti-natural killer group 2 member D monoclonal antibodies (mAbs) (1 μg/ml, 20 min; dashed line). **(B)** Co-incubation of activated NK cells and T or B cells (E/T ratios: 1:1) with (dotted line) or without (black line) 1 μg/ml TBs for 4 h (37°C, 5% CO_2). TB-mediated effects could be blocked by pre-coated T and B cells with corresponding mAbs (1 μg/ml anti-CD33 or 1 μg/ml anti-CD19, respectively) for 20 min. Data present mean ± SD from 4 to 6 independent experiments measured in duplicates. Statistically significant difference: $^{*}p \leq 0.05$ and $^{**}p \leq 0.01$. **(C)** After thawing of three primary acute myeloid leukemia samples, the viability (○, left *y*-axis) and surface expression (mean fluorescence intensity) for CD33 (□, right *y*-axis) and CD19 (◇, right *y*-axis) were monitored every 2–3 days over a time period of 2 weeks as an internal control for the followed cytotoxicity assays.

both BV-173 cells and primary blasts. This correlated with an increased CD107a$^+$ NK cell subset, but showed no alterations in the granulysin releases. This could be explained by the fact that probably the maximum time for intracellular productions of IFNγ and TNFα has long been exceeded by the long-term IL-2-driven NK cell expansion over 14 days. In contrast to our NK cell expansion protocol, Vyas et al., cultivated freshly purified NK cells only overnight (37°C, 5% CO_2) in IMDM medium supplemented with 10% heat-inactivated FCS and human IL-2 (200 IU/ml) + IL-15 (10 ng/ml) (40). We concluded that this overnight cultivation in combination with both cytokines could induce an earlier IFNγ and TNFα secretion for these short-activated NK cells. Nevertheless, Vyas and our workgroup were able to show clearly improved cytotoxic properties of activated NK cells that were consistent with an optimized anti-leukemic efficiency. However, we detected a marked decrease in the stability of activated CD33$^+$ NK cells during cytotoxic interactions against leukemia blasts confirmed by increased E/E cell contacts and higher effector cell clustering analyzed using fluorescence scanning microscope.

Future immunotherapy approaches containing primary NK cells in combination with examined TB-constructs should ensure that sufficient NK cell numbers and a strongly elevated NKG2D expression are available for an efficient receptor–ligand-(NKG2D-ULPB2)-binding as well as for complete eliminations of remaining leukemia cells, especially shown in high-risk patients. The significance of NKG2D could be also confirmed by several reports dealing with immunosurveillance and development of novel NK-based immunotherapies by using bispecific immunoligands targeting NKG2D receptors (7, 53–55). Concomitant experiments within our previous clinical phase I/II NK cell study (42) demonstrated that the NKG2D-dependent cytotoxicity against resistant neuroblastoma cells was strongly affected by immunosuppressing NKG2DLs, as one of multiple strategies to escape from immune-mediated eradication. This effect could be blocked by scavenging soluble NKG2DLs with IL-2-activated donor NK cells. As a result, NKG2D-dependent cytotoxic response was restored (41, 56). These results suggest that, in addition to a permanent characterization of NKG2D levels on NK cells, a closed monitoring of such critical immunosuppressive markers in patients' plasma appears to be necessary before TBs are administered.

Since only controversial data concerning myeloid antigen CD33 (SIGLEC-3) expression on NK cells were published so far, this expression was also closely monitored concomitantly to NCRs/NKG2D characterizations. During 14 days of NK cell expansion, we could detect a transient higher CD33 surface expression level on late-expanded NK cells compared to unstimulated and early cultured primary NK cells. In accordance with these results, several subsets of NK cells were also found in human umbilical cord blood (CB) and in diverse distributions at different development stages in the peripheral blood (PB), lymph nodes, and spleen. Because of this distributions NK cell differentiations could occur at different anatomical locations (51). Interestingly, CD56$^+$/CD33$^+$ NK cell subpopulations identified in human umbilical CB revealed only a low cytotoxic effect against K562 target cells after IL-2-triggered expansion, whereas higher cytolytic effects were observed in response to activated CD56$^+$/CD33$^-$ NK cell subsets (50, 52).

Correspondingly to our experiments, increased CD33 expression levels were proven only in a subset of IL-2-cultured NK cells compared to ubiquitous elevations of NCRs and

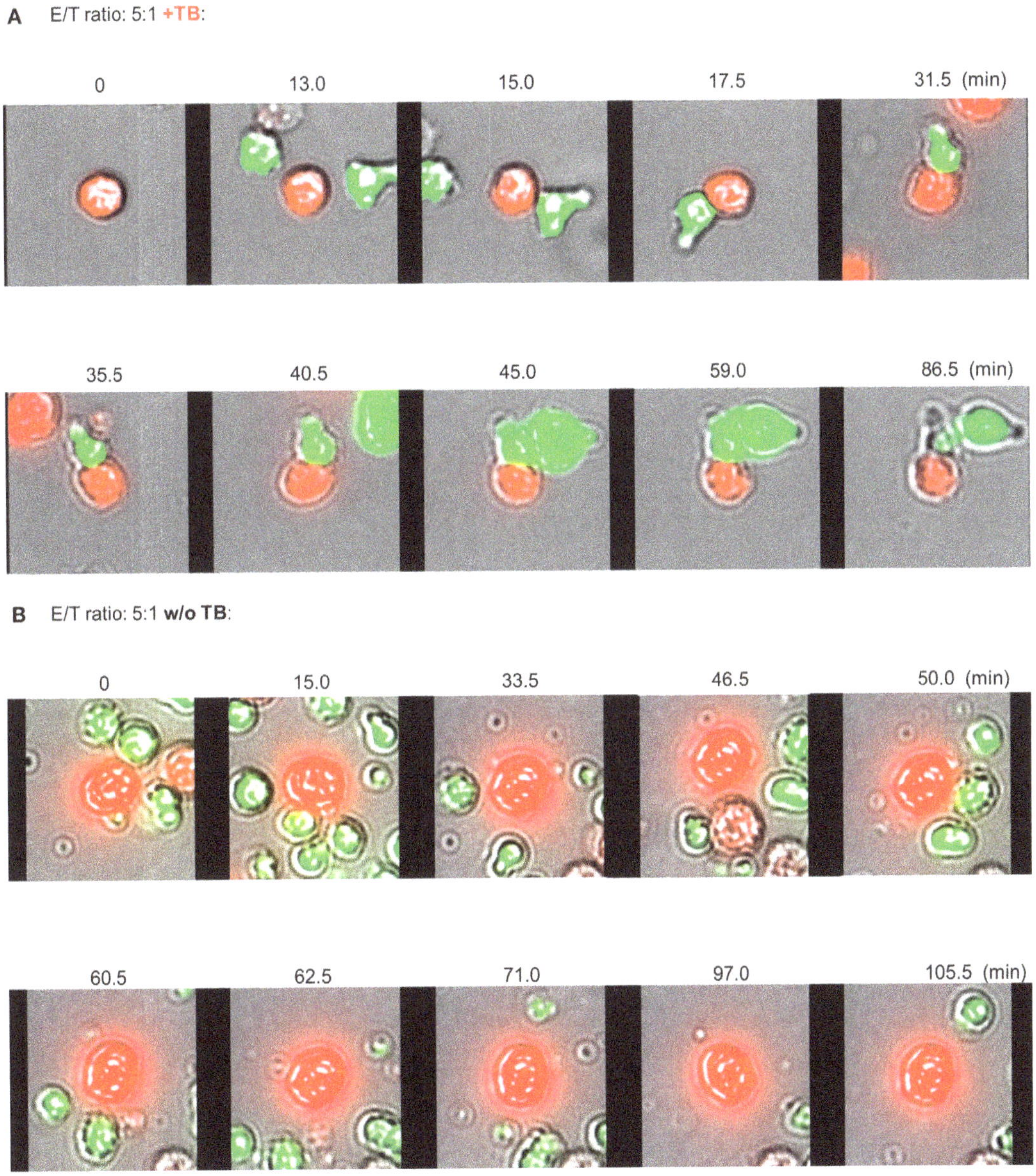

FIGURE 5 | Redirected effector–target cell interactions in response to ULBP2-aCD19-aCD33. Activated natural killer (NK) cells (expansion period: 10–14 days) and BV-173 cells were labeled with CFSE (green NK cells) or eFluor® 450 (red target cells), respectively, and co-cultured on chamber slides at the indicated E/T ratio for 8 h (37°C, 5% CO_2) in presence **(A)** or absence **(B)** of 10 µg/ml triplebodies (TBs) (ULBP2-aCD19-aCD33) exemplarily shown for three different experiments. Specific Effector-to-target (E/T)- and Effector-to-Effector (E/E)-contacts were monitored by scanR analysis allowed the time-limiting tracking of cell migrations [0–86.5 min **(A)** and 0–105.5 min **(B)**] evaluated with a fluorescence scanning microscope (IX81, Olympus, USA), visualized in response to cell morphology and fluorescence.

NKG2D surface expressions detected on all NK cells during 14 days of expansion. The viability of these CD33-expressing effector cells was adversely affected in presence of TBs (ULBP2-aCD19-aCD33) and resulted in a diminished cytotoxic response against leukemic blasts. In addition to the shown toxicity against $CD33^+$ NK cell subsets, adverse effects toward the viability of T and B lymphocytes could also be observed, which were explained according to target antigen expressions (CD19 or CD33) on those lymphocytes. Besides the well-studied CD19 surface levels that are expressed during all development stages of B cells with the exception of differentiated plasma cells (57, 58), human T cells express also a low amount of CD33. Interestingly, both T and NK cells show similarly high surface expressions of activation markers (CD25, CD28, CD38, CD45RO, or CD95) (52). This could explain the observed toxicity of activated NK cells in the presence of TBs (ULBP2-aCD19-aCD33) against T and B lymphocytes. Therefore, a close-meshed patient monitoring and examination of PB-derived immune status from AML patients should be implemented.

TABLE 1 | Quantitative evaluations of E/T cell cluster formations.

Evaluation of E/T cell cluster formations (E/T ratio: 5:1 + TB)

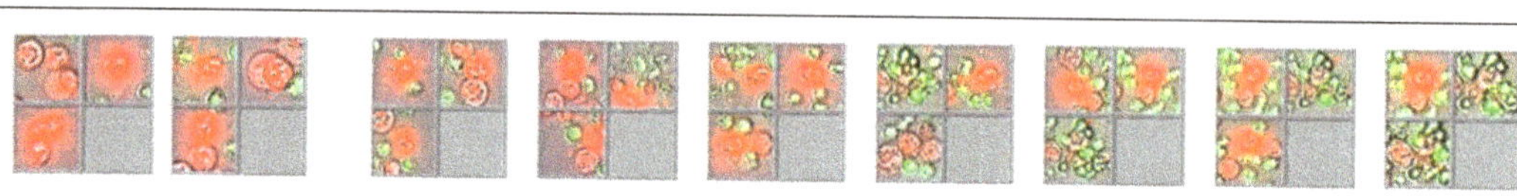

	Viable cells	Total cell clusters	T:T cell clusters	E:T cluster (1×)	E:T cluster (2×)	E:T cluster (3×)	E:T cluster (4×)	E:T cluster (5×)	E:T cluster (6×)	E:T cluster (7×)	E:T cluster (8×)
Gates	R01	R01/R02	R01/R02/ R03	R01/R02/R04	R01/R02/ R05	R01/R02/ R06	R01/R02/ R07	R01/R02/ R08	R01/R02/ R09	R01/R02/ R10	R01/R02/R11
Cell umbers	8,157	3,334	650	679	1,003	715	132	95	41	16	3
(%)	100	40.9	8.0	8.3	12.3	8.8	1.6	1.2	0.5	0.2	0.04
				32.9% (2,684 E/T cell clusters)							

Activated natural killer and BV-173 cells were co-cultured (E/T ratios: 5:1) on chamber slides over 8 h in presence of 10 µg/ml triplebodies (TBs) (37°C, 5% CO_2). Specific cell interactions between effector and target (E/T) cells were analyzed by fluorescence scanning microscope (IX81, Olympus, USA).

TABLE 2 | Quantitative analyses of E/T cell cluster formations in absence of triplebodies (TBs).

Evaluation of E/T cell cluster formations (E/T ratio: 5:1 w/o TB)

	Viable cells	Total cell clusters	T:T cell Clusters	E:T cluster (1×)	E T cluster (2×)	E:T cluster (3×)	E:T cluster (4×)	E:T cluster (5×)	E:T cluster (6×)	E:T cluster (7×)	E:T cluster (S×)
Gates	R01	R01/R02	R01/R02/ R03	R01/R02/ R04	R01/R02/ R05	R01/R02/R06	R01/R02/ R07	R01/R02/ R08	R01/R02/ R09	R01/R02/ R10	R01/R02/ R11
Cell numbers	8,588	1,287	602	589	609	44	–	–	–	–	–
(%)	100	15.0	7.0	6.9	7.1	0.5	–	–	–	–	–
				14.5% (1,242 E/T cell clusters)							

Natural killer and BV-173 cells were co-incubated (E/T ratios: 5:1) over 8 h (37°C, 5% CO_2). Clustering effector and target cells were analyzed by fluorescence scanning microscope.

TABLE 3 | Quantitative analyses of E/E cell cluster formations.

Evaluation of E/E cell cluster formations (E/T ratio: 5:1 + TB)

	Viable cells	Total E:E cell clusters	E:E cluster (1×)	E:E cluster (2×)	E:E cluster (3×)	E:E cluster (4×)
Gates	R01	R01/ R02	R01/R02/ R04	R01/R02/ R05	R01/R02/ R06	R01/R02/ R07
Cell numbers	8,157	7,537	6,394	955	157	31
(%)	100	92.4	78.4	11.7	1.9	0.4

Activated natural killer in response to BV-173 cells (E/T ratios: 5:1) were co-incubated over 8 h in presence of 10 µg/ml triplebodies (TBs) (37°C, 5% CO_2). Specific effector-to-effector cell contacts were analyzed by fluorescence scanning microscope.

TABLE 4 | Quantitative evaluations of E/E cell clusters in absence of triplebodies (TBs).

Evaluation of E/E cell cluster formations (E/T ratio: 5:1 w/o TB)

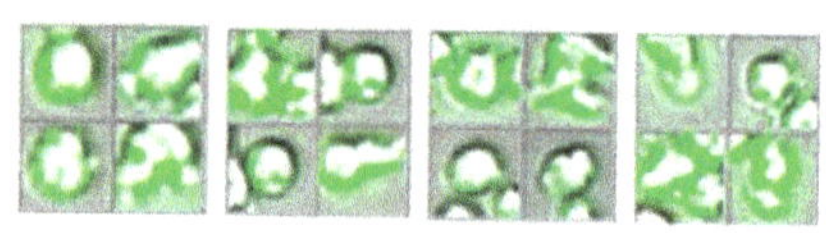

	Viable cells	Total E:E cell clusters	E:E cluster (1×)	E:E cluster (2×)	E:E cluster (3×)	E:E cluster (4×)
Gates	R01	R01/ R02	R01/R02/ R04	R01/R02/ R05	R01/R02/ R06	R01/R02/ R07
Cell numbers	8,588	2,661	2,367	229	65	3
(%)	100	31.0	27.6	2.7	0.76	0.03

Natural killer and BV-173 cells were co-cultured (E/T ratios: 5:1) over 8 h (37°C, 5% CO_2). Clustering effector cells were analyzed by fluorescence scanning microscope.

CONCLUSION

Our results indicate that TBs, especially ULBP2-aCD19-aCD33, are able to increase cytolytic properties of activated NK cells. This could be clearly demonstrated against both leukemic cell line BV-173 and primary AML blasts, but with some unfavorable toxicity effects against own effector cells and further adverse effects against T and B lymphocytes.

In summary, the experiences of our previous clinical phase I/II NK cell study for adaptive immunotherapy

(Clin-Gov-No-NCT01386619) (6, 41, 42) and our results suggest that highly activated NK cells in combination with TBs, especially ULBP2-aCD19-aCD33, might be an innovative strategy for efficient redirected eliminations of resistant AML cell. Therefore, it is necessary to monitor both CD33 and NKG2D expression levels on *ex vivo* expanded NK cells and leukemic blasts isolated from PB of AML patients to improved therapeutic benefit.

ETHICS STATEMENT

Pre-clinical development of an antibody-based triple body for NK cell-mediated immunotherapy of pediatric acute leukemia was approved and assessed by the Ethics Committee of Hannover Medical School (MHH) (ethical number: 2628-2015).

AUTHOR CONTRIBUTIONS

UK, ES, and SK designed the study, while AS, SK, and UK were mainly responsible for the performance of this study. More in detail, TG, NM, AS, and OO realized the experiments containing trispecific immunoligand ULBP2-aCD19-aCD33 (the so-called triplebodies [TBs]). SK, TG, NM, OO, and AS carried out the quality control analyses, including cell characterizations, effector cell killing activity and degranulation, time-lapse microscopy, and the enumeration of the cytokine secreting cells. MV and ES were responsible for construction, expression, and purification of the engineered trispecific immunoligand ULBP2-aCD19-aCD33. MH provided primary leukemia samples from different AML patients. SK wrote the manuscript, while UK, LA, and ES contributed to helpful discussions and the careful approval of the final manuscript. LA reworked as a specialist for native English the manuscript in word, sentence and grammar.

ACKNOWLEDGMENTS

This study was supported by grants of the Integrated Research and Treatment Center Transplantation (IFB-Tx, Ref. No. 01EO0802 and 01EO1302) and SFB738: "Optimierung konventioneller und innovativer Transplantate" financed by the German Federal Ministry of Education and Research as well as the Natural Killer Cell-Based Anti-Cancer Immunotherapies—NATURIMMUN—project (Marie Curie Initial Training Network) funded by the EC seventh Framework Programme and by a grant from the Deutsche Jose Carreras Leukämie-Stiftung to ES (DJCLS R 1408).

REFERENCES

1. Trinchieri G. Natural killer cells wear different hats: effector cells of innate resistance and regulatory cells of adaptive immunity and of hematopoiesis. *Semin Immunol* (1995) 7(2):83–8. doi:10.1006/smim.1995.0012
2. Trinchieri G. Biology of natural killer cells. *Adv Immunol* (1989) 47:187–376. doi:10.1016/S0065-2776(08)60664-1
3. Whiteside TL, Sung MW, Nagashima S, Chikamatsu K, Okada K, Vujanovic NL. Human tumor antigen-specific T lymphocytes and interleukin-2-activated natural killer cells: comparisons of antitumor effects in vitro and in vivo. *Clin Cancer Res* (1998) 4(5):1135–45.
4. Whiteside TL, Vujanovic NL, Herberman RB. Natural killer cells and tumor therapy. *Curr Top Microbiol Immunol* (1998) 230:221–44.
5. Miller JS. The biology of natural killer cells in cancer, infection, and pregnancy. *Exp Hematol* (2001) 29(10):1157–68. doi:10.1016/S0301-472X(01)00696-8
6. Koehl U, Kalberer C, Spanholtz J, Lee DA, Miller JS, Cooley S, et al. Advances in clinical NK cell studies: donor selection, manufacturing and quality control. *Oncoimmunology* (2016) 5(4):e1115178. doi:10.1080/2162402X.2015.1115178
7. Coudert JD, Held W. The role of the NKG2D receptor for tumor immunity. *Semin Cancer Biol* (2006) 16(5):333–43. doi:10.1016/j.semcancer.2006.07.008
8. Salih HR, Antropius H, Gieseke F, Lutz SZ, Kanz L, Rammensee HG, et al. Functional expression and release of ligands for the activating immunoreceptor NKG2D in leukemia. *Blood* (2003) 102(4):1389–96. doi:10.1182/blood-2003-01-0019
9. Bryceson YT, Ljunggren HG. Tumor cell recognition by the NK cell activating receptor NKG2D. *Eur J Immunol* (2008) 38(11):2957–61. doi:10.1002/eji.200838833
10. Gonzalez S, Lopez-Soto A, Suarez-Alvarez B, Lopez-Vazquez A, Lopez-Larrea C. NKG2D ligands: key targets of the immune response. *Trends Immunol* (2008) 29(8):397–403. doi:10.1016/j.it.2008.04.007
11. McGilvray RW, Eagle RA, Watson NF, Al-Attar A, Ball G, Jafferji I, et al. NKG2D ligand expression in human colorectal cancer reveals associations with prognosis and evidence for immunoediting. *Clin Cancer Res* (2009) 15(22):6993–7002. doi:10.1158/1078-0432.CCR-09-0991
12. Eagle RA, Traherne JA, Hair JR, Jafferji I, Trowsdale J. ULBP6/RAET1L is an additional human NKG2D ligand. *Eur J Immunol* (2009) 39(11):3207–16. doi:10.1002/eji.200939502
13. Voskoboinik I, Whisstock JC, Trapani JA. Perforin and granzymes: function, dysfunction and human pathology. *Nat Rev Immunol* (2015) 15(6):388–400. doi:10.1038/nri3839
14. Artis D, Spits H. The biology of innate lymphoid cells. *Nature* (2015) 517(7534):293–301. doi:10.1038/nature14189
15. Imai K, Matsuyama S, Miyake S, Suga K, Nakachi K. Natural cytotoxic activity of peripheral-blood lymphocytes and cancer incidence: an 11-year follow-up study of a general population. *Lancet* (2000) 356(9244):1795–9. doi:10.1016/S0140-6736(00)03231-1
16. Groh V, Wu J, Yee C, Spies T. Tumour-derived soluble MIC ligands impair expression of NKG2D and T-cell activation. *Nature* (2002) 419(6908):734–8. doi:10.1038/nature01112
17. Holdenrieder S, Stieber P, Peterfi A, Nagel D, Steinle A, Salih HR. Soluble MICB in malignant diseases: analysis of diagnostic significance and correlation with soluble MICA. *Cancer Immunol Immunother* (2006) 55(12):1584–9. doi:10.1007/s00262-006-0167-1
18. Song H, Kim J, Cosman D, Choi I. Soluble ULBP suppresses natural killer cell activity via down-regulating NKG2D expression. *Cell Immunol* (2006) 239(1):22–30. doi:10.1016/j.cellimm.2006.03.002
19. Chitadze G, Lettau M, Bhat J, Wesch D, Steinle A, Furst D, et al. Shedding of endogenous MHC class I-related chain molecules A and B from different human tumor entities: heterogeneous involvement of the "a disintegrin and metalloproteases" 10 and 17. *Int J Cancer* (2013) 133(7):1557–66. doi:10.1002/ijc.28174
20. Fernandez-Messina L, Ashiru O, Boutet P, Aguera-Gonzalez S, Skepper JN, Reyburn HT, et al. Differential mechanisms of shedding of the glycosylphosphatidylinositol (GPI)-anchored NKG2D ligands. *J Biol Chem* (2010) 285(12):8543–51. doi:10.1074/jbc.M109.045906
21. Baragano Raneros A, Martin-Palanco V, Fernandez AF, Rodriguez RM, Fraga MF, Lopez-Larrea C, et al. Methylation of NKG2D ligands contributes to immune system evasion in acute myeloid leukemia. *Genes Immun* (2015) 16(1):71–82. doi:10.1038/gene.2014.58

22. Nowbakht P, Ionescu MC, Rohner A, Kalberer CP, Rossy E, Mori L, et al. Ligands for natural killer cell-activating receptors are expressed upon the maturation of normal myelomonocytic cells but at low levels in acute myeloid leukemias. *Blood* (2005) 105(9):3615–22. doi:10.1182/blood-2004-07-2585
23. Pende D, Spaggiari GM, Marcenaro S, Martini S, Rivera P, Capobianco A, et al. Analysis of the receptor-ligand interactions in the natural killer-mediated lysis of freshly isolated myeloid or lymphoblastic leukemias: evidence for the involvement of the Poliovirus receptor (CD155) and Nectin-2 (CD112). *Blood* (2005) 105(5):2066–73. doi:10.1182/blood-2004-09-3548
24. Ruggeri L, Capanni M, Martelli MF, Velardi A. Cellular therapy: exploiting NK cell alloreactivity in transplantation. *Curr Opin Hematol* (2001) 8(6):355–9. doi:10.1097/00062752-200111000-00007
25. Velardi A, Ruggeri L, Moretta A, Moretta L. NK cells: a lesson from mismatched hematopoietic transplantation. *Trends Immunol* (2002) 23(9):438–44. doi:10.1016/S1471-4906(02)02284-6
26. Ruggeri L, Capanni M, Urbani E, Perruccio K, Shlomchik WD, Tosti A, et al. Effectiveness of donor natural killer cell alloreactivity in mismatched hematopoietic transplants. *Science* (2002) 295(5562):2097–100. doi:10.1126/science.1068440
27. Miller JS, Soignier Y, Panoskaltsis-Mortari A, McNearney SA, Yun GH, Fautsch SK, et al. Successful adoptive transfer and in vivo expansion of human haploidentical NK cells in patients with cancer. *Blood* (2005) 105(8):3051–7. doi:10.1182/blood-2004-07-2974
28. Bachanova V, Miller JS. NK cells in therapy of cancer. *Crit Rev Oncog* (2014) 19(1–2):133–41. doi:10.1615/CritRevOncog.2014011091
29. Bachanova V, Cooley S, Defor TE, Verneris MR, Zhang B, McKenna DH, et al. Clearance of acute myeloid leukemia by haploidentical natural killer cells is improved using IL-2 diphtheria toxin fusion protein. *Blood* (2014) 123(25):3855–63. doi:10.1182/blood-2013-10-532531
30. Bell CJ, Sun Y, Nowak UM, Clark J, Howlett S, Pekalski ML, et al. Sustained in vivo signaling by long-lived IL-2 induces prolonged increases of regulatory T cells. *J Autoimmun* (2015) 56:66–80. doi:10.1016/j.jaut.2014.10.002
31. Rossig C. Extending the chimeric receptor-based T-cell targeting strategy to solid tumors. *Oncoimmunology* (2013) 2(10):e26091. doi:10.4161/onci.26091
32. Braig F, Kriegs M, Habel B, Voigtlaender M, Grob T, Biskup K, et al. Cetuximab resistance in head and neck cancer is mediated by EGFR-K521 polymorphism. *Cancer Res* (2017) 77:1188–99. doi:10.1158/0008-5472.CAN-16-0754
33. Faderl S, Ferrajoli A, Wierda W, O'Brien S, Lerner S, Keating MJ. Alemtuzumab by continuous intravenous infusion followed by subcutaneous injection plus rituximab in the treatment of patients with chronic lymphocytic leukemia recurrence. *Cancer* (2010) 116(10):2360–5. doi:10.1002/cncr.24958
34. Tomita A. Genetic and epigenetic modulation of CD20 expression in B-cell malignancies: molecular mechanisms and significance to rituximab resistance. *J Clin Exp Hematop* (2016) 56(2):89–99. doi:10.3960/jslrt.56.89
35. Rezvani AR, Maloney DG. Rituximab resistance. *Best Pract Res Clin Haematol* (2011) 24(2):203–16. doi:10.1016/j.beha.2011.02.009
36. Sutherland MK, Yu C, Lewis TS, Miyamoto JB, Morris-Tilden CA, Jonas M, et al. Anti-leukemic activity of lintuzumab (SGN-33) in preclinical models of acute myeloid leukemia. *MAbs* (2009) 1(5):481–90. doi:10.4161/mabs.1.5.9288
37. Feldman EJ, Brandwein J, Stone R, Kalayclo M, Moore J, O'Connor J, et al. Phase III randomized multicenter study of a humanized anti-CD33 monoclonal antibody, lintuzumab, in combination with chemotherapy, versus chemotherapy alone in patients with refractory or first-relapsed acute myeloid leukemia. *J Clin Oncol* (2005) 23(18):4110–6. doi:10.1200/JCO.2005.09.133
38. Gleason MK, Ross JA, Warlick ED, Lund TC, Verneris MR, Wiernik A, et al. CD16xCD33 bispecific killer cell engager (BiKE) activates NK cells against primary MDS and MDSC CD33+ targets. *Blood* (2014) 123(19):3016–26. doi:10.1182/blood-2013-10-533398
39. Vallera DA, Felices M, McElmurry R, McCullar V, Zhou X, Schmohl JU, et al. IL15 trispecific killer engagers (TriKE) make natural killer cells specific to CD33+ targets while also inducing persistence, in vivo expansion, and enhanced function. *Clin Cancer Res* (2016) 22(14):3440–50. doi:10.1158/1078-0432.CCR-15-2710
40. Vyas M, Schneider AC, Shatnyeva O, Reiners KS, Tawadros S, Kloess S, et al. Mono- and dual-targeting triplebodies activate natural killer cells and have anti-tumor activity in vitro and in vivo against chronic lymphocytic leukemia. *Oncoimmunology* (2016) 5(9):e1211220. doi:10.1080/2162402X.2016.1211220
41. Kloess S, Huenecke S, Piechulek D, Esser R, Koch J, Brehm C, et al. IL-2-activated haploidentical NK cells restore NKG2D-mediated NK-cell cytotoxicity in neuroblastoma patients by scavenging of plasma MICA. *Eur J Immunol* (2010) 40(11):3255–67. doi:10.1002/eji.201040568
42. Koehl U, Brehm C, Huenecke S, Zimmermann SY, Kloess S, Bremm M, et al. Clinical grade purification and expansion of NK cell products for an optimized manufacturing protocol. *Front Oncol* (2013) 3:118. doi:10.3389/fonc.2013.00118
43. Kloss S, Chambron N, Gardlowski T, Weil S, Koch J, Esser R, et al. Cetuximab reconstitutes pro-inflammatory cytokine secretions and tumor-infiltrating capabilities of sMICA-inhibited NK cells in HNSCC tumor spheroids. *Front Immunol* (2015) 6:543. doi:10.3389/fimmu.2015.00543
44. Zimmermann SY, Esser R, Rohrbach E, Klingebiel T, Koehl U. A novel four-colour flow cytometric assay to determine natural killer cell or T-cell-mediated cellular cytotoxicity against leukaemic cells in peripheral or bone marrow specimens containing greater than 20% of normal cells. *J Immunol Methods* (2005) 296(1–2):63–76. doi:10.1016/j.jim.2004.10.014
45. Kloss S, Bochennek K, Huenecke S, Zimmermann SY, Kuci S, Muller T, et al. A novel five-colour flow cytometric assay to determine NK cell cytotoxicity against neuroblastoma and other adherent tumour cells. *J Immunol Methods* (2007) 325(1–2):140–7. doi:10.1016/j.jim.2007.06.013
46. Schonfeld K, Sahm C, Zhang C, Naundorf S, Brendel C, Odendahl M, et al. Selective inhibition of tumor growth by clonal NK cells expressing an ErbB2/HER2-specific chimeric antigen receptor. *Mol Ther* (2015) 23(2):330–8. doi:10.1038/mt.2014.219
47. Vitale C, Romagnani C, Puccetti A, Olive D, Costello R, Chiossone L, et al. Surface expression and function of p75/AIRM-1 or CD33 in acute myeloid leukemias: engagement of CD33 induces apoptosis of leukemic cells. *Proc Natl Acad Sci U S A* (2001) 98(10):5764–9. doi:10.1073/pnas.091097198
48. Schubert I, Kellner C, Stein C, Kugler M, Schwenkert M, Saul D, et al. A single-chain triplebody with specificity for CD19 and CD33 mediates effective lysis of mixed lineage leukemia cells by dual targeting. *MAbs* (2011) 3(1):21–30. doi:10.4161/mabs.3.1.14057
49. Kugler M, Stein C, Kellner C, Mentz K, Saul D, Schwenkert M, et al. A recombinant trispecific single-chain Fv derivative directed against CD123 and CD33 mediates effective elimination of acute myeloid leukaemia cells by dual targeting. *Br J Haematol* (2010) 150(5):574–86. doi:10.1111/j.1365-2141.2010.08300.x
50. Handgretinger R, Schafer HJ, Baur F, Frank D, Ottenlinger C, Buhring HJ, et al. Expression of an early myelopoietic antigen (CD33) on a subset of human umbilical cord blood-derived natural killer cells. *Immunol Lett* (1993) 37(2–3):223–8. doi:10.1016/0165-2478(93)90034-Y
51. Eissens DN, Spanholtz J, van der Meer A, van Cranenbroek B, Dolstra H, Kwekkeboom J, et al. Defining early human NK cell developmental stages in primary and secondary lymphoid tissues. *PLoS One* (2012) 7(2):e30930. doi:10.1371/journal.pone.0030930
52. Hernandez-Caselles T, Martinez-Esparza M, Perez-Oliva AB, Quintanilla-Cecconi AM, Garcia-Alonso A, Alvarez-Lopez DM, et al. A study of CD33 (SIGLEC-3) antigen expression and function on activated human T and NK cells: two isoforms of CD33 are generated by alternative splicing. *J Leukoc Biol* (2006) 79(1):46–58. doi:10.1189/jlb.0205096
53. Vyas M, Koehl U, Hallek M, von Strandmann EP. Natural ligands and antibody-based fusion proteins: harnessing the immune system against cancer. *Trends Mol Med* (2014) 20(2):72–82. doi:10.1016/j.molmed.2013.10.006
54. von Strandmann EP, Hansen HP, Reiners KS, Schnell R, Borchmann P, Merkert S, et al. A novel bispecific protein (ULBP2-BB4) targeting the NKG2D receptor on natural killer (NK) cells and CD138 activates NK cells and has potent antitumor activity against human multiple myeloma in vitro and in vivo. *Blood* (2006) 107(5):1955–62. doi:10.1182/blood-2005-05-2177
55. Rothe A, Jachimowicz RD, Borchmann S, Madlener M, Kessler J, Reiners KS, et al. The bispecific immunoligand ULBP2-aCEA redirects natural killer cells to tumor cells and reveals potent anti-tumor activity against colon carcinoma. *Int J Cancer* (2014) 134(12):2829–40. doi:10.1002/ijc.28609
56. Huenecke S, Zimmermann SY, Kloess S, Esser R, Brinkmann A, Tramsen L, et al. IL-2-driven regulation of NK cell receptors with regard to

the distribution of CD16+ and CD16- subpopulations and in vivo influence after haploidentical NK cell infusion. *J Immunother* (2010) 33(2):200–10. doi:10.1097/CJI.0b013e3181bb46f7

57. Robak T. Current and emerging monoclonal antibody treatments for chronic lymphocytic leukemia: state of the art. *Expert Rev Hematol* (2014) 7(6):841–57. doi:10.1586/17474086.2014.963048
58. Kellner C, Bruenke J, Horner H, Schubert J, Schwenkert M, Mentz K, et al. Heterodimeric bispecific antibody-derivatives against CD19 and CD16 induce effective antibody-dependent cellular cytotoxicity against B-lymphoid tumor cells. *Cancer Lett* (2011) 303(2):128–39. doi:10.1016/j.canlet.2011.01.020

9

Antigen Loss Variants: Catching Hold of Escaping Foes

*Maulik Vyas[1], Rolf Müller[2] and Elke Pogge von Strandmann[1]**

[1] *Experimental Tumor Research, Center for Tumor Biology and Immunology, Clinic for Hematology, Oncology and Immunology, Philipps University, Marburg, Germany,* [2] *Institute of Molecular Biology and Tumor Research, Center for Tumor Biology and Immunology, Philipps University, Marburg, Germany*

Edited by:
Antoine Toubert,
Paris Diderot University, France

Reviewed by:
Carlos Vilches,
Hospital Universitario Puerta de Hierro Majadahonda, Spain
Stephen Noel Waggoner,
Cincinnati Children's Hospital Medical Center, USA

***Correspondence:**
Elke Pogge von Strandmann
poggevon@staff.uni-marburg.de

Since mid-1990s, the field of cancer immunotherapy has seen steady growth and selected immunotherapies are now a routine and preferred therapeutic option of certain malignancies. Both active and passive cancer immunotherapies exploit the fact that tumor cells express specific antigens on the cell surface, thereby mounting an immune response specifically against malignant cells. It is well established that cancer cells typically lose surface antigens following natural or therapy-induced selective pressure and these antigen-loss variants are often the population that causes therapy-resistant relapse. CD19 and CD20 antigen loss in acute lymphocytic leukemia and chronic lymphocytic leukemia, respectively, and lineage switching in leukemia associated with mixed lineage leukemia (MLL) gene rearrangements are well-documented evidences in this regard. Although increasing number of novel immunotherapies are being developed, majority of these do not address the control of antigen loss variants. Here, we review the occurrence of antigen loss variants in leukemia and discuss the therapeutic strategies to tackle the same. We also present an approach of dual-targeting immunoligand effectively retargeting NK cells against antigen loss variants in MLL-associated leukemia. Novel immunotherapies simultaneously targeting more than one tumor antigen certainly hold promise to completely eradicate tumor and prevent therapy-resistant relapses.

Keywords: antigen loss, leukemia, NK cells, T cells, dual-targeting immunoligand

INTRODUCTION

In what is known as cancer immunoediting, the immune system not only tries to eradicate the evolving tumor but, in doing so, also shapes the immunogenicity of the tumor that may escape the immune control (1). Ultimately, the tumor cells that progress despite the immunosurveillance consist of one or more clones with lower visibility and/or higher resistance to the immune cells (1). For example, tumors often decrease the expression of components required for antigen presentation (MHC) and/or T cell activation (costimulatory molecules) as well as ligands for the NK cell-activating receptors in order to hide from the T and NK cells, respectively (1–4). Alternatively, tumor cells express ligands, which, upon binding to the respective checkpoint receptors such as CTLA-4 and PD-1 on T cells and KIR and CD94/NKG2A on NK cells, suppress their effector functions (5–9). The following sections review the current targeted therapies and the evidences of relapses associated with antigen loss variants in leukemia. Several therapeutic approaches including a dual-targeting immunoligand to manage this challenging clinical scenario are discussed.

TARGETED IMMUNOTHERAPIES—CURRENT STATUS IN LEUKEMIA

Acute leukemia represents an uncontrolled proliferation of the immature immune precursor cells and are further classified based on the lineage of the affected immune cell. Acute lymphocytic leukemia (ALL) affects the cells from the lymphoid lineage in contrast to the leukemia of myeloid cells, collectively known as acute myeloid leukemia (AML) (10). Both lymphoid and myeloid leukemia exploit the abovementioned and several other immune evasive strategies [reviewed in Ref. (6)]. However, the fact that tumor cells have to evade the immune system in order to be clinically relevant disease also supports the idea that immune system, when properly activated, can fight the cancer.

Over the last three decades, the cancer immunotherapy field has seen much progress and most of its success can be attributed to the targeted therapies against leukemia (11). The most promising immunotherapeutic options for leukemia include targeted approaches such as chimeric antigen receptor (CAR) modified T cells (CAR-T cells) and antibody-based therapies that activate T and NK cells (11). Within the CAR construct, extracellular antibody-derived scFv confers the antigen specificity, while the intracellular signaling domains (from T cell receptor and costimulatory molecule) provide the activation signal to the engineered T cells (12). Various CAR-T cells have entered the clinical studies for leukemia and the most advanced CAR is against CD19, which is being tested for ALL (13, 14). Blinatumomab, a bispecific T cell engager against CD19 and CD3 that recently got the FDA approval for the treatment of ALL, is an antibody-based molecule that also activates T cells, albeit via CD3, against the CD19-bearing target cells (15). NK cells, like T cells, have equally contributed to the clinical success of cancer immunotherapy against leukemia. For example, NK cells serve as an important effector population in chronic lymphocytic leukemia (CLL) patients who mediate antibody-dependent cell-mediated cytotoxicity through FcγRIIIa (CD16a) receptor engagement by the FDA-approved anti-CD20 antibodies (rituximab, obinutuzumab, and ofatumumab), anti-CD52 antibody alemtuzumab, and other promising anti-CD19 antibodies (MEDI-551 and XmAb5574) that are currently in clinical trials (16). In addition to the conventional antibodies, there are numerous novel approaches currently in preclinical development that aim to harness NK cell activity against cancer [reviewed in Ref. (8)].

Although many of the targeted immunotherapies have produced unprecedented responses in leukemia, especially in chemorefractory patients, the complete remissions observed following such therapies are not long-lasting and a large variety of leukemia cases are presented with relapses that are aggressive and difficult to manage. This dismal scenario emphasizes the intratumoral heterogeneity that is driven by the intrinsic factors such as accumulation of genetic and epigenetic mutations during tumor progression and extrinsic factors imposed by therapeutic pressure and tumor microenvironment (17, 18).

OCCURRENCE OF ANTIGEN LOSS VARIANTS IN LEUKEMIA

Around 30% of acute leukemia patients experience a relapse with occasional co-presentation of a phenomenon known as "lineage switch." Lineage switching occurs when acute leukemia that was initially classified as lymphoid or myeloid subtype according to the standard French–American–British guidelines shows opposite lineage when relapsed (10, 19). This phenomenon is often associated with poor prognosis and therapy resistance regardless of whether it emerged due to the lineage conversion of the original malignant clone or the selective outgrowth of a new leukemic clone (10). Out of the two possibilities, lymphoid to myeloid lineage switch is more frequently observed with more cases reported in children and often associated with the mixed lineage leukemia (MLL) gene rearrangements on chromosome 11q23 (20, 21).

Most cases of lineage switch have been reported in patients who had undergone some sort of targeted therapy. CD19-targeting immunotherapies including a bispecific antibody blinatumomab and CAR-expressing T cells have been very effective in chemorefractory B cell ALL. Anti-CD19/CD3 antibody blinatumomab redirects endogenous T cells in patients (15), while anti-CD19 CAR T cells are genetically engineered to be specifically activated against CD19 expressing target cells when infused in patients (22). Despite exceptional responses associated with these targeted therapies, some patients relapse and in many cases loss of CD19 antigen is reported. Duffner et al. reported a patient who was diagnosed with B-ALL associated with MLL-gene rearrangements but with no evidence of mixed lineage phenotype. Although blinatumomab therapy led to the complete disappearance of leukemic B cells, the patient relapsed with a more aggressive monocytic AML, which was negative for typical lymphoid markers such as CD19 (20). Similarly, CD19-specific CAR-T cell therapy could achieve complete response in all seven MLL-rearranged B-ALL patients. However, two of the seven patients relapsed with clonally related AML with no expression of B lymphoid antigens (21). Interestingly, both patients who showed lineage switch also had the presentation of cytokine release syndrome (21). Interleukin-6 (IL6), a key mediator of cytokine release syndrome, has also been shown to induce lymphoid to myeloid dedifferentiation *in vitro* (23) and *in vivo* (24). Although this is an indication of myeloid dedifferentiation of the original lymphoid blasts as an indirect effect of CAR-T cell therapy, it is also possible that myeloid clone is already present along with the lymphoid blasts, albeit below detection level, and is selected following the lymphoid-directed therapy. Ruella et al. recently described the presence of a small CD19-negative population in B-ALL patients before the administration of anti-CD19 CAR-T cell therapy (CTL019). Although there were no cases involving lineage switch, patients relapsed with CD19-negative B cell tumor following the CAR-T cell therapy (CTL019) and, as proposed by the authors, was most likely due to the selective outgrowth of the original CD19-negative subclone (13, 25). Beyond targeted immunotherapies, the phenomenon of lineage switch has also been observed following chemotherapy. As reported by Park et al., four patients of childhood B cell lineage ALL were treated with chemotherapy and were later presented with the relapse of clonally related AML (one patient) or a novel AML clone (three patients) (19).

While the link between treatment and lineage switching is not clear, the precise mechanism of antigen loss following mAb therapy is identified in several B cell malignancies. Rituximab, a chimeric antibody against CD20, has become a standard therapeutic option for various B cell ($CD20^+$) malignancies including non-Hodgkin lymphoma (NHL), follicular lymphoma, diffuse large B cell lymphoma, and CLL (26, 27). The loss of CD20 antigen following rituximab therapy has been observed for follicular lymphoma (27), B cell NHL (28), and CLL (29). Two main mechanisms have been reported for CD20 loss from the CLL cells following rituximab (anti-CD20 mAb) treatment. While CD20 internalization by malignant B cells plays a minor role, the majority of CD20, along with the bound rituximab, is removed by the Fcγ receptor-expressing monocytes and macrophages in a process called as trogocytosis or shaving (30–32). This does not only result in the rapid clearance of rituximab following the infusion but also leads to selection of CD20-negative CLL cells that are resistant to anti-CD20 therapy. Similarly, CD19 internalization is also reported by anti-CD19 antibody XmAb5574 in CLL (33). Interestingly, Jones et al. reported the loss of CD19 from the CLL cells during the shaving (trogocytosis) of anti-CD20 rituximab. It was shown that CD19 was also transferred from B cells to monocytes in Fc receptor-dependent manner (34). Moreover, antigen loss in CLL is not only associated with the mAb therapy, for example, decrease in the cell surface expression of CD20 is observed by an immune modulating agent lenalidomide (26) or following the long-term *in vitro* coculture with mesenchymal stromal cells (29).

THERAPEUTIC STRATEGIES TO COMBAT THE ANTIGEN LOSS VARIANTS

As most tumor relapses involving antigen loss have been observed following antigen-specific therapies, one plausible solution is to use therapeutic approaches that are more general in their specificity and do not depend upon a particular tumor antigen. Immunotherapy with cytokine(s) such as IL2, IL12, and IL15 act via enhancing NK and T cell-mediated immune response against tumor (35). Although side effects associated with cytokines (e.g., IL2 and interferons) greatly limit their current use in the clinics, this approach still holds promise especially at lower doses and in combination with other anti-cancer therapies (35). Alternatively, checkpoint blockade involves blocking of the inhibitory receptors on immune cells to reverse the immune suppression by tumor cells (36, 37). Recent success in blocking of inhibitory receptors on T cells such as CTLA-4 and PD-1 by FDA-approved antibodies (checkpoint inhibitors) has led to the development of novel checkpoint inhibitors blocking NK cell inhibitory receptors KIR (lirilumab, Innate Pharma) and CD94/NKG2A (IPH2201, Innate Pharma) (36–38). The advantage is that such immune-modulatory approaches aim to promote an overall antitumor environment and are predicted to be less susceptible to the limitations associated with tumor heterogeneity and antigen loss (39). However, treatment options with no specificity for tumor are less likely to be curative as mono-agents and are often associated with the systemic side effects as observed in the form of immune-related adverse events following the checkpoint blockade approach (40).

Another strategy is to broaden the specificity of the current targeted therapies that have already shown promise in the clinics. CAR-T cells with dual specificities have been developed to improve T cell targeting of tumor cells even when one of the antigens is lost from the cell surface. A prototype CAR T cell with two distinct antigen-specific scFvs in tandem (TanCAR) retained T cell activity against antigen loss variants (41). The treatment of B-ALL patients enrolled in the pediatric CTL019 trial (the University of Pennsylvania/Children's Hospital of Philadelphia) with CD19-specific CAR-T cells led to the outgrowth of CD19-negative malignant clone, which retained the expression of an IL3 receptor α chain (CD123) (25). Taking advantage of this, Ruella et al. developed CD19/CD123 CAR-T cells and proved its ability to completely eradicate the primary B-ALL blasts ($CD19^+CD123^+$ and $CD19^-CD123^+$) and to prevent the CD19 antigen loss relapse in an immunodeficient (NSG) mouse model (25). Despite the encouraging progress with the dual-specific CAR-T cell approach, major safety concerns typically associated with CAR-T cell therapy such as "on-target, off-tumor toxicity" and "cytokine release syndrome" would demand an equal attention (42).

Alternatively, NK cells, unlike T cells, express a diverse array of activating and inhibitory receptors to sense for the presence of stressed, virally infected or malignant cells. Moreover, there are multiple ligands for some of the activating receptors on NK cells (43). For example, the natural killer group 2 member D (NKG2D), an activating receptor on NK cells, can induce NK cell effector functions upon binding to any of the natural ligands such as UL16-binding proteins (ULBP1-6) and MHC-I-related chains (MICA/B) (43). This makes NK cells unlikely to succumb to the tumor heterogeneity and antigen loss provided that malignant cells remain visible to the NK cell scanning. However, the ligands for the NK cell-activating receptors, including NKG2D, are occasionally lost from the surface of leukemic cells in order to evade NK cell immunity (2, 3, 44). Of note, as shown by the recent work of Deng et al., soluble MULT1, a murine NKG2D ligand, played an indirect role in promoting NK cell immunity suggesting that soluble ligands may be more than inhibitory for overall NK cell activity (45). Our group has developed a therapeutic strategy to resensitize leukemic cells for NKG2D-dependent NK cell attack. To this end, we have developed and tested several bi- and trispecific recombinant immunoligands containing an NKG2D ligand ULBP2 fused to the various tumor antigen-specific scFvs (46–48). The idea is that these immunoligands will bind specifically to the tumor antigens and will coat the tumor cells with ULBP2 ligand. This will turn the otherwise NK cell-resistant tumor cells visible to NK cells for the attack. This was recently tested for the trispecific immunoligands (triplebodies) against CLL and MLL cells, which showed successful NK cell-mediated killing of leukemic cells in both, *in vitro* and *in vivo* settings (47).

The ability of a dual-targeting triplebody ULBP2-aCD19-aCD33 to target antigen loss variants is showed in the present report. The term "dual-targeting triplebody" represents a trispecific immunoligand targeting two distinct antigens such as CD19 and CD33 in the case of ULBP2-aCD19-aCD33 against a B-cell precursor leukemic cell line BV173. The rational of this approach is that ULBP2-aCD19-aCD33 would coat not only the CD19- and CD33-positive target cells such as leukemic cells with MLL phenotype but also any existing or newly emerging clones that lost one of the antigens (**Figure 1**). ULBP2-aCD19 and ULBP2-aCD33, the bispecific immunoligands either targeting CD19 or CD33, would fail in this regard. To mimic the antigen loss variants of BV173 cell line, CD19 and/or CD33 antigens were preblocked using molar excess of CD19- or CD33-specific scFv moieties (aCD19scFv or aCD33scFv) that lacked an ULBP2 ligand. This has previously shown to completely abolish binding of the immunoligands and subsequent killing of target cells in an antigen-specific manner (47). As shown in **Figure 2**, when both antigens were accessible on BV173 ($CD19^{+}CD33^{+}$), all three immunoligands significantly enhanced the NK-cell-dependent killing of BV173 cells, albeit depending upon the expression level of the respective antigen. Of note, the surface expression of CD19 on the BV173 is several fold higher compared to CD33 (47, 49). When CD19 antigen was blocked ($CD19^{block}CD33^{+}$) by preincubation with aCD19scFv construct, BV173 killing induced by ULBP2-aCD19 was completely abolished while ULBP2-aCD33 and ULBP2-aCD19-aCD33 retained their toxic effects. Similarly, CD33 blocking on BV173 ($CD19^{+}CD33^{block}$) by aCD33scFv could abolish killing by ULBP2-aCD33 but not by ULBP2-aCD19 and the triplebody. Only, simultaneous blocking of both CD19 and CD33 antigens could abolish the killing induced by the dual-targeting triplebody ULBP2-aCD19-aCD33. This prototype immunoligand can also be modified to target a different combination of antigens such as CD19 and CD20 in case of CLL. Theoretically, it is also possible that tumor clones that have lost the expression of both antigens preexist within the heterogeneous tumor population and can be further selected even after dual-targeting approach. Moreover, this is also relevant in the context of antigen loss following targeted therapy as simultaneous loss of CD19 and CD20 antigens has been noted following rituximab therapy (34). Therefore, clinical success of the dual-targeting strategy will require careful selection of the tumor antigen pair and combination therapies should be considered in the case of double antigen loss.

Although this study focused on NK cell-dependent effects, NKG2D is also a shared activating receptor on γ/δ T cells and a coactivating receptor on $CD8^{+}$ T cells. NKG2D-dependent antitumor effector functions of both of these T cell populations have been reported by us and others. Therefore, we believe that NKG2D targeting would facilitate a more dynamic immune

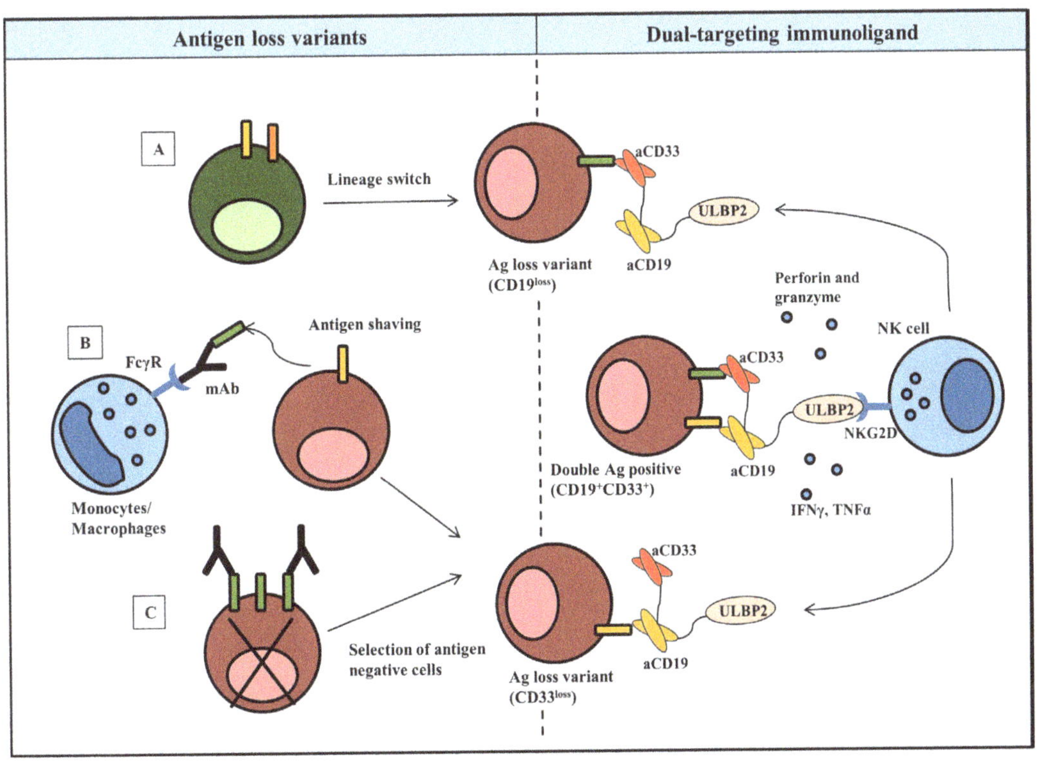

FIGURE 1 | Harnessing NK cells to control antigen loss variants: rational for the dual-targeting immunoligand approach. Emergence of antigen loss variants in most cases is seen following targeted therapy and can be associated with lineage switching **(A)**, shaving or trogocytosis of antigen–antibody complexes from the tumor cells **(B)** or selective outgrowth of antigen-negative cells **(C)**. NK cell activating dual targeting immunoligand (triplebody) consists of two scFvs against distinct antigens on tumor cells and a natural ligand to activate NK cells. As an example, ULBP2-aCD19-aCD33 (dual targeting triplebody) binds not only to the double antigen-positive ($CD19^{+}CD33^{+}$) target cells but also to the antigen loss variants. ULBP2, now coated on the target cells, activates NK cell effector functions via NKG2D receptor resulting in the killing of tumor cells by perforin and granzymes and secretion of IFNγ and TNFα. For simplicity, cross-linking is only shown between $CD19^{+}CD33^{+}$ target cells and NK cell; however, identical NK cell targeting is possible in response to antigen loss variants.

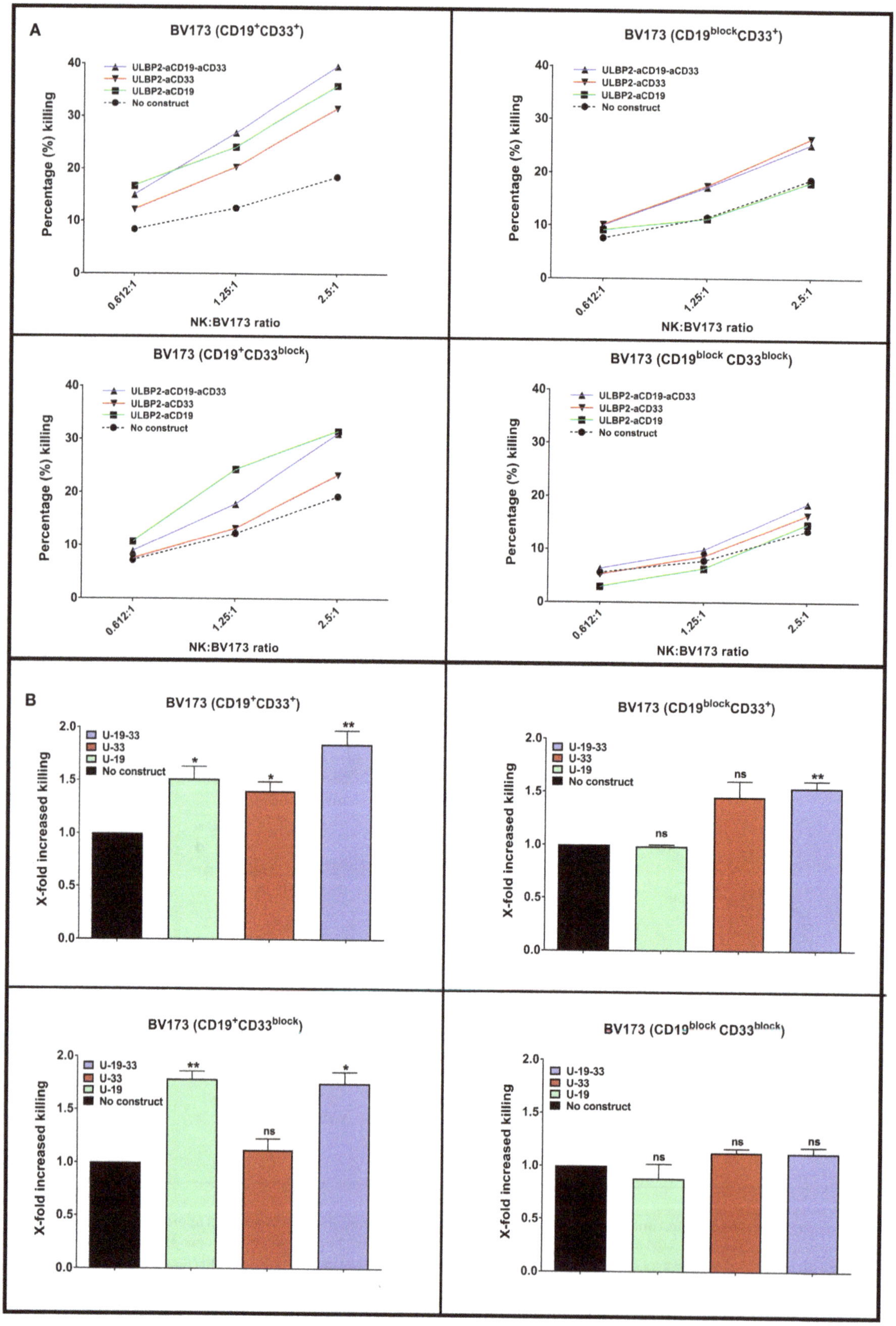

FIGURE 2 | Continued

FIGURE 2 | Continued
A dual targeting triplebody ULBP2-aCD19-aCD33 mediates NK cell-dependent killing of antigen loss variants. (A) NK cells were purified from healthy donor by negative selection and were primed by IL2 (200 U/ml) + IL15 (10 ng/ml) cytokines for 15–18 h (overnight). Next day, primed NK cells were incubated with DiR dye-labeled BV173 cells at indicated effector to target (E:T) ratio for 3 h. The incubation was continued either alone (No construct) or in the presence of 100 nM of immunoligand (U-19: ULBP2-aCD19, U-33: ULBP2-aCD33, U-19-33: ULBP2-aCD19-aCD33). After incubation, 7-AAD was added and 7-AAD-positive cells within DiR-positive gate indicated dead BV173 cells. One representative toxicity assay is shown. **(B)** Cumulative analysis of four independent toxicity assays at 2.5:1 (E:T) ratio (*N* = 4; each *N* represents an independent healthy NK cell donor). Error bars indicate SEM and statistical analysis by one-way ANOVA.

reaction involving both, innate and adaptive arms. In human and mice, chronic stimulation of NKG2D receptor by membrane bound ligands leads to the reduced surface expression of NKG2D receptor (50, 51). However, ULBP2 is not as effective as MICA in causing downmodulation of NKG2D receptor (51), and we do not anticipate that the recombinant protein will be retained in the body fluids for a relevant period to cause significant downmodulation of NKG2D receptor.

Taken together, incorporating additional tumor specificity to the current mono-targeting T and NK cell-based therapies appears to be a promising approach to prevent or treat antigen loss relapse. Their ultimate clinical benefits may be more accurately predicted by addressing whether there are any additional adverse effects that are particularly associated with dual specificities.

ETHICS STATEMENT

The collection of and the experiments with human NK cells from healthy volunteers were approved by the local ethics committee of the University of Cologne under reference number 11-140. Donors provided written consent in accordance with the Declaration of Helsinki.

AUTHOR CONTRIBUTIONS

MV and ES contributed to the design and analysis of the experiment. MV performed the experiments. MV, RM, and ES participated in writing and reviewing of the manuscript.

ACKNOWLEDGMENTS

We thank the blood donors for their generous contribution and Prof. Birgit Gathof (University of Cologne) for support.

REFERENCES

1. Dunn GP, Bruce AT, Ikeda H, Old LJ, Schreiber RD. Cancer immunoediting: from immunosurveillance to tumor escape. *Nat Immunol* (2002) 3(11):991–8. doi:10.1038/ni1102-991
2. Reiners KS, Topolar D, Henke A, Simhadri VR, Kessler J, Sauer M, et al. Soluble ligands for NK cell receptors promote evasion of chronic lymphocytic leukemia cells from NK cell anti-tumor activity. *Blood* (2013) 121(18):3658–65. doi:10.1182/blood-2013-01-476606
3. Nuckel H, Switala M, Sellmann L, Horn PA, Durig J, Duhrsen U, et al. The prognostic significance of soluble NKG2D ligands in B-cell chronic lymphocytic leukemia. *Leukemia* (2010) 24(6):1152–9. doi:10.1038/leu.2010.74
4. Salih HR, Holdenrieder S, Steinle A. Soluble NKG2D ligands: prevalence, release, and functional impact. *Front Biosci* (2008) 13:3448–56. doi:10.2741/2939
5. Yang H, Bueso-Ramos C, DiNardo C, Estecio MR, Davanlou M, Geng QR, et al. Expression of PD-L1, PD-L2, PD-1 and CTLA4 in myelodysplastic syndromes is enhanced by treatment with hypomethylating agents. *Leukemia* (2014) 28(6):1280–8. doi:10.1038/leu.2013.355
6. Ishii K, Barrett AJ. Novel immunotherapeutic approaches for the treatment of acute leukemia (myeloid and lymphoblastic). *Ther Adv Hematol* (2016) 7(1):17–39. doi:10.1177/2040620715616544
7. Sehgal A, Whiteside TL, Boyiadzis M. Programmed death-1 checkpoint blockade in acute myeloid leukemia. *Expert Opin Biol Ther* (2015) 15(8):1191–203. doi:10.1517/14712598.2015.1051028
8. Guillerey C, Huntington ND, Smyth MJ. Targeting natural killer cells in cancer immunotherapy. *Nat Immunol* (2016) 17(9):1025–36. doi:10.1038/ni.3518
9. Marin R, Ruiz-Cabello F, Pedrinaci S, Mendez R, Jimenez P, Geraghty DE, et al. Analysis of HLA-E expression in human tumors. *Immunogenetics* (2003) 54(11):767–75. doi:10.1007/s00251-002-0526-9
10. Dorantes-Acosta E, Pelayo R. Lineage switching in acute leukemias: a consequence of stem cell plasticity? *Bone Marrow Res* (2012) 2012:406796. doi:10.1155/2012/406796
11. Bachireddy P, Burkhardt UE, Rajasagi M, Wu CJ. Haematological malignancies: at the forefront of immunotherapeutic innovation. *Nat Rev Cancer* (2015) 15(4):201–15. doi:10.1038/nrc3907
12. Curran KJ, Pegram HJ, Brentjens RJ. Chimeric antigen receptors for T cell immunotherapy: current understanding and future directions. *J Gene Med* (2012) 14(6):405–15. doi:10.1002/jgm.2604
13. Maude SL, Teachey DT, Porter DL, Grupp SA. CD19-targeted chimeric antigen receptor T-cell therapy for acute lymphoblastic leukemia. *Blood* (2015) 125(26):4017–23. doi:10.1182/blood-2014-12-580068
14. Maude SL, Frey N, Shaw PA, Aplenc R, Barrett DM, Bunin NJ, et al. Chimeric antigen receptor T cells for sustained remissions in leukemia. *N Engl J Med* (2014) 371(16):1507–17. doi:10.1056/NEJMoa1407222
15. Nagorsen D, Kufer P, Baeuerle PA, Bargou R. Blinatumomab: a historical perspective. *Pharmacol Ther* (2012) 136(3):334–42. doi:10.1016/j.pharmthera.2012.07.013
16. Freeman CL, Gribben JG. Immunotherapy in chronic lymphocytic leukaemia (CLL). *Curr Hematol Malig Rep* (2016) 11(1):29–36. doi:10.1007/s11899-015-0295-9
17. McGranahan N, Swanton C. Biological and therapeutic impact of intratumor heterogeneity in cancer evolution. *Cancer Cell* (2015) 27(1):15–26. doi:10.1016/j.ccell.2014.12.001
18. Zellmer VR, Zhang S. Evolving concepts of tumor heterogeneity. *Cell Biosci* (2014) 4:69. doi:10.1186/2045-3701-4-69

19. Park M, Koh KN, Kim BE, Im HJ, Jang S, Park CJ, et al. Lineage switch at relapse of childhood acute leukemia: a report of four cases. *J Korean Med Sci* (2011) 26(6):829–31. doi:10.3346/jkms.2011.26.6.829
20. Duffner U, Abdel-Mageed A, Younge J, Tornga C, Scott K, Staddon J, et al. The possible perils of targeted therapy. *Leukemia* (2016) 30(7):1619–21. doi:10.1038/leu.2016.18
21. Gardner R, Wu D, Cherian S, Fang M, Hanafi LA, Finney O, et al. Acquisition of a CD19-negative myeloid phenotype allows immune escape of MLL-rearranged B-ALL from CD19 CAR-T-cell therapy. *Blood* (2016) 127(20):2406–10. doi:10.1182/blood-2015-08-665547
22. Ramos CA, Savoldo B, Dotti G. CD19-CAR trials. *Cancer J* (2014) 20(2):112–8. doi:10.1097/PPO.0000000000000031
23. Cohen A, Petsche D, Grunberger T, Freedman MH. Interleukin 6 induces myeloid differentiation of a human biphenotypic leukemic cell line. *Leuk Res* (1992) 16(8):751–60. doi:10.1016/0145-2126(92)90153-X
24. Maeda K, Malykhin A, Teague-Weber BN, Sun XH, Farris AD, Coggeshall KM. Interleukin-6 aborts lymphopoiesis and elevates production of myeloid cells in systemic lupus erythematosus-prone B6.Sle1. Yaa animals. *Blood* (2009) 113(19):4534–40. doi:10.1182/blood-2008-12-192559
25. Ruella M, Barrett DM, Kenderian SS, Shestova O, Hofmann TJ, Perazzelli J, et al. Dual CD19 and CD123 targeting prevents antigen-loss relapses after CD19-directed immunotherapies. *J Clin Invest* (2016) 126(10):3814–26. doi:10.1172/JCI87366
26. Lapalombella R, Yu B, Triantafillou G, Liu Q, Butchar JP, Lozanski G, et al. Lenalidomide down-regulates the CD20 antigen and antagonizes direct and antibody-dependent cellular cytotoxicity of rituximab on primary chronic lymphocytic leukemia cells. *Blood* (2008) 112(13):5180–9. doi:10.1182/blood-2008-01-133108
27. Bellesso M, Xavier FD, Costa RO, Pereira J, Siqueira SA, Chamone DA. Disease progression after R-CHOP treatment associated with the loss of CD20 antigen expression. *Rev Bras Hematol Hemoter* (2011) 33(2):148–50. doi:10.5581/1516-8484.20110036
28. Foran JM, Norton AJ, Micallef IN, Taussig DC, Amess JA, Rohatiner AZ, et al. Loss of CD20 expression following treatment with rituximab (chimaeric monoclonal anti-CD20): a retrospective cohort analysis. *Br J Haematol* (2001) 114(4):881–3. doi:10.1046/j.1365-2141.2001.03019.x
29. Pickartz T, Ringel F, Wedde M, Renz H, Klein A, von Neuhoff N, et al. Selection of B-cell chronic lymphocytic leukemia cell variants by therapy with anti-CD20 monoclonal antibody rituximab. *Exp Hematol* (2001) 29(12):1410–6. doi:10.1016/S0301-472X(01)00753-6
30. Khong HT, Restifo NP. Natural selection of tumor variants in the generation of "tumor escape" phenotypes. *Nat Immunol* (2002) 3(11):999–1005. doi:10.1038/ni1102-999
31. Williams ME, Densmore JJ, Pawluczkowycz AW, Beum PV, Kennedy AD, Lindorfer MA, et al. Thrice-weekly low-dose rituximab decreases CD20 loss via shaving and promotes enhanced targeting in chronic lymphocytic leukemia. *J Immunol* (2006) 177(10):7435–43. doi:10.4049/jimmunol.177.10.7435
32. Beum PV, Peek EM, Lindorfer MA, Beurskens FJ, Engelberts PJ, Parren PW, et al. Loss of CD20 and bound CD20 antibody from opsonized B cells occurs more rapidly because of trogocytosis mediated by Fc receptor-expressing effector cells than direct internalization by the B cells. *J Immunol* (2011) 187(6):3438–47. doi:10.4049/jimmunol.1101189
33. Awan FT, Lapalombella R, Trotta R, Butchar JP, Yu B, Benson DM Jr, et al. CD19 targeting of chronic lymphocytic leukemia with a novel Fc-domain-engineered monoclonal antibody. *Blood* (2010) 115(6):1204–13. doi:10.1182/blood-2009-06-229039
34. Jones JD, Hamilton BJ, Rigby WF. Rituximab mediates loss of CD19 on B cells in the absence of cell death. *Arthritis Rheum* (2012) 64(10):3111–8. doi:10.1002/art.34560
35. Lee S, Margolin K. Cytokines in cancer immunotherapy. *Cancers (Basel)* (2011) 3(4):3856–93. doi:10.3390/cancers3043856
36. Keir ME, Butte MJ, Freeman GJ, Sharpe AH. PD-1 and its ligands in tolerance and immunity. *Annu Rev Immunol* (2008) 26:677–704. doi:10.1146/annurev.immunol.26.021607.090331
37. Hodi FS, O'Day SJ, McDermott DF, Weber RW, Sosman JA, Haanen JB, et al. Improved survival with ipilimumab in patients with metastatic melanoma. *N Engl J Med* (2010) 363(8):711–23. doi:10.1056/NEJMoa1003466
38. Chester C, Fritsch K, Kohrt HE. Natural killer cell immunomodulation: targeting activating, inhibitory, and co-stimulatory receptor signaling for cancer immunotherapy. *Front Immunol* (2015) 6:601. doi:10.3389/fimmu.2015.00601
39. Monjazeb AM, Zamora AE, Grossenbacher SK, Mirsoian A, Sckisel GD, Murphy WJ. Immunoediting and antigen loss: overcoming the achilles heel of immunotherapy with antigen non-specific therapies. *Front Oncol* (2013) 3:197. doi:10.3389/fonc.2013.00197
40. Spain L, Diem S, Larkin J. Management of toxicities of immune checkpoint inhibitors. *Cancer Treat Rev* (2016) 44:51–60. doi:10.1016/j.ctrv.2016.02.001
41. Grada Z, Hegde M, Byrd T, Shaffer DR, Ghazi A, Brawley VS, et al. TanCAR: a novel bispecific chimeric antigen receptor for cancer immunotherapy. *Mol Ther Nucleic Acids* (2013) 2:e105. doi:10.1038/mtna.2013.32
42. Dai H, Wang Y, Lu X, Han W. Chimeric antigen receptors modified T-cells for cancer therapy. *J Natl Cancer Inst* (2016) 108(7):djv439. doi:10.1093/jnci/djv439
43. Levy EM, Roberti MP, Mordoh J. Natural killer cells in human cancer: from biological functions to clinical applications. *J Biomed Biotechnol* (2011) 2011:676198. doi:10.1155/2011/676198
44. Shatnyeva OM, Hansen HP, Reiners KS, Sauer M, Vyas M, von Strandmann EP. DNA damage response and evasion from immunosurveillance in CLL: new options for NK cell-based immunotherapies. *Front Genet* (2015) 6:11. doi:10.3389/fgene.2015.00011
45. Deng W, Gowen BG, Zhang L, Wang L, Lau S, Iannello A, et al. Antitumor immunity. A shed NKG2D ligand that promotes natural killer cell activation and tumor rejection. *Science* (2015) 348(6230):136–9. doi:10.1126/science.1258867
46. von Strandmann EP, Hansen HP, Reiners KS, Schnell R, Borchmann P, Merkert S, et al. A novel bispecific protein (ULBP2-BB4) targeting the NKG2D receptor on natural killer (NK) cells and CD138 activates NK cells and has potent antitumor activity against human multiple myeloma in vitro and in vivo. *Blood* (2006) 107(5):1955–62. doi:10.1182/blood-2005-05-2177
47. Vyas M, Schneider AC, Shatnyeva O, Reiners KS, Tawadros S, Kloess S, et al. Mono- and dual-targeting triplebodies activate natural killer cells and have anti-tumor activity in vitro and in vivo against chronic lymphocytic leukemia. *Oncoimmunology* (2016) 5(9):e1211220. doi:10.1080/2162402X.2016.1211220
48. Rothe A, Jachimowicz RD, Borchmann S, Madlener M, Kessler J, Reiners KS, et al. The bispecific immunoligand ULBP2-aCEA redirects natural killer cells to tumor cells and reveals potent anti-tumor activity against colon carcinoma. *Int J Cancer* (2014) 134(12):2829–40. doi:10.1002/ijc.28609
49. Schubert I, Kellner C, Stein C, Kugler M, Schwenkert M, Saul D, et al. A single-chain triplebody with specificity for CD19 and CD33 mediates effective lysis of mixed lineage leukemia cells by dual targeting. *mAbs* (2011) 3(1):21–30. doi:10.4161/mabs.3.1.14057
50. Molfetta R, Quatrini L, Zitti B, Capuano C, Galandrini R, Santoni A, et al. Regulation of NKG2D expression and signaling by endocytosis. *Trends Immunol* (2016) 37(11):790–802. doi:10.1016/j.it.2016.08.015
51. Molfetta R, Quatrini L, Capuano C, Gasparrini F, Zitti B, Zingoni A, et al. c-Cbl regulates MICA- but not ULBP2-induced NKG2D down-modulation in human NK cells. *Eur J Immunol* (2014) 44(9):2761–70. doi:10.1002/eji.201444512

10

Restoring Natural Killer Cell Immunity against Multiple Myeloma in the Era of New Drugs

Gianfranco Pittari[1], *Luca Vago*[2,3], *Moreno Festuccia*[4,5], *Chiara Bonini*[6,7], *Deena Mudawi*[1], *Luisa Giaccone*[4,5] *and Benedetto Bruno*[4,5]*

[1] Department of Medical Oncology, National Center for Cancer Care and Research, HMC, Doha, Qatar, [2] Unit of Immunogenetics, Leukemia Genomics and Immunobiology, IRCCS San Raffaele Scientific Institute, Milano, Italy, [3] Hematology and Bone Marrow Transplantation Unit, IRCCS San Raffaele Scientific Institute, Milano, Italy, [4] Department of Oncology/Hematology, A.O.U. Città della Salute e della Scienza di Torino, Presidio Molinette, Torino, Italy, [5] Department of Molecular Biotechnology and Health Sciences, University of Torino, Torino, Italy, [6] Experimental Hematology Unit, Division of Immunology, Transplantation and Infectious Diseases, IRCCS San Raffaele Scientific Institute, Milano, Italy, [7] Vita-Salute San Raffaele University, Milano, Italy

Edited by:
Hermann Einsele, University of Würzburg, Germany

Reviewed by:
Daniel Olive, Institut national de la santé et de la recherche médicale, France
Kerry S. Campbell, Fox Chase Cancer Center, United States

***Correspondence:**
Benedetto Bruno
benedetto.bruno@unito.it

Transformed plasma cells in multiple myeloma (MM) are susceptible to natural killer (NK) cell-mediated killing *via* engagement of tumor ligands for NK activating receptors or "missing-self" recognition. Similar to other cancers, MM targets may elude NK cell immunosurveillance by reprogramming tumor microenvironment and editing cell surface antigen repertoire. Along disease continuum, these effects collectively result in a progressive decline of NK cell immunity, a phenomenon increasingly recognized as a critical determinant of MM progression. In recent years, unprecedented efforts in drug development and experimental research have brought about emergence of novel therapeutic interventions with the potential to override MM-induced NK cell immunosuppression. These NK-cell enhancing treatment strategies may be identified in two major groups: (1) immunomodulatory biologics and small molecules, namely, immune checkpoint inhibitors, therapeutic antibodies, lenalidomide, and indoleamine 2,3-dioxygenase inhibitors and (2) NK cell therapy, namely, adoptive transfer of unmanipulated and chimeric antigen receptor-engineered NK cells. Here, we summarize the mechanisms responsible for NK cell functional suppression in the context of cancer and, specifically, myeloma. Subsequently, contemporary strategies potentially able to reverse NK dysfunction in MM are discussed.

Keywords: multiple myeloma, immunotherapy, natural killer cells, killer immunoglobulin-like receptors, cytokines, immune checkpoint inhibition, daratumumab, elotuzumab, IDO inhibitors, chimeric antigen receptor

INTRODUCTION

Multiple myeloma (MM) is a B-cell malignancy characterized by an abnormal growth of malignant plasma cells which derive from a post-germinal B-cell of the lymphoid cell lineage. The treatment paradigm for MM has undergone a dramatic evolution in the past decade given a considerable improvement in the understanding of disease pathogenesis. Despite the development of novel therapeutic agents such as proteasome inhibitors—bortezomib, carlfizomib—and immunomodulatory drugs—lenalidomide, pomalidomide—which target not only MM cells but also their interplay with the microenvironment, MM remains an incurable disease and the prognosis of patients with relapsed/refractory MM remains very poor. A number of factors concur to make MM a hard-to-treat hematologic malignancy. Drug resistance remains a major concern. MM is a highly heterogeneous disease with pathogenic processes that may greatly differ among newly diagnosed patients and others that may arise during the disease course. In recent years, several studies have focused on mechanisms of drug resistance even though many are not yet completely understood. It is widely assumed that cytogenetic and epigenetic abnormalities, deregulated signaling pathways, the MM bone marrow (BM) microenvironment, and the MM stem cell itself are all elements which play significant roles in drug resistance. Deletion 17p13 is one of the most relevant chromosomal abnormalities present in approximately 10–15% of newly diagnosed patients and observed more frequently in refractory-relapsed patients. It has been associated with resistance to new agents such as bortezomib and lenalidomide (1, 2). Aberrant drug transport processes and anti-apoptosis mechanisms have also been correlated with drug resistance (3, 4). Moreover, a pivotal role is played by the intense cell–cell crosstalk between the BM microenvironment and MM cells and their interplay with the extracellular matrix (5). All the abovementioned mechanisms make MM very challenging to eradicate with single-agent or combination modalities. Thus, an urgent need exists for new therapeutic strategies to overcome resistance to current therapies. MM is also characterized by a gradual and progressive immune dysregulation with impairs functions of B and T cell immunity, natural killer (NK) cells, and antigen-presenting/dendritic cells that allow malignant plasma cells to escape immunosurveillance. The combination of an "immunosuppressive" microenvironment and clonal evolution activate signaling pathways that invariably promote disease survival and progression. Several immunotherapies have recently been proposed and, among others, they have included monoclonal antibodies, antibody–drug conjugates, chimeric antigen receptor T cell therapy (CAR-T cells), tumor vaccines, and immune checkpoint inhibitors. This review provides an overview of the biological functions and potential clinical role of NK cells as a form of immunotherapy that may improve MM clinical outcomes.

PHYSIOLOGY OF NK CELLS AND THEIR RECEPTORS

Missing-Self Recognition and Inhibitory NK Cell Receptors

In the early 1970s, immune effectors isolated from mice and humans were found to display *in vitro* antitumor cytotoxicity without prior immunization by tumor antigens *in vivo* (6–9). These cells were functionally defined as *N-cells* or *NK cells* and were believed to belong to the lymphoid lineage, but to be distinct from B and T cells (10–13). Mechanisms regulating NK cell-mediated target recognition and killing remained obscure for more than a decade after natural cytotoxicity was first described. In 1986, Karre et al. reported that resistance of mice lymphoma cells to NK cell-mediated rejection was dependent on major histocompatibility complex (MHC) class I antigen expression on cancer surface (14). This observation led to the assumption that NK cell would possess receptors able to transduce negative signals upon MHC class I engagement, thus sparing putative targets. Lack of MHC class I would instead trigger NK cell activation, a phenomenon known as *missing-self* recognition (15).

In humans, the NK cell inhibitory receptors able to recognize HLA class I are type I transmembrane structures belonging to the immunoglobulin (Ig) superfamily, known as killer immunoglobulin-like receptors (KIR). Inhibitory KIR share a long (L) cytoplasmic tail containing immunoreceptor tyrosine-based inhibitory motifs that can process signals through the recruitment and activation of the SH2-domain-containing tyrosine phosphatase 1 protein (16–20). Three inhibitory KIR engaging HLA class I ligand groups are critical regulators of NK cell function: KIR2DL1, specific for HLA-C2 group antigens (sharing Asn at position 77 and Lys at position 80 of the HLA-Cw heavy chain); KIR2DL2/3, specific for HLA-C1 group antigens (sharing Ser at position 77 and Asn at position 80 of the HLA-Cw heavy chain) (21, 22); and KIR3DL1, specific for the HLA-Bw4 epitope (located at position 77–83 of the heavy chain of certain HLA-B and HLA-A alleles) (23–25).

In the last two decades, multiple additional inhibitory NK cells receptors have been identified, leading to the currently accepted notion that NK cell effector function is dependent on the overall balance of signals transduced by multiple inhibitory and activating receptors recognizing cognate ligands on virally infected and cancer cells. Examples of non-KIR inhibitory NK receptors include the c-type lectin-like CD94/NKG2A (CD159a) heterodimer and ILT2 (LILRB1, CD85j), respectively, engaging HLA-E and various HLA class I antigens (26, 27); NKR-P1A (CD161) recognizing the lectin-like transcript 1 (28, 29); and the carcinoembryonic antigen-related cell adhesion molecule 1 (CD66a) recognizing the CD66 ligand (30–32).

Activating NK Cell Receptors

Activating NK cell receptors are also described. Among them, NKG2D (CD314) has ligand specificity for a wide range of stress-induced cell surface ligands (NKG2D-L), including the MHC-related ligands MICA and MICB (33) and the human cytomegalovirus glycoprotein (UL16)-binding proteins ULBP1-6 (33, 34). Natural cytotoxicity receptors (NCRs) NKp46 (NCR1, CD335) (35, 36), NKp44 (NCR2, CD336) (37), and NKp30 (NCR3, CD337) (38) are potent activating receptors almost exclusively restricted to NK cells. Ligands for NCR are currently incompletely characterized. NKp46 and NKp44 are known to bind several viral hemagglutinins (39, 40), while NKp30 recognizes the HLA-B-associated transcript 3 (BAT3) (41) and B7-H6, a member of the B7 immunoreceptor family (42). CD94/NKG2C (CD159c) binds the non-classical HLA-E, similar to its inhibitory CD94/NKG2A counterpart (25). CD16 (FcγRIIIA) (43) is the low-affinity IgG receptor, strongly expressed on mature NK cells, mediating antibody-dependent cellular cytotoxicity (ADCC) (44). Other important activating receptors include the SLAM-related 2B4 (CD244) (45) engaging the pan-leukocyte surface antigen CD48 (46) and the adhesion molecule DNAM-1 (47) involved in recognition of PVR (CD155) and nectin-2 (CD112) (48).

NK CELL IMMUNITY DYSFUNCTION IN MM

Tumor-Induced Microenvironment Transformation

Accumulating evidence indicates that microenvironment transformation may significantly impair NK cell effector function in MM (49). Plasma cells and T regulatory (T_{reg}) cells from patients with MM secrete high levels of TGF-β (50, 51), a potent immunosuppressive cytokine known to downregulate multiple NK-activating receptors and to impair NK cytotoxicity (52–54). IL-10 and IL-6 are increased in MM (55–57) and independently act as powerful growth factors for malignant plasma cells (58, 59). IL-10 inhibits production of pro-inflammatory IFN-γ and TNF-α (60, 61) and promotes development of NK-resistant tumor phenotypes (62), although it may also enhance NK cytotoxicity in response to IL-15 exposure *in vitro* (63). IL-6 has been shown to impair NK cell activity in experimental models, human disease, and when administered to patients with advanced cancer (64–66). Altered levels of IFN-γ may also contribute to NK cell dysregulation in MM. In two studies, serum IFN-γ levels were found to be significantly lower in subjects with MM than in normal controls (55, 56), potentially affecting NK cell activity. Besides cytokines, other soluble factors are known to suppress NK-mediated antitumor capabilities. Prostaglandin E2 inhibits activating signals transduced by NCR, NKG2D, and CD16 (67) and has been shown to be actively produced in cultures of BM from patients with MM (68). Indoleamine 2,3-dioxygenase (IDO) promotes cancer cell immune escape through potent immunoregulatory effects on antigen-presenting cells *via* enzymatic degradation of L-tryptophan (69) (see IDO inhibitors). Della Chiesa et al. described that IDO-mediated immunosuppression also involves NK cells *via* L-kyreunine (Kyn), a L-tryptophan (Trp) degradation product impairing NKp46/NKG2D-specific lysis (70). Interestingly, interaction between CD28 on MM cells and CD80/86 stimulates IDO production by stromal dendritic cells (71), in agreement with the observation that CD28 expression on MM plasma cells is a marker correlating with poor disease outcome (72).

Additional microenvironmental factors may contribute to blunted NK cell cytotoxicity and cytokine production in MM. Among them, myeloid-derived suppressor cells (MDSCs) have been found to be expanded in MM (73, 74) and to directly contribute to downregulation of NK cell responsiveness *via* the NKp30-activating receptor (75), membrane-bound TGF-β (76), and TIGIT-mediated inhibitory signaling (77). Furthermore, reduced oxygenation described in MM BM (78, 79) may inhibit NK cell anti-myeloma responsiveness (80).

Effect of Soluble Ligands on NK Cell-Mediated Immunity in MM

MICA and MICB (collectively named MIC) are stress-inducible NKG2D ligands frequently overexpressed in response to malignant transformation (81). When bound to tumor surface, they act as markers of "abnormal self" and may trigger NK cell cytotoxicity *via* NKG2D signaling. Conversely, cleavage of membrane-bound MIC is a strategy employed by MM and other tumors to evade NK cell immunosurveillance (82–85). In individuals with MIC^+ tumors, soluble MIC (sMIC) ligands induce internalization of surface NKG2D (but also NCR and chemokine receptors) and substantial impairment of NK effector functions (86–88). In addition, sMIC has been shown to promote the accumulation of MDSC and macrophages with an immunosuppressive phenotype (89), potentially contributing to NK cell suppression. Not surprisingly, presence of sMIC is associated with poor cancer survival (90–92). In MM, shedding of MIC may result from exposure of MM cells to the genotoxic agents, doxorubicin and melphalan (93). Proteolytic cleavage by ADAMTS10 has been described to mediate this phenomenon, suggesting that the combination of metalloproteinase inhibitors with chemotherapy would exert a protective effect against escape of MM cells from NK-mediated recognition (93). Similar to NKG2D-L, NCR-specific soluble ligands may in some instances induce NK cell functional impairment. For example, circulating BAG6/BAT3 may inhibit NK cell cytotoxicity by inducing NKp30-specific hyporesponsiveness (94). Shedding of these ligands in the context of MM has not been investigated.

Effect of Cell Contact on NK Cell-Mediated Immunity in MM

Derangement of NK cell effector functions may be further amplified by tumor ligand surface expression patterns favoring dominance of inhibitory NK signals. Ligands recognized by NK-activating receptors are often poorly expressed in cancer. Downregulation of membrane-bound NKG2D-L is common in multiple tumors, resulting in impaired NKG2D-dependent NK cell cytotoxicity (95–97) and unfavorable clinical outcomes (97). In the context of monoclonal gammopathy, expression of MICA is known to decrease upon transition from pre-cancerous monoclonal gammopathy of undetermined significance (MGUS) to MM (84). Of note, various pharmacological interventions may counter NKG2D-L downregulation in MM: vincristine, *via* p38 MAPK pathway activation (98); doxorubicin, melphalan, and bortezomib as a result of oxidative stress, DNA damage, and tumor senescence (99, 100) the heat shock protein-90 (HSP90) chaperone protein inhibitors 17-allylaminogeldanamycin and radicicol (101); and inhibition or degradation of bromodomain and extra-terminal proteins (102). Exposure to therapeutic agents with activity on MM has similarly been shown to induce upregulation of PVR (an activating ligand for DNAM-1) on malignant plasma cells (98, 100, 103). Besides NKG2D-L, surface expression of the B7-H6 ligand, engaging the NKp30 NCR, has been found to be downregulated in cell lines generated from multiple cancers, including MM, resulting in NKp30-dependent NK cell functional impairment (104).

Upregulation of tumor-bound HLA class I antigens is another mechanism of protection against NK cell immunosurveillance. Malignant plasma cells obtained from the BM of early-stage myeloma patients display low HLA class I expression potentially favoring NK-mediated killing (105). In contrast, high HLA class I levels are observed on plasma cells derived from pleural effusions of patients with advanced MM (105). HLA-E is a non-classical

HLA class I antigen frequently upregulated on cancer cells, a phenomenon correlating with poor prognosis (106). In MM primary cells, high HLA-E expression results in restrained *in vitro* degranulation of NK cell subsets expressing the HLA-E-specific inhibitory NK receptor NKG2A (107).

Surface overexpression of ligands for inhibitory NK receptors is not restricted to HLA class I antigens. Notably, the CD200 glycoprotein is also commonly overexpressed on cancer surface, specifically in myeloid and lymphoid leukemias, where it is a marker of poor prognosis (108, 109). Leukemia blasts overexpressing CD200 escape NK-mediated immunosurveillance by dampening NK cell cytolytic capabilities and NKp44/NKp46 receptor expression (110), a phenomenon that can be reversed by CD200 blockade (111). CD200 is also frequently expressed in patients with MM, where it adversely affects clinical outcomes following stem cell transplantation (112).

Numerical, Phenotypic, and Functional Characteristics of NK Cells in MM

Multiple reports describe numerical, phenotypic, and functional NK cell alterations in MM. Subjects with MGUS and untreated, early-stage MM have been generally found to have higher (113–115) or similar (116–118) numbers of circulating and BM NK cells than healthy donors. Upregulation of CD57 and CD16 on NK cell surface is also observed (119, 120), suggesting the emergence of terminally differentiated subsets with high-cytotoxic potential. While these findings suggest efficient response to malignant clones subject to NK-mediated immunosurveillance, several lines of evidence favor the view that such early anti-MM effects are rather to be interpreted as a sign of immunological stress resulting in poor disease control. In fact, the effector function of expanded NK cells from MM subjects has been unexpectedly found to be similar to that of NK cells obtained from healthy donors (114), and NK cells obtained from untreated or previously treated MM patients show a lower increase in cytotoxicity to the K562 cell line in response to pre-incubation with IFN-γ (121). Moreover, NK cell effector functions positively correlate with presence of adverse prognostic factors, including anemia, low albumin, high β2-microglobulin, and renal failure (115), suggesting a "stressed" immunoresponse under the pressure of an aggressive clonal expansion (115). Notably, NK cells from patients with MM display an "exhausted" phenotype signature that includes downregulation of multiple activating receptors and upregulation of programmed death receptor-1 (PD-1). Surface expression of activating 2B4 is reduced in both PB (122) and BM (123) NK cells obtained from untreated subjects with MM, potentially preventing killing of plasma cells despite low HLA class I expression (105). NKG2D and NCR are also downregulated in MM, but preferentially in the BM (122, 123), supporting the concept that downregulation of certain activating NK cell receptors is both dependent on soluble ligands and direct cell–cell contact. Negative signaling from PD-1 is a well-established marker of exhaustion on T cells, but can also disrupt NK cell cytotoxicity and cytokine production (124). In MM, both expression of PD-1 on NK cells and of its ligand PD-L1 on plasma cells has been described (125, 126). PD-1/PD-L1 interactions may therefore promote NK cell functional exhaustion in MM, a phenomenon potentially reversible by checkpoint blockade inhibition (see Inhibitors of the PD-1/PD-L1 Pathway; **Figure 1**).

Natural killer cell-mediated immunity further deteriorates in advanced MM. Compared to MGUS and untreated MM, PB NK cell numbers are substantially reduced in advanced disease (113). Altered distribution of NK cell subsets in human BM may be likewise hypothesized based on studies in mice demonstrating selective decrease of KLRG1$^-$ NK cells during MM progression (127). Evolving phenotype editing further promotes tumor escape from NK cell-mediated immunosurveillance. Furthermore, the activating receptor DNAM-1, expressed on NK cells from healthy donors and with MM in complete remission, is downregulated on NK cells from patient with active MM (128). This phenomenon is particularly relevant for late-stage cancer immune escape, as killing of malignant plasma cells is in certain circumstances critically dependent on DNAM-1 engagement of PVR and nectin-2 (128, 129). In line with these findings, NK cell activity in advanced MM is significantly impaired (130). Taken together, these data

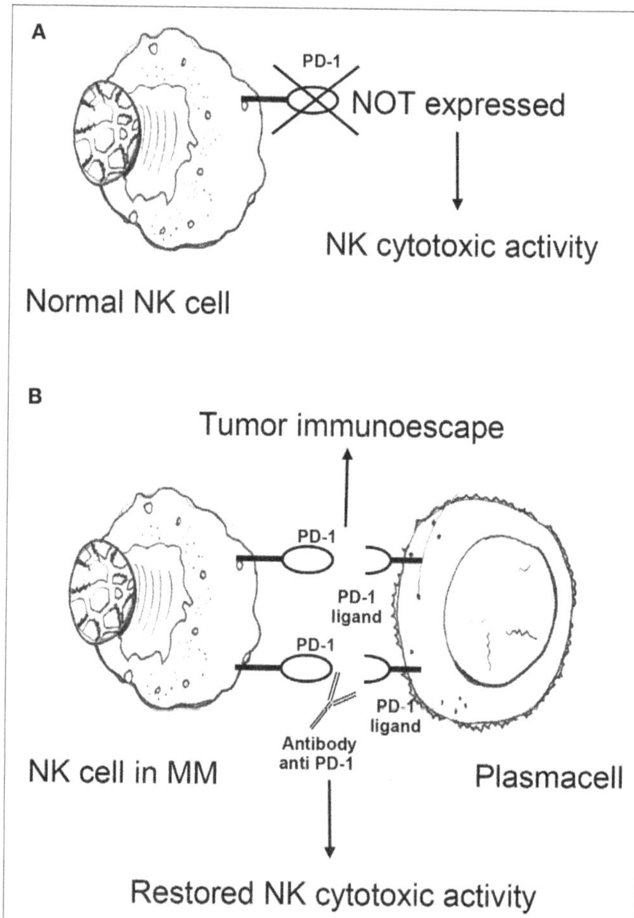

FIGURE 1 | **(A)** Antitumor cytotoxic activity of NK cells in healthy individuals is not impaired by PD-1 expression. **(B)** NK cells from MM patients express PD-1, which promotes MM escape from NK cell-mediated immunosurveillance upon engagement with cognate ligand PD-L1 on plasma cells. PD-1/PD-L1 blocking monoclonal antibodies may potentiate NK cell effector functions against MM. NK, natural killer; PD-1, programmed death receptor-1; MM, multiple myeloma.

indicate that NK cell immunity alterations, already detectable in early myeloma, progress in a clinical stage-dependent manner and that immunotherapy modalities based on efficient NK cell effector function such as (i.e., mAbs) are likely to exert a more effective anti-myeloma effect when used in early-stage disease.

Factors promoting NK immunity suppression in MM are summarized in **Table 1**.

IMMUNE CHECKPOINT BLOCKADE OF NK CELLS

Inhibitors of the PD-1/PD-L1 Pathway

Programmed death receptor-1 is a transmembrane protein expressed on the surface of antigen-activated T and B cells. It has two ligands, PD-L1 and PD-L2. PD-L1 is expressed on both antigen-presenting cells/dendritic cells and a wide spectrum of non-hematopoietic cells. PD-1/PD-L1 interactions physiologically counter T cell stimulatory signals and allow T cell homeostasis and self-tolerance by suppressing activation and proliferation of autoreactive T cells. PD-1/PD-L1 binding delivers an inhibitory costimulatory signal that induces a state of T cell exhaustion that prevents activation and proliferation of T cells. Unlike NK cells from healthy donors, NK cells from MM patients express PD-1 (**Figure 1A**), suggesting that NK cells from healthy donors do not express PD-1 (**Figure 1A**), however, NK cells from MM patients do. This may show that a functional change in NK cells in response to MM may cause an immunosuppressive microenvironment for MM to grow. In the light of these observations and the broad expression of PD-1 and its ligands in the MM microenvironment, the PD-1/PD-L1 pathway may play a pivotal role in the immune evasion of MM cells (**Figure 1B**).

A role for the PD-1/PD-L1 signaling pathway in the NK cell immunoresponse against MM and of the anti-PD1 antibody CT-011 was first shown by Benson et al. (125). CT-011 was demonstrated to enhance human NK cell function against autologous, primary MM cells by affecting NK cell trafficking, immune complex formation with MM cells, and cytotoxicity toward MM cells expressing PD-L1 while sparing normal cells (**Figure 1B**).

It was also shown that lenalidomide had the ability to down regulate PD-L1 on primary MM cells and, by so doing, increase NK cell functions against MM. Thus, targeting the PD-1/PD-L1 pathway may become a feasible clinical strategy in MM, especially in patients with persistent residual disease (131).

One preliminary phase I study reported on 17 patients treated with pembrolizumab, a PD-1 inhibitor, in combination with lenalidomide and dexamethasone (132). Overall response and very good partial response rates were 76 and 23%, respectively. Some 75% of patients achieved stable disease. Many patients were

TABLE 1 | Microenvironment alterations potentially promoting natural killer (NK) immunity suppression in multiple myeloma.

Factors	Function	Effect of TM	Impact on NK cell immunity	Reference
Soluble				
TGF-β	Anti-inflammatory cytokine	⇑	Reduced NK effector functions Downregulation of activating receptors	Castriconi et al. (52); Lee et al. (53); Mamessier et al. (54)
IL-10	Anti-inflammatory cytokine	⇑	Resistance to NK cytotoxicity Reduced NK cytokine production	Tsuruma et al. (62); Sharma et al. (55); Zheng et al. (56)
IL-6	Pro-inflammatory cytokine	⇑	Reduced NK effector functions	Bataille et al. (57); Scheid et al. (66)
IFN-γ	Pro-inflammatory cytokine	⇓	Reduced NK effector functions	Sharma et al. (55); Zheng et al. (56)
PGE2	Prostaglandin	⇑	Reduced NK effector functions Inhibition of positive intracellular signaling	Lu et al. (68); Martinet et al. (67)
sMIC	NKG2D ligand	⇑	Reduced NK effector functions Downregulation of NK activating receptors	Groh et al. (86); Jinushi et al. (84); Xiao et al. (89)
Cell bound				
mMIC	NKG2D ligand	⇓	Resistance to NK cytotoxicity	Jinushi et al. (84)
B7-H6	NKp30 ligand	⇓	Resistance to NK cytotoxicity	Fiegler et al. (104)
HLA class I	KIR/NKG2A ligands	⇑	Resistance to NK cytotoxicity	Carbone et al. (105); Bossard et al. (106); Sarkar et al. (107)
CD200	Membrane glycoprotein	⇑	Reduced NK effector functions Downregulation of NK activating receptors	Moreaux et al. (112); Coles et al. (110)
2B4	Activating receptor	⇓	Reduced NK effector functions	Fauriat et al. (122); Costello et al. (123)
NKG2D	Activating receptor	⇓	Reduced NK effector functions	Fauriat et al. (122); Costello et al. (123)
NCRs	Activating receptors	⇓	Reduced NK effector functions	Fauriat et al. (122); Costello et al. (123)
DNAM-1	Activating receptor	⇓	Reduced NK effector functions	El-Sherbiny et al. (128)
PD-1	Immune checkpoint receptor	⇑	Reduced NK effector functions	Benson et al. (125); Gorgun et al. (126); Beldi-Ferchiou et al. (124)
KLRG1	Co-inhibitory receptor	⇑	Reduced NK effector functions	Ponzetta et al. (127)

TM, tumor microenvironment; PGE2, prostaglandin E2; sMIC, soluble MIC; mMIC, membrane-bound MIC; KIRs, killer immunoglobulin-like receptors; NCRs, natural cytotoxicity receptors; PD-1, programmed cell death protein 1/programmed cell death protein ligand 1; KLRG1, killer cell lectin-like receptor subfamily G member 1.

⇑ denotes increase; ⇓ denotes decrease.

heavily pretreated with other lines of therapy. Almost all patients, however, experienced at least one adverse event with anemia, neutropenia, thrombocytopenia, fatigue, hyperglycemia, and muscle spasms being the most common. Two other recent studies with nivolumab showed acceptable toxicity but no objective responses (133, 134). Efficacy assessment of nivolumab, alone or in combination, is ongoing.

More recently, a novel subpopulation of human NK cells expressing high levels of PD-1 have been identified in ovarian cancer, characterized by low proliferative responses, and impaired antitumor activity that can be partially restored by antibody-mediated disruption of PD-1/PD-L1 interaction (135).

Future studies to evaluate the real therapeutic role of anti-PD-1 antibodies, maybe in combination with other agents with potent anti-myeloma activity such as lenalidomide, are warranted.

KIR-Specific Immune Checkpoint Inhibition

The role of NK cells as graft-vs.-myeloma effectors was first investigated in preclinical models. Frohn et al. described for the first time the killing ability of NK cells against three different MM cell lines. The mean NK cell killing ability on MM samples ranged from 23 to 34.5% (136). Moreover, KIR-ligand mismatch in T cell-depleted allogeneic stem cell transplantation reduced the relapse incidence in MM recipients. The impact of KIR-ligand mismatch was assessed in a cohort of 73 MM patients who received reduced-intensity unrelated donor transplants. KIR-ligand mismatch in the graft-vs.-host disease direction was significantly associated with lower risk of relapse (HR: 0; $p < 0.0001$) (137).

To exploit this pathway, Romagné et al. generated an IgG monoclonal antibody, 1-7F9, against three different KIRs (KIR2DL-1, KIR2DL-2, and KIR2DL-3) to enhance the NK cells antitumor effect. This checkpoint inhibitor augmented NK cell-mediated lysis of HLA-C-expressing tumor cells without interfering with normal peripheral blood (PB) mononuclear cells (138) (**Figure 2A**). The therapeutic potential of 1-7F9 was then demonstrated in preclinical mouse models, providing the platform for translational studies in humans (139).

The drug IPH2101, formerly 1-7F9, was tested in a phase I trial in 32 patients with relapsed/refractory MM. IPH2101 was administered for up to four 28-day cycles, in 7 dose-escalated cohorts (0.0003–3 mg/kg). Only one patient developed severe toxicity, characterized by grade 4 acute renal failure with hyperkalemia and hyperuricemia. From a biological point of view, the drug determined the full saturation of NK inhibitory KIRs (140). Furthermore, lenalidomide and IPH2101 were investigated as a novel, steroid-sparing, dual immunotherapy in 15 MM patients: the biological endpoint of full KIR occupancy was achieved, 5 patients had a response, and 5 severe adverse events were reported (141).

In an open-label, single arm two-stage phase II trial, IPH2101 was employed at the dose of 1 mg/kg every other month for six cycles in nine patients with smoldering MM. Despite the promising results from preclinical and phase I studies, the trial was terminated before planned second stage due to lack of patients meeting the primary objective (50% decline in M-protein) (142).

A recombinant version of IPH2101 was developed with a stabilized hinge (lirilumab). A phase I study of the safety and tolerability of lirilumab with elotuzumab in myeloma patients is currently in progress. Of note, lirilumab recognizes both the inhibitory KIR2DL1, -L2, and -L3 and the activating KIR2DS1-2. Therefore, lirilumab-mediated modulation of intracellular signals is expected to vary according to patient's HLA class I genetic background and KIR receptor repertoire.

In vitro experiments showed that KIR2D molecules are removed from NK cells surface by trogocytosis. This phenomenon culminated in a strong reduction of NK cell cytotoxic function correlating with the loss of free KIR2D surface molecules (143). These data favor future protocol designs where lirilumab is administered in combination with other NK cell-activating agents, rather than as single agent.

IMMUNOMODULATORY DRUGS AND MONOCLONAL ANTIBODIES

Lenalidomide

Lenalidomide, a thalidomide analog, is an immunomodulatory drug with multiple mechanisms of action in MM. It is currently approved in both EU and USA in association with dexamethasone for the maintenance treatment of patients with newly diagnosed MM who have undergone an autograft. Four pivotal phase III studies have associated lenalidomide with improved progression-free survival and better overall response rates (144–147). Although lenalidomide has also been associated with increased risk of a second primary cancer, the overall survival benefits outweigh the risk (148).

Due to failure of single-agent anti-KIRs in phase II studies, researchers from multiple institutions investigated possible combined therapies. *In vitro*, the immunomodulatory agent lenalidomide was responsible of NK cell expansion and activation associated with malignant cells apoptosis (149). On this platform, Benson et al. tested the cytotoxicity of IPH2101 in combination to lenalidomide against MM cell lines U266 and K562 (139, 140). Healthy donor NK cells pretreated with lenalidomide or IPH2101 alone and combined showed increased IFN-γ production against primary MM cells compared to controls ($p < 0.05$). Furthermore, NK cells pretreated with both lenalidomide and IPH2101 led to the highest IFN-γ peak. The statistical interaction of p-value was 0.0182, suggesting a synergistic effect between the two drugs. Then, healthy donor PB mononuclear cells (PBMCs) incubated as control or with lenalidomide and/or with IPH2101 were used as effectors against U266 MM cell targets. Lenalidomide increased the specific release, a surrogate for cytotoxicity, by around 1.39-fold relative to control ($p < 0.01$). IPH2101 increased the specific release by 1.48-fold ($p < 0.01$). The two drugs combined increased the specific release by 2.09-fold relative to control ($p < 0.001$), which means a significantly higher cytotoxic effect than either lenalidomide or IPH2101 alone. Patient-derived NK cell cytotoxicity against autologous MM targets was enhanced by the combination of lenalidomide plus IPH2101 (128 ± 9 spots/well) compared with control conditions (81 ± 7 spots/well). Based on *in vitro* results, the authors

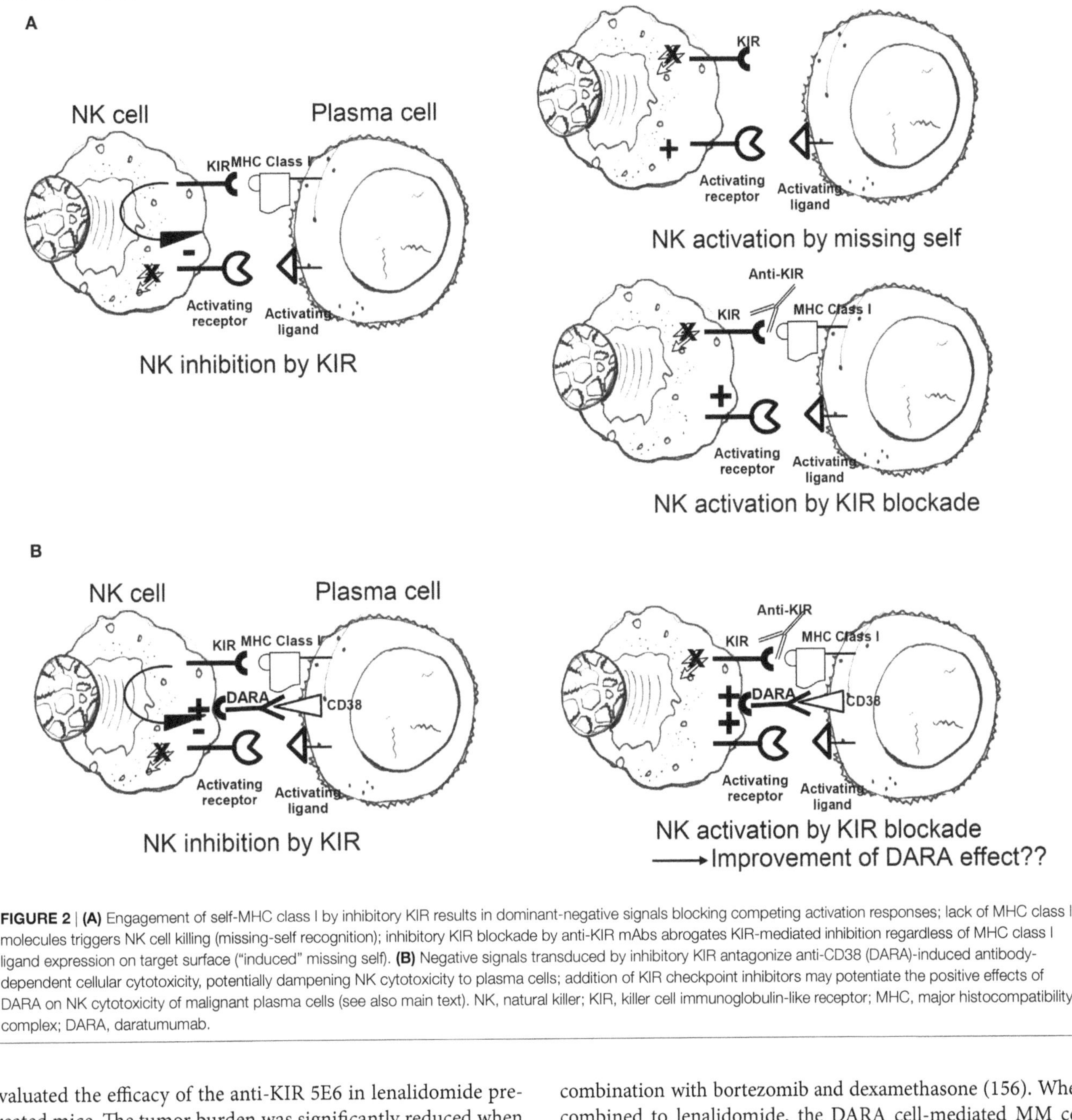

FIGURE 2 | **(A)** Engagement of self-MHC class I by inhibitory KIR results in dominant-negative signals blocking competing activation responses; lack of MHC class I molecules triggers NK cell killing (missing-self recognition); inhibitory KIR blockade by anti-KIR mAbs abrogates KIR-mediated inhibition regardless of MHC class I ligand expression on target surface ("induced" missing self). **(B)** Negative signals transduced by inhibitory KIR antagonize anti-CD38 (DARA)-induced antibody-dependent cellular cytotoxicity, potentially dampening NK cytotoxicity to plasma cells; addition of KIR checkpoint inhibitors may potentiate the positive effects of DARA on NK cytotoxicity of malignant plasma cells (see also main text). NK, natural killer; KIR, killer cell immunoglobulin-like receptor; MHC, major histocompatibility complex; DARA, daratumumab.

evaluated the efficacy of the anti-KIR 5E6 in lenalidomide pretreated mice. The tumor burden was significantly reduced when the combination of 5E6 and lenalidomide was employed, in comparison to controls ($p < 0.005$). These data provide the basis for the translation of IPH2101 and lenalidomide combination in phase I and II studies.

Daratumumab (DARA)

Daratumumab is an IgGk monoclonal antibody targeting CD38, a cell surface protein that is overexpressed on MM cells (150, 151). Preclinical studies have shown that DARA induces MM cell death through several mechanisms, including complement-dependent cytotoxicity (152), ADCC (153), antibody-dependent cellular phagocytosis (154), and apoptosis (155). The drug showed efficacy as single agent in heavily pretreated MM patients or in combination with bortezomib and dexamethasone (156). When combined to lenalidomide, the DARA cell-mediated MM cell clearance was enhanced due to lenalidomide-dependent NK cell activation. In the light of preclinical results of lenalidomide in combination with anti-KIR agents, Nijhof et al. hypothesized that the NK cell-mediated cytotoxicity induced by DARA could be enhanced by anti-KIRs (**Figure 2B**). The effect could be further improved through the association with lenalidomide which stimulates the proliferation of NK cells and activates them (157), overcoming NK cells depletion induced by DARA itself (158).

Elotuzumab

Initially, Hsi et al. described a humanized antibody, HuLuc63, which specifically targeted CS1 (CCND3 subset 1, CRACC, and SLAMF7), a cell surface glycoprotein that had not previously been associated with MM cells. By flow-cytometry, HuLuc63 showed specific staining of $CD138^+$ myeloma cells, NK cells, NK-like T cells, and $CD8^+$ T cells. HuLuc63 showed significant *in vitro* ADCC against primary myeloma cells as targets and allogeneic or autologous NK cells as effectors. The authors concluded that HuLuc63 could eliminate MM partly through NK-mediated ADCC and targeting CS1 with HuLuc63 could become a novel treatment strategy (159). Tai et al. also showed that HuLuc63 was effective in inducing ADCC against primary MM cells resistant to novel therapies such as bortezomib and HSP90 inhibitor. Moreover, pre-treatment with conventional or novel anti-MM agents enhanced HuLuc63-induced MM cell lysis (160). Collins et al. also hypothesized that elotuzumab may have other mechanisms of action. A number of findings clearly suggested that elotuzumab may enhance NK cell function beyond ADCC. Elotuzumab was shown to induce NK cell activation by binding to CS1 which promotes cytotoxicity against $CS1^+$ MM cells but not against autologous $CS1^+$ NK cells. Moreover, NK cell activation was shown to be dependent on differential expression of the signaling intermediary EAT-2 which is present in NK cells but absent in primary, human MM cells (161). Therefore, HuLu63 enhances NK cell cytotoxicity to MM *via* a dual mechanism (**Figure 3**). The synergy between current anti-CS1 antibody elotuzumab, formerly known as HuLuc63, and bortezomib was also shown by van Rhee et al. (162). Elotuzumab was approved by FDA in 2015 for the treatment of MM, specific for signaling lymphocytic activation molecule-F7 (SLAMF7, or CS1) (163). As previously mentioned, SLAMF7 is a member of the Ig gene superfamily, almost universally expressed (>95%) on the surface of marrow MM cells, but not on normal tissues, with restricted expression on specific lymphocytes including NK cells. SLAMF7 determines activating or inhibitory effects on NK cells depending on the expression or not of EAT-2, an adapter protein (**Figure 3**). Given that MM cells lack EAT-2, the molecular mechanism by which SLAMF7 mediates inhibition in NK cells was investigated by Guo et al. It was shown that the inhibitory effects of SLAMF7 in $EAT\text{-}2^-$ NK cells was mediated by a mechanism implicating lipid phosphatase SHIP-1, Src kinases, and protein tyrosine phosphatase CD45. Coupling of SLAMF7 to SHIP-1 was highly compromised in MM cells. This correlated with a lack of CD45, which is required to activate Src family kinases in hematopoietic cells and was needed to initiate SLAMF7 inhibitory signals. This defect may explain why elotuzumab eliminates MM cells by an indirect mechanism that involves NK cells activation (164, 165). An elegant preclinical model clearly showed that elotuzumab activates NK cells and promotes myeloma cell death in healthy donor PB lymphocyte (PBL)/myeloma cell cocultures (166). Moreover, the combination of elotuzumab plus lenalidomide demonstrated higher anti-myeloma activity on established *in vivo* MM xenografts and in *in vitro* PBL/myeloma cell cocultures than either agent alone. In the same study, it was interestingly shown that the increased NK cell anti-myeloma functions were also due to increased secretion of IL-2 and production of TNF-α that combined to enhance NK cell activation and MM cell killing. All these findings supported the clinical application of combination strategies. Elotuzumab initially showed activity in combination with lenalidomide and dexamethasone in a phase I and a phase Ib-2 clinical studies in relapsed/refractory MM (167, 168). In a subsequet randomized study, patients with relapsed/refractory MM received either elotuzumab with lenalidomide and dexamethasone, or lenalidomide and dexamethasone alone. Patients who received a combination of elotuzumab, lenalidomide, and dexamethasone had a significant relative reduction of 30% in the risk of disease progression or death (169). Finally, Jakubowiak et al. reported on a phase II study in relapsed/refractory MM patients where combined elotuzumab/bortezomib/dexamethasone were compared with bortezomib/dexamethasone until progression or unacceptable toxicity. Overall, elotuzumab appeared to provide clinical benefit without clinically significant toxicity when combined with bortezomib (170).

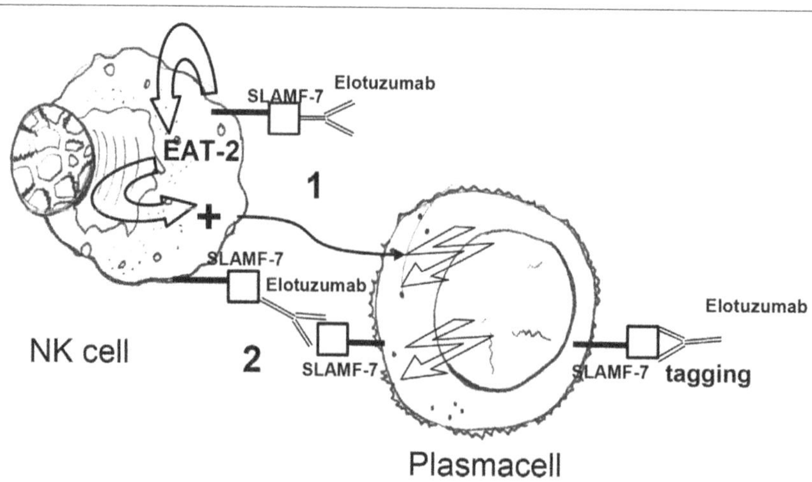

FIGURE 3 | Elotuzumab activates NK cells *via* (1) an indirect mechanism, i.e., binding of the extracellular portion of SLAMF7 and recruitment of the EAT-2 adaptor protein and (2) a direct mechanism, i.e., antibody-dependent cellular cytotoxicity in response to SLAMF7 tagging on plasma cells. Owing to the absence of EAT-2 in plasma cells, elotuzumab engagement does not cause activation of plasma cells. NK, natural killer; EAT-2 Ewing's sarcoma-associated transcript 2; Fc, fragment crystallizable; Fab, fragment antigen binding.

IDO Inhibitors

Indoleamine 2,3-dioxygenase plays a pivotal role in the metabolic cascade that converts the essential amino acid L-tryptophan (Trp) into L-kynurenine (Kyn). Moreover, IDO has also been shown to be involved in the establishment and maintenance of peripheral tolerance. This function may partly be due to IDO1 capacity to restrict the microenvironmental availability of Trp and to increase the accumulation of Kyn and derivatives. The expression of IDO on neoplastic lesions may help cancer to escape immunosurveillance. IDO inhibitors (i.e., 1-methyl-tryptophan) have therefore become a new class of anti-cancer agents. Current models imply that IDO limits both innate and adaptive immunoresponses by depleting immunoeffector cells of Trp (171, 172) and by promoting the accumulation of Kyn and its derivatives 3-dydroxykynurenine and 3-hydroxyanthranilic acid (173, 174). These derivatives have been shown not only to exert cytostatic and cytotoxic effects on several immunoeffectors including CD8 T cells, NK cells, and invariant NKT cells (173–175) but also to inhibit TH17 cells and to promote the differentiation of naıve CD4 T cells into T_{regs} and tolerogenic activity of dendritic cells (174–181). Interestingly, Bonanno et al. investigated IDO expression in 25 symptomatic MM and in 7 with either MGUS or smoldering MM (182). IDO-driven tryptophan breakdown was correlated with the release of hepatocyte growth factor (HGF) and with the frequency of T_{reg} cells and NY-ESO-1-specific CD8 T cells. Kyn was increased in 75% of MM patients and correlated with the expansion of T_{reg} cells and the contraction of NY-ESO-1-specific CD8 T cells. *In vitro*, primary MM cells promoted the differentiation of allogeneic CD4 T cells into T_{reg} cells and suppressed IFN-γ/IL-2 secretion while preserving IL-4 and IL-10 production. Both T_{reg} expansion and inhibition of Th1 differentiation were partly reverted by D,L-1-methyl-tryptophan, an inhibitor of IDO. Of note, HGF levels were higher within the marrow microenvironment of patients with $IDO^{(+)}$ MM as compared with patients with $IDO^{(-)}$ MM. The antagonism of MET receptor for HGF with SU11274, a MET inhibitor, prevented HGF-induced AKT phosphorylation and resulted in reduced IDO protein levels and functional activity. These findings suggest that IDO expression may contribute to immunosuppression in patients with MM. IDO inhibitors are currently being tested either as single agent or in combination with other anti-cancer drugs in a number of solid tumors even though this class has not yet been evaluated in MM.

NK CELLS AND THEIR ROLE IN ALLOGRAFTING: LESSONS FROM ACUTE LEUKEMIAS

Some of the most convincing proofs of the potential of NK cells as immunotherapeutic tools derive from evidences accumulated over the past two decades in the setting of allogeneic hematopoietic stem cell transplantation (allo-HSCT). The interest in NK cell immunobiology stemmed mainly from two observations: (a) NK cells are the first lymphocyte subset to recover after transplantation, often reaching percentages and absolute counts superior to those commonly observed in healthy subjects (183, 184), and (b) since KIRs and HLA ligands are encoded on different, independently inherited, chromosomes (chromosome 19 and 6, respectively), the KIR gene repertoire transferred from the donor into the host is often different posing the issue whether, and how, an efficient NK cell functionality can be achieved (185–187).

Two hallmark studies came from the Stanford group (188, 189) where it was demonstrated that after a variable number of months following HLA-matched, KIR-mismatched allo-HSCT, the NK cell repertoire is dominated by lymphocytes displaying an immature phenotype ($CD56^{bright}$ and $CD94/NKG2A^{+}$) and then it stabilizes and becomes similar to that of donor origin. Moreover, these studies highlighted significant differences in the repertoire recovery kinetics among patients, and clearly correlated impaired NK cell reconstitution with occurrence of post-transplant complications. The development of a HSCT platform which allowed to safely infuse HLA-haploidentical grafts set the stage for some of the most exciting discoveries in transplant biology and translational potential of NK cells.

Starting from preclinical studies on the tolerogenic potential of "stem cell megadoses" (190), the Perugia team developed a protocol which combined a highly immunosuppressive myeloablative conditioning regimen with the infusion of high doses of extensively T cell-depleted HLA-haploidentical hematopoietic stem cells (HSCs). Full donor engraftment of the partially incompatible HSCs was successfully achieved, and despite the absence of post-transplant pharmacological prophylaxis, neither acute nor chronic GvHD ensued (191). This elegant and technologically advanced HSCT platform offered the opportunity to investigate the metrics of NK cell reconstitution in a partially HLA-mismatched host and in the absence of confounding factors such as alloreactive T cells or immunosuppressive drugs. This highly favorable environment further boosted the early expansion of NK cells that had been already described in other transplant settings, and, importantly, led to the appearance of donor-derived NK cells with alloreactivity against the host (192).

Velardi et al. brilliantly described the principles by which NK cell alloreactivity developed and designed an algorithm to easily predict it. Based on this model—later defined as "ligand-ligand" or "KIR-ligand mismatch" model—post-transplant alloreactivity is unleashed when the donor carried one or more KIR ligands (i.e., HLA class I alleles encompassing the Bw4, C1, or C2 motifs) absent in the host. In this setting, inhibitory KIRs expressed on the surface of donor-derived NK cells—which, in the host, are continuously engaged by their respective ligands in the pre-transplant phase—do not find their cognate HLA molecules on host cells and tissues leading to a perception of "missing self" that activates an alloreactive response (193, 194).

One of the most striking observations by the Perugia group was that NK alloreactivity did not result in clinical GvHD, but, conversely, led to a potentially eradicating mechanism of residual leukemic cells reducing relapse incidence and risk of graft failure and GvHD (195, 196). Over the following years, several studies confirmed and consolidated the evidence that in T cell-depleted haploidentical HSCT NK cell alloreactivity represents the main driver of the graft-vs.-leukemia effect and a major predictor of overall clinical outcomes in both adults (197–199) and children (200, 201). In addition, these studies were a major drive for the development of cell therapy protocols in which haploidentical KIR ligand-mismatched NK cells were infused in leukemic patients after lymphodepleting chemotherapy with highly promising results (202–204).

In more recent years, several new platforms of haploidentical HSCT have been developed, mainly with the aim at improving T cell immune reconstitution and at reducing post-transplant infectious complications (205). In this "new era," studies on the impact of KIR-ligand mismatches on transplant outcomes reported some conflicting results. For instance, it was shown that infusion of unmanipulated BM grafts or of donor post-transplant T cell add-backs may mask, or blunt, the effectiveness of NK cell alloreactivity (206–208). By contrast, other recent haploidentical HSCT platforms based on the selective depletion of αβ T cells or on the infusion of balanced doses of conventional and regulatory T cells appeared to better preserve the positive effect of KIR ligand mismatches (209, 210).

In partially HLA-mismatched unrelated donor HSCT, either from adult volunteers or cord blood (CB) units, the potential role of NK alloreactivity has also been a matter of debate. Some studies supported a positive role of KIR ligand mismatches (211–213) and others found no significant advantage or even adverse effects (214–216).

To overcome these inconsistencies, several alternative immunogenetic models have been developed to better predict NK cell-driven effects on transplant outcomes. In particular, Cooley et al. focused on the donor genetic repertoire and demonstrated in a number of independent studies that donors with a KIR gene asset enriched in activating receptors—group B KIR haplotypes—can provide a superior relapse-free survival after unrelated HSCT for leukemias (217–219). Another model which takes into account both donor activating KIR asset and donor/recipient HLA typing has been proposed and validated by Venstrom et al. In an analysis on more than 1,200 unrelated HSCTs, the authors observed that the presence of donor-activating receptor KIR2DS1 and of HLA-C1 ligands provided a significant protection from relapse, further enhanced in case of recipient HLA-C1 positivity (220).

Despite the multiplicity of models proposed over the years in the setting of allo-HSCT, not all the immunogenetic mechanisms that regulate NK cell interactions and alloreactivity have fully been understood. However, it is widely assumed that NK cell alloreactivity is instrumental in control and eradication of hematological malignancies.

NK CELL THERAPIES

Expanded NK Cells for MM Treatment

Expansion of NK cells from PBMC of patients with MM has been achieved using a culture system supplemented with IL-2 and OKT3 (221). NK cells could be extensively propagated (average 1,625-fold expansion in 20 days) and displayed increased levels of activating receptors as well as cytotoxicity to the NK-susceptible K562 line and to autologous MM cells (222). Another NK cell expansion strategy for MM immunotherapy is based on the artificial feeder K562 transfected with CD137L and membrane-bound IL-15. This technique allowed extensive *in vitro* NK cell propagation (average 447-fold, range 20–10,430 on harvest day, i.e., days 10–14). Transfer of these cells into a xenogeneic model of high-risk MM resulted in myeloma growth inhibition and protection against osteolysis (223). The same group tested the safety, persistence, and activity of expanded NK cells in seven heavily pretreated patients with high-risk relapsed myeloma: no serious adverse events related to NK cell infusion was observed. Moreover, the infusion of fresh, rather than cryopreserved, cells resulted of fundamental importance for their *in vivo* expansion. Two/seven patients showed some responses which lasted for at least 6 months (224). More recently, human studies were performed with allogeneic, KIR ligand-mismatched NK cells from haploidentical family donors. NK cells were cytotoxic to K562, the myeloma line U266, and recipient primary MM cells. Fifty percent of the patients with advanced MM achieved near complete remission when these cells were infused prior to autologous SCT (225). Another phase I clinical trial (NCT02481934) evaluated safety and efficacy of multiple infusions of activated and expanded NK cells in combination with lenalidomide- or bortezomib-based regimens (226). Five heavily pretreated refractory/relapsed patients were enrolled. NK cells were activated and expanded for 3 weeks with K562mb15-41BBL cells. Patients received four cycles of new drug-based treatment with two infusions of 7.5×10^6/kg NK cells. Four patients showed stable disease while on NK cell treatment, two showed a 50% reduction in BM plasma cell infiltration and one obtained a response >1 year. No major toxicities were reported. Expanded NK cells showed a highly cytotoxic phenotype and *in vitro* killing and were detected in both BM and PB of treated patients. While efficacy and safety of multiple NK cell infusions need further assessment, these data suggest that repeated transfer of *in vitro* activated and expanded NK cells into MM patients is feasible and may result in clinical benefit when combined with anti-myeloma drugs.

CB NK Cells

Umbilical CB represents a promising source of allogeneic NK cells. However, GMP-grade large scale *ex vivo* expansion is indispensable to generate CB-derived NK (CB-NK) cell doses that may be used in the clinical setting. Shah et al. recently described a strategy for the expansion of NK cells from cryopreserved CB units (227). By co-culturing for 14 days CB units using artificial antigen-presenting feeder cells (aAPC), a highly expanded cell product (average 1,848- and 2,389-fold in 14 days from fresh and cryopreserved samples, respectively) of 95% purity for CB-NK cells and less than 1% $CD3^+$ cells was obtained. Despite differences in the expression of certain cytotoxicity receptors, aAPC-expanded CB-NK cells were phenotypically very similar to CB-NK cells expanded with IL-2 alone. Most importantly, aAPC-expanded CB-NK cells clearly showed cytotoxicity against both *in vitro* MM targets and *in vivo* anti-myeloma activity in a xenogenic mouse model. The same group investigated the mechanisms of CB-NK-mediated cytotoxicity against MM cells (228). Interestingly, a mechanism of transmissible cell death between cells induced by lipid–protein vesicles transferred from CB-NK to MM cells was described. Moreover, these vesicles were capable of migrating from recipient MM cells to neighboring MM cells enhancing cytotoxicity of CB-NK. Altogether, these findings supported the development of CB-NK-based cellular therapies for the treatment of MM. An encouraging first-in-human study of CB-NK cells for MM patients undergoing high dose chemotherapy and autologous transplantation was recently conducted (229). Patients received lenalidomide at a dose of 10 mg from day −8 through −2, standard melphalan at 200 mg/m^2 on day −7. CB-NK cells were infused on day −5 and the autograft performed on day 0. Twelve patients were treated with different dose levels. Most patients were heavily pretreated and had high-risk cytogenetics. Overall CB-NK cells with an activated phenotype (NKG2D$^+$/NKp30$^+$) were detected *in vivo* in six patients. Importantly, no signs/symptoms of GVHD were observed. Eight patients achieved at least near complete remission and two additional patients a very good partial response. After a median follow-up of 21 months, four patients relapsed or experienced progressive disease.

CAR-NK for Myeloma

The impressive clinical results obtained in patients with B cell malignancies with the infusion of T cells genetically modified to express synthetic chimeric antigen receptors (CARs) against the lineage-specific surface antigen CD19 represented a turning point in the history of cancer immunotherapy (230–236) Intriguingly, T cells engineered with an anti-CD19 CAR were capable to induce complete remission also in a patient with MM. However, given that the large majority of malignant plasma cells do not express CD19, studies to understand the mechanism that underlie this unexpected observation are currently in progress (237). More recently, a number of CARs have been developed to specifically target surface antigens expressed by pathological plasma cells, including CD38 (238, 239), CD138 (240), B cell maturation antigen (241, 242), κ light chains (243), SLAMF7

TABLE 2 | Summary of current treatments with novel agents for multiple myeloma (MM) potentially affecting natural killer (NK) cell activity.

Agent	Mechanism of action on NK cells	Clinical trials	Reference
PD-1/PD-L1 checkpoint inhibitors	*Block of the recognition of PD-L1 by PD-1 on NK cells* PD1 blockade may neutralize competitive negative signals resulting in enhanced trafficking, immune complex formation, and cytotoxicity of NK cells (**Figures 1A,B**)	Phase I trial of pembrolizumab with lenalidomide and dexamethasone. Two Phase I trials involving nivolumab showed acceptable tolerability. Efficacy assessment of nivolumab, alone or in combination is ongoing.	Benson et al. (125); San Miguel et al. (132); Suen et al. (133); Lesokhin et al. (134)
HLA-KIR checkpoint inhibitors	*Prevent inhibitory KIR recognition of cognate HLA class I ligands* Blockade of KIR-HLA interactions may neutralize negative signals transduced by inhibitory KIR2DL1/2/3 (**Figure 2A**)	Anti-KIR monoclonal antibody IPH2101 (1-7F9) determined the full saturation of NK inhibitory KIR in a phase I trial enrolling patients with RR MM. Full KIR occupancy was also achieved in a study combining lenalidomide and IPH2101. In this study, 5 (33%) patients had a response. In a single arm two-stage phase II trial, IPH2101 was employed in 9 patients with smoldering MM. The study was stopped before planned second stage due to lack of patients meeting the primary objective (50% decline in M-protein). A phase I study combining elotuzumab with lirilumab, a recombinant version of IPH2101, is currently in progress	Frohn et al. (136); Benson et al. (139); Benson et al. (140); Benson et al. (141); Korde et al. (142); Carlsten et al. (143)
Daratumumab (DARA)	*ADCC to CD38$^+$ MM cells* Cytolytic activity to MM cells triggered by CD16 signaling upon recognition of antibody tagged to CD38 antigen. NK cell-mediated cytotoxicity induced by DARA could be enhanced by lenalidomide and KIR blockade. Other mechanisms: complement-dependent cytotoxicity, antibody-dependent cellular phagocytosis, and apoptosis (**Figure 2B**)	DARA was tested in combination with bortezomib and dexamethasone in RRMM. The primary end point was progression-free survival. DARA in combination with bortezomib and dexamethasone resulted in a significantly longer progression-free survival than bortezomib and dexamethasone alone	Palumbo et al. (156)
Elotuzumab	*Direct effect: ADCC to MM cells expressing SLAMF7* Indirect effect: activation of SLAMF7$^+$ NK cells Dual mechanism of action: (1) NK cell activation *via* SLAMF7 binding and recruitment of the EAT-2 adaptor proteins; (2) NK-mediated ADCC to SLAMF7$^+$ MM cells (**Figure 3**)	Elotuzumab showed activity in combination with lenalidomide and dexamethasone in a phase I and a phase IIb-II clinical studies in RRMM. In a phase III study, patients with RRMM patients were treated with either elotuzumab with lenalidomide and dexamethasone, or lenalidomide and dexamethasone alone. Patients treated with the combination of elotuzumab, lenalidomide, and dexamethasone had a significantly reduced risk of disease progression or death. In a phase II study in RRMM patients, elotuzumab showed clinical benefit without significant toxicity when combined with bortezomib	Lonial et al. (167); Lonial et al. (169); Jakubowiak et al. (170)
IDO inhibitors	*Inhibition of L-tryptophan degradation* Reversal of NK immunosuppression by increased availability of l-tryptophan and reduced accumulation of L-kyreunine	IDO inhibitors are currently used as single agent or in combination in a number of solid tumors. This class has not yet been evaluated in clinical trials in myeloma patients	Uyttenhove et al. (172); Fallarino et al. (173); Bonanno et al. (182)

PD-1/PD-L1, programmed cell death protein 1/programmed cell death protein ligand 1; KIRs, killer immunoglobulin-like receptors; RR MM, relapsed/refractory MM; ADCC, antibody-dependent cellular cytotoxicity; SLAMF7, signaling lymphocytic activation molecule family 7; IDO, indoleamine 2,3-dioxygenase.

(244), and CD44v6 (245). However, despite their tremendous efficacy, CAR-T cells have also raised concerns on their short- and long-term toxicities, in particular the development of life-threatening cytokine release syndrome and the risks of prolonged aplasia of the healthy counterparts of the target tumor—"off tumor/on target toxicity"—and in case of allogeneic CAR-T cells the development of GvHD (246, 247).

To address these issues, genetic modifications with CARs of cells belonging to the innate immune system, and of NK cells in particular, may yield several potential advantages. For instance, most innate cells recognize and eliminate tumors by stereotyped patterns and have been infused into allogeneic recipient without excessive toxicities and with some promising intrinsic antitumor efficacy. Moreover, the short-lived persistence of innate immunocells in an allogeneic host, considered up to now one of the major limitations, may become an added value in case of CARs targeting antigens that are shared with mature cell types for which prolonged aplasia may be a concern (i.e., memory B cells, monocytes, or plasma cells) (248, 249).

Genetic modification of the human NK cell lines NKL and NK-92 by means of a lentiviral vector encoding for anti-SLAMF7 and anti-CD138 CARs has proven feasible. This did not substantially modify the expression profile of transduced cells and conferred selectivity for the target and the ability to kill human malignant plasma cells both in *ex vivo* and in an orthotopic xenograft models (250, 251). Overall, several steps to optimize and validate CAR-modified NK cells should be taken before their possible clinical use. In particular, the choice of the most appropriate NK cell source to be modified is a matter of intense debate (252). Whether freshly isolated NK cells may represent the most physiological choice to achieve sufficient cell doses and transduction efficiency remains unknown. Conversely, NK cells expanded from PB or from progenitor cells may be more easily modified even though their expression profile and functional competence may be negatively affected by prolonged *ex vivo* culture. Finally, immortalized human NK cell lines, such as NK-92, can be very efficiently transduced and expanded in desired numbers even for "off-the-shelf" use even though their cell surface expression of activating receptors is lower than in freshly isolated or expanded NK cells. Moreover, the need to irradiate the cell product before infusion would further limit their *in vivo* persistence (252, 253). A new modality that exploits the combination of the anti-CR38 monoclonal antibody DARA with $CD38^{(-)}$ NK cells armed with CS1 CAR has very recently been described by Wang et al. to treat relapsed MM (254). Given that both CS1 and CD38 are MM-associated antigens, their simultaneous targeting may prevent progression. The same authors previously showed that DARA induces apoptosis in $CD38^{(+)}$ NK cells but not in $CD38^{(-)}$ NK cells. It was then hypothesized that DARA in combination with $CD38^{(-)}$ CS1-CAR NK cells may show a synergistic effect and possibly lead to MM eradication. Long-term follow-up of clinical outcomes of this study are eagerly awaited.

FUTURE PERSPECTIVES

The potent crosstalk between malignant plasma cells and their BM microenvironment plays a central role in MM progression and resistance to current therapies. Novel forms of immunotherapy against MM represent a rapidly developing area in cancer therapy. They include treatment strategies that may be delivered either alone or in combination with currently employed therapy lines such as IMiDs and proteasome inhibitors as well as newer agents (**Table 2**). Moreover, immunotherapy may attenuate the systemic toxicity of cytotoxic chemotherapy. A robust body of evidence has clearly shown that enhancing host anti-myeloma immunity within the BM microenvironment may lead to a more efficient disease control. NK cells play a pivotal role in the intricate network of cells and signaling pathways that may prevent immune escape mechanisms. NK cells were clearly shown to have potent *in vivo* antileukemia activity in patients undergoing allografting. Recent observations on NK cell functions in MM have become promising immunotherapeutic strategies. New avenues of research have included expansion of NK cells from PB as well as CB, and the generations of specific CAR-NK cells against myeloma-specific antigens. Moreover, MM NK cells express PD-1 whereas NK cells from healthy individuals do not. This phenotypic characteristic may indicate that immunocheckpoint blockade of NK cells may be an area to fully explore given the remarkable results obtained with anti-PD 1 inhibitors in cancer treatment. Altogether, the studies reported in this review show that NK cells hold promise in changing the natural course of MM and that may help restore immunity to MM and thereby improve survival outcomes.

AUTHOR CONTRIBUTIONS

GP and BB contributed to the initial conception and designed of the manuscript. CB, MF, DM, and LG provided study materials and critically reviewed the manuscript. GP, LV, MF, and BB wrote the manuscript. All authors gave the final approval to the manuscript.

REFERENCES

1. Fonseca R, Bergsagel PL, Drach J, Shaughnessy J, Gutierrez N, Stewart AK. International Myeloma Working Group molecular classification of multiple myeloma: spotlight review. *Leukemia* (2009) 23(12):2210–21. doi:10.1038/leu.2009.174
2. Reece D, Song KW, Fu T, Roland B, Chang H, Horsman DE, et al. Influence of cytogenetics in patients with relapsed or refractory multiple myeloma treated with lenalidomide plus dexamethasone: adverse effect of deletion 17p13. *Blood* (2009) 114(3):522–5. doi:10.1182/blood-2008-12-193458
3. Cornelissen JJ, Sonneveld P, Schoester M, Raaijmakers HG, Nieuwenhuis HK, Dekker AW, et al. MDR-1 expression and response to vincristine, doxorubicin, and dexamethasone chemotherapy in multiple myeloma refractory to alkylating agents. *J Clin Oncol* (1994) 12(1):115–9. doi:10.1200/JCO.1994.12.1.115
4. Yang HH, Ma MH, Vescio RA, Berenson JR. Overcoming drug resistance in multiple myeloma: the emergence of therapeutic approaches to induce apoptosis. *J Clin Oncol* (2003) 21(22):4239–47. doi:10.1200/JCO.2003.06.001
5. Meads MB, Gatenby RA, Dalton WS. Environment-mediated drug resistance: a major contributor to minimal residual disease. *Nat Rev Cancer* (2009) 9(9):665–74. doi:10.1038/nrc2714
6. Greenberg AH, Playfair JH. Spontaneously arising cytotoxicity to the P-815-Y mastocytoma in NZB mice. *Clin Exp Immunol* (1974) 16(1):99–109.
7. Herberman RB, Nunn ME, Lavrin DH, Asofsky R. Effect of antibody to theta antigen on cell-mediated immunity induced in syngeneic mice by murine sarcoma virus. *J Natl Cancer Inst* (1973) 51(5):1509–12. doi:10.1093/jnci/51.5.1509

8. Petranyi GG, Benczur M, Onody CE, Hollan SR. Letter: HL-A 3,7 and lymphocyte cytotoxic activity. *Lancet* (1974) 1(7860):736. doi:10.1016/S0140-6736(74)92943-2
9. Takasugi M, Mickey MR, Terasaki PI. Reactivity of lymphocytes from normal persons on cultured tumor cells. *Cancer Res* (1973) 33(11):2898–902.
10. Herberman RB, Nunn ME, Holden HT, Lavrin DH. Natural cytotoxic reactivity of mouse lymphoid cells against syngeneic and allogeneic tumors. II. Characterization of effector cells. *Int J Cancer* (1975) 16(2):230–9. doi:10.1002/ijc.2910160204
11. Herberman RB, Nunn ME, Lavrin DH. Natural cytotoxic reactivity of mouse lymphoid cells against syngeneic acid allogeneic tumors. I. Distribution of reactivity and specificity. *Int J Cancer* (1975) 16(2):216–29. doi:10.1002/ijc.2910160204
12. Kiessling R, Klein E, Pross H, Wigzell H. "Natural" killer cells in the mouse. II. Cytotoxic cells with specificity for mouse Moloney leukemia cells. Characteristics of the killer cell. *Eur J Immunol* (1975) 5(2):117–21. doi:10.1002/eji.1830050209
13. Kiessling R, Klein E, Wigzell H. "Natural" killer cells in the mouse. I. Cytotoxic cells with specificity for mouse Moloney leukemia cells. Specificity and distribution according to genotype. *Eur J Immunol* (1975) 5(2):112–7. doi:10.1002/eji.1830050208
14. Karre K, Ljunggren HG, Piontek G, Kiessling R. Selective rejection of H-2-deficient lymphoma variants suggests alternative immune defence strategy. *Nature* (1986) 319(6055):675–8. doi:10.1038/319675a0
15. Ljunggren HG, Karre K. In search of the 'missing self': MHC molecules and NK cell recognition. *Immunol Today* (1990) 11(7):237–44. doi:10.1016/0167-5699(90)90097-S
16. Burshtyn DN, Scharenberg AM, Wagtmann N, Rajagopalan S, Berrada K, Yi T, et al. Recruitment of tyrosine phosphatase HCP by the killer cell inhibitor receptor. *Immunity* (1996) 4(1):77–85. doi:10.1016/S1074-7613(00)80300-3
17. Campbell KS, Dessing M, Lopez-Botet M, Cella M, Colonna M. Tyrosine phosphorylation of a human killer inhibitory receptor recruits protein tyrosine phosphatase 1C. *J Exp Med* (1996) 184(1):93–100. doi:10.1084/jem.184.1.93
18. Fry AM, Lanier LL, Weiss A. Phosphotyrosines in the killer cell inhibitory receptor motif of NKB1 are required for negative signaling and for association with protein tyrosine phosphatase 1C. *J Exp Med* (1996) 184(1):295–300. doi:10.1084/jem.184.1.295
19. Burshtyn DN, Yang W, Yi T, Long EO. A novel phosphotyrosine motif with a critical amino acid at position-2 for the SH2 domain-mediated activation of the tyrosine phosphatase SHP-1. *J Biol Chem* (1997) 272(20):13066–72. doi:10.1074/jbc.272.20.13066
20. Yusa S, Campbell KS. Src homology region 2-containing protein tyrosine phosphatase-2 (SHP-2) can play a direct role in the inhibitory function of killer cell Ig-like receptors in human NK cells. *J Immunol* (2003) 170(9):4539–47. doi:10.4049/jimmunol.170.9.4539
21. Biassoni R, Falco M, Cambiaggi A, Costa P, Verdiani S, Pende D, et al. Amino acid substitutions can influence the natural killer (NK)-mediated recognition of HLA-C molecules. Role of serine-77 and lysine-80 in the target cell protection from lysis mediated by "group 2" or "group 1" NK clones. *J Exp Med* (1995) 182(2):605–9. doi:10.1084/jem.182.2.605
22. Winter CC, Long EO. A single amino acid in the p58 killer cell inhibitory receptor controls the ability of natural killer cells to discriminate between the two groups of HLA-C allotypes. *J Immunol* (1997) 158(9):4026–8.
23. Litwin V, Gumperz J, Parham P, Phillips JH, Lanier LL. NKB1: a natural killer cell receptor involved in the recognition of polymorphic HLA-B molecules. *J Exp Med* (1994) 180(2):537–43. doi:10.1084/jem.180.2.537
24. Wagtmann N, Rajagopalan S, Winter CC, Peruzzi M, Long EO. Killer cell inhibitory receptors specific for HLA-C and HLA-B identified by direct binding and by functional transfer. *Immunity* (1995) 3(6):801–9. doi:10.1016/1074-7613(95)90069-1
25. Gumperz JE, Litwin V, Phillips JH, Lanier LL, Parham P. The Bw4 public epitope of HLA-B molecules confers reactivity with natural killer cell clones that express NKB1, a putative HLA receptor. *J Exp Med* (1995) 181(3):1133–44. doi:10.1084/jem.181.3.1133
26. Braud VM, Allan DS, O'Callaghan CA, Soderstrom K, D'Andrea A, Ogg GS, et al. HLA-E binds to natural killer cell receptors CD94/NKG2A, B and C. *Nature* (1998) 391(6669):795–9.
27. Lee N, Llano M, Carretero M, Ishitani A, Navarro F, Lopez-Botet M, et al. HLA-E is a major ligand for the natural killer inhibitory receptor CD94/NKG2A. *Proc Natl Acad Sci U S A* (1998) 95(9):5199–204. doi:10.1073/pnas.95.9.5199
28. Aldemir H, Prod'homme V, Dumaurier MJ, Retiere C, Poupon G, Cazareth J, et al. Cutting edge: lectin-like transcript 1 is a ligand for the CD161 receptor. *J Immunol* (2005) 175(12):7791–5. doi:10.4049/jimmunol.175.12.7791
29. Lanier LL, Chang C, Phillips JH. Human NKR-P1A. A disulfide-linked homodimer of the C-type lectin superfamily expressed by a subset of NK and T lymphocytes. *J Immunol* (1994) 153(6):2417–28.
30. Teixeira AM, Fawcett J, Simmons DL, Watt SM. The N-domain of the biliary glycoprotein (BGP) adhesion molecule mediates homotypic binding: domain interactions and epitope analysis of BGPc. *Blood* (1994) 84(1):211–9.
31. Watt SM, Fawcett J, Murdoch SJ, Teixeira AM, Gschmeissner SE, Hajibagheri NM, et al. CD66 identifies the biliary glycoprotein (BGP) adhesion molecule: cloning, expression, and adhesion functions of the BGPc splice variant. *Blood* (1994) 84(1):200–10.
32. Markel G, Lieberman N, Katz G, Arnon TI, Lotem M, Drize O, et al. CD66a interactions between human melanoma and NK cells: a novel class I MHC-independent inhibitory mechanism of cytotoxicity. *J Immunol* (2002) 168(6):2803–10. doi:10.4049/jimmunol.168.6.2803
33. Bauer S, Groh V, Wu J, Steinle A, Phillips JH, Lanier LL, et al. Activation of NK cells and T cells by NKG2D, a receptor for stress-inducible MICA. *Science* (1999) 285(5428):727–9. doi:10.1126/science.285.5428.727
34. Cosman D, Mullberg J, Sutherland CL, Chin W, Armitage R, Fanslow W, et al. ULBPs, novel MHC class I-related molecules, bind to CMV glycoprotein UL16 and stimulate NK cytotoxicity through the NKG2D receptor. *Immunity* (2001) 14(2):123–33. doi:10.1016/S1074-7613(01)00095-4
35. Sivori S, Vitale M, Morelli L, Sanseverino L, Augugliaro R, Bottino C, et al. p46, a novel natural killer cell-specific surface molecule that mediates cell activation. *J Exp Med* (1997) 186(7):1129–36. doi:10.1084/jem.186.7.1129
36. Pessino A, Sivori S, Bottino C, Malaspina A, Morelli L, Moretta L, et al. Molecular cloning of NKp46: a novel member of the immunoglobulin superfamily involved in triggering of natural cytotoxicity. *J Exp Med* (1998) 188(5):953–60. doi:10.1084/jem.188.5.953
37. Vitale M, Bottino C, Sivori S, Sanseverino L, Castriconi R, Marcenaro E, et al. NKp44, a novel triggering surface molecule specifically expressed by activated natural killer cells, is involved in non-major histocompatibility complex-restricted tumor cell lysis. *J Exp Med* (1998) 187(12):2065–72. doi:10.1084/jem.187.12.2065
38. Pende D, Parolini S, Pessino A, Sivori S, Augugliaro R, Morelli L, et al. Identification and molecular characterization of NKp30, a novel triggering receptor involved in natural cytotoxicity mediated by human natural killer cells. *J Exp Med* (1999) 190(10):1505–16. doi:10.1084/jem.190.10.1505
39. Mandelboim O, Lieberman N, Lev M, Paul L, Arnon TI, Bushkin Y, et al. Recognition of haemagglutinins on virus-infected cells by NKp46 activates lysis by human NK cells. *Nature* (2001) 409(6823):1055–60. doi:10.1038/35059110
40. Arnon TI, Lev M, Katz G, Chernobrov Y, Porgador A, Mandelboim O. Recognition of viral hemagglutinins by NKp44 but not by NKp30. *Eur J Immunol* (2001) 31(9):2680–9. doi:10.1002/1521-4141(200109)31:9<2680::AID-IMMU2680>3.0.CO;2-A
41. Pogge von Strandmann E, Simhadri VR, von Tresckow B, Sasse S, Reiners KS, Hansen HP, et al. Human leukocyte antigen-B-associated transcript 3 is released from tumor cells and engages the NKp30 receptor on natural killer cells. *Immunity* (2007) 27(6):965–74. doi:10.1016/j.immuni.2007.10.010
42. Brandt CS, Baratin M, Yi EC, Kennedy J, Gao Z, Fox B, et al. The B7 family member B7-H6 is a tumor cell ligand for the activating natural killer cell receptor NKp30 in humans. *J Exp Med* (2009) 206(7):1495–503. doi:10.1084/jem.20090681
43. Perussia B, Acuto O, Terhorst C, Faust J, Lazarus R, Fanning V, et al. Human natural killer cells analyzed by B73.1, a monoclonal antibody blocking Fc receptor functions. II. Studies of B73.1 antibody-antigen interaction on the lymphocyte membrane. *J Immunol* (1983) 130(5):2142–8.
44. Mandelboim O, Malik P, Davis DM, Jo CH, Boyson JE, Strominger JL. Human CD16 as a lysis receptor mediating direct natural killer cell cytotoxicity. *Proc Natl Acad Sci U S A* (1999) 96(10):5640–4. doi:10.1073/pnas.96.10.5640

45. Garni-Wagner BA, Purohit A, Mathew PA, Bennett M, Kumar V. A novel function-associated molecule related to non-MHC-restricted cytotoxicity mediated by activated natural killer cells and T cells. *J Immunol* (1993) 151(1):60–70.
46. Brown MH, Boles K, van der Merwe PA, Kumar V, Mathew PA, Barclay AN. 2B4, the natural killer and T cell immunoglobulin superfamily surface protein, is a ligand for CD48. *J Exp Med* (1998) 188(11):2083–90. doi:10.1084/jem.188.11.2083
47. Shibuya A, Campbell D, Hannum C, Yssel H, Franz-Bacon K, McClanahan T, et al. DNAM-1, a novel adhesion molecule involved in the cytolytic function of T lymphocytes. *Immunity* (1996) 4(6):573–81. doi:10.1016/S1074-7613(00)70060-4
48. Bottino C, Castriconi R, Pende D, Rivera P, Nanni M, Carnemolla B, et al. Identification of PVR (CD155) and Nectin-2 (CD112) as cell surface ligands for the human DNAM-1 (CD226) activating molecule. *J Exp Med* (2003) 198(4):557–67. doi:10.1084/jem.20030788
49. Bianchi G, Munshi NC. Pathogenesis beyond the cancer clone(s) in multiple myeloma. *Blood* (2015) 125:3049–58. doi:10.1182/blood-2014-11-568881
50. Urashima M, Ogata A, Chauhan D, Hatziyanni M, Vidriales MB, Dedera DA, et al. Transforming growth factor-beta1: differential effects on multiple myeloma versus normal B cells. *Blood* (1996) 87:1928–38.
51. Beyer M, Kochanek M, Giese T, Endl E, Weihrauch MR, Knolle PA, et al. In vivo peripheral expansion of naive CD4+CD25high FoxP3+ regulatory T cells in patients with multiple myeloma. *Blood* (2006) 107:3940–9. doi:10.1182/blood-2005-09-3671
52. Castriconi R, Cantoni C, Della Chiesa M, Vitale M, Marcenaro E, Conte R, et al. Transforming growth factor beta 1 inhibits expression of NKp30 and NKG2D receptors: consequences for the NK-mediated killing of dendritic cells. *Proc Natl Acad Sci U S A* (2003) 100:4120–5. doi:10.1073/pnas.0730640100
53. Lee JC, Lee KM, Kim DW, Heo DS. Elevated TGF-beta1 secretion and down-modulation of NKG2D underlies impaired NK cytotoxicity in cancer patients. *J Immunol* (2004) 172:7335–40. doi:10.4049/jimmunol.172.12.7335
54. Mamessier E, Sylvain A, Thibult ML, Houvenaeghel G, Jacquemier J, Castellano R, et al. Human breast cancer cells enhance self tolerance by promoting evasion from NK cell antitumor immunity. *J Clin Invest* (2011) 121:3609–22. doi:10.1172/JCI45816
55. Sharma A, Khan R, Joshi S, Kumar L, Sharma M. Dysregulation in T helper 1/T helper 2 cytokine ratios in patients with multiple myeloma. *Leuk Lymphoma* (2010) 51:920–7. doi:10.3109/10428191003699563
56. Zheng MM, Zhang Z, Bemis K, Belch AR, Pilarski LM, Shively JE, et al. The systemic cytokine environment is permanently altered in multiple myeloma. *PLoS One* (2013) 8:e58504. doi:10.1371/journal.pone.0058504
57. Bataille R, Jourdan M, Zhang XG, Klein B. Serum levels of interleukin 6, a potent myeloma cell growth factor, as a reflect of disease severity in plasma cell dyscrasias. *J Clin Invest* (1989) 84:2008–11. doi:10.1172/JCI114392
58. Zhang XG, Klein B, Bataille R. Interleukin-6 is a potent myeloma-cell growth factor in patients with aggressive multiple myeloma. *Blood* (1989) 74:11–3.
59. Kovacs E. Interleukin-6 leads to interleukin-10 production in several human multiple myeloma cell lines. Does interleukin-10 enhance the proliferation of these cells? *Leuk Res* (2010) 34:912–6. doi:10.1016/j.leukres.2009.08.012
60. Fiorentino DF, Zlotnik A, Vieira P, Mosmann TR, Howard M, Moore KW, et al. IL-10 acts on the antigen-presenting cell to inhibit cytokine production by Th1 cells. *J Immunol* (1991) 146:3444–51.
61. Conti P, Kempuraj D, Kandere K, Di Gioacchino M, Barbacane RC, Castellani ML, et al. IL-10, an inflammatory/inhibitory cytokine, but not always. *Immunol Lett* (2003) 86:123–9. doi:10.1016/S0165-2478(03)00002-6
62. Tsuruma T, Yagihashi A, Hirata K, Torigoe T, Araya J, Watanabe N, et al. Interleukin-10 reduces natural killer (NK) sensitivity of tumor cells by down-regulating NK target structure expression. *Cell Immunol* (1999) 198:103–10. doi:10.1006/cimm.1999.1586
63. Park JY, Lee SH, Yoon SR, Park YJ, Jung H, Kim TD, et al. IL-15-induced IL-10 increases the cytolytic activity of human natural killer cells. *Mol Cells* (2011) 32:265–72. doi:10.1007/s10059-011-1057-8
64. Vredevoe DL, Widawski M, Fonarow GC, Hamilton M, Martínez-Maza O, Gage JR. Interleukin-6 (IL-6) expression and natural killer (NK) cell dysfunction and anergy in heart failure. *Am J Cardiol* (2004) 93:1007–11. doi:10.1016/j.amjcard.2003.12.054
65. Cifaldi L, Prencipe G, Caiello I, Bracaglia C, Locatelli F, De Benedetti F, et al. Inhibition of natural killer cell cytotoxicity by interleukin-6: implications for the pathogenesis of macrophage activation syndrome. *Arthritis Rheumatol* (2015) 67:3037–46. doi:10.1002/art.39295
66. Scheid C, Prendiville J, Jayson G, Crowther D, Fox B, Pettit GR, et al. Immunomodulation in patients receiving intravenous Bryostatin 1 in a phase I clinical study: comparison with effects of Bryostatin 1 on lymphocyte function in vitro. *Cancer Immunol Immunother* (1994) 39:223–30. doi:10.1007/BF01525985
67. Martinet L, Jean C, Dietrich G, Fournié JJ, Poupot R. PGE2 inhibits natural killer and gamma delta T cell cytotoxicity triggered by NKR and TCR through a cAMP-mediated PKA type I-dependent signaling. *Biochem Pharmacol* (2010) 80:838–45. doi:10.1016/j.bcp.2010.05.002
68. Lu ZY, Bataille R, Poubelle P, Rapp MJ, Harousseau JL, Klein B. An interleukin 1 receptor antagonist blocks the IL-1-induced IL-6 paracrine production through a prostaglandin E2-related mechanism in multiple myeloma. *Stem Cells* (1995) 13(Suppl 2):28–34.
69. Mellor AL, Munn DH. Tryptophan catabolism and T-cell tolerance: immunosuppression by starvation? *Immunol Today* (1999) 20:469–73. doi:10.1016/S0167-5699(99)01520-0
70. Della Chiesa M, Carlomagno S, Frumento G, Balsamo M, Cantoni C, Conte R, et al. The tryptophan catabolite L-kynurenine inhibits the surface expression of NKp46- and NKG2D-activating receptors and regulates NK-cell function. *Blood* (2006) 108(13):4118–25. doi:10.1182/blood-2006-03-006700
71. Nair JR, Carlson LM, Koorella C, Rozanski CH, Byrne GE, Bergsagel PL, et al. CD28 expressed on malignant plasma cells induces a prosurvival and immunosuppressive microenvironment. *J Immunol* (2011) 187:1243–53. doi:10.4049/jimmunol.1100016
72. Mateo G, Montalbán MA, Vidriales MB, Lahuerta JJ, Mateos MV, Gutiérrez N, et al. Prognostic value of immunophenotyping in multiple myeloma: a study by the PETHEMA/GEM cooperative study groups on patients uniformly treated with high-dose therapy. *J Clin Oncol* (2008) 26:2737–44. doi:10.1200/JCO.2007.15.4120
73. Van Valckenborgh E, Schouppe E, Movahedi K, De Bruyne E, Menu E, De Baetselier P, et al. Multiple myeloma induces the immunosuppressive capacity of distinct myeloid-derived suppressor cell subpopulations in the bone marrow. *Leukemia* (2012) 26:2424–8. doi:10.1038/leu.2012.113
74. Zhuang J, Zhang J, Lwin ST, Edwards JR, Edwards CM, Mundy GR, et al. Osteoclasts in multiple myeloma are derived from Gr-1+CD11b+myeloid-derived suppressor cells. *PLoS One* (2012) 7:e48871. doi:10.1371/journal.pone.0048871
75. Hoechst B, Voigtlaender T, Ormandy L, Gamrekelashvili J, Zhao F, Wedemeyer H, et al. Myeloid derived suppressor cells inhibit natural killer cells in patients with hepatocellular carcinoma via the NKp30 receptor. *Hepatology* (2009) 50:799–807. doi:10.1002/hep.23054
76. Li H, Han Y, Guo Q, Zhang M, Cao X. Cancer-expanded myeloid-derived suppressor cells induce anergy of NK cells through membrane-bound TGF-beta 1. *J Immunol* (2009) 182:240–9. doi:10.4049/jimmunol.182.1.240
77. Sarhan D, Cichocki F, Zhang B, Yingst A, Spellman SR, Cooley S, et al. Adaptive NK cells with low TIGIT expression are inherently resistant to myeloid-derived suppressor cells. *Cancer Res* (2016) 76:5696–706. doi:10.1158/0008-5472.CAN-16-0839
78. Hu J, Handisides DR, Van Valckenborgh E, De Raeve H, Menu E, Vande Broek I, et al. Targeting the multiple myeloma hypoxic niche with TH-302, a hypoxia-activated prodrug. *Blood* (2010) 116(9):1524–7. doi:10.1182/blood-2010-02-269126
79. Martin SK, Diamond P, Gronthos S, Peet DJ, Zannettino AC. The emerging role of hypoxia, HIF-1 and HIF-2 in multiple myeloma. *Leukemia* (2011) 25(10):1533–42. doi:10.1038/leu.2011.122
80. Sarkar S, Germeraad WT, Rouschop KM, Steeghs EM, van Gelder M, Bos GM, et al. Hypoxia induced impairment of NK cell cytotoxicity against multiple myeloma can be overcome by IL-2 activation of the NK cells. *PLoS One* (2013) 8:e64835. doi:10.1371/journal.pone.0064835
81. Pende D, Rivera P, Marcenaro S, Chang CC, Biassoni R, Conte R, et al. Major histocompatibility complex class I-related chain A and UL16-binding protein expression on tumor cell lines of different histotypes: analysis of tumor susceptibility to NKG2D-dependent natural killer cell cytotoxicity. *Cancer Res* (2002) 62:6178–86.

82. Jinushi M, Takehara T, Tatsumi T, Hiramatsu N, Sakamori R, Yamaguchi S, et al. Impairment of natural killer cell and dendritic cell functions by the soluble form of MHC class I-related chain A in advanced human hepatocellular carcinomas. *J Hepatol* (2005) 43:1013–20. doi:10.1016/j.jhep.2005.05.026
83. Boissel N, Rea D, Tieng V, Dulphy N, Brun M, Cayuela JM, et al. BCR/ABL oncogene directly controls MHC class I chain-related molecule A expression in chronic myelogenous leukemia. *J Immunol* (2006) 176:5108–16. doi:10.4049/jimmunol.176.8.5108
84. Jinushi M, Vanneman M, Munshi NC, Tai YT, Prabhala RH, Ritz J, et al. MHC class I chain-related protein A antibodies and shedding are associated with the progression of multiple myeloma. *Proc Natl Acad Sci U S A* (2008) 105:1285–90. doi:10.1073/pnas.0711293105
85. Kohga K, Takehara T, Tatsumi T, Ohkawa K, Miyagi T, Hiramatsu N, et al. Serum levels of soluble major histocompatibility complex (MHC) class I-related chain A in patients with chronic liver diseases and changes during transcatheter arterial embolization for hepatocellular carcinoma. *Cancer Sci* (2008) 99:1643–9. doi:10.1111/j.1349-7006.2008.00859.x
86. Groh V, Wu J, Yee C, Spies T. Tumour-derived soluble MIC ligands impair expression of NKG2D and T-cell activation. *Nature* (2002) 419:734–8. doi:10.1038/nature01112
87. Doubrovina ES, Doubrovin MM, Vider E, Sisson RB, O'Reilly RJ, Dupont B, et al. Evasion from NK cell immunity by MHC class I chain-related molecules expressing colon adenocarcinoma. *J Immunol* (2003) 171:6891–9. doi:10.4049/jimmunol.171.12.6891
88. Wu JD, Higgins LM, Steinle A, Cosman D, Haugk K, Plymate SR. Prevalent expression of the immunostimulatory MHC class I chain-related molecule is counteracted by shedding in prostate cancer. *J Clin Invest* (2004) 114:560–8. doi:10.1172/JCI200422206
89. Xiao G, Wang X, Sheng J, Lu S, Yu X, Wu JD. Soluble NKG2D ligand promotes MDSC expansion and skews macrophage to the alternatively activated phenotype. *J Hematol Oncol* (2015) 8:13. doi:10.1186/s13045-015-0110-z
90. Paschen A, Sucker A, Hill B, Moll I, Zapatka M, Nguyen XD, et al. Differential clinical significance of individual NKG2D ligands in melanoma: soluble ULBP2 as an indicator of poor prognosis superior to S100B. *Clin Cancer Res* (2009) 15:5208–15. doi:10.1158/1078-0432.CCR-09-0886
91. Kumar V, Yi Lo PH, Sawai H, Kato N, Takahashi A, Deng Z, et al. Soluble MICA and a MICA variation as possible prognostic biomarkers for HBV-induced hepatocellular carcinoma. *PLoS One* (2012) 7:e44743. doi:10.1371/journal.pone.0044743
92. Liu G, Lu S, Wang X, Page ST, Higano CS, Plymate SR, et al. Perturbation of NK cell peripheral homeostasis accelerates prostate carcinoma metastasis. *J Clin Invest* (2013) 123:4410–22. doi:10.1172/JCI69369
93. Zingoni A, Cecere F, Vulpis E, Fionda C, Molfetta R, Soriani A, et al. Genotoxic stress induces senescence-associated ADAM10-dependent release of NKG2D MIC ligands in multiple myeloma cells. *J Immunol* (2015) 195:736–48. doi:10.4049/jimmunol.1402643
94. Reiners KS, Topolar D, Henke A, Simhadri VR, Kessler J, Sauer M, et al. Soluble ligands for NK cell receptors promote evasion of chronic lymphocytic leukemia cells from NK cell anti-tumor activity. *Blood* (2013) 121:3658–65. doi:10.1182/blood-2013-01-476606
95. Friese MA, Wischhusen J, Wick W, Weiler M, Eisele G, Steinle A, et al. RNA interference targeting transforming growth factor-beta enhances NKG2D-mediated antiglioma immune response, inhibits glioma cell migration and invasiveness, and abrogates tumorigenicity in vivo. *Cancer Res* (2004) 64:7596–603. doi:10.1158/0008-5472.CAN-04-1627
96. Eisele G, Wischhusen J, Mittelbronn M, Meyermann R, Waldhauer I, Steinle A, et al. TGF-beta and metalloproteinases differentially suppress NKG2D ligand surface expression on malignant glioma cells. *Brain* (2006) 129:2416–25. doi:10.1093/brain/awl205
97. Kamimura H, Yamagiwa S, Tsuchiya A, Takamura M, Matsuda Y, Ohkoshi S, et al. Reduced NKG2D ligand expression in hepatocellular carcinoma correlates with early recurrence. *J Hepatol* (2012) 56:381–8. doi:10.1016/j.jhep.2011.06.017
98. Soriani A, Borrelli C, Ricci B, Molfetta R, Zingoni A, Fionda C, et al. p38 MAPK differentially controls NK activating ligands at transcriptional and post-transcriptional level on multiple myeloma cells. *Oncoimmunology* (2017) 6:e1264564. doi:10.1080/2162402X.2016.1264564
99. Soriani A, Zingoni A, Cerboni C, Iannitto ML, Ricciardi MR, Di Gialleonardo V, et al. ATM-ATR-dependent up-regulation of DNAM-1 and NKG2D ligands on multiple myeloma cells by therapeutic agents results in enhanced NK-cell susceptibility and is associated with a senescent phenotype. *Blood* (2009) 113:3503–11. doi:10.1182/blood-2008-08-173914
100. Soriani A, Iannitto ML, Ricci B, Fionda C, Malgarini G, Morrone S, et al. Reactive oxygen species- and DNA damage response-dependent NK cell activating ligand upregulation occurs at transcriptional levels and requires the transcriptional factor E2F1. *J Immunol* (2014) 193:950–60. doi:10.4049/jimmunol.1400271
101. Fionda C, Soriani A, Malgarini G, Iannitto ML, Santoni A, Cippitelli M. Heat shock protein-90 inhibitors increase MHC class I-related chain A and B ligand expression on multiple myeloma cells and their ability to trigger NK cell degranulation. *J Immunol* (2009) 183:4385–94. doi:10.4049/jimmunol.0901797
102. Abruzzese MP, Bilotta MT, Fionda C, Zingoni A, Soriani A, Vulpis E, et al. Inhibition of bromodomain and extra-terminal (BET) proteins increases NKG2D ligand MICA expression and sensitivity to NK cell-mediated cytotoxicity in multiple myeloma cells: role of cMYC-IRF4-miR-125b interplay. *J Hematol Oncol* (2016) 9:134. doi:10.1186/s13045-016-0362-2
103. Fionda C, Abruzzese MP, Zingoni A, Cecere F, Vulpis E, Peruzzi G, et al. The IMiDs targets IKZF-1/3 and IRF4 as novel negative regulators of NK cell-activating ligands expression in multiple myeloma. *Oncotarget* (2015) 6:23609–30. doi:10.18632/oncotarget.4603
104. Fiegler N, Textor S, Arnold A, Rölle A, Oehme I, Breuhahn K, et al. Downregulation of the activating NKp30 ligand B7-H6 by HDAC inhibitors impairs tumor cell recognition by NK cells. *Blood* (2013) 122:684–93. doi:10.1182/blood-2013-02-482513
105. Carbone E, Neri P, Mesuraca M, Fulciniti MT, Otsuki T, Pende D, et al. HLA class I, NKG2D, and natural cytotoxicity receptors regulate multiple myeloma cell recognition by natural killer cells. *Blood* (2005) 105:251–8. doi:10.1182/blood-2004-04-1422
106. Bossard C, Bezieau S, Matysiak-Budnik T, Volteau C, Laboisse CL, Jotereau F, et al. HLA-E/beta2 microglobulin overexpression in colorectal cancer is associated with recruitment of inhibitory immune cells and tumor progression. *Int J Cancer* (2012) 131:855–63. doi:10.1002/ijc.26453
107. Sarkar S, van Gelder M, Noort W, Xu Y, Rouschop KM, Groen R, et al. Optimal selection of natural killer cells to kill myeloma: the role of HLA-E and NKG2A. *Cancer Immunol Immunother* (2015) 64:951–63. doi:10.1007/s00262-015-1694-4
108. Frey AB, Monu N. Effector-phase tolerance: another mechanism of how cancer escapes antitumor immune response. *J Leukoc Biol* (2006) 79:652–62. doi:10.1189/jlb.1105628
109. Tonks A, Hills R, White P, Rosie B, Mills KI, Burnett AK, et al. CD200 as a prognostic factor in acute myeloid leukaemia. *Leukemia* (2007) 21:566–8. doi:10.1038/sj.leu.2404559
110. Coles SJ, Wang EC, Man S, Hills RK, Burnett AK, Tonks A, et al. CD200 expression suppresses natural killer cell function and directly inhibits patient anti-tumor response in acute myeloid leukemia. *Leukemia* (2011) 25:792–9. doi:10.1038/leu.2011.1
111. Kretz-Rommel A, Qin F, Dakappagari N, Ravey EP, McWhirter J, Oltean D, et al. CD200 expression on tumor cells suppresses antitumor immunity: new approaches to cancer immunotherapy. *J Immunol* (2007) 178:5595–605. doi:10.4049/jimmunol.178.9.5595
112. Moreaux J, Hose D, Reme T, Jourdan E, Hundemer M, Legouffe E, et al. CD200 is a new prognostic factor in multiple myeloma. *Blood* (2006) 108:4194–7. doi:10.1182/blood-2006-06-029355
113. Osterborg A, Nilsson B, Björkholm M, Holm G, Mellstedt H. Natural killer cell activity in monoclonal gammopathies: relation to disease activity. *Eur J Haematol* (1990) 45:153–7. doi:10.1111/j.1600-0609.1990.tb00443.x
114. Famularo G, D'Ambrosio A, Quintieri F, Di Giovanni S, Parzanese I, Pizzuto F, et al. Natural killer cell frequency and function in patients with monoclonal gammopathies. *J Clin Lab Immunol* (1992) 37:99–109.
115. García-Sanz R, González M, Orfão A, Moro MJ, Hernández JM, Borrego D, et al. Analysis of natural killer-associated antigens in peripheral blood and bone marrow of multiple myeloma patients and prognostic implications. *Br J Haematol* (1996) 93:81–8. doi:10.1046/j.1365-2141.1996.4651006.x
116. Omedé P, Boccadoro M, Gallone G, Frieri R, Battaglio S, Redoglia V, et al. Multiple myeloma: increased circulating lymphocytes carrying plasma

cell-associated antigens as an indicator of poor survival. *Blood* (1990) 76:1375–9.
117. King MA, Radicchi-Mastroianni MA. Natural killer cells and CD56+ T cells in the blood of multiple myeloma patients: analysis by 4-colour flow cytometry. *Cytometry* (1996) 26:121–4. doi:10.1002/(SICI)1097-0320(19960615)26:2<121::AID-CYTO4>3.0.CO;2-J
118. Paiva B, Mateos MV, Sanchez-Abarca LI, Puig N, Vidriales MB, López-Corral L, et al. Immune status of high-risk smoldering multiple myeloma patients and its therapeutic modulation under LenDex: a longitudinal analysis. *Blood* (2016) 127:1151–62. doi:10.1182/blood-2015-10-662320
119. Gonzalez M, San Miguel JF, Gascon A, Moro MJ, Hernandez JM, Ortega F, et al. Increased expression of natural-killer-associated and activation antigens in multiple myeloma. *Am J Hematol* (1992) 39:84–9. doi:10.1002/ajh.2830390203
120. Van den Hove LE, Meeus P, Derom A, Demuynck H, Verhoef GE, Vandenberghe P, et al. Lymphocyte profiles in multiple myeloma and monoclonal gammopathy of undetermined significance: flow-cytometric characterization and analysis in a two-dimensional correlation biplot. *Ann Hematol* (1998) 76:249–56. doi:10.1007/s002770050397
121. Nielsen H, Nielsen HJ, Tvede N, Klarlund K, Mansa B, Moesgaard F, et al. Immune dysfunction in multiple myeloma. Reduced natural killer cell activity and increased levels of soluble interleukin-2 receptors. *APMIS* (1991) 99:340–6. doi:10.1111/j.1699-0463.1991.tb05159.x
122. Fauriat C, Mallet F, Olive D, Costello RT. Impaired activating receptor expression pattern in natural killer cells from patients with multiple myeloma. *Leukemia* (2006) 20:732–3. doi:10.1038/sj.leu.2404096
123. Costello RT, Boehrer A, Sanchez C, Mercier D, Baier C, Le Treut T, et al. Differential expression of natural killer cell activating receptors in blood versus bone marrow in patients with monoclonal gammopathy. *Immunology* (2013) 139:338–41. doi:10.1111/imm.12082
124. Beldi-Ferchiou A, Lambert M, Dogniaux S, Vély F, Vivier E, Olive D, et al. PD-1 mediates functional exhaustion of activated NK cells in patients with Kaposi sarcoma. *Oncotarget* (2016) 7:72961–77. doi:10.18632/oncotarget.12150
125. Benson DM Jr, Bakan CE, Mishra A, Hofmeister CC, Efebera Y, Becknell B, et al. The PD-1/PD-L1 axis modulates the natural killer cell versus multiple myeloma effect: a therapeutic target for CT-011, a novel monoclonal anti-PD-1 antibody. *Blood* (2010) 116(13):2286–94. doi:10.1182/blood-2010-02-271874
126. Gorgun G, Samur MK, Cowens KB, Paula S, Bianchi G, Anderson JE, et al. Lenalidomide enhances immune checkpoint blockade-induced immune response in multiple myeloma. *Clin Cancer Res* (2015) 21:4607–18. doi:10.1158/1078-0432.CCR-15-0200
127. Ponzetta A, Benigni G, Antonangeli F, Sciumè G, Sanseviero E, Zingoni A, et al. Multiple myeloma impairs bone marrow localization of effector natural killer cells by altering the chemokine microenvironment. *Cancer Res* (2015) 75:4766–77. doi:10.1158/0008-5472.CAN-15-1320
128. El-Sherbiny YM, Meade JL, Holmes TD, McGonagle D, Mackie SL, Morgan AW, et al. The requirement for DNAM-1, NKG2D, and NKp46 in the natural killer cell-mediated killing of myeloma cells. *Cancer Res* (2007) 67:8444–9. doi:10.1158/0008-5472.CAN-06-4230
129. Guillerey C, Ferrari de Andrade L, Vuckovic S, Miles K, Ngiow SF, Yong MC, et al. Immunosurveillance and therapy of multiple myeloma are CD226 dependent. *J Clin Invest* (2015) 125:2077–89. doi:10.1172/JCI77181
130. Jurisic V, Srdic T, Konjevic G, Markovic O, Colovic M. Clinical stage-depending decrease of NK cell activity in multiple myeloma patients. *Med Oncol* (2007) 24:312–7. doi:10.1007/s12032-007-0007-y
131. Paiva B, Azpilikueta A, Puig N, Ocio EM, Sharma R, Oyajobi BO, et al. PD-L1/PD-1 presence in the tumor microenvironment and activity of PD-1 blockade in multiple myeloma. *Leukemia* (2015) 29(10):2110–3. doi:10.1038/leu.2015.79
132. San Miguel J, Mateos M-V, Shah JJ, Ocio EM, Rodriguez-Otero P, Reece D, et al. Pembrolizumab in combination with lenalidomide and low-dose dexamethasone for relapsed/refractory multiple myeloma (RRMM): keynote-023 [abstract 505]. *Presented at the 57th ASH Annual Meeting and Exposition.* (Vol. 73), Dec 5–8; Orlando, FL (2015).
133. Suen H, Brown R, Yang S, Ho PJ, Gibson J, Joshua D. The failure of immune checkpoint blockade in multiple myeloma with PD-1 inhibitors in a phase 1 study. *Leukemia* (2015) 29:1621–2. doi:10.1038/leu.2015.104
134. Lesokhin AM, Ansell SM, Armand P, Scott EC, Halwani A, Gutierrez M, et al. Preliminary results of a phase I study of nivolumab (BMS-936558) in patients with relapsed or refractory lymphoid malignancies [abstract 291]. *Blood* (2014) 124(Suppl).
135. Pesce S, Greppi M, Tabellini G, Rampinelli F, Parolini S, Olive D, et al. Identification of a subset of human natural killer cells expressing high levels of programmed death 1: a phenotypic and functional characterization. *J Allergy Clin Immunol* (2017) 139(1):335–46. doi:10.1016/j.jaci.2016.04.025
136. Frohn C, Hoppner M, Schlenke P, Kirchner H, Koritke P, Luhm J. Anti-myeloma activity of natural killer lymphocytes. *Br J Haematol* (2002) 119:660–4. doi:10.1046/j.1365-2141.2002.03879.x
137. Kroger N, Shaw B, Iacobelli S, Zabelina T, Peggs K, Shimoni A, et al. Comparison between antithymocyte globulin and alemtuzumab and the possible impact of KIR-ligand mismatch after dose-reduced conditioning and unrelated stem cell transplantation in patients with multiple myeloma. *Br J Haematol* (2005) 129:631–43. doi:10.1111/j.1365-2141.2005.05513.x
138. Romagné F, Andre P, Spee P, Zahn S, Anfossi N, Gauthier L, et al. Preclinical characterization of 1-7F9, a novel human anti-KIR receptor therapeutic antibody that augments natural killer-mediated killing of tumor cells. *Blood* (2009) 114:2667–77. doi:10.1182/blood-2009-02-206532
139. Benson DM Jr, Bakan CE, Zhang S, Collins SM, Liang J, Srivastava S, et al. IPH2101, a novel anti-inhibitory KIR antibody, and lenalidomide combine to enhance the natural killer cell versus multiple myeloma effect. *Blood* (2011) 118:6387–91. doi:10.1182/blood-2011-06-360255
140. Benson DM Jr, Hofmeister CC, Padmanabhan S, Suvannasankha A, Jagannath S, Abonour R, et al. A phase 1 trial of the anti-KIR antibody IPH2101 in patients with relapsed/refractory multiple myeloma. *Blood* (2012) 120:4324–33. doi:10.1182/blood-2012-06-438028
141. Benson DM Jr, Cohen AD, Jagannath S, Munshi NC, Spitzer G, Hofmeister CC, et al. A phase I trial of the anti-KIR antibody IPH2101 and lenalidomide in patients with relapsed/refractory multiple myeloma. *Clin Cancer Res* (2015) 21(18):4055–61. doi:10.1158/1078-0432.CCR-15-0304
142. Korde N, Carlsten M, Lee MJ, Minter A, Tan E, Kwok M, et al. A phase II trial of pan-KIR2D blockade with IPH2101 in smoldering multiple myeloma. *Haematologica* (2014) 99:e81–3. doi:10.3324/haematol.2013.103085
143. Carlsten M, Korde N, Kotecha R, Reger R, Bor S, Kazandjian D, et al. Checkpoint inhibition of KIR2D with the monoclonal antibody IPH2101 induces contraction and hyporesponsiveness of NK cells in patients with myeloma. *Clin Cancer Res* (2016) 22:5211–22. doi:10.1158/1078-0432.CCR-16-1108
144. Attal M, Lauwers-Cances V, Marit G, Caillot D, Moreau P, Facon T, et al. Lenalidomide maintenance after stem-cell transplantation for multiple myeloma. *N Engl J Med* (2012) 366(19):1782–91. doi:10.1056/NEJMoa1114138
145. McCarthy PL, Owzar K, Hofmeister CC, Hurd DD, Hassoun H, Richardson PG, et al. Lenalidomide after stem cell transplantation for multiple myeloma. *N Engl J Med* (2012) 366(19):1770–81. doi:10.1056/NEJMoa1114083
146. Palumbo A, Cavallo F, Gay F, Di Raimondo F, Ben Yehuda D, Petrucci MT, et al. Autologous transplantation and maintenance therapy in multiple myeloma. *N Engl J Med* (2014) 371(10):895–905. doi:10.1056/NEJMoa1402888
147. Jackson GH, Davies FE, Pawlyn C, Cairns DA, Striha A, Collett C, et al. Lenalidomide is a highly effective maintenance therapy inmyeloma patients of all ages; results of the phase III myeloma XI study [abstract no. 1143]. *Blood* (2016) 128(22).
148. Musto P, Anderson KC, Attal M, Richardson PG, Badros A, Hou J, et al. International Myeloma Working Group. Second primary malignancies in multiple myeloma: an overview and IMWG consensus. *Ann Oncol* (2017) 28:228–45. doi:10.1093/annonc/mdx160
149. Hayashi T, Hideshima T, Akiyama M, Podar K, Yasui H, Raje N, et al. Molecular mechanisms whereby immunomodulatory drugs activate natural killer cells: clinical application. *Br J Haematol* (2005) 128:192–203. doi:10.1111/j.1365-2141.2004.05286.x
150. Malavasi F, Funaro A, Roggero S, Horenstein A, Calosso L, Mehta K. Human CD38: a glycoprotein in search of a function. *Immunol Today* (1994) 15:95–7. doi:10.1016/0167-5699(94)90148-1
151. Lin P, Owens R, Tricot G, Wilson CS. Flow cytometric immunophenotypic analysis of 306 cases of multiple myeloma. *Am J Clin Pathol* (2004) 121:482–8. doi:10.1309/74R4TB90BUWH27JX

152. Deaglio S, Mehta K, Malavasi F. Human CD38: a (r)evolutionary story of enzymes and receptors. *Leuk Res* (2001) 25:1–12. doi:10.1016/S0145-2126(00)00093-X
153. Nijhof IS, Groen RW, Lokhorst HM, van Kessel B, Bloem AC, van Velzen J, et al. Upregulation of CD38 expression on multiple myeloma cells by all-trans retinoic acid improves the efficacy of daratumumab. *Leukemia* (2015) 29:2039–49. doi:10.1038/leu.2015.123
154. Overdijk MB, Verploegen S, Bögels M, van Egmond M, van Bueren JJL, Mutis T, et al. Antibody-mediated phagocytosis contributes to the anti-tumor activity of the therapeutic antibody daratumumab in lymphoma and multiple myeloma. *MAbs* (2015) 7:311–20. doi:10.1080/19420862.2015.1007813
155. Jansen JH, Boross P, Overdijk MB, van Bueren JJ, Parren PW, Leusen JH. Daratumumab, a human CD38 antibody induces apoptosis of myeloma tumor cells via Fc receptor-mediated crosslinking [abstract]. *Blood* (2012) 120:2974.
156. Palumbo A, Chanan-Khan A, Weisel K, Nooka AK, Masszi T, Beksac M, et al. Daratumumab, bortezomib, and dexamethasone for multiple myeloma. *N Engl J Med* (2016) 375:754–66. doi:10.1056/NEJMoa1606038
157. Nijhof IS, Lammerts van Bueren JJ, van Kessel B, Andre P, Morel Y, Lokhorst HM, et al. Daratumumab-mediated lysis of primary multiple myeloma cells is enhanced in combination with the human anti-KIR antibody IPH2102 and lenalidomide. *Haematologica* (2015) 100(2):263–8. doi:10.3324/haematol.2014.117531
158. Cherkasova E, Espinoza L, Kotecha R, Reger RN, Berg M, Aue G, et al. Treatment of ex vivo expanded NK cells with daratumumab F(ab')2 fragments protects adoptively transferred NK cells from daratumumab-mediated killing and augments daratumumab-induced antibody dependent cellular toxicity (ADCC) of myeloma [abstract]. *Blood* (2015) 126:4244.
159. Hsi ED, Steinle R, Balasa B, Szmania S, Draksharapu A, Shum BP, et al. CS1, a potential new therapeutic antibody target for the treatment of multiple myeloma. *Clin Cancer Res* (2008) 14:2775–84. doi:10.1158/1078-0432.CCR-07-4246
160. Tai YT, Dillon M, Song W, Leiba M, Li XF, Burger P, et al. Anti-CS1 humanized monoclonal antibody HuLuc63 inhibits myeloma cell adhesion and induces antibody-dependent cellular cytotoxicity in the bone marrow milieu. *Blood* (2008) 112:1329–37. doi:10.1182/blood-2007-08-107292
161. Collins SM, Bakan CE, Swartzel GD, Hofmeister CC, Efebera YA, Kwon H, et al. Elotuzumab directly enhances NK cell cytotoxicity against myeloma via CS1 ligation: evidence for augmented NK cell function complementing ADCC. *Cancer Immunol Immunother* (2013) 62(12):1841–9. doi:10.1007/s00262-013-1493-8
162. van Rhee F, Szmania SM, Dillon M, van Abbema AM, Li X, Stone MK, et al. Combinatorial efficacy of anti-CS1 monoclonal antibody elotuzumab (HuLuc63) and bortezomib against multiple myeloma. *Mol Cancer Ther* (2009) 8:2616–24. doi:10.1158/1535-7163.MCT-09-0483
163. Magen H, Muchtar E. Elotuzumab: the first approved monoclonal antibody for multiple myeloma treatment. *Ther Adv Hematol* (2016) 7(4):187–95. doi:10.1177/2040620716652862
164. Guo H, Cruz-Munoz M-E, Wu N, Robbins M, Veillette A. Immune cell inhibition by SLAMF7 is mediated by a mechanism requiring src kinases, CD45, and SHIP-1 that is defective in multiple myeloma cells. *Mol Cell Biol* (2015) 35:41–51. doi:10.1128/MCB.01107-14
165. Pérez-Quintero LA, Roncagalli R, Guo H, Latour S, Davidson D, Veillette A. EAT-2, a SAP-like adaptor, controls NK cell activation through phospholipase Cγ, Ca++, and Erk, leading to granule polarization. *J Exp Med* (2014) 211:727–42. doi:10.1084/jem.20132038
166. Balasa B, Yun R, Belmar NA, Fox M, Chao DT, Robbins MD, et al. Elotuzumab enhances natural killer cell activation and myeloma cell killing through interleukin-2 and TNF-α pathways. *Cancer Immunol Immunother* (2015) 64(1):61–73. doi:10.1007/s00262-014-1610-3
167. Lonial S, Vij R, Harousseau JL, Facon T, Moreau P, Mazumder A, et al. Elotuzumab in combination with lenalidomide and low-dose dexamethasone in relapsed or refractory multiple myeloma. *J Clin Oncol* (2012) 30:1953–9. doi:10.1200/JCO.2011.37.2649
168. Richardson PG, Jagannath S, Moreau P, Jakubowiak A, Raab MS, Facon T, et al. Final results for the 1703 phase 1b/2 study of elotuzumab in combination with lenalidomide and dexamethasone in patients with relapsed/refractory multiple myeloma [abstract]. *Blood* (2014) 124:302.
169. Lonial S, Dimopoulos M, Palumbo A, White D, Grosicki S, Spicka I, et al. Elotuzumab therapy for relapsed or refractory multiple myeloma. ELOQUENT-2 investigators. *N Engl J Med* (2015) 373(7):621–31. doi:10.1056/NEJMoa1505654
170. Jakubowiak A, Offidani M, Pégourie B, De La Rubia J, Garderet L, Laribi K, et al. Randomized phase 2 study: elotuzumab plus bortezomib/dexamethasone vs bortezomib/dexamethasone for relapsed/refractory MM. *Blood* (2016) 127:2833–40. doi:10.1182/blood-2016-01-694604
171. Munn DH, Shafizadeh E, Attwood JT, Bondarev I, Pashine A, Mellor AL. Inhibition of T cell proliferation by macrophage tryptophan catabolism. *J Exp Med* (1999) 189:1363–72. doi:10.1084/jem.189.9.1363
172. Uyttenhove C, Pilotte L, Theate I, Stroobant V, Colau D, Parmentier N, et al. Evidence for a tumoral immune resistance mechanism based on tryptophan degradation by indoleamine 2,3-dioxygenase. *Nat Med* (2003) 9:1269–74. doi:10.1038/nm934
173. Fallarino F, Grohmann U, Vacca C, Orabona C, Spreca A, Fioretti MC, et al. T cell apoptosis by kynurenines. *Adv Exp Med Biol* (2003) 527:183–90. doi:10.1007/978-1-4615-0135-0_21
174. Hayashi T, Mo JH, Gong X, Rossetto C, Jang A, Beck L, et al. 3-Hydroxyanthranilic acid inhibits PDK1 activation and suppresses experimental asthma by inducing T cell apoptosis. *Proc Natl Acad Sci U S A* (2007) 104:18619–24. doi:10.1073/pnas.0709261104
175. Frumento G, Rotondo R, Tonetti M, Damonte G, Benatti U, Ferrara GB. Tryptophan-derived catabolites are responsible for inhibition of T and natural killer cell proliferation induced by indoleamine 2,3-dioxygenase. *J Exp Med* (2002) 196:459–68. doi:10.1084/jem.20020121
176. Sato N, Saga Y, Mizukami H, Wang D, Takahashi S, Nonaka H, et al. Downregulation of indoleamine-2,3-dioxygenase in cervical cancer cells suppresses tumor growth by promoting natural killer cell accumulation. *Oncol Rep* (2012) 28:1574–8. doi:10.3892/or.2012.1984
177. Molano A, Illarionov PA, Besra GS, Putterman C, Porcelli SA. Modulation of invariant natural killer T cell cytokine responses by indoleamine 2,3-dioxygenase. *Immunol Lett* (2008) 117:81–90. doi:10.1016/j.imlet.2007.12.013
178. Balachandran VP, Cavnar MJ, Zeng S, Bamboat ZM, Ocuin LM, Obaid H, et al. Imatinib potentiates antitumor T cell responses in gastrointestinal stromal tumor through the inhibition of IDO. *Nat Med* (2011) 17:1094–100. doi:10.1038/nm.2438
179. Nguyen NT, Kimura A, Nakahama T, Chinen I, Masuda K, Nohara K, et al. Aryl hydrocarbon receptor negatively regulates dendritic cell immunogenicity via a kynurenine dependent mechanism. *Proc Natl Acad Sci U S A* (2010) 107:19961–6. doi:10.1073/pnas.1014465107
180. Park MJ, Park KS, Park HS, Cho ML, Hwang SY, Min SY, et al. A distinct tolerogenic subset of splenic IDO(C)CD11b(C) dendritic cells from orally tolerized mice is responsible for induction of systemic immune tolerance and suppression of collagen-induced arthritis. *Cell Immunol* (2012) 278:45–54. doi:10.1016/j.cellimm.2012.06.009
181. Orabona C, Puccetti P, Vacca C, Bicciato S, Luchini A, Fallarino F, et al. Toward the identification of a tolerogenic signature in IDO-competent dendritic cells. *Blood* (2006) 107:2846–54. doi:10.1182/blood-2005-10-4077
182. Bonanno G, Mariotti A, Procoli A, Folgiero V, Natale D, De Rosa L, et al. Indoleamine 2,3-dioxygenase 1 (IDO1) activity correlates with immune system abnormalities in multiple myeloma. *J Transl Med* (2012) 10:247. doi:10.1186/1479-5876-10-247
183. Ault KA, Antin JH, Ginsburg D, Orkin SH, Rappeport JM, Keohan ML, et al. Phenotype of recovering lymphoid cell populations after marrow transplantation. *J Exp Med* (1985) 161(6):1483–502. doi:10.1084/jem.161.6.1483
184. Velardi A, Terenzi A, Cucciaioni S, Millo R, Grossi CE, Grignani F, et al. Imbalances within the peripheral blood T-helper (CD4+) and T-suppressor (CD8+) cell populations in the reconstitution phase after human bone marrow transplantation. *Blood* (1988) 71(5):1196–200.
185. Uhrberg M, Valiante NM, Shum BP, Shilling HG, Lienert-Weidenbach K, Corliss B, et al. Human diversity in killer cell inhibitory receptor genes. *Immunity* (1997) 7(6):753–63. doi:10.1016/S1074-7613(00)80394-5
186. Wende H, Colonna M, Ziegler A, Volz A. Organization of the leukocyte receptor cluster (LRC) on human chromosome 19q13.4. *Mamm Genome* (1999) 10(2):154–60. doi:10.1007/s003359900961
187. Shilling HG, Guethlein LA, Cheng NW, Gardiner CM, Rodriguez R, Tyan D, et al. Allelic polymorphism synergizes with variable gene content to

individualize human KIR genotype. *J Immunol* (2002) 168(5):2307–15. doi:10.4049/jimmunol.168.5.2307
188. Shilling HG, Young N, Guethlein LA, Cheng NW, Gardiner CM, Tyan D, et al. Genetic control of human NK cell repertoire. *J Immunol* (2002) 169(1):239–47. doi:10.4049/jimmunol.169.1.239
189. Shilling HG, McQueen KL, Cheng NW, Shizuru JA, Negrin RS, Parham P. Reconstitution of NK cell receptor repertoire following HLA-matched hematopoietic cell transplantation. *Blood* (2003) 101(9):3730–40. doi:10.1182/blood-2002-08-2568
190. Bachar-Lustig E, Rachamim N, Li HW, Lan F, Reisner Y. Megadose of T cell-depleted bone marrow overcomes MHC barriers in sublethally irradiated mice. *Nat Med* (1995) 1(12):1268–73. doi:10.1038/nm1295-1268
191. Aversa F, Tabilio A, Velardi A, Cunningham I, Terenzi A, Falzetti F, et al. Treatment of high-risk acute leukemia with T-cell-depleted stem cells from related donors with one fully mismatched HLA haplotype. *N Engl J Med* (1998) 339(17):1186–93. doi:10.1056/NEJM199810223391702
192. Ruggeri L, Capanni M, Casucci M, Volpi I, Tosti A, Perruccio K, et al. Role of natural killer cell alloreactivity in HLA-mismatched hematopoietic stem cell transplantation. *Blood* (1999) 94(1):333–9.
193. Farag SS, Fehniger TA, Ruggeri L, Velardi A, Caligiuri MA. Natural killer cell receptors: new biology and insights into the graft-versus-leukemia effect. *Blood* (2002) 100(6):1935–47. doi:10.1182/blood-2002-02-0350
194. Velardi A, Ruggeri L, Alessandro M, Moretta L. NK cells: a lesson from mismatched hematopoietic transplantation. *Trends Immunol* (2002) 23(9):438–44. doi:10.1016/S1471-4906(02)02284-6
195. Ruggeri L, Capanni M, Urbani E, Perruccio K, Shlomchik WD, Tosti A, et al. Effectiveness of donor natural killer cell alloreactivity in mismatched hematopoietic transplants. *Science* (2002) 295(5562):2097–100. doi:10.1126/science.1068440
196. Parham P, McQueen KL. Alloreactive killer cells: hindrance and help for haematopoietic transplants. *Nat Rev Immunol* (2003) 3(2):108–22. doi:10.1038/nri999
197. Aversa F, Terenzi A, Tabilio A, Falzetti F, Carotti A, Ballanti S, et al. Full haplotype-mismatched hematopoietic stem-cell transplantation: a phase II study in patients with acute leukemia at high risk of relapse. *J Clin Oncol* (2005) 23(15):3447–54. doi:10.1200/JCO.2005.09.117
198. Ruggeri L, Mancusi A, Capanni M, Urbani E, Carotti A, Aloisi T, et al. Donor natural killer cell allorecognition of missing self in haploidentical hematopoietic transplantation for acute myeloid leukemia: challenging its predictive value. *Blood* (2007) 110(1):433–40. doi:10.1182/blood-2006-07-038687
199. Mancusi A, Ruggeri L, Urbani E, Pierini A, Massei MS, Carotti A, et al. Haploidentical hematopoietic transplantation from KIR ligand-mismatched donors with activating KIRs reduces nonrelapse mortality. *Blood* (2015) 125(20):3173–82. doi:10.1182/blood-2014-09-599993
200. Leung W, Iyengar R, Turner V, Lang P, Bader P, Conn P, et al. Determinants of antileukemia effects of allogeneic NK cells. *J Immunol* (2004) 172(1):644–50. doi:10.4049/jimmunol.172.1.644
201. Pende D, Marcenaro S, Falco M, Martini S, Bernardo ME, Montagna D, et al. Anti-leukemia activity of alloreactive NK cells in KIR ligand-mismatched haploidentical HSCT for pediatric patients: evaluation of the functional role of activating KIR and redefinition of inhibitory KIR specificity. *Blood* (2009) 113(13):3119–29. doi:10.1182/blood-2008-06-164103
202. Miller JS, Soignier Y, Panoskaltsis-Mortari A, McNearney SA, Yun GH, Fautsch SK, et al. Successful adoptive transfer and in vivo expansion of human haploidentical NK cells in patients with cancer. *Blood* (2005) 105(8):3051–7. doi:10.1182/blood-2004-07-2974
203. Rubnitz JE, Inaba H, Ribeiro RC, Pounds S, Rooney B, Bell T, et al. NKAML: a pilot study to determine the safety and feasibility of haploidentical natural killer cell transplantation in childhood acute myeloid leukemia. *J Clin Oncol* (2010) 28(6):955–9. doi:10.1200/JCO.2009.24.4590
204. Curti A, Ruggeri L, D'Addio A, Bontadini A, Dan E, Motta MR, et al. Successful transfer of alloreactive haploidentical KIR ligand-mismatched natural killer cells after infusion in elderly high risk acute myeloid leukemia patients. *Blood* (2011) 118(12):3273–9. doi:10.1182/blood-2011-01-329508
205. Kanakry CG, Fuchs EJ, Luznik L. Modern approaches to HLA-haploidentical blood or marrow transplantation. *Nat Rev Clin Oncol* (2016) 13(1):10–24. doi:10.1038/nrclinonc.2015.128
206. Huang XJ, Zhao XY, Liu DH, Liu KY, Xu LP. Deleterious effects of KIR ligand incompatibility on clinical outcomes in haploidentical hematopoietic stem cell transplantation without in vitro T-cell depletion. *Leukemia* (2007) 21(4):848–51. doi:10.1038/sj.leu.2404566
207. Vago L, Forno B, Sormani MP, Crocchiolo R, Zino E, Di Terlizzi S, et al. Temporal, quantitative, and functional characteristics of single-KIR-positive alloreactive natural killer cell recovery account for impaired graft-versus-leukemia activity after haploidentical hematopoietic stem cell transplantation. *Blood* (2008) 112(8):3488–99. doi:10.1182/blood-2007-07-103325
208. Symons HJ, Leffell MS, Rossiter ND, Zahurak M, Jones RJ, Fuchs EJ. Improved survival with inhibitory killer immunoglobulin receptor (KIR) gene mismatches and KIR haplotype B donors after nonmyeloablative, HLA-haploidentical bone marrow transplantation. *Biol Blood Marrow Transplant* (2010) 16(4):533–42. doi:10.1016/j.bbmt.2009.11.022
209. Locatelli F, Bauquet A, Palumbo G, Moretta F, Bertaina A. Negative depletion of α/β+ T cells and of CD19+ B lymphocytes: a novel frontier to optimize the effect of innate immunity in HLA-mismatched hematopoietic stem cell transplantation. *Immunol Lett* (2013) 155(1–2):21–3. doi:10.1016/j.imlet.2013.09.027
210. Martelli MF, Di Ianni M, Ruggeri L, Falzetti F, Carotti A, Terenzi A, et al. HLA-haploidentical transplantation with regulatory and conventional T-cell adoptive immunotherapy prevents acute leukemia relapse. *Blood* (2014) 124(4):638–44. doi:10.1182/blood-2014-03-564401
211. Giebel S, Locatelli F, Lamparelli T, Velardi A, Davies S, Frumento G, et al. Survival advantage with KIR ligand incompatibility in hematopoietic stem cell transplantation from unrelated donors. *Blood* (2003) 102(3):814–9. doi:10.1182/blood-2003-01-0091
212. Beelen DW, Ottinger HD, Ferencik S, Elmaagacli AH, Peceny R, Trenschel R, et al. Genotypic inhibitory killer immunoglobulin-like receptor ligand incompatibility enhances the long-term antileukemic effect of unmodified allogeneic hematopoietic stem cell transplantation in patients with myeloid leukemias. *Blood* (2005) 105(6):2594–600. doi:10.1182/blood-2004-04-1441
213. Willemze R, Rodrigues CA, Labopin M, Sanz G, Michel G, Socié G, et al. Eurocord-Netcord and Acute Leukaemia Working Party of the EBMT. KIR-ligand incompatibility in the graft-versus-host direction improves outcomes after umbilical cord blood transplantation for acute leukemia. *Leukemia* (2009) 23(3):492–500. Erratum in: *Leukemia* (2009) 23(3):630. doi:10.1038/leu.2008.365
214. Davies SM, Ruggieri L, DeFor T, Wagner JE, Weisdorf DJ, Miller JS, et al. Evaluation of KIR ligand incompatibility in mismatched unrelated donor hematopoietic transplants. Killer immunoglobulin-like receptor. *Blood* (2002) 100(10):3825–7. doi:10.1182/blood-2002-04-1197
215. Farag SS, Bacigalupo A, Eapen M, Hurley C, Dupont B, Caligiuri MA, et al. The effect of KIR ligand incompatibility on the outcome of unrelated donor transplantation: a report from the center for international blood and marrow transplant research, the European blood and marrow transplant registry, and the Dutch registry. *Biol Blood Marrow Transplant* (2006) 12(8):876–84. doi:10.1016/j.bbmt.2006.05.007
216. Brunstein CG, Wagner JE, Weisdorf DJ, Cooley S, Noreen H, Barker JN, et al. Negative effect of KIR alloreactivity in recipients of umbilical cord blood transplant depends on transplantation conditioning intensity. *Blood* (2009) 113(22):5628–34. doi:10.1182/blood-2008-12-197467
217. Cooley S, Trachtenberg E, Bergemann TL, Saeteurn K, Klein J, Le CT, et al. Donors with group B KIR haplotypes improve relapse-free survival after unrelated hematopoietic cell transplantation for acute myelogenous leukemia. *Blood* (2009) 113(3):726–32. doi:10.1182/blood-2008-07-171926
218. Cooley S, Weisdorf DJ, Guethlein LA, Klein JP, Wang T, Le CT, et al. Donor selection for natural killer cell receptor genes leads to superior survival after unrelated transplantation for acute myelogenous leukemia. *Blood* (2010) 116(14):2411–9. doi:10.1182/blood-2010-05-283051
219. Cooley S, Weisdorf DJ, Guethlein LA, Klein JP, Wang T, Marsh SG, et al. Donor killer cell Ig-like receptor B haplotypes, recipient HLA-C1, and HLA-C mismatch enhance the clinical benefit of unrelated transplantation for acute myelogenous leukemia. *J Immunol* (2014) 192(10):4592–600. doi:10.4049/jimmunol.1302517
220. Venstrom JM, Pittari G, Gooley TA, Chewning JH, Spellman S, Haagenson M, et al. HLA-C-dependent prevention of leukemia relapse by donor activating KIR2DS1. *N Engl J Med* (2012) 367(9):805–16. doi:10.1056/NEJMoa1200503
221. Carlens S, Gilljam M, Chambers BJ, Aschan J, Guven H, Ljunggren HG, et al. A new method for in vitro expansion of cytotoxic human CD3-CD56+

natural killer cells. *Hum Immunol* (2001) 62(10):1092–8. doi:10.1016/S0198-8859(01)00313-5
222. Alici E, Sutlu T, Bjorkstrand B, Gilljam M, Stellan B, Nahi H, et al. Autologous antitumor activity by NK cells expanded from myeloma patients using GMP-compliant components. *Blood* (2008) 111(6):3155–62. doi:10.1182/blood-2007-09-110312
223. Garg TK, Szmania SM, Khan JA, Hoering A, Malbrough PA, Moreno-Bost A, et al. Highly activated and expanded natural killer cells for multiple myeloma immunotherapy. *Haematologica* (2012) 97(9):1348–56. doi:10.3324/haematol.2011.056747
224. Szmania S, Lapteva N, Garg T, Greenway A, Lingo J, Nair B, et al. Ex vivo-expanded natural killer cells demonstrate robust proliferation in vivo in high-risk relapsed multiple myeloma patients. *J Immunother* (2015) 38(1):24–36. doi:10.1097/CJI.0000000000000059
225. Shi J, Tricot G, Szmania S, Rosen N, Garg TK, Malaviarachchi PA, et al. Infusion of haplo-identical killer immunoglobulin-like receptor ligand mismatched NK cells for relapsed myeloma in the setting of autologous stem cell transplantation. *Br J Haematol* (2008) 143(5):641–53. doi:10.1111/j.1365-2141.2008.07340.x
226. Leivas A, Perez-Martinez A, Blanchard MJ, Martín-Clavero E, Fernández L, Lahuerta JJ, et al. Novel treatment strategy with autologous activated and expanded natural killer cells plus anti-myeloma drugs for multiple myeloma. *Oncoimmunology* (2016) 5(12):e1250051. doi:10.1080/2162402X.2016.1250051
227. Shah N, Martin-Antonio B, Yang H, Ku S, Lee DA, Cooper LJ, et al. Antigen presenting cell-mediated expansion of human umbilical cord blood yields log-scale expansion of natural killer cells with anti-myeloma activity. *PLoS One* (2013) 8(10):e76781. doi:10.1371/journal.pone.0076781
228. Martin-Antonio B, Najjar A, Robinson SN, Chew C, Li S, Yvon E, et al. Transmissible cytotoxicity of multiple myeloma cells by cord blood-derived NK cells is mediated by vesicle trafficking. *Cell Death Differ* (2015) 22(1):96–107. doi:10.1038/cdd.2014.120
229. Shah N, Li L, McCarty J, Kaur I, Yvon E, Shaim H, et al. Phase I study of cord blood-derived natural killer cells combined with autologous stem cell transplantation in multiple myeloma. *Br J Haematol* (2017) 177:457–66. doi:10.1111/bjh.14570
230. Porter DL, Levine BL, Kalos M, Bagg A, June CH. Chimeric antigen receptor-modified T cells in chronic lymphoid leukemia. *N Engl J Med* (2011) 365(8):725–33. doi:10.1056/NEJMoa1103849
231. Grupp SA, Kalos M, Barrett D, Aplenc R, Porter DL, Rheingold SR, et al. Chimeric antigen receptor-modified T cells for acute lymphoid leukemia. *N Engl J Med* (2013) 368(16):1509–18. doi:10.1056/NEJMoa1215134
232. Brentjens RJ, Davila ML, Riviere I, Park J, Wang X, Cowell LG, et al. CD19-targeted T cells rapidly induce molecular remissions in adults with chemotherapy-refractory acute lymphoblastic leukemia. *Sci Transl Med* (2013) 5(177):177ra38. doi:10.1126/scitranslmed.3005930
233. Kochenderfer JN, Wilson WH, Janik JE, Dudley ME, Stetler-Stevenson M, Feldman SA, et al. Eradication of B-lineage cells and regression of lymphoma in a patient treated with autologous T cells genetically engineered to recognize CD19. *Blood* (2010) 116(20):4099–102. doi:10.1182/blood-2010-04-281931
234. Maude SL, Frey N, Shaw PA, Aplenc R, Barrett DM, Bunin NJ, et al. Chimeric antigen receptor T cells for sustained remissions in leukemia. *N Engl J Med* (2014) 371(16):1507–17. doi:10.1056/NEJMoa1407222
235. Turtle CJ, Hanafi LA, Berger C, Gooley TA, Cherian S, Hudecek M, et al. CD19 CAR-T cells of defined CD4+:CD8+ composition in adult B cell ALL patients. *J Clin Invest* (2016) 126(6):2123–38. doi:10.1172/JCI85309
236. Jackson HJ, Rafiq S, Brentjens RJ. Driving CAR T-cells forward. *Nat Rev Clin Oncol* (2016) 13(6):370–83. doi:10.1038/nrclinonc.2016.36
237. Garfall AL, Maus MV, Hwang WT, Lacey SF, Mahnke YD, Melenhorst JJ, et al. Chimeric antigen receptor T cells against CD19 for multiple myeloma. *N Engl J Med* (2015) 373(11):1040–7. doi:10.1056/NEJMoa1504542
238. Mihara K, Bhattacharyya J, Kitanaka A, Yanagihara K, Kubo T, Takei Y, et al. T-cell immunotherapy with a chimeric receptor against CD38 is effective in eliminating myeloma cells. *Leukemia* (2012) 26(2):365–7. doi:10.1038/leu.2011.205
239. Drent E, Groen RW, Noort WA, Themeli M, Lammerts van Bueren JJ, Parren PW, et al. Pre-clinical evaluation of CD38 chimeric antigen receptor engineered T cells for the treatment of multiple myeloma. *Haematologica* (2016) 101(5):616–25. doi:10.3324/haematol.2015.137620
240. Guo B, Chen M, Han Q, Hui F, Dai H, Zhang W, et al. CD138-directed adoptive immunotherapy of chimeric antigen receptor (CAR)-modified T cells for multiple myeloma. *J Cell Immunother* (2016) 2(1):28–35. doi:10.1016/j.jocit.2014.11.001
241. Carpenter RO, Evbuomwan MO, Pittaluga S, Rose JJ, Raffeld M, Yang S, et al. B-cell maturation antigen is a promising target for adoptive T-cell therapy of multiple myeloma. *Clin Cancer Res* (2013) 19(8):2048–60. doi:10.1158/1078-0432.CCR-12-2422
242. Ali SA, Shi V, Maric I, Wang M, Stroncek DF, Rose JJ, et al. T cells expressing an anti-B-cell maturation antigen chimeric antigen receptor cause remissions of multiple myeloma. *Blood* (2016) 128(13):1688–700. doi:10.1182/blood-2016-04-711903
243. Ramos CA, Savoldo B, Torrano V, Ballard B, Zhang H, Dakhova O, et al. Clinical responses with T lymphocytes targeting malignancy-associated κ light chains. *J Clin Invest* (2016) 126(7):2588–96. doi:10.1172/JCI86000
244. Chu J, He S, Deng Y, Zhang J, Peng Y, Hughes T, et al. Genetic modification of T cells redirected toward CS1 enhances eradication of myeloma cells. *Clin Cancer Res* (2014) 20(15):3989–4000. doi:10.1158/1078-0432.CCR-13-2510
245. Casucci M, Nicolis di Robilant B, Falcone L, Camisa B, Norelli M, Genovese P, et al. CD44v6-*targeted T cells* mediate potent antitumor effects against acute myeloid leukemia and multiple myeloma. *Blood* (2013) 122(20):3461–72. doi:10.1182/blood-2013-04-493361
246. Casucci M, Hawkins RE, Dotti G, Bondanza A. Overcoming the toxicity hurdles of genetically targeted T cells. *Cancer Immunol Immunother* (2015) 64(1):123–30. doi:10.1007/s00262-014-1641-9
247. Brudno JN, Kochenderfer JN. Toxicities of chimeric antigen receptor T cells: recognition and management. *Blood* (2016) 127(26):3321–30. doi:10.1182/blood-2016-04-703751
248. Glienke W, Esser R, Priesner C, Suerth JD, Schambach A, Wels WS, et al. Advantages and applications of CAR-expressing natural killer cells. *Front Pharmacol* (2015) 6:21. doi:10.3389/fphar.2015.00021
249. Hermanson DL, Kaufman DS. Utilizing chimeric antigen receptors to direct natural killer cell activity. *Front Immunol* (2015) 6:195. doi:10.3389/fimmu.2015.00195
250. Chu J, Deng Y, Benson DM, He S, Hughes T, Zhang J, et al. CS1-specific chimeric antigen receptor (CAR)-engineered natural killer cells enhance in vitro and in vivo antitumor activity against human multiple myeloma. *Leukemia* (2014) 28(4):917–27. doi:10.1038/leu.2013.279
251. Jiang H, Zhang W, Shang P, Zhang H, Fu W, Ye F, et al. Transfection of chimeric anti-CD138 gene enhances natural killer cell activation and killing of multiple myeloma cells. *Mol Oncol* (2014) 8(2):297–310. doi:10.1016/j.molonc.2013.12.001
252. Tonn T, Schwabe D, Klingemann HG, Becker S, Esser R, Koehl U, et al. Treatment of patients with advanced cancer with the natural killer cell line NK-92. *Cytotherapy* (2013) 15(12):1563–70. doi:10.1016/j.jcyt.2013.06.017
253. Suck G, Odendahl M, Nowakowska P, Seidl C, Wels WS, Klingemann HG, et al. NK-92: an 'off-the-shelf therapeutic' for adoptive natural killer cell-based cancer immunotherapy. *Cancer Immunol Immunother* (2016) 65(4):485–92. doi:10.1007/s00262-015-1761-x
254. Wang Y, Yibo Zhang Y, Don Bnson D, Caligiuri M, Yu J. Daratumumab combined with CD38(-) natural killer cells armed with a CS1 chimeric antigen receptor for the treatment of relapsed multiple myeloma [abstract 4617]. *Presented at American Association for Cancer Research* (Vol. 77). Washington, DC (2017). doi:10.1158/1538-7445.AM2017-4617

11

Patient's Natural Killer Cells in the Era of Targeted Therapies: Role for Tumor Killers

Meriem Messaoudene[1,2]*, Alexandra Frazao*[3]*, Pierre Jean Gavlovsky*[3]*, Antoine Toubert*[3]*, Nicolas Dulphy*[3] *and Anne Caignard*[3]*

[1] *INSERM U1015, Gustave Roussy Cancer Campus, Villejuif, France,* [2] *Gustave Roussy Cancer Campus, Villejuif, France,* [3] *INSERM U1160, Institut Universitaire d'Hématologie, Hôpital Saint Louis, Paris, France*

Edited by:
Ulrike Koehl,
Hannover Medical School, Germany

Reviewed by:
Christian Kalberer,
University of Basel, Switzerland
Ralf Dressel,
Universitätsmedizin Göttingen, Germany

***Correspondence:**
Anne Caignard
anne.caignard@inserm.fr

Natural killer (NK) cells are potent antitumor effectors, involved in hematological malignancies and solid tumor immunosurveillance. They infiltrate various solid tumors, and their numbers are correlated with good outcome. The function of NK cells extends their lytic capacities toward tumor cells expressing stress-induced ligands, through secretion of immunoregulatory cytokines, and interactions with other immune cells. Altered NK cell function due to tumor immune escape is frequent in advanced tumors; however, strategies to release the function of NK infiltrating tumors are emerging. Recent therapies targeting specific oncogenic mutations improved the treatment of cancer patients, but patients often relapse. The actual development consists in combined therapeutic strategies including agents targeting the proliferation of tumor cells and others restorating functional antitumor immune effectors for efficient and durable efficacy of anticancer treatment. In that context, we discuss the recent results of the literature to propose hypotheses concerning the potential use of NK cells, potent antitumor cytotoxic effectors, to design novel antitumor strategies.

Keywords: tumour immunosurveillance, natural killer ligands, immune checkpoint inhibitors, BRAF inhibitor, AMLMDS, melanoma

INTRODUCTION

Natural killer cells have been known and actively studied for more than four decades. They were first described as large granular lymphocytes cytotoxic for various tumor cells without prior stimulation (1, 2). In addition to their cytolytic activity against neoplastic and virus-infected cells, NK cells also display immunomodulatory functions by their ability to release cytokines, like interferon-γ (IFNγ) and tumor necrosis factor-α (TNFα), and chemokines. NK cells represent 5–15% of blood lymphocytes. They are present in the bone marrow, liver, uterus, spleen, lungs, in mucosa-associated lymphoid tissues, thymus, and secondary lymphoid tissues (SLT) and are recruited in inflamed sites. In SLT, NK cells provide an early source of IFNγ and interact with dendritic cells to promote T helper cell type 1 responses (3).

Natural killer cells are now grouped in the system of innate lymphoid cells (4). These populations, mostly tissue resident and characterized by their capacity to produce high amounts of cytokines, constitute innate homologs of T helper cell (CD4) and cytotoxic T cell (CD8) subsets. ILCs are implicated in tissue homeostasis and autoimmune diseases. Their distribution and capacity to produce cytokines suggest that they may also be involved in the development or evolution of cancer. NK cells are considered as cytotoxic counterparts of ILC1, both depending on the T-bet transcription factor for their development.

Human NK cells, defined as $CD45^+/CD3^-/CD56^+$ cells (5), are classically subdivided in two subsets based on the relative membrane expression of CD56 and CD16, the low-affinity receptor for the Fc portion of IgG (FcγRIIIA): $CD56^{dim}$ NK cells that express high levels of CD16 mediate antibody dependent cell cytotoxicity (ADCC), whereas $CD56^{bright}$ NK cells express no or low levels of CD16. These two subsets are present in different proportions in the different tissues. $CD56^{dim}$ NK cells represent 90% of blood and splenic NK cells, while $CD56^{bright}$ NK cells predominate over $CD56^{dim}$ in the SLT [lymph nodes (LN) and tonsils] representing up to 90% of NK cells and also constitute the major NK subset in tissues. It is accepted that $CD56^{bright}$ NK cells are less mature than $CD56^{dim}$ NK cells and display an immunoregulatory function, secreting high amounts of IFNγ and TNFα. $CD56^{dim}$ NK cells represent mature NK cells with a high cytotoxic activity (6).

The activation of NK cells is tightly regulated by a balance between activating and inhibitory signals delivered through engagement of numerous activating and inhibitory receptors with ligands on the target cell. Natural cytotoxicity receptors (NCRs), such as NKp46 and NKp30, are expressed by resting NK cells while NKp44 is induced after activation by cytokines, such as IL-2 and IL-15 (7, 8). The NCRs are implicated in the lysis of various tumor cells (9). The activating NK group 2 member D (NKG2D) receptor is expressed by most circulating NK cells and binds the stress-induced MHC-class I polypeptide-related sequence (MIC)-A/B molecules and UL16-binding proteins 1–6 (ULBP1–6) (10). DNAX accessory molecule-1 (DNAM-1) binds Nectin family molecules CD155 and CD112.

Natural killer cell activation is efficiently controlled by specific inhibitory NK receptors binding human leukocyte antigen of class I (HLA-class I) molecules. The C-type lectin CD94/NKG2A receptor binds HLA-E molecules (11) sensing the global HLA-class I molecules on the target while killer Ig-like receptors (KIRs) bind classical HLA-class I molecules, including HLA-C, HLA-Bw4, and some HLA-A alleles.

NK CELLS IN TUMOR IMMUNOSURVEILLANCE

A link between NK cell function and cancer development was reported in a Japanese 11-year follow-up study including 3,625 patients in which cancer incidence was negatively correlated with blood NK-mediated cytotoxicity (12). Authors further showed that individuals with particular NKG2D haplotypes, HNK1/HNK1 haplotype (correlated with high NK activity) had a decreased risk of cancer compared to those with an LNK1/LNK1 haplotype (correlated with low NK activity) (13).

Additional results including ours showed the impact of NCR transcripts in the evolution of melanoma, lung cancers, and gastrointestinal stromal tumors (GIST) patients (14–16). High NKp46 correlated with better survival in metastatic melanoma patients and particular profiles of NKp30 isoforms was associated with better outcome and response to treatment in GIST patients.

The cancer immunoediting process (17) resumes cancer progression in three phases. In the elimination phase, immune cells and among them NK cells eradicate developing tumor cells. During the equilibrium phase, the immune system may select tumor variants with less immunogenicity gradually leading to the tumor escape phase and tumor progression. It is considered that most tumors at diagnosis are in the phase of immune escape associated with functionally altered tumor infiltrating NK cells (18). Tumor immunoediting selecting variants with decreased expression of stress-induced ligands provide tumor escape to NK cell-mediated lysis through activating receptors NKG2D or NKp46 (19, 20).

The challenge is thus to overcome tumor immunosuppression and restore NK cell activities. To this aim, understanding the mechanisms that lead to NK cell defects in tumor is required.

NK CELLS IN HEMATOLOGICAL MALIGNANCIES

Numerous studies showed that severe quantitative and qualitative alterations of NK cells are associated with different hematological malignancies, particularly in myeloid disorders. In chronic myelogenous leukemia patients, low numbers of NK cells are associated with defects in their proliferation, and weak NK cell cytolytic functions in comparison with healthy donor blood NK cells (21). Furthermore, profound alterations in the activating receptors profile have also been reported including downregulation of NKp30 and NKp46 as well as DNAM-1, 2B4, and NKG2C on NK cells from acute myeloid leukemia (AML) patients. Decreased NKp30 and NKp46 expression was correlated with reduced NK cell killing and poor leukemia prognosis (22–25). Recently, Khaznadar et al. analyzed by cell imaging the lytic NK immunological synapse following interaction with AML cells and showed defective lytic granule polarization in NK cell-AML conjugates leading to impaired NK cell cytotoxic function (26).

Importantly, the intimate relationship between immune pressure and leukemogenesis has been suggested in two recent studies. Stringaris et al. described an immunoediting process induced by AML blasts that limits NK cell control of leukemia. They showed that abnormal NKG2A expression and TNFα production predict a poor response to chemotherapy in AML patients (27). Conversely, Khaznadar et al. showed that NK cell defects in AML patients at diagnosis could be associated with a specific transcriptional program in AML blasts and with patient's outcome including relapse occurrence (28).

Furthermore, the beneficial role in the graft-versus-leukemia (GvL) of allogeneic NK cells for leukemic patients receiving allogeneic hematopoietic stem cell transplantation (HSCT) is well documented (29). Several studies showed that NK cells have a potent GvL effect in both KIR/HLA-class I-mismatched and -matched donor–recipient combinations after allogenic HSCT in AML patients (30–32). Moreover, rapid NK recovery after HSCT is also associated with a greater GvL effect and improved outcome in AML patients (33).

NK CELLS IN SOLID TUMORS

In situ detection of NK cells infiltrating various human tumors/tissues was carried out, leading sometimes to divergent results due to the disparity of NK cell markers used (CD57, CD56, NKp46, double CD3/CD56 staining). However, several reports showed that NK cells can infiltrate clear-cell renal cell carcinoma (34), melanoma (35), non-small cell lung cancer (NSCLC) (36), breast cancer (BC) (37), GIST (38), and colorectal carcinoma (CRC) (39) although NK cells were mainly localized at the tumor's periphery. In several tumors, infiltrations by NK cells were reported to have a prognostic value. Increased overall survival was associated with a high NK cell infiltrate within the tumor or tumor stroma in lung adenocarcinoma (40), metastatic renal carcinoma (41), and lung metastasis of renal cancer (42). Elevated number of NK cells was associated with reduced risk of cancer progression in prostate cancer (43), with a reduced risk of death in squamous cell lung cancer (44), and a better prognosis in gastric carcinoma (45) and CRC (46). In addition, the number of NKp46$^+$ NK cells was found inversely correlated with metastasis occurrence in patients with GIST (47). Furthermore, a positive association between a high numbers of tumor infiltrating CD56$^+$ NK cells with a regression of melanocytic lesions was observed (48).

In most tumor types studied, *ex vivo* tumor-infiltrating NK cells displayed severe phenotypic and functional alterations compared to blood NK cells and more interestingly compared to NK cells present in adjacent normal tissues. Those alterations affected the expression of activating receptors including NKp30, CD16, DNAM-1, and ILT2 on NK cells from patients with non-invasive and invasive BC (49) or NSCLC (36). A concomitant-increased expression of the inhibitory molecule NKG2A was also observed in BC (49). This deficient phenotype was associated with impaired functions including decreased cytotoxicity against tumor cells (36, 49) and reduced IFNγ production (36). Recently, Carrega et al. reported that lung and BC tissues were highly enriched in CD56^{bright}perforin^{low} NK cell subset compared to matched normal tissues (37). It is of note that comparison between NK cells from tumor and normal adjacent tissue is required for better understanding of the effect of the tumor environment on their activation.

Interestingly, our team recently identified in tumor draining LN from melanoma and BC patients, the presence of a CD56^{bright}CD16$^+$ NK-cell subset that displays higher expression of activating receptors, perforin molecules, and performs ADCC (50). We found that different NK receptors regulate the two LN-NK cell subsets in melanoma and BC (personal communication) and that NK-infiltrating LN recapitulate the alterations reported in the primary tumors. The presence of CD16$^+$ NK cells in certain tumors (51) and metastatic LN emphasizes the interest for ADCC function of such NK cells.

Alterations in Blood NK Cells from Patients with Solid Tumors

Alterations in blood NK cells from patients with solid tumors were also reported, but in a lesser extent than in tumor infiltrating NK cells. Compared to healthy donors, a downregulation of NKG2D and an increase of the inhibitory receptor CD158b expression were correlated with impaired NK cell function (52–54) in metastatic melanoma patients. Our group showed a progressive decrease of NKp46 expression on blood NK cells with the disease progression in melanoma patients (55). In BC patients with invasive tumor, blood NK cells display altered expression of activating receptors NKp30, NKG2D, DNAM-1, 2B4, and CD16 and an upregulation of the inhibitory receptors NKG2A and CD85j. This phenotypic change was correlated with decreased NK cell cytotoxicity function and cytokine production (IFNγ and TNFα) (49). Blood NK cells from soft-tissue sarcoma patients displayed reduced proportions of CD56^{dim} NK cells. Low percentages of blood NK cells associated with a reduced NKp30, NKp46, and NKG2D expression were reported in patients with invasive squamous cervical cancer (56).

NK CELLS: A POTENTIAL PARTNER FOR TARGETED THERAPIES

The advent of targeted therapies that counteract a vital cellular process within the tumor cell greatly improved cancer treatment strategies. Thus, mitogen-activated protein kinase (MAPK) inhibitors that control the mutation-driven oncogenic pathway present in most cancers are new efficient players in the arsenal of therapies for cancer patients. In addition, monoclonal antibodies (mAbs) that recognize tumor-associated antigens have been established as one of the most successful therapeutic strategies for both hematologic malignancies and solid tumors. These mAbs may activate antibody-dependent cell-mediated cytotoxicity involving NK cells.

Combining targeted therapies and methods to stimulate patient's immune players is actively evaluated and represents a promising and natural evolution in cancer treatment as this could ally immediate efficiency, specificity, and long-term antitumor efficacy.

It is of note that targeted therapies also display off-target effects, connecting oncogenesis to immunosurveillance. We discuss below the interest of NK cell-based therapies in the context of such tumor-targeted therapies (**Figure 1**).

Effect of Cancer Treatment on NK Cells

Most melanoma patients (65%) bear a *BRAF*-mutated tumor and receive specific inhibitors targeting mutated $BRAF^{V600E}$ alone or in combination with MEK inhibitors, upstream of ERK (57). These inhibitors may exert bystander effects on certain immune cells that depend on MAPK for their activation and/or proliferation. BRAF inhibitors do not affect NK cell phenotype *in vivo* and *in vitro*, but blood NK cell numbers were increased in vemurafenib-treated patients (58, 59). MEK inhibition alters the expression of the main NK receptors and the function of cytokine-activated NK cells, but the combined BRAF and MEK inhibitors did not (60).

In addition, targeted therapies may interfere with the NK/target interactions through modulation of NK ligands on cancer cells. We have shown that a BRAF inhibitor modulates the expression of MICA and ULBP2 (ligands of NKG2D), changing

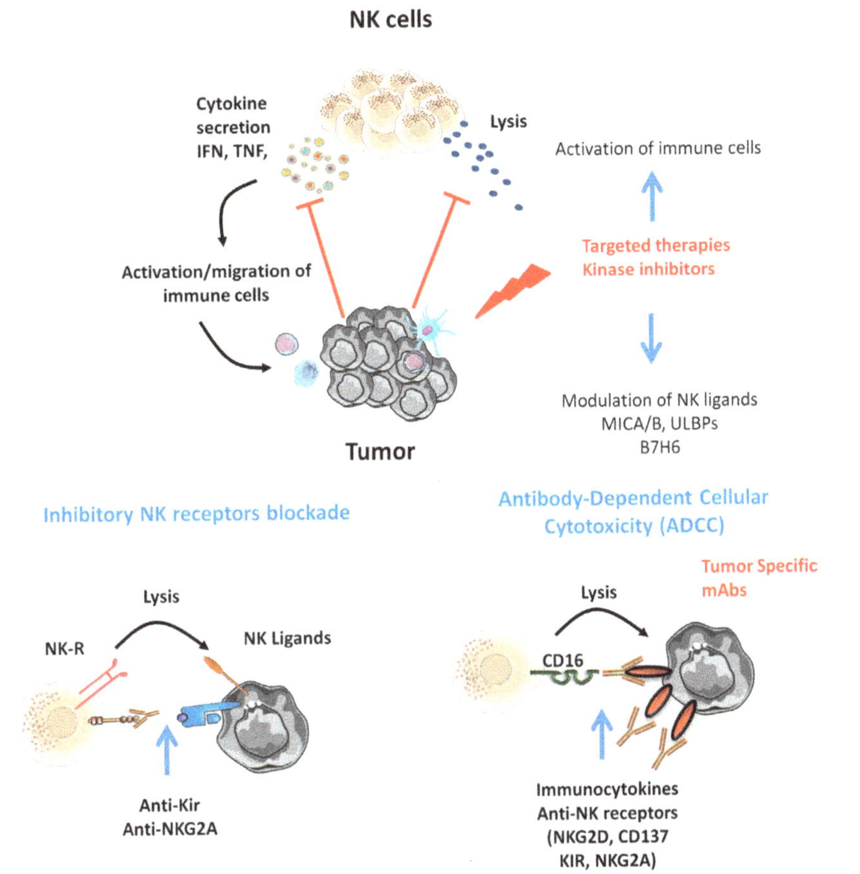

FIGURE 1 | Natural killer (NK) cells infiltrating various tumor types display altered functions. Tumor-specific therapies may potentiate NK cell activation. Inhibitors targeting MAP kinase reduce tumor growth and display off-target effect modulating NK ligands expression and immune cell activation. Tumor-specific monoclonal antibodies (mAbs) may trigger ADCC by CD16$^+$ NK cells. In that context, target NK-based immunotherapies may be proposed. Combined mitogen-activated protein kinase inhibitors with cytokines release NK function by inhibitory NK receptors [killer Ig-like receptor (KIR), NKG2A] blockade and promote NK-mediated ADCC of tumor antigen-specific mAbs and combined immunocytokines, anti-NK receptors (NKG2D, CD137, KIR, NKG2A).

the ratio between membrane expression and soluble form, and increases B7H6 (ligand of NKp30) expression and HLA-A,B,C and HLA-E molecules expression that engage inhibitory receptors (KIRs, NKG2A), thus interfering with NK cell-mediated lysis (in revision). Resistance to a BRAF inhibitor is accompanied by higher NK ligands expression (personal communications).

Our findings and recent results from the literature emphasize that therapeutics designed to limit cancer cell growth by acting through kinase inhibitors should also be considered in terms of their impact on immunosurveillance (61). In a murine model of *BRAF*-mutated melanoma, host NK cells and perforin were required for the effect of a BRAF inhibitor (62) and correlated with the reduction of tumor growth, and an increased NK and T cell infiltration of the tumors (63).

Combining specific MAPK inhibitors with immunotherapies to increase response rates is evaluated leading to yet discordant results. BRAF inhibition augments melanoma antigen expression and maintains T cell function (64). However, inhibition of BRAF in a murine model of human melanoma was associated with decreased tumor-resident lymphocytes and resistance to CTLA-4 mAb (65). MEK inhibitors increased antigen-specific T cell within the tumor sparing their cytotoxicity and combined with anti-PD-L1 mAb they exerted a synergic effect of tumor growth inhibition (66). Other kinase inhibitors such as those targeting Jak involved in the signaling cascade of cytokine receptors may influence NK (67).

A better understanding of off-target efficacy of MAPK inhibition affecting tumor–host interactions is required to develop strategies aimed at facilitating antitumor immune responses. The emerging findings indicate a potential synergy between targeted therapies, which change the balance between ligands of activating and inhibitory NK receptors, and NK-based immunotherapies, opening new interesting opportunities for the design of clinical trials.

Anti-KIR/Anti-NKG2A mAbs: Increasing NK Function by Blocking Negative Signaling

One promising approach is to release NK cell function with anti-KIR or anti-NKG2A mAbs as NK cells are strictly controlled by receptors specific for HLA-class I molecules. Fully human anti-KIR mAbs, 1-7F9 mAb, and then lirilumab (recombinant version with a stabilized hinge) were generated (68). They prevent the binding of KIR2DL1, KIR2DL2, and KIR2DL3 receptors to their HLA-C ligands and blocking their inhibitory signaling. *In vitro* and *in vivo* studies showed that anti-KIR mAbs augmented NK cell-mediated lysis of HLA-C^+ tumor cells, including autologous AML blasts and autologous $CD138^+$ multiple myeloma cells (68–71). In addition, transient increases of TNFα and MIP-1β serum concentrations and CD69 expression on NK cells were observed from treated patients (72). In a clinical trial, Benson et al. showed that 1-7F9 mAb is safe in patients with multiple myeloma and enhances *ex vivo* patient-derived NK cell cytotoxicity against tumor cells (73).

Other immune receptors highly expressed by NK cells are in development, such as anti-NKG2A (monalizumab).

Targeting inhibitory pathways in NK cell/tumor interactions may be complementary to small-molecule inhibitors for the treatment of advanced tumors such as melanoma. The prospect of combining NK cell-based immunotherapy with approaches to target the immunosuppressive tumor microenvironment or immune checkpoints, such as KIR blockade, is especially relevant to the treatment of solid tumors (74, 75) and particularly for tumors refractory to targeted therapies.

NK Cell-Mediated ADCC Using Tumor-Specific mAb

Natural killer cells express activating low-affinity FcgRIIIa (CD16) and are key mediators of antibody-dependent cellular cytotoxicity. The relevance of ADCC in tumor control using therapeutic mAbs was evaluated in several cancers. The contribution of ADCC to the clinical efficacy of a therapeutic mAb has been observed in non-Hodgkin's lymphoma patients treated by anti-CD20 (rituximab) (76). Other therapeutic mAbs likely inducing NK cell-mediated ADCC are anti-CD19 in patients with B malignancies, anti-GD2 in neuroblastoma patients, and anti-HER2 mAbs (trastuzumab) in metastatic breast and gastric cancer patients (76–78). Anti-EGFR mAb (cetuximab) was shown to increase ADCC-mediated lysis of colon tumor cells by blood NK cells from colorectal cancer patients that display altered natural cytotoxic activity (51).

Several modifications of the antibody structure, such as class switching, humanization, and point mutations to reduce complement interaction/activation, are developed to engineer mAbs with increased NK cell ADCC function and limit their toxicity. Thus, humanized anti-GD2 mAb (hu3F8-IgG1) exerts reduced toxicity compared to other anti-GD2 mAbs, by leveraging ADCC over complement-mediated cytotoxicity (79). Higher FcγRIIIA-binding affinity of anti-CD19 antibody significantly increased NK cell-mediated ADCC, leading to malignant B-cell clearing in non-human primates (78, 80). Other strategies to enhance the effect of ADCC include the coadministration of cytokines, IL-12 with anti-HER2/neu (trastuzumab) (81) to stimulate IFNγ production by NK cells and T cells and promote the $CD56^{dim}CD16^+$ NK cell differentiation to mediate ADCC (82). Co-infusion of anti-CD20 (rituximab) and TLR9 agonist (CpG) that is known to raise the membrane expression of CD20 on malignant B cells enhances ADCC (83). The infusion of immunocytokines, cytokines linked to the Fc terminus of humanized Abs, is also evaluated to potentiate ADCC. In preclinical study, Buhtoiarov et al. demonstrated that the humanized anti-GD2 immunocytokine hu14.18-IL-2 exerts higher antitumor effect than the reagents given separately (84).

Combining tumor-specific mAbs and mAbs targeting NK receptors (NKG2D, costimulatory molecule CD137) is another option. Anti-CD137 coadministred with rituximab led to a subsequent stimulation of these NK cells and enhanced rituximab-dependent cytotoxicity against the lymphoma cells (85). Furthermore, combination of rituximab with antibodies that block KIR2DL1 significantly improved NK cell-mediated lysis of tumor targets (86).

CONCLUSION

Restoring NK cell functions in addition to administration of tumor-specific therapies with kinase inhibitors or tumor-specific mAbs may benefit patients. It would increase the control of residual tumor cells, enhance mAbs efficiency, and promote the adaptive immune response necessary for long-lasting protective immunity. In that context, cytokines, blockade of inhibitory NK receptors (KIRs, NKG2A), or transfer of alloreactive NK cells are promising NK-based therapies.

ETHICS STATEMENT

The study protocol was approved by an ethic committee "Ile de France" (CPP: 2834), and the Declaration of Helsinki protocols were followed.

AUTHOR CONTRIBUTIONS

MM, ND, and AC wrote the manuscript; MM did the figure; AF, PG, and AT read and corrected the manuscript.

REFERENCES

1. Herberman RB, Nunn ME, Holden HT, Lavrin DH. Natural cytotoxic reactivity of mouse lymphoid cells against syngeneic and allogeneic tumors. II. Characterization of effector cells. *Int J Cancer* (1975) 16:230–9. doi:10.1002/ijc.2910160204
2. Kiessling R, Klein E, Wigzell H. "Natural" killer cells in the mouse. I. Cytotoxic cells with specificity for mouse Moloney leukemia cells. Specificity and distribution according to genotype. *Eur J Immunol* (1975) 5:112–7. doi:10.1002/eji.1830050209
3. Sun H, Sun C, Tian Z, Xiao W. NK cells in immunotolerant organs. *Cell Mol Immunol* (2013) 10:202–12. doi:10.1038/cmi.2013.9

4. Spits H, Artis D, Colonna M, Diefenbach A, Di Santo JP, Eberl G, et al. Innate lymphoid cells - a proposal for uniform nomenclature. *Nat Rev Immunol* (2013) 13:145–9. doi:10.1038/nri3365
5. Lanier LL, Testi R, Bindl J, Phillips JH. Identity of Leu-19 (CD56) leukocyte differentiation antigen and neural cell adhesion molecule. *J Exp Med* (1989) 169:2233–8. doi:10.1084/jem.169.6.2233
6. Caligiuri MA. Human natural killer cells. *Blood* (2008) 112:461–9. doi:10.1182/blood-2007-09-077438
7. Lanier L. Turning on natural killer cells. *J Exp Med* (2000) 191:1259–62. doi:10.1084/jem.191.8.1259
8. Moretta L, Moretta A. Unravelling natural killer cell function: triggering and inhibitory human NK receptors. *EMBO J* (2004) 23:255–9. doi:10.1038/sj.emboj.7600019
9. Moretta A, Bottino C, Vitale M, Pende D, Cantoni C, Mingari M, et al. Activating receptors and coreceptors involved in human natural killer cell-mediated cytolysis. *Annu Rev Immunol* (2001) 19:197–223. doi:10.1146/annurev.immunol.19.1.197
10. Raulet DH. Roles of the NKG2D immunoreceptor and its ligands. *Nat Rev Immunol* (2003) 3:781–90. doi:10.1038/nri1199
11. Braud V, Allan D, O'Callaghan C, Soderstrom K, D'Andrea A, Ogg G, et al. HLA-E binds to natural killer cell receptors CD94/NKG2A, B and C. *Nature* (1998) 391:795–9. doi:10.1038/35869
12. Imai K, Matsuyama S, Miyake S, Suga K, Nakachi K. Natural cytotoxic activity of peripheral-blood lymphocytes and cancer incidence: an 11-year follow-up study of a general population. *Lancet* (2000) 356:1795–9. doi:10.1016/S0140-6736(00)03231-1
13. Hayashi T, Imai K, Morishita Y, Hayashi I, Kusunoki Y, Nakachi K. Identification of the NKG2D haplotypes associated with natural cytotoxic activity of peripheral blood lymphocytes and cancer immunosurveillance. *Cancer Res* (2006) 66:563–70. doi:10.1158/0008-5472.CAN-05-2776
14. Fend L, Rusakiewicz S, Adam J, Bastien B, Caignard A, Messaoudene M, et al. Prognostic impact of the expression of NCR1 and NCR3 NK cell receptors and PD-L1 on advanced non-small cell lung cancer. *Oncoimmunology* (2017) 6:e1163456. doi:10.1080/2162402X.2016.1163456
15. Messaoudene M, Fregni G, Enot D, Jacquelot N, Neves E, Germaud N, et al. NKp30 isoforms and NKp46 transcripts in metastatic melanoma patients: unique NKp30 pattern in rare melanoma patients with favorable evolution. *Oncoimmunology* (2016) 5:e1154251. doi:10.1080/2162402X.2016.1154251
16. Rusakiewicz S, Perier A, Semeraro M, Pitt JM, Pogge von Strandmann E, Reiners KS, et al. NKp30 isoforms and NKp30 ligands are predictive biomarkers of response to imatinib mesylate in metastatic GIST patients. *Oncoimmunology* (2017) 6:e1137418. doi:10.1080/2162402X.2015.1137418
17. Dunn GP, Old LJ, Schreiber RD. The three Es of cancer immunoediting. *Annu Rev Immunol* (2004) 22:329–60. doi:10.1146/annurev.immunol.22.012703.104803
18. Vesely MD, Kershaw MH, Schreiber RD, Smyth MJ. Natural innate and adaptive immunity to cancer. *Annu Rev Immunol* (2011) 29:235–71. doi:10.1146/annurev-immunol-031210-101324
19. Elboim M, Gazit R, Gur C, Ghadially H, Betser-Cohen G, Mandelboim O. Tumor immunoediting by NKp46. *J Immunol* (2010) 184:5637–44. doi:10.4049/jimmunol.0901644
20. Guerra N, Tan YX, Joncker NT, Choy A, Gallardo F, Xiong N, et al. NKG2D-deficient mice are defective in tumor surveillance in models of spontaneous malignancy. *Immunity* (2008) 28:571–80. doi:10.1016/j.immuni.2008.02.016
21. Pierson BA, Miller JS. CD56+bright and CD56+dim natural killer cells in patients with chronic myelogenous leukemia progressively decrease in number, respond less to stimuli that recruit clonogenic natural killer cells, and exhibit decreased proliferation on a per cell basis. *Blood* (1996) 88:2279–87.
22. Costello RT, Sivori S, Marcenaro E, Lafage-Pochitaloff M, Mozziconacci MJ, Reviron D, et al. Defective expression and function of natural killer cell-triggering receptors in patients with acute myeloid leukemia. *Blood* (2002) 99:3661–7. doi:10.1182/blood.V99.10.3661
23. Fauriat C, Just-Landi S, Mallet F, Arnoulet C, Sainty D, Olive D, et al. Deficient expression of NCR in NK cells from acute myeloid leukemia: evolution during leukemia treatment and impact of leukemia cells in NCRdull phenotype induction. *Blood* (2007) 109:323–30. doi:10.1182/blood-2005-08-027979
24. Sanchez-Correa B, Gayoso I, Bergua JM, Casado JG, Morgado S, Solana R, et al. Decreased expression of DNAM-1 on NK cells from acute myeloid leukemia patients. *Immunol Cell Biol* (2012) 90:109–15. doi:10.1038/icb.2011.15
25. Lion E, Willemen Y, Berneman ZN, Van Tendeloo VF, Smits EL. Natural killer cell immune escape in acute myeloid leukemia. *Leukemia* (2012) 26:2019–26. doi:10.1038/leu.2012.87
26. Khaznadar Z, Henry G, Setterblad N, Agaugue S, Raffoux E, Boissel N, et al. Acute myeloid leukemia impairs natural killer cells through the formation of a deficient cytotoxic immunological synapse. *Eur J Immunol* (2014) 44:3068–80. doi:10.1002/eji.201444500
27. Stringaris K, Sekine T, Khoder A, Alsuliman A, Razzaghi B, Sargeant R, et al. Leukemia-induced phenotypic and functional defects in natural killer cells predict failure to achieve remission in acute myeloid leukemia. *Haematologica* (2014) 99:836–47. doi:10.3324/haematol.2013.087536
28. Khaznadar Z, Boissel N, Agaugue S, Henry G, Cheok M, Vignon M, et al. Defective NK cells in acute myeloid leukemia patients at diagnosis are associated with blast transcriptional signatures of immune evasion. *J Immunol* (2015) 195:2580–90. doi:10.4049/jimmunol.1500262
29. Ruggeri L, Capanni M, Urbani E, Perrucio K, Shlomchik W, Tosti A, et al. Effectiveness of donor natural killer cell alloreactivity in mismatched hematopoietic transplants. *Science* (2002) 295:2375–82. doi:10.1126/science.1068440
30. Beelen DW, Ottinger HD, Ferencik S, Elmaagacli AH, Peceny R, Trenschel R, et al. Genotypic inhibitory killer immunoglobulin-like receptor ligand incompatibility enhances the long-term antileukemic effect of unmodified allogeneic hematopoietic stem cell transplantation in patients with myeloid leukemias. *Blood* (2005) 105:2594–600. doi:10.1182/blood-2004-04-1441
31. Cooley S, Weisdorf DJ, Guethlein LA, Klein JP, Wang T, Le CT, et al. Donor selection for natural killer cell receptor genes leads to superior survival after unrelated transplantation for acute myelogenous leukemia. *Blood* (2010) 116:2411–9. doi:10.1182/blood-2010-05-283051
32. Stringaris K, Adams S, Uribe M, Eniafe R, Wu CO, Savani BN, et al. Donor KIR genes 2DL5A, 2DS1 and 3DS1 are associated with a reduced rate of leukemia relapse after HLA-identical sibling stem cell transplantation for acute myeloid leukemia but not other hematologic malignancies. *Biol Blood Marrow Transplant* (2010) 16:1257–64. doi:10.1016/j.bbmt.2010.03.004
33. Savani BN, Mielke S, Adams S, Uribe M, Rezvani K, Yong AS, et al. Rapid natural killer cell recovery determines outcome after T-cell-depleted HLA-identical stem cell transplantation in patients with myeloid leukemias but not with acute lymphoblastic leukemia. *Leukemia* (2007) 21:2145–52. doi:10.1038/sj.leu.2404892
34. Wittnebel S, Da Rocha S, Giron-Michel J, Jalil A, Opolon P, Escudier B, et al. Membrane-bound interleukin (IL)-15 on renal tumor cells rescues natural killer cells from IL-2 starvation-induced apoptosis. *Cancer Res* (2007) 67:5594–9. doi:10.1158/0008-5472.CAN-06-4406
35. Kornstein MJ, Stewart R, Elder DE. Natural killer cells in the host response to melanoma. *Cancer Res* (1987) 47:1411–2.
36. Platonova S, Cherfils-Vicini J, Damotte D, Crozet L, Vieillard V, Validire P, et al. Profound coordinated alterations of intratumoral NK cell phenotype and function in lung carcinoma. *Cancer Res* (2011) 71:5412–22. doi:10.1158/0008-5472.CAN-10-4179
37. Carrega P, Bonaccorsi I, Di Carlo E, Morandi B, Paul P, Rizzello V, et al. CD56(bright)perforin(low) noncytotoxic human NK cells are abundant in both healthy and neoplastic solid tissues and recirculate to secondary lymphoid organs via afferent lymph. *J Immunol* (2014) 192:3805–15. doi:10.4049/jimmunol.1301889
38. Rusakiewicz S, Semeraro M, Sarabi M, Desbois M, Locher C, Mendez R, et al. Immune infiltrates are prognostic factors in localized gastrointestinal stromal tumors. *Cancer Res* (2013) 73:3499–510. doi:10.1158/0008-5472.CAN-13-0371
39. Halama N, Braun M, Kahlert C, Spille A, Quack C, Rahbari N, et al. Natural killer cells are scarce in colorectal carcinoma tissue despite high levels of chemokines and cytokines. *Clin Cancer Res* (2011) 17:678–89. doi:10.1158/1078-0432.CCR-10-2173
40. Takanami I, Takeuchi K, Giga M. The prognostic value of natural killer cell infiltration in resected pulmonary adenocarcinoma. *J Thorac Cardiovasc Surg* (2001) 121:1058–63. doi:10.1067/mtc.2001.113026
41. Donskov F, von der Maase H. Impact of immune parameters on long-term survival in metastatic renal cell carcinoma. *J Clin Oncol* (2006) 24:1997–2005. doi:10.1200/JCO.2005.03.9594
42. Remark R, Alifano M, Cremer I, Lupo A, Dieu-Nosjean MC, Riquet M, et al. Characteristics and clinical impacts of the immune environments in colorectal and renal cell carcinoma lung metastases: influence of tumor origin. *Clin Cancer Res* (2013) 19:4079–91. doi:10.1158/1078-0432.CCR-12-3847

43. Gannon PO, Poisson AO, Delvoye N, Lapointe R, Mes-Masson AM, Saad F. Characterization of the intra-prostatic immune cell infiltration in androgen-deprived prostate cancer patients. *J Immunol Methods* (2009) 348:9–17. doi:10.1016/j.jim.2009.06.004
44. Villegas FR, Coca S, Villarrubia VG, Jimenez R, Chillon MJ, Jareno J, et al. Prognostic significance of tumor infiltrating natural killer cells subset CD57 in patients with squamous cell lung cancer. *Lung Cancer* (2002) 35:23–8. doi:10.1016/S0169-5002(01)00292-6
45. Ishigami S, Natsugoe S, Tokuda K, Nakajo A, Che X, Iwashige H, et al. Prognostic value of intratumoral natural killer cells in gastric carcinoma. *Cancer* (2000) 88:577–83. doi:10.1002/(SICI)1097-0142(20000201)88:3<577::AID-CNCR13>3.0.CO;2-V
46. Coca S, Perez-Piqueras J, Martinez D, Colmenarejo A, Saez MA, Vallejo C, et al. The prognostic significance of intratumoral natural killer cells in patients with colorectal carcinoma. *Cancer* (1997) 79:2320–8. doi:10.1002/(SICI)1097-0142(19970615)79:12<2320::AID-CNCR5>3.0.CO;2-P
47. Delahaye NF, Rusakiewicz S, Martins I, Menard C, Roux S, Lyonnet L, et al. Alternatively spliced NKp30 isoforms affect the prognosis of gastrointestinal stromal tumors. *Nat Med* (2011) 17:700–7. doi:10.1038/nm.2366
48. McKay K, Moore PC, Smoller BR, Hiatt KM. Association between natural killer cells and regression in melanocytic lesions. *Hum Pathol* (2011) 42:1960–4. doi:10.1016/j.humpath.2011.02.019
49. Mamessier E, Sylvain A, Thibult ML, Houvenaeghel G, Jacquemier J, Castellano R, et al. Human breast cancer cells enhance self tolerance by promoting evasion from NK cell antitumor immunity. *J Clin Invest* (2011) 121:3609–22. doi:10.1172/JCI45816
50. Messaoudene M, Fregni G, Fourmentraux-Neves E, Chanal J, Maubec E, Mazouz-Dorval S, et al. Mature cytotoxic CD56bright/CD16+ natural killer cells can infiltrate lymph nodes adjacent to metastatic melanoma. *Cancer Res* (2014) 74:81–92. doi:10.1158/0008-5472.CAN-13-1303
51. Rocca YS, Roberti MP, Julia EP, Pampena MB, Bruno L, Rivero S, et al. Phenotypic and functional dysregulated blood NK cells in colorectal cancer patients can be activated by cetuximab plus IL-2 or IL-15. *Front Immunol* (2016) 7:413. doi:10.3389/fimmu.2016.00413
52. Mirjacic Martinovic K, Konjevic G, Babovic N, Inic M. The stage dependent changes in NK cell activity and the expression of activating and inhibitory NK cell receptors in melanoma patients. *J Surg Res* (2011) 171:637–49. doi:10.1016/j.jss.2010.05.012
53. Konjevic G, Mirjacic Martinovic K, Vuletic A, Jovic V, Jurisic V, Babovic N, et al. Low expression of CD161 and NKG2D activating NK receptor is associated with impaired NK cell cytotoxicity in metastatic melanoma patients. *Clin Exp Metastasis* (2007) 24:1–11. doi:10.1007/s10585-006-9043-9
54. Konjevic G, Mirjacic Martinovic K, Vuletic A, Babovic N. In-vitro IL-2 or IFN-alpha-induced NKG2D and CD161 NK cell receptor expression indicates novel aspects of NK cell activation in metastatic melanoma patients. *Melanoma Res* (2010) 20:459–67. doi:10.1097/CMR.0b013e32833e3286
55. Fregni G, Messaoudene M, Fourmentraux-Neves E, Mazouz-Dorval S, Chanal J, Maubec E, et al. Phenotypic and functional characteristics of blood natural killer cells from melanoma patients at different clinical stages. *PLoS One* (2013) 8:e76928. doi:10.1371/journal.pone.0076928
56. Garcia-Iglesias T, Del Toro-Arreola A, Albarran-Somoza B, Del Toro-Arreola S, Sanchez-Hernandez PE, Ramirez-Duenas MG, et al. Low NKp30, NKp46 and NKG2D expression and reduced cytotoxic activity on NK cells in cervical cancer and precursor lesions. *BMC Cancer* (2009) 9:186. doi:10.1186/1471-2407-9-186
57. Eroglu Z, Ribas A. Combination therapy with BRAF and MEK inhibitors for melanoma: latest evidence and place in therapy. *Ther Adv Med Oncol* (2016) 8:48–56. doi:10.1177/1758834015616934
58. Comin-Anduix B, Chodon T, Sazegar H, Matsunaga D, Mock S, Jalil J, et al. The oncogenic BRAF kinase inhibitor PLX4032/RG7204 does not affect the viability or function of human lymphocytes across a wide range of concentrations. *Clin Cancer Res* (2010) 16:6040–8. doi:10.1158/1078-0432.CCR-10-1911
59. Schilling B, Sondermann W, Zhao F, Griewank KG, Livingstone E, Sucker A, et al. Differential influence of vemurafenib and dabrafenib on patients' lymphocytes despite similar clinical efficacy in melanoma. *Ann Oncol* (2014) 25:747–53. doi:10.1093/annonc/mdt587
60. Manzini C, Vene R, Cossu I, Gualco M, Zupo S, Dono M, et al. Cytokines can counteract the inhibitory effect of MEK-i on NK-cell function. *Oncotarget* (2016) 7:60858–71. doi:10.18632/oncotarget.11504
61. Chretien AS, Le Roy A, Vey N, Prebet T, Blaise D, Fauriat C, et al. Cancer-induced alterations of NK-mediated target recognition: current and investigational pharmacological strategies aiming at restoring NK-mediated anti-tumor activity. *Front Immunol* (2014) 5:122. doi:10.3389/fimmu.2014.00122
62. Ferrari de Andrade L, Ngiow SF, Stannard K, Rusakiewicz S, Kalimutho M, Khanna KK, et al. Natural killer cells are essential for the ability of BRAF inhibitors to control BRAFV600E-mutant metastatic melanoma. *Cancer Res* (2014) 74:7298–308. doi:10.1158/0008-5472.CAN-14-1339
63. Knight DA, Ngiow SF, Li M, Parmenter T, Mok S, Cass A, et al. Host immunity contributes to the anti-melanoma activity of BRAF inhibitors. *J Clin Invest* (2013) 123:1371–81. doi:10.1172/JCI66236
64. Boni A, Cogdill AP, Dang P, Udayakumar D, Njauw CN, Sloss CM, et al. Selective BRAFV600E inhibition enhances T-cell recognition of melanoma without affecting lymphocyte function. *Cancer Res* (2010) 70:5213–9. doi:10.1158/0008-5472.CAN-10-0118
65. Hooijkaas A, Gadiot J, Morrow M, Stewart R, Schumacher T, Blank CU. Selective BRAF inhibition decreases tumor-resident lymphocyte frequencies in a mouse model of human melanoma. *Oncoimmunology* (2012) 1:609–17. doi:10.4161/onci.20226
66. Ebert PJ, Cheung J, Yang Y, McNamara E, Hong R, Moskalenko M, et al. Promotes T cell and anti-tumor activity in combination with PD-L1 checkpoint blockade. *Immunity* (2016) 44:609–21. doi:10.1016/j.immuni.2016.01.024
67. Bottos A, Gotthardt D, Gill JW, Gattelli A, Frei A, Tzankov A, et al. Decreased NK-cell tumour immunosurveillance consequent to JAK inhibition enhances metastasis in breast cancer models. *Nat Commun* (2016) 7:12258. doi:10.1038/ncomms12258
68. Romagne F, Andre P, Spee P, Zahn S, Anfossi N, Gauthier L, et al. Preclinical characterization of 1-7F9, a novel human anti-KIR receptor therapeutic antibody that augments natural killer-mediated killing of tumor cells. *Blood* (2009) 114:2667–77. doi:10.1182/blood-2009-02-206532
69. Benson DM Jr, Bakan CE, Zhang S, Collins SM, Liang J, Srivastava S, et al. IPH2101, a novel anti-inhibitory KIR antibody, and lenalidomide combine to enhance the natural killer cell versus multiple myeloma effect. *Blood* (2011) 118:6387–91. doi:10.1182/blood-2011-06-360255
70. Sola C, Andre P, Lemmers C, Fuseri N, Bonnafous C, Blery M, et al. Genetic and antibody-mediated reprogramming of natural killer cell missing-self recognition in vivo. *Proc Natl Acad Sci U S A* (2009) 106:12879–84. doi:10.1073/pnas.0901653106
71. Kohrt HE, Thielens A, Marabelle A, Sagiv-Barfi I, Sola C, Chanuc F, et al. Anti-KIR antibody enhancement of anti-lymphoma activity of natural killer cells as monotherapy and in combination with anti-CD20 antibodies. *Blood* (2014) 123:678–86. doi:10.1182/blood-2013-08-519199
72. Vey N, Bourhis JH, Boissel N, Bordessoule D, Prebet T, Charbonnier A, et al. A phase 1 trial of the anti-inhibitory KIR mAb IPH2101 for AML in complete remission. *Blood* (2012) 120:4317–23. doi:10.1182/blood-2012-06-437558
73. Benson DM Jr, Hofmeister CC, Padmanabhan S, Suvannasankha A, Jagannath S, Abonour R, et al. A phase 1 trial of the anti-KIR antibody IPH2101 in patients with relapsed/refractory multiple myeloma. *Blood* (2012) 120:4324–33. doi:10.1182/blood-2012-06-438028
74. Murray S, Lundqvist A. Targeting the tumor microenvironment to improve natural killer cell-based immunotherapies: on being in the right place at the right time, with resilience. *Hum Vaccin Immunother* (2016) 12:607–11. doi:10.1080/21645515.2015.1096458
75. Navarro A, Bjorklund A, Chekenya M. Therapeutic potential and challenges of natural killer cells in treatment of solid tumors. *Front Immunol* (2015) 29:202–7. doi:10.3389/fimmu.2015.00202
76. Alderson KL, Sondel PM. Clinical cancer therapy by NK cells via antibody-dependent cell-mediated cytotoxicity. *J Biomed Biotechnol* (2011) 2011:379123. doi:10.1155/2011/379123
77. Breton CS, Nahimana A, Aubry D, Macoin J, Moretti P, Bertschinger M, et al. A novel anti-CD19 monoclonal antibody (GBR 401) with high killing activity against B cell malignancies. *J Hematol Oncol* (2014) 7:33. doi:10.1186/1756-8722-7-33
78. Ahmed M, Cheung NK. Engineering anti-GD2 monoclonal antibodies for cancer immunotherapy. *FEBS Lett* (2014) 588:288–97. doi:10.1016/j.febslet.2013.11.030
79. Cheung NK, Guo H, Hu J, Tassev DV, Cheung IY. Humanizing murine IgG3 anti-GD2 antibody m3F8 substantially improves antibody-dependent

cell-mediated cytotoxicity while retaining targeting in vivo. *Oncoimmunology* (2012) 1:477–86. doi:10.4161/onci.19864
80. Zalevsky J, Leung IW, Karki S, Chu SY, Zhukovsky EA, Desjarlais JR, et al. The impact of Fc engineering on an anti-CD19 antibody: increased Fcgamma receptor affinity enhances B-cell clearing in nonhuman primates. *Blood* (2009) 113:3735–43. doi:10.1182/blood-2008-10-182048
81. Parihar R, Dierksheide J, Hu Y, Carson WE. IL-12 enhances the natural killer cell cytokine response to Ab-coated tumor cells. *J Clin Invest* (2002) 110:983–92. doi:10.1172/JCI0215950
82. Brady J, Hayakawa Y, Smyth MJ, Nutt SL. IL-21 induces the functional maturation of murine NK cells. *J Immunol* (2004) 172:2048–58. doi:10.4049/jimmunol.172.4.2048
83. Betting DJ, Yamada RE, Kafi K, Said J, van Rooijen N, Timmerman JM. Intratumoral but not systemic delivery of CpG oligodeoxynucleotide augments the efficacy of anti-CD20 monoclonal antibody therapy against B cell lymphoma. *J Immunother* (2009) 32:622–31. doi:10.1097/CJI.0b013e3181ab23f1
84. Buhtoiarov IN, Neal ZC, Gan J, Buhtoiarova TN, Patankar MS, Gubbels JA, et al. Differential internalization of hu14.18-IL2 immunocytokine by NK and tumor cell: impact on conjugation, cytotoxicity, and targeting. *J Leukoc Biol* (2011) 89:625–38. doi:10.1189/jlb.0710422
85. Kohrt HE, Houot R, Goldstein MJ, Weiskopf K, Alizadeh AA, Brody J, et al. CD137 stimulation enhances the antilymphoma activity of anti-CD20 antibodies. *Blood* (2011) 117:2423–32. doi:10.1182/blood-2010-08-301945
86. Binyamin L, Alpaugh RK, Hughes TL, Lutz CT, Campbell KS, Weiner LM. Blocking NK cell inhibitory self-recognition promotes antibody-dependent cellular cytotoxicity in a model of anti-lymphoma therapy. *J Immunol* (2008) 180:6392–401. doi:10.4049/jimmunol.180.9.6392

12

Modeling Natural Killer Cell Targeted Immunotherapies

Silvia Lopez-Lastra[1,2,3]* *and James P. Di Santo*[1,2]

[1] *Innate Immunity Unit, Institut Pasteur, Paris, France,* [2] *Inserm U1223, Paris, France,* [3] *Université Paris-Sud (Paris-Saclay), Paris, France*

Edited by:
Ulrike Koehl,
Hannover Medical School,
Germany

Reviewed by:
Lutz Walter,
Leibniz-Institute for Primate Research, Germany
Philippe Saas,
Etablissement Français du Sang BFC, France

***Correspondence:**
Silvia Lopez-Lastra
sillolas@pasteur.fr

Animal models have extensively contributed to our understanding of human immunobiology and to uncover the underlying pathological mechanisms occurring in the development of diseases. However, mouse models do not reproduce the genetic and molecular complexity inherent in human disease conditions. Human immune system (HIS) mouse models that are susceptible to human pathogens and can recapitulate human hematopoiesis and tumor immunobiology provide one means to bridge the interspecies gap. Natural killer cells are the founding member of the innate lymphoid cell family. They exert a rapid and strong immune response against tumor and pathogen-infected cells. Their antitumor features have long been exploited for therapeutic purposes in the context of cancer. In this review, we detail the development of highly immunodeficient mouse strains and the models currently used in cancer research. We summarize the latest improvements in adoptive natural killer (NK) cell therapies and the development of novel NK cell sources. Finally, we discuss the advantages of HIS mice to study the interactions between human NK cells and human cancers and to develop new therapeutic strategies.

Keywords: humanized mouse models, innate lymphoid cell, natural killer cells, cancer immunotherapy, natural killer cell immunotherapy

INTRODUCTION

Since the generation of the first inbred mouse strains in the early 20th century, mice have served as model organisms to study mammalian biology. This approach has given birth to some of the most important scientific breakthroughs and discoveries that, in many cases, led to the development of successful treatments for previously untreatable diseases (e.g., acute promyelocytic leukemia) (1). However, *Mus musculus* and *Homo sapiens* have been evolving divergently for 85 million years, adapting to very different environments and undergoing selection for many traits, from the circadian rhythm to our body size (2). Thanks to the genome decoding, we can now appreciate that the one fifth of the genetic divergence between mice and humans is enriched in regions implicated in the immune system, metabolic processes, and stress responses (3). It is, therefore, not surprising that only less than 8% of the cancer studies in animal models reach clinical trials and that more than 80% of these eventually fail when tested in humans (4). The increasing knowledge of the molecular differences between mice and humans should allow us to evaluate the degree in which animal models may be suitable for translational research and when this is not the case, to then search for better systems.

With this aim, mice have been "humanized" by introducing human genes or genomic regions and by transferring human tissues or cells to study various aspects of human biology. The engraftment of human blood cells or blood-forming cells and organs into immunodeficient mice has opened a new era for translational immunology and the improvement of immunotherapies against human cancer and infectious diseases caused by pathogens with exclusive human tropism, such as HIV, HBV, and HCV.

DEVELOPING HUMAN IMMUNE SYSTEM (HIS) MICE

Since the discovery of the nude athymic mutations in the 1960s, our knowledge of the host immune system and its ability to reject xenografts have led to the development of several mouse strains that permit long-term "take" and function of the human tissue grafts (5). Experiments performed in the 1980s with severe combined immunodeficient (SCID) mice (that lacked functional mouse adaptive lymphocytes due to mutations in the DNA-dependent protein kinase *Prkdc*) showed that these mice could be reconstituted with human peripheral blood mononuclear cells (PBMCs) or hematopoietic stem cells (HSCs) (6, 7). However, some residual adaptive (leakiness) and an essentially intact innate immunity in SCID mice limited the complete reconstitution of all human immune subsets. Moreover, SCID mice failed to engraft human tumor xenografts, thereby limiting the development of preclinical cancer models. An alternative system with analogous immunodeficiency was obtained by mutating the recombinant activating genes (*Rag1*, *Rag2*) loci that avoided genetic "leakiness" and, in contrast to SCID mice, did not result in host radiosensitivity (8, 9). Additional genetic modifications followed to further the immunodeficiency of host mice in order to promote tolerance to human cells. Two breakthroughs have remarkably boosted the advancement of the field. First, Greiner and colleagues found that the NOD strain supported an enhanced tolerance compared to other strains and, several years later, Takenaka's team revealed that the molecular basis for this lies in the signal regulatory protein alpha (*Sirpa*) allele polymorphism (10–13). Contrarily to other strains, SIRPα from NOD mice binds to human CD47 ligand triggering a negative signal in mouse macrophages that prevents their phagocytosis (13, 14). This finding prompted the generation of transgenic mice expressing the human or NOD strain *Sirpa* allele thus conferring enhanced human cell engraftment (15–17). The second turning point for achieving a successful xenotransplantation was the common cytokine receptor gamma chain (γ_c, encoded at *Il2rg*), which leads to complete impairment of natural killer (NK) cell development and dendritic cell (DC) dysfunction (18, 19). Mice carrying *Il2rg* mutations were developed in various genetic backgrounds [NSG or NOG (both NOD $Prkdc^{SCID}Il2rg^{-/-}$) and BRG (Balb/c $Rag2^{-/-}Il2rg^{-/-}$)] allowing robust, long-lasting *de novo* multilineage development of the HIS, including human thymopoiesis, and are the basis for most of the currently used models (20–23). From that point forward, a number of model variants have been developed to address specific questions or improve particular aspects of immunity, either by genetic manipulation, engraftment of additional human tissues, or exogenous administration of human factors. This is the case of the recently described Balb/c $Rag2^{-/-}Il2rg^{-/-}Flt3^{-/-}$ (BRGF) model with specific boost of conventional and plasmacytoid DCs after exogenous Flt3 ligand treatment. This model offers a great platform for screening of immune adjuvants and DC targeting therapies (24).

HUMAN CANCER MODELS IN "HUMANIZED" MICE

Immunodeficient mice allow great flexibility for the study of human tumor immunobiology. Human tumors can be generated in NSG, NOG, BRGS, and other strains using established tumor cell lines, after transplantation of human primary tumors or following *de novo* induction of hematological neoplasms (**Figure 1**). These different models provide systems that better reflect the complexity of the disease. In order to allow human tumor to engraft and grow in mice, the host immune system is generally compromised leading to tumor kinetics that may not reflect the true patient situation. As discussed earlier, human immune components can be generated *in vivo* from human HSCs or other progenitors and "supported or potentiated" later on or infused once the tumor is established. These approaches provide "mixed" systems in which human immune cells and human tumors can co-exist allowing the dissection of immune deviation as well as studying immunotherapy.

A wide range of established tumor cell lines from different origins (brain, colon, breast, melanoma, ovarian, prostate, etc.) have been engrafted in immunocompromised mice and have greatly contributed to drug development and the preclinical assessment of potential therapies. However, the gradual accumulation of genetic and phenotypic aberrations in these cells due to their long-term culture impacts the surface markers and the tumorigenicity of the malignancy (25). These limitations have set aside these models to preliminary studies addressing specific questions like the ability of a potential therapy to target a certain molecule that has been overexpressed in the cell line. In recent years, the field has been, therefore, switching toward the engraftment of patient-derived primary tumors (PDX, patient-derived xenografts) that retain the phenotypic and genetic complexity observed in clinical samples thus better predicting drug efficacy and clinical translatability (26, 27). These include tumor stromal cells and tumor-associated lymphocytes that contribute greatly to tumor growth and metastasis and, therefore, to the therapeutic response. These PDX-HIS mouse models can engraft the tumor as efficiently as the non-humanized mice, they respond to standard chemotherapeutic drugs similarly to patients and they have proven to be responsive to newly derived immune modulators.

One of the better-characterized PDX models is the AML that has contributed to the identification of leukemia stem cells (LSC) by transplanting different stem-like cell fractions and analyzing the leukemia-initiating activity of each in SCID mice (28–31). The discovery of the concept of cancer stem cell (CSC) has been a breakthrough in cancer biology due to the clinical benefits for the long-term disease-free survival. CSC presence has been identified in numerous other malignancies through transfer into immunocompromised mice and, interestingly, markers associated to CSCs have been correlated to the tumorigenic potential (32, 33). Recent improvements in HIS mouse models by the transgenic expression of certain factors, like the NSG-SGM3 expressing human

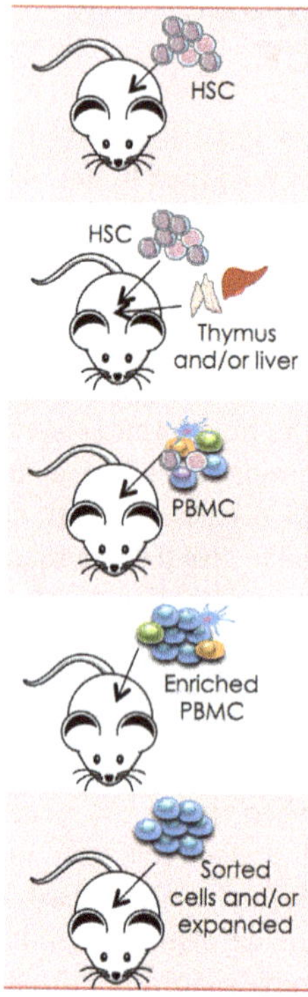

	Mouse strain	Origin human immune cells	Type of tumor xenograft	Origin of tumor xenograft	References
HSC	NSG	HSC	Rhabdomyosarcoma	Cell line	(38)
	NSG	HSC	Acute leukemia	Oncogene induced	(39)
	NOG	HSC	Pancreatic carcinoma	Cell line	(40)
	NSG	HSC+NKT	Neuroblastoma	Primary	(41)
HSC, Thymus and/or liver	NSG	HSC+FT	Leukemia	Oncogene induced	(42)
	NSG	HSC+FT	Melanoma	Cell line	(43)
	NSG	HSC+FT+liver	Oral squamous carcinoma	Primary	(44)
	NSG-SGM3	HSC+FT+liver	Melanoma	Primary	(45)
PBMC	NOG	PBMC	ATLL	Primary	(46)
	BRG	PBMC	B-cell lymphoma	EBV-induced cell line	(47)
	NSG	PBMC	Pancreatic adenocarcinoma	Cell line	(58)
	NSG	Tumor-derived	Lung tumor (non-disrupted)	Primary	(49)
Enriched PBMC	NSG	NK cells	Ovarian carcinoma	Cell line	(50)
	NSG	NK cells	Glioblastoma	Primary	(51)
	NOD/SCID	BM-T cells	Breast carcinoma	Primary	(52)
	NSG	T cells + DCs	Prostate carcinoma	Cell line	(53)
Sorted cells and/or expanded	SCID	EBV-specific T cells	B-cell lymphoma	EBV-induced cell line	(54)
	NSG	CAR-T cells	Pleural mesothelioma	Cell line	(55)
	BRG	γδ T cell	AML	Primary	(56)
	BRGS	UCB-NK cells	Colorectal carcinoma	Cell line	(37)

FIGURE 1 | Human immune system (HIS) mouse models used in cancer research. PBMC, peripheral blood mononuclear cells; NSG, NOD *Prkdc*^{SCID}*Il2rg*$^{-/-}$; BRG, Balb/c *Rag2*$^{-/-}$*Il2rg*$^{-/-}$; BRGS, Balb/c *Rag2*$^{-/-}$*Il2rg*$^{-/-}$ *Sirpa*^{NOD}; HSC, hematopoietic stem cell; FT, fetal thymus; BM, bone marrow; CAR, chimeric antigen receptor; UCB, umbilical cord blood; ATLL, T-cell leukemia/lymphoma; AML, acute myeloid leukemia; EBV, Epstein–Barr virus (37–56).

SCF, GM-SCF, and IL-3, have further ameliorate the engraftment and growth of human leukemia allowing the study of the tumor initiating cells (34). The demonstration that HIS mouse models reproduce the heterogeneity and behavior of human tumors creates great expectation on the better phenotyping of these tumor-initiating cells and the identification of drugs targeting this key population (35).

One of the advantages of modeling cancer in humanized mice is that we can study the systemic environment and the contribution of nearly all the immune cells to the pathogenesis. In this regard, a NSG model of treatment-refractory B-cell leukemia revealed that infiltration of leukemia cells into the bone marrow rewires the tumor microenvironment to inhibit engulfment of antibody-targeted tumor cells. This resistance could be overcome by combination regimens involving therapeutic antibodies and chemotherapy that lead to macrophage infiltration and phagocytic activity in the bone marrow improving the efficacy of targeted therapeutics (36).

TARGETING NK CELLS FOR CANCER IMMUNOTHERAPY

To date, most immunomodulatory strategies have focused on agents or cell therapies targeting T cell immunity. In contrast, innate immune cells, such as NK cells, have been less exploited. Nevertheless, the fundamental role for these cells has been for long justified by the higher cancer incidence in individuals with defective NK function (57). Furthermore, a number of mouse models lacking or deficient for NK cell function have corroborated their importance in tumor immunosurveillance (58).

Natural killer cells exert an immediate cytotoxicity when encountering a malignant cell and they do so without a specific antigen priming but instead, by the integrated signal of an array of activating and inhibitory receptors. Among the first group, the C-type lectin-like receptors CD94/NKG2C and NKG2D and the natural cytotoxicity receptors NKp30, NKp44, and NKp46 as well as the Ig-like receptor DNAM-1 (CD266) mediate NK cell activation when they recognize tumor cells. On the other hand, polymorphic inhibitory killer cell immunoglobulin-like receptors (KIRs) with their cognate human–leukocyte–antigen (HLA) ligands as well as CD94/NKG2A with the non-classical class I molecule HLA-E as ligand provide inhibitory signaling. In addition to the contact mediated regulation of the activity, NK cells also respond to cytokines like IL-2, IL-12, IL-15, IL-18, and IL-21, as well as toll-like receptor ligands that shape their differentiation, proliferation, and activation status (59). Cytotoxicity activity is triggered through activation of the low-affinity activating receptor FcγRIIIa (CD16) that binds the Fc portion of immunoglobulin

G1, which has been exploited in monoclonal antibody immunotherapies. NK cells kill virus-infected and tumor cells using a cargo of perforin and granzymes contained in cytotoxic granules and less efficiently by a mechanism dependent on FAS ligand, TNF, or TNF-related apoptosis-inducing ligand (60).

Given that NK cells in HSC-derived humanized mice express all the afore-mentioned receptors and respond similarly to the same cytokines (61, 62), these *in vivo* models represent a powerful platform to explore the pivotal role of NK cells in cancer immunosurveillance (63–65) (Lopez-Lastra et al., in revision). Additionally, environmental components such as inhibiting factors (TGF-β, IL-10, prostaglandin E2…) and immunosuppressive cells (Tregs, MDSCs) that influence the NK cell antitumor activity have also been described in HIS mice, enabling the evaluation of therapeutic strategies targeting the suppression of NK cells (66).

Although chemotherapy is still the core of the current clinical anticancer treatments, immunomodulators have now regained expectations after the revolutionary discovery of the CTLA-4 and PD-1 checkpoint inhibitors targeting T-cell activation (67). Humanized mice have proven to recapitulate the therapeutic effect of those antibodies as well as the side effects and have began to provide insights about the mechanism behind and possible strategies to improve them (68–70). The expression of these receptors on human NK cells suggests that they could also be targeted by checkpoint molecules and, therefore, contributes to the outcome of the therapy (71, 72). Indeed, mouse studies on a glioma model treated with activated NK cells preincubated with an anti-PD-1 blocking antibody showed an enhancement of the survival suggesting a role that must be explored in a human system (73).

ADOPTIVE TRANSFER OF NK CELLS FOR CANCER THERAPY

The potential of NK cells as innate effectors in cancer has been studied by the adoptive transfer of *ex vivo* expanded and/or activated NK cells in immunodeficient mice. Mice treated with adoptively transferred human NK cells show NK-mediated rejection of the engrafted human tumor and further administration of cytokines, such as IL-2 and IL-15 greatly improve the NK cell pool and their cytotoxic activity against transformed cell. These observations initially made in mice laid the foundation for the autologous NK cell infusion therapies started in the 1980s for metastatic cancers (74). Preclinical assessment of cytokine regimens in other cancer models, such as the low-dose IL-2 in the spontaneous EBV-associated B-cell lymphoma in PBL-SCID mice, demonstrated reduction of the tumor load and survival prolongation (75), and preceded a number of clinical trials for both hematological and solid tumors (76–79).

The discovery that inhibitory KIRs binding to MHC-I mediate inhibition of NK cells opened a new path on NK cell immunotherapies. NOD/SCID cancer models served as a platform to confirm the higher efficacy of alloreactive NK cells for the treatment of leukemia. Contrarily to T cells, NK cell do not provoke graft-versus-host disease (GVHD) in hematopoietic stem cell transplantation (HSCT) contexts but, instead, protect the patient against it and eliminate leukemia relapse and graft rejection (80). Later on, safety and efficacy of alloreactive NK cell infusion was confirmed in the clinic by Miller and colleagues in non-HSCT settings with patients suffering from metastatic melanoma, renal cell carcinoma, Hodgkin's lymphoma, and refractory AML (81). For many years, allogeneic NK cell infusions have been tested in the clinic with positive results and rare cases of mild toxicity (82). Strikingly, a recent pediatric clinical study has reported some patients suffering from acute GVHD after infusion of *ex vivo* expanded donor NK cells in HLA-matched HSCT (83), rising the necessity to perform more robust preclinical testing in humanized models. One strategy to do so was illustrated in a recent study performed in NSG mice, in which an alloreactive NK cell subpopulation expressing KIR2DS2 but lacking inhibitory KIR-HLA mismatch had dominant functional activation advantage to kill patient-derived glioblastoma cells (84). The regulation of the activity on infused NK cells has been classically based on HLA-KIR matching; however, other inhibitory receptors are implicated on the inhibition of NK cell cytotoxicity. A recent study in NSG mice engrafted with human HSC has shown that anti-NKG2A antibodies can stimulate human NK cell killing in AML and ALL models bypassing the need to search for NK cell alloreactive donors (85). *In vitro* experiments have also pointed to an increased NK cell-mediated lysis of lymphoma and myeloma cells with allogeneic NK cell infusion in combination with monoclonal antibodies blocking inhibitory KIRs but this effect need to be confirmed *in vivo* (86, 87).

Another strategy to increase NK cell activity without aggravating the side-effects is the expression of chimeric antigen receptors (CARs) directed against tumor antigens. Preclinical evaluation of CD20 targeting primary NK cell infusion in humanized mice has led to a clinical trial on B-lineage acute lymphoblastic leukemia currently undergoing (88). Other preclinical trials using CAR-engineered primary human NK cells are now being performed in lymphoma, leukemia, carcinomas, and neuroblastoma mouse models.

Natural killer cells are often infused in combination with immunomodulators that boost their antitumor effects or regulate their activity. CD16 receptor is targeted by many of those modulators since it mediates antibody-dependent cellular cytotoxicity (ADCC) when it recognizes an antibody on a tumor cell, leading to target cell lysis. This mechanism has been exploited by using monoclonal antibodies targeted tumor antigens thus stimulating the endogenous or adoptive NK cells. Evidences of NK cell-mediated ADCC and mild to moderate toxicity were observed in preclinical models and then confirmed for some cases in the clinical setting. Malignancies such as non-Hodgkin lymphoma with rituximab (anti-CD20), metastatic breast cancer with trastuzumab (anti-HER2) or metastatic colorectal, and squamous cell carcinoma of the head and neck have been treated with monoclonal antibodies together with NK cell infusions or in combination regimes extending the disease-free survival and overall survival of thousands of patients (89–91).

As mentioned before, CSCs are emerging as necessary targets to achieve cancer cures since current treatments eliminate the bulk of the tumor cells but rare resistant CSCs persist and lead to later tumor relapse (92). The upregulation of stress-induced antigens together with the ability of NK cells to target non-proliferating cells suggest that NK cells could effectively eliminate CSCs. Indeed, recent studies in pancreatic carcinoma-bearing NSG mice demonstrated the capacity of activated transferred NK cells to reduce intratumoral CSCs and tumor burden (93–95).

NOVEL NK CELL SOURCES FOR ADOPTIVE THERAPY OF CANCER

Two of the parameters to consider when evaluating the safety of NK cell products in clinical applications are the cell source and the culture conditions before the infusion. GM-CSF mobilized PBMCs, bone marrow, or umbilical cord blood (UCB) are the main sources of NK cells. With GM-CSF effects on NK function still to determine and BM being logistically difficult to obtain, UCB derived NK cells have been revealed as the best source of human material. Researchers are working on improving the expansion yield and purity as well as to enhance the activity of UCB derived NK cells before infusion in the patients. NSG mice demonstrated the capacity of these cells to migrate to BM, spleen, and liver and the inhibition of leukemia growth and prolongation of mice survival when combined with low-dose IL-15 (96). This preclinical result prompted a phase I clinical study in elderly AML patients that confirmed the safety and capacity of these cells to migrate and repopulate BM even in the absence of cytokine administration (97). This NK cell product aims at overcoming the major limitation of NK cell therapies in solid tumors, the delivery of high enough numbers of activated NK cells to the tumor site, and it is now under preclinical evaluation in the context of cervical and colorectal carcinomas (37, 98).

Alternative sources for NK cell therapy include embryonic stem cell (hESC)- or induced pluripotent stem cell (iPS)-derived NK cells, which are still under experimental development. Efficient generation of NK cells from hESC and iPS cells has been achieved, showing *in vitro* functional cytolytic activity against tumor cells, IFN-γ production, and expression of functional receptors (99). Very few reports are available regarding the *in vivo* activity of these products, with the most encouraging being in a NOD/SCID mouse model in which hESC-derived NK cells efficiently cleared a leukemia cell line tumor (100). Nevertheless, feeder-free conditions of NK cell generation need to be improved and the stability and safety of these NK cells products should be further proved in preclinical humanized models.

Finally, there is great prospect in NK cell lines as a potentially unlimited "pure" NK cell source. A clonal NK cell line NK-92 has shown the highest and most consistent cytotoxicity due to the combination of activating receptors it expresses and the absence of inhibitory KIRs (101). AML, myeloma, and melanoma are some of the numerous malignancies that have been partially eliminated from SCID mice after infusion of NK-92 (102–104). Clinical trials have further confirmed the safety and efficacy of this cell line in both solid and hematologic malignancies (105, 106). One further advantage of NK-92 is the ease of transfection with non-viral vectors allowing them to express IL-2 (required for their proliferation), thus representing a powerful "off-the-shelf" cell therapeutic (107). Additionally and inspired by the remarkable responses obtained by CAR-T cells and the early results in primary NK cells, NK-92 can be very easily transfected with a gene that expresses a tumor-CAR (108). The first preclinical tests in NSG mice have shed very optimistic results in leukemia models after CD19- or CD20-specific NK-92 infusions as well as in patient-derived glioblastoma with EGFR-specific NK-92 (109, 110). Still, these cellular therapies retain safety concerns including on-target/off-tumor effects and unregulated cytotoxicity. As such, suicide genes (including herpes-simplex-thymidine-kinase and inducible caspase-9) have been integrated into these cell products thus allowing their subsequent selective destruction (111, 112).

The latest of the NK cell therapeutic strategies was developed by Vallera and colleagues with the bi- or tri-specific killer cell engagers, BiKEs and TriKEs that are small molecules containing two or three single chain variable fragments from antibodies of different specificities (113). These are generated to bind CD16 on NK cells and one or two tumor antigens such as CD19 and CD20 (B-cell non-Hodgkin's lymphoma) (114), CD33 or CD33 and CD123 (AML) (115), CD30 (Hodgkin's lymphoma) (116), EGFR or EpCAM (EGFR/EpCAM overexpressing carcinomas) (117, 118), and many others. The initial preclinical evaluation in humanized mice proved very promising translational potential with results exceeding those of monoclonal antibodies, like in the case of CD16-CD19-CD20 TriKE versus rituximab, and also proved efficient for overpassing HLA-mediated inhibition in refractory AML blasts.

IL-15 is the master cytokine necessary for NK cell differentiation and survival and it is currently used in clinical trials alone or as an adjuvant for certain types of metastatic solid tumors to promote *in vivo* cell expansion and NK cell function (63, 119). Taking advantage of this, novel TriKE structures have been developed that use a human IL-15 as a modified cross-linker between the anti-CD16 and the antitumor antigen in order to promote *in vivo* NK cell proliferation. Assessment of the activity of a CD33 specific TriKE in an AML NSG model of NK cell adoptive transfer has shown *in vivo* persistence, high cytotoxic activity, and no toxicity to the construct (120). Clinical development is currently under progress and will probably obtain FDA approval in the upcoming months to be tested in patients.

MODELING VIRALLY INDUCED HUMAN TUMORS USING HIS MICE

While NK cell first identification was based on its antitumor activities, it is also a critical innate effector against pathogen invasions particularly viral infections. Human NK cells have been proven essential for the immune response against members of the herpesvirus, poxvirus, and papillomavirus families, as demonstrated by the predisposition of NK deficient individuals to suffer from these virus infections (121, 122). Remarkably, in one fifth of human cancers viral infection and oncogenesis are intimately linked. Viruses act on carcinogenesis either by directly promoting the initiation of the disease or by interacting in the immune response and/or immune evasion (123). Particularly, Epstein–Barr virus (EBV), hepatitis B virus (HBV), hepatitis C virus (HCV), human papillomavirus, human T-cell lymphotropic virus, Kaposi's sarcoma herpesvirus, and Merkel cell polyomavirus account for the majority of tumor cases linked to viral infection. Humanized mice offer a platform to access the molecular mechanisms behind that causal role and the receptor–ligand interactions occurring at the interface "NK cell-infected cell" that could eventually have therapeutic value. The engrafted human cells occupy relevant physiological sites, where they proliferate and function, and eventually interact with oncogenic viruses that spread and replicate to other cells or organs thus recapitulating the physiological human infection. Additionally, their unique susceptibility to infection by virus with exclusive human tropism and the possibility to manipulate the timing and dose of the infection render them indispensable for better understanding the virus–tumor interplay and disease progression as well as for developing therapeutic approaches.

Epstein–Barr virus is the most common human tumor virus worldwide (more than 200,000 associated malignancies every year) and is also the cause of infectious mononucleosis. It has been extensively studied in humanized mice modeling the different protein expression patterns of the virus that lead to latent infection 0, I, II, and III as well as low level lytic replication, although only latency III has been unequivocally demonstrated (124–129). Several studies have reported specific adaptive cellular and humoral immune response to EBV in humanized mice (128, 130, 131). Furthermore, transformation of B-cell *in vivo* has been also reported and this model has disclosed one of the viral genes (EBNA3B) responsible for tumor formation (125). Preclinical studies in HIS mice have been pivotal for the development of the therapeutic vaccines that are now undergoing clinical trials, including the EBV gp350 neutralizing antibody and infusion of EBV-specific T cells (132). The involvement of NK cells in EBV infection and disease progression was demonstrated by depletion of NK cells from EBV-infected NSG mice resulting in higher EBV DNA load in the spleen, exaggerated $CD8^+$ T-cell responses to the virus and an increased risk of EBV-induced lymphoproliferation (65). Current investigations try to deepen our understanding of the NK cell-mediated control of primary EBV infection in HIS mice and will likely provide insight on the NK cell subset responsible for that viral control (133).

About 80% of hepatocellular carcinomas (HCC) are due to HBV or HCV infections. There are more than 250,000 new cases of HCC and an estimated half a million deaths due to this disease annually (123, 134). We and others have developed mouse models harboring both the immune system and human hepatocytes, allowing the natural course of acute infection and also chronic hepatitis, characterized by advanced liver disease and hepatocellular carcinoma genesis (17, 135, 136). In addition to the immunodeficiency, these mice have liver defects that allow engraftment and expansion of transplanted human hepatocytes. Several immune system–liver humanized models have been developed, including BRGS-uPA (BALB/*c* *Rag2*$^{-/-}$*Il2rg*$^{-/-}$*Sirpa*^{NOD} *uPA*$^{tg/tg}$) (17), uPA-NOG (uPA-NOD *Prkdc*^{SCID}*Il2rg*$^{-/-}$) (135), and FRGN (*Fah*$^{-/-}$*Rag2*$^{-/-}$*Il2rg*$^{-/-}$ NOD) (136). These doubly humanized mice show high level of human liver chimerism and immune engraftment in primary and secondary lymphoid organs with reconstitution of myeloid and lymphoid populations at levels similar to the single HIS models. In BRGS-uPA mice, NK cells are present in spleen and liver in numbers even higher than BRGS mice and display the same NK receptor expression profile (unpublished data). Infection with HBV and HCV has been achieved in these mice and human immune responses have been detected as well as associated liver diseases that resemble the human pathology. Furthermore, both mimic the clinical response upon treatment with anti-HBs neutralizing antibodies and IFNα-2an, respectively, and prevented the leukocyte infiltration and liver fibrosis (137, 138). Any in depth analysis of the NK cell response against the virus or the role in tumorigenesis has been so far performed in these mice, other than the detection of $CD56^+$ cells, to our knowledge.

IMPROVING THE NK CELL COMPARTMENT IN HIS MICE

Given the central role of NK cells in immune responses in infection, malignancy and inflammation and the great therapeutic potential they hold, it is necessary to optimize the available models for understanding their biology and preclinically evaluate new therapies.

In previous sections, we discussed about two types of HIS mice for the study of NK cell biology, those in which the human immune cells develop *in vivo* from injected hematopoietic precursors and a second category that adoptively receive mature NK cells freshly isolated or derived from an *ex vivo* expansion or activation process, a cell line or an ES or iPS cell. The later have fewer requirements in terms of niche, cell–cell interactions, and soluble growth factors that are needed for NK cell development, and instead require cytokines for their survival and homeostatic proliferation. Common cytokine receptor γ_c cytokines (IL-2, IL-4, IL-7, IL-9, IL-15, and IL-21) play critical roles. In particular, IL-15 is responsible for NK cell maintenance and homeostatic proliferation through IL-15Rα presentation (139, 140), while IL-2 effect *in vivo* is oriented to the activation and induction of cytotoxicity through the regulation of the peripheral NK subsets. These humanized mice serve as platforms to understand the mechanisms underlying NK survival and function and provide preclinical information for the design of new therapeutics. Furthermore, they give valuable information about the cell migration capacity and synergistic effects with other cell types or immunomodulators.

As mentioned earlier, several immunodeficient hosts (NSG, NOG, BRG, BRGS) support multilineage development of human immune cells, including low levels of NK cells. In the BRGS model (16, 141), NK cells expressed CD56 and NKp46 as well as some level of CD16 and were able to degranulate moderately after stimulation with a cancer cell line. However, in both BRGS and NSG mice, NK cell displayed defects in maturation, functionality, and heterogeneity in comparison with the human counterparts due to a deficient cytokine signaling (142). The absence of human appropriate MHC class I expression on hematopoietic or stromal cells may result in the failure to "educate" or "license" developing NK cells in HIS mice. This could explain the abundance of immature NK cells ($CD56^{bright}CD16^-KIR^-$) and their functional defects. In line with this idea, recent publications showed improved NK cell licensing in a HIS models expressing diverse educating HLA alleles (143, 144). This approach may allow better definition of the mechanisms underlying human NK cell education *in vivo*.

Based on their cytokine requirements, IL-15 has been exogenously administered either alone or as a complex with IL-15Rα resulting in an extensive NK cell proliferation and accumulation of CD16$^+$KIR$^+$ NK cells. Also, NK cell differentiation progressed from CD56$^+$ to CD56^{low}CD16$^+$, and finally to CD56^{low}CD16$^+$KIR$^+$ mimicking the human model (63). On the other hand, the constitutive high expression of the high-affinity heterotrimeric IL-2 receptor complex in CD56^{bright}CD16$^{\pm}$ NK-cell subset and the effect of IL-2 in NK expansion and activation prompted the development of and IL-2 transgenic NOG mouse strain (145). When IL-2^{Tg}NOG mice were engrafted with human HSC, CD56$^+$ massively developed with a highly active phenotype including IFN-γ production and cytotoxicity against tumor cells. Interestingly, treatment of these mice with a therapeutic humanized anti-CCR4 Ab (mogamulizumab) suppressed the growth of a CCR4$^+$ lymphoma, suggesting that the human NK cells in the mice exerted active Ab-dependent cellular cytotoxicity *in vivo*. These cells expressed various NK receptors, including NKp30, NKp44, NKp46, NKG2D, and CD94, as well as a diverse set of killer cell Ig-like receptor molecules at levels comparable to normal human NK cells from the peripheral blood (62). Nevertheless, there are several limitations in this model due to the supra-physiological levels of IL-2 and, therefore, the high activation status of the NK cells.

It is well known that NK homeostasis and function are regulated by the interaction with other immune cells, particularly macrophages, DCs, and T cells. In addition, soluble factors released by those cells, like NKG2D ligands, IL-2, IL-12, or IL-15, signal on NK cells leading to proliferation and activation. Based on these crosstalk events, others and we have developed humanized mice that through the enhancement of the myeloid compartment, NK cell development results improved. As it happens for other lineages, human myelopoiesis is driven by soluble factors normally present in the BM niche and periphery, which are from murine origin in HIS mice. Some of these mouse cytokines cross-react to some extent with the human cells but others, the species-specific cytokines, do not. In order to circumvent this deficiency, human cytokines have been administered to HIS mice either as recombinant proteins (63) by cytokine-encoding plasmids (146) or by insertion of the cytokines either as transgenes in the mouse genome or by knocking in the human gene to replace the mouse counterpart (147, 148). As mentioned before, transgenic models provoke supra-physiological levels of the cytokine in the periphery and in the case of pro-myeloid factors, such as TPO, IL-3, GM-CSF, or M-CSF, also lead to the exacerbated mobilization and HSC exhaustion limiting the utility of the system. Swapping mouse coding exons for M-CSF, IL-3/GM-CSF, TPO, and SIRPα with their human counterparts allowed for the creation of the MISTRG strain (149). This host expresses these human cytokines under control of mouse regulatory elements and show superior human myeloid cell engraftment. Subsequently, MISTRG HIS mice showed an increased number of functional NK cells, including higher expression of KIR, CD94, and CD161 receptors (149). Nevertheless, cellular and humoral immune responses in MISTRG HIS mice are poor and these mice develop severe anemia.

The transpresentation of IL-15 occurs mainly through the IL-15Rα expressed by DCs, so efforts have been made to increase specifically this cell population in order to increase the NK pool avoiding the overdevelopment of other myeloid subsets. In our laboratory, Flt3-deficient BRG mice (BRGF) were created and after reconstitution with human HSC, human Flt3L was administered to the mice. The result was a specific increase of all the DC subsets and the promotion of NK cell hematopoiesis, with enhanced CD94, CD16, and KIR receptor expression. The combination of this system with the expression of the *Sirpa*^{NOD} protein in the BRGSF model has led to further augmentation of NK cell numbers and also an enhanced functional competency as demonstrated by their degranulation capacity and cytokine production activity (unpublished data). This HIS model provides a unique platform to study NK cell development, crosstalk mechanisms with other immune cells, and the preclinical assessment of new immunotherapies targeting innate cells.

The combination of the protocols detailed in the previous sections for modeling human cancer or infection with the abovementioned strategies to boost the NK cells in HIS mice will raise the potential to understand how NK cell interact with malignant or infected cells. Moreover, HSC-HIS mice offer the possibility to study the tissue specific interactions, the reservoirs, the migration patterns, and the crosstalks within the immune compartment that may be important to develop combinatorial therapies that avoid metastasis, tumor relapse, and "relocation" of the viruses.

CONCLUDING REMARKS AND FUTURE PERSPECTIVES

Therapies designed to induce or potentiate the immune response against tumors are an appealing strategy to control tumor growth and have been the object of intense research since their discovery in the 1970s. Despite representing the most promising cancer treatment since the emergence of chemotherapy, several cases of side effects or disappointing clinical results have downshifted the development of new immunotherapies. The better understanding of the tumor heterogeneity, the mechanisms of the immune response, and the interaction with the tumor microenvironment is a required step for the development of safe and effective therapies. Humanized mice have the potential to reproduce the HIS, the tumor growth and immune evasion, and the response to treatments targeting immune effector cells or immunomodulators. One of the most challenging aspects of tumor research has been to understand the variability within he same type of cancer among individuals and, therefore, the disparate responses and outcomes after therapy. Current efforts are being made to overcome these limitations by creating truly personalized HIS-PDX mouse models in which both the immune system and the tumor are derived form the same individual. These models will provide an invaluable bridge between immunotherapy discovery and the clinic, increasing the success rate of new therapies in human trials and improving the chances to beat cancer.

Natural killer cells have been for long time considered the only innate effectors of the lymphoid system but nowadays we appreciate that they belong to a larger family, the "innate lymphoid cells" (ILCs) (150). These recently described populations lack cytotoxic capacity but instead, they exert very potent cytokine production.

In recent years, there has been a rapid advance in our understanding of their development, phenotypic, and functional diversity, which has been nicely reviewed elsewhere (151–155). ILCs come in three groups mirroring the cytokine and transcriptional profile of CD4+ helper T cells (Th1, Th2, and Th17/22). Given the myriad of cytokines they produce, ILCs have been involved in the early orchestration of immune responses against a number of pathogens, in tumor immunosurveillance and in inflammatory diseases. Recent works in mice have proven the antitumor effects of ILC1s whereas ILC3s have been found to exert both beneficial and tumor-promoting effects depending on the circumstances (156–158). The multifaceted functions of ILCs suggest new alternatives for immunotherapeutic approaches against tumors that need to be explored in *in vivo* humanized models. The improvement of human helper ILCs in HIS mice could open new avenues for harnessing innate immunity to treat cancer and inflammatory diseases.

AUTHOR CONTRIBUTIONS

Both authors have made substantial, direct, and intellectual contribution to the work and approved it for publication.

REFERENCES

1. Bernardi R, Grisendi S, Pandolfi PP. Modelling haematopoietic malignancies in the mouse and therapeutical implications. *Oncogene* (2002) 21:3445–58. doi:10.1038/sj.onc.1205313
2. Springer MS, Murphy WJ. Mammalian evolution and biomedicine: new views from phylogeny. *Biol Rev Camb Philos Soc* (2007) 82:375–92. doi:10.1111/j.1469-185X.2007.00016.x
3. Yue F, Cheng Y, Breschi A, Vierstra J, Wu W, Ryba T, et al. A comparative encyclopedia of DNA elements in the mouse genome. *Nature* (2014) 515:355–64. doi:10.1038/nature13992
4. Mak IW, Evaniew N, Ghert M. Lost in translation: animal models and clinical trials in cancer treatment. *Am J Transl Res* (2014) 6(2):114–8.
5. Rygaard J, Povsen CO. Heterotransplantation of a human malignant tumour to "nude" mice. *APMIS* (2007) 115:604–6. doi:10.1111/j.1600-0463.2007.apm_689a.x
6. Mosier DE, Gulizia RJ, Baird SM, Wilson DB. Transfer of a functional human immune system to mice with severe combined immunodeficiency. *Nature* (1988) 335:256–9. doi:10.1038/335256a0
7. Lapidot T, Pflumio F, Doedens M, Murdoch B, Williams DE, Dick JE. Cytokine stimulation of multilineage hematopoiesis from immature human cells engrafted in SCID mice. *Science* (1992) 255:1137–41. doi:10.1126/science.1372131
8. Mombaerts P, Iacomini J, Johnson RS, Herrup K, Tonegawa S, Papaioannou VE. RAG-1-deficient mice have no mature B and T lymphocytes. *Cell* (1992) 68:869–77. doi:10.1016/0092-8674(92)90030-G
9. Shinkai Y, Rathbun G, Lam KP, Oltz EM, Stewart V, Mendelsohn M, et al. RAG-2-deficient mice lack mature lymphocytes owing to inability to initiate V(D)J rearrangement. *Cell* (1992) 68:855–67. doi:10.1016/0092-8674(92)90029-C
10. Greiner DL, Shultz LD, Yates J, Appel MC, Perdrizet G, Hesselton RM, et al. Improved engraftment of human spleen cells in NOD/LtSz-scid/scid mice as compared with C.B-17-scid/scid mice. *Am J Pathol* (1995) 146:888–902.
11. Hesselton RM, Greiner DL, Mordes JP, Rajan TV, Sullivan JL, Shultz LD. High levels of human peripheral blood mononuclear cell engraftment and enhanced susceptibility to human immunodeficiency virus type 1 infection in NOD/LtSz-scid/scid mice. *J Infect Dis* (1995) 172:974–82. doi:10.1093/infdis/172.4.974
12. Pflumio F, Izac B, Katz A, Shultz LD, Vainchenker W, Coulombel L. Phenotype and function of human hematopoietic cells engrafting immune-deficient CB17-severe combined immunodeficiency mice and nonobese diabetic-severe combined immunodeficiency mice after transplantation of human cord blood mononuclear cells. *Blood* (1996) 88:3731–40.
13. Takenaka K, Prasolava TK, Wang JCY, Mortin-Toth SM, Khalouei S, Gan OI, et al. Polymorphism in Sirpa modulates engraftment of human hematopoietic stem cells. *Nat Immunol* (2007) 8:1313–23. doi:10.1038/ni1527
14. Yamauchi T, Takenaka K, Urata S, Shima T, Kikushige Y, Tokuyama T, et al. Polymorphic *Sirpa* is the genetic determinant for NOD-based mouse lines to achieve efficient human cell engraftment. *Blood* (2013) 121(8):1316–25. doi:10.1182/blood-2012-06-440354
15. Legrand N, Huntington ND, Nagasawa M, Bakker AQ, Schotte R, Strick-Marchand H, et al. Functional CD47/ signal regulatory protein alpha (SIRP α) interaction is required for optimal human T- and natural killer- (NK) cell homeostasis in vivo. *Proc Natl Acad Sci U S A* (2011) 108:13224–9. doi:10.1073/pnas.1101398108
16. Strowig T, Rongvaux A, Rathinam C, Takizawa H, Borsotti C. Transgenic expression of human signal regulatory protein of human hematopoietic cells in humanized mice. *Proc Natl Acad Sci U S A* (2011) 108(32):13218–23. doi:10.1073/pnas.1109769108
17. Strick-Marchand H, Dusséaux M, Darche S, Huntington ND, Legrand N, Masse-Ranson G, et al. A novel mouse model for stable engraftment of a human immune system and human hepatocytes. *PLoS One* (2015) 10(3):e0119820. doi:10.1371/journal.pone.0119820
18. Di Santo JP, Vosshenrich CA. Bone marrow versus thymic pathways of natural killer cell development. *Immunol Rev* (2006) 214:35–46. doi:10.1111/j.1600-065X.2006.00461.x
19. Cao X, Shores EW, Hu-Li J, Anver MR, Kelsall BL, Russell SM, et al. Defective lymphoid development in mice lacking expression of the common cytokine receptor gamma chain. *Immunity* (1995) 2:223–38. doi:10.1016/1074-7613(95)90047-0
20. Ito M, Hiramatsu H, Kobayashi K, Suzue K, Kawahata M, Hioki K, et al. NOD/SCID/gamma(c)(null) mouse: an excellent recipient mouse model for engraftment of human cells. *Blood* (2002) 100:3175–82. doi:10.1182/blood-2001-12-0207
21. Traggiai E, Chicha L, Mazzucchelli L, Bronz L, Piffaretti J-C, Lanzavecchia A, et al. Development of a human adaptive immune system in cord blood cell-transplanted mice. *Science* (2004) 304:104–7. doi:10.1126/science.1093933
22. González-Clemente JM, Mauricio D, Richart C, Broch M, Caixàs A, Megia A, et al. Diabetic neuropathy is associated with activation of the TNF-alpha system in subjects with type 1 diabetes mellitus. *Clin Endocrinol (Oxf)* (2005) 63:525–9. doi:10.1111/j.1365-2265.2005.02376.x
23. Gimeno R, Weijer K, Voordouw A, Uittenbogaart CH, Legrand N, Alves NL, et al. Monitoring the effect of gene silencing by RNA interference in human CD34+ cells injected into newborn RAG2-/- γc-/- mice: functional inactivation of p53 in developing T cells. *Blood* (2004) 104:3886–93. doi:10.1182/blood-2004-02-0656
24. Li Y, Mention J-J, Court N, Masse-Ranson G, Toubert A, Spits H, et al. A novel Flt3-deficient HIS mouse model with selective enhancement of human DC development. *Eur J Immunol* (2016) 46:1291–9. doi:10.1002/eji.201546132
25. Yu F, Yao H, Zhu P, Zhang X, Pan Q, Gong C, et al. let-7 regulates self renewal and tumorigenicity of breast cancer cells. *Cell* (2007) 131:1109–23. doi:10.1016/j.cell.2007.10.054
26. Townsend EC, Murakami MA, Christodoulou A, Christie AL, Köster J, DeSouza TA, et al. The public repository of xenografts enables discovery and randomized phase II-like trials in mice. *Cancer Cell* (2016) 29:574–86. doi:10.1016/j.ccell.2016.03.008

27. Gao H, Korn JM, Ferretti S, Monahan JE, Wang Y, Singh M, et al. High-throughput screening using patient-derived tumor xenografts to predict clinical trial drug response. *Nat Med* (2015) 21:1318–25. doi:10.1038/nm.3954
28. Eppert K, Takenaka K, Lechman ER, Waldron L, Nilsson B, van Galen P, et al. Stem cell gene expression programs influence clinical outcome in human leukemia. *Nat Med* (2011) 17:1086–93. doi:10.1038/nm.2415
29. Gal H, Amariglio N, Trakhtenbrot L, Jacob-Hirsh J, Margalit O, Avigdor A, et al. Gene expression profiles of AML derived stem cells; similarity to hematopoietic stem cells. *Leukemia* (2006) 20:2147–54. doi:10.1038/sj.leu.2404401
30. Bonnet D, Dick JE. Human acute myeloid leukemia is organized as a hierarchy that originates from a primitive hematopoietic cell. *Nat Med* (1997) 3:730–7. doi:10.1038/nm0797-730
31. Lapidot T, Sirard C, Vormoor J, Murdoch B, Hoang T, Caceres-Cortes J, et al. A cell initiating human acute myeloid leukaemia after transplantation into SCID mice. *Nature* (1994) 367:645–8. doi:10.1038/367645a0
32. Lohberger B, Rinner B, Stuendl N, Absenger M, Liegl-Atzwanger B, Walzer SM, et al. Aldehyde dehydrogenase 1, a potential marker for cancer stem cells in human sarcoma. *PLoS One* (2012) 7:e43664. doi:10.1371/journal.pone.0043664
33. Rasheed Z, Wang Q, Matsui W. Isolation of stem cells from human pancreatic cancer xenografts. *J Vis Exp* (2010) 43:2169. doi:10.3791/2169
34. Billerbeck E, Barry WT, Mu K, Dorner M, Rice CM, Ploss A. Development of human CD4+FoxP3+ regulatory T cells in human stem cell factor-, granulocyte-macrophage colony-stimulating factor-, and interleukin-3-expressing NOD-SCID IL2Rγ(null) humanized mice. *Blood* (2011) 117:3076–86. doi:10.1182/blood-2010-08-301507
35. Goyama S, Wunderlich M, Mulloy JC. Xenograft models for normal and malignant stem cells. *Blood* (2015) 125(17):2630–40. doi:10.1182/blood-2014-11-570218
36. Pallasch CP, Leskov I, Braun CJ, Vorholt D, Drake A, Soto-Feliciano YM, et al. Sensitizing protective tumor microenvironments to antibody-mediated therapy. *Cell* (2014) 156:590–602. doi:10.1016/j.cell.2013.12.041
37. Veluchamy JP, Lopez-Lastra S, Spanholtz J, Bohme F, Kok N, Heideman DA, et al. In vivo efficacy of umbilical cord blood stem cell derived NK cells in the treatment of metastatic colorectal cancer. *Front Immunol* (2017) 8:87. doi:10.3389/fimmu.2017.00087
38. Seitz G, Pfeiffer M, Fuchs J, Warmann SW, Leuschner I, Vokuhl C, et al. Establishment of a rhabdomyosarcoma xenograft model in human-adapted mice. *Oncol Rep* (2010) 24:1067–72. doi:10.3892/or.2010.1067
39. Barabé F, Kennedy JA, Hope KJ, Dick JE. Modeling the initiation and progression of human acute leukemia in mice. *Science* (2007) 316(5824):600–4. doi:10.1126/science.1139851
40. Suemizu H, Monnai M, Ohnishi Y, Ito M, Tamaoki N, Nakamura M. Identification of a key molecular regulator of liver metastasis in human pancreatic carcinoma using a novel quantitative model of metastasis in NOD/SCID/gammacnull (NOG) mice. *Int J Oncol* (2007) 31(4):741–51. doi:10.3892/ijo.31.4.741
41. Liu D, Song L, Wei J, Courtney AN, Gao X, Marinova E, et al. IL-15 protects NKT cells from inhibition by tumor-associated macrophages and enhances antimetastatic activity. *J Clin Invest* (2012) 122:2221–33. doi:10.1172/JCI59535
42. Xia J, Hu Z, Yoshihara S, Li Y, Jin C-H, Tan S, et al. Modeling human leukemia immunotherapy in humanized mice. *EBioMedicine* (2016) 10:101–8. doi:10.1016/j.ebiom.2016.06.028
43. Hu Z, Xia J, Fan W, Wargo J, Yang Y-G, Hu Z, et al. Human melanoma immunotherapy using tumor antigen-specific T cells generated in humanized mice. *Oncotarget* (2016) 7:6448–59. doi:10.18632/oncotarget.7044
44. Kaur K, Cook J, Park S-H, Topchyan P, Kozlowska A, Ohanian N, et al. Novel strategy to expand super-charged NK cells with significant potential to lyse and differentiate cancer stem cells; differences in NK expansion and function between healthy and cancer patients. *Front Oncol* (2017) 7:9. doi:10.3389/fimmu.2017.00297
45. Aryee K-E, Brehm MA, Shultz LD, Jurczyk A. Modeling immune system-tumor interactions using humanized mice. *J Immunol* (2016) 196:212.12.
46. Ito A, Ishida T, Utsunomiya A, Sato F, Mori F, Yano H, et al. Defucosylated anti-CCR4 monoclonal antibody exerts potent ADCC against primary ATLL cells mediated by autologous human immune cells in NOD/Shi-scid, IL-2Rγnull mice in vivo. *J Immunol* (2009) 183(7):4782–91. doi:10.4049/jimmunol.0900699
47. Xiang Z, Liu Y, Zheng J, Liu M, Lv A, Gao Y, et al. Targeted activation of human Vγ9Vd2-T cells controls Epstein–Barr virus-induced B cell lymphoproliferative disease. *Cancer Cell* (2014) 26:565–76. doi:10.1016/j.ccr.2014.07.026
48. Abate-Daga D, Lagisetty KH, Tran E, Zheng Z, Gattinoni L, Yu Z, et al. A novel chimeric antigen receptor against prostate stem cell antigen mediates tumor destruction in a humanized mouse model of pancreatic cancer. *Hum Gene Ther* (2014) 25:1003–12. doi:10.1089/hum.2013.209
49. Simpson-Abelson MR, Sonnenberg GF, Takita H, Yokota SJ, Conway TF, Kelleher RJ, et al. Long-term engraftment and expansion of tumor-derived memory T cells following the implantation of non-disrupted pieces of human lung tumor into NOD-scid IL2Rgamma(null) mice. *J Immunol* (2008) 180:7009–18. doi:10.4049/jimmunol.180.10.7009
50. Geller MA, Knorr DA, Hermanson DA, Pribyl L, Bendzick L, McCullar V, et al. Intraperitoneal delivery of human natural killer cells for treatment of ovarian cancer in a mouse xenograft model. *Cytotherapy* (2013) 15:1297–306. doi:10.1016/j.jcyt.2013.05.022
51. Lee SJ, Kang WY, Yoon Y, Jin JY, Song HJ, Her JH, et al. Natural killer (NK) cells inhibit systemic metastasis of glioblastoma cells and have therapeutic effects against glioblastomas in the brain. *BMC Cancer* (2015) 15:1011. doi:10.1186/s12885-015-2034-y
52. Feuerer M, Beckhove P, Bai L, Solomayer E-F, Bastert G, Diel IJ, et al. Therapy of human tumors in NOD/SCID mice with patient-derived reactivated memory T cells from bone marrow. *Nat Med* (2001) 7:452–8. doi:10.1038/86523
53. Roth MD, Harui A. Human tumor infiltrating lymphocytes cooperatively regulate prostate tumor growth in a humanized mouse model. *J Immunother cancer* (2015) 3:12. doi:10.1186/s40425-015-0056-2
54. Koehne G, Doubrovin M, Doubrovina E, Zanzonico P, Gallardo HF, Ivanova A, et al. Serial in vivo imaging of the targeted migration of human HSV-TK-transduced antigen-specific lymphocytes. *Nat Biotechnol* (2003) 21:405–13. doi:10.1038/nbt805
55. Cherkassky L, Morello A, Villena-Vargas J, Feng Y, Dimitrov DS, Jones DR, et al. Human CAR T cells with cell-intrinsic PD-1 checkpoint blockade resist tumor-mediated inhibition. *J Clin Invest* (2016) 126:3130–44. doi:10.1172/JCI83092
56. Gründer C, van Dorp S, Hol S, Drent E, Straetemans T, Heijhuurs S, et al. γ9 and d2CDR3 domains regulate functional avidity of T cells harboring γ9d2TCRs. *Blood* (2012) 120(26):5153–62. doi:10.1182/blood-2012-05-432427
57. Orange JS. Natural killer cell deficiency. *J Allergy Clin Immunol* (2013) 132:515–25; quiz 526. doi:10.1016/j.jaci.2013.07.020
58. Kim S, Iizuka K, Aguila HL, Weissman IL, Yokoyama WM. In vivo natural killer cell activities revealed by natural killer cell-deficient mice. *Proc Natl Acad Sci U S A* (2000) 97:2731–6. doi:10.1073/pnas.050588297
59. Colucci F, Caligiuri MA, Di Santo JP. What does it take to make a natural killer? *Nat Rev Immunol* (2003) 3:413–25. doi:10.1038/nri1088
60. Briercheck EL, Freud AG, Caligiuri MA. *Natural Killer Cells: Basic Science and Clinical Application*. Amsterdam, Boston, London: Elsevier/Academic Press (2010). p. 113–22.
61. Huntington ND, Alves NL, Legrand N, Lim A, Strick-Marchand H, Mention J-J, et al. IL-15 transpresentation promotes both human T-cell reconstitution and T-cell-dependent antibody responses in vivo. *Proc Natl Acad Sci U S A* (2011) 108:6217–22. doi:10.1073/pnas.1019167108
62. Katano I, Takahashi T, Ito R, Kamisako T, Mizusawa T, Ka Y, et al. Predominant development of mature and functional human NK cells in a novel human IL-2-producing transgenic NOG mouse. *J Immunol* (2015) 194:3513–25. doi:10.4049/jimmunol.1401323
63. Huntington ND, Legrand N, Alves NL, Jaron B, Weijer K, Plet A, et al. IL-15 trans-presentation promotes human NK cell development and differentiation in vivo. *J Exp Med* (2009) 206:25–34. doi:10.1084/jem.20082013
64. Jost S, Chang J, Rands K, Tomezsko P, Doyle E, Daillere R, et al. Impact of NK cells on control of HIV infection in humanized BLT mice (P6207). *J Immunol* (2013) 190:118.27–118.27.
65. Chijioke O, Müller A, Feederle R, Barros MHM, Krieg C, Emmel V, et al. Human natural killer cells prevent infectious mononucleosis features by targeting lytic Epstein–Barr virus infection. *Cell Rep* (2013) 5:1489–98. doi:10.1016/j.celrep.2013.11.041

66. Wilson EB, El-Jawhari JJ, Neilson AL, Hall GD, Melcher AA, Meade JL, et al. Human tumour immune evasion via TGF-β blocks NK cell activation but not survival allowing therapeutic restoration of anti-tumour activity. *PLoS One* (2011) 6:e22842. doi:10.1371/journal.pone.0022842
67. Intlekofer AM, Thompson CB. At the bench: preclinical rationale for CTLA-4 and PD-1 blockade as cancer immunotherapy. *J Leukoc Biol* (2013) 94:25–39. doi:10.1189/jlb.1212621
68. Ma S-D, Xu X, Jones R, Delecluse H-J, Zumwalde NA, Sharma A, et al. PD-1/CTLA-4 blockade inhibits Epstein–Barr virus-induced lymphoma growth in a cord blood humanized-mouse model. *PLoS Pathog* (2016) 12:e1005642. doi:10.1371/journal.ppat.1005642
69. Vudattu NK, Waldron-Lynch F, Truman LA, Deng S, Preston-Hurlburt P, Torres R, et al. Humanized mice as a model for aberrant responses in human T cell immunotherapy. *J Immunol* (2014) 193:587–96. doi:10.4049/jimmunol.1302455
70. Singh M, Singh P, Vaira D, Amand M, Rahmouni S, Moutschen M. Minocycline attenuates HIV-1 infection and suppresses chronic immune activation in humanized NOD/LtsZ-scidIL-2Rγ(null) mice. *Immunology* (2014) 142:562–72. doi:10.1111/imm.12246
71. Stojanovic A, Fiegler N, Brunner-Weinzierl M, Cerwenka A. CTLA-4 is expressed by activated mouse NK cells and inhibits NK cell IFN-γ production in response to mature dendritic cells. *J Immunol* (2014) 192:4184–91. doi:10.4049/jimmunol.1302091
72. Baumeister SH, Freeman GJ, Dranoff G, Sharpe AH. Coinhibitory pathways in immunotherapy for cancer. *Annu Rev Immunol* (2016) 34:539–73. doi:10.1146/annurev-immunol-032414-112049
73. Huang BY, Zhan YP, Zong WJ, Yu CJ, Li JF, Qu YM, et al. The PD-1/B7-H1 pathway modulates the natural killer cells versus mouse glioma stem cells. *PLoS One* (2015) 10:e0134715. doi:10.1371/journal.pone.0134715
74. Rosenberg SA, Lotze MT, Muul LM, Leitman S, Chang AE, Ettinghausen SE, et al. Observations on the systemic administration of autologous lymphokine-activated killer cells and recombinant interleukin-2 to patients with metastatic cancer. *N Engl J Med* (1985) 313:1485–92. doi:10.1056/NEJM198512053132327
75. Baiocchi RA, Caligiuri MA. Low-dose interleukin 2 prevents the development of Epstein–Barr virus (EBV)-associated lymphoproliferative disease in scid/scid mice reconstituted i.p. with EBV-seropositive human peripheral blood lymphocytes. *Proc Natl Acad Sci U S A* (1994) 91:5577–81. doi:10.1073/pnas.91.12.5577
76. Boiardi A, Silvani A, Ruffini PA, Rivoltini L, Parmiani G, Broggi G, et al. Loco-regional immunotherapy with recombinant interleukin-2 and adherent lymphokine-activated killer cells (A-LAK) in recurrent glioblastoma patients. *Cancer Immunol Immunother* (1994) 39:193–7. doi:10.1007/BF01533386
77. Hayes RL, Koslow M, Hiesiger EM, Hymes KB, Moore EJ, Pierz DM, et al. Improved long term survival after intracavitary interleukin-2 and lymphokine-activated killer cells for adults with recurrent malignant glioma. *Cancer* (1995) 76:840–52. doi:10.1002/1097-0142(19950901)76:5<840::AID-CNCR2820760519>3.0.CO;2-R
78. Keilholz U, Scheibenbogen C, Brado M, Georgi P, Maclachlan D, Brado B, et al. Regional adoptive immunotherapy with interleukin-2 and lymphokine-activated killer (LAK) cells for liver metastases. *Eur J Cancer* (1994) 30A:103–5. doi:10.1016/S0959-8049(05)80028-0
79. Sutlu T, Alici E. Natural killer cell-based immunotherapy in cancer: current insights and future prospects. *J Intern Med* (2009) 266:154–81. doi:10.1111/j.1365-2796.2009.02121.x
80. Ruggeri L, Mancusi A, Burchielli E, Aversa F, Martelli MF, Velardi A. Natural killer cell alloreactivity in allogeneic hematopoietic transplantation. *Curr Opin Oncol* (2007) 19:142–7. doi:10.1097/CCO.0b013e3280148a1a
81. Miller JS, Soignier Y, Panoskaltsis-Mortari A, McNearney SA, Yun GH, Fautsch SK, et al. Successful adoptive transfer and in vivo expansion of human haploidentical NK cells in patients with cancer. *Blood* (2005) 105:3051–7. doi:10.1182/blood-2004-07-2974
82. Curti A, Ruggeri L, Parisi S, Bontadini A, Dan E, Motta MR, et al. Larger size of donor alloreactive NK cell repertoire correlates with better response to NK cell immunotherapy in elderly acute myeloid leukemia patients. *Clin Cancer Res* (2016) 22:1914–21. doi:10.1158/1078-0432.CCR-15-1604
83. Shah NN, Baird K, Delbrook CP, Fleisher TA, Kohler ME, Rampertaap S, et al. Acute GVHD in patients receiving IL-15/4-1BBL activated NK cells following T-cell-depleted stem cell transplantation. *Blood* (2015) 125:784–92. doi:10.1182/blood-2014-07-592881
84. Gras Navarro A, Kmiecik J, Leiss L, Zelkowski M, Engelsen A, Bruserud Ø, et al. NK cells with KIR2DS2 immunogenotype have a functional activation advantage to efficiently kill glioblastoma and prolong animal survival. *J Immunol* (2014) 193:6192–206. doi:10.4049/jimmunol.1400859
85. Ruggeri L, Urbani E, Andre P, Mancusi A, Tosti A, Topini F, et al. Effects of anti-NKG2A antibody administration on leukemia and normal hematopoietic cells. *Haematologica* (2016) 101:626–33. doi:10.3324/haematol.2015.135301
86. Binyamin L, Alpaugh RK, Hughes TL, Lutz CT, Campbell KS, Weiner LM. Blocking NK cell inhibitory self-recognition promotes antibody-dependent cellular cytotoxicity in a model of anti-lymphoma therapy. *J Immunol* (2008) 180:6392–401. doi:10.4049/jimmunol.180.9.6392
87. Nijhof IS, van Bueren JJL, van Kessel B, Andre P, Morel Y, Lokhorst HM, et al. Daratumumab-mediated lysis of primary multiple myeloma cells is enhanced in combination with the human anti-KIR antibody IPH2102 and lenalidomide. *Haematologica* (2015) 100(2):263–8. doi:10.3324/haematol.2014.117531
88. Glienke W, Esser R, Priesner C, Suerth JD, Schambach A, Wels WS, et al. Advantages and applications of CAR-expressing natural killer cells. *Front Pharmacol* (2015) 6:21. doi:10.3389/fphar.2015.00021
89. Du J, Lopez-Verges S, Pitcher BN, Johnson J, Jung S-H, Zhou L, et al. CALGB 150905 (alliance): rituximab broadens the antilymphoma response by activating unlicensed NK cells. *Cancer Immunol Res* (2014) 2:878–89. doi:10.1158/2326-6066.CIR-13-0158
90. Daldrup-Link HE, Meier R, Rudelius M, Piontek G, Piert M, Metz S, et al. In vivo tracking of genetically engineered, anti-HER2/neu directed natural killer cells to HER2/neu positive mammary tumors with magnetic resonance imaging. *Eur Radiol* (2005) 15:4–13. doi:10.1007/s00330-004-2526-7
91. Monteverde M, Milano G, Strola G, Maffi M, Lattanzio L, Vivenza D, et al. The relevance of ADCC for EGFR targeting: a review of the literature and a clinically-applicable method of assessment in patients. *Crit Rev Oncol Hematol* (2015) 95:179–90. doi:10.1016/j.critrevonc.2015.02.014
92. Valent P, Bonnet D, De Maria R, Lapidot T, Copland M, Melo JV, et al. Cancer stem cell definitions and terminology: the devil is in the details. *Nat Rev Cancer* (2012) 12:767–75. doi:10.1038/nrc3368
93. Ames E, Canter RJ, Grossenbacher SK, Mac S, Chen M, Smith RC, et al. NK cells preferentially target tumor cells with a cancer stem cell phenotype. *J Immunol* (2015) 195:4010–9. doi:10.4049/jimmunol.1500447
94. Tallerico R, Todaro M, Di Franco S, Maccalli C, Garofalo C, Sottile R, et al. Human NK cells selective targeting of colon cancer-initiating cells: a role for natural cytotoxicity receptors and MHC class I molecules. *J Immunol* (2013) 190:2381–90. doi:10.4049/jimmunol.1201542
95. Dionne LK, Driver ER, Wang XJ. Head and neck cancer stem cells: from identification to tumor immune network. *J Dent Res* (2015) 94:1524–31. doi:10.1177/0022034515599766
96. Cany J, van der Waart AB, Tordoir M, Franssen GM, Hangalapura BN, de Vries J, et al. Natural killer cells generated from cord blood hematopoietic progenitor cells efficiently target bone marrow-residing human leukemia cells in NOD/SCID/IL2Rgnull mice. *PLoS One* (2013) 8(6):e64384. doi:10.1371/journal.pone.0064384
97. Dolstra H, Roeven MWH, Spanholtz J, Hangalapura B, Tordoir M, Maas F, et al. A phase I study of allogeneic natural Killer cell therapy generated from cord blood hematopoietic stem and progenitor cells in elderly acute myeloid leukemia patients. *Blood* (2015) 126:1357.
98. Veluchamy JP, Heeren AM, Spanholtz J, van Eendenburg JDH, Heideman DAM, Kenter GG, et al. High-efficiency lysis of cervical cancer by allogeneic NK cells derived from umbilical cord progenitors is independent of HLA status. *Cancer Immunol Immunother* (2016) 66(1):51–61. doi:10.1007/s00262-016-1919-1
99. Woll PS, Martin CH, Miller JS, Kaufman DS. Human embryonic stem cell-derived NK cells acquire functional receptors and cytolytic activity. *J Immunol* (2005) 175:5095–103. doi:10.4049/jimmunol.175.8.5095
100. Woll PS, Grzywacz B, Tian X, Marcus RK, Knorr DA, Verneris MR, et al. Human embryonic stem cells differentiate into a homogeneous population of natural killer cells with potent in vivo antitumor activity. *Blood* (2009) 113(24):6094–101. doi:10.1182/blood-2008-06-165225

101. Gong J, Maki G, Klingemann H. Characterization of a human cell line (NK-92) with phenotypical and functional characteristics of activated natural killer cells. *Leukemia* (1994) 8:652–8.
102. Tam YK, Miyagawa B, Ho VC, Klingemann H-G. Immunotherapy of malignant melanoma in a SCID mouse model using the highly cytotoxic natural killer cell line NK-92. *J Hematother* (1999) 8:281–90. doi:10.1089/106161299320316
103. Yan Y, Steinherz P, Klingemann HG, Dennig D, Childs BH, McGuirk J, et al. Antileukemia activity of a natural killer cell line against human leukemias. *Clin Cancer Res* (1998) 4:2859–68.
104. Swift BE, Williams BA, Kosaka Y, Wang X-H, Medin JA, Viswanathan S, et al. Natural killer cell lines preferentially kill clonogenic multiple myeloma cells and decrease myeloma engraftment in a bioluminescent xenograft mouse model. *Haematologica* (2012) 97:1020–8. doi:10.3324/haematol.2011.054254
105. Arai S, Meagher R, Swearingen M, Myint H, Rich E, Martinson J, et al. Infusion of the allogeneic cell line NK-92 in patients with advanced renal cell cancer or melanoma: a phase I trial. *Cytotherapy* (2008) 10:625–32. doi:10.1080/14653240802301872
106. Tonn T, Schwabe D, Klingemann HG, Becker S, Esser R, Koehl U, et al. Treatment of patients with advanced cancer with the natural killer cell line NK-92. *Cytotherapy* (2013) 15:1563–70. doi:10.1016/j.jcyt.2013.06.017
107. Boissel L, Campbell K, Toneguzzo F, Nichols K, Klingemann H. NK-92: an "off the shelf" target-specific cytotoxic cell therapeutic. *Cytotherapy* (2015) 17(6):S19. doi:10.1016/j.jcyt.2015.03.367
108. Klingemann H. Are natural killer cells superior CAR drivers? *Oncoimmunology* (2014) 3:e28147. doi:10.4161/onci.28147
109. Han J, Chu J, Keung Chan W, Zhang J, Wang Y, Cohen JB, et al. CAR-engineered NK cells targeting wild-type EGFR and EGFRvIII enhance killing of glioblastoma and patient-derived glioblastoma stem cells. *Sci Rep* (2015) 5:11483. doi:10.1038/srep11483
110. Boissel L, Betancur-Boissel M, Lu W, Krause DS, Van Etten RA, Wels WS, et al. Retargeting NK-92 cells by means of CD19- and CD20-specific chimeric antigen receptors compares favorably with antibody-dependent cellular cytotoxicity. *Oncoimmunology* (2013) 2:e26527. doi:10.4161/onci.26527
111. Jones BS, Lamb LS, Goldman F, Di Stasi A. Improving the safety of cell therapy products by suicide gene transfer. *Front Pharmacol* (2014) 5:254. doi:10.3389/fphar.2014.00254
112. Zhou X, Di Stasi A, Brenner MK. iCaspase 9 suicide gene system. *Methods Mol Biol* (2015) 1317:87–105. doi:10.1007/978-1-4939-2727-2_6
113. Miller JS. Therapeutic applications: natural killer cells in the clinic. *Hematol Am Soc Hematol Educ Program* (2013) 2013:247–53. doi:10.1182/asheducation-2013.1.247
114. Kipriyanov SM, Cochlovius B, Schäfer HJ, Moldenhauer G, Bähre A, Le Gall F, et al. Synergistic antitumor effect of bispecific CD19 x CD3 and CD19 x CD16 diabodies in a preclinical model of non-Hodgkin's lymphoma. *J Immunol* (2002) 169:137–44. doi:10.4049/jimmunol.169.1.137
115. Pizzitola I, Anjos-Afonso F, Rouault-Pierre K, Lassailly F, Tettamanti S, Spinelli O, et al. Chimeric antigen receptors against CD33/CD123 antigens efficiently target primary acute myeloid leukemia cells in vivo. *Leukemia* (2014) 28:1–10. doi:10.1038/leu.2014.62
116. Hartmann F, Renner C, Jung W, Deisting C, Juwana M, Eichentopf B, et al. Treatment of refractory Hodgkin's disease with an anti-CD16/CD30 bispecific antibody. *Blood* (1997) 89:2042–7.
117. Elsässer D, Stadick H, Stark S, Van de Winkel JG, Gramatzki M, Schrott KM, et al. Preclinical studies combining bispecific antibodies with cytokine-stimulated effector cells for immunotherapy of renal cell carcinoma. *Anticancer Res* (1999) 19:1525–8.
118. Vallera DA, Zhang B, Gleason MK, Oh S, Weiner LM, Kaufman DS, et al. Heterodimeric bispecific single-chain variable-fragment antibodies against EpCAM and CD16 induce effective antibody-dependent cellular cytotoxicity against human carcinoma cells. *Cancer Biother Radiopharm* (2013) 28:274–82. doi:10.1089/cbr.2012.1329
119. Carson WE, Giri JG, Lindemann MJ, Linett ML, Ahdieh M, Paxton R, et al. Interleukin (IL)-15 is a novel cytokine that activates human natural-killer cells via component of the IL-2 receptor. *J Exp Med* (1994) 180:1395–403. doi:10.1084/jem.180.4.1395
120. Miller JS, Felice M, McElmurry R, McCullar V, Zhou X, Tolar J, et al. Trispecific killer engagers (TriKEs) that contain IL-15 to make NK cells antigen specific and to sustain their persistence and expansion. *Blood* (2015) 126:232.
121. Biron CA, Byron KS, Sullivan JL. Severe herpesvirus infections in an adolescent without natural killer cells. *N Engl J Med* (1989) 320:1731–5. doi:10.1056/NEJM198906293202605
122. Orange JS. Human natural killer cell deficiencies and susceptibility to infection. *Microbes Infect* (2002) 4:1545–58. doi:10.1016/S1286-4579(02)00038-2
123. Venook AP, Papandreou C, Furuse J, de Guevara LL. The incidence and epidemiology of hepatocellular carcinoma: a global and regional perspective. *Oncologist* (2010) 15(Suppl 4):5–13. doi:10.1634/theoncologist.2010-S4-05
124. Islas-Ohlmayer M, Padgett-Thomas A, Domiati-Saad R, Melkus MW, Cravens PD, Martin Mdel P, et al. Experimental infection of NOD/SCID mice reconstituted with human CD34+ cells with Epstein–Barr virus. *J Virol* (2004) 78:13891–900. doi:10.1128/JVI.78.24.13891-13900.2004
125. White RE, Rämer PC, Naresh KN, Meixlsperger S, Pinaud L, Rooney C, et al. EBNA3B-deficient EBV promotes B cell lymphomagenesis in humanized mice and is found in human tumors. *J Clin Invest* (2012) 122(4):1487–502. doi:10.1172/JCI58092
126. Ma S-D, Hegde S, Young KH, Sullivan R, Rajesh D, Zhou Y, et al. A new model of Epstein–Barr virus infection reveals an important role for early lytic viral protein expression in the development of lymphomas. *J Virol* (2011) 85:165–77. doi:10.1128/JVI.01512-10
127. Ma S-D, Yu X, Mertz JE, Gumperz JE, Reinheim E, Zhou Y, et al. An Epstein–Barr Virus (EBV) mutant with enhanced BZLF1 expression causes lymphomas with abortive lytic EBV infection in a humanized mouse model. *J Virol* (2012) 86:7976–87. doi:10.1128/JVI.00770-12
128. Strowig T, Gurer C, Ploss A, Liu Y-F, Arrey F, Sashihara J, et al. Priming of protective T cell responses against virus-induced tumors in mice with human immune system components. *J Exp Med* (2009) 206:1423–34. doi:10.1084/jem.20081720
129. Kutok JL, Wang F. Spectrum of Epstein–Barr virus-associated diseases. *Annu Rev Pathol* (2006) 1:375–404. doi:10.1146/annurev.pathol.1.110304.100209
130. Yajima M, Imadome K-I, Nakagawa A, Watanabe S, Terashima K, Nakamura H, et al. A new humanized mouse model of Epstein–Barr virus infection that reproduces persistent infection, lymphoproliferative disorder, and cell-mediated and humoral immune responses. *J Infect Dis* (2008) 198:673–82. doi:10.1086/590502
131. Shultz LD, Saito Y, Najima Y, Tanaka S, Ochi T, Tomizawa M, et al. Generation of functional human T-cell subsets with HLA-restricted immune responses in HLA class I expressing NOD/SCID/IL2r null humanized mice. *Proc Natl Acad Sci U S A* (2010) 107:13022–7. doi:10.1073/pnas.1000475107
132. Cohen JI. Epstein–barr virus vaccines. *Clin Transl Immunol* (2015) 4:e32. doi:10.1038/cti.2015.4
133. Fujiwara S, Imadome K-I, Takei M. Modeling EBV infection and pathogenesis in new-generation humanized mice. *Exp Mol Med* (2015) 47:e135. doi:10.1038/emm.2014.88
134. El-Serag HB, Rudolph KL. Hepatocellular carcinoma: epidemiology and molecular carcinogenesis. *Gastroenterology* (2007) 132:2557–76. doi:10.1053/j.gastro.2007.04.061
135. Gutti TL, Knibbe JS, Makarov E, Zhang J, Yannam GR, Gorantla S, et al. Human hepatocytes and hematolymphoid dual reconstitution in treosulfan-conditioned uPA-NOG mice. *Am J Pathol* (2014) 184:101–9. doi:10.1016/j.ajpath.2013.09.008
136. Wilson EM, Bial J, Tarlow B, Bial G, Jensen B, Greiner DL, et al. Extensive double humanization of both liver and hematopoiesis in FRGN mice. *Stem Cell Res* (2014) 13:404–12. doi:10.1016/j.scr.2014.08.006
137. Bility MT, Cheng L, Zhang Z, Luan Y, Li F, Chi L, et al. Hepatitis B virus infection and immunopathogenesis in a humanized mouse model: induction of human-specific liver fibrosis and M2-like macrophages. *PLoS Pathog* (2014) 10:e1004032. doi:10.1371/journal.ppat.1004032
138. Keng CT, Sze CW, Zheng D, Zheng Z, Yong KSM, Tan SQ, et al. Characterisation of liver pathogenesis, human immune responses and drug testing in a humanised mouse model of HCV infection. *Gut* (2016) 65:1744–53. doi:10.1136/gutjnl-2014-307856
139. Prlic M, Blazar BR, Farrar MA, Jameson SC. In vivo survival and homeostatic proliferation of natural killer cells. *J Exp Med* (2003) 197:967–76. doi:10.1084/jem.20021847
140. Marçais A, Viel S, Grau M, Henry T, Marvel J, Walzer T. Regulation of mouse NK cell development and function by cytokines. *Front Immunol* (2013) 4:450. doi:10.3389/fimmu.2013.00450

141. Legrand N, Huntington ND, Nagasawa M, Bakker AQ, Schotte R, Strick-Marchand H, et al. Functional CD47/signal regulatory protein alpha (SIRP(alpha)) interaction is required for optimal human T- and natural killer- (NK) cell homeostasis in vivo. *Proc Natl Acad Sci U S A* (2011) 108:13224–9. doi:10.1073/pnas.1101398108
142. Strowig T, Chijioke O, Carrega P, Arrey F, Meixlsperger S, Rämer PC, et al. Human NK cells of mice with reconstituted human immune system components require preactivation to acquire functional competence Human NK cells of mice with reconstituted human immune system components require preactivation to acquire functional competence. *Blood* (2010) 116:4158–67. doi:10.1182/blood-2010-02-270678
143. Landtwing V, Raykova A, Pezzino G, Béziat V, Marcenaro E, Graf C, et al. Cognate HLA absence in *trans* diminishes human NK cell education. *J Clin Invest* (2016) 126(10):3772–82. doi:10.1172/JCI86923
144. Boudreau JE, Liu XR, Zhao Z, Zhang A, Shultz LD, Greiner DL, et al. Cell-extrinsic MHC Class I molecule engagement augments human NK cell education programmed by cell-intrinsic MHC Class I. *Immunity* (2016) 45:280–91. doi:10.1016/j.immuni.2016.07.005
145. Caligiuri MA. Human natural killer cells. *Blood* (2008) 112:461–9. doi:10.1182/blood-2007-09-077438
146. Li Y, Chen Q, Zheng D, Yin L, Chionh YH, Wong LH, et al. Induction of functional human macrophages from bone marrow promonocytes by M-CSF in humanized mice. *J Immunol* (2013) 191:3192–9. doi:10.4049/jimmunol.1300742
147. Willinger T, Rongvaux A, Strowig T, Manz MG, Flavell RA. Improving human hemato-lymphoid-system mice by cytokine knock-in gene replacement. *Trends Immunol* (2011) 32:321–7. doi:10.1016/j.it.2011.04.005
148. Ito R, Takahashi T, Katano I, Kawai K, Kamisako T, Ogura T, et al. Establishment of a human allergy model using human IL-3/GM-CSF-transgenic NOG mice. *J Immunol* (2013) 191:2890–9. doi:10.4049/jimmunol.1203543
149. Rongvaux A, Willinger T, Martinek J, Strowig T, Gearty SV, Teichmann LL, et al. Development and function of human innate immune cells in a humanized mouse model. *Nat Biotechnol* (2014) 32:364–72. doi:10.1038/nbt.2858
150. Spits H, Artis D, Colonna M, Diefenbach A, Di Santo JP, Eberl G, et al. Innate lymphoid cells – a proposal for uniform nomenclature. *Nat Rev Immunol* (2013) 13:145–9. doi:10.1038/nri3365
151. Tait Wojno ED, Artis D. Emerging concepts and future challenges in innate lymphoid cell biology. *J Exp Med* (2016) 213:2229–48. doi:10.1084/jem.20160525
152. Serafini N, Vosshenrich CA, Di Santo JP. Transcriptional regulation of innate lymphoid cell fate. *Nat Rev Immunol* (2015) 15(7):415–28. doi:10.1038/nri3855
153. Artis D, Spits H. The biology of innate lymphoid cells. *Nature* (2015) 517:293–301. doi:10.1038/nature14189
154. Eberl G, Di Santo JP, Vivier E. The brave new world of innate lymphoid cells. *Nat Immunol* (2015) 16:1–5. doi:10.1038/ni.3059
155. Sonnenberg GF, Artis D. Innate lymphoid cells in the initiation, regulation and resolution of inflammation. *Nat Med* (2015) 21:698–708. doi:10.1038/nm.3892
156. Klose CSN, Artis D. Innate lymphoid cells as regulators of immunity, inflammation and tissue homeostasis. *Nat Immunol* (2016) 17:765–74. doi:10.1038/ni.3489
157. Cording S, Medvedovic J, Cherrier M, Eberl G. Development and regulation of RORγt+ innate lymphoid cells. *FEBS Lett* (2014) 588:4176–81. doi:10.1016/j.febslet.2014.03.034
158. Sanos SL, Diefenbach A. Innate lymphoid cells: from border protection to the initiation of inflammatory diseases. *Immunol Cell Biol* (2013) 91:215–24. doi:10.1038/icb.2013.3

13

Natural Killer Cell-Based Cancer Immunotherapies: From Immune Evasion to Promising Targeted Cellular Therapies

Erhard Hofer[1] and Ulrike Koehl[2]**

[1] Department of Vascular Biology, Medical University of Vienna, Vienna, Austria, [2] Institute of Cellular Therapeutics, IFB-Tx, Hannover Medical School, Hannover, Germany

Edited by:
Marcelo Anibal Fernandez-Vina, University of Texas MD Anderson Cancer Center, United States

Reviewed by:
Eric Spierings, Utrecht University, Netherlands
Guido Moll, Charité Universitätsmedizin Berlin, Germany

***Correspondence:**
Erhard Hofer
erhard.hofer@meduniwien.ac.at, erhard.hofer@gmx.at;
Ulrike Koehl
koehl.ulrike@mh-hannover.de

Immunotherapies based on natural killer (NK) cells are among the most promising therapies under development for the treatment of so far incurable forms of leukemia and other types of cancer. The importance of NK cells for the control of viral infections and cancer is supported among others by the findings that viruses and tumors use a multitude of mechanisms to subvert and evade the NK cell system. Infections and malignant diseases can further lead to the shaping of NK cell populations with altered reactivity. Counter measures of potential therapeutic impact include the blocking of inhibitory interactions between NK cell receptors and their cellular ligands, the enhancement of activating receptor signals, and the infusion of large numbers of *ex vivo* generated and selected NK cells. Moreover, the specific cross-linking of NK cells to their target cells using chimeric antigen receptors or therapeutic bi-/trispecific antibody reagents is a promising approach. In this context, NK cells stand out by their positive effects and safety demonstrated in most clinical trials so far. Based in part on results of the recent EC-sponsored project "NATURIMMUN" and considering additional published work in the field, we discuss below new developments and future directions that have the potential to further advance and establish NK cell-based therapies at the clinics on a broader scale.

Keywords: immunotherapy, natural killer cells, immune evasion, cell therapy, checkpoint inhibitors, chimeric antigen receptors, bispecific antibodies

INTRODUCTION

Natural killer (NK) cells have been classically defined as part of the innate immune system providing immediate reactivity against their main targets, virally infected and tumor cells (1). This view has been substantially extended over the recent years based on the findings that NK cells are calibrated to provide self-tolerance, can develop a memory, and play a role in the regulation of the adaptive immune response (2–5). Furthermore, NK cells have turned out to be part of a larger family of innate lymphoid cells (ILCs) that include ILC1–3 (6).

Natural killer reactivity, including cytokine secretion and cytotoxicity, is controlled by a balance of several germ-line encoded inhibitory and activating receptors such as killer immunoglobulin-like receptors (KIRs) and natural cytotoxicity receptors (NCRs) (1, 5, 7, 8). Evidence for the anticancer efficacy of NK cells comes from allogeneic or haploidentical hematopoietic stem cell

(HSC) transplantations that have been used in combination with chemotherapy in the treatment of different forms of leukemia (9). This has shown that NK cells formed from the transplant not only are efficient in killing of allogeneic leukemia cells but are also instrumental in reducing the incidence of graft versus host disease due to their killing activity for dendritic cells (10). Taken together with clinical NK infusion trials in leukemia patients, which have shown exciting antitumor activities and generally safety of the procedure (11–13), it appears that NK cells could be the cells of choice in cellular therapies of leukemia not displaying the critical graft versus host activities of T lymphocytes. Although it is currently less clear whether NK cells will be similarly active in solid cancers, this is a further important area of interest.

IMMUNE EVASION MECHANISMS AND SHAPING OF THE NK CELL COMPARTMENT

Given the importance of NK cells, it is not astonishing that viruses and tumors use a wide array of mechanisms to avoid recognition by NK cells. A paradigm is represented by the Herpes virus family. Many mechanisms such as expression of viral ligands for inhibitory receptors have been described (14). Important is further the downregulation of human stress-induced ligands recognized by the activating NKG2D receptor present on the majority of NK cells. Normally, these stress ligands appear on the cell surface whenever a cell is virally infected or undergoes oncogenic transformation. Whereas internalization and miRNA-mediated downregulation of several stress ligands have been shown previously (15), additional novel mechanisms have been recently identified within the EC-funded project NATURIMMUN. For example, in the case of HHV-6B the expression of stress ligands is suppressed by proteasomal degradation induced by the virus. Consequently, HHV-6B-infected cells can evade immune surveillance by NK cells (16). These various evasion mechanisms of Herpes viruses are reviewed (17) within this research topic ("Tailoring NK Cell Receptor-Ligand Interactions: an Art in Evolution").

Extending the importance of NKG2D ligands to tumors, Schmiedel et al. have shown within the NATURIMMUN project that the stress ligand ULBP2 can be suppressed by an RNA-binding protein that is frequently overexpressed in tumor cells. By binding of this oncogenic protein to ULBP2 mRNA the stability of the mRNA is reduced and ULBP2 levels on the cell surface are downregulated. In consequence, the tumor cells are protected from NK cell recognition (18). This strongly supports that modulation of stress ligands is an important escape mechanism used by cancer cells to diminish NK cell recognition. Involving a different inhibitory receptor, another unexpected novel evasion mechanism could be shown by the same group for colon cancer. NK cell killing was inhibited by the presence of fecal bacteria in the tumor environment. Bacterial proteins interacted with the inhibitory TIGIT receptor on NK cells leading to the inhibition of NK cell cytotoxicity (19). Inhibition of NK cells can also occur by blocking of NKG2D *via* soluble forms of the stress ligand MICA as shown for neuroblastoma as well as head and neck carcinoma. This tumor escape can be overcome in part by highly activated NK cells with upregulated NKG2D (20, 21).

Viruses and human cancers can further have profound effects on and shape the NK cell compartment. Human cytomegalovirus (HCMV), a herpes family member, can trigger an adaptive NK cell response leading to the expansion of NK cell subsets with specific receptor expression (22–24), e.g., the activating NKG2C receptor. The adaptive NKG2C NK cells have been implicated in improved survival of leukemia patients receiving a HSC transplant from HCMV-positive donors (23, 25). Given the potential higher antitumor reactivity of the NKG2C NK cells, this subset is of therapeutic interest and was investigated within the frame of the NATURIMMUN project. Obtained results support that different adaptive NK cell subsets develop in response to viral infection and this is influenced by the copy number of the NKG2C gene (26).

It has been established that certain forms of leukemia display a defective NK cell compartment (27) rendering these forms priority cases for the exploration of NK cell-based therapies. In regard of acute myeloid leukemia (AML), we investigated within the NATURIMMUN project NK cells in patients receiving a novel maintenance therapy with histamine plus IL-2. In this study, AML patients displayed diminished and partly defective NK cells. The therapy strongly induced the immunomodulatory $CD56^{bright}CD16^{-}$ and $CD56^{bright}CD16^{low}$ NK cell subtypes and contributed to the restoration of the NK cell compartment (28). This is in line with the described positive effects of the therapy on disease-free survival of AML patients (29, 30). In addition, our cooperation partner S. Huenecke describes in this research topic that during immune reconstitution after HSC transplantation the degree of development of the two $CD56^{bright}$ and the $CD56^{dim}$ NK cell subpopulations can serve as prognostic marker for both graft versus host disease and viral infections (31).

MODULATION OF INHIBITORY NK RECEPTOR–LIGAND INTERACTIONS AND NOVEL LIGANDS OF ACTIVATING RECEPTORS

Unprecedented rates and durations of clinical responses have been recently achieved in cancer patients by the treatment with antibody reagents that block inhibitory "checkpoint receptors" (32). Whereas these therapies have so far been restricted to the blockade of inhibitory pathways acting on T lymphocytes, the inhibition of NK cells by the interaction of inhibitory NK cell receptors with MHC class I ligands can be regarded as typical checkpoint inhibition. In fact, efforts are currently been undertaken to evaluate blockade of the inhibitory NKG2A/CD94 receptor and of inhibitory KIRs to elicit NK reactivity to cancer cells. The company Innate Pharma has developed first-in-class monoclonal antibodies that target inhibitory NK cell receptors and these are currently in preclinical and clinical evaluation (33).

While the ligands for inhibitory NK cell receptors are well established, ligands bound by important activating receptors are still incompletely identified. This is the case for the activating

NKG2C/CD94 receptor, several activating KIRs, and the NCRs. In this regard, a group participating in NATURIMMUN has studied how HCMV stimulates NK cells *via* the activating KIR2DS1 receptor. The ligand was identified as a specific class I molecule, HLA-C2, which in its normal form is recognized by the related inhibitory KIR2DL1 receptor. Possibly, a conformational change in normal HLA-C2 triggered by HCMV was required for KIR2DS1-mediated NK cell activation (34). Other participants in NATURIMMUN have developed assay systems and have work in progress to identify virally induced and potentially tumor ligands for the activating NKG2C receptor (Pupuleku et al., manuscript in preparation for this research topic) and the NCRs The clarification of the molecular nature and mechanism of action of the corresponding activating ligands on virally infected and tumor cells will allow novel pathways of NK cell activation to be triggered.

GENERATION OF LARGE-SCALE THERAPEUTIC NK CELLS AND TECHNOLOGY TO TARGET AND CROSS-LINK NK CELLS TO CANCER CELLS

Exploiting and strengthening the NK cell response is a highly promising approach for future successful immunotherapies of cancer. This could be achieved by infusion of *ex vivo* expanded and activated NK cells, by genetic modification of NK cells with chimeric antigen receptors (CAR), by multivalent reagents cross-linking NK cells to cancer cells, or by a combination of these methods (**Figure 1**).

In regard of *ex vivo* expansion of peripheral donor NK cells several groups have developed corresponding technologies and some were or are being applied in clinical trials of NK cell infusions (11, 12, 35). Important for broader availability of these therapies are commercial sources of the necessary equipment and reagents and further development of automated systems for production of GMP-compliant clinical-grade NK cells. A pioneer in this regard is the company Miltenyi Biotec. In part as participant of NATURIMMUN, this company has further developed a protocol to expand peripheral NK cells using irradiated autologous peripheral blood mononuclear cells as feeder cells. NK cell isolation and expansion were further fully automated for future clinical applications (36, 37). NK cells generated by this procedure have been evaluated in detail (Delso-Vallejo et al., submitted to this research topic).

Another possibility is the generation of therapeutic NK cells from umbilical cord blood stem cells (UCBSC), which was pioneered by the company Glycostem (38). Within NATURIMMUN, NK cells differentiated in this system were characterized in detail and the procedure improved to yield more mature NK cells (39). Furthermore, an important role of the transcription factor ZNF683/HOBIT for NK cell differentiation could be shown supporting that the factor could be used to modulate NK cell generation [(40), this research topic]. This research topic. UCBSC-derived NK cells have been evaluated in a phase I clinical trial in elderly AML patients and found to be safe (41). Furthermore, recent evidence obtained in NATURIMMUN supports that the cells possess high cytotoxicity against metastatic colorectal cancer cells (42, 43) and could be used in the therapy of solid cancers [(44), this research topic].

An important topic in the field is to harmonize the manufacturing of GMP-compliant therapeutic NK cell products, which was initiated within NATURIMMUN and has been described in a summary of the worldwide experience obtained so far with allogeneic adaptive NK cell therapies (12). It is conceivable that expanded therapeutic NK cells could be stored frozen and be shipped on demand. These NK cells could, therefore, qualify as off-the-shelf-products, and to what extent this will be possible is a relevant question for future research.

It has been shown that expanded and cytokine-activated NK cells can be functional in certain cancer types. However, evidence suggests that specific targeting and cross-linking of

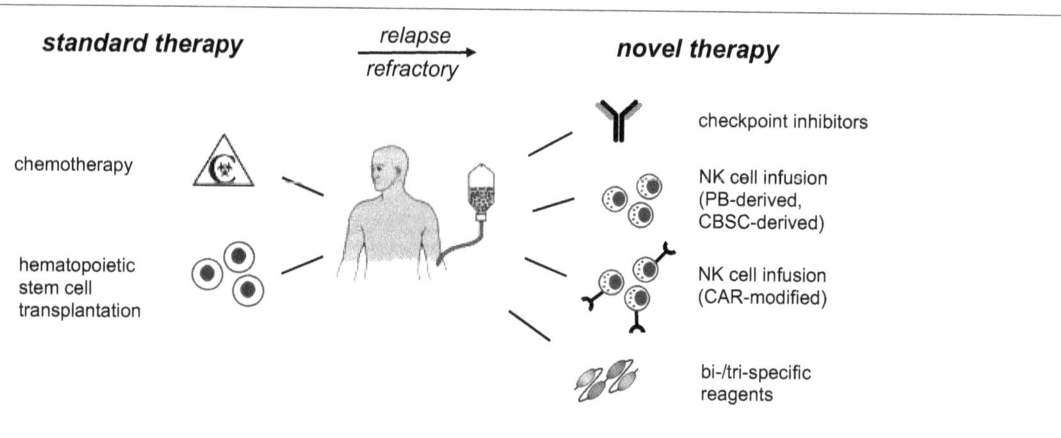

FIGURE 1 | NK cell-based immunotherapies. Standard therapy for high-risk leukemia includes high-dose chemotherapy, followed by hematopoietic stem cell transplantation. Patients, who do not reach remission or suffer from early relapse thereafter, have a poor prognosis and are in urgent medical need for advanced therapies. Current immunotherapeutic developments and phase I/II trials include checkpoint inhibitors for inhibitory NK receptors, infusion of expanded and activated autologous or allogeneic NK cells, and targeting of NK cells to cancer cells. The latter can be done by modification of NK cells with CAR or by application of multispecific reagents to cross-link NK cells with cancer cells. These immunotherapies should reduce relapse rates and constitute promising additional treatment options for high-risk patients. PB, peripheral blood; CBSC, cord blood stem cell; CAR, chimeric antigen receptor; NK, natural killer.

NK cells to cancer cells would strongly enhance their reactivity and the applicability of NK cell therapies. A paradigm in the field is currently the exciting successes of targeting of T lymphocytes to CD19 *via* genetic CAR modification (45, 46) or corresponding bispecific reagents (46). We believe that NK cells will provide important advantages to the use of T lymphocytes based on their comparable reactivity but much higher safety. We have achieved increased NK cell cytotoxicity against leukemia cells using transduction of NK cells with CAR constructs (47, 48) or by cross-linking with trispecific reagents (Kloess et al., submitted to this research topic). Furthermore, it is conceivable that procedures to achieve redirected primary human NK cells as an "off-the-shelf-immunotherapy" can be developed. For this, optimizing both the respective antigen binding and the triggering of the intracellular signaling cascade by the CAR will be desirable (49).

A possibility to target NK cells to cancer cells can be the use of monoclonal antibody therapeutics already approved for clinical application. Examples of these are the anti-CD20 antibody rituximab (50) for B cell leukemia and the anti-EGF receptor antibody cetuximab (51–53). The latter is in use for the therapy of colon carcinoma and head and neck cancer. It displays limited efficacy in colon with better activities in head and neck cancer. It is possible that synergistic activities could be gained by coapplication of NK cell infusions as these antibodies trigger ADCC *via* binding to the low-affinity Fcγ receptor present on NK cells. It could be shown within NATURIMMUN that NK cytotoxicity toward $EGFR^+$ colon and cervical cancer cells was strongly enhanced by cetuximab (42, 43). This provides a rationale to strengthen NK cell immunotherapy through a combination with cetuximab for metastatic colorectal cancer patients [(44), this research topic].

PRECLINICAL MODELS FOR EVALUATION OF HUMAN NK CELL-BASED CANCER THERAPIES

The preclinical evaluation of NK cell-based therapies in mouse models is hampered by the inherent problem that reagents designed to trigger human immune cell would not react at all or only partially with murine NK cells. Similarly, the evaluation of human NK cell infusions in mice does not provide a human immune cell compartment necessary for full functioning. This problem can be partly circumvented by mouse models with humanized immune system (HIS) in combination with xenotransplantation models of human cancers.

In this regard, a novel method to boost the inefficient human NK cell development in mice observed after engraftment of human HSC was recently developed. Normally, the differentiation of NK cells depends on the interplay with myeloid cells, and human myeloid cells are poorly reconstituted in available HIS mice due to competition with the murine cells (54). Therefore, a new model was developed in the NATURIMMUN project using mice that lack the Flt3 receptor (55) and display reduced murine myeloid differentiation. In these mice, human dendritic cells and consequently human NK cells could be successfully boosted by human Flt3 ligand providing a novel mouse model with increased NK cell numbers [(56), this research topic]. This will be valuable for future evaluations of immunotherapies involving reagents designed for human cells as well as human NK cell infusions.

As an exemplary preclinical evaluation, we tested within NATURIMMUN the efficacy of NK cell infusions alone or in combination with the clinically approved cetuximab against human colon cancer. HIS mice were engrafted with a human colorectal carcinoma cell line and treated with cetuximab and infusions of PB-derived and UCBSC-derived NK cells. Then the tumor load and survival rate were monitored. Significant inhibition of tumor growth and improvement of survival rates were observed. These results provide a rationale for NK infusion therapies not only for leukemia but also for solid cancer treatment [(44), this research topic].

MAIN FUTURE DIRECTIONS TO ACHIEVE NK CELL-BASED CANCER IMMUNOTHERAPIES ON A BROADER SCALE

Collectively, the basic work on NK cells, their receptors, and NK evasion mechanisms have provided evidence for the importance of the NK cell system in the control of human cancers. Clinical trials of NK infusion therapies, performed mostly in different forms of leukemia, have uniformly shown safety of infused NK cells and in certain cases exciting effects on disease-free survival (11). This together underlines the feasibility and potential efficacy of NK cell-based immunotherapies. However, based on the currently available data a number of questions and major routes should be further explored in order for NK cell therapies to become clinically used on a broader scale. Among those are improved methods for the selection of the best donor NK cells to be able to optimally exploit the antitumor alloreactivity of NK cells (12). Then the question of best activation of NK cells by cytokines such as IL-2, IL-12, IL-15, IL-18, and IL-21 needs to be settled as reviewed within this research topic (57). In addition, the best expansion time points of clinical-scale NK cells have to be evaluated regarding both safety and efficacy with the overall goal to allow multiple adaptive NK cell application to the respective patients. The optimal application of the newly developed NK cell-directed checkpoint inhibitors needs to be explored. Further additional reagents for targeting and cross-linking of NK cells to cancer cells using bi-/trispecific antibody-based reagents should be developed to extend the range of targeted cancer cells. Similarly, additional CAR constructs for wider targeting should be derived and corresponding standard "off-the-shelf-procedures" developed for genetic modification of NK cells. Of special importance for NK infusion therapies, available technologies for NK cell generation need to be fully automated and harmonized protocols developed for large-scale GMP-compliant generation of clinical-grade therapeutic NK cells that have been recently classified as advanced therapy medicinal products in Europe. They are regulated accordingly either centralized or under hospital exemption by the member states [Regulation (EC) No 1394/2007; Directive 2001/83/EC and Regulation (EC) No 726/2004]. Given the accessibility of the tumor cells the primary focus should be on leukemia as it is to be expected that progress will be more rapid in this area. But in light of the high need of new therapies for solid cancers these should also be pursued.

CONCLUDING REMARKS

The recent years have seen significant progress in immunotherapies of cancer based on novel checkpoint inhibitors and reagents and technology to boost T and NK lymphocytes. We propose that based on the available knowledge of NK cells, these cells will be much more amenable for therapeutic purposes based on their high cytotoxicity and generally demonstrated safety. Therefore, we suggest that a concerted effort in the development of NK cell-based immunotherapies has high potential to achieve novel therapies of hitherto untreatable and relapsed forms of leukemia and potentially also solid cancers. The development of broadly applicable NK cell-based therapies should extend the currently more restricted available T cell-based therapies and could thus boost the long-standing promise of cellular cancer therapies.

ETHICS STATEMENT

Described work at the Medical School Hannover was carried out in accordance with the recommendations of the Ethics Committee of the Medical School Hannover with written informed consent from all subjects. All subjects gave written informed consent in accordance with the Declaration of Helsinki. The protocol was approved by the Ethics Committee of the Medical School Hannover (No 2159-2014).

AUTHOR CONTRIBUTIONS

EH prepared the outline, the article parts were jointly written with EH focusing on the more basic immunology parts and UK on the therapeutic aspects.

ACKNOWLEDGMENTS

We thank all partners in the project NATURIMMUN for contributions to the project and continuous discussion.

REFERENCES

1. Caligiuri MA. Human natural killer cells. *Blood* (2008) 112(3):461–9. doi:10.1182/blood-2007-09-077438
2. Moretta A, Marcenaro E, Parolini S, Ferlazzo G, Moretta L. NK cells at the interface between innate and adaptive immunity. *Cell Death Differ* (2008) 15(2):226–33. doi:10.1038/sj.cdd.4402170
3. Vivier E, Raulet DH, Moretta A, Caligiuri MA, Zitvogel L, Lanier LL, et al. Innate or adaptive immunity? The example of natural killer cells. *Science* (2011) 331(6013):44–9. doi:10.1126/science.1198687
4. Cerwenka A, Lanier LL. Natural killer cell memory in infection, inflammation and cancer. *Nat Rev Immunol* (2016) 16(2):112–23. doi:10.1038/nri.2015.9
5. Hoglund P, Brodin P. Current perspectives of natural killer cell education by MHC class I molecules. *Nat Rev Immunol* (2010) 10(10):724–34. doi:10.1038/nri2835
6. Diefenbach A, Colonna M, Koyasu S. Development, differentiation, and diversity of innate lymphoid cells. *Immunity* (2014) 41(3):354–65. doi:10.1016/j.immuni.2014.09.005
7. Pende D, Marcenaro S, Falco M, Martini S, Bernardo ME, Montagna D, et al. Anti-leukemia activity of alloreactive NK cells in KIR ligand-mismatched haploidentical HSCT for pediatric patients: evaluation of the functional role of activating KIR and redefinition of inhibitory KIR specificity. *Blood* (2009) 113(13):3119–29. doi:10.1182/blood-2008-06-164103
8. Heidenreich S, Kroger N. Reduction of relapse after unrelated donor stem cell transplantation by KIR-based graft selection. *Front Immunol* (2017) 8:41. doi:10.3389/fimmu.2017.00041
9. Mancusi A, Ruggeri L, Velardi A. Haploidentical hematopoietic transplantation for the cure of leukemia: from its biology to clinical translation. *Blood* (2016) 128(23):2616–23. doi:10.1182/blood-2016-07-730564
10. Morandi B, Mortara L, Chiossone L, Accolla RS, Mingari MC, Moretta L, et al. Dendritic cell editing by activated natural killer cells results in a more protective cancer-specific immune response. *PLoS One* (2012) 7(6):e39170. doi:10.1371/journal.pone.0039170
11. Bachanova V, Miller JS. NK cells in therapy of cancer. *Crit Rev Oncog* (2014) 19(1–2):133–41. doi:10.1615/CritRevOncog.2014011091
12. Koehl U, Kalberer C, Spanholtz J, Lee DA, Miller JS, Cooley S, et al. Advances in clinical NK cell studies: donor selection, manufacturing and quality control. *Oncoimmunology* (2016) 5(4):e1115178. doi:10.1080/2162402X.2015.1115178
13. Rubnitz JE, Inaba H, Ribeiro RC, Pounds S, Rooney B, Bell T, et al. NKAML: a pilot study to determine the safety and feasibility of haploidentical natural killer cell transplantation in childhood acute myeloid leukemia. *J Clin Oncol* (2010) 28(6):955–9. doi:10.1200/JCO.2009.24.4590
14. Jonjic S, Babic M, Polic B, Krmpotic A. Immune evasion of natural killer cells by viruses. *Curr Opin Immunol* (2008) 20(1):30–8. doi:10.1016/j.coi.2007.11.002
15. Stern-Ginossar N, Elefant N, Zimmermann A, Wolf DG, Saleh N, Biton M, et al. Host immune system gene targeting by a viral miRNA. *Science* (2007) 317(5836):376–81. doi:10.1126/science.1140956
16. Schmiedel D, Tai J, Levi-Schaffer F, Dovrat S, Mandelboim O. Human herpesvirus 6B downregulates expression of activating ligands during lytic infection to escape elimination by natural killer cells. *J Virol* (2016) 90(21):9608–17. doi:10.1128/JVI.01164-16
17. Schmiedel D, Mandelboim O. Disarming cellular alarm systems-manipulation of stress-induced NKG2D ligands by human herpesviruses. *Front Immunol* (2017) 8:390. doi:10.3389/fimmu.2017.00390
18. Schmiedel D, Tai J, Yamin R, Berhani O, Bauman Y, Mandelboim O. The RNA binding protein IMP3 facilitates tumor immune escape by downregulating the stress-induced ligands ULPB2 and MICB. *Elife* (2016) 5:e13426. doi:10.7554/eLife.13426
19. Gur C, Ibrahim Y, Isaacson B, Yamin R, Abed J, Gamliel M, et al. Binding of the Fap2 protein of *Fusobacterium nucleatum* to human inhibitory receptor TIGIT protects tumors from immune cell attack. *Immunity* (2015) 42(2):344–55. doi:10.1016/j.immuni.2015.01.010
20. Kloess S, Huenecke S, Piechulek D, Esser R, Koch J, Brehm C, et al. IL-2-activated haploidentical NK cells restore NKG2D-mediated NK-cell cytotoxicity in neuroblastoma patients by scavenging of plasma MICA. *Eur J Immunol* (2010) 40(11):3255–67. doi:10.1002/eji.201040568
21. Kloss S, Chambron N, Gardlowski T, Arseniev L, Koch J, Esser R, et al. Increased sMICA and TGFbeta1 levels in HNSCC patients impair NKG2D-dependent functionality of activated NK cells. *Oncoimmunology* (2015) 4(11):e1055993. doi:10.1080/2162402X.2015.1055993
22. Guma M, Angulo A, Vilches C, Gomez-Lozano N, Malats N, Lopez-Botet M. Imprint of human cytomegalovirus infection on the NK cell receptor repertoire. *Blood* (2004) 104(12):3664–71. doi:10.1182/blood- 40.
2004-05-2058

23. Muntasell A, Vilches C, Angulo A, Lopez-Botet M. Adaptive reconfiguration of the human NK-cell compartment in response to cytomegalovirus: a different perspective of the host-pathogen interaction. *Eur J Immunol* (2013) 43(5):1133–41. doi:10.1002/eji.201243117
24. Liu LL, Pfefferle A, Yi Sheng VO, Bjorklund AT, Beziat V, Goodridge JP, et al. Harnessing adaptive natural killer cells in cancer immunotherapy. *Mol Oncol* (2015) 9(10):1904–17. doi:10.1016/j.molonc.2015.10.001
25. Cichocki F, Cooley S, Davis Z, DeFor TE, Schlums H, Zhang B, et al. CD56dimCD57+NKG2C+ NK cell expansion is associated with reduced leukemia relapse after reduced intensity HCT. *Leukemia* (2016) 30(2):456–63. doi:10.1038/leu.2015.260
26. Muntasell A, Pupuleku A, Cisneros E, Vera A, Moraru M, Vilches C, et al. Relationship of NKG2C copy number with the distribution of distinct cytomegalovirus-induced adaptive NK cell subsets. *J Immunol* (2016) 196(9): 3818–27. doi:10.4049/jimmunol.1502438
27. Khaznadar Z, Boissel N, Agaugue S, Henry G, Cheok M, Vignon M, et al. Defective NK cells in acute myeloid leukemia patients at diagnosis are associated with blast transcriptional signatures of immune evasion. *J Immunol* (2015) 195(6):2580–90. doi:10.4049/jimmunol.1500262
28. Cuapio A, Post M, Cerny-Reiterer S, Gleixner KV, Stefanzl G, Basilio J, et al. Maintenance therapy with histamine plus IL-2 induces a striking expansion of two CD56bright NK cell subpopulations in patients with acute myeloid leukemia and supports their activation. *Oncotarget* (2016) 7(29):46466–81. doi:10.18632/oncotarget.10191
29. Brune M, Castaigne S, Catalano J, Gehlsen K, Ho AD, Hofmann WK, et al. Improved leukemia-free survival after postconsolidation immunotherapy with histamine dihydrochloride and interleukin-2 in acute myeloid leukemia: results of a randomized phase 3 trial. *Blood* (2006) 108(1):88–96. doi:10.1182/blood-2005-10-4073
30. Martner A, Rydstrom A, Riise RE, Aurelius J, Anderson H, Brune M, et al. Role of natural killer cell subsets and natural cytotoxicity receptors for the outcome of immunotherapy in acute myeloid leukemia. *Oncoimmunology* (2016) 5(1):e1041701. doi:10.1080/2162402X.2015.1041701
31. Huenecke S, Cappel C, Esser R, Pfirrmann V, Salzmann-Manrique E, Betz S, et al. Development of three different NK cell subpopulations during immune reconstitution after pediatric allogeneic hematopoietic stem cell transplantation: prognostic markers in GvHD and viral infections. *Front Immunol* (2017) 8:109. doi:10.3389/fimmu.2017.00109
32. Sharma P, Allison JP. The future of immune checkpoint therapy. *Science* (2015) 348(6230):56–61. doi:10.1126/science.aaa8172
33. Ruggeri L, Urbani E, Andre P, Mancusi A, Tosti A, Topini F, et al. Effects of anti-NKG2A antibody administration on leukemia and normal hematopoietic cells. *Haematologica* (2016) 101(5):626–33. doi:10.3324/haematol.2015.135301
34. van der Ploeg K, Chang C, Ivarsson MA, Moffett A, Wills MR, Trowsdale J. Modulation of human leukocyte antigen-C by human cytomegalovirus stimulates KIR2DS1 recognition by natural killer cells. *Front Immunol* (2017) 8:298. doi:10.3389/fimmu.2017.00298
35. Bachanova V, Cooley S, Defor TE, Verneris MR, Zhang B, McKenna DH, et al. Clearance of acute myeloid leukemia by haploidentical natural killer cells is improved using IL-2 diphtheria toxin fusion protein. *Blood* (2014) 123(25):3855–63. doi:10.1182/blood-2013-10-532531
36. Granzin M, Soltenborn S, Muller S, Kollet J, Berg M, Cerwenka A, et al. Fully automated expansion and activation of clinical-grade natural killer cells for adoptive immunotherapy. *Cytotherapy* (2015) 17(5):621–32. doi:10.1016/j.jcyt.2015.03.611
37. Granzin M, Stojanovic A, Miller M, Childs R, Huppert V, Cerwenka A. Highly efficient IL-21 and feeder cell-driven ex vivo expansion of human NK cells with therapeutic activity in a xenograft mouse model of melanoma. *Oncoimmunology* (2016) 5(9):e1219007. doi:10.1080/2162402X.2016.1219007
38. Spanholtz J, Preijers F, Tordoir M, Trilsbeek C, Paardekooper J, de Witte T, et al. Clinical-grade generation of active NK cells from cord blood hematopoietic progenitor cells for immunotherapy using a closed-system culture process. *PLoS One* (2011) 6(6):e20740. doi:10.1371/journal.pone.0020740
39. Lehmann D, Spanholtz J, Sturtzel C, Tordoir M, Schlechta B, Groenewegen D, et al. IL-12 directs further maturation of ex vivo differentiated NK cells with improved therapeutic potential. *PLoS One* (2014) 9(1):e87131. doi:10.1371/journal.pone.0087131
Post M, Cuapio A, Osl M, Lehmann D, Resch U, Davies DM, et al. The transcription factor ZNF683/HOBIT regulates human NK-cell development. *Front Immunol* (2017) 8:535. doi:10.3389/fimmu.2017.00535
41. Dolstra H, Roeven MW, Spanholtz J, Hangalapura BN, Tordoir M, Maas F, et al. Successful transfer of umbilical cord blood CD34$^+$ hematopoietic stem and progenitor-derived NK cells in older acute myeloid leukemia patients. *Clin Canc Res* (2017). doi:10.1158/1078-0432.CCR-16-2981
42. Veluchamy JP, Heeren AM, Spanholtz J, van Eendenburg JD, Heideman DA, Kenter GG, et al. High-efficiency lysis of cervical cancer by allogeneic NK cells derived from umbilical cord progenitors is independent of HLA status. *Cancer Immunol Immunother* (2017) 66(1):51–61. doi:10.1007/s00262-016-1919-1
43. Veluchamy JP, Spanholtz J, Tordoir M, Thijssen VL, Heideman DA, Verheul HM, et al. Combination of NK cells and cetuximab to enhance anti-tumor responses in RAS mutant metastatic colorectal cancer. *PLoS One* (2016) 11(6):e0157830. doi:10.1371/journal.pone.0157830
44. Veluchamy JP, Lopez-Lastra S, Spanholtz J, Bohme F, Kok N, Heideman DA, et al. In vivo efficacy of umbilical cord blood stem cell-derived NK cells in the treatment of metastatic colorectal cancer. *Front Immunol* (2017) 8:87. doi:10.3389/fimmu.2017.00087
45. Ramos CA, Savoldo B, Dotti G. CD19-CAR trials. *Cancer J* (2014) 20(2):112–8. doi:10.1097/PPO.0000000000000031
46. Newman MJ, Benani DJ. A review of blinatumomab, a novel immunotherapy. *J Oncol Pharm Pract* (2016) 22(4):639–45. doi:10.1177/1078155215618770
47. Suerth JD, Morgan MA, Kloess S, Heckl D, Neudorfl C, Falk CS, et al. Efficient generation of gene-modified human natural killer cells via alpharetroviral vectors. *J Mol Med (Berl)* (2016) 94(1):83–93. doi:10.1007/s00109-015-1327-6
48. Glienke W, Esser R, Priesner C, Suerth JD, Schambach A, Wels WS, et al. Advantages and applications of CAR-expressing natural killer cells. *Front Pharmacol* (2015) 6:21. doi:10.3389/fphar.2015.00021
49. Oberschmidt O, Kloess S, Koehl U. Redirected primary human chimeric antigen receptor natural killer cells as an "off-the-shelf immunotherapy" for improvement in cancer treatment. *Front Immunol* (2017) 8:654. doi:10.3389/fimmu.2017.00654
50. Jabbour E, O'Brien S, Ravandi F, Kantarjian H. Monoclonal antibodies in acute lymphoblastic leukemia. *Blood* (2015) 125(26):4010–6. doi:10.1182/blood-2014-08-596403
51. Graham J, Muhsin M, Kirkpatrick P. Cetuximab. *Nat Rev Drug Discov* (2004) 3(7):549–50. doi:10.1038/nrd1445
52. Silvestris N, Vincenzi B, Brunetti AE, Loupakis F, Dell'Aquila E, Russo A, et al. Pharmacogenomics of cetuximab in metastatic colorectal carcinoma. *Pharmacogenomics* (2014) 15(13):1701–15. doi:10.2217/pgs.14.124
53. Kloss S, Chambron N, Gardlowski T, Weil S, Koch J, Esser R, et al. Cetuximab reconstitutes pro-inflammatory cytokine secretions and tumor-infiltrating capabilities of sMICA-inhibited NK cells in HNSCC tumor spheroids. *Front Immunol* (2015) 6:543. doi:10.3389/fimmu.2015.00543
54. Li Y, Di Santo JP. Probing human NK cell biology using human immune system (HIS) mice. *Curr Top Microbiol Immunol* (2016) 395:191–208. doi:10.1007/82_2015_488
55. Li Y, Mention JJ, Court N, Masse-Ranson G, Toubert A, Spits H, et al. A novel Flt3-deficient HIS mouse model with selective enhancement of human DC development. *Eur J Immunol* (2016) 46(5):1291–9. doi:10.1002/eji.201546132
56. Lopez-Lastra S, Di Santo JP. Modeling natural killer cell targeted immunotherapies. *Front Immunol* (2017) 8:370. doi:10.3389/fimmu.2017.00370
57. Granzin M, Wagner J, Koehl U, Cerwenka A, Huppert V, Ullrich E. Shaping of natural killer cell antitumor activity by ex vivo cultivation. *Front Immunol* (2017) 8:458. doi:10.3389/fimmu.2017.00458

Conflict of Interest Statement: The authors declare that the work was conducted in the absence of any commercial or financial relationships that could be construed as a potential conflict of interest.

14

Cetuximab Reconstitutes Pro-Inflammatory Cytokine Secretions and Tumor-Infiltrating Capabilities of sMICA-Inhibited NK Cells in HNSCC Tumor Spheroids

Stephan Klöss[1]*, Nicole Chambron[2], Tanja Gardlowski[1], Sandra Weil[3], Joachim Koch[3], Ruth Esser[1], Elke Pogge von Strandmann[4], Michael A. Morgan[5], Lubomir Arseniev[1], Oliver Seitz[2] and Ulrike Köhl[1]

[1] Institute of Cellular Therapeutics, Integrated Research and Treatment Center Transplantation (IFB-Tx), Hannover Medical School, Hannover, Germany, [2] Department of Oral, Cranio-Maxillofacial and Facial Plastic Surgery, Klinikum Hanau GmbH, Hanau, Germany, [3] Georg-Speyer-Haus Institute for Tumor Biology and Experimental Therapy, Frankfurt, Germany, [4] Klinik I für Innere Medizin, Uniklinik Köln, Cologne, Germany, [5] Institute of Experimental Haematology, Hannover Medical School, Hannover, Germany

Edited by:
Susana Larrucea, BioCruces Health Research Institute, Spain

Reviewed by:
Francisco Borrego, BioCruces Health Research Institute and Cruces University Hospital, Spain
Frank Momburg, Deutsches Krebsforschungszentrum, Germany

***Correspondence:**
Stephan Klöss
kloess.stephan@mh-hannover.de

Immunosuppressive factors, such as soluble major histocompatibility complex class I chain-related peptide A (sMICA) and transforming growth factor beta 1 (TGF-β_1), are involved in tumor immune escape mechanisms (TIEMs) exhibited by head and neck squamous cell carcinomas (HNSCCs) and may represent opportunities for therapeutic intervention. In order to overcome TIEMs, we investigated the antibody-dependent cellular cytotoxicity (ADCC), cytokine release and retargeted tumor infiltration of sMICA-inhibited patient NK cells expressing Fcγ receptor IIIa (FcγRIIIa, CD16a) in the presence of cetuximab, an anti-epidermal growth factor receptor (HER1) monoclonal antibody (mAb). Compared to healthy controls, relapsed HNSCC patients ($n = 5$), not currently in treatment revealed decreased levels of circulating regulatory NK cell subsets in relation to increased cytotoxic NK cell subpopulations. Elevated sMICA and TGF-β_1 plasma levels correlated with diminished TNFα and IFN-γ release and decreased NKG2D (natural killer group 2 member D)-dependent killing of HNSCC cells by NK cells. Incubation of IL-2-activated patient NK cells with patient plasma containing elevated sMICA or sMICA analogs (shed MICA and recombinant MICA) significantly impaired NKG2D-mediated killing by down-regulation of NKG2D surface expression. Of note, CD16 surface expression levels, pro-apoptotic and activation markers, and viability of patient and healthy donor NK cell subpopulations were not affected by this treatment. Accordingly, cetuximab restored killing activity of sMICA-inhibited patient NK cells against cetuximab-coated primary HNSCC cells via ADCC in a dose-dependent manner. Rapid reconstitution of anti-tumor recognition and enhanced tumor infiltration of treated NK cells was monitored by 24 h co-incubation of HNSCC tumor spheroids with cetuximab (1 μg/ml) and was characterized by increased IFN-γ and TNFα secretion. This data show that the impaired NK cell-dependent tumor surveillance in relapsed HNSCC patients could be reversed by the re-establishment of ADCC-mediated effector cell activity, thus supporting NK cell-based immunotherapy in combination with antineoplastic monoclonal mAbs.

Keywords: ADCC, cetuximab-activated NK cells, HNSCC tumor spheroids, soluble MICA, TGF-β_1

INTRODUCTION

Natural killer cells are lymphoid effector cells important for the innate immune response against virally infected and malignant cells (1, 2). NK cells eliminate transformed target cells during cytotoxic interactions by releasing pro-inflammatory cytokines, especially IFN-γ and TNF-α (3). Similarly, tumor-infiltrating NK cells can trigger stimulating interactions via "cell cross-talk" with dendritic cells (DCs), possibly facilitating tumor antigen presentation and induction of tumor antigen-directed T-cell responses. This demonstrates the constitutive role of NK cells as mediators between the innate and acquired immune systems (4–6). In this respect, NK cell killing activity is regulated by both stimulatory and inhibitory receptors. Interaction between NK cell inhibitory receptors in the presence or absence of MHC class I molecules on normal and possible target cells was described as the "missing self" hypothesis (7, 8). Activating receptors include the natural cytotoxicity receptors (NCRs) NKp30, NKp44, and NKp46 with poorly characterized ligands, and the NKG2D receptor, which recognizes a variety of well-defined ligands expressed by transformed cells (9–12). Several studies confirmed the predominant relevance of NKG2D in efficient recognition and elimination of tumor- and "stressed" cells by targeted binding of MICA and MICB (9, 10, 12). Interestingly, elevated levels of soluble forms of these NKG2D ligands (sMICA and sMICB), generated by matrix metalloproteinase (MMP)-dependent proteolytic cleavage ("shedding") were detected in plasma/serum of cancer patients (13). These soluble NKG2D ligands are responsible for systemic reduction of NKG2D expression on the surface of various circulating blood lymphocytes, especially cytotoxic NK cells, NK-like T (NKT) cells, and $CD8^+$ $\alpha\beta^+$- and $\gamma\delta^+$-T cells. Thus, these immune modulating effects resulted in decreased tumor surveillance by attenuated recognition and elimination of malignant cancer cells (14–16).

Some reports described NK cell dysfunction in patients with head and neck squamous cell carcinoma (HNSCC). These highly aggressive solid tumors originate from the epithelial lining of the upper aero-digestive tract and are able to escape NK cell-mediated immunosurveillance. Tumor progression is accomplished by significantly reduced expression levels of NKG2D on effector cells (17–20). Detection of increased sMICA plasma levels monitored in HNSCC patients at advanced disease stages (stage IV) and poor clinical prognosis further supports the importance of diminished tumor surveillance in HNSCC progression (21, 22). Indeed, high sMICA levels coincide with increased frequencies of lymph node (LN) metastasis. Additionally, decreased survival rates in high-risk cancer patients are potentiated by high sMICB levels (19, 20, 23). Interestingly, the multiple dysfunctions of NK cells can be largely reversed by cancer antigen-targeted antibodies, which stimulate the antibody-dependent cellular cytotoxicity (ADCC)-mediated cytotoxicity of activated NK cells to selectively eliminate malignant cells. Therefore, multiple monoclonal antibodies (mAbs) have been designed to target diverse tumor surface molecules. Based on the assumption that mAbs interact specifically with tumor target molecules, mAbs could affect tumor cells by directly inhibiting essential signaling pathways initiated by target molecules and/or by stimulating effector cell cytotoxicity, resulting in tumor elimination. One highly investigated cancer antigen for treatment of solid tumors is the epidermal growth factor receptor (EGFR). EGFR is a member of the ErbB protein family, which consists of four transmembrane receptor proteins, including HER1 [EGFR, ErbB1: avian erythroblastic leukemia viral (v-erb-b) oncogene homolog, receptor for EGF], HER2 (ErbB2), HER3 (ErbB3), and HER4 (ErbB4) (24–26). EGFR overexpression is associated with poor clinical prognosis in multiple solid tumors and EGFR signaling plays an important role in malignant cell migration, evasion, and proliferation (27). Therefore, mAbs designed to target EGFR were developed. In the last decade, cetuximab, an anti-EGFR human-mouse chimeric IgG_1 monoclonal antibody (mAb) was approved by the Food and Drug Administration (FDA) for treatment of metastatic colorectal cancer, metastatic non-small cell lung cancer, and HNSCC patients (28, 29). However, only a weak to moderate (10–20%) benefit was observed in clinical trials with high-risk cancer patients (30–35). In this context, HNSCC cells may have evaded NK cell immunosurveillance due to polymorphisms in the FcγRIIIa (CD16a) on effector cells that impact interaction with the IgG_1 Fc, heavy-chain, portion of cetuximab. This partially elucidates patient-specific responses to cetuximab and underscores the essential importance of modulating immunological synapses (36–38). In addition, the tumor microenvironment can impact lymphocyte-dependent immunosurveillance, which correlated strongly with tumor infiltration as well as NK cell-mediated killing activity. The ability to control these factors could contribute to improved prognosis in some malignant diseases (39–42). Thus, NK cells as key players in ADCC-related cetuximab activity were able to infiltrate primary colorectal adenocarcinomas and NK cell infiltration was an independent predictor for response and progression-free survival in patients receiving cetuximab treatment (43).

Recently, we described a decreased anti-tumor recognition, cytokine release and a reduced NKG2D expression on NK cells from untreated HNSCC patients. *In vitro* blocking experiments revealed a synergistic negative effect of sMICA potentiated by TGF-β_1 on the killing activity of patient NK cells (22). In the current study, cetuximab treatment reconstituted the tumor surveillance capacity of sMICA-inhibited NK cells from HNSCC patients ($n = 5$), thus demonstrating the potential usefulness of cetuximab in retargeted ADCC. In order to investigate specific NK cell-dependent tumor infiltrations and ADCC-related cetuximab response, we developed HNSCC tumor-like cell clusters and a tumor spheroid model derived from primary, singularized tumor cells from these HNSCC patients. Our results indicate a crucial relevance of enhanced sMICA levels in tumor surveillance and infiltrations of inhibited patient NK cells. Finally, we demonstrate that these immunosuppressive effects on NK cell-mediated killing activity could be bypassed using cetuximab-coated HNSCC cells.

PATIENTS AND METHODS

HNSCC Patients

We analyzed five HNSCC patients (three male and two female, age range: 24–76 years) and five age-matched healthy individuals (three male and two female, age range: 26–58 years) served as controls. Histopathology confirmed that patients had stage II–IV HNSCC. Patients were included in this study after tumor recidivism but before initiation of any clinical treatment (**Table 1**). Corresponding patient blood samples (80–100 ml) were received shortly before the tumor surgery and associated tumor fragments were collected during tumor extractions from all patients. Informed consent was obtained from patients, caretakers, and healthy controls (HCs). Patient characteristics are summarized in **Table 1**. Blood samples were collected from HNSCC patients and healthy individuals in Heparin- and EDTA-coated tubes. Total leukocytes and the resultant subpopulations were counted by five-color flow cytometry (FCM) analysis as described previously (44). Immunocompetent cell subpopulation distributions were compared among patients and HCs (**Figure 1**).

Target Cell Line

The human HNSCC cell line SCC-4 (ATCC: CRL-1624) (45, 46) was used to compare the cytotoxic activities of freshly purified patient and healthy NK cells and served as an internal control for scored intensities of comparable fluorescence staining's from different patient-derived primary HNSCC cells. Therefore, the SCC-4 was cultured in DMEM and GlutaMAX™ medium (GIBCO, Invitrogen, Germany) supplemented with 10% (v/v) heat-inactivated fetal bovine serum (FBS) and 2 mM L-glutamine (PAA Laboratories GmbH, Austria).

Preparation of Single Cell Suspension from Primary Tumor Samples

Tumor samples from untreated HNSCC patients ($n = 5$) were collected post-surgery and washed twice in serum-free DMEM/F12HAM/Glutamax supplemented with 100 U/ml penicillin, 100 µg/ml streptomycin, 0.25 µg/ml amphotericin B (antibiotic–antimycotic 100×, all purchased from Life Technologies, Gibco®, Darmstadt, Germany). After dissociation with 0.05% trypsin/EDTA (Life Technologies, Gibco®, Darmstadt, Germany), tumor pieces were minced with scissors and scalpels in a sterile dish. The digestion was stopped with DMEM/F12HAM/Glutamax containing 10% AB-Serum (former: PAA, Linz, Austria), antibiotic–antimycotic and the sample was passed through a 70 or 100-µm nylon mesh cell strainer (BD Biosciences, Heidelberg, Germany) to achieve single cell suspensions. Cells were collected in a 50-ml conical tube and subsequently centrifuged. Suspended cells were counted with trypan blue, characterized with FCM and cultivated in DMEM/F12HAM/Glutamax/10% AB serum/antibiotic–antimycotic in an incubator (37°C, 5% CO_2, 90% humidity). Cultured tumor cells formed small tumor cell clusters after a few days, which resulted in tightly arranged HNSCC tumor spheroids (diameter: 1–3 mm) after cell cultivation of 1–2 weeks. Tumor cluster and spheroids derived from our HNSCC patients were used for NK cell-based cytotoxicity and tumor-infiltration assays monitored by fluorescence microscopy and time-lapsed transmitted imaging.

Cytokine Analysis

The BD CBA Kit was utilized for scavenging soluble cytokines, especially IFN-γ and TNFα with beads of known size and fluorescence, allowing identification of soluble molecules in blood or supernatants from cell culture medium by FCM as described previously (44).

TABLE 1 | The clinical parameters and immune status of HNSCC ($n = 5$) patients summarized after tumor (TU) and lymph node (LN) surgery but before any acute clinical therapeutic regimens.

Patient characteristics		1	2	3	4	5
Age (years)		24	69	66	76	29
TNM classification		T2N2M1	T3N2M1	T3N2M1	T1N1M0	T2N2M0
Grading		2	2	2	1	2
Treatment/chemotherapy		Surgery/–	Surgery/–	Surgery/–	Surgery/–	Surgery/–
TU/LN locations		Neck/ear	Upper gingiva/lower jaw	Neck/ear	Jaw angle	Tongue
TU/LN material for research		+/–	+/+	+/+	+/+	+/–
Immune status	Leukocytes (cells/µl)	11,000	6250	8000	4900	8200
	Lymphocytes (% of leukocytes)	13.6	19.4	14.6	11.5	43.2
	Monocytes (% of leukocytes)	10.5	4.5	8.9	6.6	6.2
	T cells (% of lymphocytes)	74.7	70.1	82.7	82.2	54.9
	B cells (% of lymphocytes)	11.2	1.9	12.1	6.2	5.0
	NKT cells (% of lymphocytes)	1.1	7.5	1.7	1.3	3.4
	NK cells (% of lymphocytes)	12.8	19.5	2.7	9.3	33.2

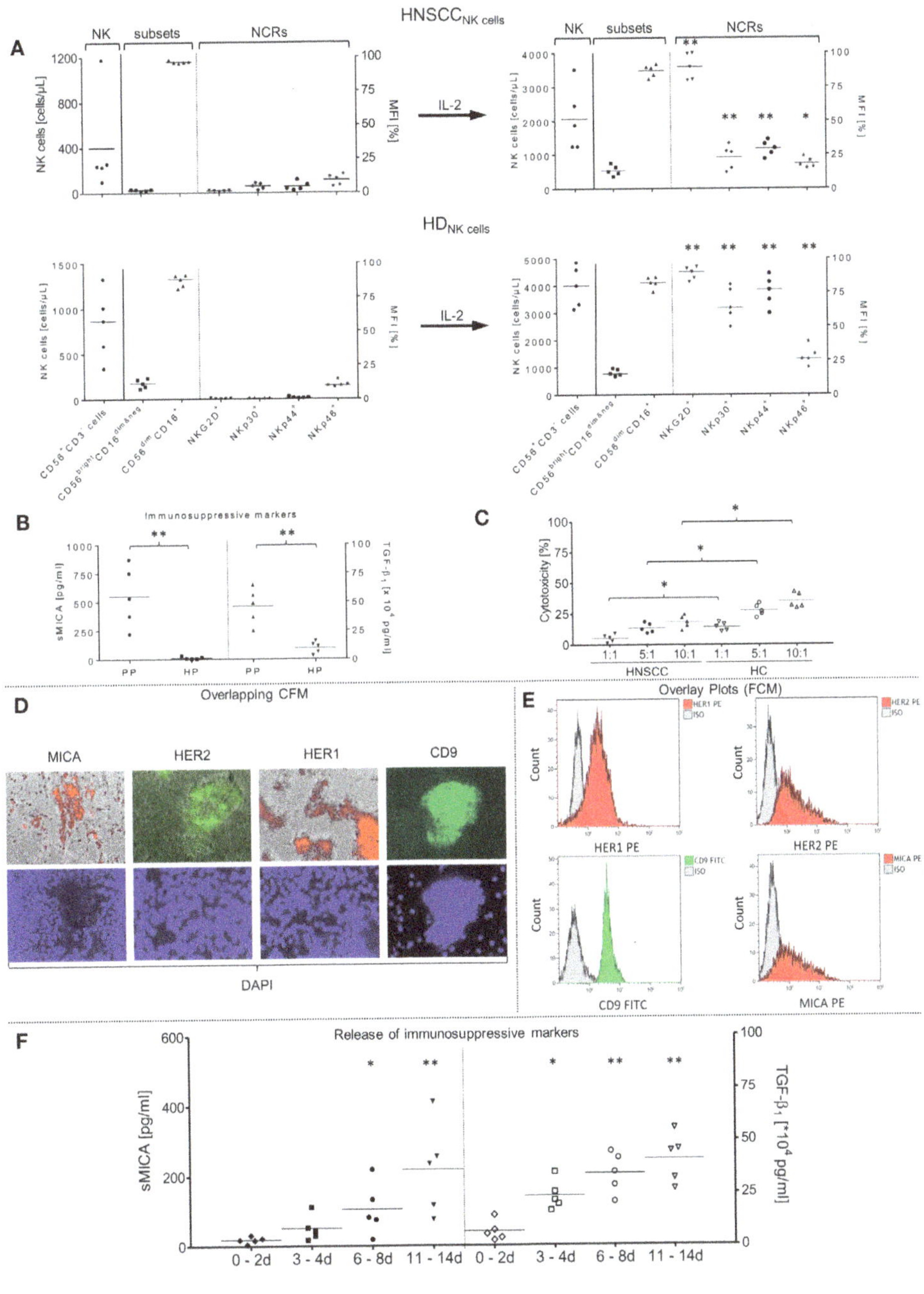

FIGURE 1 | Phenotypical and functional characterizations of patient NK cell subsets and primary tumor cells from non-treated HNSCC patients (*n* = 5) after tumor relapse. (A) *In vitro* rearrangement of the NK cell phenotype was quantified in the PB before separation of NK cells and after IL-2 expansion (1000 IU/ml IL-2; 9–12 days). Shown are the absolute numbers of patient ($HNSCC_{NK\ cells}$) and healthy donor ($HD_{NK\ cells}$) $CD56^{+}/CD3^{-}$ NK cells [cells/μl] [left graph area (NK)], the mean fluorescence intensity [MFI (%)] of distribution of resultant $CD56^{bright}/CD16^{dim\&neg}$ and $CD56^{dim}/CD16^{+}$ NK subpopulations [middle graph area (subsets)] and co-expressed NCRs [MFI (%) right graph area (NCRs)] among total NK cells. **(B)** SMICA and TGF-β_1 levels were analyzed in blood plasma from corresponding HNSCC patients (PP) and compared to age-matched healthy donor plasma controls (HP). **(C)** Assessment of the basic killing activities between effector cells isolated from patient and healthy donor NK cells against SCC-4 target cells. Freshly isolated, non-stimulated NK cells from patients (HNSCC), and healthy controls (HC) were treated with corresponding HNSCC patient plasma (high sMICA/TGF-β_1) or associated healthy control plasma (low sMICA/TGF-β_1) and co-incubated for 4 h (37°C, 5% CO_2, 250 rpm) with SCC-4 cells at the indicated E:T ratios and cytotoxicity (%) was measured by FCM. **(D,E)** Immunofluorescence staining and FCM-based characterization of relevant tumor antigen expression from primary tumor samples derived from corresponding HNSCC patients (*n* = 5). After preparation of single cell suspensions from primary tumor samples, tumor cells were cultured (1–2 days, 37°C, 5% CO_2) on chamber slides and characterized phenotypically for CD9, MICA, HER1, and HER2 expression profiles by immunofluorescence microscopy. Depicted is the staining for one representative HNSCC sample. **(D)** The HNSCC cells were also analyzed for CD9 (FITC), MICA (PE), HER1 (PE), and HER2 (PE) surface expression levels by FCM **(E)**. Nuclei were stained with DAPI (4′,6-diamidino-2-phenylindole, blue fluorescence). **(F)** Secretion of soluble immunosuppressive factors derived from five tumor samples. SMICA and TGF-β_1 levels in expanded primary HNSCC cell cultures were analyzed by ELISAs in the collected cell medium supernatants at the indicated time frames. **(A–F)** Data are shown as mean ± SD from two to four experiments for the five patients. Range of statistically significant differences: from $^{*}p \leq 0.01$ to $^{**}p \leq 0.001$.

Quantification of sMICA and TGF-β_1 in HNSCC Patients

The *BAMOMAB* MICA-Sandwich ELISA kit for sMICA (AXXORA GmbH, Germany) was designed for quantification of soluble MICA (sMICA). The kit was utilized for detection and monitoring of immunosuppressive molecules in HNSCC patient blood plasma ($n = 5$), HCs ($n = 5$), and supernatants of cell culture medium during adherent growth phase of tumor cell clusters and generation of tumor spheroids as described previously (47). TGF-β_1 levels in human blood plasma samples and cell culture supernatants were quantified by MTPL ELISA (Milenia Biotec, Version 3.0, Germany).

Immunomagnetic Separation of CD56$^+$CD3$^-$ NK Cells

Up to 100 ml heparinized blood from HNSCC patients ($n = 5$) was used to isolate viable mononuclear cells (MNC) in high yield and purity by Ficoll-Paque density gradient. Primary NK cells were separated from purified MNCs via "non-touched" depletion using the EasySep® Human NK Cell Enrichment Kit (STEMCELL Technologies SARL, Germany). Other leukocyte subsets were labeled with tetrameric antibody complexes against CD3, CD4, CD14, CD19, CD20, CD36, CD66b, CD123, HLA-DR, glycophorin A, and dextran-coated magnetic particles. The non-labeled cells were scavenged by an EasySep® hand magnet according to the manufacturer's recommendations. Freshly purified NK cells (purity: 95.1 ± 2.8%) were expanded and activated with 1000 IU/ml IL-2 for 9–12 days as described previously (47).

Production of Shed MICA

To generate shMICA, the DNA sequence encoding for full length MICA engineered with an N-terminal histidine-tag was cloned into a tet-on vector system and transfected into the UKF-NB3 tumor cell line using the Neon transfection system (Life Technologies, USA). Cells were selected with 0.25 mg/ml G418 and 1 µg/ml puromycin (Life Technologies, USA). MICA expression was induced by addition of 2 µg/ml doxycycline (Sigma, Germany). Further on cells were stressed by serum starvation for 72 h to induce MICA shedding. Finally shMICA was purified with Protino Ni-NTA agarose (Macherey-Nagel, Germany) from cell culture supernatants and concentrated using Amicon centrifugal filter units (Merckmillipore, Germany).

Cytotoxicity Assays

To analyze the NK cell-mediated killing activity in presence and absence of cetuximab, we developed a matched effector-target cell system based on a FCM-based cytotoxicity assay. Therefore, we utilized only concordant patient NK cells, patient plasma (PP, high sMICA) and primary patient HNSCC tumor cells ($n = 5$). To demonstrate the sMICA-mediated inhibition, IL-2 expanded (9–12 days) primary patient NK cells were co-incubated overnight (24 h, 37°C, 5% CO_2, 250 rpm) with 500 pg/ml shMICA, PP containing high MICA levels (PP, range: 220.9–870.7 pg/ml) and, as a comparative control, with HC plasma (HP, range: 2.8–22.0 pg/ml) diluted 1:2 in X-VIVO™10 medium (Biowhittacker™Cambex Bioscience, Belgium). Phenotypical cell characterizations were accomplished to detect altered expression patterns of CD16 and NKG2D on treated NK cells (**Figures 2** and **3**). Singularized HNSCC cells were also tested for MICA, HER2, and HER1 surface expression, the latter as a target for restored ADCC before initiation of described cytotoxicity assays. To determine the effect of cetuximab on NK cell-dependent killing activity, various cetuximab doses (Cetmab: 1 pg/ml–1 µg/ml) were used to coat the corresponding HNSCC target cells (E:T ratio: 10:1). To inhibit putative effects of cetuximab toward NK cell-mediated cytotoxicity, the NK cells were pre-incubated for 20 min with anti-CD16 mAb (20 µg/ml). The effector-based cytotoxicity of these treated NK cell samples were analyzed against corresponding patient HNSCC cells in the indicated effector-to-target ratios (**Figures 3** and **4**). To avoid effector and target cell sedimentations or insufficient stirring of our co-incubated approaches during the cytotoxic reactions, the co-cultured cell suspensions were shaken in an CO_2-incubator (CO2cell, 170-400 Plus, RS Biotech, Scotland) for 4 h (37°C, 5% CO_2, 250 rpm). An optimized gating panel (Figure S1 in Supplementary Material) based on a no-wash single platform FCM procedure (FC500, Beckman Coulter, Germany) was applied. Treated NK cells were stained with several monoclonal antibodies (mAbs): CD45 FITC (fluorescein isothiocyanate), CD56 PE (phycoerythrin) or NKG2D PE, CD16 PC-7 (phycoerythrin-cyanin-7) in order to exclude the effector cells from primary HNSCC target cells stained with CD9 FITC, CD9 PE, HER1 PE, MICA PE, or CD81 PE. Effector and target cells were stained with mAbs as described previously (47, 48). Target cell elimination by effector cells was calculated as the total loss of viable HNSCC target cells as follows (48, 49):

$$\text{Killing activity} = (1 - \text{concentration}_{\text{co-cultured HNSCC cells/}\mu\text{L}} / \text{concentration}_{\text{HNSCC control cells/}\mu\text{L}}) \times 100\%$$

Fluorescence Microscopy

IL-2 activated patient NK cells were cultured (24 h) with PP [(high sMICA); 1:2 diluted with X-VIVO™10] or healthy plasma [HP (low sMICA); 1:2 diluted with X-VIVO™10]. Afterward, these treated NK cells were co-incubated (1 h, 37°C, 5% CO_2, approximate E:T ratios of 5:1) on 4-well chamber slides (1.7 cm^2 growth area/well, 0.5–1.0 ml working volume, Nunc™, USA) with corresponding adherent HNSCC cell clusters derived from HNSCC patients ($n = 5$, time of cultivation: 3–12 days) in presence or absence of 1 µg/ml cetuximab for indicated periods of time (**Figure 5A**). The capability of tumor infiltration from those treated patients NK cells (see above) in corresponding patient HNSCC tumor spheroids ($n = 5$) were assessed with and without 1 µg/ml cetuximab and analyzed after 24 h co-cultivation (37°C, 5% CO_2) (**Figure 6**). 2D/3D confocal fluorescence microscopy (CFM) using FITC- and PE-conjugated mAbs was used to separate stained patient NK cells (CD45 FITC or NKG2D PE) and corresponding HNSCC cell clusters or tumor spheroids (CD9 FITC, HER1 PE, HER2 PE, or MICA PE), as described before (47).

Time-Lapse Microscopy

Transmission microscopy was used to monitor infiltration of sMICA-inhibited NK cells into corresponding HNSCC tumor spheroids. Low numbers of tumor spheroids were grown scattered on chamber slides for 16 h (37°C, 5% CO_2). Subsequently,

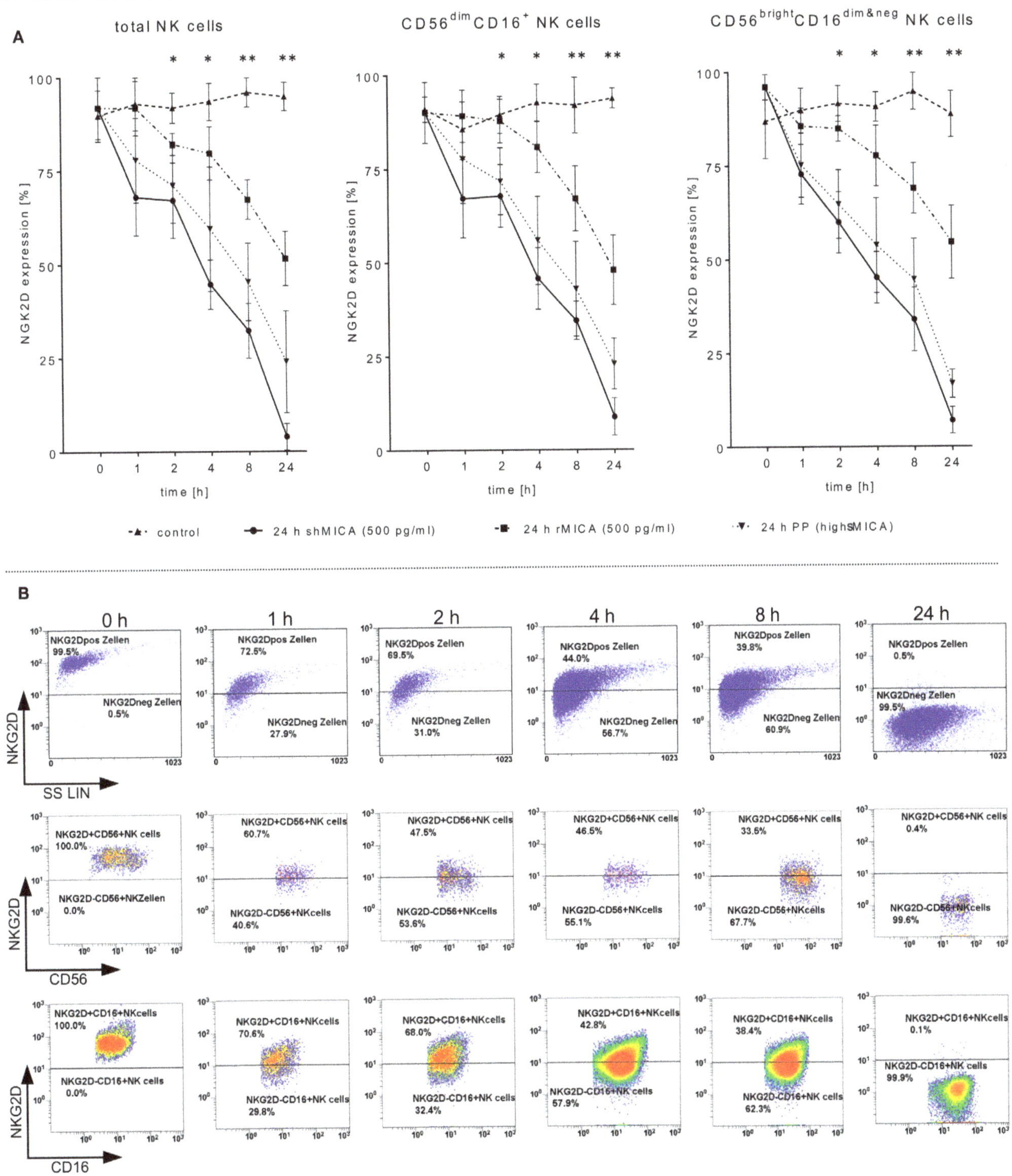

FIGURE 2 | Impact of sMICA on NKG2D surface expression and NKG2D-mediated NK cell cytotoxicity against HNSCC cells. Isolated CD56+/CD3− NK cells from untreated HNSCC patients ($n = 5$) were stimulated for 9–12 days with 1000 IU/ml IL-2. **(A)** Effect of high sMICA levels on NKG2D surface expression. IL-2 activated patient NK cells were incubated (37°C, 5% CO_2) at the indicated points of time with shMICA (500 pg/ml), rMICA (500 pg/ml) and patient plasma containing high MICA levels (PP, range: 220.9–870.7 pg/ml). The NKG2D expression levels on total NK cells and both $CD56^{dim}/CD16^{+}$ and $CD56^{bright}/CD16^{dim\&neg}$ NK subsets were compared to non-treated control NK cells. **(B)** Phenotypical analyses on time-dependent impact of shMICA on NKG2D surface expression were determined by FCM. Exemplarily depicted here are representative dot plots for one time-dependent experiment.

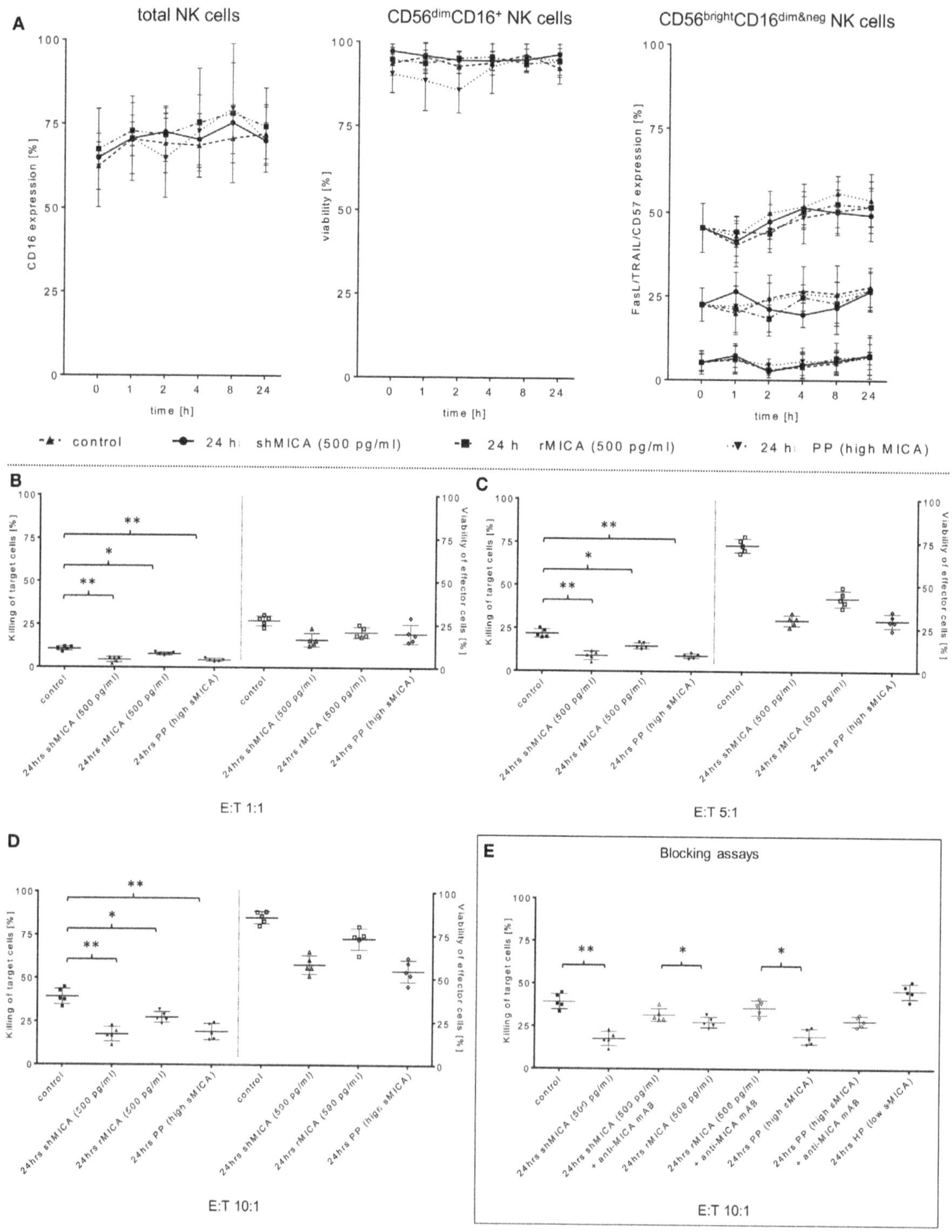

FIGURE 3 | Determination of the sMICA impact on ADCC-stimulating and pro-apoptotic NK cell receptors. (A) Surface expression levels of CD16, TRAIL, FasL, and CD57 on the same effector cell samples were measured at the indicated time points and compared to non-treated control cells. As an additional control, effector cell viability was monitored over a 24-h period **(B–D)**. The impact of sMICA on the NKG2D-mediated NK cell cytotoxicity and NK cell stability during effector-target interactions against primary HNSCC cells ($n = 5$) was analyzed by FCM. Therefore, the same overnight-treated NK cell samples (see above, section **Figure 2A**) were subsequently co-incubated for 4 h (37°C, 5% CO_2) with corresponding patient HNSCC target cells at the indicated E:T ratios. **(E)** Inhibition of sMICA effects on NKG2D-mediated NK cell cytotoxicity. SMICA molecules in all treatment samples (shMICA, rMICA, and PP) were blocked by pre-incubation (20 min) with MICA-specific mABs (20 μg/ml, MAB13001). NKG2D-dependent killing rates of those incubated NK cells against primary HNSCC cells were measured after 4 h ("Blocking assays," ratio: 10:1, 37°C, 5% CO_2) as described above. Data are shown as mean ± SD from three to four experiments for each patient. Statistically significant difference: $^*p \leq 0.01$ and $^{**}p \leq 0.001$.

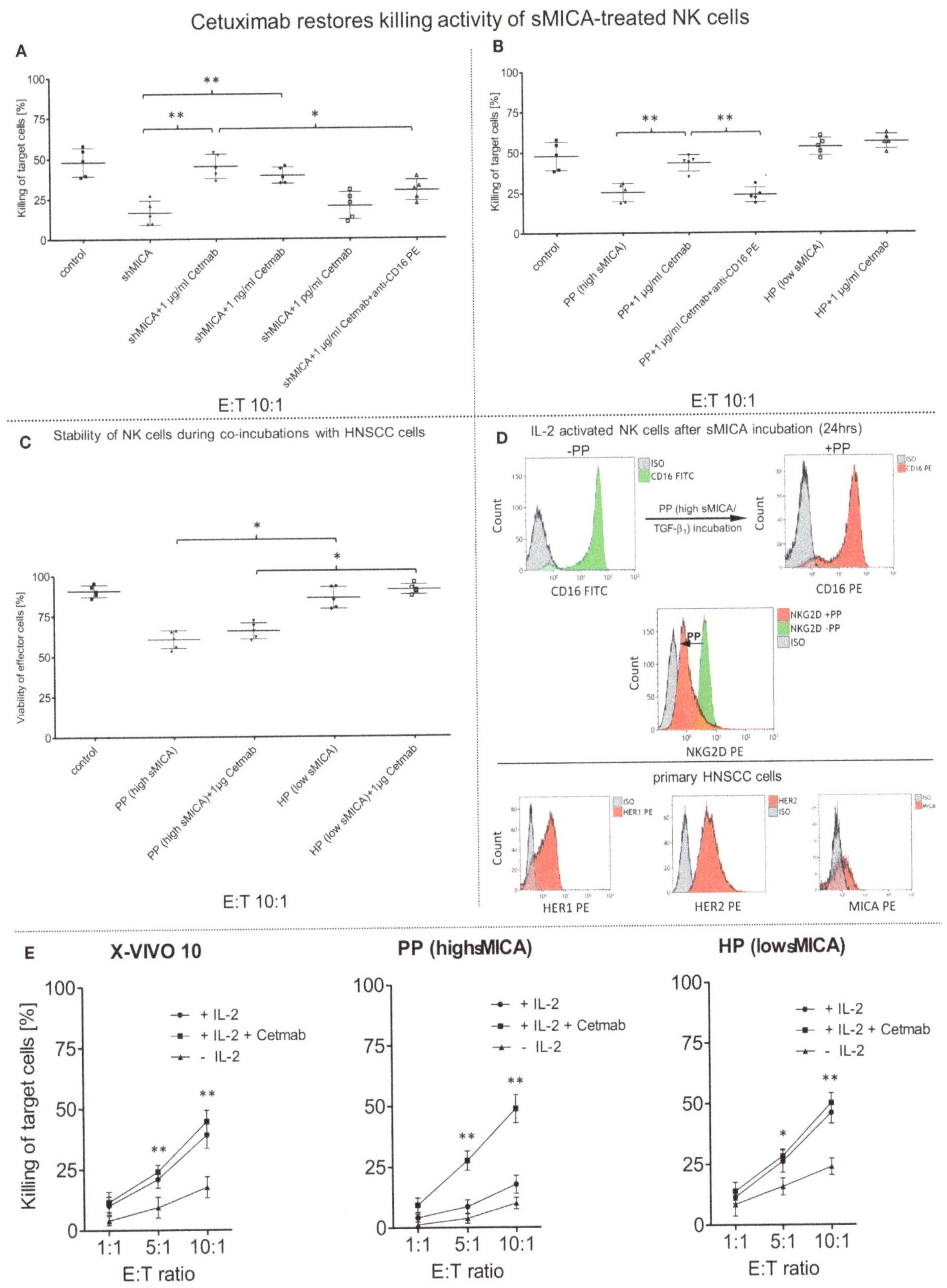

FIGURE 4 | Restored killing activity against primary HNSCC cells mediated by cetuximab.

(Continued)

FIGURE 4 | Continued
IL-2 activated patient NK cells (n = 5) were incubated for 24 h (37°C, 5% CO_2) with **(A)** shed MICA (shMICA, 500 pg/ml), **(B)** with patient plasma containing high MICA levels (PP, n = 5, range: 220.9–870.7 pg/ml) and healthy plasma (HP, n = 5, range: 1.9–28.5 pg/ml) served as a control. **(A)** CD16-mediated ADCC of those treated patient NK cells was assessed by FCM-based cytotoxicity assay using cetuximab (Cetmab, range of dose titration: 1 µg/ml–1 pg/ml)-coated primary HNSCC cells and compared to killing activity of shMICA-treated NK cells against non-coated HNSCC cells. **(B)** To overcome sMICA effects against NK cell cytotoxicity, 1 µg/ml cetuximab was chosen based on the earlier titration experiments (see above). Inhibition of cetuximab-dependent ADCC was achieved by blocking CD16 epitopes on treated patient NK cells with anti-CD16 mAb [20 µg/ml, 20 min pre-incubation, graphs **(A,B)**]. **(C)** The effector cell stability of PP- and HP-treated viable NK cells during tumor cell lysis was analyzed in presence and absence of 1 µg/ml cetuximab and compared to non-treated effector cell controls quantified by single platform functionality assays. **(D)** FCM-based characterization was utilized to monitor CD16 and NKG2D surface expression patterns on treated NK cells (CD16/NKG2D, see overlay histograms) before (−PP) and after (+PP) overnight incubation with patient plasma (PP) containing high sMICA and TGF-β_1 levels. Analogously, HNSCC cell clusters were singularized for FCM-based characterization of target cell antigens (HER1/HER2/MICA) to assess alterations in the killing activity from different treated patient NK cells. **(E)** Effect of IL-2 on the killing activity of patient NK cells (n = 5) in presence (+IL-2/+Cetmab) and absence (+IL-2) of 1 µg/ml cetuximab (Cetmab). IL-2 expanded NK cells were incubated with patient plasma containing high MICA levels (PP, n = 5, range: 220.9–870.7 pg/ml), healthy plasma (HP, n = 5, range: 1.9–28.5 pg/ml) and control medium. Cytotoxicity of treated NK cells was analyzed at the indicated ratios with and without Cetmab (1 µg/ml)-coated primary patient HNSCC cells and compared to killing activity of unstimulated NK cells against non-coated HNSCC cells (−IL-2). **(A–D)** Data represent the mean ± SD of three experiments for each patient. Statistically significant difference: $^*p \leq 0.01$ and $^{**}p \leq 0.001$.

HNSCC spheroids were co-incubated with freshly isolated, non-stimulated patient NK cells (1 × 10^6 effector cells/ml). Putative effector cell migration and tumor chemoattraction were monitored by time-lapse microscopy and imaging (Video S1 in Supplementary Material) as described previously (47).

Blocking Assays

To inhibit the restored NK cell-based cytotoxicity by cetuximab, patient NK cells were pre-incubated for 20 min with anti-CD16 mAb (clone 3G8, 20 µg/ml, A07766, Beckman Coulter, Germany). Afterward, killing activity of treated NK cells was analyzed with cetuximab-coated primary patient HNSCC cells over a time period of 4 h (37°C, 5% CO_2, 250 rpm) at an E:T ratio of 10:1 and compared to corresponding controls (**Figure 4**) as described previously (47). To analyze the direct role of sMICA on NKG2D-mediated cytotoxicity, we incubated IL-2-activated patient NK cells with shMICA, rMICA, and PP containing high sMICA levels overnight in the presence or absence of specific MICA antibodies, respectively (20 µg/ml, MAB13001, R&D systems, Germany). Additionally, plasma from five healthy donors (low sMICA, see above) was utilized to incubate patient NK cells as a control.

Statistical Analyses

The Mann–Whitney non-parametric U-test was utilized to compare clinical and pathological parameters of plasma sMICA and TGF-β_1 levels from HNSCC patients (n = 5) with healthy individuals (n = 5). The Student's t test was used to assess the significance of the killing activity of patient NK cells incubated under various conditions. A p level ≥0.01 was considered statistically as non-significant. Unless otherwise declared, results of statistical evaluations from functional assays are indicated as mean ± SD and represent three to four experiments for each patient.

RESULTS

Characterization of Altered NK Cell Subsets and Expression of NCRs in HNSCC Patients

Compared to age-matched healthy individuals (50), HNSCC patients showed a broad range of leukocyte subpopulations and absolute numbers of lymphocytes and leukocytes (**Table 1**). Although median NK cell amounts (12.8%; range: 2.7–33.2%) did not differ from HCs (**Table 1**), the absolute NK cell numbers (cells/µl) differed widely in the peripheral blood (PB) of patients and healthy donors (left graph sector, **Figure 1A**). Moreover, the proportion of immunoregulatory NK cells ($CD56^{bright}$/$CD16^{dim\&neg}$) was markedly reduced in all patients [median: 2.4% ($HNSCC_{NK\ cells}$) versus 11.8% in healthy donors ($HD_{NK\ cells}$), middle graph sectors, **Figure 1A**]. In contrast, the cytotoxic NK cell subpopulation ($CD56^{dim}$/$CD16^{+}$) was strongly increased for all investigated HNSCC patients [median ($HNSCC_{NK\ cells}$): 96.2% versus 86.8% ($HD_{NK\ cells}$), middle graph sector, **Figure 1A**]. Moreover, freshly isolated patient NK cells revealed low to moderate expression levels of the NCRs, NKp30, NKp44, NKp46, and NKG2D compared to higher frequencies of IL-2 stimulated NK cells from HCs (right graph sector, **Figure 1A**). Nevertheless, the percentage of NK cells expressing NCRs increased (~4.7-fold, 3.8-fold, and 2-fold for NKp30, NKp44, and NKp46, respectively) during IL-2 activation over 9–12 days and was accompanied by ~60.7-fold higher expression levels of NKG2D (**Figure 1A**, right graphs) for all stimulated patient NK cells. The distribution of NK cell subpopulations shifted to higher $CD56^{bright}$/$CD16^{dim\&neg}$ NK cell subsets (median before IL-2: 2.4% versus median after IL-2: 12.5%) and consequently lower percentages of $CD56^{dim}$/$CD16^{+}$ NK cells (median before IL-2: 96.2% versus median after IL-2: 88.9%) (**Figure 1A**).

Reduced NK Cell-Dependent Cytotoxicity and Increased Immunosuppressive Factors in HNSCC Patients

The well-defined immunosuppressive factors sMICA and TGF-β_1, which are responsible for impaired immunosurveillance, were quantified in PB from our HNSCC patients. Higher levels of both soluble factors were detected in HNSCC patients compared to HCs (sMICA, median: 532.8 versus 5.9 pg/ml; TGF-β_1, median: 48.9 × 10^4 versus 10.9 × 10^4 pg/ml, respectively) (**Figure 1B**). Healthy plasma samples showed sMICA and TGF-β_1 levels close to the detection limits of this assay [sMICA (mean ± SD): 10.8 ± 11.2 pg/ml; TGF-β_1: 9.5 ± 5.2 × 10^4 pg/ml; **Figure 1B**, HP], whereas sMICA (TGF-β_1) in PP ranged between 220.9 and

TABLE 2 | Antigen expression levels measured on singularized patient HNSCC cells (n = 5) after preparation from tumors.

	SCC-4	Patient 1	Patient 2	Patient 3	Patient 4	Patient 5
HER1	+++	++	+++	++	+++	++
HER2	+++	+/−	+/−	+	+	+
MICA	++	+/−	+	+/−	+	+
CD9	+++	++	++	++	+	++

Fluorescence labeling of these tumor cells were compared to the SCC-4 cell line and intensities were scored as follows: weak (+/−), moderate (+), good (++), and strong (+++).

870.7 pg/ml (25–64.8 × 10^4 pg/ml) (**Figure 1B**, PP). To compare the basic killing activity between patient NK cells and NK cells from healthy individuals, freshly isolated, non-stimulated NK cells from both, patients and HCs were co-incubated overnight (37°C, 5% CO_2) with corresponding HNSCC PP (high sMICA/TGF-β_1) or associated HC plasma (low sMICA/TGF-β_1). In both cases, the NK cell-mediated cytotoxicity was analyzed against the target cell line SCC-4 at the indicated ratios (4 h, 37°C, 5% CO_2, 250 rpm) by FCM (**Figure 1C**). HNSCC plasma-treated patient NK cells (HNSCC, **Figure 1C**) exhibited significantly reduced cytotoxicity in all prepared E:T ratios when compared to NK cells from healthy donors pre-incubated with the corresponding HC plasma (**Figure 1C**).

Phenotypical Analysis of Tumor-Relevant Expression Markers on Primary Patient HNSCC Cells

Expression levels of relevant surface antigens, especially HER1 and HER2, were examined on single tumor cell suspensions prepared from primary tumors surgically removed from HNSCC patients (n = 5, see above). Pre-characterized SCC-4 cells were utilized as a positive control for HER1 and HER2 expression because it was previously shown that these HNSCC cells displayed high amplification rates of HER1 genes in combination with enhanced HER2 and MICA expression levels (51). In addition to the HER1/2 levels, we analyzed expression levels of surface markers CD9 and MICA on the mono-dispersed patient tumor cells (**Figure 1E**). Representative overlay plots (FCM) were shown exemplarily in **Figure 1E** for prepared tumor cells derived from one HNSCC patient and were compared with qualitative evaluations determined by CFM. In **Figure 1D**, representative results of overlapping CFM exemplarily illustrated for cultured (1–2 days) adherent patient tumor cells after fluorescence labeling with similar stimulation energy and duration are shown. For counterstaining, we utilized DAPI as a fluorescent stain that binds strongly to A–T rich DNA regions of nuclei and chromosomes and emits blue fluorescence (**Figure 1D**). All tumor samples showed higher HER1 and CD9 levels in contrast to low-to-moderate expression of membrane-associated MICA and HER2 molecules. In summary, the antigens exhibited variable expression levels on all tumor samples of HNSCC patients and were slightly lower than on the SCC-4 control line as presented in **Table 2**.

Primary patient HNSCC cells (n = 5) were cultured over 2 weeks (37°C, 5% CO_2) to investigate time-dependent accumulation of immunosuppressive factors (sMICA and TGF-β_1), and cell culture supernatants were collected at the indicated time periods (**Figure 1F**). A time-dependent increase of both immunosuppressive molecules was identified in cell culture supernatants in which the mean of sMICA levels increased from 18.7 ± 9.5 to 218.7 ± 131.9 pg/ml, and the mean of TGF-β_1 levels increased from 6.8 ± 4.8 × 10^4 to 40.9 ± 11.6 × 10^4 pg/ml between the first and last time points monitored (**Figure 1F**).

SMICA Affects NKG2D Expression and Cytotoxicity of IL-2-Activated Patient NK Cells

Tumor-derived TGF-β_1 potentiated the sMICA-mediated down-regulation of NKG2D surface expression on various effector cells, especially NK cells, and resulted in decreased NKG2D-dependent immunity, thus reflecting the predominant role of the sMICA-NKG2D system (20, 52). Therefore, we co-cultured IL-2-activated patient NK cells overnight (24 h) with sMICA analogs (shMICA and rMICA, each in 500 pg/ml) and PP containing high sMICA levels (range: 220.9–870.7 pg/ml). All incubated NK cell samples from the five HNSCC patients exhibited a time-dependent down-regulation of NKG2D expression on total NK cells and both NK cell subsets (**Figures 2A,B**), whereas surface expression of CD16, pro-apoptotic FasL, and TRAIL receptors as well as activation marker CD57 were largely unaffected and stable over the indicated time period on all NK cell fractions by the described sMICA analogs (**Figure 3A**). We detected a distinct effect of shMICA on NKG2D expression on total NK cells and resultant NK subpopulations analyzed during the indicated time frame of 24 h (37°C, 5% CO_2) by FCM. The time-dependent NKG2D down-regulation is displayed in plots exemplarily shown for one experiment (**Figure 2B**). Although surface phenotype, cell proportions, and viability of NK cells were not affected by co-incubation with PP (24 h, high MICA, **Figure 3A**), the observed decrease of NKG2D surface expression was accompanied by reduced IFN-γ and TNFα secretion compared to IL-2 activated patient NK cells [IL-2 (9–12 days), **Figure 5D**].

Additionally, the impact of sMICA on the NKG2D-mediated NK cell cytotoxicity of overnight co-incubated NK cells was analyzed against corresponding primary tumor cells derived from HNSCC patients (n = 5) to assess the degree of killing activity. Effector and target cells were co-incubated for 4 h (37°C, 5% CO_2, 250 rpm) at indicated E:T ratios and killing activity was subsequently compared to X-VIVO™10-incubated controls without sMICA analogs. The cytotoxicity of shMICA- and PP-treated NK cells was strongly suppressed compared to low inhibition frequencies of rMICA-incubated effector cells at all E:T ratios (**Figures 3B–E**). During the cytotoxicity assays of co-cultured effector and target cells, the NK cell viability was decreased in shMICA-, rMICA-, and PP-treated NK cells as compared to

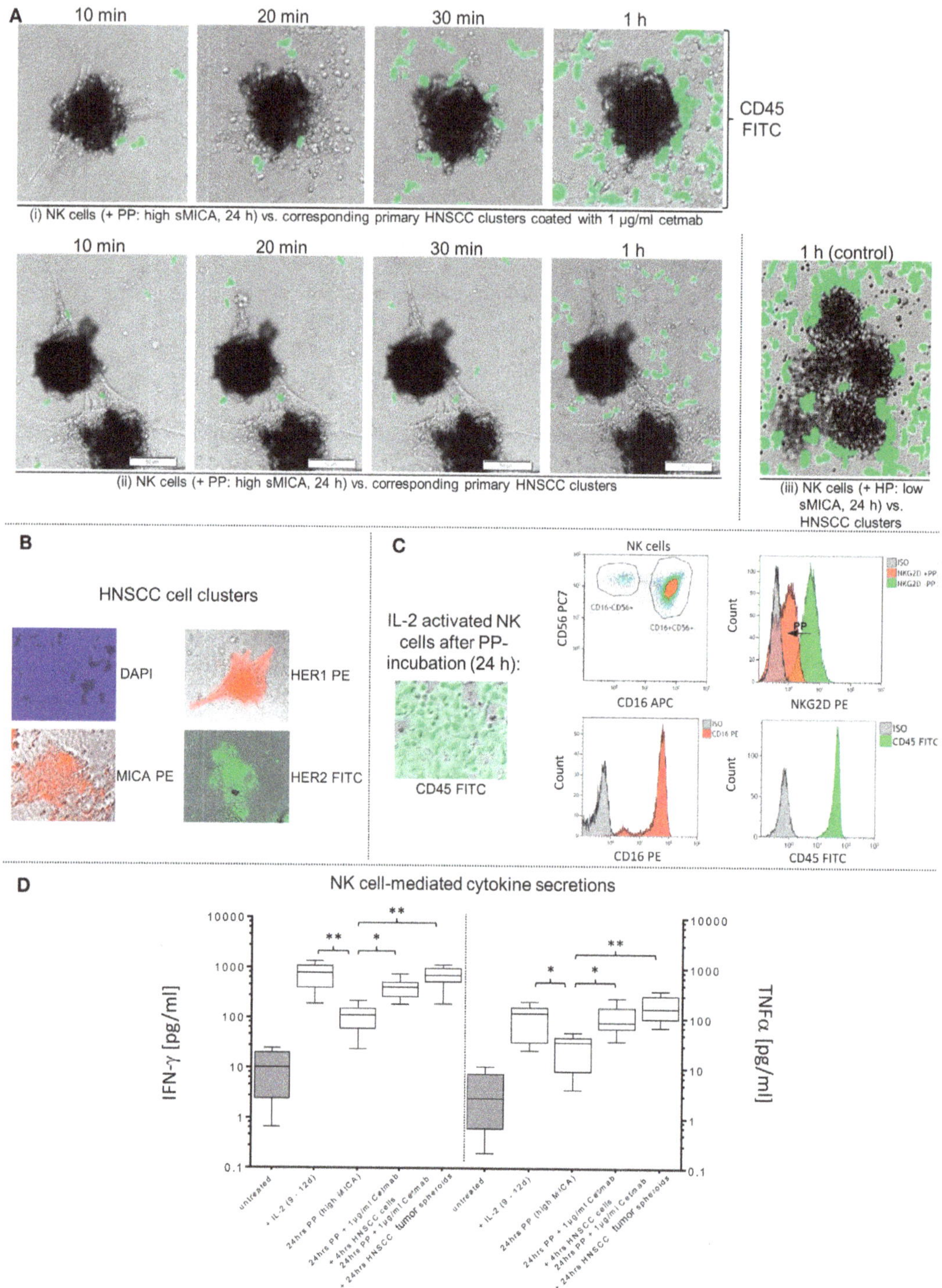

FIGURE 5 | Restored effector-target cell interactions via cetuximab-dependent ADCC of sMICA-inhibited NK cells. (A) Activated patient NK cells were cultured (24 h) with patient plasma (PP; 1:2 diluted with X-VIVO™10) containing high sMICA levels and co-incubated (1 h, approximate E:T ratios of 5:1) with corresponding HNSCC cell clusters (time of cultivation: 3–12 days) in (i) presence and (ii) absence of 1 µg/ml cetuximab. (iii) Activated patient NK cells treated for 24 h with healthy plasma (HP [low sMICA]; 1:2 diluted with X-VIVO™10) were used to monitor early effector-target cell interactions, served as a positive control. CD45 (green) for PP- and HP-treated patient NK cells were analyzed by staining with FITC-conjugated mAb, whereas co-incubated adherent HNSCC clusters were unlabeled. Before initiating these functional assays, effector and target cells were tested flow cytometrically for relevant surface markers, especially MICA, HER2 and HER1 on HNSCC cells **(B)**, and CD16 and NKG2D levels on PP (high sMICA/TGF-β_1)-treated NK cells **(C)**. Variable IFN-γ and TNFα secretions levels during co-incubations of treated NK cells with HNSCC cell clusters or tumor spheroids in presence or absence of cetuximab were quantified and compared with untreated controls **(D)**. Statistically significant difference: $^*p \leq 0.01$ and $^{**}p \leq 0.001$.

more stable effector cells in X-VIVO™10-incubated controls (**Figures 3B–E**). To confirm the direct sMICA impact on the NKG2D-mediated NK cell cytotoxicity, MICA-specific mAbs (20 μg/ml, MAB13001) were used to deplete sMICA molecules in different treatment mixtures (shMICA, rMICA, and PP) (**Figure 3E**). NK cell cytotoxicity of these treated effector cells co-incubated with corresponding primary HNSCC cells were compared with sMICA analog-treated NK cells co-cultured without sMICA-specific mAbs as a negative control. The diminished killing activity of sMICA-affected NK cells was partially restored when compared to untreated (X-VIVO™10) control effector cells and to NK cells pre-treated with healthy plasma (HP, low sMICA) (**Figure 3E**).

Cetuximab Restores NK Cell-Dependent Cytotoxicity Against Primary HNSCC Cells via ADCC

Cetuximab is a therapeutic mAb directed against the HER1 epitopes on several types of high-malignant tumors (53). Consequently, cetuximab is a powerful stimulus of NK cell-mediated ADCC via activation of FcγRIIIa against cetuximab-coated tumor cells and for induction of cytokine release, especially IFN-γ and TNFα secretion. **Figure 4A** demonstrates significantly restored NK cell-based cytotoxicity against cetuximab-coated HNSCC cells with mAb concentrations of 1 μg/ml (mean: 43.8 ± 10.8%) and 1 ng/ml (mean: 38.4 ± 6.8%) as compared to reduced killing activity of shMICA-incubated NK cells (mean: 13.4 ± 11.2%) against non-coated HNSCC cells and anti-CD16 mAb-blocked NK cells (mean: 28.8 ± 8.9%). Based on these titration experiments, only 1 μg/ml cetuximab were applied in subsequent experiments to restore the decreased NK cell cytotoxicity (mean: 41.5 ± 6.3%) after overnight incubation with PP (high sMICA, mean of cytotoxicity: 26.1 ± 6.3%) and blocking experiments with anti-CD16 mAb-treated NK cells (mean of cytotoxicity: 24.8 ± 5.7%) (**Figure 4B**). Untreated (X-VIVO™10) and healthy plasma (HP, low sMICA)-incubated patient NK cells were defined as unaffected control cells. NK cell viability was affected exclusively by PP-treatment and HP-incubated NK cells exhibited similar viability levels as untreated effector cells (**Figure 4C**). As internal effector cell controls, we assessed the activation marker NKG2D and IgG Fc receptor (CD16) of PP-treated (high sMICA) NK cells after 24 h overnight incubation by FCM. As shown in **Figure 4D** (upper row), the expression of NKG2D was significantly decreased but the CD16 expression levels were not changed on these effector cells. Otherwise, the phenotypical characterizations of target cell parameters showed higher levels of HER1 and HER2 antigens and only low-to-moderate MICA expressions on corresponding patient HNSCC cells (**Figure 4D**, lower row).

To determine whether IL-2 potentiates the cetuximab-mediated outcome on restored effector cell cytotoxicity against HNSCC cells, we analyzed the degree of ADCC from X-VIVO™10-, PP-, and HP-treated patient NK cells against cetuximab-coated and non-labeled HNSCC cells in the presence or absence of IL-2 (**Figure 4E**). Reconstituted NK cell cytotoxicity via ADCC was detected against cetuximab-coated HNSCC cells (IL-2 and Cetmab) compared to decreased cytotoxicity of PP-incubated NK cells against non-labeled HNSCC cells independently from the presence (+IL-2) or absence of IL-2 (**Figure 4E**, middle graph). Combination of IL-2 with cetuximab-coated or non-coated HNSCC cells (IL-2 with Cetmab or IL-2 alone) revealed no improvement of ADCC from X-VIVO™10- and HP-treated NK cells, but inclusion of IL-2 significantly enhanced NK cell cytotoxicity compared to untreated NK cells (-IL-2) (**Figure 4E**, left/right graphs). The cetuximab-mediated reconstitution of NK cell cytotoxicity correlated with higher IFN-γ and TNFα secretion levels (24 h PP + 1 μg/ml Cetmab + 4 h HNSCC cells) compared to PP-treated NK cells [24 h PP (high MICA) **Figure 5D**].

Cetuximab Reconstitutes NK Cell Infiltration into HNSCC Clusters and Tumor Spheroids

To assess the capability of tumor infiltrations from sMICA-affected patient NK cells in absence or presence of 1 μg/ml cetuximab, we developed different *in vitro* models by establishing co-cultures (37°C, 5% CO_2). Therefore, activated patient NK cells were pre-incubated overnight with corresponding PP (high sMICA) or healthy plasma (HP, low sMICA) served as a positive control. Afterward, the early recognition and tumor-infiltration capabilities of these treated effector cells were monitored after 1 h co-incubation with primary HNSCC cell clusters (**Figure 5**) or after 24 h in HNSCC tumor spheroids derived from five HNSCC patients to identify specific tumor-infiltrated NK cells (**Figure 6**).

Early "effector-to-target" affinities of treated NK cells against adherent HNSCC cell clusters coated or not coated with 1 μg/ml cetuximab were monitored (10 min–1 h) by CFM. The experiments showed clearly impaired and disordered "effector-to-target" interactions and decreased HNSCC cell cluster infiltrations from PP-treated NK cells in absence of cetuximab (**Figure 5A**, one row) compared to normal tumor infiltration capabilities of HP-treated (low sMICA) NK cells, served as a positive control [**Figure 5A**, lower row, right picture (control)]. In contrast, early effector cell-dependent infiltration within 20–30 min and restored anti-tumor reaction was observed by PP-treated NK cells against cetuximab-coated HNSCC clusters after 1 h (**Figure 5A**, upper row) accompanied by raised IFN-γ and TNFα medium levels (24 h PP + 1 μg/ml Cetmab + 4 h HNSCC cells, **Figure 5D**). Otherwise, the cytokine secretions of PP-inhibited NK cells against non-coated HNSCC cell clusters (without cetuximab) revealed significantly degraded IFN-γ and TNFα concentrations analyzed in medium supernatants [24 h PP (high sMICA), **Figure 5D**].

In addition, phenotypical analysis of relevant effector and target cell parameters were analyzed before co-cultivations. HNSCC cells revealed high HER1 and CD9 expression levels (not shown) and low-to-moderate MICA levels on the target cell clusters (**Figure 5B**). Otherwise, treated NK cells with corresponding PP (high sMICA, 24 h) showed a decreased NKG2D surface expression as detected by FCM analysis, while CD16 surface expression was unaffected (**Figure 5C**). In this context, sMICA and TGF-β_1 accumulated time-dependently in the cell culture medium as summarized in **Figure 1F** analyzed for different time periods (0–2 up to 11–14 days).

Analysis of tumor-infiltration assays were also performed with PP-treated (high sMICA) NK cells co-cultured (24 h, 37°C, 5% CO_2) against non-coated HNSCC tumor spheroids. As monitored for one spheroid by time-lapse imaging, effector-to-target cell interactions and tumor infiltrations of NK cells were strongly abolished and resulted in a lack of recognition of the HNSCC spheroid by non-functional NK cells, even though they were in close proximity to the spheroids (24 h, Video S1 in Supplementary Material). However, the PP-mediated inhibition of NK cell functions was also clearly restored by cetuximab-coated HNSCC spheroids (1 μg/ml cetuximab) via multiple effector-target interactions exemplarily shown for one representative overlay photograph (**Figure 6B**, right overlay plot containing yellow regions) generated by 2D CFM. Similarly, 3D fluorescence microscopy allowed visualization of the reconstituted specific tumor infiltrations by PP-treated (high sMICA) NK cells via putative ADCC in HNSCC tumor spheroids over 24 h in the presence of 1 μg/ml cetuximab (green-yellow areas, **Figure 6C**). Moreover, the medium supernatants of those incubated samples (PP-cultured NK cells and cetuximab-coated HNSCC tumor spheroids) revealed significantly increased IFN-γ and TNFα secretion levels (24 h PP + 1 μg/ml Cetmab + 24 h HNSCC tumor spheroids, **Figure 5D**). However, the tumor infiltration [$CD45^+$ NK cells (green signals), **Figure 6D**] and cytokine release [24 h PP (high sMICA), **Figure 5D**] of same PP-inhibited NK cells were strongly reduced in 24 h-incubated samples in absence of cetuximab and consequently non-coated HNSCC tumor spheroids.

DISCUSSION

Dysfunctional tumor surveillance in HNSCC patients is further hampered by tumor immune escape mechanisms, which may induce dysregulation of immunocompetent cell profiles. In contrast to literature reports describing decreased NK cell numbers in HNSCC MNCs (54), we did not observe marked alterations in total NK cells among the total lymphocytes in our cohort of five relapsed HNSCC patients not currently in treatment. Previously, we demonstrated disbalances in NK cell subpopulations in 55 patients with initial and relapsed HNSCC (22). However, the immunoregulatory NK cell population responsible for stimulation of immature DCs by TNFα and IFN-γ secretion (5, 55) was significantly decreased in comparison with age-matched healthy individuals. Accordingly, Wulff et al. reported reduced regulatory NK cells in many HNSCC patients at different tumor stages (56). We detected higher levels of cytotoxic NK cell subpopulations compared to HCs, which may reflect suppressed cytotoxic interactions against HNSCC cells and limited tumor-infiltration capacities of these inhibited NK cell subsets. Additionally, we demonstrated that alterations in NK cell subtypes in our HNSCC patients were accompanied by disrupted TNFα and IFN-γ secretion. Enhanced sMICA and TGF-β_1 plasma levels in HNSCC patients also correlated strongly with NKG2D-dependent dysfunction of patient NK cells, resulting in suppressed killing activity against HNSCC cells and decreased NK cell viability during cytotoxic effector-target interactions. In accordance, Bose et al. described marked alterations in the Th1/Th2 cytokine ratios and significantly increased suppressor regulatory T cells in cultured MNCs from HNSCC patients resulted in decreased cytotoxicity of HNSCC effector cells (54). According to our cytotoxicity experiments, high sMICA/TGF-β_1 levels derived from HNSCC PP also contribute to diminished effector cell stability. This was also confirmed by Rossi et al. via correlation between impaired NK cell viability, effector cell cytotoxicity and decreased NKG2D and NKp46 surface expression (57). Increased NK cell susceptibility also reflects the ability of the tumor to release apoptosis-promoting factors (programmed death receptor ligand, PDL-1), which can abolish several effector cell functions within the tumor microenvironment as detected on $PD\text{-}1^+$ NK cells in cancer patients (58, 59).

We translated an activation protocol from a previous phase I/II trial (Clin-Gov-No-NCT01386619), which described stimulation of allogeneic NK cells (1000 IU/ml IL-2) with resultant increased distribution of NK cell subsets and high NCR expression levels (11), to NK cells isolated from our HNSCC patients. After IL-2 stimulation, the activated patient NK cells revealed improved distribution of increased immunoregulatory NK cells ($CD56^{bright}$/$CD16^{dim\&neg}$), enhanced TNFα and IFN-γ secretion and up-regulated NKG2D expression levels that resulted in enhanced cytotoxicity against associated HNSCC cells. Increased expression of CD56 was coincident with higher levels of NCRs as was also detected in other studies, suggesting that tumor-infiltrating NK cells were activated effector cells, but characterized by poor functionality (39, 60, 61). Indeed, our incubations (24 h) of activated NK cells with sMICA analogs or PP containing increased sMICA resulted in decreased NKG2D expression followed by impaired NKG2D-dependent killing activity against associated HNSCC cells. However, NK cell viability and expression levels of CD16, pro-apoptotic FasL and TRAIL receptors as well as for the activation marker CD57 were not affected by sMICA. In contrast, the significantly decreased levels of FasL on cultured HNSCC MNC NK cells shown by Bose et al. might be due to marked up-regulation of suppressor regulatory T cells followed by enhanced TGF-β_1 secretion levels (54).

Our observations regarding inhibited cytotoxic functions of sMICA-treated NK cells due to suppressed HNSCC tumor infiltration indicate suppressed migratory capacity of NK cells toward HNSCC tumors as detected by transmitted time-lapse imaging (Video S1 in Supplementary Material). This is one of the numerous effector cell functionalities, impairment of which can result in TIEM allowing unhindered tumor growth. NKG2D is an activating receptor for NK, NKT, $CD8^+$, and $\gamma\delta^+$ T effector cells, and down-regulation of NKG2D in HNSCC patients seems to be a crucial mechanism of immune evasion. Soluble NKG2D ligands (NKG2DL) in association with growth factors, such as TGF-β_1, released from mesothelioma cell-generated exosomes were described to potentiate the down-modulation of NKG2D surface expression on activated NK cells (62). Indeed, we observed significant correlations between time-dependent high sMICA and TGF-β_1 secretion levels and increased HNSCC cell growth of our tumor clusters or spheroids, which also resulted in disrupted NKG2D-mediated immunosurveillance of blocked patient NK cells as shown previously in our HNSCC patient study (22). However, the negative impact of increased TGF-β_1 levels detected in these HNSCC patients could be inhibited by

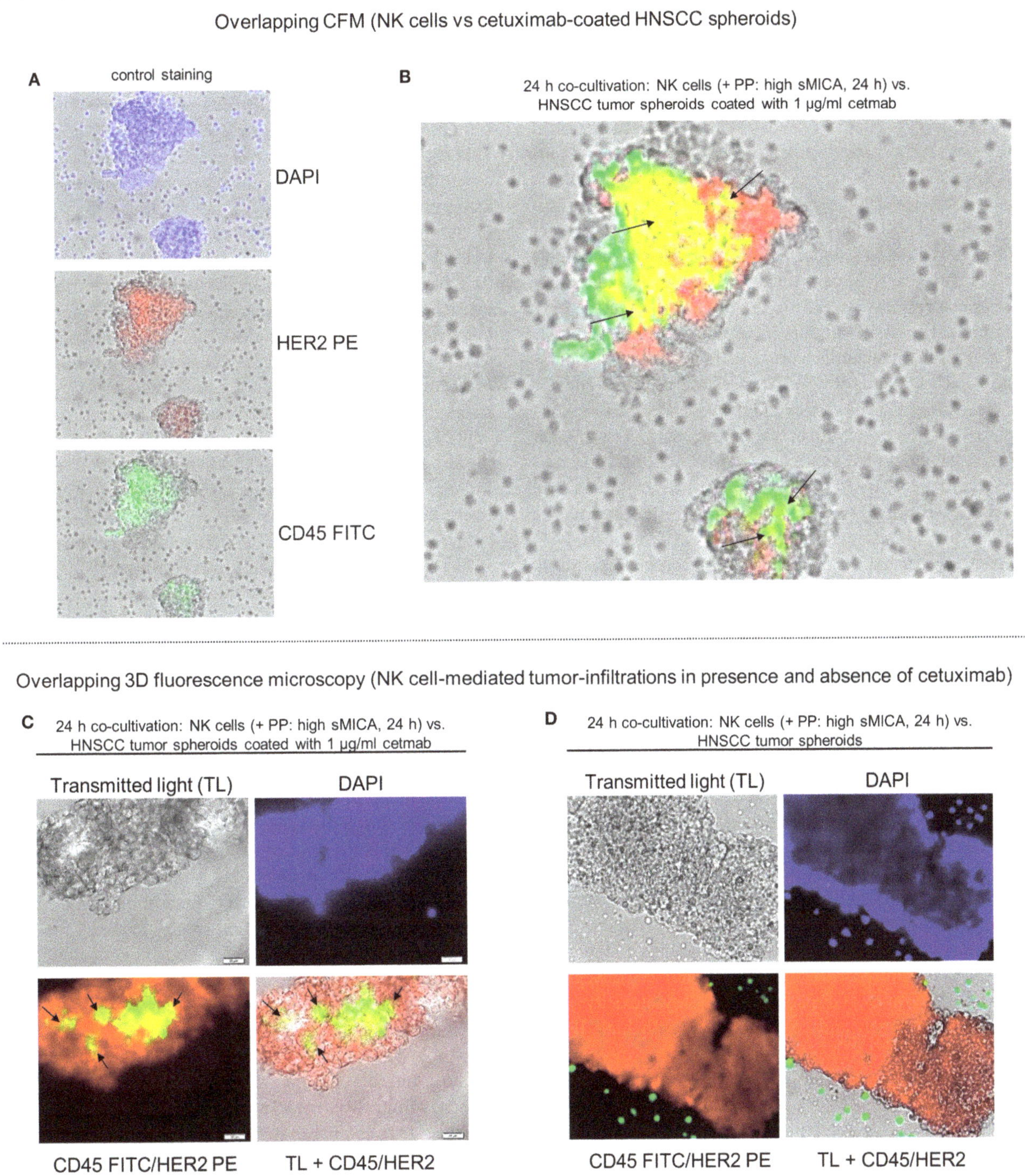

FIGURE 6 | Reconstitution of NK cell-dependent tumor infiltration by cetuximab. Tumor recognition and infiltration of sMICA-affected NK cells in absence or presence of cetuximab (1 μg/ml) were analyzed by co-culturing (24 h, 37°C, 5% CO_2) of corresponding patient NK cells, plasma (high sMICA) and HNSCC tumor spheroids derived from HNSCC patient tumors ($n = 5$). Activated NK cells were pre-incubated (24 h) with corresponding patient plasma containing high sMICA (PP: > 500 pg/ml; diluted 1:2 with X-VIVO™10) and co-incubated (24 h) with primary HNSCC spheroids (time of cultivation: 11–14 days) in the presence of 1 μg/ml cetuximab. CD45 (green) for PP-treated patient NK cells were identified by staining with FITC-conjugated mAb, whereas tumor spheroids were labeled with anti-HER2 PE-conjugated mABs **(A)**. Tumor-infiltrating NK cells were illustrated by overlay plots (CFM). Exactly the same orientation of $CD45^+$ NK cells (green) and $HER2^+$ tumor spheroids (red) showed a positive, overlapping signal in yellow [see black arrows **(B)**] for ADCC-related effector-target cell interactions, whereas single positive, HNSCC cells ($CD45^-/HER2^+$) are shown as red signals and single positive NK cells are depicted by green signals in the corresponding photographs **(A)**. 3D fluorescence microscopy visualized tumor-infiltrating patient NK cells (green-yellow areas) in corresponding HNSCC tumor spheroids in the presence **(C)** and absence **(D)** of 1 μg/ml cetuximab. DAPI [blue signals **(A,C,D)**] was used to stain DNA from methanol-fixed effector and target cells for analyses of the nuclear morphology.

neutralization antibodies against this immunosuppressive factor, indicating that TGF-β_1 seemed to potentiate the sMICA-induced decrease of patient NK cell cytotoxicity and diminished NKG2D expression. According to our results, others have demonstrated restored NKG2D expression levels on NK and CD8$^+$ T cells after tumor resection in glioma patients, which was accompanied by enhanced killing activities of those effector cells against NKG2DL-positive tumor targets (63). Interestingly, NKG2D was markedly down-regulated on activated CD8$^+$ T cells but only if CD4$^+$ T cells and NKG2DLs, such as soluble MICB, were present (64). This observation supports the hypothesis that soluble NKG2DLs played a secondary role and the down-regulation of NKG2D was primarily caused by tumor-derived TGF-β_1 (63). The immunosuppressive tumor environment was responsible for the high diversity of antigen presentation patterns in stroma-infiltrating NK cells and tumor-infiltrating NK cells. The effector cells showed lower CD56 levels and higher CD16 expression during cytotoxic interactions with breast cancer cells and resulted in altered NK cell phenotypes with decreased functional capacities (65). In accordance with this report, we demonstrate stable CD16 expression levels, which were not affected by sMICA analogs or HNSCC plasma containing high amounts of sMICA and TGF-β_1.

NK cell subsets have a stimulating Fc receptor for binding IgG (Fcγ RIIIa), which induces ADCC and may trigger TNFα and IFN-γ secretion to finally recognize and kill antibody-coated targets (66). Since several ErbB family members, including HER1 (EGFR), HER2 and HER3, seem to be strong predictors for the outcome of HNSCC (67), we stained different tumor samples post-surgery from corresponding relapsed patients for HER1/2 and collected PB for NK cell separation as well as blood plasma for quantification of sMICA and TGF-β_1. The chimeric (human-murine) IgG$_1$ cetuximab directly affects HER1-positive tumor cells by activation immunocompetent cells (66). Therefore, we assessed our sMICA-affected NK cells regarding to cytotoxicity and tumor infiltrations. We found improved killing activity via ADCC for sMICA-inhibited NK cells against cetuximab-coated HNSCC cells. Additionally, enhanced cytokine (TNFα and IFN-γ) release was observed compared to low cell lysis of the same effector cells co-cultured with untreated tumor cells. Negative effects of sMICA and TGF-β_1 on NK cell cytotoxicity were overcome by cetuximab and correlated well with high IFN-γ and TNFα secretion levels described previously in HNSCC and other cancers (53, 68). In this context, it was reported that cetuximab-induced NK cells are able to activate DC maturation markers and antigen presentation machinery via IFN-γ secretion, thus allowing initiation of adaptive immune responses by NK cell stimulated DC maturation (69). We demonstrated that combining cetuximab with IL-2 (1000 IU/ml) re-established NK cell cytotoxicity and reconstituted TNFα and IFN-γ secretion. This is in agreement with other reports that described increased ADCC activity and cytokine release in the presence of cetuximab and several cytokines (IL-2, IL-12, IL-15, and IL-21) compared to cetuximab-mediated ADCC in absence of these cytokines (53, 65, 68). Prospective studies should clarify how to improve the cetuximab effect combined with pro-inflammatory cytokines, especially IL-2. Interestingly, it was demonstrated that ADCC- and IL-2-activated NK cells were less susceptible to immunosuppression by chemotherapy or other immunosuppressive drugs, especially mycophenolate mofetil (MMF), than non-stimulated NK cells in cancer patients (68, 70).

To investigate the potency of cetuximab on cytotoxic functionalities of sMICA-inhibited NK cells, we generated HER1-positive tumor spheroids to simulate the *in vivo* tumor microenvironment. This enabled us to analyze the restoration of tumor-infiltrating capability of NK cells with down-regulated NKG2D expression levels in a well-defined system. Tumor spheroids were utilized to monitor specific migratory capability and tumor chemoattraction of effector cells, especially NK cells, over well-defined time periods (71). However, only few reports describe the effector cell-mediated immunosurveillance toward tumor spheroids. Inactivation of NK cells by clustered Ewing's sarcoma cells and cytotoxicity of $\gamma\delta^+$ T cells against pediatric liver tumor spheroids was demonstrated (72, 73). In our tumor models, we were able to assess the cetuximab-mediated tumor infiltration from sMICA-inhibited NK cells toward corresponding primary HNSCC cell clusters and tumor spheroids that expressed high levels of HER1 and HER2. HNSCC spheroids expressed only low-to-moderate levels of membrane-bound MICA, but sMICA and TGF-β_1 release steadily increased in supernatants of cultured HNSCC tumor spheroids and reached saturation levels after a few days of cultivation. The accumulation of immunosuppressive factors in HNSCC spheroid supernatants supported the assumption that NKG2D-mediated cytotoxicity of co-cultured patient NK cells was diminished, which corroborates the negative effect toward NK cell-mediated immunosurveillance by these tumor spheroids as described previously (71). Importantly, we detected early effector-to-target interactions displayed in small HNSCC tumor cell clusters during the first hours of co-cultivation in the presence of cetuximab. Furthermore, we successfully demonstrated cetuximab-mediated tumor infiltrations and increased TNFα and IFN-γ secretions of sMICA-inhibited NK cells in associated HNSCC spheroids using 2D- and 3D-microscopy techniques. This supported results from monolayer cultures of different effector-to-target cell ratios co-incubated for previous cytotoxicity assays. Correspondingly, others evaluated the direct localization of fluorescently stained therapeutic antibody, cetuximab-IRDye800CW, in histologic sections with tonsil, tongue, and cutaneous squamous cell carcinoma (SCC) by fluorescence immunohistochemistry (74). Interestingly, the heterogeneous composition and tumor architecture in short-term culture of HNSCC tumor slices demonstrated a high diversity of individual responses to cetuximab, but the absence of any effector cell subset (75). It was recently suggested that drug resistance to anti-EGFR therapies in HNSCC is not affected by the hypoxic tumor microenvironment within the investigated tumors (76). Indeed, the inhibition of EGFR via cetuximab reduces angiogenesis via hypoxia-inducible factor-1α and Notch1 in HNSCC (77).

CONCLUSION

In conclusion, our current results emphasize that cetuximab is able to neutralize negative effects of TIEMs. This was shown in a corresponding effector-target system of patient NK cells, PP containing immunosuppressive factors (sMICA and TGF-β_1), and associated HNSCC tumors/tumor cells as targets. This is of clinical relevance as the shedding of soluble NKG2DLs combined with secretion of immunosuppressive cytokines may promote tumor progression and is predictive for a negative prognosis in tumor patients (78).

In summary, observations of our clinical phase I/II haploidentical NK cell study for adaptive immunotherapy (Clin-Gov-No-NCT01386619) (47, 79, 80) and our current results indicate that allogeneic NK cells in combination with tumor antigen-specific mAbs, especially cetuximab, might be an innovative approach to circumvent TIEM-derived limitations against patient NK cells. Therefore, it is necessary to monitor overexpressed tumor antigens, such as HER1, and immunosuppressive ligands and the polymorphic Fcγ receptor IIIa to regenerate the tumoricidal properties of NK cells for improved therapeutic benefit.

ACKNOWLEDGMENTS

This project was supported by the Wilhelm Sander-Stiftung für Krebsforschung, the Frankfurter Stiftung für krebskranke Kinder, the Kind-Philipp-Stiftung, and the Integrated Research and Treatment Center Transplantation (IFB-Tx, Ref. No. 01EO1302) financed by the German Federal Ministry of Education and Research.

SUPPLEMENTARY MATERIAL

Figure S1 | Gating strategy. Flow cytometry-based cytotoxicity assay of NK cells against singularized HNSCC cells. NK and HNSCC cells were stained with CD9 FITC, CD56 PC7, and CD45 KO mAbs. 7-AAD (7-amino-actinomycin D) was used to discriminate non-viable cells in FCM. Blue: CD45+ NK cells; red: CD9+ HNSSC cells; and grey: 7-AAD+ effector and target cells. Dot plot areas were defined by the analysis of mono-cultured NK and HNSCC cells incubated for the identical time periods. Dot plot (WBC/MNC) displays an overview of all scatter events properties, and shows differentiations to non-specifically stained debris by low forward scatter signals. Plot (CAL) illustrates the events of the region "beads" along the time course to calibrate the events for "cells/μl" and detect even sample flow. Region (Target cells) is defined to include viable CD9+ HNSCC cells, and region (Effector cells) is defined to include all CD45+ NK cells. Plot ("Cell clusters") represents the viable effector-target-cell interactions. Plot ("Viability") is defined as a region to exclude the 7-AAD+ cells. These 7-AAD+ cells are not further presented in the following dot plots by using the characteristic signal when representing the 7-AAD fluorescence against the side scatter (SSC) properties.

Video S1 | Time-lapse imaging of PP (high sMICA and TGF-beta1 levels)-treated NK cells against corresponding HNSCC tumor spheroids over a time period of 24 h. These NK cells (smaller rounded effector cells) isolated from the same HNSCC patient showed a decreased tumor recognition against associated tumor spheroids via reduced migratory capability and a decreased cytotoxicity.

REFERENCES

1. Trinchieri G. Biology of natural killer cells. *Adv Immunol* (1989) **47**:187–376. doi:10.1016/S0065-2776(08)60664-1
2. Farag SS, Caligiuri MA. Human natural killer cell development and biology. *Blood Rev* (2006) **20**(3):123–37. doi:10.1016/j.blre.2005.10.001
3. Trinchieri G. Natural killer cells wear different hats: effector cells of innate resistance and regulatory cells of adaptive immunity and of hematopoiesis. *Semin Immunol* (1995) **7**(2):83–8. doi:10.1006/smim.1995.0012
4. Zwirner NW, Fuertes MB, Girart MV, Domaica CI, Rossi LE. Cytokine-driven regulation of NK cell functions in tumor immunity: role of the MICA-NKG2D system. *Cytokine Growth Factor Rev* (2007) **18**(1–2):159–70. doi:10.1016/j.cytogfr.2007.01.013
5. Walzer T, Dalod M, Robbins SH, Zitvogel L, Vivier E. Natural-killer cells and dendritic cells: "l'union fait la force". *Blood* (2005) **106**(7):2252–8. doi:10.1182/blood-2005-03-1154
6. Lee SC, Srivastava RM, Lopez-Albaitero A, Ferrone S, Ferris RL. Natural killer (NK): dendritic cell (DC) cross talk induced by therapeutic monoclonal antibody triggers tumor antigen-specific T cell immunity. *Immunol Res* (2011) **50**(2–3):248–54. doi:10.1007/s12026-011-8231-0
7. Lanier LL. Missing self, NK cells, and the white album. *J Immunol* (2005) **174**(11):6565. doi:10.4049/jimmunol.174.11.6565
8. Lanier LL. NK cell recognition. *Annu Rev Immunol* (2005) **23**:225–74. doi:10.1146/annurev.immunol.23.021704.115526
9. Coudert JD, Held W. The role of the NKG2D receptor for tumor immunity. *Semin Cancer Biol* (2006) **16**(5):333–43. doi:10.1016/j.semcancer.2006.07.008
10. Salih HR, Antropius H, Gieseke F, Lutz SZ, Kanz L, Rammensee HG, et al. Functional expression and release of ligands for the activating immunoreceptor NKG2D in leukemia. *Blood* (2003) **102**(4):1389–96. doi:10.1182/blood-2003-01-0019
11. Huenecke S, Zimmermann SY, Kloess S, Esser R, Brinkmann A, Tramsen L, et al. IL-2-driven regulation of NK cell receptors with regard to the distribution of CD16+ and CD16- subpopulations and in vivo influence after haploidentical NK cell infusion. *J Immunother* (2010) **33**(2):200–10. doi:10.1097/CJI.0b013e3181bb46f7
12. Bryceson YT, Ljunggren HG. Tumor cell recognition by the NK cell activating receptor NKG2D. *Eur J Immunol* (2008) **38**(11):2957–61. doi:10.1002/eji.200838833
13. Franchi A, Santucci M, Masini E, Sardi I, Paglierani M, Gallo O. Expression of matrix metalloproteinase 1, matrix metalloproteinase 2, and matrix metalloproteinase 9 in carcinoma of the head and neck. *Cancer* (2002) **95**(9):1902–10. doi:10.1002/cncr.10916
14. Groh V, Wu J, Yee C, Spies T. Tumour-derived soluble MIC ligands impair expression of NKG2D and T-cell activation. *Nature* (2002) **419**(6908):734–8. doi:10.1038/nature01112
15. Holdenrieder S, Stieber P, Peterfi A, Nagel D, Steinle A, Salih HR. Soluble MICB in malignant diseases: analysis of diagnostic significance and correlation with soluble MICA. *Cancer Immunol Immunother* (2006) **55**(12):1584–9. doi:10.1007/s00262-006-0167-1
16. Holdenrieder S, Stieber P, Peterfi A, Nagel D, Steinle A, Salih HR. Soluble MICA in malignant diseases. *Int J Cancer* (2006) **118**(3):684–7. doi:10.1002/ijc.21382
17. Shibuya K, Mathers CD, Boschi-Pinto C, Lopez AD, Murray CJ. Global and regional estimates of cancer mortality and incidence by site: II. Results for the global burden of disease 2000. *BMC Cancer* (2002) **2**:37. doi:10.1186/1471-2407-2-37
18. Ragin CC, Modugno F, Gollin SM. The epidemiology and risk factors of head and neck cancer: a focus on human papillomavirus. *J Dent Res* (2007) **86**(2):104–14. doi:10.1177/154405910708600202
19. Duray A, Demoulin S, Hubert P, Delvenne P, Saussez S. Immune suppression in head and neck cancers: a review. *Clin Dev Immunol* (2010) **2010**:701657. doi:10.1155/2010/701657
20. Dasgupta S, Bhattacharya-Chatterjee M, O'Malley BW Jr, Chatterjee SK. Inhibition of NK cell activity through TGF-beta 1 by down-regulation of NKG2D in a murine model of head and neck cancer. *J Immunol* (2005) **175**(8):5541–50. doi:10.4049/jimmunol.175.8.5541
21. Tamaki S, Sanefuzi N, Kawakami M, Aoki K, Imai Y, Yamanaka Y, et al. Association between soluble MICA levels and disease stage IV oral squamous cell carcinoma in Japanese patients. *Hum Immunol* (2008) **69**(2):88–93. doi:10.1016/j.humimm.2008.01.010

22. Kloess S, et al. Increased sMICA and TGF-β1 levels in HNSCC patients impair NKG2D-dependent functionality of activated NK cells. *Oncoimmunology* (2015) **4**(11):e1055993. doi:10.1080/2162402X.2015.1055993
23. Tamaki S, Kawakami M, Ishitani A, Kawashima W, Kasuda S, Yamanaka Y, et al. Soluble MICB serum levels correlate with disease stage and survival rate in patients with oral squamous cell carcinoma. *Anticancer Res* (2010) **30**(10):4097–101.
24. Roskoski R Jr. ErbB/HER protein-tyrosine kinases: structures and small molecule inhibitors. *Pharmacol Res* (2014) **87**:42–59. doi:10.1016/j.phrs.2014.06.001
25. Roskoski R Jr. The ErbB/HER family of protein-tyrosine kinases and cancer. *Pharmacol Res* (2014) **79**:34–74. doi:10.1016/j.phrs.2013.11.002
26. Montemurro F, Scaltriti M. Biomarkers of drugs targeting HER-family signalling in cancer. *J Pathol* (2014) **232**(2):219–29. doi:10.1002/path.4269
27. Zuo JH, Zhu W, Li MY, Li XH, Yi H, Zeng GQ, et al. Activation of EGFR promotes squamous carcinoma SCC10A cell migration and invasion via inducing EMT-like phenotype change and MMP-9-mediated degradation of E-cadherin. *J Cell Biochem* (2011) **112**(9):2508–17. doi:10.1002/jcb.23175
28. Lee CM, Tannock IF. The distribution of the therapeutic monoclonal antibodies cetuximab and trastuzumab within solid tumors. *BMC Cancer* (2010) **10**:255. doi:10.1186/1471-2407-10-255
29. Hong KW, Kim CG, Lee SH, Chang KH, Shin YW, Ryoo KH, et al. A novel anti-EGFR monoclonal antibody inhibiting tumor cell growth by recognizing different epitopes from cetuximab. *J Biotechnol* (2010) **145**(1):84–91. doi:10.1016/j.jbiotec.2009.09.023
30. Campoli M, Ferris R, Ferrone S, Wang X. Immunotherapy of malignant disease with tumor antigen-specific monoclonal antibodies. *Clin Cancer Res* (2010) **16**(1):11–20. doi:10.1158/1078-0432.CCR-09-2345
31. Ferris RL, Jaffee EM, Ferrone S. Tumor antigen-targeted, monoclonal antibody-based immunotherapy: clinical response, cellular immunity, and immunoescape. *J Clin Oncol* (2010) **28**(28):4390–9. doi:10.1200/JCO.2009.27.6360
32. William WN Jr, Kim ES, Herbst RS. Cetuximab therapy for patients with advanced squamous cell carcinomas of the head and neck. *Nat Clin Pract Oncol* (2009) **6**(3):132–3. doi:10.1038/ncponc1321
33. Blumenschein GR Jr, Paulus R, Curran WJ, Robert F, Fossella F, Werner-Wasik M, et al. Phase II study of cetuximab in combination with chemoradiation in patients with stage IIIA/B non-small-cell lung cancer: RTOG 0324. *J Clin Oncol* (2011) **29**(17):2312–8. doi:10.1200/JCO.2010.31.7875
34. Lenz HJ, Van Cutsem E, Khambata-Ford S, Mayer RJ, Gold P, Stella P, et al. Multicenter phase II and translational study of cetuximab in metastatic colorectal carcinoma refractory to irinotecan, oxaliplatin, and fluoropyrimidines. *J Clin Oncol* (2006) **24**(30):4914–21. doi:10.1200/JCO.2006.06.7595
35. Bonner JA, Keene KS. Is cetuximab active in patients with cisplatin-refractory squamous cell carcinoma of the head and neck? *Nat Clin Pract Oncol* (2007) **4**(12):690–1. doi:10.1038/ncponc0962
36. Taylor RJ, Chan SL, Wood A, Voskens CJ, Wolf JS, Lin W, et al. FcgammaRIIIa polymorphisms and cetuximab induced cytotoxicity in squamous cell carcinoma of the head and neck. *Cancer Immunol Immunother* (2009) **58**(7):997–1006. doi:10.1007/s00262-008-0613-3
37. López-Albaitero A, Lee SC, Morgan S, Grandis JR, Gooding WE, Ferrone S, et al. Role of polymorphic Fc gamma receptor IIIa and EGFR expression level in cetuximab mediated, NK cell dependent in vitro cytotoxicity of head and neck squamous cell carcinoma cells. *Cancer Immunol Immunother* (2009) **58**(11):1853–64. doi:10.1007/s00262-009-0697-4
38. López-Albaitero A, Nayak JV, Ogino T, Machandia A, Gooding W, DeLeo AB, et al. Role of antigen-processing machinery in the in vitro resistance of squamous cell carcinoma of the head and neck cells to recognition by CTL. *J Immunol* (2006) **176**(6):3402–9. doi:10.4049/jimmunol.176.6.3402
39. Carrega P, Morandi B, Costa R, Frumento G, Forte G, Altavilla G, et al. Natural killer cells infiltrating human nonsmall-cell lung cancer are enriched in CD56 bright CD16(-) cells and display an impaired capability to kill tumor cells. *Cancer* (2008) **112**(4):863–75. doi:10.1002/cncr.23239
40. Ohtani H. Focus on TILs: prognostic significance of tumor infiltrating lymphocytes in human colorectal cancer. *Cancer Immun* (2007) **7**:4.
41. Bruno A, Ferlazzo G, Albini A, Noonan DM. A think tank of TINK/TANKs: tumor-infiltrating/tumor-associated natural killer cells in tumor progression and angiogenesis. *J Natl Cancer Inst* (2014) **106**(8):dju200. doi:10.1093/jnci/dju200
42. Vitale M, Cantoni C, Pietra G, Mingari MC, Moretta L. Effect of tumor cells and tumor microenvironment on NK-cell function. *Eur J Immunol* (2014) **44**(6):1582–92. doi:10.1002/eji.201344272
43. Maréchal R, De Schutter J, Nagy N, Demetter P, Lemmers A, Devière J, et al. Putative contribution of CD56 positive cells in cetuximab treatment efficacy in first-line metastatic colorectal cancer patients. *BMC Cancer* (2010) **10**:340. doi:10.1186/1471-2407-10-340
44. Huenecke S, Behl M, Fadler C, Zimmermann SY, Bochennek K, Tramsen L, et al. Age-matched lymphocyte subpopulation reference values in childhood and adolescence: application of exponential regression analysis. *Eur J Haematol* (2008) **80**(6):532–9. doi:10.1111/j.1600-0609.2008.01052.x
45. Rheinwald JG, Beckett MA. Defective terminal differentiation in culture as a consistent and selectable character of malignant human keratinocytes. *Cell* (1980) **22**(2 Pt 2):629–32. doi:10.1016/0092-8674(80)90373-6
46. Rheinwald JG, Beckett MA. Tumorigenic keratinocyte lines requiring anchorage and fibroblast support cultured from human squamous cell carcinomas. *Cancer Res* (1981) **41**(5):1657–63.
47. Kloess S, Huenecke S, Piechulek D, Esser R, Koch J, Brehm C, et al. IL-2-activated haploidentical NK cells restore NKG2D-mediated NK-cell cytotoxicity in neuroblastoma patients by scavenging of plasma MICA. *Eur J Immunol* (2010) **40**(11):3255–67. doi:10.1002/eji.201040568
48. Klöss S, Bochennek K, Huenecke S, Zimmermann SY, Kuçi S, Müller T, et al. A novel five-colour flow cytometric assay to determine NK cell cytotoxicity against neuroblastoma and other adherent tumour cells. *J Immunol Methods* (2007) **325**(1–2):140–7. doi:10.1016/j.jim.2007.06.013
49. Zimmermann SY, Esser R, Rohrbach E, Klingebiel T, Koehl U. A novel four-colour flow cytometric assay to determine natural killer cell or T-cell-mediated cellular cytotoxicity against leukaemic cells in peripheral or bone marrow specimens containing greater than 20% of normal cells. *J Immunol Methods* (2005) **296**(1–2):63–76. doi:10.1016/j.jim.2004.10.014
50. Comans-Bitter WM, de Groot R, van den Beemd R, Neijens HJ, Hop WC, Groeneveld K, et al. Immunophenotyping of blood lymphocytes in childhood. Reference values for lymphocyte subpopulations. *J Pediatr* (1997) **130**(3):388–93.
51. Li H, Wawrose JS, Gooding WE, Garraway LA, Lui VW, Peyser ND, et al. Genomic analysis of head and neck squamous cell carcinoma cell lines and human tumors: a rational approach to preclinical model selection. *Mol Cancer Res* (2014) **12**(4):571–82. doi:10.1158/1541-7786.MCR-13-0396
52. Marten A, von Lilienfeld-Toal M, Buchler MW, Schmidt J. Soluble MIC is elevated in the serum of patients with pancreatic carcinoma diminishing gammadelta T cell cytotoxicity. *Int J Cancer* (2006) **119**(10):2359–65. doi:10.1002/ijc.22186
53. Roda JM, Joshi T, Butchar JP, McAlees JW, Lehman A, Tridandapani S, et al. The activation of natural killer cell effector functions by cetuximab-coated, epidermal growth factor receptor positive tumor cells is enhanced by cytokines. *Clin Cancer Res* (2007) **13**(21):6419–28. doi:10.1158/1078-0432.CCR-07-0865
54. Bose A, Chakraborty T, Chakraborty K, Pal S, Baral R. Dysregulation in immune functions is reflected in tumor cell cytotoxicity by peripheral blood mononuclear cells from head and neck squamous cell carcinoma patients. *Cancer Immun* (2008) **8**:10.
55. Walzer T, Dalod M, Vivier E, Zitvogel L. Natural killer cell-dendritic cell crosstalk in the initiation of immune responses. *Expert Opin Biol Ther* (2005) **5**(Suppl 1):S49–59. doi:10.1517/14712598.5.1.S49
56. Wulff S, Pries R, Borngen K, Trenkle T, Wollenberg B. Decreased levels of circulating regulatory NK cells in patients with head and neck cancer throughout all tumor stages. *Anticancer Res* (2009) **29**(8):3053–7.
57. Rossi LE, Avila DE, Spallanzani RG, Ziblat A, Fuertes MB, Lapyckyj L, et al. Histone deacetylase inhibitors impair NK cell viability and effector functions through inhibition of activation and receptor expression. *J Leukoc Biol* (2012) **91**(2):321–31. doi:10.1189/jlb.0711339
58. Lyford-Pike S, Peng S, Young GD, Taube JM, Westra WH, Akpeng B, et al. Evidence for a role of the PD-1:PD-L1 pathway in immune resistance of HPV-associated head and neck squamous cell carcinoma. *Cancer Res* (2013) **73**(6):1733–41. doi:10.1158/0008-5472.CAN-12-2384
59. Benson DM Jr, Bakan CE, Mishra A, Hofmeister CC, Efebera Y, Becknell B, et al. The PD-1/PD-L1 axis modulates the natural killer cell versus

multiple myeloma effect: a therapeutic target for CT-011, a novel monoclonal anti-PD-1 antibody. *Blood* (2010) **116**(13):2286–94. doi:10.1182/blood-2010-02-271874

60. Platonova S, Cherfils-Vicini J, Damotte D, Crozet L, Vieillard V, Validire P, et al. Profound coordinated alterations of intratumoral NK cell phenotype and function in lung carcinoma. *Cancer Res* (2011) **71**(16):5412–22. doi:10.1158/0008-5472.CAN-10-4179
61. Mamessier E, Sylvain A, Bertucci F, Castellano R, Finetti P, Houvenaeghel G, et al. Human breast tumor cells induce self-tolerance mechanisms to avoid NKG2D-mediated and DNAM-mediated NK cell recognition. *Cancer Res* (2011) **71**(21):6621–32. doi:10.1158/0008-5472.CAN-11-0792
62. Clayton A, Mitchell JP, Court J, Linnane S, Mason MD, Tabi Z. Human tumor-derived exosomes down-modulate NKG2D expression. *J Immunol* (2008) **180**(11):7249–58. doi:10.4049/jimmunol.180.11.7249
63. Crane CA, Han SJ, Barry JJ, Ahn BJ, Lanier LL, Parsa AT. TGF-beta downregulates the activating receptor NKG2D on NK cells and CD8+ T cells in glioma patients. *Neuro Oncol* (2010) **12**(1):7–13. doi:10.1093/neuonc/nop009
64. Cerboni C, Ardolino M, Santoni A, Zingoni A. Detuning CD8+ T lymphocytes by down-regulation of the activating receptor NKG2D: role of NKG2D ligands released by activated T cells. *Blood* (2009) **113**(13):2955–64. doi:10.1182/blood-2008-06-165944
65. Roberti MP, Rocca YS, Amat M, Pampena MB, Loza J, Coló F, et al. IL-2- or IL-15-activated NK cells enhance cetuximab-mediated activity against triple-negative breast cancer in xenografts and in breast cancer patients. *Breast Cancer Res Treat* (2012) **136**(3):659–71. doi:10.1007/s10549-012-2287-y
66. Varilla V, Atienza J, Dasanu CA. Immune alterations and immunotherapy prospects in head and neck cancer. *Expert Opin Biol Ther* (2013) **13**(9):1241–56. doi:10.1517/14712598.2013.810716
67. Xia W, Lau YK, Zhang HZ, Xiao FY, Johnston DA, Liu AR, et al. Combination of EGFR, HER-2/neu, and HER-3 is a stronger predictor for the outcome of oral squamous cell carcinoma than any individual family members. *Clin Cancer Res* (1999) **5**(12):4164–74.
68. Kurai J, Chikumi H, Hashimoto K, Yamaguchi K, Yamasaki A, Sako T, et al. Antibody-dependent cellular cytotoxicity mediated by cetuximab against lung cancer cell lines. *Clin Cancer Res* (2007) **13**(5):1552–61. doi:10.1158/1078-0432.CCR-06-1726
69. Srivastava RM, Lee SC, Andrade Filho PA, Lord CA, Jie HB, Davidson HC, et al. Cetuximab-activated natural killer and dendritic cells collaborate to trigger tumor antigen-specific T-cell immunity in head and neck cancer patients. *Clin Cancer Res* (2013) **19**(7):1858–72. doi:10.1158/1078-0432.CCR-12-2426
70. Brehm C, Huenecke S, Esser R, Kloess S, Quaiser A, Betz S, et al. Interleukin-2-stimulated natural killer cells are less susceptible to mycophenolate mofetil than non-activated NK cells: possible consequences for immunotherapy. *Cancer Immunol Immunother* (2014) **63**(8):821–33. doi:10.1007/s00262-014-1556-5
71. Giannattasio A, Weil S, Kloess S, Ansari N, Stelzer EH, Cerwenka A, et al. Cytotoxicity and infiltration of human NK cells in in vivo-like tumor spheroids. *BMC Cancer* (2015) **15**:351. doi:10.1186/s12885-015-1321-y
72. Hoh A, Dewerth A, Vogt F, Wenz J, Baeuerle PA, Warmann SW, et al. The activity of gammadelta T cells against paediatric liver tumour cells and spheroids in cell culture. *Liver Int* (2013) **33**(1):127–36. doi:10.1111/liv.12011
73. Holmes TD, El-Sherbiny YM, Davison A, Clough SL, Blair GE, Cook GP. A human NK cell activation/inhibition threshold allows small changes in the target cell surface phenotype to dramatically alter susceptibility to NK cells. *J Immunol* (2011) **186**(3):1538–45. doi:10.4049/jimmunol.1000951
74. de Boer E, Warram JM, Tucker MD, Hartman YE, Moore LS, de Jong JS, et al. In vivo fluorescence immunohistochemistry: localization of fluorescently labeled cetuximab in squamous cell carcinomas. *Sci Rep* (2015) **5**:10169. doi:10.1038/srep10169
75. Peria M, Donnadieu J, Racz C, Ikoli JF, Galmiche A, Chauffert B, et al. Evaluation of individual sensitivity of head and neck squamous cell carcinoma to cetuximab by short-term culture of tumor slices. *Head Neck* (2015). doi:10.1002/hed.24126
76. Boeckx C, Van den Bossche J, De Pauw I, Peeters M, Lardon F, Baay M, et al. The hypoxic tumor microenvironment and drug resistance against EGFR inhibitors: preclinical study in cetuximab-sensitive head and neck squamous cell carcinoma cell lines. *BMC Res Notes* (2015) **8**:203. doi:10.1186/s13104-015-1197-6
77. Wang WM, Zhao ZL, Ma SR, Yu GT, Liu B, Zhang L, et al. Epidermal growth factor receptor inhibition reduces angiogenesis via hypoxia-inducible factor-1alpha and Notch1 in head neck squamous cell carcinoma. *PLoS One* (2015) **10**(2):e0119723. doi:10.1371/journal.pone.0119723
78. Baragano Raneros A, Suarez-Alvarez B, Lopez-Larrea C. Secretory pathways generating immunosuppressive NKG2D ligands: new targets for therapeutic intervention. *Oncoimmunology* (2014) **3**:e28497. doi:10.4161/onci.28497
79. Koehl U, Brehm C, Huenecke S, Zimmermann SY, Kloess S, Bremm M, et al. Clinical grade purification and expansion of NK cell products for an optimized manufacturing protocol. *Front Oncol* (2013) **3**:118. doi:10.3389/fonc.2013.00118
80. Stern M, Passweg JR, Meyer-Monard S, Esser R, Tonn T, Soerensen J, et al. Pre-emptive immunotherapy with purified natural killer cells after haploidentical SCT: a prospective phase II study in two centers. *Bone Marrow Transplant* (2013) **48**(3):433–8. doi:10.1038/bmt.2012.162

15

The Application of Natural Killer Cell Immunotherapy for the Treatment of Cancer

Katayoun Rezvani[1] and Rayne H. Rouce[2,3]*

[1] Department of Stem Cell Transplantation, Division of Cancer Medicine, The University of Texas MD Anderson Cancer Center, Houston, TX, USA, [2] Department of Pediatrics, Texas Children's Cancer and Hematology Centers, Baylor College of Medicine, Houston, TX, USA, [3] Center for Cell and Gene Therapy, Baylor College of Medicine Houston Methodist Hospital and Texas Children's Hospital, Houston, TX, USA

Edited by:
Rafael Solana,
University of Cordoba, Spain

Reviewed by:
Kamalakannan Rajasekaran,
Blood Research Institute, USA
Alejandra Pera,
Instituto Maimonides de Investigaciones Biomédicas de Córdoba, Spain

****Correspondence:***
Katayoun Rezvani
krezvani@mdanderson.org

Natural killer (NK) cells are essential components of the innate immune system and play a critical role in host immunity against cancer. Recent progress in our understanding of NK cell immunobiology has paved the way for novel NK cell-based therapeutic strategies for the treatment of cancer. In this review, we will focus on recent advances in the field of NK cell immunotherapy, including augmentation of antibody-dependent cellular cytotoxicity, manipulation of receptor-mediated activation, and adoptive immunotherapy with *ex vivo*-expanded, chimeric antigen receptor (CAR)-engineered, or engager-modified NK cells. In contrast to T lymphocytes, donor NK cells do not attack non-hematopoietic tissues, suggesting that an NK-mediated antitumor effect can be achieved in the absence of graft-vs.-host disease. Despite reports of clinical efficacy, a number of factors limit the application of NK cell immunotherapy for the treatment of cancer, such as the failure of infused NK cells to expand and persist *in vivo*. Therefore, efforts to enhance the therapeutic benefit of NK cell-based immunotherapy by developing strategies to manipulate the NK cell product, host factors, and tumor targets are the subject of intense research. In the preclinical setting, genetic engineering of NK cells to express CARs to redirect their antitumor specificity has shown significant promise. Given the short lifespan and potent cytolytic function of mature NK cells, they are attractive candidate effector cells to express CARs for adoptive immunotherapies. Another innovative approach to redirect NK cytotoxicity towards tumor cells is to create either bispecific or trispecific antibodies, thus augmenting cytotoxicity against tumor-associated antigens. These are exciting times for the study of NK cells; with recent advances in the field of NK cell biology and translational research, it is likely that NK cell immunotherapy will move to the forefront of cancer immunotherapy over the next few years.

Keywords: natural killer cells, adoptive immunotherapy, CAR NK cells, ADCC, anti-KIR antibody, NK-92, transplantation

INTRODUCTION

Natural killer (NK) cell-mediated cytotoxicity contributes to the innate immune response against various malignancies, including leukemia (1, 2). The antitumor effect of NK cells is a subject of intense investigation in the field of cancer immunotherapy. In this review, we will focus on recent advances in NK cell immunotherapy, including augmentation of antibody-dependent cytotoxicity, manipulation of receptor-mediated activation, and adoptive immunotherapy with *ex vivo*-expanded, chimeric antigen receptor (CAR)-engineered, or engager-modified NK cells.

BIOLOGY OF NK CELLS RELEVANT TO ADOPTIVE IMMUNOTHERAPY

Natural killer cells are characterized by the lack of CD3/TCR molecules and by the expression of CD16 and CD56 surface antigens. Around 90% of circulating NK cells are $CD56^{dim}$, characterized by their distinct ability to mediate cytotoxicity in response to target cell stimulation (3, 4). This subset includes the alloreactive NK cells that play a central role in targeting leukemia cells in the setting of allogeneic hematopoietic stem cell transplant (HSCT) (5). The remaining NK cells, predominantly housed in lymphoid organs, are $CD56^{bright}$, and although less mature ("unlicensed") (3, 6, 7), they have a greater capability to secrete and respond to cytokines (8, 9). $CD56^{bright}$ and $CD56^{dim}$ NK cells are also distinguished by their differential expression of FcγRIII (CD16), an integral determinant of NK-mediated antibody-dependent cellular cytotoxicity (ADCC), with $CD56^{dim}$ NK cells expressing high levels of the receptor, while $CD56^{bright}$ NK cells are CD16 dim or negative (6). In contrast to T and B lymphocytes, NK cells do not express rearranged, antigen-specific receptors; rather, NK effector function is dictated by the integration of signals received through germ-line-encoded receptors that can recognize ligands on their cellular targets. Functionally, NK cell receptors are classified as activating or inhibitory. NK cell function, including cytotoxicity and cytokine release, is governed by a balance between signals received from inhibitory receptors, notably the killer Ig-like receptors (KIRs) and the heterodimeric C-type lectin receptor (NKG2A), and activating receptors, in particular the natural cytotoxicity receptors (NCRs) NKp46, NKp30, NKp44, and the C-type lectin-like activating immunoreceptor NKG2D (9).

The inhibitory KIRs (iKIRs) with known HLA ligands include KIR2DL2 and KIR2DL3, which recognize the HLA-C group 1-related alleles characterized by an asparagine residue at position 80 of the α-1 helix (HLA-CAsn80); KIR2DL1, which recognizes the HLA-C group 2-related alleles characterized by a lysine residue at position 80 (HLA-CLys80); and KIR3DL1, which recognizes the HLA-Bw4 alleles (9, 10). NK cells also express several activating receptors that are potentially specific for self-molecules. KIR2DS1 has been shown to interact with group 2 HLA-C molecules (HLA-C2), while KIR2DS2 was recently shown to recognize HLA-A*11 (10, 11). Hence, these receptors require mechanisms to prevent inadvertent activation against normal tissues, processes referred to as "tolerance to self." Engagement of iKIR receptors by HLA class I leads to signals that block NK-cell triggering during effector responses. These receptors explain the "missing self" hypothesis, which postulates that NK cells survey tissues for normal levels of the ubiquitously expressed MHC class I molecules (12, 13). Upon cellular transformation or viral infection, surface MHC class I expression on the cell surface is often reduced or lost to evade recognition by antitumor T cells. When a mature NK cell encounters transformed cells lacking MHC class I, their inhibitory receptors are not engaged, and the unsuppressed activating signals, in turn, can trigger cytokine secretion and targeted attack of the virus-infected or transformed cell (13, 14). In parallel, cellular stress and DNA damage (occurring in cells during viral or malignant transformation) results in upregulation of "stress ligands" that can be recognized by activating NK receptors. Thus, human tumor cells that have lost self-MHC class I expression or bear "altered-self" stress-inducible proteins are ideal targets for NK recognition and killing (14–16). NK cells directly kill tumor cells through several mechanisms, including release of cytoplasmic granules containing perforin and granzyme (16–18), expression of tumor necrosis factor (TNF) family members, such as FasL or TNF-related apoptosis-inducing ligand (TRAIL), which induce tumor cell apoptosis by interacting with their respective receptors Fas and TRAIL receptor (TRAILR) (16–19) as well as ADCC (9).

INTERACTION BETWEEN NATURAL KILLER CELLS AND OTHER IMMUNE SUBSETS

Increasing understanding of NK cell biology and their interaction with other cells of the immune system has led to several novel immunotherapeutic approaches as discussed in this review. NK cells produce cytokines that can exert regulatory control of downstream adaptive immune responses by influencing the magnitude of T cell responses, specifically T helper-1 (TH1) function (20). NK cell function, in turn, is regulated by cytokines, such as IL-2, IL-15, IL-12, and IL-18 (21), as well as by interactions with other cell types, such as dendritic cells, macrophages, and mesenchymal stromal cells (10, 22, 23). IL-15 has emerged as a pivotal cytokine required for NK cell development and maintenance. Whereas mice deficient in IL-2 (historically the cytokine of choice to expand and activate NK cells) have normal NK cells, IL-15-deficient mice lack NK cells (24).

Several cytokines are also known to inhibit NK cell activation and function, thus playing a crucial role in tumor escape from NK immune surveillance. Recently, considerable attention has been paid to the inhibitory effects of transforming growth factor-beta (TGF-β) and IL-10 on NK cell cytotoxicity (12, 25, 26). Several groups have shown that secretion of TGF-β by tumor cells results in downregulation of activating receptors, such as NKp30 and NKG2D, with resultant NK dysfunction (25, 26). Similarly, IL-10 production by acute myeloid leukemia (AML) blasts induces upregulation of NKG2A with significant impairment in NK function (3).

MODULATION OF ANTIBODY-DEPENDENT CELLULAR CYTOTOXICITY

The CD56^{dim} subset of NK cells expresses the Fcγ receptor CD16, through which NK cells mount ADCC, providing opportunities for its modulation to augment NK effector function (27, 28). In fact, a number of clinically approved therapeutic antibodies targeting tumor-associated antigens (such as rituximab or cetuximab) function at least partially through triggering NK cell-mediated ADCC. Several studies using mouse tumor models have established that efficient antibody–Fc receptor (FcR) interactions are essential for the efficacy of monoclonal antibody (mAb) therapy, a mainstay of cancer therapy (28, 29). Based on this premise, Romain et al. successfully engineered the Fc region of the IgG mAb, HuM195 targeting the AML leukemia antigen CD33, by introducing the triple mutation S293D/A330L/I332E (DLE). Using timelapse imaging microscopy in nanowell grids (TIMING, a method of analyzing kinetics of thousands of NK cells and mAb-coated targets), they demonstrated that the DLE-HuM195 antibody increased both the quality and quantity of NK cell-mediated ADCC by recruiting NK cells to participate in cytotoxicity via CD16-mediated signaling. NK cells encountering DLE-HuM195-coated targets induced rapid target cell apoptosis by promoting conjugation to multiple target cells (leading to increased "serial killing" of targets), thus inducing apoptosis in twice the number of targets as the wild-type mAb (27).

Additional approaches under investigation to enhance NK cell-mediated ADCC include antibody engineering and therapeutic combination of antibodies predicted to have synergistic activity. For example, mogamulizumab (an anti-CCR4 mAb recently approved in Japan) is defucosylated to increase binding by FcγRIIIA and thereby enhances ADCC. Mogamulizumab successfully induced ADCC activity against CCR4-positive cell lines and inhibited the growth of EBV-positive NK-cell lymphomas in a murine xenograft model (30). These findings suggest that mogamulizumab may be a therapeutic option against EBV-associated T and NK-lymphoproliferative diseases (30). Obinutuzumab (GA101) is a novel type II glycoengineered mAb against CD20 with increased FcγRIII binding and ADCC activity. In contrast to rituximab, GA101 induces activation of NK cells irrespective of their inhibitory KIR expression, and its activity is not negatively affected by KIR/HLA interactions (31). These data show that modification of the Fc fragment to enhance NK-mediated ADCC can be an effective strategy to augment the efficacy of therapeutic mAbs (31).

Although enhanced NK-mediated ADCC occurs in the presence of certain mAbs, in the case of non-engineered mAbs (such as rituximab), this NK-mediated cytotoxicity is typically still under the jurisdiction of KIR-mediated inhibition. However, ADCC responses can be potentiated *in vitro* in the presence of antibodies that block NK cell inhibitory receptor interaction with MHC class I ligands (32). These include the use of anti-KIR Abs to block the interaction of iKIRs with their cognate HLA class I ligands. To exploit this pathway pharmacologically, a fully humanized anti-KIR mAb 1-7F9 (IPH2101) (33) with the ability to block KIR2DL1/L2/L3 and KIR2DS1/S2 was generated. *In vitro*, anti-KIR mAbs can augment NK cell-mediated lysis of HLA-C-expressing tumor cells, including autologous AML blasts and autologous CD138$^+$ multiple myeloma (MM) cells (34). Additionally, in a dose-escalation phase 1 clinical trial in elderly patients with AML, 1-7F9 mAb was reported to be safe and could block KIRs for prolonged periods (35). A recombinant version of this mAb with a stabilized hinge (lirilumab) was recently developed. Lirilumab is a fully humanized IgG4 anti-KIR2DL1, -L2, -L3, -S1, and -S2 mAb. The iKIRs targeted by lirilumab collectively recognize virtually all HLA-C alleles, and the blockade of the three KIR2DLs allows targeting of every patient without the need for prior HLA or KIR typing (33, 34). Furthermore, the combination of an anti-KIR mAb with the immunomodulatory drug lenalidomide was shown to potentiate ADCC and is being tested in a phase 1 clinical trial in patients with MM [NCT01217203 (35)]. A potential concern is related to how inhibitory KIR blockade may impact on the ability of NK cells to discriminate self, healthy cells from abnormal virally infected or cancerous cells. Preliminary *in vitro* data suggest that Ab blockade of iKIRs will preferentially augment the ADCC response, without increasing cytotoxicity against self healthy cells (32). It is reassuring that in the IPH2101 phase 1 studies, no alterations in the expression of major inhibitory or activating NK receptors or frequencies of circulating peripheral lymphocytes were reported, indicating that the Ab does not induce clinically significant targeting of normal cells by NK cells (35). Lin et al. recently reported on the application of an agonistic NK cell-targeted mAb to augment ADCC (36). Following FcR triggering during ADCC, expression of the activation marker CD137 is increased. Agonistic antibodies targeting CD137 have been reported to augment NK-cell function, including degranulation, secretion of IFN-γ, and antitumor cytotoxicity in *in vitro* and *in vivo* preclinical models of tumor (36–39). The combination of the agonistic anti-CD137 antibody with rituximab is currently being evaluated in a phase 1 trial in patients with lymphoma [NCT01307267 (35–37)].

Other factors, such as specific CD16 polymorphisms and NKG2D engagement, can also influence ADCC, with certain polymorphisms (such as FcγRIIIa-V158F polymorphism) resulting in a stronger IgG binding (40). These findings are clinically relevant, as supported by the observation that patients with non-Hodgkin lymphoma (NHL) with the FcγRIIIa-V158F polymorphism experienced improved clinical response to rituximab (41, 42). In summary, several antibody combinations designed to boost ADCC have shown promising results in preclinical and early clinical trials, thus warranting further study of this strategy to enhance NK cell activity against tumor cells.

ADOPTIVE TRANSFER OF AUTOLOGOUS NK CELLS

The early studies of adoptive NK cell therapy focused on enhancing the antitumor activity of endogenous NK cells (43). Initial trials of adoptive NK therapy in the autologous setting involved using CD56 beads to select NK cells from a leukapheresis product and subsequently infusing the bead-selected autologous NK cells into patients (43, 44). Infusions were followed by administration of systemic cytokines (most commonly IL-2) to provide additional *in vivo* stimulation and support their expansion. This strategy met with limited success due to a combination of factors (44). Although cytokine stimulation promoted NK cell activation and resulted in greater cytotoxicity against malignant targets *in vitro*, only limited *in vivo* antitumor activity was observed (43–45). Similar findings were observed when autologous NK cells and systemic IL-2 were given as consolidation treatment to patients with lymphoma who underwent autologous BMT (46). The poor clinical outcomes observed with adoptive transfer of *ex vivo* activated autologous NK cells followed by systemic IL-2 were attributed to three factors: (1) development of severe life-threatening side effects, such as vascular leak syndrome as a result of IL-2 therapy; (2) IL-2-induced expansion of regulatory T cells known to directly inhibit NK cell function and induce activation-induced cell death (47–49); and (3) lack of antitumor effect related to the inhibition of autologous NK cells by self-HLA molecules. Strategies to overcome this autologous "checkpoint," thus redirecting autologous NK cells to target and kill leukemic blasts are the subject of intense investigation (33–35). These include the use of anti-KIR Abs (such as the aforementioned lirilumab) to block the interaction of inhibitory receptors on the surface of NK cells with their cognate HLA class I ligand.

EXPLOITING THE ALLOREACTIVITY OF ALLOGENEIC NK CELLS – ADOPTIVE IMMUNOTHERAPY AND BEYOND

An alternative strategy is to use allogeneic instead of autologous NK cells, thus taking advantage of the inherent alloreactivity afforded by the "missing self" concept (13). Over the past decade, adoptive transfer of *ex vivo*-activated or -expanded allogeneic NK cells has emerged as a promising immunotherapeutic strategy for cancer (24, 50–52). Allogeneic NK cells are less likely to be subject to the inhibitory response resulting from NK cell recognition of self-MHC molecules as seen with autologous NK cells. A number of studies have shown that infusion of haploidentical NK cells to exploit KIR/HLA alloreactivity is safe and can mediate impressive clinical activity in some patients with AML (50–52). In fact, algorithms have been developed to ensure selection of stem cell donors with the greatest potential for NK cell alloreactivity for allogeneic HSCT (50).

Promising results in the HSCT setting suggest that the application of this strategy in the non-transplant setting may be a plausible option. Miller et al. were among the first to show that adoptive transfer of *ex vivo*-expanded haploidentical NK cells after lymphodepleting chemotherapy is safe, and can result in expansion of NK cells *in vivo* without inducing graft-vs.-host disease (GVHD) (50). In a phase I dose-escalation trial, 43 patients with either hematologic malignancies (poor prognosis AML or Hodgkin lymphoma) or solid tumor (metastatic melanoma or renal cell carcinoma) received up to 2×10^7cells/kg of haploidentical NK cells following either low intensity [low-dose cyclophosphamide (Cy) and methylprednisolone or fludarabine (Flu)] or high intensity regimens (Hi-Cy/Flu). All patients received subcutaneous IL-2 after NK cell infusion. Whereas adoptively infused NK cells persisted only transiently following low intensity regimens, AML patients who received the more intense Hi-Cy/Flu regimen had a marked rise in endogenous IL-15 associated with expansion of donor NK cells and induction of complete remission (CR) in five of 19 very high-risk patients. The superior NK expansion observed after high-dose compared to low-dose chemotherapy was attributed to a combination of factors including prevention of host T cell-mediated rejection and higher levels of cytokines, such as IL-15. These findings provided the first evidence that haploidentical NK cells are safe and can persist and expand *in vivo*, supporting the proof of concept that NK cells may be applied for the treatment of selected malignancies either alone or as an adjunct to HSCT (50).

Another pivotal pilot study, the NKAML trial (Pilot Study of Haploidentical NK Transplantation for AML), reported that infusion of KIR-HLA-mismatched donor NK cells can reduce the risk of relapse in childhood AML (51). Ten pediatric patients with favorable or intermediate risk AML in first CR were enrolled following completion of 4–5 cycles of chemotherapy. All patients received a low-dose conditioning regimen consisting of Cy/Flu prior to infusion of NK cells (median, 29×10^6/kg NK cells) from a haploidentical donor, followed by six doses of IL-2. NK infusions were well tolerated with limited non-hematologic toxicity. All patients had transient engraftment of NK cells for a median of 10 days (range 2–189 days) with significant expansion of KIR-mismatched NK cells. With a median follow-up of 964 days, all patients remained in remission, suggesting that donor-recipient HLA-mismatched NK cells may reduce the risk of relapse in childhood AML (51).

Other strategies currently under investigation include the infusion of KIR-ligand-mismatched haploidentical NK cells as part of the pre-HSCT conditioning regimen (NCT00402558), and NK cell infusion to prevent relapse or as therapy for minimal residual disease in patients after haploidentical HSCT (NCT01386619).

ADOPTIVE NK CELL THERAPY IN SOLID MALIGNANCIES

Natural killer cell-based immunotherapies are also a promising therapeutic option for solid tumors. A number of studies have shown that the presence of intratumoral NK cells correlates with delayed tumor progression and improved outcomes (53–55). However, the successful application of NK cell-based therapies in the solid tumor setting poses a special challenge. In addition to the immune evasion strategies common to hematologic malignancies, such as secretion of immunosuppressive cytokines and downregulation of activating ligands (55–57), additional challenges specific to solid tumors exist; NK cells must not only traffic to sites of disease, but also penetrate the tumor capsule in order to exert their effector function. Furthermore, tumor targets must be inherently susceptible to NK-mediated cytotoxicity (58). Several groups have focused on strategies to alter the tumor microenvironment by targeting myeloid-derived suppressor cells or regulatory T cells (Treg) rather than the tumors themselves (58, 59). In fact, the prospect of combining NK cell-based immunotherapy with approaches to target the immunosuppressive tumor microenvironment or immune checkpoints, such as KIR blockade, is especially relevant to the treatment of solid tumors (55, 58). Several early phase clinical trials have demonstrated the feasibility of adoptive therapy with autologous or allogeneic *ex vivo* activated/expanded NK cells in patients with refractory solid malignancies [NCT01875601 (60)]; however, outside of the post-HSCT setting (namely in neuroblastoma), limited data on the clinical efficacy of NK cells in eradicating solid tumors exist. Currently, several trials are actively recruiting patients with refractory solid tumors for adoptive NK therapy (including NCT01807468, NCT02130869, and NCT0210089).

THE IDEAL MANUFACTURING STRATEGY FOR *EX VIVO* ACTIVATION OF NK CELLS

Recent approaches to adoptive NK therapy focused on infusion of NK cells that have undergone a process of *ex vivo* cytokine activation and expansion (61). A number of cytokines (IL-2, IL-12, IL-15, IL-18, IL-21, and type I IFNs) have been studied to activate and expand NK cells *ex vivo* (62–65). The most extensively studied cytokine is IL-2 (62, 63). This is not surprising, considering IL-2 was the only cytokine available in clinical grade until recently. Nevertheless, NK cells expanded in the presence of IL-12, IL-15, and IL-18, either alone or in combination, have shown remarkable activity against tumor targets in experimental models and offer an attractive strategy for clinical expansion of NK cells (64, 65). IL-15, in particular, is appealing as it does not stimulate Tregs (65). IL-15 has been tested in preclinical models with promising results; however, very high doses were necessary to observe any meaningful *in vivo* antitumor effects, and toxicity of systemic cytokine administration and cytokine-induced NK-cell apoptosis remained major issues (65). Recently, Miller et al. compared the persistence and *in vivo* efficacy of adoptively infused freshly activated NK cells (FA-NK) and *ex vivo*-expanded NK cells (Ex-NK) in a xenotransplantation model. They showed that *in vivo* NK cell persistence is cytokine dependent, with IL-15 being superior to IL-2. They also reported that cryopreservation of FA-NK or Ex-NK was detrimental to NK cell function, and that culture conditions influence homing, persistence, and expansion of NK cells *in vivo* (66).

Although the results from the abovementioned trials proved that transient persistence of adoptively transferred NK cells obtained via apheresis is feasible and safe, the requirement of a willing, available donor precludes the widespread applicability of this approach. Hence, more recent efforts have focused on optimizing methods for *ex vivo* expansion of NK cells from peripheral blood mononuclear cells (PBMCs) collected by a simple blood draw, with a goal of producing large quantities of purified, functionally active NK cells for clinical use. These expansion strategies include the use of "feeder cells," such as monocytes in the form of irradiated PBMCs, EBV-transformed lymphoblastoid cell lines (EBV-LCLs) or gene-modified, irradiated K562 cells expressing membrane-bound IL-15 or IL-21 and 41BB ligand for costimulation (61, 66–69) in gas-permeable large-scale expansion flasks. These techniques have dramatically increased the yield and activation status of NK cells, potentially overcoming the need for leukapheresis. Because the feeder cells used in these manufacture methods are lethally irradiated prior to use in culture (leaving the remaining feeder cells to be lysed by the expanding NK cells), the risk of infusing viable feeder cells is negligible. However, a number of safeguards have also been incorporated that include monitoring the growth rate of feeder cells and testing for the presence of viable feeder cells at the end of the culture period. Clinical products are, therefore, only released if no viable gene-modified K562 cells or transformed LCLs are present, with strict cutoff values for contaminating B cells and monocytes at the end of the culture period as well (67).

Although these expansion methods can produce large numbers of functionally active NK cells, concomitant expansion of competing cells with immunosuppressive properties, such as Tregs remains problematic. Early studies reported that NK cell infusions from haploidentical donors are able to induce remissions in some patients with AML, but not others (50–52). Several groups, therefore, set out to explore factors that may contribute to the failure of NK expansion *in vivo*. Bachanova explored the effect of competition between Tregs and NK cells in 57 patients with refractory AML who received lymphodepleting chemotherapy followed by NK cell infusion and IL-2 administration [NCT00274846 and NCT01106950 (70)]. Fifteen patients also received the IL-2-diphtheria toxin fusion protein (IL2DT) to deplete Tregs prior to NK cell infusion. IL2DT treatment was associated with increased donor NK cell persistence and improved CR and disease-free survival at 6 months (33 vs. 5% in patients not receiving IL2DT; $P < 0.01$). In the IL2DT cohort, NK cell expansion correlated with higher post-chemotherapy serum IL-15 levels ($P = 0.002$) and effective peripheral blood (PB) Treg depletion (<5%) at day 7 ($P < 0.01$). This study shed light on the importance of optimizing the cytokine milieu to facilitate the *in vivo* expansion of adoptively transferred NK cells and identifying ways to abrogate the immunosuppressive elements, such as regulatory T cells.

Although these data are encouraging, adoptive transfer of NK cells under good manufacturing practices (GMP) requires significant infrastructure and specialized processing equipment, thus limiting the availability and scalability of these NK cell therapies to a few specialized institutions (61). Nonetheless, the feasibility of centralized processing and safe delivery of *ex vivo*-manufactured NK cells for infusion at remote clinics have been demonstrated, suggesting that the practice might become more widespread as procedures are optimized (71). For example, in order to improve access to *ex vivo* activated NK cells and ease the burden associated with producing cellular products at individual treatment centers, the National Heart, Lung, and Blood Institute (NHLBI, Bethesda, MD, USA) sponsored the Production Assistance for Cellular Therapies (PACT) program. Using this approach, activated NK cells have been sent to other centers for infusion into patients (72, 73).

Since the initial reports of successful adoptive transfer of NK cells (50–52), many groups continue to perform extensive preclinical exploration of the ideal manufacturing strategy for *ex vivo* activation and expansion of NK cells. Several expansion methods optimized in the preclinical setting have been successfully scaled up for the clinic (61, 67–70). In addition to the six clinical trials of adoptive NK cell therapy for leukemia that have reported their data (48–52, 70), there are currently 12 active clinical trials enrolling patients with hematologic malignancies for NK cell adoptive therapy, a number that is steadily rising.

ALTERNATIVE SOURCES OF NK CELLS FOR ADOPTIVE TRANSFER – IS CORD BLOOD THE ANSWER?

Although the majority of clinical studies of NK cell immunotherapy have used PB NK cells, several alternative sources of NK cells exist. These include bone marrow, human embryonic stem cells (hESCs), induced pluripotent stem cells [iPCSs (74, 75)], and umbilical cord blood (CB). While the generation of NK cells from hESCs or iPCS has been largely experimental to date, clinical-grade generation and expansion of NK cells from CB-derived $CD34^+$ cells has been successfully achieved (76).

Umbilical CB as a source for NK cells lends additional clinical advantages. CB contains a high percentage of NK cells (77, 78) and serves as an immediate "off-the-shelf" source of NK cells, with less stringent requirements for HLA matching, and lower risk of causing GVHD following infusion due to the naivety of the cord T cell repertoire (77, 78). Although no direct comparison of PB- and CB-derived NKs has been performed in the clinical setting, *in vitro* studies have identified a number of differences between CB and PB NK cells. CB NK cells form weaker conjugates with target cells due to the lower membrane expression of adhesion molecules on their surface (79, 80). CB NK cells also express higher levels of lectin-like inhibitory receptors (CD94/NKG2A) and lower levels of KIRs, indicating an immature phenotype (81). CB NK cells are similarly sensitive to cytokines for *in vivo* expansion and persistence (82). However, it appears that the requirements for *in vitro* expansion of CB NK cells may be different to those required for PB NKs. CB NKs are less responsive to IL-2 stimulation, which may be related to the lower expression of IL-2Rα and reduced activation of the STAT5 signaling pathway as compared with PB NK cells (83). The combination of IL-15 and IL-18, however, can induce significant proliferation and cytokine production by CB NK cells, while the killing capacity of CB NK cells is significantly enhanced after stimulation with IL-15 (83). As with PB-derived NK cells, T-cell contamination is a concern, but can be ameliorated by CD3 depletion. T-cell contamination should be limited to $<1–5 \times 10^5$/kg (61) to minimize the risk of GVHD. In addition, $CD56^+$ selection reduces B-cell contamination to <1%, which minimizes passenger B lymphocyte-mediated complications, such as EBV-related post-transplant lymphoproliferative disorder (PTLD) and acute hemolytic anemia.

More recently, efforts have focused on optimizing the large-scale expansion of purified CB-derived NK cells. Shah et al. were the first to describe a strategy for expanding NK cells from cryopreserved CB units in which they employed K562-based artificial antigen-presenting cells (aAPCs) expressing membrane-bound IL-21 (clone 9.mbIL21) (77, 84). The clone 9.mbIL21 cell line is GMP-grade and expresses membrane-bound IL-21, 4-1BB ligand, CD64 (FcγRI), and CD86. After only 14 days of culture in a gas-permeable culture system, mean-fold expansion of CB-NK cells was 1848-fold from fresh and 2389-fold from cryopreserved CB with >95% purity for $CD56^+CD3^-$ NK cells. aAPC-expanded CB-NK cells displayed a phenotype similar to that of expanded PB-NK cells and maintained strong expression of the transcription factors eomesodermin and T-bet. Furthermore, CB-NK cells formed functional immune synapses and efficiently killed various MM targets *in vitro*. Finally, aAPC-expanded CB-NK cells showed significant *in vivo* activity against MM in a xenogenic mouse model. These findings highlight a clinically applicable strategy for the generation of highly functional CB-NK cells using an aAPC platform, which can be potentially extended to other hematologic malignancies and solid tumors (77). A number of phase I/II clinical trials are underway to test the feasibility and efficacy of CB-NK cell adoptive therapy in patients with hematologic malignancies (NCT01619761, NCT01729091 NCT02280525, NCT01914263, and NCT00412360) (summarized in **Table 1**).

HUMAN NK CELL LINES AS A SOURCE OF NK IMMUNOTHERAPY

The adoptive transfer of NK cell lines has several theoretical advantages over the use of patient- or donor-derived NK cells. These are primarily related to the lack of expression of iKIRs, presumed lack of immunogenicity, ease of expansion and availability as an "off-the-shelf" product (85). Several human NK cell lines, such as NK-92 and KHYG-1, have been documented to exert antitumor activity in both preclinical and clinical settings (86–88). NK-92, the most extensively characterized NK-cell line, was established in 1994 from the PB of a male Caucasian patient with NHL. NK-92 cells are IL-2-dependent, harbor a $CD2^+CD56^+CD57^+$ phenotype and exert potent *in vitro* cytotoxicity (86). Infusion of up to 10^{10} cells/m^2 NK-92 cells into patients with advanced lung cancer and other advanced malignancies was well tolerated and the cells persisted for a minimum of 48 h with encouraging clinical responses (86, 88–91). However, potential limitations of using NK cell lines, such as NK-92 cells, include the requirement for irradiation to reduce the risk of engrafting cells with potential *in vivo* tumorigenicity, and the need for pre-infusion conditioning to avoid host rejection. Furthermore, infusion of allogeneic NK cell lines may induce T and B cell alloimmune responses, limiting their *in vivo* persistence and precluding multiple infusions. A number of studies are testing NK-92 cells (Neukoplast®) in patients with solid tumors, such as Merkel cell cancer and renal cell carcinoma, as well as in hematological malignancies (85).

While results from clinical studies of NK cell adoptive therapy are encouraging (48–52, 70), significant gaps remain in our understanding of the optimal conditions for NK cell infusion. Based on the pioneering work from Rosenberg et al. demonstrating the importance of lymphodepletion to support the expansion of tumor-infiltrating T cells (92) and given its emergence as a key determinant of efficacy with CAR therapy, several groups are actively investigating the ideal preparative regimen to promote the expansion and persistence of adoptively infused NK cells (53, 69, 70, 75). Available data support the use of high-dose Cy/Flu regimen as the frontrunner, considering it is reasonably well tolerated and shown to support the *in vivo* expansion of NK cells

TABLE 1 | Published results of NK adoptive immunotherapy trials in hematologic malignancies.

Reference	Approach	Disease	NK source of cells	Conditioning regimen	Dose of cells	Outcome
Burns et al. (43)	*Ex vivo* IL-2 activated autologous NKs or bolus IL-2	Relapsed lymphoma *N* = 29	Autologous	None	4 × 10^7–8 × 10^7 cells/kg	1° endpoint safety/feasibility; no change in outcome compared to historical controls
[a]Miller et al. (50)	IL-2 activated NK cells	HR AML (adults) *N* = 19	Haplo-related donors	Hi-Cy/Flu	1 × 10^6–2 × 10^7 cells/kg followed by 14 days IL-2	5/19 (26%) CR
[a]Rubnitz and Inaba (51)	Fresh-activated NK (FA-NK)	LR/IR AML (pedi) *N* = 10	Haplo-related donors	Hi-Cy/Flu	Median 29 × 10^6 cells/kg followed by IL-2 × 6 doses	10/10 (100%) CR at 964 days
Yoon et al. (49)	IL-7/15/21 *ex vivo* cultured NKs	HR ALL/AML/MDS (adults) *N* = 14	Haplo-related HSCT donors (from CD34⁺ fraction)	Pre-SCT conditioning regimen (Bu/Flu/thymo)	Median 2.2 × 10^6 cells/kg	1° endpoint safety/feasibility; (no toxicity; low-grade GVHD); 4/14 (28%) alive and well
[a]Curti and Ruggeri (52)	CD56⁺ selected NKs	AML-CR and relapsed (adult) *N* = 13	Haplo-related donors	Hi-Cy/Flu	5 × 10^6 cells/kg followed by IL-2 × 6 doses	6/13 (46%) remain in CR
Stern et al. (48)	1–3 doses positively selected NKs	ALL, AML (adult and pedi) *N* = 15	Haplo donors	Pre-SCT conditioning regimen	Median 1.2 × 10^7 cells/kg	4/16 (25%) alive
Klingemann and Grodman (71)	Apheresis-mobilized CD56 selection	HL, NHL, MM *N* = 13	Haplo donors	None	1 × 10^5–2 × 10^7 cells/kg	1° endpoint safety/feasibility; 7/13 in remission
[a]Bachanova (70)	NK infusion w/IL-2 ± IL2DT Treg depletion	AML *N* = 42 (IL-2 alone) *N* = 15 (+IL2DT)	Haplo donors	Hi-Cy/Flu	Mean 2.6 ± 1.5 × 10^7 cells/kg	IL-2 alone: 9/42 (21%) CR/CRi IL2DT: 8/15 CR/CRi (53%)
Choi et al. (116)	Apheresis-mobilized, *ex vivo* IL-15/21 induced NK cells	*N* = 41	Haplo donors	Bu/Flu/ATG	Median 1 × 10^8 cells/kg	Reduced leukemia progression 46 vs. 74%

HR, high risk; haplo, haploidentical; LR, low risk; IR, intermediate risk; Hi-Cy/Flu, high-dose cyclophosphamide and fludarabine; CR, complete remission; ALL, acute lymphoblastic leukemia; AML, acute myeloid leukemia; MDS, myelodysplastic syndromes; SCT, hematopoietic stem cell transplant; HL, Hodgkin lymphoma; NHL, non-Hodgkin lymphoma; pedi, pediatric; CRi, complete remission with incomplete platelet recovery; IL2DT, IL-2-diphtheria fusion protein.

[a]NK cells infused outside of the setting of hematopoietic cell transplantation.

(51, 70). IL-15 is an ideal candidate cytokine for the expansion of NK cells *in vivo*, especially since it does not promote expansion of regulatory T cells (66), which have been shown to suppress NK cell effector function in IL-2-based trials (69, 70). In a recent phase 1 study in patients with metastatic melanoma or renal cell carcinoma, rhIL-15 was shown to activate NK cells, monocytes, γδ, and CD8 T cells (93). However, as an intravenous bolus dose, rhIL-15 proved too difficult to administer because of significant clinical toxicities (93). Based on these promising data, alternative dosing strategies are being investigated, including continuous intravenous infusions. To this effect, systemic IL-15 along with infusion of donor NK cells are currently being tested in a phase I clinical trial for AML (NCT01385423).

CHIMERIC ANTIGEN RECEPTOR-MODIFIED NK CELLS

Chimeric antigen receptors have been used extensively to redirect the specificity of T cells against leukemia (94). Recently, use of an anti-CD19-BB-ζ receptor transduced into autologous or allogeneic T cells produced dramatic clinical responses in patients with acute lymphoblastic leukemia (95, 96); however, infusions of activated T cells from an allogeneic source are likely to increase the risk of GVHD. T cell-depleted allogeneic NK cells, by contrast, should not cause GVHD, as predicted by observations in murine models, as well as in patients with leukemia and solid malignancies treated with haploidentical NK cells (50–52). Given their shorter lifespan and potent cytolytic function, mature NK cells provide attractive candidate effector cells to express CARs and, provide an excellent source of off-the-shelf cellular therapy for patients with cancer.

The feasibility of genetically engineering NK cells to express CARs has been shown in the preclinical setting (97, 98). Primary human NK cells, as well as NK-92 cells, have been successfully engineered to express CARs against a number of targets including CD19, CD20, CD244, and HER2 (97). CAR-transduced NK cells mediate efficient *in vitro* and *in vivo* killing of tumor targets (97, 98) although to date, no clinical data of CAR NK cell therapy have been reported. Shimasaki et al. recently tested the expression of a receptor containing CD3ζ and 4-1BB signaling molecules (anti-CD19-BB-ζ) in human NK cells after mRNA electroporation using a clinical-grade electroporator. The authors reported adequate transfection efficiency 24 h after electroporation, with median anti-CD19-BB-ζ expression of 40.3% in freshly purified and 61.3% in expanded NK cells. NK cells expressing anti-CD19-BB-ζ secreted IFN-γ in response to CD19-positive target cells. Interestingly, the levels of CAR expression in NK cells after mRNA transfection were comparable to those achieved by retroviral transduction. A large-scale protocol was developed to transfect expanded NK cells, achieving excellent receptor expression and considerable cytotoxicity of CAR-transduced NK cells in xenograft models of B-cell leukemia (99). Another interesting strategy is the development of CAR-modified NK cells that target NKG2D ligands on the surface of tumor cells, rendering NK more cytotoxic against a variety of hematologic and solid malignancies (100). NK cells have also been successfully engineered to target antigens on a variety of solid tumors. For example, an NK-CAR targeting the ganglioside GD2 (present on neuroblastoma cells) has been tested in preclinical studies (101, 102). GD2 is also expressed on breast cancer stem cells, thus raising the potential for its widespread use as a target for immunotherapy (103). Additional antigens targeted by NK CARs include HER2 (overexpressed in a number of solid tumors), CD138, and CS1 (overexpressed in MM) (104, 105).

Although these data support the use of CAR engineering to redirect the specificity of NK cells to augment their cytotoxicity, a number of challenges remain. These include the relative difficulty in expressing exogenous genes in primary human NK cells and the need to expand NK cells in culture to achieve adequate numbers for clinical studies of immunotherapy. To counteract this difficulty, some groups have expressed CARs in the human NK-like cell line NK-92, in an attempt to engineer a uniformly cytolytic effector cell population (106). As previously mentioned, NK-92 cells can be easily expanded in culture and their safety has been shown in phase I clinical trials in human subjects. Thus, CAR-expressing NK-92 cells may offer a practical source of cells for NK cell-based immunotherapeutic trials. In order to prevent the risk of engrafting cells with potential *in vivo* tumorigenicity, however, NK-92 cells must be irradiated prior to infusion, which may in turn significantly impact their *in vivo* persistence and long-term antitumor efficacy. Although limited *in vivo* persistence could prove beneficial once the alloreactive NK cells have eradicated the tumors, a number of studies of adoptive therapy with NK cells and CAR-modified T cells have reported the importance of cell persistence in inducing long-term antitumor response (50, 95, 96).

As with CAR-modified T cell therapy, a number of variables can affect the activation, antitumor efficacy, and persistence of CAR-NK cells. Second and third generation CAR constructs incorporating additional costimulatory domains (e.g., CD28, OX-40, or 4-1BB) have been shown to enhance both *in vitro* and *in vivo* activation, and the persistence of CAR T cells. Further studies exploring the optimal vector, construct and transduction method are necessary to identify the "perfect NK CAR."

SAFETY CONCERNS RELATED TO ADOPTIVE TRANSFER OF CAR-MODIFIED NK CELLS

When considering the use of CAR-modified effector cells, one must take into account their safety profile. Many of the same concerns raised with CAR-modified T cells may be relevant to CAR-NK cells. These include on-target/off-tumor effects, GVHD, cytokine release syndrome, tumor lysis syndrome, and toxicity to normal tissues due to limited selectivity of the target antigen (107–109). Thus, the necessity of equipping CAR-modified NK cells with a "safety switch" or suicide gene is an important question to explore. While mature allogeneic CAR-engineered NK cells are expected to be short lived, data on the persistence of more immature NK cells, such as those derived from CB, are lacking. Interestingly, a recent study reported that IL15/4-1BBL-activated NK cells infused early after T-depleted allogeneic stem cell transplantation in patients not receiving immunosuppressive prophylaxis could contribute to acute GVHD (110). To this effect, the insertion of a suicide safety switch system, as employed with CAR-modified T cells (111, 112), can provide an efficient means for depletion of these cells if needed. Inducible suicide systems have safely and effectively eradicated GVHD in patients receiving adoptively transferred T cells without causing deleterious effects (112). However, these systems have not been extensively studied in NK cells, and in the absence of clinical data on the *in vivo* persistence of CAR-modified NK cells, the necessity of a suicide switch in this setting remains unknown.

Despite the growing wealth of preclinical experience with CAR-engineered NK cells, to date, only two clinical studies (both targeting CD19$^+$ malignancies using a retroviral transduced anti-CD19-BB-ζ NK-CAR) have obtained regulatory approval: one is a recently completed pediatric study at St. Jude Children's Research Hospital, where haploidentical NK cells modified with anti-CD19-BB-ζ CAR were infused into patients with B-ALL (ClinicalTrials.gov.NCT00995137) and the other is an ongoing study at the National University Hospital in Singapore (ClinicalTrials.gov. NCT01974479) using IL-2-activated haploidentical CAR-modified NK cells in pediatric and adult patients with refractory B-ALL (99). The results of these studies have not been reported to date.

BISPECIFIC AND TRISPECIFIC ENGAGERS

An innovative immunoglobulin-based strategy to redirect NK cytotoxicity towards tumor cells is to create either bispecific or trispecific antibodies (BiKE, TriKE) (113). BiKEs are constructed by joining a single-chain Fv against CD16 and a single-chain Fv against a tumor-associated antigen (BiKE), or two tumor-associated antigens (TriKE). Gleason et al. showed that bispecific (bscFv) CD16/CD19 and trispecific (tscFv) CD16/CD19/CD22 engagers directly trigger NK cell activation through CD16, significantly increasing NK cell cytolytic activity and cytokine production against various CD19-expressing B cell lines. The same group also developed and tested a CD16 × 33 BiKE in refractory

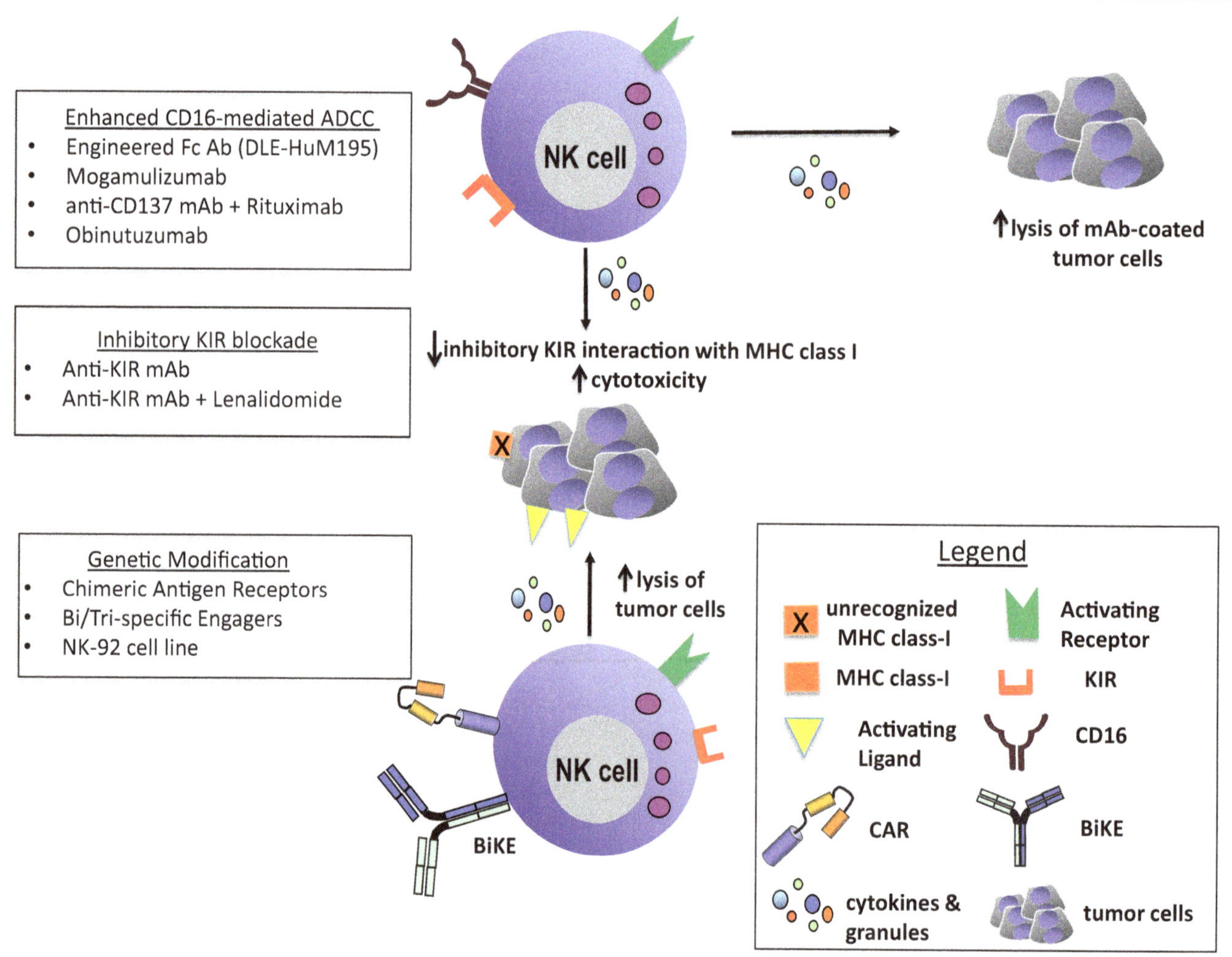

FIGURE 1 | Enhanced CD16-mediated ADCC: engineered Fc Ab (DLE-HuM195) (27), mogamulizumab (30), anti-CD137 mAb + rituximab (36–39), and obinutuzumab (31). Inhibitory KIR blockade: anti-KIRmAb (33–35) and anti-KIR mAb + lenalidomide (35). Genetic modification: chimeric antigen receptors (97–105), bi/trispecific engagers (113, 114), and NK-92 cell line (86–91).

AML and demonstrated that the potent killing by NK cells could overcome the inhibitory effect of KIR signaling (113, 114).

Notably, activated NK cells lose CD16 (FcRγIII) and CD62L through a metalloprotease called ADAM17, which is expressed on NK cells, which may in turn impact on the efficacy of Fc-mediated cytotoxicity (115). Romee et al. recently showed that selective inhibition of ADAM17 enhances CD16-mediated NK cell function by preserving CD16 on the NK cell surface, thus enhancing ADCC (115). Additionally, Fc-induced production of cytokines by NK cells exposed to rituximab-coated B cell targets can be further enhanced by ADAM17 inhibition. These findings support a role for targeting ADAM17 to prevent CD16 shedding and to improve the efficacy of therapeutic mAbs. The same group subsequently discovered that ADAM17 inhibition enhances CD16 × 33 BiKE responses against primary AML targets (114).

NK CELLS – WHAT DOES THE FUTURE HOLD?

Recent advances in the understanding of NK cell immunobiology have paved the way for novel and innovative anti-cancer therapies. Here, we have discussed a representation of these novel immunotherapeutic strategies to potentiate NK cell function and enhance antitumor activity including ADCC-inducing mAbs, *ex vivo* activated or genetically modified NK cells and bi- or trispecific engagers (**Figure 1**).

Although experience has shown that adoptive immunotherapy with allogeneic NK cells may be more efficacious than with autologous NK cells, to date, their long-term antitumor benefits have been modest (3). Expansion and persistence of NK cells following infusion appear to be the main determinants of clinical response (50–52, 70), thus underscoring the importance of identifying ways to enhance their persistence and antitumor activity. It is likely that the combination of high-dose lymphodepleting chemotherapy with additional modifications (such as Treg depletion, *in vivo* administration of cytokines, such as IL-15 or enhancement of CD16-mediated antigen targeting) may maximize NK persistence and efficacy.

In addition, the possibility of third-party "off-the-shelf" products with partially HLA-matched NK cells from CB, third-party donors, or NK cell lines allow the advantage of unlimited sources of cells to improve the practicality of cell therapy. With increasing focus on genetically modifying NK cells to redirect their specificity or engager-modified NK cells, it is likely that NK cells will move to the forefront of cancer therapy over the next few years.

AUTHOR CONTRIBUTIONS

RR and KR wrote and edited the manuscript.

REFERENCES

1. Lotzova E, Savary CA, Herberman RB. Inhibition of clonogenic growth of fresh leukemia cells by unstimulated and IL-2 stimulated NK cells of normal donors. *Leuk Res* (1987) **11**(12):1059–66. doi:10.1016/0145-2126(87)90158-5
2. Lowdell MW, Craston R, Samuel D, Wood ME, O'Neill E, Saha V, et al. Evidence that continued remission in patients treated for acute leukaemia is dependent upon autologous natural killer cells. *Br J Haematol* (2002) **117**(4):821–7. doi:10.1046/j.1365-2141.2002.03495.x
3. Davies J, Stringaris K, Barrett AJ, Rezvani K. Opportunities and limitations of NK cells as adoptive therapy for malignant disease. *Cytotherapy* (2014) **16**(11):1453–66. doi:10.1016/j.jcyt.2014.03.009
4. Lanier LL, Le AM, Civin CI, Loken MR, Phillips JH. The relationship of CD16 (Leu-11) and Leu-19 (NKH-1) antigen expression on human peripheral blood NK cells and cytotoxic T lymphocytes. *J Immunol* (1986) **136**(12):4480–6.
5. Locatelli F, Moretta F, Brescia L, Merli P. Natural killer cells in the treatment of high-risk acute leukaemia. *Semin Immunol* (2014) **26**(2):173–9. doi:10.1016/j.smim.2014.02.004
6. Cooper MA, Fehniger TA, Turner SC, Chen KS, Ghaheri BA, Ghayur T, et al. Human natural killer cells: a unique innate immunoregulatory role for the CD56(bright) subset. *Blood* (2001) **97**(10):3146–51. doi:10.1182/blood.V97.10.3146
7. Campbell JJ, Qin S, Unutmaz D, Soler D, Murphy KE, Hodge MR, et al. Unique subpopulations of CD56+ NK and NK-T peripheral blood lymphocytes identified by chemokine receptor expression repertoire. *J Immunol* (2001) **166**(11):6477–82. doi:10.4049/jimmunol.166.11.6477
8. De Maria A, Bozzano F, Cantoni C, Moretta L. Revisiting human natural killer cell subset function revealed cytolytic CD56(dim)CD16+ NK cells as rapid producers of abundant IFN-gamma on activation. *Proc Natl Acad Sci U S A* (2011) **108**:728–32. doi:10.1073/pnas.1012356108
9. Farag S, Caligiuri M. Human natural killer cell development and biology. *Blood Rev* (2006) **20**(3):123–37. doi:10.1016/j.blre.2005.10.001
10. Moretta L, Pietra G, Montaldo E, Vacca P, Pende D, Falco M, et al. Human NK cells: from surface receptors to the therapy of leukemias and solid tumors. *Front Immunol* (2014) **5**:87. doi:10.3389/fimmu.2014.00087
11. Liu J, Xiao Z. Activating killer immunoglobulin-like receptor 2DSs2 binds to HLA A*11. *Proc Natl Acad Sci U S A* (2014) **111**(7):2662–7. doi:10.1073/pnas.1322052111
12. Fauriat C, Long EO, Ljunggren HG, Bryceson YT. Regulation of human NK-cell cytokine and chemokine production by target cell recognition. *Blood* (2010) **115**:2167–76. doi:10.1182/blood-2009-08-238469
13. Karre K. Immunology. A perfect mismatch. *Science* (2002) **295**:2029–31.
14. Campbell K, Hasegawa J. Natural killer cell biology: an update and future directions. *J Allergy Clin Immuol.* (2013) **32**(3):536–44. doi:10.1016/j.jaci.2013.07.006
15. Caligiuri MA. Human natural killer cells. *Blood* (2008) **112**:461–9. doi:10.1182/blood-2007-09-077438
16. Bradley M, Zeytun A, Rafi-Janajreh A, Nagarkatti PS, Nagarkatti M. Role of spontaneous and interleukin-2-induced natural killer cell activity in the cytotoxicity and rejection of Fasl and Fas- tumor cells. *Blood* (1998) **92**:4248–55.
17. Screpanti V, Wallin RP, Ljunggren HG, Grandien A. A central role for death receptor-mediated apoptosis in the rejection of tumors by NK cells. *J Immunol* (2001) **167**:2068–73. doi:10.4049/jimmunol.167.4.2068
18. Kayagaki N, Yamaguchi N, Nakayama M, Takeda K, Akiba H, Tsutsui H, et al. Expression and function of TNF-related apoptosis-inducing ligand on murine activated NK cells. *J Immunol* (1999) **163**:1906–13.
19. Joncker NT, Shifrin N, Delebecque F, Raulet DH. Mature natural killer cells reset their responsiveness when exposed to an altered MHC environment. *J Exp Med* (2010) **207**(10):2065–72. doi:10.1084/jem.20100570
20. Martin-Fontecha A, Thomsen L, Brett S, Gerard C, Lipp M, Lanzavecchia A, et al. Induced recruitment of NK cells to lymph nodes provides IFN-γ for TH1 priming. *Nat Immunol* (2004) **5**:1260–5. doi:10.1038/ni1138
21. Agaugue S, Marcenaro E. Human natural killer cells exposed to IL-2, IL-12, IL-18 or IL-4 differently modulate priming of naïve T cells by

monocyte-derived dendritic cells. *Blood* (2008) **112**(5):1776–83. doi:10.1182/blood-2008-02-135871
22. Ferlazzo G, Tsang ML, Moretta L, Melioli G, Steinman RM, Munz C. Human dendritic cells activate resting natural killer (NK) cells and are recognized via the NKp30 receptor by activated NK cells. *J Exp Med* (2002) **195**(3):343–511. doi:10.1084/jem.20011149
23. Bellora F, Castriconi R, Dondero A, Reggiardo G, Moretta L, Mantovani A, et al. The interaction of human natural killer cells with either unpolarized or polarized macrophages results in different functional outcomes. *Proc Natl Acad Sci U S A* (2010) **107**(50):21641–59. doi:10.1073/pnas.1007654108
24. Murphy WJ, Parham P, Miller JS. NK cells – from bench to clinic. *Biol Blood Marrow Transplant* (2012) **18**(1 Suppl):S2–7. doi:10.1016/j.bbmt.2011.10.033
25. Castriconi R, Dondero A, Bellora F, Moretta L, Castellano A, Locatelli F, et al. Neuroblastoma-derived TGF-beta1 modulates the chemokine receptor repertoire of human resting NK cells. *J Immunol.* (2013) **190**:5321–8. doi:10.4049/jimmunol.1202693
26. Lee HM, Kim KS, Kim J. A comparative study of the effects of inhibitory cytokines on human natural killer cells and the mechanistic features of transforming growth factor-beta. *Cell Immunol* (2014) **290**:52–61. doi:10.1016/j.cellimm.2014.05.001
27. Romain G, Senyukov V, Rey-Villamizar N, Merouane A, Kelton W, Liadi I, et al. Antibody Fc-engineering improves frequency and promotes kinetic boosting of serial killing mediated by NK cells. *Blood* (2014) **124**(22):3241–9. doi:10.1182/blood-2014-04-569061
28. Clynes RA, Towers TL, Presta LG, Ravetch JV. Inhibitory Fc receptors modulate in vivo cytotoxicity against tumor targets. *Nat Med* (2000) **6**(4):443–6. doi:10.1038/74704
29. de Haij S, Jansen JH, Boross P, Beurskens FJ, Bakema JE, Bos DL, et al. In vivo cytotoxicity of type I CD20 antibodies critically depends on Fc receptor ITAM signaling. *Cancer Res* (2010) **70**(8):3209–17. doi:10.1158/0008-5472.CAN-09-4109
30. Kanazawa T, Hiramatsu Y, Iwata S, Siddiquey M, Sato Y, Suzuki M, et al. Anti CCR-4 monoclonal antibody mogamulizumab for the treatment of EBV-associated T-and NK-cell lymphoproliferative diseases. *Clin Cancer Res* (2014) **20**(19):5075–84. doi:10.1158/1078-0432.CCR-14-0580
31. Terszowski G, Klein CKIR. /HLA interactions negatively affect rituximab-but not GA101 (obinutuzumab)-induced antibody-dependent cellular cytotoxicity. *J Immunol* (2014) **192**(12):5618–24. doi:10.4049/jimmunol.1400288
32. Binyamin L, Alpaugh RK, Hughes TL, Lutz CT, Campbell KS, Weiner LM. Blocking NK cell inhibitory self-recognition promotes antibody-dependent cellular cytotoxicity in a model of anti-lymphoma therapy. *J Immunol* (2008) **180**(9):6392–401. doi:10.4049/jimmunol.180.9.6392
33. Kohrt H, Thielens A. Anti-KIR antibody enhancement of anti-lymphoma activity of natural killer cells as monotherapy and in combination with anti-CD20 antibodies. *Blood* (2014) **123**(5):678–86. doi:10.1182/blood-2013-08-519199
34. Romagne F, Andre P, Spee P, Zahn S, Anfossi N, Gauthier L, et al. Preclinical characterization of I-7F9, a novel human anti-KIR receptor therapeutic antibody that augments natural killer-mediated killing of tumor cells. *Blood* (2009) **114**(13):2667–77. doi:10.1182/blood-2009-02-206532
35. Vey N, Bourhis JH, Boissel N, Bordessoule D, Prebet T, Charbonnier A, et al. A phase 1 trial of the anti-inhibitory KIR mAb IPH2101 for AML in complete remission. *Blood* (2012) **120**(22):4317–23. doi:10.1182/blood-2012-06-437558
36. Lin W, Voskens C, Zhang X, Schindler DG, Wood A, Burch E, et al. Fc-dependent expression of CD137 on human NK cells: insights into "agonistic" effects of anti-CD137 monoclonal antibodies. *Blood* (2008) **112**(3):699–707. doi:10.1182/blood-2007-11-122465
37. Benson DM Jr, Bakan CE, Zhang S, Collins SM, Liang J, Srivastava S, et al. IPH2101, a novel anti-inhibitory KIR antibody, and lenalidomide combine to enhance the natural killer cell versus multiple myeloma effect. *Blood* (2011) **118**(24):6387–91. doi:10.1182/blood-2011-06-360255
38. Kohrt HE, Houot R, Goldstein MJ, Weiskopf K, Alizadeh AA, Brody J, et al. CD137 stimulation enhances the antilymphoma activity of anti-CD20 antibodies. *Blood* (2011) **117**(8):2423–32. doi:10.1182/blood-2010-08-301945 Epub 2011/01/05.,
39. Wilcox RA, Tamada K, Strome SE, Chen L. Signaling through NK cell-associated CD137 promotes both helper function for CD8+ cytolytic T cells and responsiveness to IL-2 but not cytolytic activity. *J Immunol* (2002) **169**(8):4230–6. doi:10.4049/jimmunol.169.8.4230
40. Seidel UJ, Schlegel P, Lang P. Natural killer cell mediated antibody-dependent cellular cytotoxicity in tumor immunotherapy with therapeutic antibodies. *Front Immunol* (2013) **4**:76. doi:10.3389/fimmu.2013.00076
41. Cartron G, Dacheux L. Therapeutic activity of humanized anti-CD20 monoclonal antibody and polymorphism in IgG Fc receptor FcgammaRIIIa gene. *Blood* (2002) **99**(3):754–8. doi:10.1182/blood.V99.3.754
42. Weng WK, Levy R. Two immunoglobulin G fragment C receptor polymorphisms independently predict response to rituximab in patients with follicular lymphoma. *J Clin Oncol* (2012) **21**(21):3940–7. doi:10.1200/JCO.2003.05.013
43. Burns LJ, Weisdorf DJ, DeFor TE, Vesole DH, Repka TL, Blazar BR, et al. IL-2-based immunotherapy after autologous transplantation for lymphoma and breast cancer induces immune activation and cytokine release: a phase I/II trial. *Bone Marrow Transplant* (2003) **32**:177–86. doi:10.1038/sj.bmt.1704086
44. Rosenberg SA, Lotze MT, Muul LM, Leitman S, Chang AE, Ettinghausen SE, et al. Observations on the systemic administration of autologous lymphokine-activated killer cells and recombinant interleukin-2 to patients with metastatic cancer. *N Engl J Med* (1985) **313**:1485–92. doi:10.1056/NEJM198512053132327
45. Parkhurst MR, Riley JP, Dudley ME, Rosenberg SA. Adoptive transfer of autologous natural killer cells leads to high levels of circulating natural killer cells but does not mediate tumor regression. *Clin Cancer Res* (2011) **17**:6287–97. doi:10.1158/1078-0432.CCR-11-1347
46. Benyunes MC, Higuchi C. Immunotherapy with interleukin 2 with or without lymphokine-activated killer cells after autologous bone marrow transplantation for malignant lymphoma: a feasibility trial. *Bone Marrow Transplant* (1995) **16**(2):283–8.
47. Cheng M, Chen Y, Xiao W, Sun R, Tian Z. NK cell-based immunotherapy for malignant diseases. *Cell Mol Immunology* (2013) **10**(3):230–52. doi:10.1038/cmi.2013.10
48. Stern M, Passweg JR, Meyer-Monard S, Esser R, Tonn T, Soerensen J, et al. Pre-emptive immunotherapy with purified natural killer cells after haploidentical SCT: a prospective phase II study in two centers. *Bone Marrow Transplant* (2013) **48**:433–8. doi:10.1038/bmt.2012.162
49. Yoon SR, Lee YS, Yang SH, Ahn KH, Lee JH, Lee JH, et al. Generation of donor natural killer cells from CD34+progenitor cells and subsequent infusion after HLA-mismatched allogeneic hematopoietic cell transplantation: a feasibility study. *Bone Marrow Transplant* (2010) **45**(6):1038–46. doi:10.1038/bmt.2009.304
50. Miller JS, Soignier Y. Successful adoptive transfer and in vivo expansion of human haploidentical NK cells in patients with cancer. *Blood* (2005) **105**(8):3051–7. doi:10.1182/blood-2004-07-2974
51. Rubnitz JE, Inaba H. NKAML: a pilot study to determine the safety and feasibility of haploidentical natural killer cell transplantation in childhood acute myeloid leukemia. *J Clin Oncol* (2010) **28**(6):955–9. doi:10.1200/JCO.2009.24.4590
52. Curti A, Ruggeri L. Succesful transfer of alloreactive haploidentical KIR ligand-mismatched natural killer cells after infusion in elderly high risk acute myeloid leukemia patients. *Blood* (2011) **118**(12):3273–9. doi:10.1182/blood-2011-01-329508
53. Rusakiewicz S, Semeraro M, Sarabi M, Desbois M, Locher C, Mendez R, et al. Immune infiltrates are prognostic factors in localized gastrointestinal stromal tumors. *Cancer Res* (2013) **73**(12):3499–510. doi:10.1158/0008-5472.CAN-13-0371
54. Mamessier E, Sylvain A, Thibult ML, Houvenaeghel G, Jacquemier J, Castellano R, et al. Human breast cancer cells enhance self tolerance by promoting evasion from NK cell antitumor immunity. *J. Clin. Invest* (2011) **121**:3609–22. doi:10.1172/JCI45816
55. Navarro A, Bjorklund A, Chekenya M. Therapeutic potential and challenges of natural killer cells in treatment of solid tumors. *Front Immunol* (2015) **29**(6):202. doi:10.3389/fimmu.2015.00202
56. Stringaris K, Sekine T, Khoder A, Alsuliman A, Razzaghi B, Sargeant R, et al. Leukemia-induced phenotypic and functional defects in natural killer cells predict failure to achieve remission in acute myeloid leukemia. *Haematologica* (2014) **99**(5):836–47. doi:10.3324/haematol.2013.087536
57. Rouce R, Shaim H, Sekine T, et al. The TGF-β/SMAD pathway is an important mechanism for NK cell immune evasion in childhood B acute lymphoblastic leukemia. *Leukemia* (in press).

58. Murray S, Lundqvist A. Targeting the tumor microenvironment to improve natural killer cell-based immunotherapies: on being in the right place at the right time, with resilience. *Hum Vaccin Immunother* (2015) **24**:0. doi:10.1080/21645515.2015.1096458
59. Nausch N, Galani I, Schlecker E, Cerwenka A. Mononuclear myeloid-derived "suppressor" cells express RAE-1 and activate natural killer cells. *Blood* (2008) **112**(10):4080–9. doi:10.1182/blood-2008-03-143776
60. Geller MA, Cooley S, Judson PL, Ghebre R, Carson LF, Argenta PA, et al. A phase II study of allogeneic natural killer cell therapy to treat patients with recurrent ovarian and breast cancer. *Cytotherapy* (2011) **13**(1):98–107. doi:10.3109/14653249.2010.515582
61. Koepsell SA, Miller JS. Natural killer cells: a review of manufacturing and clinical utility. *Transfusion* (2013) **53**(2):404–10. doi:10.1111/j.1537-2995.2012.03724.x
62. Lang P, Pfeiffer M, Teltschik HM, Schlegel P, Feuchtinger T, Ebinger M, et al. Natural killer cell activity influences outcome after T cell depleted stem cell transplantation from matched unrelated and haploidentical donors. *Best Pract Res Clin Haematol* (2011) **24**:403–11. doi:10.1016/j.beha.2011.04.009
63. Pfeiffer M, Schumm M, Feuchtinger T, Dietz K, Handgretinger R, Lang P. Intensity of HLA class I expression and KIR-mismatch determine NK-cell mediated lysis of leukaemic blasts from children with acute lymphatic leukaemia. *Br J Haematol* (2007) **138**:97–100. doi:10.1111/j.1365-2141.2007.06631.x
64. Leong W, Chase JM, Romee R, Schneider SE, Sullivan RP, Cooper MA, et al. Pre-activation with IL-12, IL-15, and IL-18 induces CD25 and a functional high affinity IL-2 receptor on human cytokine-induced memory-like NK cells. *Biol Blood Marrow Transplant* (2014) **20**(4):463–73. doi:10.1016/j.bbmt.2014.01.006
65. Ni J, Miller M, Stojanovic A, Garbi N, Cerwenka A. Sustained effector function of IL-12/15/18-preactivated NK cells against established tumors. *J Exp Med* (2012) **209**:2351–65. doi:10.1084/jem.20120944
66. Miller J, Rooney C, Curtsinger J, McElmurry R, McCullar V, Verneris MR, et al. Expansion and homing of adoptively transferred human natural killer cells in immunodeficient mice varies with product preparation and in vivo cytokine administration: implications for clinical therapy. *Biol Blood Marrow Transplant* (2014) **20**(8):1252–7. doi:10.1016/j.bbmt.2014.05.004
67. Fujisaki H, Kakuda H, Shimasaki N, Imai C, Ma J, Lockey T, et al. Expansion of highly cytotoxic human natural killer cells for cancer cell therapy. *Cancer Res* (2009) **69**:4010–7. doi:10.1158/0008-5472.CAN-08-3712
68. Berg M, Lundqvist A, McCoy P Jr, Samsel L, Fan Y, Tawab A, et al. Clinical-grade ex vivo-expanded human natural killer cells up-regulate activating receptors and death receptor ligands and have enhanced cytolytic activity against tumor cells. *Cytotherapy* (2009) **11**:341–55. doi:10.1080/14653240902807034
69. Lundqvist A, Yokoyama H, Smith A, Berg M, Childs R. Bortezomib treatment and regulatory T-cell depletion enhance the antitumor effects of adoptively infused NK cells. *Blood* (2009) **113**:6120–7. doi:10.1182/blood-2008-11-190421
70. Bachanova C. Clearance of acute myeloid leukemia by haploidentical natural killer cells is improved using IL-2 diphtheria toxin fusion protein. *Blood* (2014) **123**(25):3855–63. doi:10.1182/blood-2013-10-532531
71. Klingemann H, Grodman C. Autologous stem cell transplant recipients tolerate haploidentical related-donor natural killer cell-enriched infusions. *Transfusion* (2013) **53**:412–8. doi:10.1111/j.1537-2995.2012.03764.x
72. Reed W, Noga SJ, Gee AP, Rooney CM, Wagner JE, McCullough J, et al. Production assistance for cellular therapies (PACT): four-year experience from the United States National Heart, Lung, and Blood Institute (NHLBI) contract research program in cell and tissue therapies. *Transfusion* (2009) **49**:786–96. doi:10.1111/j.1537-2995.2008.02027.x
73. Koepsell SA, Kadidlo DM, Fautsch S, McCullough J, Klingemann H, Wagner JE, et al. Successful "in-flight" activation of natural killer cells during long-distance shipping. *Transfusion* (2013) **53**:398–403. doi:10.1111/j.1537-2995.2012.03695.x
74. Woll P, Grzywacz B, Tian X, Marcus RK, Knorr DA, Verneris MR, et al. Human embryonic stem cells differentiate into a homogenous population of natural killer cells with potent in vivo antitumor activity. *Blood* (2009) **113**(24):6094–101. doi:10.1182/blood-2008-06-165225
75. Chouaib S, Pittari G, Nanbakhsh A, ElAyoubi H, Amsellem S, Bourhis JH, et al. Improving the outcome of leukemia by natural killer cell-based immunotherapeutic strategies. *Front Immunol* (2014) **5**:95. doi:10.3389/fimmu.2014.00095
76. Spanholtz J, Preijers F, Tordoir M, Trilsbeek C, Paardekooper J, de Witte T, et al. Clinical-grade generation of active NK cells from cord blood hematopoietic progenitor cells for immunotherapy using a closed-system culture process. *PLoS One* (2011) **6**:e20740. doi:10.1371/journal.pone.0020740
77. Shah N, Martin-Antonio B, Yang H, Ku S, Lee DA, Cooper LJ, et al. Antigen presenting cell-mediated expansion of human umbilical cord blood yields log-scale expansion of natural killer cells with anti-myeloma activity. *PLoS One* (2013) **8**(10):e76781. doi:10.1371/journal.pone.0076781
78. Shaim H, Yvon E. Cord blood: a promising source of allogeneic natural killer cells for immunotherapy. *Cytotherapy* (2015) **17**(1):1–2. doi:10.1016/j.jcyt.2014.12.001
79. Dalle JH, Menezes J, Wagner E, Blagdon M, Champagne J, Champagne MA, et al. Characterization of cord blood natural killer cells: implications for transplantation and neonatal infections. *Pediatr Res* (2005):57.
80. Tomchuck S, Leung W, Dallas M. Enhanced cytotoxic function of natural killer and CD3+CD56+ cells in cord blood after culture. *Biol Blood Marrow Transplant* (2015) **21**(1):39–49. doi:10.1016/j.bbmt.2014.10.014
81. Della C, Falco M, Podesta M, Locatelli F, Moretta L, Frassoni F, et al. Phenotypic and functional heterogeneity of human NK cells developing after umbilical cord blood transplantation: a role for human cytomegalovirus. *Blood* (2012) **119**(2):399–410. doi:10.1182/blood-2011-08-372003
82. Xing R. Cord blood natural killer cells exhibit impaired lytic immunological synapse formation that is reversed with IL-2 ex vivo expansion. *J Immunother* (2010) **33**(7):684–96. doi:10.1097/CJI.0b013e3181e475e9
83. Alnabhan R, Madrigal A, Saudemont A. Differential activation of cord blood and peripheral blood NK cells by cytokines. *Cytotherapy* (2015) **17**(1):73–85. doi:10.1016/j.jcyt.2014.08.003
84. Denman CJ, Senyukov VV, Somanchi SS, Phatarpekar PV, Kopp LM, Johnson JL, et al. Membrane-bound IL-21 promotes sustained ex vivo proliferation of human natural killer cells. *PLoS One* (2012) **7**:e30264. doi:10.1371/journal.pone.0030264
85. Pittari G, Filippini P, Gentilcore G, Grivel JC, Rutella S. Revving up natural killer cells and cytokine-induced killer cells against hematological malignancies. *Front Immunol* (2015) **6**:230. doi:10.3389/fimmu.2015.00230
86. Klingemann HG, Wong E, Maki GA. cytotoxic NK-cell line (NK-92) for ex vivo purging of leukemia from blood. *Biol Blood Marrow Transplant* (1996) **2**:68–75.
87. Yagita M, Huang C, Umehara H, Matsuo Y, Tabata R, Miyake M, et al. A novel natural killer cell line (KHYG-1) from a patient with aggressive natural killer cell leukemia carrying a p53 point mutation. *Leukemia* (2000) **14**(5):922–30. doi:10.1038/sj.leu.2401769
88. Tonn T, Schwabe D, Klingemann H, Becker S, Esser R, Koehl U, et al. Treatment of patients with advanced cancer with the natural killer cell line NK-92. *Cytotherapy* (2013) **15**(12):1563–70. doi:10.1016/j.jcyt.2013.06.017
89. Swift B, Williams B. Natural killer cell lines preferentially kill clonogenic multiple myeloma cells and decrease myeloma engraftment in a bioluminescent xenograft mouse model. *Haematologica* (2012) **97**(7):1020–8. doi:10.3324/haematol.2011.054254
90. Yan Y, Steinherz P, Klingemann HG, Dennig D, Childs BH, McGuirk J, et al. Antileukemia activity of a natural killer cell line against human leukemias. *Clin Cancer Res* (1998) **4**:2859–68.
91. Tam YK, Miyagawa B, Ho VC, Klingemann HG. Immunotherapy of malignant melanoma in a SCID mouse model using the highly cytotoxic natural killer cell line NK-92. *J Hematother* (1999) **8**:281–90. doi:10.1089/106161299320316
92. Dudley ME, Wunderlich JR, Yang JC, Sherry RM, Topalian SL, Restifo NP, et al. Adoptive cell transfer therapy following non-myeloablative but lymphodepleting chemotherapy for the treatment of patients with refractory metastatic melanoma. *Journal of Clinical Oncology* (2005) **23**:2346–57. doi:10.1200/JCO.2005.00.240
93. Conlon K, Lugli E, Welles H. Redistribution, hyperproliferation, activation of natural killer cells and CD8 T cells, and cytokine production during first-in-human clinical trial of recombinant human interleukin-15 in patients with cancer. *J Clin Oncol* (2015) **33**(1):74–82. doi:10.1200/JCO.2014.57.3329
94. Pegram H, Smith E, Rafiq S, Brentgens R. CAR therapy for hematological cancers: can success seen in the treatment of B-cell acute lymphoblastic leukemia be applied to other hematological malignancies? *Immunotherapy* (2015) **7**(5):545–61. doi:10.2217/imt.15.6

95. Grupp S, Kalos M, Barrett D, Aplenc R, Porter D, Rheingold S, et al. Chimeric antigen receptor-modified T cells for acute lymphoid leukemia. *N Engl J Med* (2013) **368**(16):1509–18. doi:10.1056/NEJMoa1215134
96. Lee DW, Kochenderfer JN, Stetler-Stevenson M, Cui YK, Delbrook C, Feldman SA, et al. T cells expressing CD19 chimeric antigen receptors for acute lymphoblastic leukaemia in children and young adults: a phase 1 dose-escalation trial. *Lancet* (2015) **385**(9967):517–28. doi:10.1016/S0140-6736(14)61403-3
97. Glienke W, Esser R, Priesner C, Suerth J, Schambach A, Wels WS, et al. Advantages and applications of CAR-expressing NK cells. *Front Pharmacol* (2015) **6**:21. doi:10.3389/fphar.2015.00021
98. Tettamanti S, Marin V, Pizzitola I, Magnani CF, Giordano Attianese GM, Cribioli E, et al. Targeting of acute myeloid leukaemia by cytokine-induced killer cells redirected with a novel CD123-specific chimeric antigen receptor. *Br J Haematol* (2013) **161**:389–401. doi:10.1111/bjh.12282
99. Shimasaki N, Fujisaki H, Cho D, Masselli M, Lockey T, Eldridge P, et al. A clinically adaptable method to enhance the cytotoxicity of natural killer cells against B-cell malignancies. *Cytotherapy* (2012) **14**(7):830–40. doi:10.3109/14653249.2012.671519
100. Chang YH, Connolly J, Shimasaki N, Mimura K, Kono K, Campana D. A chimeric receptor with NKG2D specificity enhances natural killer cell activation and killing of tumor cells. *Cancer Res* (2013) **73**(6):1777–86. doi:10.1158/0008-5472.CAN-12-3558
101. Altvater B, Landmeier S, Pscherer S, Temme J, Schweer K, Kailayangiri S, et al. 2B4(CD244) signaling by recombinant antigen-specific chimeric receptors costimulates natural killer cell activation to leukemia and neuroblastoma cells. *Clin. Cancer Res* (2009) **15**(15):4857–66. doi:10.1158/1078-0432.CCR-08-2810
102. Esser R, Muller T, Stefes D, Kloess S, Seidel D, Gillies S, et al. NK cells engineered to express a GD2-specific antigen receptor display built-in ADCC-like activity against tumour cells of neuroectodermal origin. *J. Cell Mol. Med* (2012) **16**(3):569–81. doi:10.1111/j.1582-4934.2011.01343.x
103. Battula VL, Shi Y, Evans KW, Wang RY, Spaeth EL, Jacamo RO, et al. Ganglioside GD2 identifies breast cancer stem cells and promotes tumorigenesis. *J Clin Invest* (2012) **122**(6):2066–78. doi:10.1172/JCI59735
104. Jiang H, Zhang W, Shang P, Zhang H, Fu W, Ye F, et al. Transfection of chimeric anti-CD138 gene enhances natural killer cell activation and killing of multiple myeloma cells. *Mol.Oncol* (2014) **8**(2):297–310. doi:10.1016/j.molonc.2013.12.001
105. Chu J, Deng Y, Benson DM, He S, Hughes T, Zhang J, et al. CS1-specific chimeric antigen receptor (CAR)-engineered natural killer cells enhance in vitro and in vivo antitumor activity against human multiple myeloma. *Leukemia* (2014) **28**(4):917–27. doi:10.1038/leu.2013.279
106. Boissel L, Betancur-Boissel M, Lu W, Krause DS, Van Etten RA, Wels WS, et al. Retargeting NK-92 cells by means of CD19- and CD20-specific chimeric antigen receptors compares favorably with antibody-dependent cellular cytotoxicity. *OncoImmunology* (2013) **2**(10):e26527. doi:10.4161/onci.26527
107. Ferrara JL, Levine JE, Reddy P, Holler E. Graft-versus-host disease. *Lancet* (2009) **373**(9674):1550–61. doi:10.1016/S0140-6736(09)60237-3
108. Stauss H, Morris E. Immunotherapy with gene-modified T cells: limiting side effects provides new challenges. *Gene Ther* (2013) **20**(11):1029–32. doi:10.1038/gt.2013.34
109. Maus M, Grupp S, Porter D, June C. Antibody-modified T cells: CARs take the front seat for hematologic malignancies. *Blood* (2014) **123**(17):2625–35. doi:10.1182/blood-2013-11-492231
110. Shah N, Baird K, Delbrook C, Fleisher T, Kohler M, Rampertaap S, et al. Acute GVHD in patients receiving IL-15/4-1BBL activated NK cells following T cell depleted stem cell transplantation. *Blood* (2015) **125**(5):784–92. doi:10.1182/blood-2014-07-592881
111. Jones B, Lamb L, Goldman F, DiStasi A. Improving the safety of cell therapy products by suicide gene transfer. *Front Pharmacol* (2014) **5**:254. doi:10.3389/fphar.2014.00254
112. Zhou X, Dotti G, Krance R, Martinez C, Naik S, Kamble R, et al. Inducible caspase-9 suicide gene controls adverse effects from alloreplete T cells after haploidentical stem cell transplantation. *Blood* (2015) **125**(26):4103–13. doi:10.1182/blood-2015-02-628354
113. Gleason R. CD16xCD33 bispecific killer cell engager (BiKE) activates NK cells against primary MDS and MDSC CD33+ targets. *Blood* (2014) **123**(19):3016–26. doi:10.1182/blood-2013-10-533398
114. Wiernik A, Foley B, Zhang B, Verneris MR, Warlick E, Gleason MK, et al. Targeting natural killer cells to AML in vitro with a CD16x33 engager. *Clin Cancer Res* (2013) **19**(14):3844–55. doi:10.1158/1078-0432.CCR-13-0505
115. Romee R, Foley B, Lenvik T, Wang Y, Zhang B, Ankarlo D, et al. NK cell CD16 surface expression is regulated by a disintegrin and metalloprotease-17 (ADAM17). *Blood* (2013) **121**(18):3599–608. doi:10.1182/blood-2012-04-425397
116. Choi I, Yoon SR, Park SY, Kim H, Jung SJ, et al. Donor-derived natural killer cells infused after human leukocyte antigen-haploidentical hematopoietic cell transplantation: a dose-escalation study. *Biol Blood Marrow Transplant* (2014) **20**(5):696–704. doi:10.1016/j.bbmt.2014.01.031

16

NK Cell-Mediated Antibody-Dependent Cellular Cytotoxicity in Cancer Immunotherapy

*Wei Wang[1], Amy K. Erbe[1], Jacquelyn A. Hank[1], Zachary S. Morris[1] and Paul M. Sondel[1,2]**

[1] *Department of Human Oncology, University of Wisconsin-Madison, Madison, WI, USA,* [2] *Department of Pediatrics, University of Wisconsin-Madison, Madison, WI, USA*

Edited by:
Susana Larrucea, BioCruces Health Research Institute, Spain

Reviewed by:
John T. Vaage, Oslo University Hospital and University of Oslo, Norway
Todd A. Fehniger, Washington University School of Medicine, USA

***Correspondence:**
Paul M. Sondel, University of Wisconsin-Madison, Department of Human Oncology & Department of Pediatrics, 1111 Highland Avenue, 4159 WIMR Building, Madison, WI 53705, USA
pmsondel@humonc.wisc.edu

Natural killer (NK) cells play a major role in cancer immunotherapies that involve tumor-antigen targeting by monoclonal antibodies (mAbs). NK cells express a variety of activating and inhibitory receptors that serve to regulate the function and activity of the cells. In the context of targeting cells, NK cells can be "specifically activated" through certain Fc receptors that are expressed on their cell surface. NK cells can express FcγRIIIA and/or FcγRIIC, which can bind to the Fc portion of immunoglobulins, transmitting activating signals within NK cells. Once activated through Fc receptors by antibodies bound to target cells, NK cells are able to lyse target cells without priming, and secrete cytokines like interferon gamma to recruit adaptive immune cells. This antibody-dependent cell-mediated cytotoxicity (ADCC) of tumor cells is utilized in the treatment of various cancers overexpressing unique antigens, such as neuroblastoma, breast cancer, B cell lymphoma, and others. NK cells also express a family of receptors called killer immunoglobulin-like receptors (KIRs), which regulate the function and response of NK cells toward target cells through their interaction with their cognate ligands that are expressed on tumor cells. Genetic polymorphisms in KIR and KIR-ligands, as well as FcγRs may influence NK cell responsiveness in conjunction with mAb immunotherapies. This review focuses on current therapeutic mAbs, different strategies to augment the anti-tumor efficacy of ADCC, and genotypic factors that may influence patient responses to antibody-dependent immunotherapies.

Keywords: natural killer cell, therapeutic monoclonal antibody, antibody-dependent cellular cytotoxicity, cancer, immunotherapy

Introduction

Natural killer (NK) cells have been described throughout the literature for their ability to kill virally infected and malignant cells without priming. Unlike B and T cells, NK cells do not require somatic gene rearrangements to produce highly specific receptors that recognize target cells (1, 2). Instead, mature NK cells reserve large amounts of cytotoxic granules containing perforin and granzymes, as well as the mRNA of IFNγ that is ready for translation if stimulated. As soon as the balance between inhibitory and activating signals within NK cells are skewed toward activation, NK cells are capable of forming synapses with target cells, allowing the release of the perforin and granzyme to lyse the target cells, as well as for IFNγ production (3). In addition, NK cells can initiate the transduction of death signals within target cells through death receptor/ligand ligation (4). Their capabilities of tumor cytotoxicity and inflammatory cytokine production enable NK cells to play an important role in different settings of cancer immunotherapy.

NK Cell Recognition and "Missing-Self" Hypothesis

Mature NK cells express a series of transmembrane receptors. The activating receptors allow them to recognize stress-induced ligands, while their inhibitory receptors prevent them from attacking normal cells. Activating receptors are often associated with adaptor proteins that have activation motifs in their cytoplasmic domains. Upon ligand binding, followed by phosphorylation, these activating receptors can activate down-stream kinases, leading to NK cell degranulation and cytokine secretion (5). Inhibitory receptors, on the other hand, have one or more inhibitory motifs in their cytoplasmic tails. Once phosphorylated, they recruit phosphatases and deactivate signaling kinases, resulting in NK cell inhibition (6). NK cell activity is tightly regulated through the balance between inhibitory and activating signals transduced by these receptors.

One family of receptors on human NK cells is that of the killer immunoglobulin-like receptors (KIR), which recognize HLA as their ligands (7). Some inhibitory and activating KIRs share the same ligand (8); however, most inhibitory KIRs have stronger binding affinity to their shared ligand (9, 10). One way to shift the activating-inhibitory balance toward NK cell activation is by decreasing the inhibitory KIR signaling. When cells are under stress or virally infected, the HLA expression on their surface is often downregulated in order to escape from T cell recognition. When an NK cell encounters these "target" cells, an immune mediated synapse can occur. If the target cell is missing the expression of the HLA ligand for the inhibitory receptors on that NK cell, and expresses ligands for the activating receptors, the interaction could lead to NK cell activation due to lack of inhibitory signals ("missing-self" hypothesis) (11, 12).

However, within an individual, not all NK cells have the same receptor expression profile, and each individual has a different KIR expression profile. As a result, not all NK cells express inhibitory KIR receptors (13, 14). The NK cells without any inhibitory KIR receptors have been shown to be hyporesponsive to HLA-null targets compared to NK cells expressing inhibitory KIR receptors that can recognize self-HLA molecules (15). NK cells that express no inhibitory KIR receptors, or KIR receptors that only recognize allogeneic HLA (i.e., which do not recognize or bind self-HLA) do not go through a "licensing" process during NK cell differentiation. Licensing plays a role in enabling the licensed NK cells to be more capable of killing targets that do not express inhibitory HLA ligands (such as HLA-null targets). Unlicensed NK cells are less potent in their activation by, or killing of, HLA-null targets, than are licensed NK cells (16). Interestingly, in mouse models, such unlicensed (hyporesponsive) NK cells, which express Ly49 receptors (the inhibitory receptors found on mouse NK cells that are the functional counterparts of human inhibitory KIRs) but have not seen "self-ligand," can be made to be functional by transferring these NK cells to an environment where cognate ligand is expressed (17, 18).

Besides the expression of HLA ligands for the inhibitory receptors on NK cells, there are other mechanisms by which self-normal cells are protected from being attacked by NK cells. Studies in both mice (19–23) and humans (24, 25) have shown that after continuous exposure to the ligand for NK-activating KIR receptors, NK cells expressing these activating receptors can become hyporesponsive to target cells that express the ligand. This is evidence that the NK cells can be desensitized through their continual receptor contact with activating ligands.

ADCC Mechanism

In the setting of tumor-targeting monoclonal antibody (mAb) therapies, the anti-tumor efficacy of many mAb are shown to be NK cell-dependent (26). Human NK cells can express both FcγRIIC/CD32c (27) and FcγRIIIA/CD16a (28), which bind to the Fc portion of human immunoglobulins. FcγRIIIA often associates with FcεRI-γ chains or CD3-ζ chains within the cell membrane, or with a heterodimer of these two chains (5). Both FcεRI-γ and CD3-ζ chains have immune tyrosine-based activating motifs (ITAM) in their cytoplasmic tails. Unlike most activating receptors on NK cells, FcγRIIC has an ITAM in its own cytoplasmic tail. Upon FcγR binding, these ITAMs are phosphorylated, and through signal transduction mechanisms (binding to tyrosine kinases ZAP-70 and Syk and activation of PI3K, NF-κb and ERK pathways) NK cell degranulation, cytokine secretion, and finally tumor cell lysis occur (29).

Antibody-dependent NK-mediated tumor killing occurs through several different pathways, including: (1) exocytosis of cytotoxic granules; (2) TNF family death receptors signaling; (3) pro-inflammatory cytokine release, such as IFNγ. Both the uptake of perforin and granzymes by target cells and TNF family death receptor signaling cause target cell apoptosis (29), while IFNγ released by NK cells activate nearby immune cells to promote antigen presentation and adaptive immune responses (30). IFNγ production and cytotoxicity have been considered two distinctive functions of different NK subsets (31, 32), but growing evidence shows that the main cytotoxic NK subset, $CD56^{dim}CD16+$ NK cells, that are responsible for mAb-mediated tumor killing, are also able to produce IFNγ following activation (33, 34). In addition to inhibiting cell proliferation, angiogenesis, and increasing MHC surface expression (35), IFNγ was also shown to contribute to upregulation of TRAIL expression on NK cells (36), which suggests that one mechanism may interact with another to synergistically enhance tumor killing. A recent study indicates that NK-insensitive targets can become NK-sensitive via treatment with IFNγ, which induces lysis through ICAM-1 upregulation and increasing conjugate formation with NK cells (37). These mechanisms might work together to eliminate tumor targets through engagement of both innate and adaptive immunity; whether one is predominant over the others in tumor killing is still unknown (29–37).

NK Cell FcγRs in ADCC

Genotypic variations (polymorphisms) exist in humans in both FcγRIIIA and FcγRIIC that influence FcR function. Thus, FcγRIIIA and FcγRIIC genotype can influence the interaction of these receptors with immunoglobulin, resulting in differential effectiveness of mAb therapy depending on an individual's genotype. In addition, immunoglobulin isotypes (IgG1, IgG2, IgG3, and IgG4), as well as fucosylation and glucosylation patterns, have varying influence on the affinities of these IgG molecules for both FcγRIIIA and FcγRIIC (38–41). Such factors (patient FcγR genotype and antibody Fc backbone) create the opportunity for considering therapeutic treatment options that may optimize the degree to which a patient will respond when administering mAb therapy. Selection of an optimized regimen may lead to a more effective trigger of the proper immune response.

A number of studies have shown that anti-tumor activity of certain tumor-specific mAbs is associated with higher affinity FcRs, based on the FcR genotype. These results suggest that these mAbs are acting through antibody-dependent cell-mediated cytotoxicity (ADCC) by the cells that express those FcRs. In particular, FcγRIIIA expressed on NK cells has a single nucleotide polymorphism (SNP) that results in FcγRIIIA polymorphic variants [FcγRIIIA with phenylalanine (F) at amino acid position 158, or FcγRIIIA with valine (V) at amino acid 158], which vary in their strength of binding to immunoglobulins. Several studies have shown that those individuals that are homozygous for V (FcγRIIIA-158-V/V) have improved clinical outcome (after treatment with ADCC-inducing tumor-reactive mAb) over those that are either heterozygous (FcγRIIIA-158-V/F) or homozygous (FcγRIIIA-158-F/F) for the lower affinity FcγRIIIA isoform (42–46). While not all such studies confirm these findings, several clinical analyses are consistent with this result. These are presented in the section below regarding a few of the mAbs that appear to be acting, at least in part, via ADCC.

Therapeutic Monoclonal Antibodies for Cancer Treatment

Tumor-specific mAbs that recognize tumor-selective antigens on the surface of tumor cells are being used as cancer therapy. These therapeutic mAbs target and attack tumor cells through various mechanisms, including directing toxic molecules to target cells, inhibiting target cell proliferation, blocking inhibitory signals for immune cells, and directing immune cells to kill targets through ADCC (47). Some newer antibodies, such as bi-specific antibodies (bsAbs), work through promoting conjugate formation between target cells and NK or T cells (48). A more comprehensive list of anti-cancer mAbs has been summarized in other reviews (47, 49, 50). This review focuses on NK cell-mediated anti-tumor activity, via ADCC, and thus includes discussion of representative tumor-specific mAbs that have been shown in pre-clinical or clinical models to function, at least in part, via ADCC (**Table 1**).

Anti-GD2 mAb for Melanoma and Neuroblastoma Treatment

GD2 is a disialoganglioside expressed on human melanoma and neuroblastoma cells with restricted expression on normal tissues, which makes it a suitable target for mAb immunotherapy (**Table 1**). The first anti-GD2 antibody, 3F8, is a murine IgG3 mAb that was produced in 1985 from a mouse hybridoma (62). 3F8 is able to elicit complement activation and ADCC against human neuroblastoma cells. However, most patients in early clinical trials that received 3F8 developed human anti-mouse antibody (HAMA) response (63). HAMA may compete for the binding site of the therapeutic antibodies resulting in decreased binding to GD2, therefore dampening the anti-tumor efficacy and leading to acceleration of clearance of the therapeutic antibody from circulation (64).

Another murine anti-GD2 mAb, 14.18, which is also an IgG3, was generated separately by Mujoo et al. in 1987 (65). This antibody also displayed the capability to induce efficient *in vitro* ADCC and *in vivo* anti-tumor effects. An isotype variant of this murine anti-human GD2 antibody, 14.G2a (66), was tested clinically and showed some anti-tumor activity (67, 68), but HAMA response was still present in a significant portion of patients. While effective in targeting tumor and reducing tumor size in occasional patients, it became evident that it was necessary to improve the backbone of these initial mAb to increase efficacy and decrease the immunogenicity of this immunotherapeutic option.

In order to reduce the HAMA response and lengthen the antibody half-life in patients, efforts were made to create chimeric anti-GD2 antibodies, containing human constant regions with murine variable regions. Since a chimeric antibody has a majority of human epitopes, these epitopes should not be recognized by the immune system as foreign, and thus be less immunogenic than the fully murine antibodies. Dinituximab (formerly known as ch14.18) is a chimeric mAb comprising a fusion protein of the human constant portion of IgG1 and the GD2-reactive variable portion of the murine 14.18 mAb (69). Dinituximab has been shown to induce stronger ADCC than 14.G2a *in vitro* against GD2-positive neuroblastoma cells (70), and have anti-tumor activity against GD2-positive melanoma cells *in vivo* (71). In the initial published phase I clinical study of dinituximab treatment for pediatric neuroblastoma (72), no human anti-chimeric antibody (HACA) response was detected. Four out of nine children had anti-tumor response and one had a minor response. Thus, by modifying the backbone of the antibody, improved clinical outcome was observed.

To further improve antibodies, a fully human antibody was "grafted" with murine complementarity determining regions (CDRs), which confer antigen specificity. These humanized antibodies are considered less immunogenic than chimeric antibodies (73). However, even with humanized antibodies specific for GD2, pain and capillary leak were seen as significant toxicities. These toxicities limit the dose that can be administered, which restrains the possible anti-tumor effect that one would expect if a higher dose could be given. The toxicities are mainly attributed to complement activation (74), which is elicited by the CH_2 domain on antibodies (75). Therefore, by reducing complement activation via a point mutation at amino acid position 322 in the CH2 domain of humanized antibody, complement activation is greatly reduced. Such reduction in complement activation, and thus reduced toxicities (76), allowed for higher treatment-dose to be administered to patients, while at the same time maintaining the anti-tumor ADCC effect (77). Both humanized 14.18K322A and humanized 3F8 are under clinical investigation (**Table 1**) (73, 78).

TABLE 1 | Representative tumor-antigen targeting monoclonal antibodies and immunocytokines functioning through ADCC.

	Target	Status	Reference or clinical trial#
MAb			
Rituximab	CD20	FDA approved for non-Hodgkin's lymphoma	Cartron et al. (42)
		Phase I relapsed indolent B-cell non-Hodgkin Lymphoma, combined with ALT-803[a]	NCT02384954
		Combination with matrix metalloproteases inhibitor in pre-clinical models	Romee et al. (51)
Obinutuzumab	CD20	FDA approved for chronic lymphocytic leukemia	Goede et al. (52)
Hul4.18K322A	GD2	Phase I neuroblastoma, melanoma,	NCT01576692
		Osteosarcoma, ewing sarcoma	NCT00743496
		Phase II neuroblastoma	NCT01857934
Hu3F8	GD2	Phase I GD2+ tumors	NCT01419834
		Phase I high-risk neuroblastoma and GD2+ solid Tumors, combined with IL2	NCT01662804
		Phase I refractory high-risk neuroblastoma, combined with GM-CSF	NCT01757626
Dinituximab	GD2	FDA approved for high-risk neuroblastoma, combined with IL2 and GM-CSF	Yu et al. (53)
Trastuzumab	HER2	FDA approved for HER2+ breast cancer and HER2+ metastatic gastric adenocarcinoma	Junttila et al. (54)
Cetuximab	EGFR	FDA approved for metastatic colorectal cancer and head and neck cancer	Messersmith and Ahnen (55)
IMMUNOCYTOKINE			
Rituximab-RLI[b]	CD20	Tested for human B lymphoma in SCID mouse	Vincent et al. (56)
c.60C3-RLI[c]	GD2	Tested in mouse GD2+ cell lines EL4 (subcutaneous) and NXS2 (metastatic)-grafted mouse models	Vincent et al. (57)
Hul4.18-IL2	GD2	Completed phase II refractory neuroblastoma	Delgado et al. (58), Shusterman et al. (59)
KM2812	PSMA	Tested in human prostate cancer cell LNCaP-xenografted mouse model	Sugimoto et al. (60)
BI-SPECIFIC ANTIBODY AND SINGLE CHAIN VARIABLE FRAGMENT			
AFM13	CD30/CD16	Phase II relapsed Hodgkin lymphoma	NCT02321592
$(CD20)_2$xCD16	CD20/CD16	Tested in humanized mouse grafted with autologous human B cells	Glorius et al. (61)

[a]ALT-803 is a fusion protein consisting of mutated IL15 and IL15Rα/Fc complex.
[b]RLI (IL15Rα-linker-IL15) is a fusion protein linking the NH_2-terminal domains of IL15Rα to IL15 through a 20-amino acid linker.
[c]c.60C3 is a chimeric anti-GD2 mAb.
This table is a selected (not complete) list of therapeutic mAbs that are capable of inducing antibody-dependent cellular cytotoxicity.

Herceptin/Trastuzumab

Trastuzumab is a humanized anti-HER2 mAb used to treat HER2-positive breast carcinoma (**Table 1**), as well as many other types of cancers that overexpress HER2, a member of the human epidermal growth factor receptor (EGFR) family. HER2 is a transmembrane tyrosine kinase with no known ligand. Dimerization of HER2 with certain EGFR family members leads to activation of signaling pathways that promote cell proliferation and survival (79). HER2 is overexpressed on a variety of tumors with limited expression on normal tissues, thus it is an ideal target for treatment of HER2-positive cancers.

Trastuzumab was first approved by the FDA in 1998 to treat HER2-positive metastatic breast cancer. Besides preventing HER2 from dimerization, trastuzumab was also shown to mediate ADCC against HER2-positive tumor cells *in vitro*, and the major effector cells were NK cells expressing FcγRIIIA (80, 81). A mutant trastuzumab that lost the ability to bind to FcγR lost anti-tumor activity *in vivo* in a xenograft breast tumor model, suggesting that ADCC is involved in the anti-tumor effect of anti-HER2 mAb therapy *in vivo* (54). In addition, Clynes et al. showed that the anti-tumor response to trastuzumab in a breast carcinoma xenograft mouse model was decreased in mice lacking the activating receptor FcγRIIIA, but enhanced in mice lacking the inhibitory receptor FcγRIIB (82). These experimental data demonstrate that Fc receptor recognition is responsible for at least part of anti-tumor efficacy of trastuzumab.

Cetuximab

Cetuximab is an FDA-approved chimeric mAb for treatment of EGFR-expressing metastatic colorectal cancer (mCRC) (55), metastatic non-small cell lung cancer, and head and neck cancer. It reacts against the human EGFR, and can interfere with tumor growth via receptor blockade from growth factor activation. *In vitro* studies indicate that some of the anti-tumor activity of cetuximab is mediated via ADCC (83, 84), and cetuximab-mediated *in vitro* ADCC is correlated with NK cell FcγR polymorphisms of the effector donors (46, 85). In addition, in mCRC patients treated with cetuximab and irinotecan, those who have higher affinity FcγR polymorphisms had longer progression-free survival (45, 86). In other studies of mCRC patients, the opposite association has been found, namely low affinity FcγR polymorphism was associated with better clinical outcomes (87–89). However, the patients in these studies either had different percentage of KRAS mutation or received other antibody concurrently with cetuximab, which indicates patient mutation profile as well as treatment regimen could also influence the impact of FcγR on cetuximab response (90). Nevertheless, both *in vitro* and some clinical data suggest that cetuximab-mediated ADCC through NK cells may contribute to its anti-tumor activity.

Rituximab and Obinutuzumab

Rituximab is a chimeric IgG1 mAb targeting CD20, a B cell differentiation antigen (**Table 1**). There are various mechanisms that may account for the anti-tumor effect of rituximab: complement-dependent cytotoxicity (CDC), direct target cell apoptosis, antibody-dependent phagocytosis and ADCC (91). In xenograft mouse models of B-cell lymphoma, the anti-tumor effect of rituximab was greatly reduced in $FcR\gamma^{-/-}$ nude mice (82), or in mice treated with FcγR block (92). Interestingly, clinical evidence gathered by Cartron et al. suggested that FcγRIIIA on NK cells plays an important role in anti-tumor effect of rituximab. Cartron et al. evaluated FcγRIIIA polymorphisms in follicular lymphoma patients treated with rituximab, and this was the first time that better response to rituximab was associated with higher affinity FcγRIIIA genotype (42). However, there have been mixed results as to the influence of FcR genotype on clinical response to rituximab. Some groups have found in B cell lymphoma patients with no association between FcγR polymorphism and outcome (93–95). There are various factors that could contribute to the differences in these findings, including patient population, genotyping methodology, rituximab treatment strategy, and whether or not patients received concurrent chemotherapy (96–99).

Modifications have been made to improve the binding affinity of therapeutic anti-CD20 mAb to activating Fc receptors, in hopes of maximizing the ADCC function of this mAb therapy. One such effort consists of using glycoengineered antibodies, which are produced in CHO cells that overexpress β-1,4-N-acetyl-glucosaminyltransferase III and Golgi α-mannosidase II, and will increase the binding affinity to both the higher affinity and to the lower affinity isoforms of FcγRIIIA (39, 41, 100). The first Fc-glycoengineered anti-CD20 humanized mAb, obinutuzumab, was shown to induce stronger ADCC and direct target cell death *in vitro*, and it also elicited better anti-tumor activity in a lymphoma xenograft mouse model compared to rituximab (101, 102). In a phase III clinical trial comparing obinutuzumab vs. rituximab, combined with chemotherapy, for treating chronic lymphocytic leukemia (CLL) patients (52), obinutuzumab plus chlorambucil significantly prolonged progression-free survival compared to rituximab plus chlorambucil. Obinutuzumab alone, or in combination with chemotherapy, is currently under clinical investigations for other B-cell malignancies as well (103–106).

How to Augment Anti-Tumor Effects of ADCC

Since ADCC is an important contributor to the anti-tumor activity of many mAb therapies, enhanced immune activation of the effector cells may be an ideal adjuvant therapy to augment ADCC activity of mAb. In addition to the ADCC capabilities of NK cells, they can also stimulate the activity of other immune processes through their release of cytokines (such as IFNγ), and thus can provide a link to initiate additional immune responses to attack target tumors.

mAb + Radiation Therapy

Approximately 60% of oncology patients receive radiation therapy as part of their cancer treatment. Radiation elicits an anti-tumor effect through the induction of DNA damage but may also increase tumor susceptibility to immune response (107). Consequently, the effect of radiation may be modulated by immune response (108, 109) and radiation may augment the efficacy of immunotherapies (110). The mechanisms by which radiation may interact with the immune system include radiation-induced production of inflammatory cytokines, release of tumor-specific antigens, phenotypic changes in tumor cell expression of immune susceptibility markers, and effects on vascular architecture that enhance immune surveillance (107, 110). Included among the phenotypic changes induced by radiation are the upregulation of MHC class I, NKG2D ligand, and the Fas death receptor; all of which may potentiate the ADCC response (111–114). In addition, radiation may impact the expression of antigens targeted by tumor-specific antibodies and this has been show to enhanced ADCC response *in vitro* (115, 116). The potential interaction of radiation and ADCC has not yet been clarified *in vivo*. Interestingly, however, a number of tumor-specific antibodies that are known to elicit ADCC (including cetuximab, trastuzumab, and dinituximab) are commonly administered to patients that also receive radiation therapy. The role of ADCC and NK cells in the clinical response to such combined modality treatment has not been defined. Further pre-clinical investigation is needed to evaluate whether the effects of radiation on tumor immune susceptibility may be leveraged to enhance ADCC response in the clinical setting.

mAb + Matrix Metalloproteases Inhibitor

Upon activation, NK cells have been shown to shed FcγRIIIA (also known as CD16) from their cell membrane, a process that is mediated by matrix metalloproteases (MMPs) activation (117, 118). *In vitro* treatment of NK cells with the MMP inhibitor, GM6001, rescues FcγRIIIA loss stimulated by K562 tumor cells, but does not interfere with NK cell degranulation (118), which indicates that FcγRIIIA shedding and degranulation of NK cells are dependent upon separate pathways. It is plausible, therefore, that maintaining FcγRIIIA expression on NK cell surface via an MMP inhibitor could enhance the ADCC function of NK cells without interrupting NK cell degranulation (119). Recently, one specific MMP, ADAM17, was identified by Romee et al. as the key MMP responsible for FcγRIIIA expression loss after NK cell activation (51). According to their report, an ADAM17-specific inhibitor not only rescues FcγRIIIA shedding stimulated by tumor targets but also improves NK cell degranulation as well as IFNγ production in the presence of tumor-specific mAb (51). These *in vitro* studies suggest that limiting the loss of FcγRIIIA on the cell surface is important for enhanced NK cell-mediated ADCC, and opens new possibilities for combination of both MMP and mAb therapies to further promote ADCC.

Anti-GD2 mAb + IL2 + GM-CSF

While anti-GD2 mAb alone produced some anti-tumor activity in neuroblastoma patients, combining the mAb with GM-CSF and interleukin 2 (IL2) to further activate immune effector cells enabled potent clinical anti-tumor efficacy of anti-GD2 mAb.

Our lab previously showed that peripheral blood mononuclear cells (PBMCs) obtained from cancer patients pre-IL2 treatment had low levels of ADCC against a neuroblastoma cell line, even in the presence of both anti-GD2 mAb and IL2. However, PBMCs obtained from the same patient following 4 weeks of IL2 infusions mediated much higher ADCC of neuroblastoma cells, and the addition of IL2 *in vitro* dramatically boosted the anti-GD2 mAb-mediated ADCC (120, 121). Moreover, depletion of FcγRIIIA-positive cells eliminated ADCC completely (120). These data suggest that *in vivo* infusion of IL2 could overcome the immune suppression seen in some cancer patients; when using NK cells obtained from patients following therapy with IL2, the *in vitro* combination of IL2 and anti-GD2 mAb greatly boosted ADCC of neuroblastoma cells.

Besides IL2, GM-CSF also acts as an immune stimulator, especially following immune suppressive chemotherapy, to rescue bone marrow myeloid function (122–125). The combination of GM-CSF and the murine anti-GD2 mAb, 3F8, resulted in complete responses of GD2-positive cancer patients, particularly in patients with minimal residual disease (MRD) (126). In a phase III clinical trial for high-risk neuroblastoma patients following induction therapy and autologous transplant, patients were randomized into standard therapy (isotretinoin) or into a group that received the combined immunotherapy regimen of dinituximab + IL2 + GM-CSF in addition to isotretinoin. Patients in the immunotherapy treatment group had significantly improved event-free survival and overall survival as compared to the patients that received the standard of care alone (53). Other mAb therapies, such as rituximab and trastuzumab (81), are also being combined with IL2 in clinical trials to evaluate the anti-tumor efficacy (127–129).

mAb + IL15

IL15 is another cytokine that can activate immune cells, such as NK and T cells. IL15 receptors (IL15R) share the same β chain and common γ chain as IL2 receptors, but IL15R have a unique α chain that is specific for IL15, which is used for IL15 presentation to β/γ receptors on immune cells. Soluble IL15 has been shown to activate NK cells, enhance NK-mediated cytotoxicity, and cytokine production *in vitro* (130), and administration of recombinant human IL15 (rhIL15) to cancer patients resulted in *in vivo* NK cell proliferation and activation (131). While IL2 has the potential for T regulatory cell (Treg) maintenance, which could dampen activating immune response, IL15 does not stimulate Tregs (132). Moreover, IL15 does not trigger T cell death after activation, or lead to vascular leak syndrome (VLS), both of which are significant toxicities seen in pre-clinical models using IL2 (133–135). Therefore, in addition to IL2, IL15 is a potential candidate that can be used clinically combined with mAb, to improve NK cell-mediated ADCC without severe toxicity.

Indeed, the Caligiuri group reported enhanced NK-mediated ADCC by IL15 in 1994 (130). More recently, Moga et al. showed that *in vitro* activation of PBMCs by IL15 significantly improved rituximab-mediated ADCC against B lymphoma cells, and the major effector cells were NK cells (136). They also showed this enhancement of rituximab-mediated ADCC using PMBCs from CLL patients against a CLL cell line (137). In addition, they showed that IL15 treatment resulted in similar ADCC levels between individuals with the lower affinity FcγRIIIA and individuals that have the higher affinity FcγRIIIA (137). These findings suggested that the difference between the capabilities of the high and low affinity FcγRIIIA to mediate ADCC might be overcome by IL15 treatment. Such findings suggest that IL15 administration may compensate for mAb-affinity differences that are dependent upon FcγRIIIA polymorphisms.

Different from IL2, the activation of NK and T cells by IL15 is through trans-presentation by the IL15Rα chain, which is expressed independent of the β/γ chains. With efforts to increase the stimulation activity of soluble IL15, several different IL15 agonists have been generated and tested in pre-clinical models in order to boost NK-mediated ADCC efficacy. A fusion protein linking IL15Rα chain sushi domain and human IL15 (RLI) has been generated and shown superior stimulation potential *in vitro* and better anti-tumor effects *in vivo* than soluble IL15 (138, 139). In addition, an IL15 mutant has been identified as a better surrogate for soluble IL15 due to its enhanced activity to stimulate proliferation of cells expressing IL15R (140). Later on, the same group generated a fusion protein consisting of the IL15 mutant and an IL15Rα/Fc complex (141). This IL15 superagonist, ALT-803, could improve rituximab-mediated ADCC against B cell lymphoma both *in vitro* and *in vivo* [Maximillian (142)]. Thus, IL15 combined with tumor-specific mAb appears to merit clinical testing as a potential immunotherapy, which is currently underway (NCT02384954) (**Table 1**).

Immunocytokines

Anti-GD2 Immunocytokines

Immunocytokines (IC) are fusion proteins made by linking immune-activating cytokine to a tumor-specific mAb. The initially described IC (ch14.18-IL2) linked IL2 to the C-terminus of the ch14.18 chimeric anti-GD2 mAb (143). The IL2 component on these IC has the same ability as soluble IL2 in terms of stimulating cell proliferation via IL2 receptors (IL2R) (144), and have been shown to mediate the conjugation between IL2R-positive cells and tumor target cells (145). IC also preserve FcR binding ability, and thus are capable of mediating *in vitro* ADCC (144). By targeting IL2 to the tumor microenvironment, IC might be superior at activating tumor infiltrating immune cells resulting in improved ADCC while causing less toxicity than soluble IL2. Lode et al. showed that ch14.18-IL2 had improved anti-tumor efficacy than the combination of ch14.18 and soluble IL2 in a spontaneous neuroblastoma metastases mouse model (146). They also showed that the mechanism behind this anti-tumor effect strongly depended on NK cells, since NK cell depletion completely abrogated the anti-tumor effect in this model (147).

A separate humanized IC was created by linking human IL2 to the humanized 14.18 mAb (hu14.18-IL2; **Table 1**). This IC was assessed for anti-tumor activity in a phase II clinical trial of refractory neuroblastoma patients divided into two strata of patients depending on disease burden. Stratum 1 included patients with disease measurable by computed tomography and/or magnetic resonance imaging using standard radiographic criteria, while stratum 2 included patients with disease evaluable only by ^{123}I-MIBG scintigraphy and/or BM histology. Of 36 patients evaluable for response, no responses were seen in the 13 patients in stratum 1 while 5 complete responses were noted in the 23 patients in stratum II (59). In another phase II clinical trial of hu14.18-IL2 in 14 patients with measurable metastatic melanoma, 1 patient had a partial response to the immunotherapy (148). Both of these trials suggest that hu14.18-IL2 works better in clinical cases where MRD is present than in individuals that have bulky disease. Furthermore, in the phase II neuroblastoma study of hu14.18-IL2 noted above where 21.7% of stratum II patients with MRD responded, all of the responders had a favorable KIR/KIR-ligand genotype (i.e., KIR-ligand missing) (58). This study indicates that NK cells play an important role in response to the hu14.18-IL2 treatment in neuroblastoma patients.

Anti-PSMA Immunocytokine

Prostate-specific membrane antigen (PSMA) is a surface antigen highly expressed by poorly differentiated prostate cancers, which makes it a suitable target for mAb therapies. Several anti-PSMA mAb and antibody-drug conjugates have been tested clinically, and a small fraction of patients showed anti-tumor response, which in part was due to ADCC (149). To improve ADCC function of anti-PSMA mAb, an anti-PSMA IC was generated by fusing IL2 to a mouse/human chimeric anti-PSMA mAb (60). This IC showed enhanced *in vitro* ADCC, and superior *in vivo* anti-tumor activity compared to a naked anti-PMSA mAb (60). Continued research is underway in both pre-clinical models and clinical testing to determine if superior anti-tumor efficacy is noted with this novel IC.

IL15-Linked Immunocytokine

One limitation of IL2-linked ICs is IL2-mediated toxicity that restricts the maximum-tolerated dose (MTD) that can be administered. In contrast to IL2, IL15 does not appear to elicit as severe adverse side effects. In addition, IL15 may have improved anti-tumor efficacy (133). Since IL15 is most active in a trans-presentation form, a fusion protein of human IL15Rα and human IL15, called RLI, was generated. This fusion protein showed superior activation function than soluble IL15 both *in vitro* and *in vivo* (138, 150). Vincent et al. linked RLI to the end of chimeric anti-GD2 antibody (c.60C3) and generated an RLI-linked anti-GD2 IC (57). This IC had similar IL15-induced proliferative activity as RLI, and similar ADCC-inducing ability as anti-GD2 antibody *in vitro*. Moreover, anti-GD2-RLI exhibited better anti-tumor activity than either RLI or anti-GD2, alone or in combination, in an NXS2 mouse neuroblastoma model (57). The same group also showed that RLI-linked rituximab significantly prolonged survival as compared to administration of RLI and rituximab at the same time in a residual lymphoma mouse model (56). These pre-clinical data suggest that IL15-linked IC merit clinical investigation to evaluate both efficacy and toxicity relative to IL2-linked IC.

Novel Bi-Specific Antibodies or Single Chain Variable Fragment Targeting NK Cells

With evolving genetic engineering technologies, non-conventional antibodies that have dual or tri specificity have been constructed. They either target two different antigens on tumor cells, or facilitate conjugate formation between immune cells and tumor targets (151). There are two classes of bsAbs: Fc-containing bsAbs that are of similar size as conventional mAbs and those bsAbs without an Fc domain, which are of much smaller size. They are further divided according to their structures, specificities, or how they are constructed (152). In efforts to bring NK cells and tumor cells in proximity, several bsAbs or single chain variable fragment (scFv) have been constructed to bind both tumor antigen and FcγRIIIA (CD16). AFM13 is a bi-specific tetravalent antibody construct that targets both CD30 and CD16 (153), which is currently in a phase II clinical trial for relapsed Hodgkin lymphoma patients (NCT02321592) (**Table 1**). It has been shown to exhibit superior cytotoxicity than other CD30-targeting antibodies *in vitro* and its ADCC activity is independent of the FcγRIIIA allotypes (154). A CD20/CD16 bsAb also showed improved clearance of B cell malignancy compared to rituximab both *in vitro* and *in vivo*. Again, the efficacy of this antibody construct is not influenced by FcγRIIIA polymorphism (61). Two different CD33/CD16 bsAbs have been shown to efficiently kill acute myeloid leukemia (AML) cells or stem cells from myelodysplastic syndrome, a precursor of AML (155–157). A few CD19/CD16 bsAb or derivatives have also been generated to target leukemia and lymphoma cells and showed promising ADCC activity *in vitro* (158–161).

For solid tumors, a bsAb-targeting HER2/CD16 has been tested in mouse model against HER2-positive tumor cells and showed enhanced anti-tumor efficacy than trastuzumab against HER2-low expressing tumor. Interestingly, the efficacy of this bsAb is also FcγRIIIA polymorphism independent (162). In addition, an EpCAM/CD16 bi-specific scFv antibody fragment showed enhanced *in vitro* killing of human carcinomas with a broad range of origins, as well as efficient killing of NK-resistant carcinoma targets (163). The fact that a lot of these bsAbs discussed above have proficient anti-tumor effect regardless of FcγRIIIA affinity on NK cells enlarges the patient population that will potentially benefit from such type of immunotherapy.

Adoptive Transfer of *Ex Vivo*-Activated NK Cells

A separate approach to augment NK cell function and ADCC is to infuse NK cells that have been activated and expanded in cell number *ex vivo*. During the *ex vivo* expansion of NK cells, these activated effectors are primed to kill tumors more effectively, and they can then be infused into cancer patients. Autologous NK cells may be relatively tolerant to self-tumors and, in certain settings, may have less anti-tumor potential than allogeneic NK cells (164). There are several approaches being pursued in order to expand and activate NK cells *ex vivo*. In the presence of IL2 and irradiated feeder cells, such as EBV-LCL cells or genetically modified K562 cells, NK cells are preferentially expanded to 200- to 400-fold within a 21-day period (165, 166). NK cells can be potently expanded and activated by culturing them with K562 cells that have been modified to be far more stimulatory by expressing membrane-bound IL15 and 41BBL. NK cells that have been *ex vivo* expanded using these modified K562 cells have better anti-tumor activity *in vitro* and *in vivo* in mouse models (166). In addition, *ex vivo*-expanded NK cells have also been shown to be able to mediate ADCC, and in combination with tumor antigen-specific antibodies, they exerted better anti-tumor efficacy (167, 168). Different protocols using expanded NK cells are currently under clinical testing, either alone or in combination with tumor-specific mAb (169–171). These ongoing clinical trials may answer the questions of how long these NK cells may survive *in vivo*, their homing capacity, and tumor-targeting specificity.

Genotypic Factors that Affect Antibody/ NK Cell Based Immunotherapy

Despite the success noted in some immunotherapies involving NK cells, some patients do not benefit from these immunotherapies. Therefore, it would be advantageous to be able to identify those patients who would likely respond, and those who would be less likely to benefit from immunotherapies that require NK cell response. There are many different genotypic factors that could potentially affect the efficacy of different immunotherapies in various cancer types. Here, we focus mainly on those genotypes that could affect NK cell function and ADCC efficacy.

Killer-Immunoglobulin Receptors on NK Cells and Their Ligands

Killer-immunoglobulin receptors are expressed on a subset of NK cells and on some T cell subsets. KIRs are a family of receptors with high polymorphisms inherited on chromosome 19 (172). There are 15 functional KIR genes and 2 KIR pseudogenes in the human genome. Of the functional KIR genes, some function as inhibitory receptors and others are activating receptors (173). For some KIR genes, the ligands are well defined (7). KIRs with defined ligands generally recognize distinct HLA class I molecules, but do so with less specificity than T cell receptors. Furthermore, and again in distinction to T cell receptors, KIRs appear to recognize class I HLA molecules without regard to peptides that may be presented in the cleft (pocket) of the HLA structure. Some inhibitory and activating KIRs with high homology in extra cellular domains share the same HLA ligand, with the inhibitory KIR/KIR-ligand interaction being a much stronger binding pair. The interaction between KIRs and KIR-ligands influence NK cell education and function (15, 174). Since KIR haplotype, as well as KIR expression, is highly diverse among the population, and since KIR-ligands are inherited independently from KIRs, an individual's KIR/KIR-ligand genotype can affect NK cell function and ADCC in different settings. These differences in genotypes for KIR/KIR-L between different individuals have been associated with the individual's outcome in response to certain forms of immunotherapy, particularly those immunotherapies that involve NK cells.

The KIR haplotype of NK cell donors (175–177) as well as the relationship between the donor's inhibitory KIR and the recipient's KIR-ligand genotypes can predict clinical outcomes in allogeneic hematopoietic transplantation for AML patients; the impact of KIR and KIR-L genotypes may also play a role in some, but not all, settings of hematopoietic transplantation for acute lymphoid leukemia (ALL) patients (178–180). Recipients that are missing the KIR-ligands for the KIR genes that are present on the donor NK cells are predicted to have longer progression-free survival post-transplantation than those who have the KIR-ligand genes present for the donor KIR genes (178–180). In an autologous hematopoietic transplantation setting for high-risk neuroblastoma patients, Venstrom et al. showed that patients who were missing any self-inhibitory KIR-ligand had improved progression-free survival and improved overall survival (181). Even though direct ADCC was not involved in these transplants for cancer patients, these data involving KIR/KIR-L relationships indicate that NK cells likely contribute to the anti-tumor response post-transplantation.

In some immunotherapies where ADCC is involved, patients missing their self-inhibitory KIR-ligand also had better clinical outcome. In neuroblastoma patients treated with anti-GD2 mAb or IC, and in lymphoma patients treated with rituximab, inhibitory KIR-ligand missing was associated with improved clinical outcome (58, 181–183). These findings suggest that self KIR/KIR-ligand genotypes not only affect NK cell function but also affect NK-mediated ADCC effects in the clinical setting.

The effects of inhibitory KIR/KIR-ligand interactions on NK cell function have been investigated for over a decade, but the role of activating KIR/KIR-ligand interactions in immunotherapies involving NK cells has only recently been reported. In 2011, Scquizzato et al. evaluated the impact of recipient KIR-ligand missing for both inhibitory and activating donor KIR genes, in a group of patients with various hematopoietic malignancies who received allogeneic hematopoietic stem cell transplantation (184). They found that the cohort of recipient patients missing the KIR-ligand for donor KIR genes had better disease-free survival, but only when patients who had KIR-ligands missing for their activating KIR were excluded from the analysis (184). This finding suggests that activating KIR/KIR-ligand interactions may have differential impact on patient outcome than inhibitory KIR/KIR-ligand interactions, thus the KIR-ligand missing model to predict patient outcome appears to be applicable only to inhibitory KIRs, with exclusion of activating KIRs.

In 2012, Venstrom et al. reported that in a large group of AML patients, those who received allografts from donors that had the activating KIR gene, KIR2DS1, and two copies of the ligand for KIR2DS1, HLA-C2, exhibited a higher relapse rate than patients who received grafts from KIR2DS1-positive donors and either one or no copies of HLA-C2 (24). Consistent with this clinical observation, Pittari et al. showed that KIR2DS1-positive NK cells from healthy donors that have two copies of HLA-C2 were hyporesponsive to tumor targets as compared to KIR2DS1-positive NK cells from donors with one or no copy of HLA-C2 (25). Both clinical and *in vitro* data suggest that NK cells that express activating KIR2DS1 are subjected to hyporesponsiveness via long-term contact of the activating KIR with its cognate ligand when two copies of the ligand are present. Whether or not NK cell hyporesponsiveness affects patient response in the context of ADCC, where mAb is administered to trigger NK cell activation through FcγRs, is still under investigation. As the interactions of activating KIRs and their ligands have not been studied as extensively as the interactions of the inhibitory KIRs and their ligands, much more characterization is required. Inhibitory KIR/KIR-ligand genotypes and activating KIR/KIR-ligand interactions may also serve as potential predictors of clinical outcome in the future.

NK Cell Fc Receptor Genetic Variations

FcγRs function as receptors for the Fc portion of IgG immunoglobulins, and in doing so serve as a link of the innate immune system to the humoral system. In humans, there are three classes of FcγR, including variations of FcγRI, FcγRII, and FcγRIII. These three receptor classes are characteristically expressed on various immune cells. NK cells express both FcγRIIC and FcγRIIIA, which have low to intermediate affinity for IgGs (depending on FcR polymorphisms and IgG subclasses). As noted above, individuals with intact immune systems all express FcγRIIIA on most of their NK cells, less than half the population expresses FcγRIIC on their NK cells. Genetic variability exists within both FcγRIIC and FcγRIIIA, and this genotypic variation can vary the expression and avidity of these FcγRs for IgG molecules (185, 186).

As was mentioned above, several groups have found associations between SNP genotype of FcγRIII3A-158-V/F and patient response; those with the higher affinity V/V genotype respond better to some antibody therapies than those with V/F or F/F (42–46). In fact, FcγRIIIA-158-V has improved binding affinity for IgG1 subclasses as compared to FcγRIIIA-158-F, which is the most common IgG subclass used in mAb cancer immunotherapeutics. Interestingly, FcγRIIIA-158-V and FcγRIIIA-158-F have increased binding strength for IgG3 over IgG1 (187). This suggests that antibody IgG subclass variations may improve their interaction with NK cells, via improved engagement of FcγRIIIA, and thus improve tumor destruction. However, other FcRs (such as Fcwever) are expressed on other immune cells, such as neutrophils and monocytes/macrophages, that also play an important role in tumor killing, and have improved binding to IgG1 isotypes over IgG3 (188).

Another FcγR that is expressed on NK cells, that also has improved binding to IgG3 over IgG1, is FcγRIIC. FcγRIIC is the result of the unequal crossover of FcγRIIA (an activating FcR expressed on myeloid immune cell lineages) and FcγRIIB (an inhibitory FcR expressed on B cells, monocytes, and macrophages) (185). FcγRIIC also has SNPs within its nucleotide sequence that governs its expression (or non-expression) on the cell surface. SNP sites within exon 3 of FcγRIIC result in either an open reading frame, hence protein expression on the cell surface (FcγRIIC-ORF), or a stop codon, thus non-expression of FcγRIIC on the cell surface (FcγRIIC-STOP). As such, FcγRIIC is expressed in ~20–40% of the population, and co-expression of FcγRIIC with FcγRIIIA may result in enhanced ADCC capabilities of the NK cells (186).

Some of the FcγRs have been shown to be subject to copy number variability (CNV), and this CNV can result in variable expression of these FcRs on the cell surface. Both FcγRIIC and FcγRIIIA can be CNV, and CNV in these receptors correlates with differences in protein expression levels (189, 190), as well as increased ADCC function through enhanced NK cell activation (186, 190).

Given that the isotypes of FcγRIIIA have been shown to have differential binding affinity to IgG subclasses, that FcγRIIC expression on NK cells is variable within the population, and that these FcγRs can be CNV, using genotypic measures to pre-select patients based on their FcγR genotype for therapeutics that require NK cell may be of critical importance for future clinical investigations. Detailed simultaneous testing of polymorphisms and CNV in all three of these genes (FcγRIIA, FcγRIIC, and FcγRIIIA) has yet to be evaluated for associations with clinical outcome in clinical trials of ADCC-inducing tumor-reactive mAbs. However, it is likely that these factors, which should influence *in vivo* ADCC function, will be found to play a role in the clinical activity of ADCC-inducing mAb therapies.

Concluding Remarks

Monoclonal antibodies utilize different mechanisms to destroy cancer cells, one of which is ADCC. As these treatments have continued to evolve from original mouse antibodies to chimeric antibodies to humanized and fully human antibodies, therapeutic mAbs for cancer treatment are still being engineered to achieve improved anti-tumor efficacy. Besides optimizing the characteristics of the mAb itself, there are other ways to improve anti-tumor effect, one of which is to improve ADCC by combining immune stimulatory therapies with the mAb. Since NK cells are considered a major player in mAb-mediated ADCC against tumor cells, reagents that can enhance NK cell activation, such as IL2, may be combined with mAb to improve the ADCC effect. However, due to IL2-associated toxicity in patients (191), novel IC therapeutics are being generated in an attempt to reduce toxicity and to allow for an increased MTD. Another cytokine that helps NK cell activation is IL15, and it is currently being tested clinically in various types of cancer. Besides cytokines or IC, newly discovered mechanisms that could potentially improve ADCC include MMP inhibitors to prevent FcR shedding, and ionizing radiation to make tumors more immunogenic and vulnerable to immune-mediated destruction. These approaches combined with tumor-specific mAb are still being explored in pre-clinical models to determine efficacy and optimize dosing regimens. Finally, genotypic profiles of NK cells also may contribute to our understanding of the magnitude of ADCC responses of NK cells in any given patient. Specifically, KIRs and FcR genotypes may help to predict clinical outcome to ADCC-inducing mAb therapy, allowing for more personalized treatment. More detailed analyses of associations of clinical outcome with NK cell genotype profiles are needed in order to determine the predictive value of this form of genotyping for distinct types of cancer and for different immunotherapies that involve NK cells.

Acknowledgments

This work was supported, in part, by National Institutes of Health Grants CA032685, CA87025, CA166105, CA14520, a Stand Up To Cancer – St. Baldrick's Pediatric Dream Team Translational Research Grant (SU2C-AACR-DT1113), the University of Wisconsin-Madison Institute for Clinical and Translational Research Grant 1TL1RR025013-01, RSNA Research Resident Seed Grant, ASTRO Resident Seed Grant, and grants from the Midwest Athletes for Childhood Cancer Fund, The Crawdaddy Foundation, and Hyundai Hope on Wheels.

References

1. Dorshkind K, Pollack SB, Bosma MJ, Phillips RA. Natural killer (NK) cells are present in mice with severe combined immunodeficiency (SCID). *J Immunol* (1985) **134**:3798–801.
2. Sun JC, Lanier LL. NK cell development, homeostasis and function: parallels with CD8+ T cells. *Nat Rev Immunol* (2011) **11**:645–57. doi:10.1038/nri3044
3. Bryceson YT, March ME, Ljunggren HG, Long EO. Synergy among receptors on resting NK cells for the activation of natural cytotoxicity and cytokine secretion. *Blood* (2006) **107**:159–66. doi:10.1182/blood-2005-04-1351
4. Chua HL, Serov Y, Brahmi Z. Regulation of FasL expression in natural killer cells. *Hum Immunol* (2004) **65**:317–27. doi:10.1016/j.humimm.2004.01.004
5. Lanier LL. Up on the tightrope: natural killer cell activation and inhibition. *Nat Immunol* (2008) **9**:495–502. doi:10.1038/ni1581
6. Lanier LL. NK cell recognition. *Annu Rev Immunol* (2005) **23**:225–74. doi:10.1146/annurev.immunol.23.021704.115526
7. Parham P. MHC class I molecules and KIRs in human history, health and survival. *Nat Rev Immunol* (2005) **5**:201–14. doi:10.1038/nri1570
8. David G, Djaoud Z, Willem C, Legrand N, Rettman P, Gagne K, et al. Large spectrum of HLA-C recognition by killer Ig-like receptor (KIR)2DL2 and KIR2DL3 and restricted C1 specificity of KIR2DS2: dominant impact of KIR2DL2/KIR2DS2 on KIR2D NK cell repertoire formation. *J Immunol* (2013) **191**:4778–88. doi:10.4049/jimmunol.1301580
9. Stewart CA, Laugier-Anfossi F, Vely F, Saulquin X, Riedmuller J, Tisserant A, et al. Recognition of peptide-MHC class I complexes by activating killer immunoglobulin-like receptors. *Proc Natl Acad Sci U S A* (2005) **102**:13224–9. doi:10.1073/pnas.0503594102
10. Moesta AK, Parham P. Diverse functionality among human NK cell receptors for the C1 epitope of HLA-C: KIR2DS2, KIR2DL2, and KIR2DL3. *Front Immunol* (2012) **3**:336. doi:10.3389/fimmu.2012.00336
11. Ljunggren HG, Kärre K. In search of the 'missing self': MHC molecules and NK cell recognition. *Immunol Today* (1990) **11**:237–44. doi:10.1016/0167-5699(90)90097-S
12. Draghi M. Single-cell analysis of the human NK cell response to missing self and its inhibition by HLA class I. *Blood* (2005) **105**:2028–35. doi:10.1182/blood-2004-08-3174
13. Li HC, Pascal V, Martin MP, Carrington M, Anderson SK. Genetic control of variegated KIR gene expression: polymorphisms of the bi-directional KIR3DL1 promoter are associated with distinct frequencies of gene expression. *PLoS Genet* (2008) **4**(11):e1000254. doi:10.1371/journal.pgen.1000254
14. Cichocki F, Miller JS, Anderson SK. Killer immunoglobulin-like receptor transcriptional regulation: a fascinating dance of multiple promoters. *J Innate Immun* (2011) **3**:242–8. doi:10.1159/000323929
15. Anfossi N, André P, Guia S, Falk CS, Roetynck S, Stewart CA, et al. Human NK cell education by inhibitory receptors for MHC class I. *Immunity* (2006) **25**:331–42. doi:10.1016/j.immuni.2006.06.013
16. Kim S, Poursine-Laurent J, Truscott SM, Lybarger L, Song YJ, Yang L, et al. Licensing of natural killer cells by host major histocompatibility complex class I molecules. *Nature* (2005) **436**:709–13. doi:10.1038/nature03847
17. Elliott JM, Wahle JA, Yokoyama WM. MHC class I-deficient natural killer cells acquire a licensed phenotype after transfer into an MHC class I-sufficient environment. *J Exp Med* (2010) **207**:2073–9. doi:10.1084/jem.20100986
18. Joncker NT, Shifrin N, Delebecque F, Raulet DH. Mature natural killer cells reset their responsiveness when exposed to an altered MHC environment. *J Exp Med* (2010) **207**:2065–72. doi:10.1084/jem.20100570
19. Cerwenka A, Baron JL, Lanier LL. Ectopic expression of retinoic acid early inducible-1 gene (RAE-1) permits natural killer cell-mediated rejection of a MHC class I-bearing tumor in vivo. *Proc Natl Acad Sci USA* (2001) **98**:11521–6. doi:10.1073/Pnas.201238598
20. Ogasawara K, Hamerman JA, Hsin H, Chikuma S, Bour-Jordan H, Chen T, et al. Impairment of NK cell function by NKG2D modulation in NOD mice. *Immunity* (2003) **18**:41–51. doi:10.1016/S1074-7613(02)00505-8
21. Coudert JD, Zimmer J, Tomasello E, Cebecauer M, Colonna M, Vivier E, et al. Altered NKG2D function in NK cells induced by chronic exposure to NKG2D ligand-expressing tumor cells. *Blood* (2005) **106**:1711–7. doi:10.1182/blood-2005-03-0918
22. Sun JC, Lanier LL. Tolerance of NK cells encountering their viral ligand during development. *J Exp Med* (2008) **205**:1819–28. doi:10.1084/Jem.20072448
23. Tripathy SK, Keyel PA, Yang LP, Pingel JT, Cheng TP, Schneeberger A, et al. Continuous engagement of a self-specific activation receptor induces NK cell tolerance. *J Exp Med* (2008) **205**:1829–41. doi:10.1084/Jem.20072446
24. Venstrom JM, Pittari G, Gooley TA, Chewning JH, Spellman S, Haagenson M, et al. HLA-C-dependent prevention of leukemia relapse by donor activating KIR2DS1. *N Engl J Med* (2012) **367**:805–16. doi:10.1056/NEJMoa1200503
25. Pittari G, Liu XR, Selvakumar A, Zhao Z, Merino E, Huse M, et al. NK Cell tolerance of self-specific activating receptor KIR2DS1 in individuals with cognate HLA-C2 ligand. *J Immunol* (2013) **190**:4650–60. doi:10.4049/jimmunol.1202120
26. Alderson KL, Sondel PM. Clinical cancer therapy by NK cells via antibody-dependent cell-mediated cytotoxicity. *J Biomed Biotechnol* (2011) **2011**:379123. doi:10.1155/2011/379123
27. Morel PA, Ernst LK, Metes D. Functional CD32 molecules on human NK cells. *Leuk Lymphoma* (1999) **35**:47–56. doi:10.3109/10428199909145704
28. Lanier LL, Ruitenberg JJ, Phillips JH. Functional and biochemical-analysis of Cd16 antigen on natural-killer cells and granulocytes. *J Immunol* (1988) **141**:3478–85.
29. Smyth MJ, Cretney E, Kelly JM, Westwood JA, Street SE, Yagita H, et al. Activation of NK cell cytotoxicity. *Mol Immunol* (2005) **42**:501–10. doi:10.1016/j.molimm.2004.07.034
30. Srivastava RM, Lee SC, Andrade Filho PA, Lord CA, Jie HB, Davidson HC, et al. Cetuximab-activated natural killer and dendritic cells collaborate to trigger tumor antigen-specific T-cell immunity in head and neck cancer patients. *Clin Cancer Res* (2013) **19**:1858–72. doi:10.1158/1078-0432.CCR-12-2426
31. Poli A, Michel T, Thérésine M, Andrès E, Hentges F, Zimmer J. CD56bright natural killer (NK) cells: an important NK cell subset. *Immunology* (2009) **126**:458–65. doi:10.1111/j.1365-2567.2008.03027.x
32. Kinder M, Greenplate AR, Strohl WR, Jordan RE, Brezski RJ. An Fc engineering approach that modulates antibody-dependent cytokine release without altering cell-killing functions. *MAbs* (2015) **7**:494–504. doi:10.1080/19420862.2015.1022692
33. Juelke K, Killig M, Luetke-Eversloh M, Parente E, Gruen J, Morandi B, et al. CD62L expression identifies a unique subset of polyfunctional CD56dim NK cells. *Blood* (2010) **116**:1299–307. doi:10.1182/blood-2009-11-253286
34. De Maria A, Bozzano F, Cantoni C, Moretta L. Revisiting human natural killer cell subset function revealed cytolytic CD56(dim)CD16+ NK cells as rapid producers of abundant IFN-gamma on activation. *Proc Natl Acad Sci U S A* (2011) **108**:728–32. doi:10.1073/pnas.1012356108
35. Schroder K, Hertzog PJ, Ravasi T, Hume DA. Interferon-gamma: an overview of signals, mechanisms and functions. *J Leukoc Biol* (2004) **75**:163–89. doi:10.1189/jlb.0603252
36. Smyth MJ, Takeda K, Hayakawa Y, Peschon JJ, Van Den Brink MR, Yagita H. Nature's TRAIL – on a path to cancer immunotherapy. *Immunity* (2003) **18**:1–6. doi:10.1016/S1074-7613(02)00502-2
37. Wang R, Jaw JJ, Stutzman NC, Zou Z, Sun PD. Natural killer cell-produced IFN-gamma and TNF-alpha induce target cell cytolysis through up-regulation of ICAM-1. *J Leukoc Biol* (2012) **91**:299–309. doi:10.1189/jlb.0611308
38. Koene HR, Kleijer M, Algra J, Roos D, Vondemborne AEGK, Dehaas M. Fc gamma RIIIa-158V/F polymorphism influences the binding of IgG by natural killer cell Fc gamma RIIIa, independently of the Fc gamma IIIa-48L/R/H phenotype. *Blood* (1997) **90**:1109–14.
39. Shields RL, Lai J, Keck R, O'Connell LY, Hong K, Meng YG, et al. Lack of fucose on human IgG1 N-linked oligosaccharide improves binding to human Fcgamma RIII and antibody-dependent cellular toxicity. *J Biol Chem* (2002) **277**:26733–40. doi:10.1074/jbc.M202069200
40. Shinkawa T, Nakamura K, Yamane N, Shoji-Hosaka E, Kanda Y, Sakurada M, et al. The absence of fucose but not the presence of galactose or bisecting N-acetylglucosamine of human IgG1 complex-type oligosaccharides shows the critical role of enhancing antibody-dependent cellular cytotoxicity. *J Biol Chem* (2003) **278**:3466–73. doi:10.1074/jbc.M210665200
41. Niwa R, Hatanaka S, Shoji-Hosaka E, Sakurada M, Kobayashi Y, Uehara A, et al. Enhancement of the antibody-dependent cellular cytotoxicity of low-fucose IgG1 Is independent of FcgammaRIIIa functional polymorphism. *Clin Cancer Res* (2004) **10**:6248–55. doi:10.1158/1078-0432.CCR-04-0850
42. Cartron G, Dacheux L, Salles G, Solal-Celigny P, Bardos P, Colombat P, et al. Therapeutic activity of humanized anti-CD20 monoclonal antibody and polymorphism in IgG Fc receptor FcgammaRIIIa gene. *Blood* (2002) **99**:754–8. doi:10.1182/blood.V99.3.754

43. Weng WK, Levy R. Two immunoglobulin G fragment C receptor polymorphisms independently predict response to rituximab in patients with follicular lymphoma. *J Clin Oncol* (2003) **21**:3940–7. doi:10.1200/jco.2003.05.013
44. Musolino A, Naldi N, Bortesi B, Pezzuolo D, Capelletti M, Missale G, et al. Immunoglobulin G fragment C receptor polymorphisms and clinical efficacy of trastuzumab-based therapy in patients with HER-2/neu-positive metastatic breast cancer. *J Clin Oncol* (2008) **26**:1789–96. doi:10.1200/Jco.2007.14.8957
45. Bibeau F, Lopez-Crapez E, Di Fiore F, Thezenas S, Ychou M, Blanchard F, et al. Impact of Fc gamma RIIa-Fc gamma RIIIa polymorphisms and KRAS mutations on the clinical outcome of patients with metastatic colorectal cancer treated with cetuximab plus irinotecan. *J Clin Oncol* (2009) **27**:1122–9. doi:10.1200/Jco.2008.18.0463
46. Taylor RJ, Chan SL, Wood A, Voskens CJ, Wolf JS, Lin W, et al. FcgammaRIIIa polymorphisms and cetuximab induced cytotoxicity in squamous cell carcinoma of the head and neck. *Cancer Immunol Immunother* (2009) **58**:997–1006. doi:10.1007/s00262-008-0613-3
47. Scott AM, Wolchok JD, Old LJ. Antibody therapy of cancer. *Nat Rev Cancer* (2012) **12**:278–87. doi:10.1038/nrc3236
48. May C, Sapra P, Gerber HP. Advances in bispecific biotherapeutics for the treatment of cancer. *Biochem Pharmacol* (2012) **84**:1105–12. doi:10.1016/J.Bcp.2012.07.011
49. Ross JS, Gray K, Gray GS, Worland PJ, Rolfe M. Anticancer antibodies. *Am J Clin Pathol* (2003) **119**:472–85. doi:10.1309/Y6LP-C0LR-726L-9DX9
50. Weidle UH, Kontermann RE, Brinkmann U. Tumor-antigen-binding bispecific antibodies for cancer treatment. *Semin Oncol* (2014) **41**:653–60. doi:10.1053/J.Seminoncol.2014.08.004
51. Romee R, Foley B, Lenvik T, Wang Y, Zhang B, Ankarlo D, et al. NK cell CD16 surface expression and function is regulated by a disintegrin and metalloprotease-17 (ADAM17). *Blood* (2013) **121**:3599–608. doi:10.1182/blood-2012-04-425397
52. Goede V, Fischer K, Busch R, Engelke A, Eichhorst B, Wendtner CM, et al. Obinutuzumab plus chlorambucil in patients with CLL and coexisting conditions. *N Engl J Med* (2014) **370**:1101–10. doi:10.1056/NEJMoa1313984
53. Yu AL, Gilman AL, Ozkaynak MF, London WB, Kreissman SG, Chen HX, et al. Anti-GD2 antibody with GM-CSF, interleukin-2, and isotretinoin for neuroblastoma. *N Engl J Med* (2010) **363**:1324–34. doi:10.1056/NEJMoa0911123
54. Junttila TT, Parsons K, Olsson C, Lu Y, Xin Y, Theriault J, et al. Superior in vivo efficacy of afucosylated trastuzumab in the treatment of HER2-amplified breast cancer. *Cancer Res* (2010) **70**:4481–9. doi:10.1158/0008-5472.CAN-09-3704
55. Messersmith WA, Ahnen DJ. Targeting EGFR in colorectal cancer. *N Engl J Med* (2008) **359**:1834–6. doi:10.1056/NEJMe0806778
56. Vincent M, Teppaz G, Lajoie L, Sole V, Bessard A, Maillasson M, et al. Highly potent anti-CD20-RLI immunocytokine targeting established human B lymphoma in SCID mouse. *MAbs* (2014) **6**:1026–37. doi:10.4161/mabs.28699
57. Vincent M, Bessard A, Cochonneau D, Teppaz G, Sole V, Maillasson M, et al. Tumor targeting of the IL-15 superagonist RLI by an anti-GD2 antibody strongly enhances its antitumor potency. *Int J Cancer* (2013) **133**:757–65. doi:10.1002/ijc.28059
58. Delgado DC, Hank JA, Kolesar J, Lorentzen D, Gan J, Seo S, et al. Genotypes of NK cell KIR receptors, their ligands, and Fc receptors in the response of neuroblastoma patients to Hu14.18-IL2 immunotherapy. *Cancer Res* (2010) **70**:9554–61. doi:10.1158/0008-5472.can-10-2211
59. Shusterman S, London WB, Gillies SD, Hank JA, Voss SD, Seeger RC, et al. Antitumor activity of hu14.18-IL2 in patients with relapsed/refractory neuroblastoma: a Children's Oncology Group (COG) phase II study. *J Clin Oncol* (2010) **28**:4969–75. doi:10.1200/JCO.2009.27.8861
60. Sugimoto Y, Hirota M, Yoshikawa K, Sumitomo M, Nakamura K, Ueda R, et al. The therapeutic potential of a novel PSMA antibody and its IL-2 conjugate in prostate cancer. *Anticancer Res* (2014) **34**:89–97.
61. Glorius P, Baerenwaldt A, Kellner C, Staudinger M, Dechant M, Stauch M, et al. The novel tribody [(CD20)(2)xCD16] efficiently triggers effector cell-mediated lysis of malignant B cells. *Leukemia* (2013) **27**:190–201. doi:10.1038/leu.2012.150
62. Cheung NKV, Saarinen UM, Neely JE, Landmeier B, Donovan D, Coccia PF. Monoclonal-antibodies to a glycolipid antigen on human neuro-blastoma cells. *Cancer Res* (1985) **45**:2642–9.
63. Cheung NK, Lazarus H, Miraldi FD, Abramowsky CR, Kallick S, Saarinen UM, et al. Ganglioside GD2 specific monoclonal antibody 3F8: a phase I study in patients with neuroblastoma and malignant melanoma. *J Clin Oncol* (1987) **5**:1430–40.
64. Klee GG. Human anti-mouse antibodies. *Arch Pathol Lab Med* (2000) **124**:921–3.
65. Mujoo K, Cheresh DA, Yang HM, Reisfeld RA. Disialoganglioside GD2 on human neuroblastoma cells: target antigen for monoclonal antibody-mediated cytolysis and suppression of tumor growth. *Cancer Res* (1987) **47**:1098–104.
66. Mujoo K, Kipps TJ, Yang HM, Cheresh DA, Wargalla U, Sander DJ, et al. Functional-properties and effect on growth suppression of human neuro-blastoma tumors by isotype switch variants of monoclonal antiganglioside Gd2 antibody 14.18. *Cancer Res* (1989) **49**:2857–61.
67. Handgretinger R, Baader P, Dopfer R, Klingebiel T, Reuland P, Treuner J, et al. A phase-I study of neuroblastoma with the antiganglioside Gd2 antibody 14.G2a. *Cancer Immunol Immunother* (1992) **35**:199–204. doi:10.1007/Bf01756188
68. Frost JD, Hank JA, Reaman GH, Frierdich S, Seeger RC, Gan J, et al. A phase I/IB trial of murine monoclonal anti-GD2 antibody 14.G2a plus interleukin-2 in children with refractory neuroblastoma: a report of the Children's Cancer Group. *Cancer* (1997) **80**:317–33. doi:10.1002/(SICI)1097-0142(19970715)80:2<317::AID-CNCR21>3.0.CO;2-W
69. Mueller BM, Romerdahl CA, Gillies SD, Reisfeld RA. Enhancement of antibody-dependent cytotoxicity with a chimeric anti-GD2 antibody. *J Immunol* (1990) **144**:1382–6.
70. Barker E, Mueller BM, Handgretinger R, Herter M, Yu AL, Reisfeld RA. Effect of a chimeric anti-ganglioside GD2 antibody on cell-mediated lysis of human neuroblastoma cells. *Cancer Res* (1991) **51**:144–9.
71. Kendra K, Malkovska V, Allen M, Guzman J, Albertini M. In vivo binding and antitumor activity of Ch14.18. *J Immunother* (1999) **22**:423–30. doi:10.1097/00002371-199909000-00006
72. Handgretinger R, Anderson K, Lang P, Dopfer R, Klingebiel T, Schrappe M, et al. A phase I study of human/mouse chimeric antiganglioside GD2 antibody ch14.18 in patients with neuroblastoma. *Eur J Cancer* (1995) **31A**:261–7. doi:10.1016/0959-8049(94)00413-Y
73. Cheung NKV, Guo HF, Hu J, Tassev DV, Cheung IY. Humanizing murine IgG3 anti-GD2 antibody m3F8 substantially improves antibody-dependent cell-mediated cytotoxicity while retaining targeting in vivo. *Oncoimmunology* (2012) **1**(4):477–86. doi:10.4161/Onci.19864
74. Imai M, Landen C, Ohta R, Cheung NK, Tomlinson S. Complement-mediated mechanisms in anti-GD2 monoclonal antibody therapy of murine metastatic cancer. *Cancer Res* (2005) **65**:10562–8. doi:10.1158/0008-5472.CAN-05-1894
75. Thommesen JE, Michaelsen TE, Loset GA, Sandlie I, Brekke OH. Lysine 322 in the human IgG3 C(H)2 domain is crucial for antibody dependent complement activation. *Mol Immunol* (2000) **37**:995–1004. doi:10.1016/S0161-5890(01)00010-4
76. Sorkin LS, Otto M, Baldwin WM III, Vail E, Gillies SD, Handgretinger R, et al. Anti-GD(2) with an FC point mutation reduces complement fixation and decreases antibody-induced allodynia. *Pain* (2010) **149**:135–42. doi:10.1016/j.pain.2010.01.024
77. Leung K. 64Cu-1,4,7,10-tetraazacyclododecane-N,N',N'',N'''-tetraacetic acid-p-benzyl-NH-hu14.18K322A. *Molecular Imaging and Contrast Agent Database (MICAD)*. Bethesda, MD: National Center for Biotechnology Information (US) (2004-2013).
78. Navid F, Sondel PM, Barfield R, Shulkin BL, Kaufman RA, Allay JA, et al. Phase I trial of a novel anti-GD2 monoclonal antibody, Hu14.18K322A, designed to decrease toxicity in children with refractory or recurrent neuroblastoma. *J Clin Oncol* (2014) **32**:1445–52. doi:10.1200/JCO.2013.50.4423
79. Graus-Porta D, Beerli RR, Daly JM, Hynes NE. ErbB-2, the preferred heterodimerization partner of all ErbB receptors, is a mediator of lateral signaling. *EMBO J* (1997) **16**:1647–55. doi:10.1093/emboj/16.7.1647
80. Sliwkowski MX, Lofgren JA, Lewis GD, Hotaling TE, Fendly BM, Fox JA. Nonclinical studies addressing the mechanism of action of trastuzumab (Herceptin). *Semin Oncol* (1999) **26**:60–70.
81. Carson WE, Parihar R, Lindemann MJ, Personeni N, Dierksheide J, Meropol NJ, et al. Interleukin-2 enhances the natural killer cell response to Herceptin-coated Her2/neu-positive breast cancer cells. *Eur J Immunol* (2001) **31**:3016–25. doi:10.1002/1521-4141(2001010)31:10<3016::AID-IMMU3016>3.0.CO;2-J

82. Clynes RA, Towers TL, Presta LG, Ravetch JV. Inhibitory Fc receptors modulate in vivo cytotoxicity against tumor targets. *Nat Med* (2000) **6**:443–6. doi:10.1038/74704
83. Kurai J, Chikumi H, Hashimoto K, Yamaguchi K, Yamasaki A, Sako T, et al. Antibody-dependent cellular cytotoxicity mediated by cetuximab against lung cancer cell lines. *Clin Cancer Res* (2007) **13**:1552–61. doi:10.1158/1078-0432.CCR-06-1726
84. Monteverde M, Milano G, Strola G, Maffi M, Lattanzio L, Vivenza D, et al. The relevance of ADCC for EGFR targeting: a review of the literature and a clinically-applicable method of assessment in patients. *Crit Rev Oncol Hematol* (2015). doi:10.1016/j.critrevonc.2015.02.014
85. Lopez-Albaitero A, Lee SC, Morgan S, Grandis JR, Gooding WE, Ferrone S, et al. Role of polymorphic Fc gamma receptor IIIa and EGFR expression level in cetuximab mediated, NK cell dependent in vitro cytotoxicity of head and neck squamous cell carcinoma cells. *Cancer Immunol Immunother* (2009) **58**:1853–64. doi:10.1007/s00262-009-0697-4
86. Etienne-Grimaldi MC, Bennouna J, Formento JL, Douillard JY, Francoual M, Hennebelle I, et al. Multifactorial pharmacogenetic analysis in colorectal cancer patients receiving 5-fluorouracil-based therapy together with cetuximab-irinotecan. *Br J Clin Pharmacol* (2012) **73**:776–85. doi:10.1111/j.1365-2125.2011.04141.x
87. Zhang W, Gordon M, Schultheis AM, Yang DY, Nagashima F, Azuma M, et al. FCGR2A and FCGR3A polymorphisms associated with clinical outcome of epidermal growth factor receptor expressing metastatic colorectal cancer patients treated with single-agent cetuximab. *J Clin Oncol* (2007) **25**:3712–8. doi:10.1200/JCO.2006.08.8021
88. Zhang W, Azuma M, Lurje G, Gordon MA, Yang D, Pohl A, et al. Molecular predictors of combination targeted therapies (cetuximab, bevacizumab) in irinotecan-refractory colorectal cancer (BOND-2 study). *Anticancer Res* (2010) **30**:4209–17.
89. Dahan L, Norguet E, Etienne-Grimaldi MC, Formento JL, Gasmi M, Nanni I, et al. Pharmacogenetic profiling and cetuximab outcome in patients with advanced colorectal cancer. *BMC Cancer* (2011) **11**:496. doi:10.1186/1471-2407-11-496
90. Lievre A, Bachet JB, Le Corre D, Boige V, Landi B, Emile JF, et al. KRAS mutation status is predictive of response to cetuximab therapy in colorectal cancer. *Cancer Res* (2006) **66**:3992–5. doi:10.1158/0008-5472.CAN-06-0191
91. Manches O, Lui G, Chaperot L, Gressin R, Molens JP, Jacob MC, et al. In vitro mechanisms of action of rituximab on primary non-Hodgkin lymphomas. *Blood* (2003) **101**:949–54. doi:10.1182/blood-2002-02-0469
92. Funakoshi S, Longo DL, Murphy WJ. Differential in vitro and in vivo antitumor effects mediated by anti-CD40 and anti-CD20 monoclonal antibodies against human B-cell lymphomas. *J Immunother Emphasis Tumor Immunol* (1996) **19**:93–101. doi:10.1097/00002371-199603000-00002
93. Carlotti E, Palumbo GA, Oldani E, Tibullo D, Salmoiraghi S, Rossi A, et al. FcgammaRIIIA and FcgammaRIIA polymorphisms do not predict clinical outcome of follicular non-Hodgkin's lymphoma patients treated with sequential CHOP and rituximab. *Haematologica* (2007) **92**:1127–30. doi:10.3324/haematol.11288
94. Mitrovic Z, Aurer I, Radman I, Ajdukovic R, Sertic J, Labar B. FCgammaRIIIA and FCgammaRIIA polymorphisms are not associated with response to rituximab and CHOP in patients with diffuse large B-cell lymphoma. *Haematologica* (2007) **92**:998–9. doi:10.3324/haematol.10327
95. Prochazka V, Papajik T, Gazdova J, Divoka M, Rozmanova S, Faber E, et al. FcgammaRIIIA receptor genotype does not influence an outcome in patients with follicular lymphoma treated with risk-adapted immunochemotherapy. *Neoplasma* (2011) **58**:263–70. doi:10.4149/neo_2011_03_263
96. Ahlgrimm M, Pfreundschuh M, Kreuz M, Regitz E, Preuss KD, Bittenbring J. The impact of Fc-gamma receptor polymorphisms in elderly patients with diffuse large B-cell lymphoma treated with CHOP with or without rituximab. *Blood* (2011) **118**:4657–62. doi:10.1182/blood-2011-04-346411
97. Ghesquieres H, Cartron G, Seymour JF, Delfau-Larue MH, Offner F, Soubeyran P, et al. Clinical outcome of patients with follicular lymphoma receiving chemoimmunotherapy in the PRIMA study is not affected by FCGR3A and FCGR2A polymorphisms. *Blood* (2012) **120**:2650–7. doi:10.1182/blood-2012-05-431825
98. Persky DO, Dornan D, Goldman BH, Braziel RM, Fisher RI, Leblanc M, et al. Fc gamma receptor 3a genotype predicts overall survival in follicular lymphoma patients treated on SWOG trials with combined monoclonal antibody plus chemotherapy but not chemotherapy alone. *Haematologica* (2012) **97**:937–42. doi:10.3324/haematol.2011.050419
99. Burchard PR, Malhotra S, Kaur P, Tsongalis GJ. Detection of the FCGR3a polymorphism using a real-time polymerase chain reaction assay. *Cancer Genet* (2013) **206**:130–4. doi:10.1016/j.cancergen.2013.03.002
100. Ferrara C, Brunker P, Suter T, Moser S, Puntener U, Umana P. Modulation of therapeutic antibody effector functions by glycosylation engineering: influence of Golgi enzyme localization domain and co-expression of heterologous beta1, 4-N-acetylglucosaminyltransferase III and Golgi alpha-mannosidase II. *Biotechnol Bioeng* (2006) **93**:851–61. doi:10.1002/bit.20777
101. Mossner E, Brunker P, Moser S, Puntener U, Schmidt C, Herter S, et al. Increasing the efficacy of CD20 antibody therapy through the engineering of a new type II anti-CD20 antibody with enhanced direct and immune effector cell-mediated B-cell cytotoxicity. *Blood* (2010) **115**:4393–402. doi:10.1182/blood-2009-06-225979
102. Herter S, Herting F, Mundigl O, Waldhauer I, Weinzierl T, Fauti T, et al. Preclinical activity of the type II CD20 antibody GA101 (obinutuzumab) compared with rituximab and ofatumumab in vitro and in xenograft models. *Mol Cancer Ther* (2013) **12**:2031–42. doi:10.1158/1535-7163.MCT-12-1182
103. Sehn LH, Assouline SE, Stewart DA, Mangel J, Gascoyne RD, Fine G, et al. A phase 1 study of obinutuzumab induction followed by 2 years of maintenance in patients with relapsed CD20-positive B-cell malignancies. *Blood* (2012) **119**:5118–25. doi:10.1182/blood-2012-02-408773
104. Morschhauser FA, Cartron G, Thieblemont C, Solal-Celigny P, Haioun C, Bouabdallah R, et al. Obinutuzumab (GA101) monotherapy in relapsed/refractory diffuse large b-cell lymphoma or mantle-cell lymphoma: results from the phase II GAUGUIN study. *J Clin Oncol* (2013) **31**:2912–9. doi:10.1200/JCO.2012.46.9585
105. Ogura M, Tobinai K, Hatake K, Uchida T, Suzuki T, Kobayashi Y, et al. Phase I study of obinutuzumab (GA101) in Japanese patients with relapsed or refractory B-cell non-Hodgkin lymphoma. *Cancer Sci* (2013) **104**:105–10. doi:10.1111/cas.12040
106. Salles GA, Morschhauser F, Solal-Celigny P, Thieblemont C, Lamy T, Tilly H, et al. Obinutuzumab (GA101) in patients with relapsed/refractory indolent non-Hodgkin lymphoma: results from the phase II GAUGUIN study. *J Clin Oncol* (2013) **31**:2920–6. doi:10.1200/JCO.2012.46.9718
107. Demaria S, Bhardwaj N, McBride WH, Formenti SC. Combining radiotherapy and immunotherapy: a revived partnership. *Int J Radiat Oncol Biol Phys* (2005) **63**:655–66. doi:10.1016/j.ijrobp.2005.06.032
108. Stone HB, Peters LJ, Milas L. Effect of host immune capability on radiocurability and subsequent transplantability of a murine fibrosarcoma. *J Natl Cancer Inst* (1979) **63**:1229–35.
109. Budach W, Taghian A, Freeman J, Gioioso D, Suit HD. Impact of stromal sensitivity on radiation response of tumors. *J Natl Cancer Inst* (1993) **85**:988–93. doi:10.1093/jnci/85.12.988
110. Formenti SC, Demaria S. Combining radiotherapy and cancer immunotherapy: a paradigm shift. *J Natl Cancer Inst* (2013) **105**:256–65. doi:10.1093/jnci/djs629
111. Reap EA, Roof K, Maynor K, Borrero M, Booker J, Cohen PL. Radiation and stress-induced apoptosis: a role for Fas/Fas ligand interactions. *Proc Natl Acad Sci U S A* (1997) **94**:5750–5. doi:10.1073/pnas.94.11.5750
112. Chakraborty M, Abrams SI, Coleman CN, Camphausen K, Schlom J, Hodge JW. External beam radiation of tumors alters phenotype of tumor cells to render them susceptible to vaccine-mediated T-cell killing. *Cancer Res* (2004) **64**:4328–37. doi:10.1158/0008-5472.CAN-04-0073
113. Garnett CT, Palena C, Chakraborty M, Tsang KY, Schlom J, Hodge JW. Sublethal irradiation of human tumor cells modulates phenotype resulting in enhanced killing by cytotoxic T lymphocytes. *Cancer Res* (2004) **64**:7985–94. doi:10.1158/0008-5472.CAN-04-1525
114. Reits EA, Hodge JW, Herberts CA, Groothuis TA, Chakraborty M, Wansley EK, et al. Radiation modulates the peptide repertoire, enhances MHC class I expression, and induces successful antitumor immunotherapy. *J Exp Med* (2006) **203**:1259–71. doi:10.1084/jem.20052494
115. Voutsas IF, Mahaira LG, Fotopoulou K, Kapranos N, Reclos JG, Gritzapis AD, et al. Gamma-irradiation induces HER-2/neu overexpression in breast cancer cell lines and sensitivity to treatment with trastuzumab. *Int J Radiat Biol* (2013) **89**:319–25. doi:10.3109/09553002.2013.765617
116. Wattenberg MM, Kwilas AR, Gameiro SR, Dicker AP, Hodge JW. Expanding the use of monoclonal antibody therapy of cancer by using ionising radiation

to upregulate antibody targets. *Br J Cancer* (2014) **110**:1472–80. doi:10.1038/bjc.2014.79
117. Harrison D, Phillips JH, Lanier LL. Involvement of a metalloprotease in spontaneous and phorbol ester-induced release of natural killer cell-associated Fc gamma RIII (CD16-II). *J Immunol* (1991) **147**:3459–65.
118. Grzywacz B, Kataria N, Verneris MR. CD56dimCD16+ NK cells downregulate CD16 following target cell induced activation of matrix metalloproteinases. *Leukemia* (2007) **21**:356–9. doi:10.1038/sj.leu.2404499
119. Liu Q, Sun Y, Rihn S, Nolting A, Tsoukas PN, Jost S, et al. Matrix metalloprotease inhibitors restore impaired NK cell-mediated antibody-dependent cellular cytotoxicity in human immunodeficiency virus type 1 infection. *J Virol* (2009) **83**:8705–12. doi:10.1128/jvi.02666-08
120. Hank JA, Robinson RR, Surfus J, Mueller BM, Reisfeld RA, Cheung NK, et al. Augmentation of antibody dependent cell mediated cytotoxicity following in vivo therapy with recombinant interleukin 2. *Cancer Res* (1990) **50**:5234–9.
121. Hank JA, Surfus J, Gan J, Chew TL, Hong R, Tans K, et al. Treatment of neuroblastoma patients with antiganglioside GD2 antibody plus interleukin-2 induces antibody-dependent cellular cytotoxicity against neuroblastoma detected in vitro. *J Immunother Emphasis Tumor Immunol* (1994) **15**:29–37. doi:10.1097/00002371-199401000-00004
122. Metcalf D, Begley CG, Johnson GR, Nicola NA, Vadas MA, Lopez AF, et al. Biologic properties in vitro of a recombinant human granulocyte-macrophage colony-stimulating factor. *Blood* (1986) **67**:37–45.
123. Giebel S, Thomas X, Hallbook H, Geissler K, Boiron JM, Huguet F, et al. The prophylactic use of granulocyte-colony stimulating factor during remission induction is associated with increased leukaemia-free survival of adults with acute lymphoblastic leukaemia: a joint analysis of five randomised trials on behalf of the EWALL. *Eur J Cancer* (2012) **48**:360–7. doi:10.1016/j.ejca.2011.11.023
124. Renner P, Milazzo S, Liu JP, Zwahlen M, Birkmann J, Horneber M. Primary prophylactic colony-stimulating factors for the prevention of chemotherapy-induced febrile neutropenia in breast cancer patients. *Cochrane Database Syst Rev* (2012) **10**:CD007913. doi:10.1002/14651858.CD007913.pub2
125. Mhaskar R, Clark OA, Lyman G, Engel Ayer Botrel T, Morganti Paladini L, Djulbegovic B. Colony-stimulating factors for chemotherapy-induced febrile neutropenia. *Cochrane Database Syst Rev* (2014) **10**:CD003039. doi:10.1002/14651858.CD003039.pub2
126. Kushner BH, Kramer K, Cheung NK. Phase II trial of the anti-G(D2) monoclonal antibody 3F8 and granulocyte-macrophage colony-stimulating factor for neuroblastoma. *J Clin Oncol* (2001) **19**:4189–94.
127. Repka T, Chiorean EG, Gay J, Herwig KE, Kohl VK, Yee D, et al. Trastuzumab and interleukin-2 in HER2-positive metastatic breast cancer: a pilot study. *Clin Cancer Res* (2003) **9**:2440–6.
128. Holmberg LA, Maloney D, Bensinger W. Immunotherapy with rituximab/interleukin-2 after autologous stem cell transplantation as treatment for CD20+ non-Hodgkin's lymphoma. *Clin Lymphoma Myeloma* (2006) **7**:135–9. doi:10.3816/CLM.2006.n.051
129. Lopes De Menezes DE, Denis-Mize K, Tang Y, Ye H, Kunich JC, Garrett EN, et al. Recombinant interleukin-2 significantly augments activity of rituximab in human tumor xenograft models of B-cell non-Hodgkin lymphoma. *J Immunother* (2007) **30**:64–74. doi:10.1097/01.cji.0000211315.21116.07
130. Carson WE, Giri JG, Lindemann MJ, Linett ML, Ahdieh M, Paxton R, et al. Interleukin (IL) 15 is a novel cytokine that activates human natural killer cells via components of the IL-2 receptor. *J Exp Med* (1994) **180**:1395–403. doi:10.1084/jem.180.4.1395
131. Conlon KC, Lugli E, Welles HC, Rosenberg SA, Fojo AT, Morris JC, et al. Redistribution, hyperproliferation, activation of natural killer cells and CD8 T cells, and cytokine production during first-in-human clinical trial of recombinant human interleukin-15 in patients with cancer. *J Clin Oncol* (2015) **33**:74–82. doi:10.1200/JCO.2014.57.3329
132. Berger C, Berger M, Hackman RC, Gough M, Elliott C, Jensen MC, et al. Safety and immunologic effects of IL-15 administration in nonhuman primates. *Blood* (2009) **114**:2417–26. doi:10.1182/blood-2008-12-189266
133. Munger W, Dejoy SQ, Jeyaseelan R Sr, Torley LW, Grabstein KH, Eisenmann J, et al. Studies evaluating the antitumor activity and toxicity of interleukin-15, a new T cell growth factor: comparison with interleukin-2. *Cell Immunol* (1995) **165**:289–93. doi:10.1006/cimm.1995.1216
134. Marks-Konczalik J, Dubois S, Losi JM, Sabzevari H, Yamada N, Feigenbaum L, et al. IL-2-induced activation-induced cell death is inhibited in IL-15 transgenic mice. *Proc Natl Acad Sci U S A* (2000) **97**:11445–50. doi:10.1073/pnas.200363097
135. Waldmann TA, Lugli E, Roederer M, Perera LP, Smedley JV, Macallister RP, et al. Safety (toxicity), pharmacokinetics, immunogenicity, and impact on elements of the normal immune system of recombinant human IL-15 in rhesus macaques. *Blood* (2011) **117**:4787–95. doi:10.1182/blood-2010-10-311456
136. Moga E, Alvarez E, Canto E, Vidal S, Rodriguez-Sanchez JL, Sierra J, et al. NK cells stimulated with IL-15 or CpG ODN enhance rituximab-dependent cellular cytotoxicity against B-cell lymphoma. *Exp Hematol* (2008) **36**:69–77. doi:10.1016/j.exphem.2007.08.012
137. Moga E, Canto E, Vidal S, Juarez C, Sierra J, Briones J. Interleukin-15 enhances rituximab-dependent cytotoxicity against chronic lymphocytic leukemia cells and overcomes transforming growth factor beta-mediated immunosuppression. *Exp Hematol* (2011) **39**:1064–71. doi:10.1016/j.exphem.2011.08.006
138. Mortier E, Quemener A, Vusio P, Lorenzen I, Boublik Y, Grotzinger J, et al. Soluble interleukin-15 receptor alpha (IL-15R alpha)-sushi as a selective and potent agonist of IL-15 action through IL-15R beta/gamma. Hyperagonist IL-15 x IL-15R alpha fusion proteins. *J Biol Chem* (2006) **281**:1612–9. doi:10.1074/jbc.M508624200
139. Bessard A, Sole V, Bouchaud G, Quemener A, Jacques Y. High antitumor activity of RLI, an interleukin-15 (IL-15)-IL-15 receptor alpha fusion protein, in metastatic melanoma and colorectal cancer. *Mol Cancer Ther* (2009) **8**:2736–45. doi:10.1158/1535-7163.MCT-09-0275
140. Zhu X, Marcus WD, Xu W, Lee HI, Han K, Egan JO, et al. Novel human interleukin-15 agonists. *J Immunol* (2009) **183**:3598–607. doi:10.4049/jimmunol.0901244
141. Xu W, Jones M, Liu B, Zhu X, Johnson CB, Edwards AC, et al. Efficacy and mechanism-of-action of a novel superagonist interleukin-15: interleukin-15 receptor alphaSu/Fc fusion complex in syngeneic murine models of multiple myeloma. *Cancer Res* (2013) **73**:3075–86. doi:10.1158/0008-5472.CAN-12-2357
142. Rosario M, Liu B, Kong L, Schneider SE, Jeng EK, Rhode PR, et al. The IL-15 superagonist ALT-803 enhances NK cell ADCC and in vivo clearance of B cell lymphomas directed by an anti-CD20 monoclonal antibody. *Blood* (2014) **124**:807–807.
143. Gillies SD, Reilly EB, Lo KM, Reisfeld RA. Antibody-targeted interleukin 2 stimulates T-cell killing of autologous tumor cells. *Proc Natl Acad Sci U S A* (1992) **89**:1428–32. doi:10.1073/pnas.89.4.1428
144. Hank JA, Surfus JE, Gan J, Jaeger P, Gillies SD, Reisfeld RA, et al. Activation of human effector cells by a tumor reactive recombinant anti-ganglioside GD2 interleukin-2 fusion protein (ch14.18-IL2). *Clin Cancer Res* (1996) **2**:1951–9.
145. Gubbels JAA, Gadbaw B, Buhtoiarov IN, Horibata S, Kapur AK, Patel D, et al. Ab-IL2 fusion proteins mediate NK cell immune synapse formation by polarizing CD25 to the target cell-effector cell interface. *Cancer Immunol Immunother* (2011) **60**:1789–800. doi:10.1007/S00262-011-1072-9
146. Lode HN, Xiang R, Varki NM, Dolman CS, Gillies SD, Reisfeld RA. Targeted interleukin-2 therapy for spontaneous neuroblastoma metastases to bone marrow. *J Natl Cancer Inst* (1997) **89**:1586–94. doi:10.1093/jnci/89.21.1586
147. Lode HN, Xiang R, Dreier T, Varki NM, Gillies SD, Reisfeld RA. Natural killer cell-mediated eradication of neuroblastoma metastases to bone marrow by targeted interleukin-2 therapy. *Blood* (1998) **91**:1706–15.
148. Albertini MR, Hank JA, Gadbaw B, Kostlevy J, Haldeman J, Schalch H, et al. Phase II trial of hu14.18-IL2 for patients with metastatic melanoma. *Cancer Immunol Immunother* (2012) **61**:2261–71. doi:10.1007/s00262-012-1286-5
149. Olson WC, Heston WD, Rajasekaran AK. Clinical trials of cancer therapies targeting prostate-specific membrane antigen. *Rev Recent Clin Trials* (2007) **2**:182–90. doi:10.2174/157488707781662724
150. Huntington ND, Legrand N, Alves NL, Jaron B, Weijer K, Plet A, et al. IL-15 trans-presentation promotes human NK cell development and differentiation in vivo. *J Exp Med* (2009) **206**:25–34. doi:10.1084/jem.20082013
151. Riethmuller G. Symmetry breaking: bispecific antibodies, the beginnings, and 50 years on. *Cancer Immun* (2012) **12**:12.
152. Spiess C, Zhai Q, Carter PJ. Alternative molecular formats and therapeutic applications for bispecific antibodies. *Mol Immunol* (2015). doi:10.1016/j.molimm.2015.01.003
153. Rothe A, Sasse S, Topp MS, Eichenauer DA, Hummel H, Reiners KS, et al. A phase I study of the bispecific anti-CD30/CD16A antibody construct AFM13

in patients with relapsed or refractory Hodgkin lymphoma. *Blood* (2015) **125**(26):4024–31. doi:10.1182/blood-2014-12-614636

154. Reusch U, Duell J, Ellwanger K, Herbrecht C, Knackmuss SH, Fucek I, et al. A tetravalent bispecific T and Ab (CD19/CD3), AFM11, efficiently recruits T cells for the potent lysis of CD19(+) tumor cells. *MAbs* (2015) **7**:584–604. doi:10.1080/19420862.2015.1029216
155. Silla LM, Chen J, Zhong RK, Whiteside TL, Ball ED. Potentiation of lysis of leukaemia cells by a bispecific antibody to CD33 and CD16 (Fc gamma RIII) expressed by human natural killer (NK) cells. *Br J Haematol* (1995) **89**:712–8. doi:10.1111/j.1365-2141.1995.tb08406.x
156. Wiernik A, Foley B, Zhang B, Verneris MR, Warlick E, Gleason MK, et al. Targeting natural killer cells to acute myeloid leukemia in vitro with a CD16 x 33 bispecific killer cell engager and ADAM17 inhibition. *Clin Cancer Res* (2013) **19**:3844–55. doi:10.1158/1078-0432.CCR-13-0505
157. Gleason MK, Ross JA, Warlick ED, Lund TC, Verneris MR, Wiernik A, et al. CD16xCD33 bispecific killer cell engager (BiKE) activates NK cells against primary MDS and MDSC CD33+ targets. *Blood* (2014) **123**:3016–26. doi:10.1182/blood-2013-10-533398
158. Bruenke J, Barbin K, Kunert S, Lang P, Pfeiffer M, Stieglmaier K, et al. Effective lysis of lymphoma cells with a stabilised bispecific single-chain Fv antibody against CD19 and FcgammaRIII (CD16). *Br J Haematol* (2005) **130**:218–28. doi:10.1111/j.1365-2141.2005.05414.x
159. Kellner C, Bruenke J, Horner H, Schubert J, Schwenkert M, Mentz K, et al. Heterodimeric bispecific antibody-derivatives against CD19 and CD16 induce effective antibody-dependent cellular cytotoxicity against B-lymphoid tumor cells. *Cancer Lett* (2011) **303**:128–39. doi:10.1016/j.canlet.2011.01.020
160. Gleason MK, Verneris MR, Todhunter DA, Zhang B, McCullar V, Zhou SX, et al. Bispecific and trispecific killer cell engagers directly activate human NK cells through CD16 signaling and induce cytotoxicity and cytokine production. *Mol Cancer Ther* (2012) **11**:2674–84. doi:10.1158/1535-7163.MCT-12-0692
161. Schubert I, Saul D, Nowecki S, Mackensen A, Fey GH, Oduncu FS. A dual-targeting triplebody mediates preferential redirected lysis of antigen double-positive over single-positive leukemic cells. *MAbs* (2014) **6**:286–96. doi:10.4161/mabs.26768
162. Turini M, Chames P, Bruhns P, Baty D, Kerfelec B. A FcgammaRIII-engaging bispecific antibody expands the range of HER2-expressing breast tumors eligible to antibody therapy. *Oncotarget* (2014) **5**:5304–19.
163. Vallera DA, Zhang B, Gleason MK, Oh S, Weiner LM, Kaufman DS, et al. Heterodimeric bispecific single-chain variable-fragment antibodies against EpCAM and CD16 induce effective antibody-dependent cellular cytotoxicity against human carcinoma cells. *Cancer Biother Radiopharm* (2013) **28**:274–82. doi:10.1089/cbr.2012.1329
164. Burns LJ, Weisdorf DJ, Defor TE, Vesole DH, Repka TL, Blazar BR, et al. IL-2-based immunotherapy after autologous transplantation for lymphoma and breast cancer induces immune activation and cytokine release: a phase I/II trial. *Bone Marrow Transplant* (2003) **32**:177–86. doi:10.1038/sj.bmt.1704086
165. Berg M, Lundqvist A, McCoy P Jr, Samsel L, Fan Y, Tawab A, et al. Clinical-grade ex vivo-expanded human natural killer cells up-regulate activating receptors and death receptor ligands and have enhanced cytolytic activity against tumor cells. *Cytotherapy* (2009) **11**:341–55. doi:10.1080/14653240902807034
166. Fujisaki H, Kakuda H, Shimasaki N, Imai C, Ma J, Lockey T, et al. Expansion of highly cytotoxic human natural killer cells for cancer cell therapy. *Cancer Res* (2009) **69**:4010–7. doi:10.1158/0008-5472.CAN-08-3712
167. Voskens CJ, Watanabe R, Rollins S, Campana D, Hasumi K, Mann DL. Ex-vivo expanded human NK cells express activating receptors that mediate cytotoxicity of allogeneic and autologous cancer cell lines by direct recognition and antibody directed cellular cytotoxicity. *J Exp Clin Cancer Res* (2010) **29**:134. doi:10.1186/1756-9966-29-134
168. Deng X, Terunuma H, Nieda M, Xiao W, Nicol A. Synergistic cytotoxicity of ex vivo expanded natural killer cells in combination with monoclonal antibody drugs against cancer cells. *Int Immunopharmacol* (2012) **14**:593–605. doi:10.1016/j.intimp.2012.09.014
169. Arai S, Meagher R, Swearingen M, Myint H, Rich E, Martinson J, et al. Infusion of the allogeneic cell line NK-92 in patients with advanced renal cell cancer or melanoma: a phase I trial. *Cytotherapy* (2008) **10**:625–32. doi:10.1080/14653240802301872
170. Barkholt L, Alici E, Conrad R, Sutlu T, Gilljam M, Stellan B, et al. Safety analysis of ex vivo-expanded NK and NK-like T cells administered to cancer patients: a phase I clinical study. *Immunotherapy* (2009) **1**:753–64. doi:10.2217/imt.09.47
171. Linn YC, Niam M, Chu S, Choong A, Yong HX, Heng KK, et al. The anti-tumour activity of allogeneic cytokine-induced killer cells in patients who relapse after allogeneic transplant for haematological malignancies. *Bone Marrow Transplant* (2012) **47**:957–66. doi:10.1038/bmt.2011.202
172. Dupont B, Selvakumar A, Steffens U. The killer cell inhibitory receptor genomic region on human chromosome 19q13.4. *Tissue Antigens* (1997) **49**:557–63. doi:10.1111/j.1399-0039.1997.tb02802.x
173. Barten R, Torkar M, Haude A, Trowsdale J, Wilson MJ. Divergent and convergent evolution of NK-cell receptors. *Trends Immunol* (2001) **22**:52–7. doi:10.1016/S1471-4906(00)01802-0
174. Yawata M. Roles for HLA and KIR polymorphisms in natural killer cell repertoire selection and modulation of effector function. *J Exp Med* (2006) **203**:633–45. doi:10.1084/jem.20051884
175. Cooley S, Trachtenberg E, Bergemann TL, Saeteurn K, Klein J, Le CT, et al. Donors with group B KIR haplotypes improve relapse-free survival after unrelated hematopoietic cell transplantation for acute myelogenous leukemia. *Blood* (2008) **113**:726–32. doi:10.1182/blood-2008-07-171926
176. Cooley S, Weisdorf DJ, Guethlein LA, Klein JP, Wang T, Le CT, et al. Donor selection for natural killer cell receptor genes leads to superior survival after unrelated transplantation for acute myelogenous leukemia. *Blood* (2010) **116**:2411–9. doi:10.1182/blood-2010-05-283051
177. Cooley S, Weisdorf DJ, Guethlein LA, Klein JP, Wang T, Marsh SGE, et al. Donor killer cell Ig-like receptor B haplotypes, recipient HLA-C1, and HLA-C mismatch enhance the clinical benefit of unrelated transplantation for acute myelogenous leukemia. *J Immunol* (2014) **192**:4592–600. doi:10.4049/jimmunol.1302517
178. Ruggeri L. Effectiveness of donor natural killer cell alloreactivity in mismatched hematopoietic transplants. *Science* (2002) **295**:2097–100. doi:10.1126/science.1068440
179. Hsu KC, Keever-Taylor CA, Wilton A, Pinto C, Heller G, Arkun K, et al. Improved outcome in HLA-identical sibling hematopoietic stem-cell transplantation for acute myelogenous leukemia predicted by KIR and HLA genotypes. *Blood* (2005) **105**:4878–84. doi:10.1182/Blood-2004-12-4825
180. Leung W, Handgretinger R, Iyengar R, Turner V, Holladay MS, Hale GA. Inhibitory KIR-HLA receptor-ligand mismatch in autologous haematopoietic stem cell transplantation for solid tumour and lymphoma. *Br J Cancer* (2007) **97**:539–42. doi:10.1038/sj.bjc.6603913
181. Venstrom JM, Zheng J, Noor N, Danis KE, Yeh AW, Cheung IY, et al. KIR and HLA genotypes are associated with disease progression and survival following autologous hematopoietic stem cell transplantation for high-risk neuroblastoma. *Clin Cancer Res* (2009) **15**:7330–4. doi:10.1158/1078-0432.ccr-09-1720
182. Tarek N, Luduec JL, Gallagher MM, Zheng J, Venstrom JM, Chamberlain E, et al. Unlicensed NK cells target neuroblastoma following anti-GD2 antibody treatment. *J Clin Investig* (2012) **122**:3260–70. doi:10.1172/jci62749
183. Du J, Lopez-Verges S, Pitcher BN, Johnson J, Jung SH, Zhou L, et al. CALGB 150905 (alliance): rituximab broadens the antilymphoma response by activating unlicensed NK cells. *Cancer Immunol Res* (2014) **2**:878–89. doi:10.1158/2326-6066.cir-13-0158
184. Scquizzato E, Zambello R, Teramo A, Baesso I, Varotto S, Albergoni MP, et al. KIR/HLA-I mismatching and risk of relapse in paediatric patients undergoing non-haploidentical allogeneic haematopoietic stem cell transplantation. *Pediatr Transplant* (2011) **15**:198–204. doi:10.1111/j.1399-3046.2010.01447.x
185. Bournazos S, Woof JM, Hart SP, Dransfield I. Functional and clinical consequences of Fc receptor polymorphic and copy number variants. *Clin Exp Immunol* (2009) **157**:244–54. doi:10.1111/j.1365-2249.2009.03980.x
186. Breunis WB, Van Mirre E, Geissler J, Laddach N, Wolbink G, Van Der Schoot E, et al. Copy number variation at the FCGR locus includes FCGR3A, FCGR2C and FCGR3B but not FCGR2A and FCGR2B. *Hum Mutat* (2009) **30**:E640–50. doi:10.1002/humu.20997
187. Bruhns P, Iannascoli B, England P, Mancardi DA, Fernandez N, Jorieux S, et al. Specificity and affinity of human Fc gamma receptors and their polymorphic variants for human IgG subclasses. *Blood* (2009) **113**:3716–25. doi:10.1182/Blood-2008-09-179754
188. Mellor JD, Brown MP, Irving HR, Zalcberg JR, Dobrovic A. A critical review of the role of Fc gamma receptor polymorphisms in the response to

monoclonal antibodies in cancer. *J Hematol Oncol* (2013) **6**:1. doi:10.1186/1756-8722-6-1

189. Breunis WB, Van Mirre E, Bruin M, Geissler J, De Boer M, Peters M, et al. Copy number variation of the activating FCGR2C gene predisposes to idiopathic thrombocytopenic purpura. *Blood* (2008) **111**:1029–38. doi:10.1182/blood-2007-03-079913
190. Van Der Heijden J, Breunis WB, Geissler J, De Boer M, Van Den Berg TK, Kuijpers TW. Phenotypic variation in IgG receptors by nonclassical FCGR2C alleles. *J Immunol* (2012) **188**:1318–24. doi:10.4049/jimmunol.1003945
191. Tartour E, Mathiot C, Fridman WH. Current status of interleukin-2 therapy in cancer. *Biomed Pharmacother* (1992) **46**:473–84. doi:10.1016/0753-3322(92)90005-R

17

NK Cell Subgroups, Phenotype, and Functions after Autologous Stem Cell Transplantation

*Benedikt Jacobs[1,2,3], Sara Tognarelli[4,5], Kerstin Poller[3], Peter Bader[4,5], Andreas Mackensen[3] and Evelyn Ullrich[3,4,5]**

[1] Department of Cancer Immunology, Institute for Cancer Research, Oslo University Hospital, Radiumhospital, Oslo, Norway,
[2] The KG Jebsen Center for Cancer Immunotherapy, Institute of Clinical Medicine, University of Oslo, Oslo, Norway,
[3] Department of Haematology and Oncology, University Hospital Erlangen, Erlangen, Germany, [4] Department of Pediatric Stem Cell Transplantation and Immunology, Children's Hospital, Johann Wolfgang Goethe-University, Frankfurt, Germany,
[5] LOEWE Center for Cell and Gene Therapy, Johann Wolfgang Goethe-University, Frankfurt, Germany

Edited by:
Francisco Borrego, BioCruces Health Research Institute and Cruces University Hospital, Spain

Reviewed by:
Roberto Biassoni, Istituto Giannina Gaslini, Italy
Jacques Zimmer, Luxembourg Institute of Health, Luxembourg

***Correspondence:**
Evelyn Ullrich
evelyn.ullrich@kgu.de

High-dose chemotherapy with consecutive autologous stem cell transplantation (autoSCT) is a well-established treatment option for patients suffering from malignant lymphoma or multiple myeloma. Natural killer (NK) cells are an important part of the immune surveillance, and their cell number after autoSCT is predictive for progression-free and overall survival. To improve knowledge about the role of NK cells after autoSCT, we investigated different NK cell subgroups, their phenotype, and their functions in patients treated with autoSCT. Directly after leukocyte regeneration (>1000 leukocytes/μl) following autoSCT, $CD56^{++}$ NK cells were the major NK cell subset. Surprisingly, these cells showed unusually high surface expression levels of CD57 and killer Ig-like receptors (KIRs) compared to expression levels before or at later time points after autoSCT. Moreover, these NK cells strongly upregulated KIR2DL2/3/S2 and KIR3DL1, whereas KIR2DL1/S1 remained constant, indicating that this cell population arose from more immature NK cells instead of from activated mature ones. Remarkably, NK cells were already able to degranulate and produce IFN-γ and MIP-1β upon tumor interaction early after leukocyte regeneration. In conclusion, we describe an unusual upregulation of CD57 and KIRs on $CD56^{++}$ NK cells shortly after autoSCT. Importantly, these NK cells were functionally competent upon tumor interaction at this early time point.

Keywords: NK cells, CD57, KIR, autologous stem cell transplantation, CD107a expression, IFN-γ production

INTRODUCTION

Natural killer (NK) cells are an important part of the innate immune system and are able to kill virus-infected or malignantly transformed cells (1). Their important role in tumor surveillance has been demonstrated in many different tumor models (1). NK cell cytotoxicity is regulated by a diverse repertoire of inhibitory and activating receptors. Inhibitory receptors, such as killer Ig-like receptors (KIRs) and the C-type lectin-like receptor NKG2A, recognize

---

Abbreviations: HDC, high-dose chemotherapy; (auto)SCT, (autologous) stem cell transplantation; MM, multiple myeloma; NHL, non-Hodgkin lymphoma; TP1–3, time point 1–3; KIRs, killer Ig-like receptors; NK cell, natural killer cell.

different alleles of HLA molecules (HLA-A, B, and C by KIRs and HLA-E by NKG2A) on healthy cells. In contrast, many tumor cells downregulate their HLA molecules to evade T cell recognition, making them more susceptible to NK cell killing (2). Additionally, tumor cells may express stress-induced molecules, such as MHC I chain-related molecule A/B or UL-16-binding proteins, which are ligands for the activating NK cell receptor NKG2D (3, 4).

High-dose chemotherapy (HDC) with consecutive autologous stem cell transplantation (autoSCT) is an effective and well-established treatment option for patients suffering from multiple myeloma (MM) (5) or malignant lymphoma (6–9). Before treatment with the myeloablative chemotherapy, hematopoietic stem cells are collected from peripheral blood and frozen. Following HDC, these cells are thawed and given back to the patient in order to shorten the time of aplasia, thereby reducing the infection and blood transfusion rates.

Many reports have demonstrated the important role of the absolute lymphocyte count after HDC/autoSCT (10). It has been shown that an absolute lymphocyte count >500/μl is associated with improved overall and progression-free survival in patients with Hodgkin lymphoma (11), non-Hodgkin lymphoma (NHL) (12), acute myeloid leukemia (13), MM (12), and metastatic breast cancer (14). By analyzing the different lymphocyte subsets at day 15 following autoSCT, a clear correlation between improved overall survival and progression-free survival could only be found for NK cell counts >80/μl. No correlation was found for any other lymphocyte subset (15). In a more recent study, improved median overall and progression-free survival as well as the NK cell count at day 15 after HDC/autoSCT were all associated with an increased IL-15 concentration at day 15 of ≥76.5 pg/ml for NHL patients receiving HDC/autoSCT (16).

Because there is no information available regarding the detailed analysis of NK cell subsets or function early after HDC/autoSCT, in our study, we prospectively investigated the major NK cell subsets directly after leukocyte recovery (leukocytes >1000/μl) and also at later time points after HDC/autoSCT in patients with different lymphoproliferative diseases. Moreover, we further analyzed the different NK cell subsets, evaluating their education and differentiation markers, as well as their functional properties, such as cytokine/chemokine production and degranulation capacity.

MATERIALS AND METHODS

Patients' Characteristics and Study Design

This study was carried out in accordance with the recommendations of the local ethics committee of the University of Erlangen, and all patients gave written informed consent in accordance with the Declaration of Helsinki. Patients who suffered from MM or malignant lymphoma and received HDC/autoSCT were included. Blood was taken from these patients at three different time points. Time point 1 (TP1) was before the start of the HDC and at least 3 weeks after the last chemotherapy. The second time point (TP2) was 1–2 days after leukocyte regeneration (>1000 leukocytes/μl) following autoSCT, and the third time point (TP3) was after at least 2 weeks following leukocyte recovery.

Reagents

For NK and K562 cell culture, we used full media containing RPMI 1640 media (Gibco®) supplemented with 10% FBS, MEM non-essential amino acids (1%), sodium pyruvate (1%), L-glutamine (1%; all from PanBiotech), and penicillin/streptomycin (1%; Thermo Fischer Scientific). For the washing steps, we used Dulbecco's phosphate-buffered saline (DPBS; Gibco®).

To analyze the different leukocyte subsets, CD3, CD14, CD16, CD19, CD45, and CD56 antibodies with different fluorochromes from Becton Dickinson (BD) were used. For detailed NK cell subset analyses, we used anti-KIR2D-, KIR3DL1/2-, and KIR2DL1/S1-PE (Miltenyi), KIR2DL1-PerCP (R&D), KIR2DL2/3/S2-APC (Beckman Coulter), KIR3DL1 PE-Vio770 (Miltenyi), NKG2A FITC (Miltenyi), and CD57 APC (BD). For the KIR staining, the clones of the antibodies were selected according to Czaja et al. (17), and a sequential staining protocol was used as described by Beziat et al. (18). For intracellular staining, we used IFN-γ PE-Cy-7 and MIP-1β APC-H7 (BD). To exclude dead cells, 7-AAD (BD) for extracellular and Fixable Viability Dye eFluor® 520 (eBioscience) for intracellular staining were used.

PBMC Preparation, Freezing, and Thawing

Blood samples were obtained from the patients at the indicated time points. PBMCs were isolated by performing a Ficoll density centrifugation of whole blood samples, and then the PBMCs were frozen (5×10^6 PBMCs/ml freezing media containing 90% FCS + 10% DMSO; from Sigma) until they were used.

For thawing, tubes were incubated at room temperature and gently thawed by re-suspending the cells in prewarmed full media (without FCS). The cells were washed twice and counted before being used for further analysis.

Extra- and Intracellular Antibody Staining

For surface staining, the cells were incubated with different antibody cocktails for 10′ at 4°C; then, they were washed and either fixed in BD CytoFix™ solution or further processed for intracellular staining using the BD Cytofix/Cytoperm™ kit. Briefly, the cells were first incubated for 30′ at 4°C with Fixable Viability Dye eFluor® 520 (eBioscience), and then they were washed and fixed for 20′ at 4°C in 100 μl BD Cytofix™ solution. Subsequently, the cells were washed in BD Cytoperm™ solution and incubated with the indicated intracellular antibodies for 20′ at 4°C. Finally, the cells were washed, placed into BD CytoFix™ solution, and analyzed with a BD FACSCanto II™ or Canto10c™ using the FlowJo® software (FlowJo, LLC) was used to analyze the FACS data.

Functional Assays

After thawing, the cells were placed into full media supplemented with 100 IU IL-2/ml (Proleukin®, Novartis) in a 96-U well plate overnight. The cells were harvested, washed, counted, and incubated with K562 cells (ratio 1:1) in full media in a 96-U well plate for 4 h. CD107a APC (BD) was added at the start of the coculturing period and BD GolgiStop™ (BD) was added after 1 h for the rest of the incubation time.

Statistics

For statistical analysis, we used GraphPad Prism® software. In all graphs, the mean and SD were calculated and plotted. For comparison between matched samples, we used a Wilcoxon test, whereas for non-matched samples, we performed a Mann–Whitney test. Statistical significance is indicated with the *p*-values (*<0.05; **<0.01; ***<0.001; *****<0.0001).

RESULTS

Patients' Characteristics and Leukocyte Subsets

Peripheral blood samples from 32 different patients collected at three specific time points (TP1–3, as described in Section "Materials and Methods") were available for the analysis of leukocyte subsets. The basic patient characteristics are summarized in **Table 1**. The ratio between male and female patients was approximately 2:1. Half of the patients suffered from MM. The average age at HDC/autoSCT was 56.7 years. The average time between SCT and TP2 was 11.6 days, whereas the time between collecting samples at TP2 and TP3 was 38.8 days. All three values were normally distributed.

Before the start of HDC/autoSCT at TP1, $CD3^+CD56^-$ T cells and $CD14^+$ monocytes were the two major leukocyte subsets in all patients (**Figure 1A**; for gating strategy, see Figure S1 in Supplementary Material). While $CD14^+$ monocytes were the major subset early after leukocyte recovery after HDC/autoSCT at TP2 (**Figures 1A,B**), $CD3^+CD56^-$ T cells were the major subset at TP3 (**Figures 1A,B**). In contrast, the NK cell percentages within the leukocyte population significantly decreased from TP1 to TP2 (*p*-value: 0.0175) but recovered to the initial value at TP3 (*p*-value TP2/TP3: 0.0263; TP1/3: 0.19; **Figures 1A,B**).

By correlating the NK cell dynamics at the three different time points with clinical data, we observed that the fold change of the NK cell percentage within the leukocyte population between TP1 and TP2 (ratio TP2/TP1) significantly differed between patients having a time period of ≤11 days between SCT and TP2 and those having a period of >11 days (*p*-value: 0.04). When the time period was >11 days, no decrease in the NK cell percentage within the leukocyte population at TP2 was observed. Moreover, another significant difference was observed when comparing the fold change in the NK cell percentage between TP1 and TP2 in patients who were refractory/recurrent or not at 1 year after SCT (*p*-value: 0.0258). Patients with recurrent or refractory disease did not have a decrease in their fold change ratio (TP2/TP1), while patients without recurrent/refractory disease did have a decrease at 1 year after SCT. Additionally, the fold increase in the NK cell percentage between TP2 and TP3 (TP3/TP2 ratio) was more pronounced when the time period between TP2 and TP3 was ≤38 days (*p*-value: 0.12; **Figure 1C**).

No differences were observed when analyzing the patients' age or hematological malignancies in relation to the fold changes of the NK cell percentages within the leukocyte population between the three different time points (Figure S2 in Supplementary Material).

TABLE 1 | Patient characteristics.

Gender	Male (23), female (9)
Mean age (range)	56.7 years (30–74 years)
Malignancy	
Multiple myeloma	16
Diffuse large-cell B cell lymphoma	7
Mantle cell lymphoma	3
T cell lymphoma	5
Hodgkin lymphoma	1
Recurrent/refractory disease at 1 year after SCT (yes/no)	6/21 (5 patients were lost during follow-up)
Time period SCT–TP2 (mean, days)	11.6
Time period TP2–TP3 (mean, days)	38.8

$CD56^{++}CD16^{-/+}$ NK Cells are the Major Subset at Leukocyte Recovery

Next, we analyzed the different NK cell subsets based on their CD56 and CD16 expression. NK cells were divided into $CD56^{++}CD16^-$ or $CD16^+$ and $CD56^+CD16^{++}$ NK cells (see Figure S1 in Supplementary Material). The $CD56^{++}CD16^+$ population has been reported to be an intermediate state between $CD56^{++}CD16^-$ and $CD56^+CD16^{++}$ NK cells (19, 20). The $CD56^+CD16^-$ population was excluded, as it was shown that this population could be induced by cryopreservation (21).

At TP1, the major NK cell subset was the $CD56^+CD16^{++}$ NK cell population (71.86%), followed by the $CD56^{++}CD16^+$ (17.71%) and the $CD56^{++}CD16^-$ (10%) populations (**Figure 2A**). After leukocyte regeneration (TP2), the $CD56^+CD16^{++}$ cells significantly decreased (39.98%; *p*-value: <0.0001), whereas both $CD56^{++}$ NK cell subsets significantly increased ($CD16^-$: 22.85%, *p*-value: <0.0001; $CD16^+$: 36.51%, *p*-value: <0.0001). At TP3, the levels of $CD56^+CD16^{++}$ NK cells increased again (64.85%; *p*-value: <0.0001) but remained reduced in contrast to the starting levels at TP1 (*p*-value: 0.0064). Conversely, the levels of $CD56^{++}CD16^-$ (14.41%; *p*-value: 0.0006) and $CD16^+$ (20.66%; *p*-value: <0.0001) cells decreased but remained significantly elevated compared to the starting values [*p*-values: 0.0184 ($CD16^-$) and 0.0428 ($CD16^+$); **Figure 2B**].

The ratio between mature $CD56^+$ and more immature $CD56^{++}$ NK cells, which is approximately 10 within healthy donors (data not shown), was already significantly reduced within our patients before HDC/autoSCT (TP1; ratio: 3.707) because all patients had previously received chemotherapy. Nevertheless, the ratio significantly dropped at TP2 (ratio: 1.157; *p*-value: <0.0001) and did not recover to initial values at TP3 [ratio: 2.49; *p*-value: 0.0001 (TP2–3); 0.0111 (TP1–3); **Figure 2C**].

The $CD56^+/CD56^{++}$ ratios at TP2 and TP3 were both independent of the time period between SCT and TP2 (≤11 days: 1.247 vs. >11 days: 1.041; *p*-value: 0.392) and between TP2 and TP3 (≤38 days: 2.59 vs. >38 days: 2.32; *p*-value: 0.504; **Figure 2D**). Within the group of MM patients, the $CD56^+/CD56^{++}$ ratio was lower compared to lymphoma patients, but

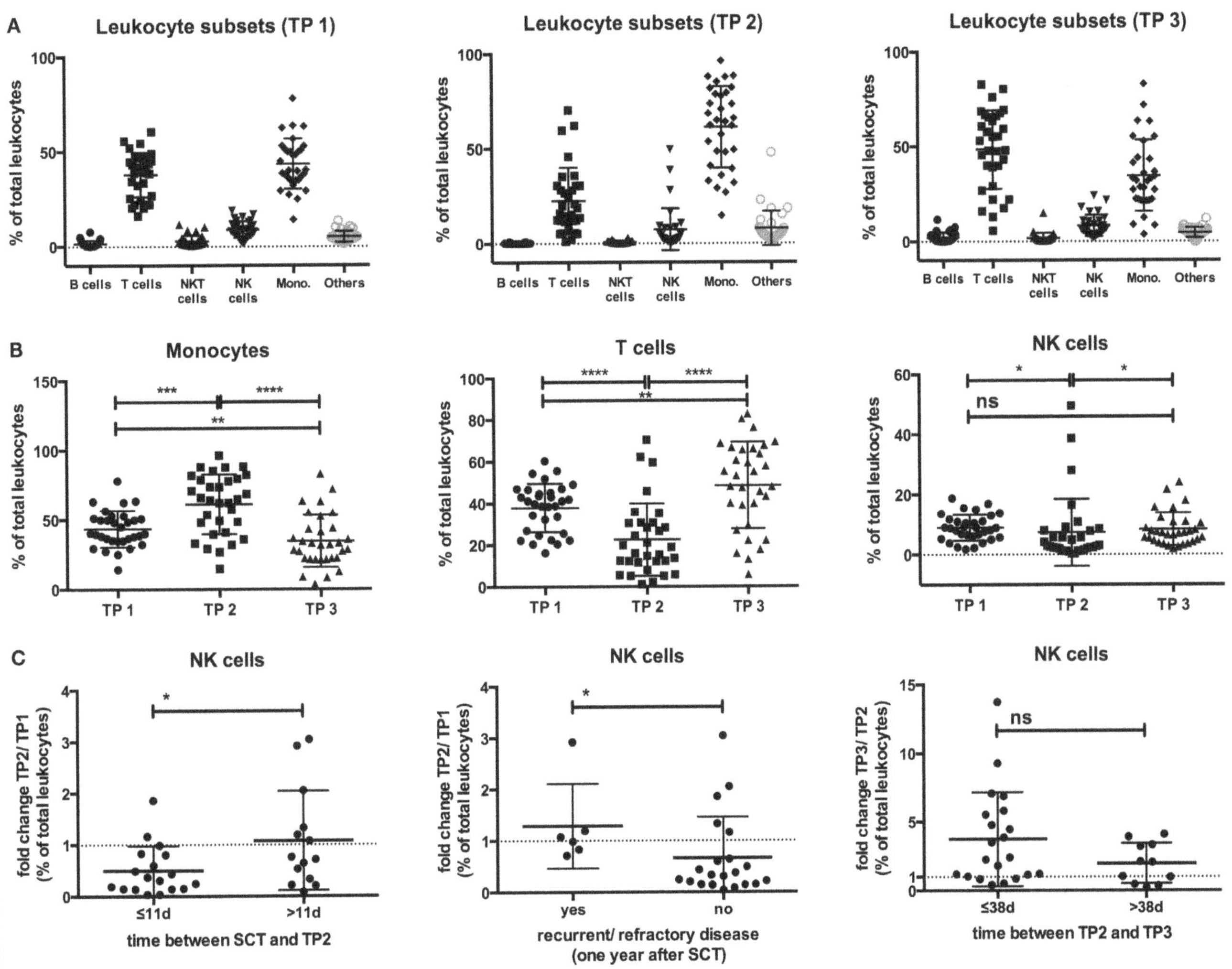

FIGURE 1 | (A) The percentages of the different leukocyte subsets within the leukocyte population at all three time points are indicated. **(B)** The percentage of CD14⁺ monocytes within the leukocyte population increased from TP1 (43.49%) to TP2 (61.22%; *p*-value: 0.0003), becoming the major leukocyte subset at TP2 but decreased below the starting levels at TP3 (34.6%; *p*-value TP2/TP3: <0.0001; TP1/3: 0.0088). In contrast, the T cell percentages decreased from TP1 (37.89%) to TP2 (22.5%; *p*-value: <0.0001), but they increased again at TP3 (48.49%) above the initial values to become the major leukocyte subset at TP3 (*p*-value TP2/TP3: <0.0001; TP1/3: 0.0093). The NK cell percentages decreased from TP1 (8.94%) to TP2 (7.32%) but reached the initial levels again at TP3 (8.29%). **(C)** Patients with a time period of ≤11 days between SCT and TP2 had a decrease from TP1 to TP2 within their NK cell percentage (ratio TP2/TP1: 0.49), in contrast to patients with a time period >11 days (ratio TP2/TP1: 1.07). Similarly, patients with no recurrent/refractory disease 1 year after SCT had a decrease of their NK cell percentage at TP2 (ratio TP2/TP1: 0.66), as opposed to patients who were recurrent/refractory at 1 year (ratio TP2/TP1: 1.28). The increase of the NK cell percentage from TP2 to TP3 was more pronounced in patients with a time period of ≤38 days between TP2 and TP3 (ratio TP3/TP2: 3.73) than in patients with a time period of >38 days (ratio TP3/TP2: 1.97).

it was only significantly lower at TP2 (1.656 vs. 0.6578; *p*-value: 0.0019; **Figure 2D**). In contrast, the patients' age and relapse/refractory status at 1 year after autoSCT seemed to have no impact on the $CD56^{+}/CD56^{++}$ ratio at any time point (Figure S3 in Supplementary Material).

Increased Levels of CD57 and KIR Expression After Leukocyte Regeneration

Next, we analyzed the expression of markers for NK cell education and differentiation at the indicated time points.

As expected, the NKG2A expression on all NK cells increased from TP1 (67.63%) to TP2 (76.51%; *p*-value: 0.0179), and the percentage of $NKG2A^{+}$ NK cells remained elevated above the starting values until TP3 (77.67%; *p*-value: 0.0009). Further NK cell subset analyses revealed that the observed early increase in NKG2A at TP2 was potentially an effect of the elevated $CD56^{++}$ population expressing higher levels of NKG2A than the $CD56^{+}CD16^{++}$ population because the NKG2A expression did not significantly differ within the distinct subsets between TP1 and TP2. In contrast, the percentage of NKG2A-expressing NK cells was significantly elevated in all subsets at TP3 compared to

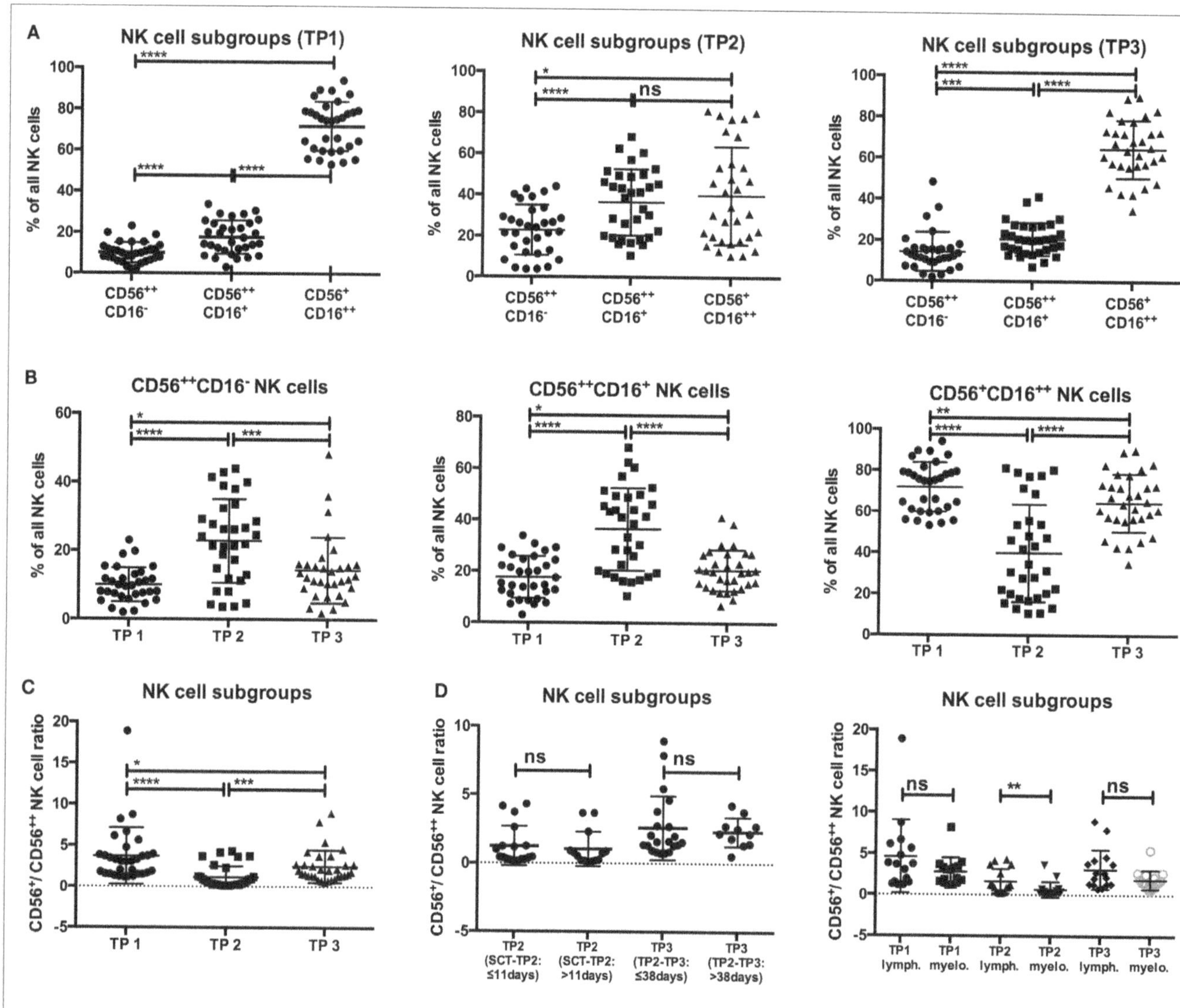

FIGURE 2 | (A,B) The level of $CD56^{+}CD16^{++}$ NK cells significantly dropped at TP2 (39.98%) and recovered at TP3 (64.58%), but it remained lower in contrast to the starting levels (TP1: 71.86%). Moreover, the percentages of $CD56^{++}CD16^{-/+}$ NK cells were markedly increased at TP2 ($CD16^{-}$: 22.85%; $CD16^{+}$: 36.51%) and remained elevated at TP3 ($CD16^{-}$: 14.41%; $CD16^{+}$: 20.66%) compared to the TP1 values ($CD16^{-}$: 10%; $CD16^{+}$: 17.71%). **(C)** The ratio of $CD56^{+}/CD56^{++}$ NK cells at TP1 was already lower than in healthy controls (approx 10; data not shown), with a ratio of 3.707, and decreased further at TP2 (1.157). Although the ratio increased at TP3 (2.49), it was still lower compared to TP1 and healthy control samples. **(D)** There were no significant differences when analyzing the $CD56^{+}/CD56^{++}$ ratio at TP2 with regard to the time period between SCT and TP2 (≤11 days: 1.247; >11 days: 1.041) or at TP3 with regard to the period between TP2 and TP3 (≤38 days: 2.59; >38 days: 2.32). A significant difference of the $CD56^{+}/CD56^{++}$ ratio between lymphoma and myeloma patients was only observed at TP2 (lymphoma patients: 1.656; myeloma patients: 0.6578).

TP1 [*p*-value: 0.0054 ($CD56^{++}CD16^{-}$); 0.0259 ($CD56^{++}CD16^{+}$); 0.0031 ($CD56^{+}CD16^{++}$); **Figure 3A**].

$CD57^{+}$ NK cells, known to define terminally differentiated NK cells, significantly increased from TP1 (43.53%) to TP2 (56.66%; *p*-value: 0.0163) in all NK cells but decreased to the initial values at TP3 (39.13%; *p*-value: 0.4274). Surprisingly, CD57 expression was significantly increased within the $CD56^{++}CD16^{+/-}$ population at TP2 ($CD16^{-}$: 32.98%; $CD16^{+}$: 57.78%, *p*-value: <0.0001) but decreased again from TP2 to TP3 ($CD16^{-}$: 5.592%; $CD16^{+}$: 19.06%, *p*-value: <0.0001). Nevertheless, the percentage of $CD57^{+}$ cells was still elevated in contrast to the starting values at TP1 [$CD16^{-}$: 4.1%; $CD16^{+}$: 15.58%, *p*-values: 0.0102 ($CD16^{-}$); 0.0214 ($CD16^{+}$)]. CD57 expression within the $CD56^{+}CD16^{++}$ population was also elevated at TP2 (72.15%; *p*-value: 0.0006) but to a much lesser extent (**Figure 3B**).

Strikingly, the percentage of KIR^{+} NK cells remained constant over time (TP1: 42.45%; TP3: 44.09%; *p*-value: 0.463), even shortly after leukocyte regeneration (TP2: 40.77%; *p*-value: 0.4106). Of note, the KIR expression within the $CD56^{++}CD16^{+/-}$ NK cell population was markedly increased at TP2 ($CD16^{-}$:

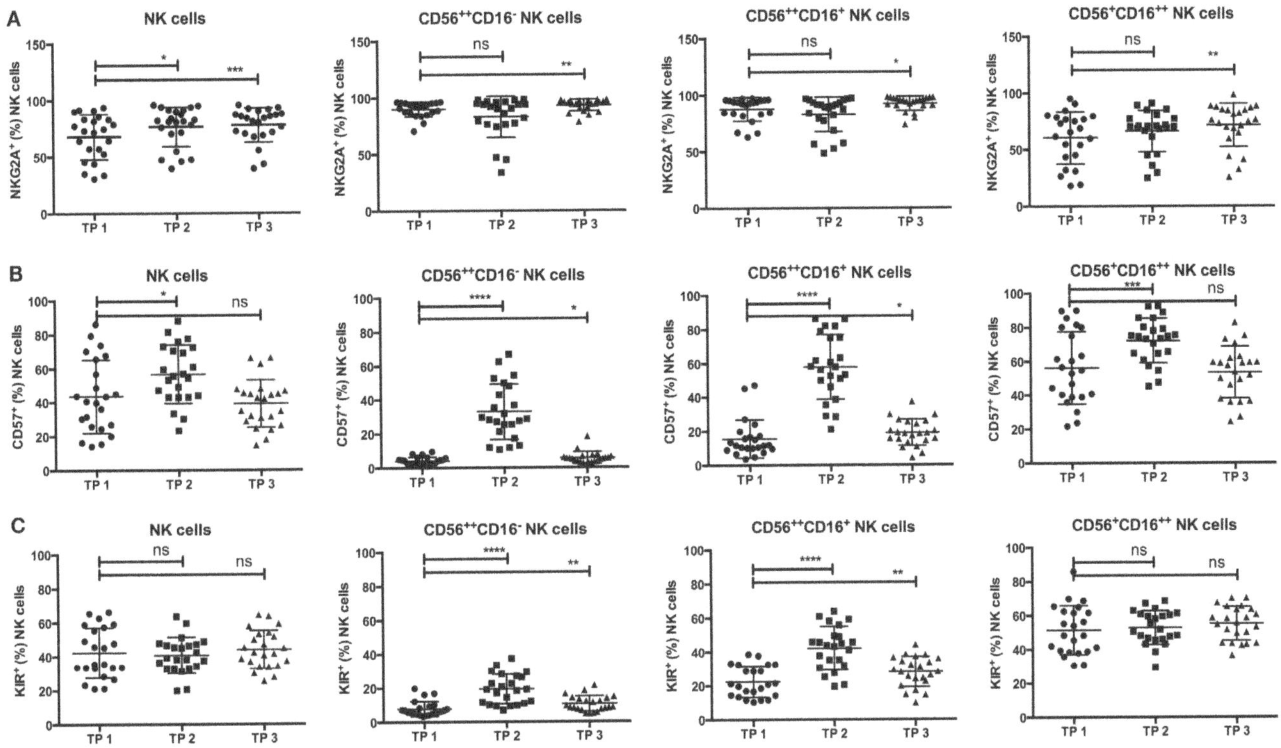

FIGURE 3 | (A) NKG2A expression was significantly increased at TP2 (76.41%) and TP3 (77.67%) within all NK cells compared to the starting values (TP1: 67.63). Moreover, the percentage of NKG2A$^+$ cells was significantly increased at TP3 within the CD56^{++}CD16$^-$ (TP1: 89.94%; TP3: 93.1%) and CD16$^+$ (TP1: 87.6%; TP3: 92.27%) subsets as well as within the CD56$^+$CD16^{++} subset (TP1: 60.39%; TP3: 71.42%). **(B)** Similarly, the percentage of CD57$^+$ NK cells increased at TP2 (56.66%) but returned to the initial values at TP3 (TP1: 43.53%; TP3: 39.13%). Consistently, within the CD56^{++}CD16$^-$ and CD16$^+$ as well as within the CD56$^+$CD16^{++} NK cell subsets, there was a highly significant increase of CD57 expression at TP2 (CD56^{++}CD16$^-$: 32.98%; CD56^{++}CD16$^+$: 57.78%; and CD56$^+$CD16^{++}: 72.15%), which decreased at TP3 again but remained above the starting values within the CD56^{++} subsets (CD16$^-$: TP1 4.1% and TP3 5.592% and CD16$^+$: TP1 15.58% and TP3 19.06%). **(C)** In parallel, the KIR expression was significantly elevated at TP2 (CD16$^-$: 19.35%; CD16$^+$: 42.14%) and TP3 (CD16$^-$: 10.43%; CD16$^+$: 28.14%) compared to TP1 (CD16$^-$: 7.82%; CD16$^+$: 22.44%) within the CD56^{++} subsets but not within the CD56$^+$CD16^{++} cells (TP1: 51.42%; TP2: 52.89%; and TP3: 55.23%).

19.35%; CD16$^+$: 42.14%; p-value: <0.0001 for both populations) and remained elevated at TP3 (CD16$^-$: 10.43%; CD16$^+$: 28.14%) in comparison to the starting values at TP1 [CD16$^-$: 7.82%; CD16$^+$: 22.44%; p-values: 0.0083 (CD16$^-$); 0.0015 (CD16$^+$)]. KIR expression within the CD56$^+$CD16^{++} population remained stable throughout all time points (TP1: 51.42%, TP2: 52.89%, and TP3: 55.23%; **Figure 3C**).

Elevated CD57 and KIR Expression is Age Dependent

Additionally, we analyzed the influence of different clinical factors on the expression levels of NKG2A, CD57, and KIRs within the different NK cell subsets. As CD57 expression levels within healthy individuals are known to be age dependent (22), we compared CD57 expression within the younger (≤56 years) and older (>56 years) patient populations. Notably, there was a significant difference in the percentage of CD57$^+$ cells within the CD56^{++}CD16$^-$ population at TP2 (≤56 years: 23.98%; >56 years: 42.8%; p-value: 0.0056), whereas this effect was not observed within the CD56^{++}CD16$^+$ or the CD56$^+$CD16^{++} population (**Figure 4A**).

Similar, by analyzing KIR expression within the CD56^{++}CD16$^{+/-}$ populations, we observed a significant difference in their expression at TP2 between the two age groups [≤56 years: 14.74% (CD16$^-$) and 37.29% (CD16$^+$); >56 years: 24.39% (CD16$^-$) and 47.44% (CD16$^+$); p-values: 0.0088 (CD16$^-$), 0.0317 (CD16$^+$)]. No age-dependent difference was found for KIR expression within the CD56$^+$CD16^{++} population at TP2 but at TP3 (≤56 years: 50.51%; >56 years: 60.37%; p-value: 0.0225; **Figure 4B**).

In contrast, there was no difference in NKG2A expression throughout all age groups, time points, or NK cell populations (**Figure 4C**).

We also addressed CD57, KIR, and NKG2A expressions in different subsets with regard to recurrent/refractory disease at 1 year after SCT and hematological malignancy. There were no differences in the expression of the three markers regarding the rate of relapsed/refractory disease throughout all NK cell subsets and time points, except for CD57 expression at TP1 within the CD56^{++}CD16$^-$ NK cell population, which was higher in patients

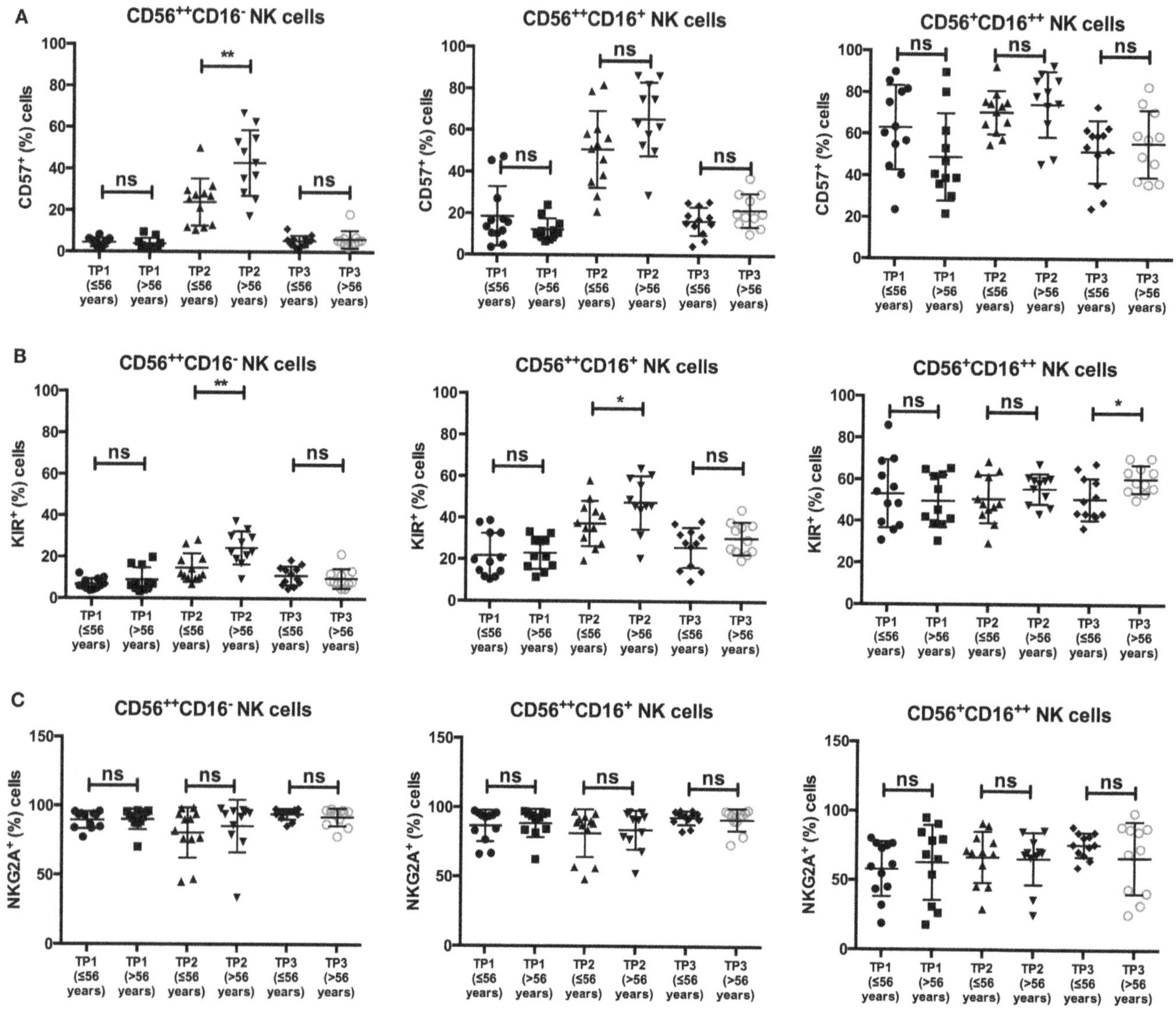

FIGURE 4 | (A) There was a significant difference in the average of CD57^{+} cells within the CD56^{++}CD16^{-} NK cell subsets at TP2 in patients ≤56 years (23.98%) vs. >56 years (42.8%). A significant difference was not observed within the CD56^{++}CD16^{+} (≤56 years: 50.83% vs. >56 years: 65.36%) or the CD56^{+}CD16^{++} subset (≤56 years: 70.25% vs. >56 years: 74.23%). **(B)** The percentage of KIR^{+} NK cells at TP2 differed significantly between the two age groups within the CD56^{++} subset (≤56 years: CD56^{++}CD16^{-}: 14.74% and CD56^{++}CD16^{+}: 37.29% vs. >56 years: CD56^{++}CD16^{-}: 24.39% and CD56^{++}CD16^{+}: 47.44%). A significant difference within the CD56^{+}CD16^{++} subset was observed at TP3 (≤56 years: 50.51% vs. >56 years: 60.37%). **(C)** No difference was observed in NKG2A expression within the different NK subsets and age groups.

with no recurrent/refractory disease 1 year after SCT (*p*-value: 0.0083; Figure S4A in Supplementary Material).

Furthermore, patients with MM had higher NKG2A expression within the CD56^{++}CD16$^{-/+}$ subset at TP2 compared to lymphoma patients. No further differences were observed (Figure S4B in Supplementary Material).

Detailed KIR Expression Analysis

We next performed an extended KIR analysis within the samples of five additional patients, who were not included into the original analysis group because we did not have a sample from TP3. We analyzed the expression levels and distribution of global KIR expression (anti-KIR2D and anti-KIR3DL1/2), as well as KIR2DL1/S1, KIR2DL2/3/S2, and KIR3DL1, within the different subsets at TP1 and TP2. Notably, global KIR expression levels were upregulated within both CD56^{++} NK cell populations at TP2 compared to TP1, whereas they remained stable within the CD56^{+}CD16^{++} NK cell population, confirming the results from the former analyzed patient cohort (**Figures 5A–C**). Within the CD56^{+}CD16^{++} population, no clear differences between the expression levels of the different KIR subsets were observed comparing TP1 and TP2 (**Figure 5C**). In contrast, both CD56^{++} NK cell subsets upregulated their KIR2DL2/3/S2 and KIR3DL1 expression levels from TP1 to TP2, whereas the KIR2DL1/S1 levels

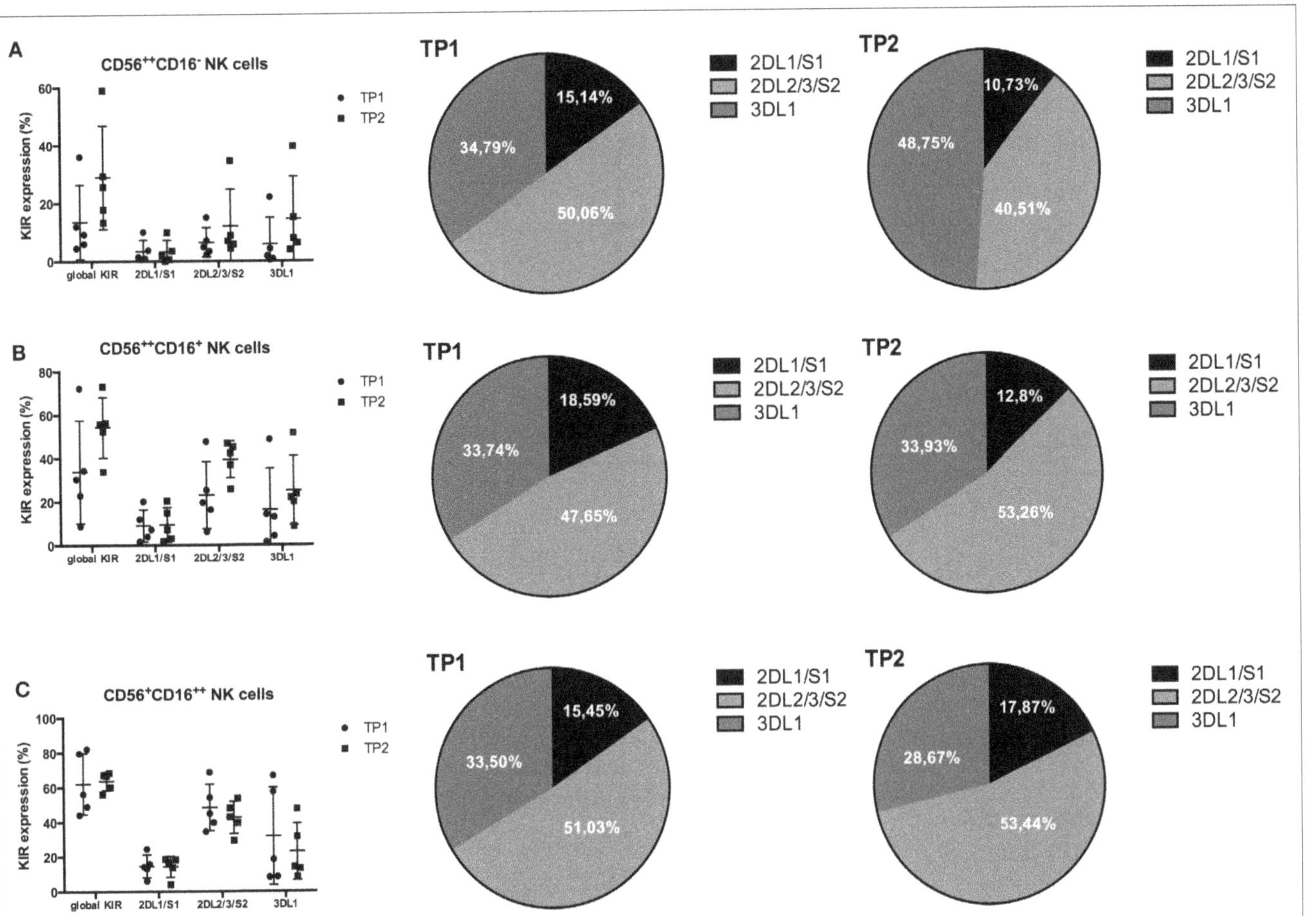

FIGURE 5 | (A) Global KIR expression (anti-KIR2D and anti-KIR3DL1/2) was upregulated within the CD56^{++}CD16^{-} NK cell population from TP1 to TP2 (TP1: 13.56%; TP2: 29.04%). KIR2DL1/S1 remained stable (TP1: 3.33%; TP2: 3.2%), whereas KIR2DL2/3/S2 (TP1: 6.49%; TP2: 12.08%) and KIR3DL1 (TP1: 5.85%; TP2: 14.54%) were both upregulated. **(B)** Within the CD56^{++}CD16^{+} NK cells, the global KIR expression was upregulated from TP1 to TP2 (TP1: 33.74%; TP2: 54.24%) in addition to the KIR2DL2/3/S2 (TP1: 22.95%; TP2: 39.28%) and KIR3DL1 expression (TP1: 16.25%; TP2: 25.02%). In contrast, the KIR2DL1/S1 levels remained stable (TP1: 8.95%; TP2: 9.44%). **(C)** The expression levels of global KIR expression (TP1: 62.24%; TP2: 63.66%) as well as that of KIR2DL1/S1 (TP1: 14.65%; TP2: 14.26%) remained stable, whereas the two other KIR subsets revealed a slight decrease from TP1 to TP2 within the CD56^{+}CD16^{++} NK cell population [2DL2/3/S2: 48.4% (TP1) and 42.65% (TP2); 3DL1: 31.77% (TP1) and 22.88% (TP2)]. The proportions of the different KIR subsets within the whole KIR population are indicated within the pie charts **(A–C)**.

remained stable between the two time points (**Figures 5A,B**). Moreover, the proportion of the different KIR subsets within the global KIR population changed within the CD56^{++}CD16^{-} population, with KIR2DL2/3/S2 being the dominant KIR subset at TP1 and KIR3DL1 being the dominant one at TP2 (**Figure 5A**; pie chart). Within the CD56^{++}CD16^{+} population, KIR2DL2/3/S2 was the dominant KIR population at TP1 and was the only one to increase from TP1 to TP2 (**Figure 5B**; pie chart).

NK Cell Function is Preserved After Leukocyte Recovery

Finally, we analyzed the functions of the different NK cell subsets at the three time points before and after HDC/autoSCT. After an overnight incubation with low-dose IL-2 (100 IU/ml) and 4 h of coculture with K562 cells, we investigated the cytokine (IFN-γ) and chemokine (MIP-1β) productions, as well as the NK cell degranulation (CD107a expression) capacity (for gating strategy see Figure S5 in Supplementary Material). Due to very low cell numbers at TP2, functional analysis was only possible in a subset of all included patients ($n = 17$).

As expected, CD56^{++}CD16^{-} NK cells were the main subset to produce IFN-γ upon interaction with K562 cells at all three time points (**Figure 6A**). The percentage of IFN-γ-positive CD56^{++}CD16^{-} NK cells was slightly decreased at TP2 compared to TP1 but significantly increased from TP2 to TP3 (p-value: 0.0008). Similarly, MIP-1β- and CD107a-positive CD56^{++}CD16^{-} cells remained constant between TP1 and TP2, whereas their percentages increased from TP2 to TP3 [p-values: 0.0056 (MIP-1β) and 0.0232 (CD107a)].

Whereas IFN-γ production was only marginal at all three time points within the CD56^{++}CD16^{+} NK cell population (**Figure 6B**), MIP-1β- and CD107a-positive cells had similar percentages at TP1 and TP2. Both percentages significantly increased from TP2 to TP3 (p-value: 0.0079 for both).

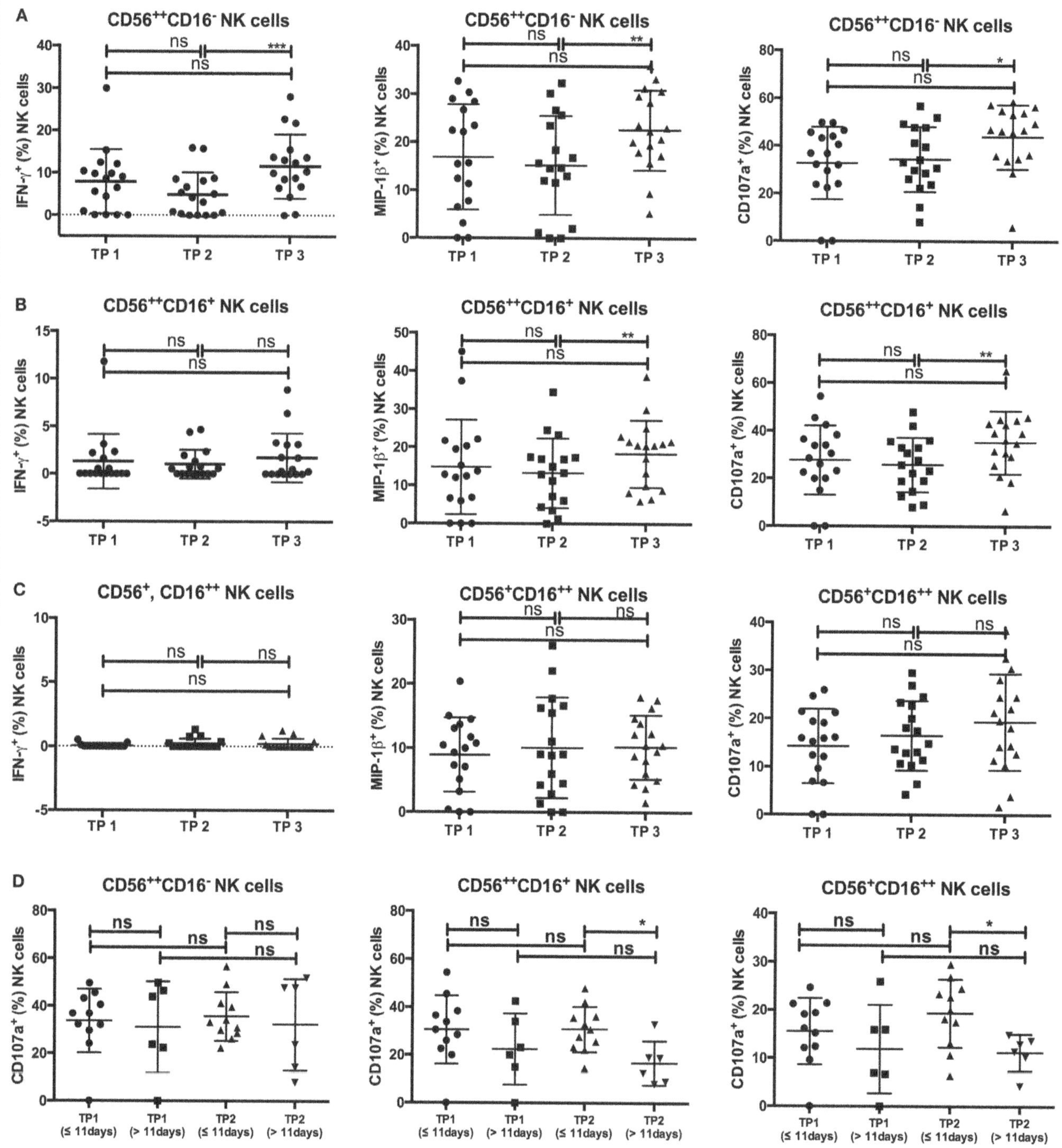

FIGURE 6 | (A) The CD56⁺⁺CD16⁻ subset revealed an increase of IFN-γ-, MIP-1β-, and CD107a-positive cells between TP2 and TP3 [TP2: 4.9% (IFN-γ), 15.17% (MIP-1β), 34.37% (CD107a), TP3: 11.58% (IFN-γ), 22.58% (MIP-1β), 43.82% (CD107a)], while the percentages between TP1 and TP2 were similar [TP1: 7.92% (IFN-γ), 16.89% (MIP-1β), and 32.7% (CD107a)]. **(B)** IFN-γ-positive cells within the CD56⁺⁺CD16⁺ population were only marginal at all three time points (TP1: 1.29%; TP2: 1.01%; TP3: 1.72%), whereas MIP-1β- and CD107a-positive cells increased from TP2 to TP3 [TP2: 13.26% (MIP-1β), 25.78% (CD107a); TP3: 18.37% (MIP-1β), 35.26% (CD107a)] but not from TP1 to TP2 [TP1: 14.79% (MIP-1β), 27.64% (CD107a)]. **(C)** No IFN-γ-positive cells were detectable within the CD56⁺CD16⁺⁺ population at any of the three time points. MIP-1β-positive cells remained stable throughout all three time points (TP1: 8.93%; TP2: 10.06%; TP3: 10.18%) which is similar to CD107a-expressing cells (TP1: 14.26%; TP2: 16.47%; TP3: 19.33%). **(D)** Patients were grouped according to the duration between SCT and TP2 (≤11 vs. >11 days). While there were no differences between the percentage of CD107a-expressing cells within the CD56⁺⁺CD16⁻ population at TP1 and TP2, the percentage of CD107a-positive NK cells at TP2 was lower within patients with a longer duration between SCT and TP2 within the CD56⁺⁺CD16⁺ (≤11 days: 30.74%; >11 days: 16.68%) and the CD56⁺CD16⁺⁺ (≤11 days: 19.34%; >11 days: 11.21%) populations.

Within the $CD56^{+}CD16^{++}$ NK cell subsets, the percentage of MIP-1β- and CD107a-positive NK cells after coincubation with K562 cells remained constant at all three time points, whereas no IFN-γ-positive NK cells were detected at any time point (**Figure 6C**).

Although the number of available patient samples was low, we tried to correlate the NK cell function values with clinical data. Remarkably, there was an impact of the duration from SCT to TP2 (≤11 vs. >11 days), as the percentage of CD107a-positive cells within the $CD56^{++}CD16^{+}$ and $CD56^{+}CD16^{++}$ populations was significantly lower when the time period between SCT and TP2 exceeded 11 days [*p*-values: 0.0111 ($CD56^{++}CD16^{+}$); 0.027 ($CD56^{+}CD16^{++}$); **Figure 6D**]. In investigating age-dependent differences (≤56 vs. >56 years), we observed that patients older than 56 years tended to have slightly higher percentages of IFN-γ- and MIP-1β-positive $CD56^{++}CD16^{-}$ cells at TP1, although the difference was not significant. Furthermore, no significant differences in the presence of MIP-1β- or CD107a-positive $CD56^{++}CD16^{+}$ or $CD56^{+}CD16^{++}$ NK cells between the two age groups were observed at all three time points, although older patients tended to have lower CD107a expression within the $CD56^{+}CD16^{++}$ subset at TP3 (*p*-value: 0.074). Notably, the observed increase of CD107a-positive $CD56^{+}CD16^{++}$ cells from TP2 to TP3 was only present in younger but not older patients [≤56 years: 15.69% (TP2) and 24.05% (TP3), *p*-value: 0.1; >56 years: 17.16% (TP2) and 15.14% (TP3), p-value: 0.82; Figure S6 in Supplementary Material].

DISCUSSION

In the setting of HDC/autoSCT, it has been demonstrated that a rapid NK cell recovery at 1 month after HDC/autoSCT is associated with a prolonged progression-free survival in MM (23) and NHL patients (16). In those studies, the absolute NK cell count (cells/μl) at 1 month or 15 days after HDC/autoSCT was investigated, whereas in our study, we analyzed the NK cell percentage within the leukocyte population in correlation with the day of leukocyte recovery following autoSCT.

Our data demonstrate that the percentage of NK cells within the leukocyte population decreased after leukocyte recovery but increased to the initial levels over time. Notably, when the time period between SCT and TP2 was >11 days, indicating a delay in leukocyte recovery, the decrease within the NK cell percentage was lower. This may be explained by the fact that leukocyte recovery (white blood cell count >1000/μl) after SCT is mainly due to the recovery of neutrophil granulocytes (10), and their recovery can be delayed in contrast to NK cell recovery (24). Moreover, the higher increase of the NK cell percentage from TP2 to TP3 in patients from whom the third blood sample was collected ≤38 days after TP2 indicates that the NK cell percentage within the leukocytes increases much more within the first month after leukocyte recovery and then decreases again. Similar results have been demonstrated by Rueff et al. by analyzing the absolute NK cell count numbers (cells/μl) 1, 3, 6, 12, and 24 months after SCT. Here, the NK cell numbers first increased 1 month after SCT, but then they decreased again until 6 months after SCT (23). The absence of a decrease in the NK cell percentage at TP2 within patients with recurrent/refractory disease at 1 year after SCT could be explained by the fact that most of these patients (4/6; 66.6%) had a time period of >11 days from SCT to TP2, unlike non-relapsing/refractory patients (8/21; 38%).

In line with other NK cell reconstitution studies after SCT (24–26), we observed elevated percentages of the more immature $CD56^{++}CD16^{+/-}$ NK cell subsets shortly after SCT, decreasing only slowly at later time points. As the ratio between $CD56^{+}$/$CD56^{++}$ NK cells did not differ with shorter (≤38 days) or longer (>38 days) time periods between TP2 and TP3, we assume that normalization of the NK cell subset distribution takes much longer than the recovery of the NK cell numbers. Similar observations have been made within patients receiving an allogeneic SCT after reduced-intensity conditioning (26). The conditioning and former treatment regimens could explain the different $NKG2A^{+}$ NK cells ratios between myeloma and lymphoma patients at TP2.

Moreover, the percentage of $NKG2A^{+}$ NK cells was increased after HDC/autoSCT and remained high even after several months, as it has been recently demonstrated by Pical-Izard et al. after allogeneic SCT (26). In contrast, we could demonstrate a highly significant increase of $CD57^{+}$ and KIR^{+} NK cells, mainly within the $CD56^{++}CD16^{+/-}$ subsets at TP2. This effect has not been described thus far within the literature because most of the studies have evaluated NK cell subsets 1 month after SCT (23–27). At this time point, the percentage of $CD57^{+}$ and KIR^{+} NK cells had already decreased back toward normal levels within our study group. The prolonged immature phenotype in the Pical-Izard study may be attributed to GVHD prophylaxis, especially cyclosporine A (CSA). Vukicevic et al. investigate the NK cell phenotype at an equally early time point after allogeneic stem cell transplantation as we did but did not detect an upregulation of KIRs within the $CD56^{++}$ subsets (28), which might be due to the allogeneic transplantation setting. Acquisition of CD57 and KIRs as well as downregulation of NKG2A has been demonstrated as signs of NK cell differentiation and maturation (22). It is known that CD57 expression increases with age within the $CD56^{+}CD16^{++}$ population (29), whereas we did not observe any age-related differences within this NK cell subset at any time point. Nevertheless, an age-dependent difference was observed at TP2 within the $CD56^{++}CD16^{-}$ population.

Therefore, the question arises whether these $CD56^{++}$ cells described here represent more mature new NK cells or are just activated "old" NK cells that might have increased their CD56 expression and lost CD16 on their surface. As this phenotype is reset after at least 2 weeks after leukocyte regeneration, one could argue that the phenotype shift is due to the cytokine milieu during HDC/autoSCT. Indeed, there have been several reports about increased cytokine concentrations during allogeneic and autologous SCT (16, 25, 30) shaping the NK cell phenotype. It has been reported that $CD56^{+}CD16^{++}$ NK cells are capable of up-regulating CD56 expression upon IL-15 (28) or IL-12 stimulation (31), of which IL-15 is known to be increased during HDC/autoSCT (16). Furthermore, different groups demonstrated a downregulation of CD16 by metalloproteinases, which can be induced by IL-2 (32, 33). In general, the combined effect of CD56 upregulation upon IL-15 stimulation and the loss of CD16 through IL-2-stimulated upregulation of metalloproteinases might result in the observed $CD56^{++}CD16^{+/-}$ NK cell phenotype at TP2. Nevertheless, the $CD56^{++}$ NK cells described in our study upregulated KIR3DL1 and KIR2DL2/3/S2, while KIR2DL1/S1 remained stable. KIR3DL1 and KIR2DL2/3 are the first KIRs expressed after SCT, whereas KIR2DL1 is upregulated quite late (34, 35). Therefore, we could assume that the KIR upregulation was due to the generation of "fresh/new" NK cells and was not due to a shift of "old" NK cells to a $CD56^{++}$ phenotype because we should have observed no changes within the KIR subtype distribution between TP1 and TP2 if the cells were derived from the "old" NK population. Moreover, we observed a slight upregulation of CX3CR1 at TP2 (data not shown). CX3CR1 expression is associated with a more mature and differentiated NK cell phenotype within healthy donors (36). In contrast to the observed CD56 upregulation, which can be explained by IL-15 stimulation (28), CX3CR1 is known to be downregulated upon IL-15 stimulation (36), which contradicts the idea that $CD56^{++}$ NK cells with a mature phenotype ($CD57^{+}$, $CX3CR1^{+}$, and KIR^{+}) have arisen from $CD56^{dim}$ NK cells. In future studies, it would be very interesting to investigate which factors are responsible for this NK cell phenotype because protocols for inducing NK cell maturation and differentiation have yet to be optimized.

Most importantly, we analyzed NK cell functional activity directly after leukocyte recovery after HDC/autoSCT. Upon interaction with K562 tumor cells, the percentage of IFN-γ- and MIP-1β-positive $CD56^{++}CD16^{+/-}$ NK cells did not differ between TP1 and TP2. This result demonstrates that NK cells are capable of recognizing tumor cells and inducing cytokine and chemokine production, even at a very early time point after HDC/autoSCT. Moreover, the degranulation capacity of the $CD56^{+}CD16^{++}$ NK cell subset, known to be mainly responsible for NK cell cytotoxicity (37), remained stable throughout the whole time period until TP3, indicating that these NK cells were able to kill tumor cells at an early time point after SCT. This finding is consistent with other studies in which patients received allogeneic SCT. For example, in the setting of HLA-matched SCT after reduced-intensity conditioning, it has been demonstrated that the NK cell degranulation and chemokine production capacity was similar to healthy controls as early as 1 month after SCT. In contrast to our data, IFN-γ production upon interaction with K562 cells was significantly reduced after SCT compared to healthy donors (26). These differences might be due to the use of immunosuppressive drugs such as CSA because it has been demonstrated that CSA is able to reduce IFN-γ production upon target-cell interaction (38), although a recent report has failed to demonstrate such an effect (39). Furthermore, we compared IFN-γ production upon interaction with K562 cells before and after SCT and not directly with healthy control samples. Therefore, although we did not observe a significant decrease in IFN-γ-positive cells between TP1 and TP2, their percentage might be still significantly lower than in healthy controls. Additionally, we observed that the degranulation capacity was influenced by the time period between SCT and TP2 because patients with a time period >11 days had significantly reduced CD107a-positive $CD56^{+}CD16^{++}$ cells at TP2. Because the prolonged time from SCT to TP2 indicates a longer period for leukocyte recovery, this might give an explanation for the reduced degranulation capacity. Nevertheless, we could not discover a correlation between NK cell function at TP2 and the rate of recurrent/refractory disease at 1 year after SCT, which might be due to the low number of recurrent/refractory patients and the short follow-up period.

To the best of our knowledge, this is the first study investigating NK cell function at such an early time point after HDC/autoSCT. We were able to demonstrate that NK cells were capable of cytokine/chemokine production and degranulation upon tumor cell interaction. Furthermore, we describe an unusual $CD56^{++}$ NK cell population expressing high levels of CD57 and KIRs shortly after SCT. Further analysis and characterization of this population might reveal more details about how NK cell maturation and differentiation are regulated.

AUTHOR CONTRIBUTIONS

All authors critically revised the work for important intellectual content and approved the final version of the manuscript. They all agreed to be accountable for all aspects of the work in ensuring that questions related to the accuracy or integrity of any part of the work are appropriately investigated and resolved. BJ, ST, and KP were responsible for acquisition, analysis, and interpretation of the data. EU, PB, AM, and BJ were responsible for the concept and design of the work.

ACKNOWLEDGMENTS

We thank Franziska Ganss for working with the patient samples during the first phase of the study. We are grateful to the Department of Haematology and Oncology physician team for collecting blood samples from the patients. The authors thank Becton Dickinson (BD) for providing the FACSCanto II™ and Canto10c™ Flow Cytometry Analyzers used in this study.

FUNDING

Authors were supported by the German Cancer Aid (Max Eder Nachwuchsgruppe, Deutsche Krebshilfe; EU), the LOEWE Center for Cell and Gene Therapy Frankfurt (EU, ST, and PB) funded by the Hessian Ministry of Higher Education, Research and the Arts, Germany (III L 4-518/17.004) and by the "Alfred- und Angelika Gutermuth-Stiftung," Frankfurt, Germany (EU). BJ was funded by a Mildred Scheel postdoctoral scholarship from the Dr. Mildred Scheel Foundation for Cancer Research of the German Cancer Aid Organization. ST was founded by a GO-IN postdoctoral fellowship (PCOFUND-GA-2011-291776).

REFERENCES

1. Vivier E, Tomasello E, Baratin M, Walzer T, Ugolini S. Functions of natural killer cells. *Nat Immunol* (2008) **9**(5):503–10. doi:10.1038/ni1582
2. Ljunggren HG, Karre K. Host resistance directed selectively against H-2-deficient lymphoma variants. Analysis of the mechanism. *J Exp Med* (1985) **162**(6):1745–59. doi:10.1084/jem.162.6.1745
3. Nash WT, Teoh J, Wei H, Gamache A, Brown MG. Know thyself: NK-cell inhibitory receptors prompt self-tolerance, education, and viral control. *Front Immunol* (2014) **5**:175. doi:10.3389/fimmu.2014.00175
4. Ullrich E, Koch J, Cerwenka A, Steinle A. New prospects on the NKG2D/NKG2DL system for oncology. *Oncoimmunology* (2013) **2**(10):e26097. doi:10.4161/onci.26097
5. Shah N, Callander N, Ganguly S, Gul Z, Hamadani M, Costa L, et al. Hematopoietic stem cell transplantation for multiple myeloma: guidelines from the American Society for blood and marrow transplantation. *Biol Blood Marrow Transplant* (2015) **21**(7):1155–66. doi:10.1016/j.bbmt.2015.03.002
6. Perales MA, Ceberio I, Armand P, Burns LJ, Chen R, Cole PD, et al. Role of cytotoxic therapy with hematopoietic cell transplantation in the treatment of Hodgkin lymphoma: guidelines from the American Society for blood and marrow transplantation. *Biol Blood Marrow Transplant* (2015) **21**(6):971–83. doi:10.1016/j.bbmt.2015.02.022
7. Chen Y, Wang M, Romaguera J. Current regimens and novel agents for mantle cell lymphoma. *Br J Haematol* (2014) **167**(1):3–18. doi:10.1111/bjh.13000
8. Habermann TM. New developments in the management of diffuse large B-cell lymphoma. *Hematology* (2012) **17**(Suppl 1):S93–7. doi:10.1179/102453312X13336169156014
9. Perrone G, Corradini P. Autologous stem cell transplantation for T-cell lymphomas. *Semin Hematol* (2014) **51**(1):59–66. doi:10.1053/j.seminhematol.2013.11.002
10. Visani G, Lemoli R, Tosi P, Martinelli G, Testoni N, Ricci P, et al. Use of peripheral blood stem cells for autologous transplantation in acute myeloid leukemia patients allows faster engraftment and equivalent disease-free survival compared with bone marrow cells. *Bone Marrow Transplant* (1999) **24**(5):467–72. doi:10.1038/sj.bmt.1701920
11. Porrata LF, Inwards DJ, Micallef IN, Ansell SM, Geyer SM, Markovic SN. Early lymphocyte recovery post-autologous haematopoietic stem cell transplantation is associated with better survival in Hodgkin's disease. *Br J Haematol* (2002) **117**(3):629–33. doi:10.1046/j.1365-2141.2002.03478.x
12. Porrata LF, Gertz MA, Inwards DJ, Litzow MR, Lacy MQ, Tefferi A, et al. Early lymphocyte recovery predicts superior survival after autologous hematopoietic stem cell transplantation in multiple myeloma or non-Hodgkin lymphoma. *Blood* (2001) **98**(3):579–85. doi:10.1182/blood.V98.3.579
13. Porrata LF, Litzow MR, Tefferi A, Letendre L, Kumar S, Geyer SM, et al. Early lymphocyte recovery is a predictive factor for prolonged survival after autologous hematopoietic stem cell transplantation for acute myelogenous leukemia. *Leukemia* (2002) **16**(7):1311–8. doi:10.1038/sj.leu.2402503
14. Porrata LF, Ingle JN, Litzow MR, Geyer S, Markovic SN. Prolonged survival associated with early lymphocyte recovery after autologous hematopoietic stem cell transplantation for patients with metastatic breast cancer. *Bone Marrow Transplant* (2001) **28**(9):865–71. doi:10.1038/sj.bmt.1703236
15. Porrata LF, Inwards DJ, Ansell SM, Micallef IN, Johnston PB, Gastineau DA, et al. Early lymphocyte recovery predicts superior survival after autologous stem cell transplantation in non-Hodgkin lymphoma: a prospective study. *Biol Blood Marrow Transplant* (2008) **14**(7):807–16. doi:10.1016/j.bbmt.2008.04.013
16. Porrata LF, Inwards DJ, Micallef IN, Johnston PB, Ansell SM, Hogan WJ, et al. Interleukin-15 affects patient survival through natural killer cell recovery after autologous hematopoietic stem cell transplantation for non-Hodgkin lymphomas. *Clin Dev Immunol* (2010) **2010**:914945. doi:10.1155/2010/914945
17. Czaja K, Borer AS, Schmied L, Terszowski G, Stern M, Gonzalez A. A comprehensive analysis of the binding of anti-KIR antibodies to activating KIRs. *Genes Immun* (2014) **15**(1):33–7. doi:10.1038/gene.2013.58
18. Beziat V, Traherne J, Malmberg JA, Ivarsson MA, Bjorkstrom NK, Retiere C, et al. Tracing dynamic expansion of human NK-cell subsets by high-resolution analysis of KIR repertoires and cellular differentiation. *Eur J Immunol* (2014) **44**(7):2192–6. doi:10.1002/eji.201444464
19. Beziat V, Duffy D, Quoc SN, Le Garff-Tavernier M, Decocq J, Combadiere B, et al. CD56brightCD16+ NK cells: a functional intermediate stage of NK cell differentiation. *J Immunol* (2011) **186**(12):6753–61. doi:10.4049/jimmunol.1100330
20. Romagnani C, Juelke K, Falco M, Morandi B, D'Agostino A, Costa R, et al. CD56brightCD16- killer Ig-like receptor- NK cells display longer telomeres and acquire features of CD56dim NK cells upon activation. *J Immunol* (2007) **178**(8):4947–55. doi:10.4049/jimmunol.178.8.4947
21. Lugthart G, van Ostaijen-ten Dam MM, van Tol MJ, Lankester AC, Schilham MW. CD56(dim)CD16(-) NK cell phenotype can be induced by cryopreservation. *Blood* (2015) **125**(11):1842–3. doi:10.1182/blood-2014-11-610311
22. Bjorkstrom NK, Riese P, Heuts F, Andersson S, Fauriat C, Ivarsson MA, et al. Expression patterns of NKG2A, KIR, and CD57 define a process of CD56dim NK-cell differentiation uncoupled from NK-cell education. *Blood* (2010) **116**(19):3853–64. doi:10.1182/blood-2010-04-281675
23. Rueff J, Medinger M, Heim D, Passweg J, Stern M. Lymphocyte subset recovery and outcome after autologous hematopoietic stem cell transplantation for plasma cell myeloma. *Biol Blood Marrow Transplant* (2014) **20**(6):896–9. doi:10.1016/j.bbmt.2014.03.007
24. Bosch M, Dhadda M, Hoegh-Petersen M, Liu Y, Hagel LM, Podgorny P, et al. Immune reconstitution after anti-thymocyte globulin-conditioned hematopoietic cell transplantation. *Cytotherapy* (2012) **14**(10):1258–75. doi:10.3109/14653249.2012.715243
25. Dulphy N, Haas P, Busson M, Belhadj S, Peffault de Latour R, Robin M, et al. An unusual CD56(bright) CD16(low) NK cell subset dominates the early posttransplant period following HLA-matched hematopoietic stem cell transplantation. *J Immunol* (2008) **181**(3):2227–37. doi:10.4049/jimmunol.181.3.2227
26. Pical-Izard C, Crocchiolo R, Granjeaud S, Kochbati E, Just-Landi S, Chabannon C, et al. Reconstitution of natural killer cells in HLA-matched HSCT after reduced-intensity conditioning: impact on clinical outcome. *Biol Blood Marrow Transplant* (2015) **21**(3):429–39. doi:10.1016/j.bbmt.2014.11.681
27. Foley B, Cooley S, Verneris MR, Curtsinger J, Luo X, Waller EK, et al. NK cell education after allogeneic transplantation: dissociation between recovery of cytokine-producing and cytotoxic functions. *Blood* (2011) **118**(10):2784–92. doi:10.1182/blood-2011-04-347070
28. Vukicevic M, Chalandon Y, Helg C, Matthes T, Dantin C, Huard B, et al. CD56bright NK cells after hematopoietic stem cell transplantation are activated mature NK cells that expand in patients with low numbers of T cells. *Eur J Immunol* (2010) **40**(11):3246–54. doi:10.1002/eji.200940016
29. Nielsen CM, White MJ, Goodier MR, Riley EM. Functional significance of CD57 expression on human NK cells and relevance to disease. *Front Immunol* (2013) **4**:422. doi:10.3389/fimmu.2013.00422
30. Boyiadzis M, Memon S, Carson J, Allen K, Szczepanski MJ, Vance BA, et al. Up-regulation of NK cell activating receptors following allogeneic hematopoietic stem cell transplantation under a lymphodepleting reduced intensity regimen is associated with elevated IL-15 levels. *Biol Blood Marrow Transplant* (2008) **14**(3):290–300. doi:10.1016/j.bbmt.2007.12.490
31. Loza MJ, Perussia B. The IL-12 signature: NK cell terminal CD56+high stage and effector functions. *J Immunol* (2004) **172**(1):88–96. doi:10.4049/jimmunol.172.1.88
32. Harrison D, Phillips JH, Lanier LL. Involvement of a metalloprotease in spontaneous and phorbol ester-induced release of natural killer cell-associated Fc gamma RIII (CD16-II). *J Immunol* (1991) **147**(10):3459–65.
33. Peruzzi G, Femnou L, Gil-Krzewska A, Borrego F, Weck J, Krzewski K, et al. Membrane-type 6 matrix metalloproteinase regulates the activation-induced downmodulation of CD16 in human primary NK cells. *J Immunol* (2013) **191**(4):1883–94. doi:10.4049/jimmunol.1300313
34. Giebel S, Dziaczkowska J, Czerw T, Wojnar J, Krawczyk-Kulis M, Nowak I, et al. Sequential recovery of NK cell receptor repertoire after allogeneic hematopoietic SCT. *Bone Marrow Transplant* (2010) **45**(6):1022–30. doi:10.1038/bmt.2009.384
35. Fischer JC, Ottinger H, Ferencik S, Sribar M, Punzel M, Beelen DW, et al. Relevance of C1 and C2 epitopes for hemopoietic stem cell transplantation: role

for sequential acquisition of HLA-C-specific inhibitory killer Ig-like receptor. *J Immunol* (2007) **178**(6):3918–23. doi:10.4049/jimmunol.178.6.3918

36. Hamann I, Unterwalder N, Cardona AE, Meisel C, Zipp F, Ransohoff RM, et al. Analyses of phenotypic and functional characteristics of CX3CR1-expressing natural killer cells. *Immunology* (2011) **133**(1):62–73. doi:10.1111/j.1365-2567.2011.03409.x
37. Freud AG, Caligiuri MA. Human natural killer cell development. *Immunol Rev* (2006) **214**:56–72. doi:10.1111/j.1600-065X.2006.00451.x
38. Eissens DN, Van Der Meer A, Van Cranenbroek B, Preijers FW, Joosten I. Rapamycin and MPA, but not CsA, impair human NK cell cytotoxicity due to differential effects on NK cell phenotype. *Am J Transplant* (2010) **10**(9):1981–90. doi:10.1111/j.1600-6143.2010.03242.x
39. Derniame S, Perazzo J, Lee F, Domogala A, Escobedo-Cousin M, Alnabhan R, et al. Differential effects of mycophenolate mofetil and cyclosporine A on peripheral blood and cord blood natural killer cells activated with interleukin-2. *Cytotherapy* (2014) **16**(10):1409–18. doi:10.1016/j.jcyt.2014.05.010

18

Natural Killer Cell-Based Therapies Targeting Cancer: Possible Strategies to Gain and Sustain Anti-Tumor Activity

Carin I. M. Dahlberg[1,2], *Dhifaf Sarhan*[3,4], *Michael Chrobok*[1,2], *Adil D. Duru*[1,2] and *Evren Alici*[1,2,5]*

[1] *Cell Therapies Institute, Nova Southeastern University, Fort Lauderdale, FL, USA,* [2] *Cell and Gene Therapy Group, Center for Hematology and Regenerative Medicine (HERM), Karolinska University Hospital Huddinge, NOVUM, Stockholm, Sweden,* [3] *Oncology-Pathology, Cancer Center Karolinska, Karolinska Institutet, Stockholm, Sweden,* [4] *Division of Hematology, Oncology and Transplantation, Masonic Cancer Research Center, University of Minnesota, Minnesota, MN, USA,* [5] *Hematology Center, Karolinska University Hospital Huddinge, Stockholm, Sweden*

Edited by:
Francisco Borrego, Cruces University Hospital, Spain

Reviewed by:
Subramaniam Malarkannan, Medical College of Wisconsin, USA
Vincent Vieillard, Institut National de la Santé et de la Recherche Scientifique (INSERM), France

***Correspondence:**
Evren Alici
evren.alici@ki.se

Natural killer (NK) cells were discovered 40 years ago, by their ability to recognize and kill tumor cells without the requirement of prior antigen exposure. Since then, NK cells have been seen as promising agents for cell-based cancer therapies. However, NK cells represent only a minor fraction of the human lymphocyte population. Their skewed phenotype and impaired functionality during cancer progression necessitates the development of clinical protocols to activate and expand to high numbers *ex vivo* to be able to infuse sufficient numbers of functional NK cells to the cancer patients. Initial NK cell-based clinical trials suggested that NK cell-infusion is safe and feasible with almost no NK cell-related toxicity, including graft-versus-host disease. Complete remission and increased disease-free survival is shown in a small number of patients with hematological malignances. Furthermore, successful adoptive NK cell-based therapies from haploidentical donors have been demonstrated. Disappointingly, only limited anti-tumor effects have been demonstrated following NK cell infusion in patients with solid tumors. While NK cells have great potential in targeting tumor cells, the efficiency of NK cell functions in the tumor microenvironment is yet unclear. The failure of immune surveillance may in part be due to sustained immunological pressure on tumor cells resulting in the development of tumor escape variants that are invisible to the immune system. Alternatively, this could be due to the complex network of immune-suppressive compartments in the tumor microenvironment, including myeloid-derived suppressor cells, tumor-associated macrophages, and regulatory T cells. Although the negative effect of the tumor microenvironment on NK cells can be transiently reverted by *ex vivo* expansion and long-term activation, the aforementioned NK cell/tumor microenvironment interactions upon reinfusion are not fully elucidated. Within this context, genetic modification of NK cells may provide new possibilities for developing effective cancer immunotherapies by improving NK cell responses and making them less susceptible to the tumor microenvironment. Within this review, we will discuss clinical trials using NK cells with a specific reflection on novel potential strategies, such as genetic modification of NK cells and complementary therapies aimed at improving the clinical outcome of NK cell-based immune therapies.

Keywords: natural killer cells, adoptive cell therapy, immunotherapy, cancer, clinical trials, expansion, tumor microenvironment, genetic modifications

INTRODUCTION

Natural killer (NK) cells are lymphocytes of the innate immune system. They are cytokine producing and have cytotoxic ability to kill both viral infected and tumor cells. Tumor-killing lymphocytes were first reported in 1968 by Hellström et al. (1). Kiessling and colleagues, in parallel with Ronald Herberman's research laboratory, defined a novel lymphocyte population named NK cells that are able to target tumor cells in 1975 (2–5). Unlike T cells and B cells, NK cell recognition is not governed by high-resolution antigen specificity. Target cell recognition is mediated by the signals delivered through several activating and inhibitory receptors. The balance between activating and inhibitory signals decides the response of NK cells. When there is a mismatch between an inhibitory subgroup of killer immunoglobulin-like receptors (KIRs) on NK cells and self-human leukocyte antigen (HLA) class I proteins on the surface of target cells the NK cells can get activated due to lack of inhibitory signals leading to lysis of the host cell. This mismatch mediates alloreactivity and is the strategy behind the missing-self concept (6). KIRs can be divided into two haplotypes; the A haplotype with predominantly inhibitory KIRs plus only one activating KIR and the B haplotype containing inhibitory and activating receptors (7). During NK cell education, KIRs go through a random sequential acquisition process where they get functionally competent after they encounter self-MHC class I molecules. Consequently, mature NK cell function is inhibited by self-MHC class I and KIR interaction (8). When a NK cell confronts a target cell without expression of self-MHC class I molecules, the inhibitory signals are not active and the NK cell gets activated.

The majority of NK cells, as well as certain T cell subpopulations, may express the receptor family NKG2. One of the ligands for most NKG2 receptors is HLA-E, which is expressed on all nucleated cells. NKG2-family consists of seven members: NKG2A, B, C, D, E, F, and H in which NKG2A and B are inhibitory receptors. NK cells also express activation receptors on the surface, such as natural cytotoxicity receptors (NCRs), DNAM-1, and receptor members of the 2B4 family. NCRs, including NKp30, NKp44, and NKp46, are one of the main and initial groups of NK cell-activating receptors identified and they recognize viral ligands, heat shock-associated proteins, or tumor antigens (9). NK cells can also get activated by crosslinking of Fc receptor CD16 to target cell leading to antibody-dependent cellular cytotoxicity (ADCC) and lysis of the target cell (10, 11).

Natural killer cells perform their cytotoxic activity through granzyme B- and perforin-mediated apoptosis or by expression of death receptor ligands such as FasL and TNF-related apoptosis-inducing ligand (TRAIL). While the release of cytolytic granules is one of the essential cytotoxic responses, perforin deficient NK cells can still kill tumor cells through Fas-mediated apoptosis (12). Moreover, TRAIL-TRAILR mediated cytotoxicity also plays an important role in eliminating the target cells. Various tumor cells express TRAIL death receptors, which could be upregulated by proteasome inhibitors such as bortezomib (13). Additionally, immunomodulatory drugs (IMiDs) such as lenalidomide upregulates TRAIL expression on NK cells that potentially enhance the TRAIL-mediated elimination of tumor cells (14, 15).

Natural killer cells are derived from hematopoietic stem cells (HSC) in the bone marrow. The differentiation from HSC can be divided into five stages based on surface markers [detailed review in Ref. (16)]. The stages can be identified by the following surface markers, CD34, CD117, CD94, and CD16 among the Lin^- events, where stage 1 is $CD34^+CD117^-CD94^-CD16^-$. First at stage 2, the cells are able to respond to IL-15, which is necessary for NK cell development (17, 18). In the transition between stage 2 and 3, they lose their CD34 expression. At stage 4, the NK cells are $CD56^{bright}$, produce IFNγ, and are capable of cytotoxic killing of K562 cells *in vitro* (19). NK cells in stage 5 are $CD56^{dim}$ and express CD16.

The majority of human NK cells are $CD14^-CD19^-CD3^-CD56^+$. While most of the $CD56^+$ cells express lower levels of CD56 (~90% $CD56^{dim}$), they are potent cytotoxic killers of target cells and secrete cytokines such as IFNγ. Approximately 10% of peripheral NK cells express high levels of CD56 ($CD56^{bright}$), have low cytolytic activity, and have the capacity to produce high titers of immunoregulatory cytokines. The cell surface phenotypes of these two subpopulations also differ in respect to the receptors they express: the $CD56^{bright}$ population expresses the inhibitory receptor NKG2A that could also be expressed on $CD56^{dim}$ NK cells. While the $CD56^{dim}$ population expresses FcγRIIIa (CD16a) as well as the inhibitory receptors KIRs (20).

NK CELLS IN CANCER

Natural killer cells recognize tumor cells by the activating receptors like NCRs, which detect the altered expression of their ligands on the tumor cell surface. Additionally, downregulation or lack of MHC class I molecules on the cell surface of tumor cells can trigger NK cell activation since it diminishes the inhibitory signals transduced through KIR-MHC interactions. Moreover, since NK cells' target recognition and activation are mainly through NCRs and missing-self, this engagement could induce upregulation of FasL on the NK cell surface leading to an alternative pathway inducing apoptosis in tumor cells. Nevertheless, both IL-2 stimulation and NK cell activation through NCRs also upregulate Fas on NK cells that may initiate regulation of the NK cell activation and expansion (21, 22).

Many tumors have gained methods to evade the surveillance by NK cells and other members of the immune system. For example, 16 of 18 patients with acute myeloid leukemia (AML) had reduced NCR surface expression compared to healthy donor NK cells, resulting in reduced cytotoxic capacity against target cells (23). Another way for tumor cells to escape recognition by NK cells is upregulation of the non-classical MHC class I molecule HLA-G, which dampens NK cell responses (24, 25). In numerous malignancies, there are also abnormalities found in the NK cell population. Examples of this include defective expression of activating receptors found in hepatocellular carcinoma (26), metastatic melanoma (27), AML (23), chronic lymphocytic leukemia (CLL) (28), and multiple myeloma (29, 30) or defective NK cell proliferation in metastatic renal cell carcinoma (31) and chronic myelogenous leukemia (CML) (32).

In renal cell carcinoma, infiltrating NK cells have, compared to peripheral blood NK cells, increased expression of NKG2A receptor contributing to decreased NK cell activity (33). NKG2D is a well-studied activating receptor on NK cells. Membrane-bound NKG2D ligand has a stimulatory effect on immunity, while soluble NKG2D ligands have the opposite effect on immune system leading to metastatic cancer progression (34). Patients with colorectal cancer have increased serum titers of the soluble NKG2D ligand, MHC class I chain-related protein A (sMICA), compared to healthy controls, leading to downmodulation of activating and cytokine receptors on the NK cells (35). A potential way to reduce the risk of soluble NKG2D ligand is to give the patients neutralizing antibody treatment. Clinical observations demonstrate that patients treated with cytotoxic T lymphocyte-associated antigen 4 (CTLA-4) antibody blockade have reduced sMICA in a close correlation with increased titers of autoantibodies against MICA (36). Interestingly, a new report from Deng et al. shows that the soluble high-affinity ligand MUL1 causes NK cell activation and stimulates tumor rejection in mice, instead of inhibition of NK cells as earlier reported (37).

The potential benefits of NK cell-based cancer immunotherapy products have led to the design of *in vitro* methods aiming to cultivate NK cells in cGMP conditions. Some of these methods have already been tested in clinical trials, which will be discussed later in this review.

CLINICAL-GRADE NK CELL PRODUCTS

It is possible to activate NK cells and increase their anti-tumor activity through short-term cytokine exposure *in vitro* prior to adoptive transfer (38). However, to achieve clinically relevant numbers of NK cells, there also needs to be development of long-term NK cell expansion protocols (**Table 1**; **Figure 1**) (39–47). Yet, there are concerns when expanding NK cells *in vitro*, such as potential phenotypic changes, selective expansion, and reduced cytotoxic killing. When expanded *in vitro* with IL-2, there is a chance of $CD3^+$ cell expansion as well (48, 49). Thus, there is still room for improvement to achieve optimum clinically relevant NK cell numbers, *in vivo* NK cell persistence and survival, and most importantly, anti-tumor activity. There are numerous parameters affecting the clinical-grade NK cell manufacturing such as source of the NK cells, cytokine stimulation, cell culture medium, and expansion platform. Here, in this section, we will address these parameters.

Source of the NK Cells

The majority of clinical NK cell products or pre-clinical research on efficient NK cell manufacturing platforms are making use of peripheral blood mononuclear cells (PBMC), umbilical cord blood (UCB), cell lines, and human embryonic stem cells (hESC), as well as induced pluripotent stem cells (iPSC) as a source of start material.

Peripheral Blood Mononuclear Cells

The majority of NK cell products are generated through utilization of PBMCs either by apheresis or ficoll separation under cGMP conditions. An advantage of using PBMCs is the ability to collect cells in a closed aseptic system. Although PBMC consists of 5–20% NK cells, it is not possible to achieve sufficient numbers of potent NK cells. Thus, various techniques to expand NK cells *ex vivo* have been developed. For example, we have designed a feeder-free NK cell expansion system where it is possible to expand and activate tumor-reactive NK cells in a clinically compatible manner (45). These cells have a high cytotoxic effect specifically against autologous and allogeneic tumors *in vitro* and *in vivo* (42, 45). We have also completed a first-in-man clinical trial using donor-derived *ex vivo* expanded NK cells in terminal cancer patients that had CLL, kidney cancer, colorectal cancer, and hepatocellular carcinoma with promising results (43). Having optimized the procedure for NK cell expansion in a closed-automated bioreactor using clinical-grade GMP-compliant components, we have initiated a first-in-man phase I/II clinical trial to expand and restore the function of patients' own NK cells (45, 62). To our knowledge, this is the first advanced therapy investigational medicinal product trial performed using autologous NK cells in Sweden.

Sakamoto et al. have established another similar approach that generates large numbers of activated NK cells from peripheral blood without prior purification of the cells. The PBMCs are cultured with autologous plasma, IL-2, OK-432, and γ-irradiated autologous FN-CH296 stimulated T cells, reaching up to a median purity of 90.96% of NK cells at day 21 or 22. Many of the NK expansion protocols are based on enrichment of NK cells either prior to NK cell activation and expansion through cell selection or sorting in order to achieve pure cell therapy product and avoid unwanted side effects stemming from T cells especially in allogeneic NK cell transfusions.

One of the main methods of enriching the purity and the number of initial NK cells is the clinical-grade immuno-magnetic depletion of other lymphocyte subsets such as T cells and/or B cells as well as myeloid cells (60). Depletion of $CD3^+$ cells followed by $CD56^+$ cell enrichment can lead to highly pure NK cells which could be supplemented by $CD19^+$ cell depletion before infusion in order to prevent passenger lymphocyte syndrome in allogeneic transplantation (63). Nguyen et al. have shown that a partial depletion of T cells could get a more beneficial clinical outcome compared to a complete T cell depletion after hematopoietic stem cell transplantation, suggesting that T cells may have a positive role in *in vivo* NK cell function (64).

TABLE 1 | Clinical-grade NK cell products.

Cell source	Medium	Serum	Feeder cell	Other	System	Time (days)	Purity (% NK cells)
UCB CD34+ cells (46)	GBGM	10% HS	–	High-dose cytokine cocktail (SCF, Flt3L, TPO, IL-7), low-dose cytokine cocktail (GM-CSF, G-CSF, IL-6), IL-15, low molecular weight heparin, high-dose cytokine cocktail (IL-7, SCF, IL-15, IL-2)	Vuelife bags Wave bioreactor system Biostat CultiBag system	42	>90
UCB CD34+ cells (50)	GBGM SCGM	2% HS	–	250 pg/mL G-CSF, 10 pg/mL GM-CSF, 50 pg/mL IL-6, high-dose cytokine cocktail (20 ng/mL IL-7, SCF, IL-15), 1000 U/mL IL-2, 200 pg/mL IL-12	Vuelife bags	21–28	>80
NK-92 cell line (51)	X-Vivo 10, 15, 20 Aim VR TCM QBSF-56	HS Human HP HSA	–	450 IU/mL IL-2, 0.2 mM I-inositol, 2 mM L-glutamine, 20 mM folic acid, 10^{-4} M 2-mercaptoethanole	Flaske Vuelife bags X-Fold culture bags	15–17	–
NK-92 cell line (52)	X-Vivo 10	2.5% HP	–	500 U/mL IL-2, 0.6 mM L-asparagine, 3 mM L-glutamine, 1.8 mM L-serine	Vuelife culture bags	15–17	–
Total PBMC (48, 53)	CellGro SCGM	5% HS	–	500 U/mL IL-2, 10 ng/mL OKT3	Flasks	21	55–74
Total PBMC (54)	CellGro SCGM or RPMI-1640	10% FBS	K562-mb15-41BBL	10 U/mL IL-2	Flasks Teflon bags	7–14	96.8 83.1
Total PBMC (55)	RHAMα	5%AP	HFWT	100 U/mL IL-2	24-well plates	6–7	86
Total PBMC (42)	CellGro SCGM	5% HS	–	500 U/mL IL-2, 10 ng/mL OKT3	Flasks	20	65%
Total PBMC (45)	CellGro SCGM	5% HS	–	500 U/mL IL-2, 10 ng/mL OKT3	Wave bioreactor system Flasks Vuelife bags	20	Relative: 64% 74% 47%
Total PBMC (47)	GT-T510	1% HP	Autologous FN-CH296 induced T cells	IL-2, OK-432	CultiLife bag	20–21	90%
CD56 enriched PBMC (49)	X-VIVO 10	10% HS	–	500 U/mL IL-2, 10 ng/mL IL-15, 200 mM L-glutamine	NR	14	NR
CD5 and CD8 depleted PBMC (56)	RPMI-1640	10% HS	–	1000 U/mL IL-2, 2 mM L-glutamine, 1000 U/mL penicillin, 100 U/mL streptomycin	Polystyrene Cell Factories Teflon bags Polyolefin bags	21	88
CD5 and CD8 depleted PBMC (41)	2:1 DMEM:Ham's F12-based NK medium	10% HS	–	1000 U/mL IL-2, 20 μM 2-mercaptoethanole, 50 μM ethanolamine, 20 mg/mL L-ascorbic acid, 5 μg/L sodium selenite, 100 U/mL penicillin, and streptomycin	Stirred-tank bioreactor Spinner flasks 24-well plates	33	95–96
Non-adherent PBMC (57)	RPMI-1640	10% FBS	RPMI 8866	50 U/mL IL-2	24-well plates	10–12	80
Non-adherent PBMC (58)	RPMI-1640	10% FBS	RPMI 8866	50 U/mL IL-2	24-well plates	10–12	90
CD3 depleted non-adherent PBMC (39, 59)	DMEM	8% HS	LAZ 388	200 U/mL IL-2, 2 mM L-glutamine, 1 mM sodium pyruvate, 0.2% NaOH, 100 U/mL penicillin, 0.1 mg/mL streptomycin	V-bottom microplates	13–21	>90
Purified NK cells (60)	X-VIVO 20	–	Allogeneic mononuclear cells	100 U/mL IL-2, 10 U/mL IL-15, 100 μg/mL PHA, 1 μmol/mL ionomycin	Teflon bags	14–21	92
Purified NK cells (61)	X-VIVO 20	10% HS	EBV-TM-LCL	500 U/mL IL-2, 2 mM GlutaMAX-1 at 6.5% CO_2	Flasks or Baxter bags	28	99
Adherent activated NK cells (40)	RPMI-1640	10% HS	Allogeneic mononuclear cells	6000 U/mL IL-2	Flasks	14–18	85

PBMC, peripheral blood mononuclear cells; HS, human serum; FBS, fetal bovine serum; HP, human plasma; HAS, human serum albumin.

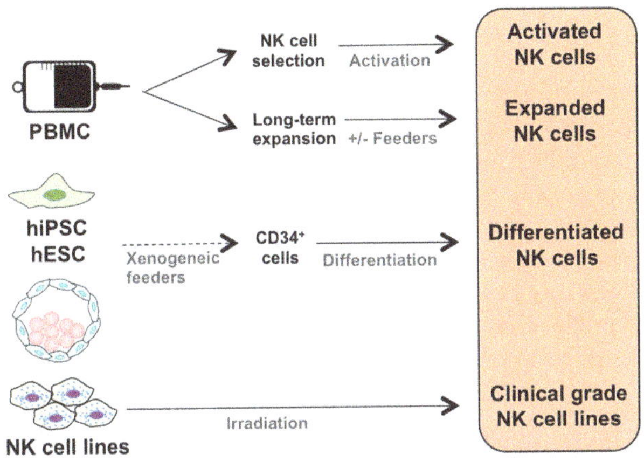

FIGURE 1 | Clinical NK cell therapy products.

Additionally, direct enrichment of CD56$^+$ NK cells through immuno-magnetic selection is an option to achieve high purity initial NK cell product. Nevertheless, NK cells might require physical and cytokine-dependent communication with other cells such as monocytes (65) in order to activate and expand. Thus, it is essential to fine-tune the enrichment of NK cells by making use of feeder cells and/or optimizing the cytokine cocktail used in *ex vivo* NK cell expansion protocols.

Furthermore, using feeder cells and cell lines is another approach in expanding NK cells *ex vivo* since feeder cells can provide essential stimulatory signals for NK cells proliferation. Monocytes, irradiated PBMC, feeder cell lines, and engineered feeder cell lines are the most commonly used sources for stimulation of NK cell expansion through humoral signals and cell-to-cell contact. Example of feeder cells that have been used in clinical trials are irradiated autologous PBMCs (60, 66), irradiated Epstein–Barr virus-transformed lymphoblastoid cells (61), and K562 engineered cells expressing 4-1BB ligand (67) or membrane-bound IL-21 (68, 69) on cell surface.

Stem Cells

While PBMC is one of the major sources for achieving clinically relevant doses of tumor-reactive NK cells, HSC and potentially hESC as well as iPSC are likewise essential sources for achieving clinically relevant doses of NK cells.

One of the potential sources to accomplish clinically relevant doses of tumor-reactive NK cells is making use of HSC (CD34$^+$) through differentiation and expansion of CD34$^+$ cells isolated from bone marrow, peripheral blood, or UCB into functional NK cells. It was recently demonstrated that it is possible to expand activated, tumor cytotoxic and pure NK cells by differentiating UCB CD34$^+$ HSC under cGMP condition (46). Furthermore, NK cells derived from CD34$^+$ UCB cells lack expression of KIRs such as KIR2DL1 (CD158a), KIR2DL2/DL3 (CD158b), and NKB1, as well as diminished CD16 expression in the CD56^{dim} population (70). Even though NK cells derived from UCB have reduced cytotoxicity, this could be restored by *ex vivo* cytokine stimulation such as IL-2, IL-12, and IL-15 (50, 71–73). Infusion of UCB-derived NK cells supplemented with IL-15 has shown to inhibit growth of human bone marrow resident leukemia cells *in vivo* (74). Recently, it was demonstrated that frozen UCB CD34$^+$ cells differentiate into NK cells with better expansion than freshly isolated UCB CD34$^+$ cells, and more importantly, UCB CD34$^+$ cells gave more NK cell product than peripheral blood HSC without jeopardizing NK cell functionality (75). Thus, UCB CD34$^+$ cells are one of the essential sources for manufacturing NK cell therapy protocols, providing an option to create NK cell biobanks.

Another potential source of NK cells is hESC and iPSC, with the advantage of potential usage of iPSCs in autologous settings with reduced risk of immune rejection. The first step is to generate CD34$^+$ hematopoietic precursor cells from the hESCs and iPSCs and then differentiate these cells into NK cells, which could be efficiently achieved through growing hESCs and iPSCs on murine stromal cells (76, 77). Yet, the involvement of xenogeneic cells could limit the potential clinical usage of hESCs and iPSCs. Addressing this potential problem, Knorr et al. developed a two-stage culture method where hESCs and iPSCs are first differentiated to CD34$^+$ hematopoietic cells by spin-EB system in xeno-free and serum-free conditions followed by stroma-free NK cell differentiation, which enables generation of cytotoxic NK cells without involvement of xenogeneic cells taking a step forward toward clinical-scale production (78). Since IL-2-activated NK cells are potent killers of both allogeneic and autologous iPSCs (79), it is possible to manufacture a pure NK cell therapy product. This sticks out as one of the advantages of using *in vitro* NK cell differentiation from iPSCs.

Cell Lines

Cell lines derived from NK cells with similar biological functions (NK-92, NKL, KYHG-1, and NKG) are potential candidates for NK cell-based products enabling design and development of off-the-shelf anti-cancer cell therapy products. Furthermore, it is more feasible to generate genetically modified NK cell lines expressing intracellular IL-2 for activation or cell surface molecules such as CD16, NCRs, and chimeric antigen receptors (CARs). To our knowledge, the NK-92 cell line is the most clinically studied one. The IL-2-dependent NK-92 cell line is cytotoxic to a wide range of malignant cells (80–83). It has also been used as a source of NK cells for cGMP-grade cellular therapy products (51) as well as in clinical trials (52, 84). The NK-92 cell line expresses several activating receptors but lacks most of the inhibitory KIRs, NKp44, and CD16 (80, 85). NK-92 cells require irradiation to prevent proliferation prior to being used effectively in immunotherapeutic approaches without compromising hematopoietic cell function. For example, recently, clinical-grade NK-92 cells have been manufactured and were safely used as anti-tumor therapy for patients with a variety of tumors (84) with promising results (52). As of today's date, two phase I clinical trials (NCT00900809 and NCT00990717) are recruiting patients with hematological malignancies for treatment with NK-92 cells. The first clinical phase II study (NCT02465957) with NK-92 cells has recently been initiated.

KHYG-1 is the first NK cell line derived from NK leukemia and has higher cytotoxicity than NK-92 cell line (86). Likewise

NK-92 cells, these cells can also be irradiated to inhibit proliferation and can still efficiently kill tumor targets. Furthermore, NKL cell line, which is the most biologically and functionally similar to primary NK cells, is more cytotoxic to certain tumor cells than NK-92 cell line and, additionally, it has the ADCC capacity whereas NK-92 cells lack CD16 expression. Thus, both KHYG-1 and NKL cell lines have the potential to be used as anti-cancer NK cell products.

Additionally, one of the advantages of using such master cell bank is an appealing opportunity in the manufacture of cellular therapy products since it is possible to establish a comprehensive standardization and characterization of the cell source. It is also possible to genetically modify these cell lines to exert more tumor specificity and cytotoxicity. For example, NK-92 cell lines are dependent on external IL-2 stimulation, which increases manufacturing costs as well as potentially reducing the long-term cytotoxic capacity of these cells unless they are supported by IL-2 infusions. Thus, constitutive expression of IL-2 in NK-92 cells through genetic modification leads to auto-activated and -proliferating cells, which reduces the manufacturing costs as well as potentially increases the *in vivo* tumor reactivity (87, 88).

Cytokines

Ex vivo manufacturing of NK cell-based products is dependent on extensive use of cytokines to stimulate, differentiate, activate, and expand NK cells in order to get clinically relevant doses and enhanced anti-tumor reactivity. Historically, one of the most popular cytokines in NK cell research is IL-2 since it was the first cytokine to be injected to patients to treat metastatic melanoma (89). Thirty years ago, Rosenberg et al. published the first report where they treated 25 metastatic cancer patients, who did not respond to standard therapy, with autologous lymphokine-activated killer (LAK) cells together with recombinant-derived IL-2. LAK cells are generated from mononuclear cells collected from IL-2 injected patients. In 11 patients, the cancer regression was observed with >50% of tumor volume (90). This adoptive immunotherapy was followed by a larger scale study, where 157 patients with advanced metastatic cancer were treated with successful results (91). In the same year, it was shown that it was the NK cells that mediated the cytotoxic activity in response to systemic administered recombinant IL-2 (92). These reports were followed by many years of IL-2 and NK cell research. In a dose-dependent manner, IL-2 is important for NK cell infiltration and killing of the tumor. For example, in the bone marrow, there are hypoxic regions leading to reduced NK cell killing of plasma cells in multiple myeloma. IL-2-activated NK cells *ex vivo* have increased NKG2D expression resulting in increased targeting of multiple myeloma upon infusion (93). Cytokine-activated NK cells *in vitro* are dependent on constant stimulation both *in vitro* and *in vivo*. Basse et al. reported that when no exogenous IL-2 is present the amount of injected NK cells found in tumors were very low (94). The half-life of IL-2 in serum is not more than 10 min, which makes the administration of IL-2-dependent cells difficult (95). By transducing NK cells to produce IL-2 prior to transplantation, the activated NK cells would have a constant source of IL-2 *in vivo* (87, 96). One of the disadvantages of using IL-2 to activate NK cell *in vivo* is the competition over IL-2 by regulatory T cells, which express high levels of the high-affinity receptor for IL-2, IL-2Rα (CD25). By treating patients with lympho-depleting agents (fludarabine and cyclophosphamide) followed by NK cell infusion and IL-2 fused with diphtheria toxin (IL-2DT), $CD25^+$ cells are selectively depleted, leading to increased NK cell expansion and complete remission rate for patients with AML compared to regular IL-2 treatment (97). Overall, the majority of cGMP-grade NK cell therapy protocols include IL-2 as a main cytokine to stimulate NK cell activation and proliferation.

Another important cytokine is IL-15 which is required for both NK cell maturation and survival (98). IL-2 and IL-15 share the same receptor components: IL-2/15Rβ and common γ chain (also shared with IL-4, IL-7, IL-9, and IL-21). Recent advances in the production of cGMP quality cytokines enabled further optimization of cytokine supplementation during NK cell expansion. For example, use of IL-15 in combination with IL-2 has a synergetic effect on product viability and NK cell proliferation (66). This highlights the necessity of other cytokines to achieve NK cell product potency especially when it comes to the NK cell expansion protocols that are not using feeder cell support. Additionally, IL-21, primarily described in 2000 (99), has significant homology with IL-2 and IL-15. Compared to IL-2 and IL-15, IL-21 promotes maturation and survival but does not promote proliferation of NK cells alone. However, IL-21 does have synergetic effects with IL-2 and IL-15 (100). Interestingly, it has been suggested that IL-21 does not drive proliferation of regulatory T cells *in vivo* and might be a good candidate to substitute for IL-2 in CLL (101).

Other Factors

Besides NK cell source, feeder support, and cytokine stimulation, other parameters such as expansion platform, cell culture media, and serum supplementation are also very important in achieving clinically relevant cell numbers, viability, and tumor cytotoxicity. More specifically, we have recently investigated the importance of the culture vessels on the quality and efficacy of the NK cell product. Briefly, PBMCs from healthy donors and myeloma patients were cultured for 21 days using flasks, cell culture bags, and bioreactors. Even though we have achieved high yield in NK cell expansions in all systems, NK cells expanded in the bioreactor displayed significantly higher cytotoxic capacity. These results demonstrate that highly active NK cells can be produced in a closed, automated, large-scale bioreactor under feeder-free current GMP conditions facilitating adoptive immunotherapy clinical trials (45).

Additionally, cell culture media is another important factor to consider in the manufacturing of cellular therapy products. There are very few cGMP quality medias that work optimally for *ex vivo* NK cell expansion protocols. The most commonly preferred media in the generation of NK cell products are stem cell growth medium (SCGM; CellGenix, Freiburg, Germany), X-VIVO serum-free media (BioWhittaker, Verviers, Belgium), or AIM V (Life Technologies, Grand Island, NY, USA) (49, 102, 103). Generally, medium is supplemented by human AB serum or fetal bovine serum from certified sources.

Finally, there are numerous variables that may impact quality and quantity of NK cell products. Future pre-clinical research and results from more clinical trials will evaluate the contribution of each factor to the product purity, potency, and safety, as well as assist in acquiring NK cell products that can be manufactured reproducibly with the optimal safety and anti-tumor responses.

CLINICAL USE OF NK CELL-BASED ANTICANCER PRODUCTS

Autologous NK Cells

Several clinical studies have been performed with adoptive autologous NK cells in an attempt to target tumors, such as breast cancer, lymphoma, glioma renal cell carcinoma, non-small cell lung cancer, and adenocarcinoma (**Table 2**) (39, 40, 55, 103–107). In general, autologous NK cell trials are safe with no toxic side effects (39, 40, 55, 105). For example, *ex vivo* activated autologous peripheral blood lymphocytes get enhanced cytolytic activity against heat shock protein 70 (Hsp70) membrane-positive tumors *in vivo* if pre-incubated with Hsp70 peptide and IL-2 (105, 107). However, some clinical trials with autologous NK cells have only partial effect on tumors, such as glioma (55). While other tumors, such as metastatic carcinoma or relapsed lymphoma, do not demonstrate any improvement (103, 104, 108). Moreover, a recent clinical trial used *ex vivo* FN-CH296 stimulated T cells and OK-432 expanded, autologous NK cells with enrolled patients diagnosed with rectal, esophageal, gastric, or colon cancer that was either recurrent or at metastatic disease stage. The NK cell therapy in these patients was well tolerated with no severe adverse events and the cytotoxicity of peripheral blood was elevated approximately twofold up to 4 weeks post the last transfer (47).

Allogeneic NK Cells

Allogeneic NK cell products have been used in the treatment of a range of malignancies, such as leukemia, renal cell carcinoma, leukemia, colorectal cancer, hepatocellular cancer, lymphoma, and melanoma (**Table 3**) (38, 109–113). The major risk with allogeneic NK cell transplantation is the development of graft-versus-host disease (GvHD). Several precautions can be taken to reduce the risk of GvHD, for example, immunosuppression,

TABLE 2 | Clinical trials with infusion of autologous NK cells.

Malignancy	n	NK cell source	Depletion	Product	*Ex vivo* handling	Purity	Dose	Outcome
Colorectal carcinoma/ NSCLC (105)	11/1	PBMC	–	IL-2 + Hsp70 peptide	4 days	Mean: 14% (range: 8–20%)	Range: 0.1–1.5 × 10^9 NK cells	Cytotoxic activity of NK cells. No significant tumor response
Colon carcinoma (107)	1	PBMC	–	IL-2 + Hsp70 peptide	4 days	Mean: 22.4% (range: 16–25%)	Mean: 1.48 × 10^9 NK cells (range: 0.9–1.9 × 10^9)	Anti-tumor activity by NK cells
Glioma (55)	9	PBMC	–	Irradiated feeder cell line (HFWT) + autologous plasma + IL-2	14 days	82.2 ± 10.5%	i.c. 0.4–2.3 × 10^9 cells i.v. 0.2–6.5 × 10^9 cells	3 partial responses, 2 minimal responses
RCC (39)	10	PBMC	CD3+ depletion or Immunorosette depletion	Cultured on LAZ388 with allogeneic irradiated PBMNC as feeder cells + IL-2	13–21 days	>90% except 1 patient (33%)	Mean: 5.8 × 10^9 total cells (range: 1.8–15.1 × 10^9)	All patients improved, 4 complete response, 2 partial response
Melanoma/ RCC (103)	7/1	PBMC	CD3+ depletion	Autologous irradiated PBMNC as feeder cells + IL-2 and OKT3	21 days	96% ± 2%	Range: 4.7 × 10^{10} (±2.1 × 10^{10}) NK cells	No tumor lysis by NK cells. No tumor response
Rectal/ esophageal/ gastric/colon cancer (47)	4/4/3/3	PBMC	–	Autologous FN-CH296 stimulated T cells + autologous plasma + IL-2 and OK-432	21–22 days	Median: 90.96% (range: 65.94–99.45%)	0.5–2.0 × 10^9 cells	No tumor response
Lymphoma/ breast cancer (104)	20/14	*In vivo* IL-2 activated NK cell	–	IL-2	Over night	Not reported	Range: 0.33–2.09 × 10^8 cells/kg	No improvement of survival
Breast cancer (108)	5	*In vivo* IL-2 activated NK cell	Monocyte depletion	Allogeneic irradiated PBMNC as feeder cells + IL-2	14 days	Mean: 83.2% (range: 67–93%)	Mean: 3.97 × 10^9 total cells (range: 1.55–9.1 × 10^9)	1 complete response, 1 partial response, 2 had stable disease, 1 disease progression
Lymphoma/ breast cancer (40)	10/1	*In vivo* IL-2 activated NK cell progenitors	Monocyte depletion	Allogeneic irradiated PBMNC as feeder cells + IL-2	14–18 days	Mean: 85% (range: 64–98%)	Range: 6.8 × 10^8–4 × 10^{10} total cells	Increased NK cell numbers and activity in 4 patients

NSCLC, non-small cell lung cancer; PBMC, peripheral blood mononuclear cell; RCC, renal cell carcinoma.

TABLE 3 | Clinical trials with infusion of allogeneic NK cells.

Malignancy	n	NK cell source	Depletion	Product	*Ex vivo* handling	Purity	Dose	Outcome
		Cell line						
RCC/MM (52)	11/1	NK-92	–	IL-2	3 weeks	Clonal cell line	1×10^8–3×10^9 cells/kg	1 mixed response, 4 stable disease, 6 progressive disease
Solid tumor/CLL/B-NHL (84)	13/1/1	NK-92	–	IL-2	2–2.5 weeks	Clonal cell line	1×10^9, 3×10^9, 1×10^{10} cells/m^2	2 mixed response, 1 stable disease, 12 progressive disease
		Progenitor cells						
AML/ALL/high-grade MDS (114)	11/1/2	Related CD34+ progenitors	CD34+ selection	IL-15, IL-21+ hydrocortisone	42 days	Not reported	Mean: 3.49×10^8 NK cells/kg (range: 1.8–6.3×10^8)	2 with active leukemia had no response
		Adult cells						
AML/CML (110)	4/1	Haploidentical PBMC	CD3+ depleted, CD56+ enrichment	–	Overnight storage in +4°C	Median: 97.35% (range: 77.9–98.9%)	Median: 0.93×10^7 cells/kg (range: 0.21–1.41×10^7)	3 donor chimerism, 1 relapse
AML (112)	10	Haploidentical PBMC	CD3+ depleted, CD56+ enrichment	–	Overnight storage	Not stated	Mean: 29×10^6 NK cells/kg (range: 5–81×10^6)	*In vivo* expansion of NK cells. 2 years event-free remission in 100%
AML (113)	13	Haploidentical PBMC	CD3+ depleted, CD56+ enrichment	–	–	Median: 93.5% (range: 66.4–99.2%)	Median: 2.74×10^6 cells/kg (range: 1.11–5×10^6)	3 disease-free. 4 complete remissions, 5 with active disease had no clinical benefit
AML (116)	1	Haploidentical PBMC	CD3+ depleted	–	–	Not stated	3×10^7 NK cells/kg	Complete response, relapse on day 80
Melanoma/RCC/HD/AML (38)	10/13/1/19	Haploidentical PBMC	CD3+ depleted	IL-2	Over night	Mean: 40% (range: 18–68%)	1×10^5–2×10^7 cells/kg	*In vivo* expansion of NK cells. 5 complete remission (AML)
Breast/ovarian carcinoma (115)	6/14	Haploidentical PBMC	CD3+ depleted	IL-2	Overnight	25.0 ± 0.3%	Mean: 2.15×10^7 NK cells/kg 8.33×10^6–3.94×10^7 cells	4 partial responses, 12 and 3 stable or progressive diseases, respectively
Neurobalstoma/AML/ALL/RMS/HD (117)	4/5/5/1/1	Haploidentical PBMC	CD3+ depleted, CD56+ enrichment	Group 1: – Group 2: IL-2	9–14 days	Median: 95% (range: 84.4–98.6%)	Range Group 1: 3.2–38.3×10^6 cells/kg, Range Group 2: 6–45.1×10^6 cells/kg	Group 1: 3 complete remissions (1 NB, 2 ALL) Group 2: 2 complete remissions (NB)
ALL/AML (109)	2/1	Haploidentical PBMC	CD3+ depleted, CD56+ enrichment	IL-2	14 days	Mean: 95%	Mean: 11.9×10^6 cells/kg (range: 3.3–29.5×10^6)	3 complete remissions, AML patient got early relapse
Neuroblastoma (118)	2	Haploidentical PBMC	CD3+ depleted, CD56+ enrichment	IL-2	14 days	>95%	7.8–45.1×10^6 cells/kg	Initially enhanced NK cells cytotoxicity

(Continued)

TABLE 3 | Continued

Malignancy	n	NK cell source	Depletion	Product	*Ex vivo* handling	Purity	Dose	Outcome
CRC/HCC/RCC/CLL (43)	1/1/2/1	Haploidentical PBMC	–	IL-2 + OKT3	20 days	Not stated	1.0–10 × 10^6 NK cells/kg	Signs of response in HCC. No tumor response
NSCLC (44)	15	Haploidentical PBMC	CD56+ enrichment	IL-15 + hydrocortisone	20–23 days	Median: 97.9% (range: 82.7–899.6%)	Median 4.15 × 10^6 NK cells/kg (range: 0.2–29 × 10^6)	15 months median overall survival. 56% 1-year-survival, 19% 2-year survival
MM (119)	8	Haploidentical PBMC (3) or autologous PBMC (5)	CD3+ depleted	IL-2 + irradiated K562-mb15-41BBL	7–9 days	Median: 78% (range: 52–90%)	2 × 10^7–1 × 10^8 NK cells/kg	1 partial response, 1 reduced disease progression, 5 no clinical benefit
Solid tumor (67)	9	Haploidentical PBMC	CD34+ deleted, CD3+ depleted, CD56+ enrichment	IL-15 + irradiated KT32.A2.41BBL.64	Cryopreserved + culture 9–11 days	≥90%	1–10 × 10^5 NK cells/kg	5 aGvHD

NSCLC, non-small cell lung cancer; PBMC, peripheral blood mononuclear cell; CRC, colorectal cancer; HCC, hepatocellular carcinoma; CLL, chronic lymphocytic lymphoma; ALL, acute lymphoblastic leukemia; EBV, Epstein-Barr virus; MDS, myelodysplastic syndrome; CML, chronic myelogenous leukemia; RCC, renal cell carcinoma; HD, Hodgkin disease; RMS, rhabdomyosarcoma; MM, multiple myeloma; B-NHL, non-Hodgkin lymphoma.

infusion of CD3 depleted high purity NK cells and if available, selecting the donor that matches the host HLA (44, 114, 115).

In the first phase I clinical trial using the feeder-free *ex vivo* expansion platform, adoptive transfer of NK cells from HLA identical siblings into patients with leukemia or carcinoma was well tolerated and safe alongside *in vivo* NK cell expansion, with only some infusion-related complications (43).

If no HLA identical donor is available, host cells from a receptor–ligand-mismatched donor can be used. If the donor is HLA matched, it is preferentially better if the donor cells are KIR B haplotype. Also, to further improve the outcome, T cell depletion is performed (120). In haploidentical transplantation, at least one KIR ligand is not expressed on the host cells leading to reduced inhibition of donor NK cells. Less inhibited NK cells could lead to better prognosis and might be the best treatment for a good clinical outcome if GvHD can be avoided (38, 121). When haploidentical transplantation is performed, it is strictly necessary to make extensive T cell depletion to avoid GvHD. In most clinical trials, NK cells are collected from leukapheresis followed by a two-step purification procedure, with depletion of $CD3^+$ T cells followed by enrichment of $CD56^+$ cell (109, 110, 117, 118).

Completed clinical trials with haploidentical donors are safe with only a few reports of infusion-related complications such as dyspnea, nausea, hypertension, stroke, febrile reaction, and vomiting (38, 115). So far, allogeneic NK cell transplantations derived from PBMCs or $CD34^+$ cells have shown promising results with engraftment, *in vivo* expansion of NK cells, complete remission, and a 100% 2-year event-free survival in one clinical trial by Rubnitz et al. (109, 112–114, 116).

IMMUNE SUPPRESSION OF NK CELLS IN THE TUMOR MICROENVIRONMENT

Natural killer cells can recognize and kill tumor cells *in vitro*. However, their efficiency in targeting solid tumors has not yet been fully acknowledged in the clinical setting even though endogenous and adoptively transferred activated NK cells can be detected in various tumors (122, 123). Nevertheless, not all tumors are equally well infiltrated by NK cells, and many of the infiltrating cells are dysfunctional (124–127). The failure of immune surveillance may in part be due to sustained immunological selection pressure on tumor cells resulting in the development of tumor escape variants that are in fact invisible to the immune system (**Figure 2**). In addition, cytotoxic function of immune effector cells is also largely suppressed in the tumor microenvironment (128), which could be explained by suppressive tumor-secreted factors as well as suppressive immune compartments, such as myeloid-derived suppressor cells (MDSCs), tumor-associated macrophages (TAM), and regulatory T cells (**Figure 2**). One of the most studied immune-suppressive cell types associated with tumor progression is regulatory T cells (Treg), characterized by their expression of CD4, high CD25 ($CD4^+CD25^+CD127^{low/neg}$) as well as the transcription factor forkhead box P3 (FoxP3) (129). The expansion of Treg population is promoted in different cancers and their accumulation correlates with impaired immune cell function and poor prognosis (130–135). *In vitro*, NK cells are suppressed by Treg cells in a cell contact-dependent manner where membrane-bound TGF-β is utilized to attenuate their cytotoxicity (136). In line with this, inverse correlation between NK cell activity and Treg cell expansion has been observed in patients with gastrointestinal stromal tumor (GIST) (136) as well as in hepatocellular carcinoma patients (137). Treg cells express the high-affinity IL-2 receptor alpha (CD25, IL-2Rα) and need IL-2 for their full function. Recent studies have indicated that NK cell proliferation, accumulation, and activation can be limited by Treg cells through hampering the availability of IL-2 released by activated $CD4^+$ T cells (138, 139). Consequently, inadequate IL-2 levels in the tumor microenvironment limits the extent of NK cell-mediated tumor rejection.

Another group of immunosuppressive cells in the tumor is the MDSCs. MDSCs are heterogeneous precursors of the myeloid cells, granulocytes, macrophages, and immature dendritic cells with immunosuppressive activity (140). Recently, MDSCs have been proposed as a key immunoregulator in various solid and hematologic malignancies (141, 142). MDSCs are divided into two groups that can originate from granulocytic (grMDSCs) and monocytic precursors (moMDSCs) (143). In human beings, distinct phenotypes of MDSCs are associated with different types of cancers (144–148). Their suppressive function is mediated by a few different mechanisms such as production of suppressive cytokines including IL-10 and TGF-β, depletion of arginine in the tumor or production of reactive oxygen species (ROS) (144, 149–151). Additionally, recent studies investigated the induction mechanism of MDSCs and how they suppress T cells *in vitro* (152–154). Furthermore, several studies have characterized cytokines that can induce MDSCs from healthy human PBMCs. We found that prostaglandin E2 treated healthy monocytes resemble patient-derived moMDSCs and suppress NK cell responses through TGF-β-dependent mechanism (155). In patients with hepatocellular carcinoma, NK cells were shown to be suppressed by monocytic MDSC in a cell contact-dependent manner, but did not rely on the arginase activity of MDSCs, which is a hallmark function of these cells; however, MDSC-mediated inhibition of NK cell function was revealed to be mainly dependent on the NKp30 on NK cells (146). Moreover, a negative correlation between increased $CD33^+$-MDSC accumulation and functional loss of NK cells has been demonstrated in patients with myelodysplastic syndromes (156).

Macrophages are the dominant myeloid-derived population that is found in the tumor microenvironment. TAM has been identified as regulators of solid tumor development based on their capacity to enhance angiogenic, invasive, and metastatic programing of neoplastic tissue (157–160). TAMs could be found in several types of human cancer correlating with poor clinical outcome (161, 162). The immune-suppressive mechanisms applied by TAMs on NK cells in the tumor microenvironment can be different, such as recruitment of Treg, prostaglandin E2-mediated inactivation, and production of IL-10 (163–165). Furthermore, tumors are able to escape NK cells by releasing indoleamine 2,3-dioxygenase and prostaglandin E2, which inhibit the expression of activating receptors of NCRs and NKG2D (166). These molecules are also released by mesenchymal stem cells to inhibit NK cell function in the tumor microenvironment (167). There is a direct association between the surface density of NCRs (NKp46) and the intensity of anti-tumor cytolytic activity of the NK cells (168).

As mentioned earlier, the tumor microenvironment plays a significant role in suppressing NK cell responses against cancer. Therefore, therapies aim to target immunosuppressive cell

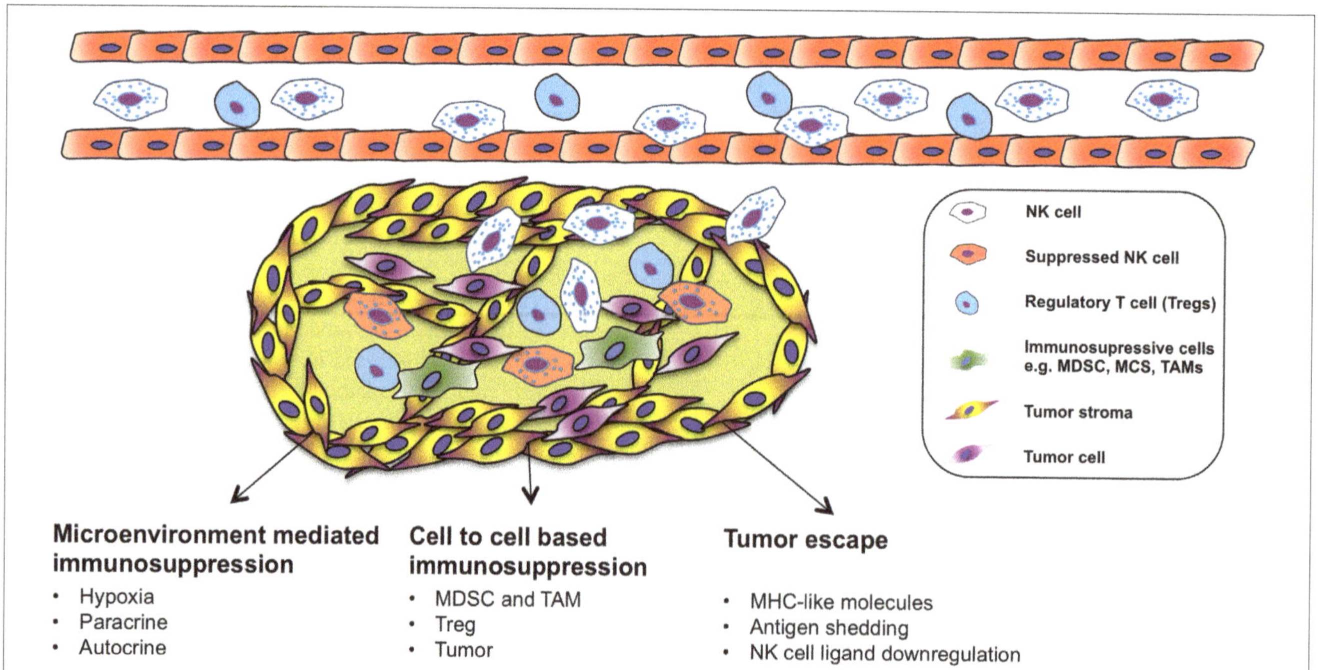

FIGURE 2 | Immune evasion and immunosuppressive in the tumor microenvironment.

populations are emerging (169–174). In the next section, some of the alternative ways aiming to enhance tumor-specific targeting and NK cell survival in order to overcome immunosuppressive effect of the tumor microenvironment on NK cells and to improve intra-tumoral NK cell responses will be discussed.

FUTURE PERSPECTIVES

Genetically Modified NK Cells

In the last decade, several NK cell based anti-cancer products have been taken to clinical trial stage with promising clinical outcomes. However, in order to manufacture more efficient NK cell therapy products, it is essential to develop novel potential strategies such as genetic modification of NK cells (**Figure 3**). Although NK cells are inherently resistant to retroviral infections (96, 175–177), our group has significantly enhanced retroviral and lentiviral gene delivery to NK cells through enhanced proliferation and targeting intracellular viral defense mechanism by small molecule inhibitors (96). Therefore, it is easier to design genetically modified NK cells expressing cytokine transgenes, silenced inhibitory receptors, overexpressing activating receptors, or retargeting NK cells by expression of CARs on the cell surface. By genetically modifying NK cells to produce cytokines such as IL-2 or IL-15, their survival capacity and proliferation increase and their activation and anti-tumor activity *in vivo* are enhanced (83, 87, 88, 178, 179). To enhance the specificity for the target cells, NK cells can be modified to recognize antigens specifically expressed on the tumor cells.

Furthermore, another approach aiming to enhance tumor specificity is to make use of ADCC. The constant region of the tumor-specific monoclonal antibodies (mAbs) targeting the tumor cells can engage to the FcγRIIIa receptor (CD16a) on the NK cell, activating the NK cell. However, NK-92 cell line cannot perform ADCC since they lack CD16a expression (80, 85). This defect on NK-92 cells can be reverted by the introduction of CD16a through genetic modification so that they are able to perform ADCC in antibody combination treatments (180). Finally, CAR-modified NK cell lines can also function as tumor-specific standardized and characterized NK cell-based therapy products. Most of the NK cell lines require further *in vivo* characterization with a potential to become standard NK cell-based products for certain tumors.

Monoclonal Antibodies

When the antigen-binding fraction (Fab) of the antibody binds to the tumor target cell and the constant region (Fc) of the antibody binds to CD16 on the NK cells, NK cells get activated and ADCC is triggered. Several different mAbs have been developed for targeting specific tumor antigens, such as anti-CD20 (retuximab), anti-Her2 (trastuzumab), anti-CD52 (alemtuzumab), anti-EGFR (certuximab), and anti-CD38 (daratumumab) (181). Daratumumab treatment of patients with relapsed myeloma has mild infusion-related reactivity, complete or very good partial responses with reduced bone marrow plasma cell levels (182). mAbs bind to the target tumor cell plus engaging CD16 on NK cells and other cell types resulting in killing of tumor cell by ADCC both *in vivo* and *in vitro* [reviewed in Ref. (183)]. New generations of mAbs have been developed to increase ADCC and complement-dependent cytotoxicity. Second-generation anti-CD20 mAbs, such as veltuzumab (hA20) (184, 185) and ofatumumab (HuMax-CD20) (186–193), have the advantage of being humanized or of fully human origin. Both veltuzumab and ofatumumab had promising preliminary outcomes in various studies (184, 186, 187, 189, 190, 193). The benefit of third-generation anti-CD20 mAbs, ublituximab (TG-1101), ocaratuzumab (AME-133) (194, 195), and obinutuzumab (GA-101) (196–200), is that they are both humanized and that their Fc regions have been modified for increased binding affinity to CD16a. So far, the most studied third-generation anti-CD20 mAbs is obinutuzumab. The overall response rate for obinutuzumab is 44.6%, which is higher than the overall response rate for rituximab treatment which is 33.7% (200). In the same study, the progression-free survival did not promote obinutuzumab over rituximab. By increased affinity between CD16a and mAb better NK cell cytolysis can be induced by ADCC. Ublituximab, ocaratuzumab, or obinutuzumab-treated NK cells from CLL patients or healthy donors have more efficient ADCC compared to same cells treated with first- or second-generation anti-CD20 mAb *in vitro* (201–203).

Monoclonal antibody therapies in combination with already existing treatments can potentially enhance NK cell activity in anti-tumor therapy. The completely human IgG4 anti-KIR antibody, IPH-2102, has been tested in several clinical trials for hematological diseases both as single treatment and as combination (204, 205). Some clinical trials for combination treatment of

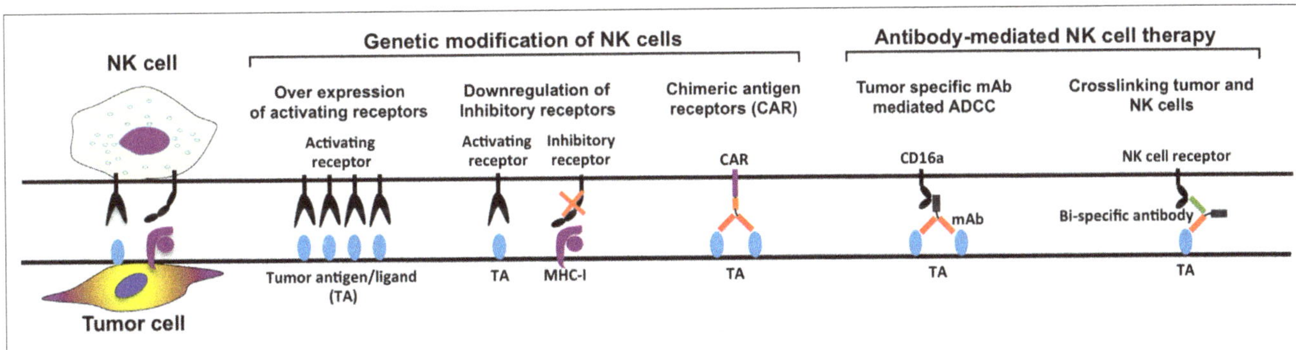

FIGURE 3 | NK cell therapy approaches.

advanced solid tumors with anti-KIR antibodies are done as well, for example, in combination with anti-CTLA antibody or anti-PD1 antibody (NCT01750580 and NCT01714739, respectively). Thus, use of mAbs enhancing ADCC and stimulation of NK cells as well as blocking NK cell inhibition could potentially improve outcome of clinical anti-cancer NK cell products (**Figure 3**).

Bi- and Trispecific Antibodies

Likewise designing CARs through tumor-specific mAbs can be used to engineer bi- and trispecific antibodies crosslinking CD16 with tumor-specific mAbs in order to enhance NK cell tumor reactivity (**Figure 3**). Briefly, the design of bi- and trispecific antibodies, fusing the Fab region of the antibody targeting the tumor cell antigen, such as CD19, CD20, and CD33, in combination with another Fab region recognizing CD16 on NK cell leads to stimulation of the NK cells followed by tumor cell killing. This technology makes it possible to select the amount of NK cells that should be activated as well as it is possible to add more Fab regions targeting other tumor-associated antigens. These Fab regions can be exchanged to other tumor-associated antigen-recognizing antibody parts, as long as the part crosslinking CD16 on the NK cell is present (206, 207).

Chimeric Antigen Receptors (CARs)

Design of CARs using antigen-specific variable part of these tumor antigen antibodies fused with intracellular lymphocyte stimulatory molecules (CD3ξ, CD28, 4-1BB) enables high-affinity specific recognition of tumor antigens and tumors. CAR modifications of T cells have been studied extensively and have led to several phase I and phase II clinical trials (208–211). NK cells are less explored and so far only two clinical trials using CAR NK cells have been approved. The first study (NCT00995137) at St. Jude Children's Research Hospital is completed and was a phase I clinical trial with 14 relapsed or refractory B-lineage ALL patients below 18 years. Haploidentical NK cells were expanded by co-culture with irradiated K562 cell line expressing IL-15 and 4-1BB ligand on the surface to be transduced with a signaling receptor binding CD19 (anti-CD19 CAR). The second study (NCT01974479) is a phase II pilot study, which is still recruiting refractory B-lineage ALL patients in all ages. NK cells are expanded by co-culture with K562 cells as the previous trial, together with IL-2 before transduction with the same construct. The patients will also receive IL-2 after NK cell administration to support NK cell viability and expansion. Although CAR T cell studies have been extremely promising, CARs designed for T cell therapies are still suboptimal for NK cells. Thus, it is essential to further optimize the construct design, especially the intracellular stimulatory adapter molecules, in order to trigger most efficient NK cell responses.

Immunomodulatory Drugs (IMiDs)

Immunomodulatory drugs (IMiDs) such as thalidomide, lenalidomide, and pomalidomide, can stimulate both NK cells and T cells, potentially resulting in better targeting cancer cells (212). Lenalidomide upregulates TRAIL molecules on NK cells and enhances anti-tumor activity (14, 15). So far, several different malignancies, both solid and hematological, have been treated with IMiDs. A large part of the nearly 100 clinical trials with IMiDs that has been reported with results to clinicaltrials.gov is treatment of myeloma, lymphoma, and leukemia. IMiDs can be used as combination treatment, such as lenalidomide in combination with IPH-2102, anti-inhibitory KIR antibody therapy (205). Lenalidomide expands and activates the NK cells, while anti-inhibitory KIR antibody (IPH2101) promotes NK cell recognition and lysis of tumor cells. This combination could give a better therapeutic outcome.

Combination Treatments

It is possible that NK cell products cannot fully eliminate tumor cells due to several immunosuppressive effects of tumor microenvironment as well as reduced *in vivo* expansion and cytotoxicity. These obstacles could be overcome by combination treatments using NK cell therapy products together with other drugs either directly targeting tumor cells or modulating cytotoxic activity of NK cells. As mentioned earlier, use of mAbs and IMiDs together with appropriate NK cell products could enhance tumor targeting and elimination. Another way to enhance NK cell-mediated killing is to combine drug therapy with NK cell stimulating cytokines such as IL-2, IL-12, IL-15, and IL-21 (213).

Furthermore, chemotherapy in combination with NK cell infusions is an alternative way to overcome tumor-induced dysfunctions. NK cells from haploidentical donor require combination treatments with the intense chemotherapy drugs high-dose fludarabin and cyclophosphamide (Hi-Cy/Flu) plus daily infusion of IL-2 to be able to expand *in vivo* (38). Total body irradiation could help to create immunological space for expanding NK cells in addition to chemotherapy after short-term *ex vivo* activation of NK cells (214).

CONCLUSION

In this review, we have summarized current NK cell-based therapy strategies as well as some of the challenges that need to be addressed. Even though NK cell-based therapies represent one of the most promising strategies to combat cancer, to our knowledge, no clinical trial has clearly demonstrated a significant benefit in patients with malignancies. This is in part due to the lack of prospective large-scale clinical trials and partly due to a lack of consensus in which NK cell product preparation would show the best effect. Further comparative clinical studies are definitely warranted; however, the design of such clinical trials is challenging due to the advanced therapy regulations in major countries such as European Union member states and the United States of America. Although cell therapy clinical trials are reaching a log-linear expansion, the number of NK cell-based therapies is not aligned with this increase. Nevertheless, there is a lot of promise in early clinical and pre-clinical data that cannot be omitted. In the near future, different NK cell-based products will reach multicenter clinical trial stage and we will start to see efficacy data.

Separately, NK cell-based therapies are in theory complementary to many different upfront, maintenance, and late-line therapies. Further studies clarifying the complementary efficacies and synergies have to be initiated to conclusively state if there is any place for these intriguing cells in search for an effective treatment of cancer.

AUTHOR CONTRIBUTIONS

All the authors performed the review of the literature, wrote, and edited the manuscript.

ACKNOWLEDGMENTS

This research was supported by generous grants from the Swedish Cancer Fund (Cancerfonden) (Reference number 141022), Swedish Cancer Society in Stockholm (Cancerföreningen/Radiumhemmets forskningsfonder), VINNOVA (Project number 2010-00501), and State of Florida Department of Health. The authors acknowledge the linguistic comments from Kim Kusser at Nova Southeastern University and Elizabeth Henry Alici.

REFERENCES

1. Hellström I, Hellström KE, Pierce GE, Yang JP. Cellular and humoral immunity to different types of human neoplasms. *Nature* (1968) **220**:1352–4. doi:10.1038/2201352a0
2. Herberman RB, Nunn ME, Holden HT, Lavrin DH. Natural cytotoxic reactivity of mouse lymphoid cells against syngeneic and allogeneic tumors. II. Characterization of effector cells. *Int J Cancer* (1975) **16**:230–9.
3. Herberman RB, Nunn ME, Lavrin DH. Natural cytotoxic reactivity of mouse lymphoid cells against syngeneic acid allogeneic tumors. I. Distribution of reactivity and specificity. *Int J Cancer* (1975) **16**:216–29.
4. Kiessling R, Klein E, Pross H, Wigzell H. "Natural" killer cells in the mouse. II. Cytotoxic cells with specificity for mouse Moloney leukemia cells. Characteristics of the killer cell. *Eur J Immunol* (1975) **5**:117–21. doi:10.1002/eji.1830050208
5. Kiessling R, Klein E, Wigzell H. "Natural" killer cells in the mouse. I. Cytotoxic cells with specificity for mouse Moloney leukemia cells. Specificity and distribution according to genotype. *Eur J Immunol* (1975) **5**:112–7. doi:10.1002/eji.1830050208
6. Karre K, Ljunggren HG, Piontek G, Kiessling R. Selective rejection of H-2-deficient lymphoma variants suggests alternative immune defence strategy. *Nature* (1986) **319**:675–8. doi:10.1038/319675a0
7. Pyo CW, Guethlein LA, Vu Q, Wang R, Abi-Rached L, Norman PJ, et al. Different patterns of evolution in the centromeric and telomeric regions of group A and B haplotypes of the human killer cell Ig-like receptor locus. *PLoS One* (2010) **5**:e15115. doi:10.1371/journal.pone.0015115
8. Fauriat C, Ivarsson MA, Ljunggren HG, Malmberg KJ, Michaelsson J. Education of human natural killer cells by activating killer cell immunoglobulin-like receptors. *Blood* (2010) **115**:1166–74. doi:10.1182/blood-2009-09-245746
9. Koch J, Steinle A, Watzl C, Mandelboim O. Activating natural cytotoxicity receptors of natural killer cells in cancer and infection. *Trends Immunol* (2013) **34**:182–91. doi:10.1016/j.it.2013.01.003
10. Titus JA, Perez P, Kaubisch A, Garrido MA, Segal DM. Human K/natural killer cells targeted with hetero-cross-linked antibodies specifically lyse tumor cells in vitro and prevent tumor growth in vivo. *J Immunol* (1987) **139**:3153–8.
11. Garrido MA, Perez P, Titus JA, Valdayo MJ, Winkler DF, Barbieri SA, et al. Targeted cytotoxic cells in human peripheral blood lymphocytes. *J Immunol* (1990) **144**:2891–8.
12. Wallin RP, Screpanti V, Michaelsson J, Grandien A, Ljunggren HG. Regulation of perforin-independent NK cell-mediated cytotoxicity. *Eur J Immunol* (2003) **33**:2727–35. doi:10.1002/eji.200324070
13. De Wilt LH, Kroon J, Jansen G, De Jong S, Peters GJ, Kruyt FA. Bortezomib and TRAIL: a perfect match for apoptotic elimination of tumour cells? *Crit Rev Oncol Hematol* (2013) **85**:363–72. doi:10.1016/j.critrevonc.2012.08.001
14. Wu L, Adams M, Carter T, Chen R, Muller G, Stirling D, et al. Lenalidomide enhances natural killer cell and monocyte-mediated antibody-dependent cellular cytotoxicity of rituximab-treated CD20+ tumor cells. *Clin Cancer Res* (2008) **14**:4650–7. doi:10.1158/1078-0432.CCR-07-4405
15. Jungkunz-Stier I, Zekl M, Stuhmer T, Einsele H, Seggewiss-Bernhardt R. Modulation of natural killer cell effector functions through lenalidomide/dasatinib and their combined effects against multiple myeloma cells. *Leuk Lymphoma* (2014) **55**:168–76. doi:10.3109/10428194.2013.794270
16. Freud AG, Yu J, Caligiuri MA. Human natural killer cell development in secondary lymphoid tissues. *Semin Immunol* (2014) **26**:132–7. doi:10.1016/j.smim.2014.02.008
17. Suzuki H, Duncan GS, Takimoto H, Mak TW. Abnormal development of intestinal intraepithelial lymphocytes and peripheral natural killer cells in mice lacking the IL-2 receptor beta chain. *J Exp Med* (1997) **185**:499–505. doi:10.1084/jem.185.3.499
18. Carotta S, Pang SH, Nutt SL, Belz GT. Identification of the earliest NK-cell precursor in the mouse BM. *Blood* (2011) **117**:5449–52. doi:10.1182/blood-2010-11-318956
19. Freud AG, Yokohama A, Becknell B, Lee MT, Mao HC, Ferketich AK, et al. Evidence for discrete stages of human natural killer cell differentiation in vivo. *J Exp Med* (2006) **203**:1033–43. doi:10.1084/jem.20052507
20. Bruno A, Ferlazzo G, Albini A, Noonan DM. A think tank of TINK/TANKs: tumor-infiltrating/tumor-associated natural killer cells in tumor progression and angiogenesis. *J Natl Cancer Inst* (2014) **106**:dju200. doi:10.1093/jnci/dju200
21. Medvedev AE, Johnsen AC, Haux J, Steinkjer B, Egeberg K, Lynch DH, et al. Regulation of Fas and Fas-ligand expression in NK cells by cytokines and the involvement of Fas-ligand in NK/LAK cell-mediated cytotoxicity. *Cytokine* (1997) **9**:394–404. doi:10.1006/cyto.1996.0181
22. Poggi A, Massaro AM, Negrini S, Contini P, Zocchi MR. Tumor-induced apoptosis of human IL-2-activated NK cells: role of natural cytotoxicity receptors. *J Immunol* (2005) **174**:2653–60. doi:10.4049/jimmunol.174.5.2653
23. Costello RT, Sivori S, Marcenaro E, Lafage-Pochitaloff M, Mozziconacci MJ, Reviron D, et al. Defective expression and function of natural killer cell-triggering receptors in patients with acute myeloid leukemia. *Blood* (2002) **99**:3661–7. doi:10.1182/blood.V99.10.3661
24. Rouas-Freiss N, Moreau P, Ferrone S, Carosella ED. HLA-G proteins in cancer: do they provide tumor cells with an escape mechanism? *Cancer Res* (2005) **65**:10139–44. doi:10.1158/0008-5472.CAN-05-0097
25. Urosevic M, Dummer R. Human leukocyte antigen-G and cancer immunoediting. *Cancer Res* (2008) **68**:627–30. doi:10.1158/0008-5472.CAN-07-2704
26. Jinushi M, Takehara T, Tatsumi T, Hiramatsu N, Sakamori R, Yamaguchi S, et al. Impairment of natural killer cell and dendritic cell functions by the soluble form of MHC class I-related chain A in advanced human hepatocellular carcinomas. *J Hepatol* (2005) **43**:1013–20. doi:10.1016/j.jhep.2005.05.026
27. Konjevic G, Mirjacic Martinovic K, Vuletic A, Jovic V, Jurisic V, Babovic N, et al. Low expression of CD161 and NKG2D activating NK receptor is associated with impaired NK cell cytotoxicity in metastatic melanoma patients. *Clin Exp Metastasis* (2007) **24**:1–11. doi:10.1007/s10585-006-9043-9
28. Veuillen C, Aurran-Schleinitz T, Castellano R, Rey J, Mallet F, Orlanducci F, et al. Primary B-CLL resistance to NK cell cytotoxicity can be overcome in vitro and in vivo by priming NK cells and monoclonal antibody therapy. *J Clin Immunol* (2012) **32**:632–46. doi:10.1007/s10875-011-9624-5
29. Fauriat C, Mallet F, Olive D, Costello RT. Impaired activating receptor expression pattern in natural killer cells from patients with multiple myeloma. *Leukemia* (2006) **20**:732–3. doi:10.1038/sj.leu.2404096
30. El-Sherbiny YM, Meade JL, Holmes TD, Mcgonagle D, Mackie SL, Morgan AW, et al. The requirement for DNAM-1, NKG2D, and NKp46 in the natural killer cell-mediated killing of myeloma cells. *Cancer Res* (2007) **67**:8444–9. doi:10.1158/0008-5472.CAN-06-4230
31. Gati A, Da Rocha S, Guerra N, Escudier B, Moretta A, Chouaib S, et al. Analysis of the natural killer mediated immune response in metastatic

renal cell carcinoma patients. *Int J Cancer* (2004) **109**:393–401. doi:10.1002/ijc.11730

32. Pierson BA, Miller JS. CD56+bright and CD56+dim natural killer cells in patients with chronic myelogenous leukemia progressively decrease in number, respond less to stimuli that recruit clonogenic natural killer cells, and exhibit decreased proliferation on a per cell basis. *Blood* (1996) **88**:2279–87.
33. Schleypen JS, Von Geldern M, Weiss EH, Kotzias N, Rohrmann K, Schendel DJ, et al. Renal cell carcinoma-infiltrating natural killer cells express differential repertoires of activating and inhibitory receptors and are inhibited by specific HLA class I allotypes. *Int J Cancer* (2003) **106**:905–12. doi:10.1002/ijc.11321
34. Liu G, Lu S, Wang X, Page ST, Higano CS, Plymate SR, et al. Perturbation of NK cell peripheral homeostasis accelerates prostate carcinoma metastasis. *J Clin Invest* (2013) **123**:4410–22. doi:10.1172/JCI69369
35. Doubrovina ES, Doubrovin MM, Vider E, Sisson RB, O'Reilly RJ, Dupont B, et al. Evasion from NK cell immunity by MHC class I chain-related molecules expressing colon adenocarcinoma. *J Immunol* (2003) **171**:6891–9. doi:10.4049/jimmunol.171.12.6891
36. Jinushi M, Hodi FS, Dranoff G. Therapy-induced antibodies to MHC class I chain-related protein A antagonize immune suppression and stimulate antitumor cytotoxicity. *Proc Natl Acad Sci U S A* (2006) **103**:9190–5. doi:10.1073/pnas.0603503103
37. Deng W, Gowen BG, Zhang L, Wang L, Lau S, Iannello A, et al. Antitumor immunity. A shed NKG2D ligand that promotes natural killer cell activation and tumor rejection. *Science* (2015) **348**:136–9. doi:10.1126/science.1258867
38. Miller JS, Soignier Y, Panoskaltsis-Mortari A, Mcnearney SA, Yun GH, Fautsch SK, et al. Successful adoptive transfer and in vivo expansion of human haploidentical NK cells in patients with cancer. *Blood* (2005) **105**:3051–7. doi:10.1182/blood-2004-07-2974
39. Escudier B, Farace F, Angevin E, Charpentier F, Nitenberg G, Triebel F, et al. Immunotherapy with interleukin-2 (IL2) and lymphokine-activated natural killer cells: improvement of clinical responses in metastatic renal cell carcinoma patients previously treated with IL2. *Eur J Cancer* (1994) **30A**:1078–83. doi:10.1016/0959-8049(94)90460-X
40. Lister J, Rybka WB, Donnenberg AD, Demagalhaes-Silverman M, Pincus SM, Bloom EJ, et al. Autologous peripheral blood stem cell transplantation and adoptive immunotherapy with activated natural killer cells in the immediate post-transplant period. *Clin Cancer Res* (1995) **1**:607–14.
41. Pierson BA, Europa AF, Hu WS, Miller JS. Production of human natural killer cells for adoptive immunotherapy using a computer-controlled stirred-tank bioreactor. *J Hematother* (1996) **5**:475–83. doi:10.1089/scd.1.1996.5.475
42. Alici E, Sutlu T, Bjorkstrand B, Gilljam M, Stellan B, Nahi H, et al. Autologous antitumor activity by NK cells expanded from myeloma patients using GMP-compliant components. *Blood* (2008) **111**:3155–62. doi:10.1182/blood-2007-09-110312
43. Barkholt L, Alici E, Conrad R, Sutlu T, Gilljam M, Stellan B, et al. Safety analysis of ex vivo-expanded NK and NK-like T cells administered to cancer patients: a phase I clinical study. *Immunotherapy* (2009) **1**:753–64. doi:10.2217/imt.09.47
44. Iliopoulou EG, Kountourakis P, Karamouzis MV, Doufexis D, Ardavanis A, Baxevanis CN, et al. A phase I trial of adoptive transfer of allogeneic natural killer cells in patients with advanced non-small cell lung cancer. *Cancer Immunol Immunother* (2010) **59**:1781–9. doi:10.1007/s00262-010-0904-3
45. Sutlu T, Stellan B, Gilljam M, Quezada HC, Nahi H, Gahrton G, et al. Clinical-grade, large-scale, feeder-free expansion of highly active human natural killer cells for adoptive immunotherapy using an automated bioreactor. *Cytotherapy* (2010) **12**:1044–55. doi:10.3109/14653249.2010.504770
46. Spanholtz J, Preijers F, Tordoir M, Trilsbeek C, Paardekooper J, De Witte T, et al. Clinical-grade generation of active NK cells from cord blood hematopoietic progenitor cells for immunotherapy using a closed-system culture process. *PLoS One* (2011) **6**:e20740. doi:10.1371/journal.pone.0020740
47. Sakamoto N, Ishikawa T, Kokura S, Okayama T, Oka K, Ideno M, et al. Phase I clinical trial of autologous NK cell therapy using novel expansion method in patients with advanced digestive cancer. *J Transl Med* (2015) **13**:277. doi:10.1186/s12967-015-0632-8
48. Carlens S, Gilljam M, Chambers BJ, Aschan J, Guven H, Ljunggren HG, et al. A new method for in vitro expansion of cytotoxic human CD3-CD56+ natural killer cells. *Hum Immunol* (2001) **62**:1092–8. doi:10.1016/S0198-8859(01)00313-5
49. Klingemann HG, Martinson J. Ex vivo expansion of natural killer cells for clinical applications. *Cytotherapy* (2004) **6**:15–22. doi:10.1080/14653240310004548
50. Cany J, Van Der Waart AB, Spanholtz J, Tordoir M, Jansen JH, Van Der Voort R, et al. Combined IL-15 and IL-12 drives the generation of CD34-derived natural killer cells with superior maturation and alloreactivity potential following adoptive transfer. *Oncoimmunology* (2015) **4**:e1017701. doi:10.1080/2162402X.2015.1017701
51. Tam YK, Martinson JA, Doligosa K, Klingemann HG. Ex vivo expansion of the highly cytotoxic human natural killer-92 cell-line under current good manufacturing practice conditions for clinical adoptive cellular immunotherapy. *Cytotherapy* (2003) **5**:259–72. doi:10.1080/14653240310001523
52. Arai S, Meagher R, Swearingen M, Myint H, Rich E, Martinson J, et al. Infusion of the allogeneic cell line NK-92 in patients with advanced renal cell cancer or melanoma: a phase I trial. *Cytotherapy* (2008) **10**:625–32. doi:10.1080/14653240802301872
53. Guven H, Gilljam M, Chambers BJ, Ljunggren HG, Christensson B, Kimby E, et al. Expansion of natural killer (NK) and natural killer-like T (NKT)-cell populations derived from patients with B-chronic lymphocytic leukemia (B-CLL): a potential source for cellular immunotherapy. *Leukemia* (2003) **17**:1973–80. doi:10.1038/sj.leu.2403083
54. Fujisaki H, Kakuda H, Shimasaki N, Imai C, Ma J, Lockey T, et al. Expansion of highly cytotoxic human natural killer cells for cancer cell therapy. *Cancer Res* (2009) **69**:4010–7. doi:10.1158/0008-5472.CAN-08-3712
55. Ishikawa E, Tsuboi K, Saijo K, Harada H, Takano S, Nose T, et al. Autologous natural killer cell therapy for human recurrent malignant glioma. *Anticancer Res* (2004) **24**:1861–71.
56. Miller JS, Klingsporn S, Lund J, Perry EH, Verfaillie C, Mcglave P. Large scale ex vivo expansion and activation of human natural killer cells for autologous therapy. *Bone Marrow Transplant* (1994) **14**:555–62.
57. Torelli GF, Guarini A, Palmieri G, Breccia M, Vitale A, Santoni A, et al. Expansion of cytotoxic effectors with lytic activity against autologous blasts from acute myeloid leukaemia patients in complete haematological remission. *Br J Haematol* (2002) **116**:299–307. doi:10.1046/j.1365-2141.2002.03277.x
58. Torelli GF, Guarini A, Maggio R, Alfieri C, Vitale A, Foa R. Expansion of natural killer cells with lytic activity against autologous blasts from adult and pediatric acute lymphoid leukemia patients in complete hematologic remission. *Haematologica* (2005) **90**:785–92.
59. Hercend T, Farace F, Baume D, Charpentier F, Droz JP, Triebel F, et al. Immunotherapy with lymphokine-activated natural killer cells and recombinant interleukin-2: a feasibility trial in metastatic renal cell carcinoma. *J Biol Response Mod* (1990) **9**:546–55.
60. Luhm J, Brand JM, Koritke P, Hoppner M, Kirchner H, Frohn C. Large-scale generation of natural killer lymphocytes for clinical application. *J Hematother Stem Cell Res* (2002) **11**:651–7. doi:10.1089/15258160260194794
61. Berg M, Lundqvist A, Mccoy P Jr., Samsel L, Fan Y, Tawab A, et al. Clinical-grade ex vivo-expanded human natural killer cells up-regulate activating receptors and death receptor ligands and have enhanced cytolytic activity against tumor cells. *Cytotherapy* (2009) **11**:341–55. doi:10.1080/14653240902807034
62. Sutlu T, Alici E. Ex vivo expansion of natural killer cells: a question of function. *Cytotherapy* (2011) **13**:767–8. doi:10.3109/14653249.2011.563295
63. Skeate R, Singh C, Cooley S, Geller M, Northouse J, Welbig J, et al. Hemolytic anemia due to passenger lymphocyte syndrome in solid malignancy patients treated with allogeneic natural killer cell products. *Transfusion* (2013) **53**:419–23. doi:10.1111/j.1537-2995.2012.03942.x
64. Nguyen S, Kuentz M, Vernant JP, Dhedin N, Bories D, Debre P, et al. Involvement of mature donor T cells in the NK cell reconstitution after haploidentical hematopoietic stem-cell transplantation. *Leukemia* (2008) **22**:344–52. doi:10.1038/sj.leu.2405041
65. Miller JS, Oelkers S, Verfaillie C, Mcglave P. Role of monocytes in the expansion of human activated natural killer cells. *Blood* (1992) **80**:2221–9.
66. Siegler U, Meyer-Monard S, Jorger S, Stern M, Tichelli A, Gratwohl A, et al. Good manufacturing practice-compliant cell sorting and large-scale expansion of single KIR-positive alloreactive human natural killer cells for multiple infusions to leukemia patients. *Cytotherapy* (2010) **12**:750–63. doi:10.3109/14653241003786155
67. Shah NN, Baird K, Delbrook CP, Fleisher TA, Kohler ME, Rampertaap S, et al. Acute GVHD in patients receiving IL-15/4-1BBL activated NK cells

following T-cell-depleted stem cell transplantation. *Blood* (2015) **125**:784–92. doi:10.1182/blood-2014-07-592881

68. Somanchi SS, Senyukov VV, Denman CJ, Lee DA. Expansion, purification, and functional assessment of human peripheral blood NK cells. *J Vis Exp* (2011):doi:10.3791/2540
69. Denman CJ, Senyukov VV, Somanchi SS, Phatarpekar PV, Kopp LM, Johnson JL, et al. Membrane-bound IL-21 promotes sustained ex vivo proliferation of human natural killer cells. *PLoS One* (2012) **7**:e30264. doi:10.1371/journal.pone.0030264
70. Kalberer CP, Siegler U, Wodnar-Filipowicz A. Human NK cell development in NOD/SCID mice receiving grafts of cord blood CD34+ cells. *Blood* (2003) **102**:127–35. doi:10.1182/blood-2002-07-2024
71. Tanaka H, Kai S, Yamaguchi M, Misawa M, Fujimori Y, Yamamoto M, et al. Analysis of natural killer (NK) cell activity and adhesion molecules on NK cells from umbilical cord blood. *Eur J Haematol* (2003) **71**:29–38. doi:10.1034/j.1600-0609.2003.00081.x
72. Xing D, Ramsay AG, Gribben JG, Decker WK, Burks JK, Munsell M, et al. Cord blood natural killer cells exhibit impaired lytic immunological synapse formation that is reversed with IL-2 ex vivo expansion. *J Immunother* (2010) **33**:684–96. doi:10.1097/CJI.0b013e3181e475e9
73. Lehmann D, Spanholtz J, Sturtzel C, Tordoir M, Schlechta B, Groenewegen D, et al. IL-12 directs further maturation of ex vivo differentiated NK cells with improved therapeutic potential. *PLoS One* (2014) **9**:e87131. doi:10.1371/journal.pone.0087131
74. Cany J, Van Der Waart AB, Tordoir M, Franssen GM, Hangalapura BN, De Vries J, et al. Natural killer cells generated from cord blood hematopoietic progenitor cells efficiently target bone marrow-residing human leukemia cells in NOD/SCID/IL2Rg(null) mice. *PLoS One* (2013) **8**:e64384. doi:10.1371/journal.pone.0064384
75. Luevano M, Domogala A, Blundell M, Jackson N, Pedroza-Pacheco I, Derniame S, et al. Frozen cord blood hematopoietic stem cells differentiate into higher numbers of functional natural killer cells in vitro than mobilized hematopoietic stem cells or freshly isolated cord blood hematopoietic stem cells. *PLoS One* (2014) **9**:e87086. doi:10.1371/journal.pone.0087086
76. Woll PS, Martin CH, Miller JS, Kaufman DS. Human embryonic stem cell-derived NK cells acquire functional receptors and cytolytic activity. *J Immunol* (2005) **175**:5095–103. doi:10.4049/jimmunol.175.8.5095
77. Woll PS, Grzywacz B, Tian X, Marcus RK, Knorr DA, Verneris MR, et al. Human embryonic stem cells differentiate into a homogeneous population of natural killer cells with potent in vivo antitumor activity. *Blood* (2009) **113**:6094–101. doi:10.1182/blood-2008-06-165225
78. Knorr DA, Ni Z, Hermanson D, Hexum MK, Bendzick L, Cooper LJ, et al. Clinical-scale derivation of natural killer cells from human pluripotent stem cells for cancer therapy. *Stem Cells Transl Med* (2013) **2**:274–83. doi:10.5966/sctm.2012-0084
79. Kruse V, Hamann C, Monecke S, Cyganek L, Elsner L, Hubscher D, et al. Human induced pluripotent stem cells are targets for allogeneic and autologous natural killer (NK) cells and killing is partly mediated by the activating NK receptor DNAM-1. *PLoS One* (2015) **10**:e0125544. doi:10.1371/journal.pone.0125544
80. Gong JH, Maki G, Klingemann HG. Characterization of a human cell line (NK-92) with phenotypical and functional characteristics of activated natural killer cells. *Leukemia* (1994) **8**:652–8.
81. Klingemann HG, Miyagawa B. Purging of malignant cells from blood after short ex vivo incubation with NK-92 cells. *Blood* (1996) **87**:4913–4.
82. Yan Y, Steinherz P, Klingemann HG, Dennig D, Childs BH, Mcguirk J, et al. Antileukemia activity of a natural killer cell line against human leukemias. *Clin Cancer Res* (1998) **4**:2859–68.
83. Tam YK, Maki G, Miyagawa B, Hennemann B, Tonn T, Klingemann HG. Characterization of genetically altered, interleukin 2-independent natural killer cell lines suitable for adoptive cellular immunotherapy. *Hum Gene Ther* (1999) **10**:1359–73. doi:10.1089/10430349950018030
84. Tonn T, Schwabe D, Klingemann HG, Becker S, Esser R, Koehl U, et al. Treatment of patients with advanced cancer with the natural killer cell line NK-92. *Cytotherapy* (2013) **15**:1563–70. doi:10.1016/j.jcyt.2013.06.017
85. Maki G, Klingemann HG, Martinson JA, Tam YK. Factors regulating the cytotoxic activity of the human natural killer cell line, NK-92. *J Hematother Stem Cell Res* (2001) **10**:369–83. doi:10.1089/152581601750288975
86. Suck G, Branch DR, Smyth MJ, Miller RG, Vergidis J, Fahim S, et al. KHYG-1, a model for the study of enhanced natural killer cell cytotoxicity. *Exp Hematol* (2005) **33**:1160–71. doi:10.1016/j.exphem.2005.06.024
87. Nagashima S, Mailliard R, Kashii Y, Reichert TE, Herberman RB, Robbins P, et al. Stable transduction of the interleukin-2 gene into human natural killer cell lines and their phenotypic and functional characterization in vitro and in vivo. *Blood* (1998) **91**:3850–61.
88. Konstantinidis KV, Alici E, Aints A, Christensson B, Ljunggren HG, Dilber MS. Targeting IL-2 to the endoplasmic reticulum confines autocrine growth stimulation to NK-92 cells. *Exp Hematol* (2005) **33**:159–64. doi:10.1016/j.exphem.2004.11.003
89. Atkins MB, Kunkel L, Sznol M, Rosenberg SA. High-dose recombinant interleukin-2 therapy in patients with metastatic melanoma: long-term survival update. *Cancer J Sci Am* (2000) **1**(6 Suppl):S11–4.
90. Rosenberg SA, Lotze MT, Muul LM, Leitman S, Chang AE, Ettinghausen SE, et al. Observations on the systemic administration of autologous lymphokine-activated killer cells and recombinant interleukin-2 to patients with metastatic cancer. *N Engl J Med* (1985) **313**:1485–92. doi:10.1056/NEJM198512053132327
91. Rosenberg SA, Lotze MT, Muul LM, Chang AE, Avis FP, Leitman S, et al. *N Engl J Med* (1987) **316**:889–97. doi:10.1056/NEJM198704093161501
92. Phillips JH, Gemlo BT, Myers WW, Rayner AA, Lanier LL. In vivo and in vitro activation of natural killer cells in advanced cancer patients undergoing combined recombinant interleukin-2 and LAK cell therapy. *J Clin Oncol* (1987) **5**:1933–41.
93. Sarkar S, Germeraad WT, Rouschop KM, Steeghs EM, Van Gelder M, Bos GM, et al. Hypoxia induced impairment of NK cell cytotoxicity against multiple myeloma can be overcome by IL-2 activation of the NK cells. *PLoS One* (2013) **8**:e64835. doi:10.1371/journal.pone.0064835
94. Basse PH, Goldfarb RH, Herberman RB, Hokland ME. Accumulation of adoptively transferred A-NK cells in murine metastases: kinetics and role of interleukin-2. *In vivo* (1994) **8**:17–24.
95. Donohue JH, Rosenberg SA. The fate of interleukin-2 after in vivo administration. *J Immunol* (1983) **130**:2203–8.
96. Sutlu T, Nystrom S, Gilljam M, Stellan B, Applequist SE, Alici E. Inhibition of intracellular antiviral defense mechanisms augments lentiviral transduction of human natural killer cells: implications for gene therapy. *Hum Gene Ther* (2012) **23**:1090–100. doi:10.1089/hum.2012.080
97. Bachanova V, Cooley S, Defor TE, Verneris MR, Zhang B, Mckenna DH, et al. Clearance of acute myeloid leukemia by haploidentical natural killer cells is improved using IL-2 diphtheria toxin fusion protein. *Blood* (2014) **123**:3855–63. doi:10.1182/blood-2013-10-532531
98. Cooper MA, Bush JE, Fehniger TA, Vandeusen JB, Waite RE, Liu Y, et al. In vivo evidence for a dependence on interleukin 15 for survival of natural killer cells. *Blood* (2002) **100**:3633–8. doi:10.1182/blood-2001-12-0293
99. Ozaki K, Kikly K, Michalovich D, Young PR, Leonard WJ. Cloning of a type I cytokine receptor most related to the IL-2 receptor beta chain. *Proc Natl Acad Sci U S A* (2000) **97**:11439–44. doi:10.1073/pnas.200360997
100. De Rham C, Ferrari-Lacraz S, Jendly S, Schneiter G, Dayer JM, Villard J. The proinflammatory cytokines IL-2, IL-15 and IL-21 modulate the repertoire of mature human natural killer cell receptors. *Arthritis Res Ther* (2007) **9**:R125. doi:10.1186/ar2336
101. Gowda A, Ramanunni A, Cheney C, Rozewski D, Kindsvogel W, Lehman A, et al. Differential effects of IL-2 and IL-21 on expansion of the CD4+ CD25+ Foxp3+ T regulatory cells with redundant roles in natural killer cell mediated antibody dependent cellular cytotoxicity in chronic lymphocytic leukemia. *MAbs* (2010) **2**:35–41. doi:10.4161/mabs.2.1.10561
102. Mckenna DH Jr., Sumstad D, Bostrom N, Kadidlo DM, Fautsch S, Mcnearney S, et al. Good manufacturing practices production of natural killer cells for immunotherapy: a six-year single-institution experience. *Transfusion* (2007) **47**:520–8. doi:10.1111/j.1537-2995.2006.01145.x
103. Parkhurst MR, Riley JP, Dudley ME, Rosenberg SA. Adoptive transfer of autologous natural killer cells leads to high levels of circulating natural killer cells but does not mediate tumor regression. *Clin Cancer Res* (2011) **17**:6287–97. doi:10.1158/1078-0432.CCR-11-1347
104. Burns LJ, Weisdorf DJ, Defor TE, Vesole DH, Repka TL, Blazar BR, et al. IL-2-based immunotherapy after autologous transplantation for lymphoma and

breast cancer induces immune activation and cytokine release: a phase I/II trial. *Bone Marrow Transplant* (2003) **32**:177–86. doi:10.1038/sj.bmt.1704086
105. Krause SW, Gastpar R, Andreesen R, Gross C, Ullrich H, Thonigs G, et al. Treatment of colon and lung cancer patients with ex vivo heat shock protein 70-peptide-activated, autologous natural killer cells: a clinical phase I trial. *Clin Cancer Res* (2004) **10**:3699–707. doi:10.1158/1078-0432.CCR-03-0683
106. Motohashi S, Ishikawa A, Ishikawa E, Otsuji M, Iizasa T, Hanaoka H, et al. A phase I study of in vitro expanded natural killer T cells in patients with advanced and recurrent non-small cell lung cancer. *Clin Cancer Res* (2006) **12**:6079–86. doi:10.1158/1078-0432.CCR-06-0114
107. Milani V, Stangl S, Issels R, Gehrmann M, Wagner B, Hube K, et al. Anti-tumor activity of patient-derived NK cells after cell-based immunotherapy – a case report. *J Transl Med* (2009) **7**:50. doi:10.1186/1479-5876-7-50
108. Demagalhaes-Silverman M, Donnenberg A, Lembersky B, Elder E, Lister J, Rybka W, et al. Posttransplant adoptive immunotherapy with activated natural killer cells in patients with metastatic breast cancer. *J Immunother* (2000) **23**:154–60. doi:10.1097/00002371-200001000-00018
109. Koehl U, Sorensen J, Esser R, Zimmermann S, Gruttner HP, Tonn T, et al. IL-2 activated NK cell immunotherapy of three children after haploidentical stem cell transplantation. *Blood Cells Mol Dis* (2004) **33**:261–6. doi:10.1016/j.bcmd.2004.08.013
110. Passweg JR, Tichelli A, Meyer-Monard S, Heim D, Stern M, Kuhne T, et al. Purified donor NK-lymphocyte infusion to consolidate engraftment after haploidentical stem cell transplantation. *Leukemia* (2004) **18**:1835–8. doi:10.1038/sj.leu.2403524
111. Rizzieri DA, Storms R, Chen DF, Long G, Yang Y, Nikcevich DA, et al. Natural killer cell-enriched donor lymphocyte infusions from A 3-6/6 HLA matched family member following nonmyeloablative allogeneic stem cell transplantation. *Biol Blood Marrow Transplant* (2010) **16**:1107–14. doi:10.1016/j.bbmt.2010.02.018
112. Rubnitz JE, Inaba H, Ribeiro RC, Pounds S, Rooney B, Bell T, et al. NKAML: a pilot study to determine the safety and feasibility of haploidentical natural killer cell transplantation in childhood acute myeloid leukemia. *J Clin Oncol* (2010) **28**:955–9. doi:10.1200/JCO.2009.24.4590
113. Curti A, Ruggeri L, D'addio A, Bontadini A, Dan E, Motta MR, et al. Successful transfer of alloreactive haploidentical KIR ligand-mismatched natural killer cells after infusion in elderly high risk acute myeloid leukemia patients. *Blood* (2011) **118**:3273–9. doi:10.1182/blood-2011-01-329508
114. Yoon SR, Lee YS, Yang SH, Ahn KH, Lee JH, Lee JH, et al. Generation of donor natural killer cells from CD34(+) progenitor cells and subsequent infusion after HLA-mismatched allogeneic hematopoietic cell transplantation: a feasibility study. *Bone Marrow Transplant* (2010) **45**:1038–46. doi:10.1038/bmt.2009.304
115. Geller MA, Cooley S, Judson PL, Ghebre R, Carson LF, Argenta PA, et al. A phase II study of allogeneic natural killer cell therapy to treat patients with recurrent ovarian and breast cancer. *Cytotherapy* (2011) **13**:98–107. doi:10.3109/14653249.2010.515582
116. Nguyen S, Beziat V, Norol F, Uzunov M, Trebeden-Negre H, Azar N, et al. Infusion of allogeneic natural killer cells in a patient with acute myeloid leukemia in relapse after haploidentical hematopoietic stem cell transplantation. *Transfusion* (2011) **51**:1769–78. doi:10.1111/j.1537-2995.2010.03058.x
117. Brehm C, Huenecke S, Quaiser A, Esser R, Bremm M, Kloess S, et al. IL-2 stimulated but not unstimulated NK cells induce selective disappearance of peripheral blood cells: concomitant results to a phase I/II study. *PLoS One* (2011) **6**:e27351. doi:10.1371/journal.pone.0027351
118. Huenecke S, Zimmermann SY, Kloess S, Esser R, Brinkmann A, Tramsen L, et al. IL-2-driven regulation of NK cell receptors with regard to the distribution of CD16+ and CD16- subpopulations and in vivo influence after haploidentical NK cell infusion. *J Immunother* (2010) **33**:200–10. doi:10.1097/CJI.0b013e3181bb46f7
119. Szmania S, Lapteva N, Garg T, Greenway A, Lingo J, Nair B, et al. Ex vivo-expanded natural killer cells demonstrate robust proliferation in vivo in high-risk relapsed multiple myeloma patients. *J Immunother* (2015) **38**:24–36. doi:10.1097/CJI.0000000000000059
120. Ruggeri L, Capanni M, Casucci M, Volpi I, Tosti A, Perruccio K, et al. Role of natural killer cell alloreactivity in HLA-mismatched hematopoietic stem cell transplantation. *Blood* (1999) **94**:333–9.
121. Ruggeri L, Capanni M, Urbani E, Perruccio K, Shlomchik WD, Tosti A, et al. Effectiveness of donor natural killer cell alloreactivity in mismatched hematopoietic transplants. *Science* (2002) **295**:2097–100. doi:10.1126/science.1068440
122. Kumano M, Hara I, Furukawa J, Oniki S, Nagai H, Miyake H, et al. Interleukin-21 activates cytotoxic T lymphocytes and natural killer cells to generate antitumor response in mouse renal cell carcinoma. *J Urol* (2007) **178**:1504–9. doi:10.1016/j.juro.2007.05.115
123. Rocha FG, Calvo FB, Chaves KC, Peron JP, Marques RF, De Borba TR, et al. Endostatin- and interleukin-2-expressing retroviral bicistronic vector for gene therapy of metastatic renal cell carcinoma. *J Gene Med* (2011) **13**:148–57. doi:10.1002/jgm.1547
124. Watanabe M, Kono K, Kawaguchi Y, Mizukami Y, Mimura K, Maruyama T, et al. NK cell dysfunction with down-regulated CD16 and up-regulated CD56 molecules in patients with esophageal squamous cell carcinoma. *Dis Esophagus* (2010) **23**:675–81. doi:10.1111/j.1442-2050.2010.01073.x
125. Izawa S, Kono K, Mimura K, Kawaguchi Y, Watanabe M, Maruyama T, et al. H(2)O(2) production within tumor microenvironment inversely correlated with infiltration of CD56(dim) NK cells in gastric and esophageal cancer: possible mechanisms of NK cell dysfunction. *Cancer Immunol Immunother* (2011) **60**:1801–10. doi:10.1007/s00262-011-1082-7
126. Koo KC, Shim DH, Yang CM, Lee SB, Kim SM, Shin TY, et al. Reduction of the CD16(-)CD56bright NK cell subset precedes NK cell dysfunction in prostate cancer. *PLoS One* (2013) **8**:e78049. doi:10.1371/journal.pone.0078049
127. Sun C, Sun H, Zhang C, Tian Z. NK cell receptor imbalance and NK cell dysfunction in HBV infection and hepatocellular carcinoma. *Cell Mol Immunol* (2015) **12**(3):292–302. doi:10.1038/cmi.2014.91
128. Hasmim M, Messai Y, Ziani L, Thiery J, Bouhris JH, Noman MZ, et al. Critical role of tumor microenvironment in shaping NK cell functions: implication of hypoxic stress. *Front Immunol* (2015) **6**:482. doi:10.3389/fimmu.2015.00482
129. Hartigan-O'Connor DJ, Poon C, Sinclair E, Mccune JM. Human CD4+ regulatory T cells express lower levels of the IL-7 receptor alpha chain (CD127), allowing consistent identification and sorting of live cells. *J Immunol Methods* (2007) **319**:41–52. doi:10.1016/j.jim.2006.10.008
130. Berendt MJ, North RJ. T-cell-mediated suppression of anti-tumor immunity. An explanation for progressive growth of an immunogenic tumor. *J Exp Med* (1980) **151**:69–80.
131. Woo EY, Chu CS, Goletz TJ, Schlienger K, Yeh H, Coukos G, et al. Regulatory CD4(+)CD25(+) T cells in tumors from patients with early-stage non-small cell lung cancer and late-stage ovarian cancer. *Cancer Res* (2001) **61**:4766–72.
132. Ichihara F, Kono K, Takahashi A, Kawaida H, Sugai H, Fujii H. Increased populations of regulatory T cells in peripheral blood and tumor-infiltrating lymphocytes in patients with gastric and esophageal cancers. *Clin Cancer Res* (2003) **9**:4404–8.
133. Ormandy LA, Hillemann T, Wedemeyer H, Manns MP, Greten TF, Korangy F. Increased populations of regulatory T cells in peripheral blood of patients with hepatocellular carcinoma. *Cancer Res* (2005) **65**:2457–64. doi:10.1158/0008-5472.CAN-04-3232
134. Fu J, Xu D, Liu Z, Shi M, Zhao P, Fu B, et al. Increased regulatory T cells correlate with CD8 T-cell impairment and poor survival in hepatocellular carcinoma patients. *Gastroenterology* (2007) **132**:2328–39. doi:10.1053/j.gastro.2007.03.102
135. Sinicrope FA, Rego RL, Ansell SM, Knutson KL, Foster NR, Sargent DJ. Intraepithelial effector (CD3+)/regulatory (FoxP3+) T-cell ratio predicts a clinical outcome of human colon carcinoma. *Gastroenterology* (2009) **137**:1270–9. doi:10.1053/j.gastro.2009.06.053
136. Ghiringhelli F, Menard C, Terme M, Flament C, Taieb J, Chaput N, et al. CD4+CD25+ regulatory T cells inhibit natural killer cell functions in a transforming growth factor-beta-dependent manner. *J Exp Med* (2005) **202**:1075–85. doi:10.1084/jem.20051511
137. Cai L, Zhang Z, Zhou L, Wang H, Fu J, Zhang S, et al. Functional impairment in circulating and intrahepatic NK cells and relative mechanism in hepatocellular carcinoma patients. *Clin Immunol* (2008) **129**:428–37. doi:10.1016/j.clim.2008.08.012
138. Pandiyan P, Zheng L, Ishihara S, Reed J, Lenardo MJ. CD4+CD25+Foxp3+ regulatory T cells induce cytokine deprivation-mediated apoptosis of effector CD4+ T cells. *Nat Immunol* (2007) **8**:1353–62. doi:10.1038/ni1536
139. Gasteiger G, Hemmers S, Firth MA, Le Floc'h A, Huse M, Sun JC, et al. IL-2-dependent tuning of NK cell sensitivity for target cells is controlled by regulatory T cells. *J Exp Med* (2013) **210**:1167–78. doi:10.1084/jem.20122462

140. Gabrilovich DI, Bronte V, Chen SH, Colombo MP, Ochoa A, Ostrand-Rosenberg S, et al. The terminology issue for myeloid-derived suppressor cells. *Cancer Res* (2007) **67**:425. doi:10.1158/0008-5472.CAN-06-3037
141. Arihara F, Mizukoshi E, Kitahara M, Takata Y, Arai K, Yamashita T, et al. Increase in CD14+HLA-DR -/low myeloid-derived suppressor cells in hepatocellular carcinoma patients and its impact on prognosis. *Cancer Immunol Immunother* (2013) **62**:1421–30. doi:10.1007/s00262-013-1447-1
142. Gorgun GT, Whitehill G, Anderson JL, Hideshima T, Maguire C, Laubach J, et al. Tumor-promoting immune-suppressive myeloid-derived suppressor cells in the multiple myeloma microenvironment in humans. *Blood* (2013) **121**:2975–87. doi:10.1182/blood-2012-08-448548
143. Youn JI, Nagaraj S, Collazo M, Gabrilovich DI. Subsets of myeloid-derived suppressor cells in tumor-bearing mice. *J Immunol* (2008) **181**:5791–802. doi:10.4049/jimmunol.181.8.5791
144. Loercher AE, Nash MA, Kavanagh JJ, Platsoucas CD, Freedman RS. Identification of an IL-10-producing HLA-DR-negative monocyte subset in the malignant ascites of patients with ovarian carcinoma that inhibits cytokine protein expression and proliferation of autologous T cells. *J Immunol* (1999) **163**:6251–60.
145. Serafini P, Meckel K, Kelso M, Noonan K, Califano J, Koch W, et al. Phosphodiesterase-5 inhibition augments endogenous antitumor immunity by reducing myeloid-derived suppressor cell function. *J Exp Med* (2006) **203**:2691–702. doi:10.1084/jem.20061104
146. Hoechst B, Voigtlaender T, Ormandy L, Gamrekelashvili J, Zhao F, Wedemeyer H, et al. Myeloid derived suppressor cells inhibit natural killer cells in patients with hepatocellular carcinoma via the NKp30 receptor. *Hepatology* (2009) **50**:799–807. doi:10.1002/hep.23054
147. Poschke I, Mougiakakos D, Hansson J, Masucci GV, Kiessling R. Immature immunosuppressive CD14+HLA-DR-/low cells in melanoma patients are Stat3hi and overexpress CD80, CD83, and DC-sign. *Cancer Res* (2010) **70**:4335–45. doi:10.1158/0008-5472.CAN-09-3767
148. Lin Y, Gustafson MP, Bulur PA, Gastineau DA, Witzig TE, Dietz AB. Immunosuppressive CD14+HLA-DR(low)/- monocytes in B-cell non-Hodgkin lymphoma. *Blood* (2011) **117**:872–81. doi:10.1182/blood-2010-05-283820
149. Bronte V, Serafini P, Mazzoni A, Segal DM, Zanovello P. L-arginine metabolism in myeloid cells controls T-lymphocyte functions. *Trends Immunol* (2003) **24**:302–6. doi:10.1016/S1471-4906(03)00132-7
150. Kusmartsev S, Gabrilovich DI. Inhibition of myeloid cell differentiation in cancer: the role of reactive oxygen species. *J Leukoc Biol* (2003) **74**:186–96. doi:10.1189/jlb.0103010
151. Filipazzi P, Valenti R, Huber V, Pilla L, Canese P, Iero M, et al. Identification of a new subset of myeloid suppressor cells in peripheral blood of melanoma patients with modulation by a granulocyte-macrophage colony-stimulation factor-based antitumor vaccine. *J Clin Oncol* (2007) **25**:2546–53. doi:10.1200/JCO.2006.08.5829
152. Lechner MG, Liebertz DJ, Epstein AL. Characterization of cytokine-induced myeloid-derived suppressor cells from normal human peripheral blood mononuclear cells. *J Immunol* (2010) **185**:2273–84. doi:10.4049/jimmunol.1000901
153. Kohanbash G, Mckaveney K, Sakaki M, Ueda R, Mintz AH, Amankulor N, et al. GM-CSF promotes the immunosuppressive activity of glioma-infiltrating myeloid cells through interleukin-4 receptor-alpha. *Cancer Res* (2013) **73**:6413–23. doi:10.1158/0008-5472.CAN-12-4124
154. Mao Y, Poschke I, Wennerberg E, Pico De Coana Y, Egyhazi Brage S, Schultz I, et al. Melanoma-educated CD14+ cells acquire a myeloid-derived suppressor cell phenotype through COX-2-dependent mechanisms. *Cancer Res* (2013) **73**:3877–87. doi:10.1158/0008-5472.CAN-12-4115
155. Mao Y, Sarhan D, Steven A, Seliger B, Kiessling R, Lundqvist A. Inhibition of tumor-derived prostaglandin-e2 blocks the induction of myeloid-derived suppressor cells and recovers natural killer cell activity. *Clin Cancer Res* (2014) **20**:4096–106. doi:10.1158/1078-0432.CCR-14-0635
156. Gleason MK, Ross JA, Warlick ED, Lund TC, Verneris MR, Wiernik A, et al. CD16xCD33 bispecific killer cell engager (BiKE) activates NK cells against primary MDS and MDSC CD33+ targets. *Blood* (2014) **123**:3016–26. doi:10.1182/blood-2013-10-533398
157. Pollard JW. Trophic macrophages in development and disease. *Nat Rev Immunol* (2009) **9**:259–70. doi:10.1038/nri2528
158. Qian BZ, Pollard JW. Macrophage diversity enhances tumor progression and metastasis. *Cell* (2010) **141**:39–51. doi:10.1016/j.cell.2010.03.014
159. Capece D, Fischietti M, Verzella D, Gaggiano A, Cicciarelli G, Tessitore A, et al. The inflammatory microenvironment in hepatocellular carcinoma: a pivotal role for tumor-associated macrophages. *Biomed Res Int* (2013) **2013**:187204. doi:10.1155/2013/187204
160. Wyler L, Napoli CU, Ingold B, Sulser T, Heikenwalder M, Schraml P, et al. Brain metastasis in renal cancer patients: metastatic pattern, tumour-associated macrophages and chemokine/chemoreceptor expression. *Br J Cancer* (2014) **110**:686–94. doi:10.1038/bjc.2013.755
161. Steidl C, Lee T, Shah SP, Farinha P, Han G, Nayar T, et al. Tumor-associated macrophages and survival in classic Hodgkin's lymphoma. *N Engl J Med* (2010) **362**:875–85. doi:10.1056/NEJMoa0905680
162. Campbell MJ, Tonlaar NY, Garwood ER, Huo D, Moore DH, Khramtsov AI, et al. Proliferating macrophages associated with high grade, hormone receptor negative breast cancer and poor clinical outcome. *Breast Cancer Res Treat* (2011) **128**:703–11. doi:10.1007/s10549-010-1154-y
163. Liu J, Zhang N, Li Q, Zhang W, Ke F, Leng Q, et al. Tumor-associated macrophages recruit CCR6+ regulatory T cells and promote the development of colorectal cancer via enhancing CCL20 production in mice. *PLoS One* (2011) **6**:e19495. doi:10.1371/journal.pone.0019495
164. Chen C, Qu QX, Shen Y, Mu CY, Zhu YB, Zhang XG, et al. Induced expression of B7-H4 on the surface of lung cancer cell by the tumor-associated macrophages: a potential mechanism of immune escape. *Cancer Lett* (2012) **317**:99–105. doi:10.1016/j.canlet.2011.11.017
165. Helm O, Held-Feindt J, Grage-Griebenow E, Reiling N, Ungefroren H, Vogel I, et al. Tumor-associated macrophages exhibit pro- and anti-inflammatory properties by which they impact on pancreatic tumorigenesis. *Int J Cancer* (2014) **135**(4):843–61. doi:10.1002/ijc.28736
166. Pietra G, Manzini C, Rivara S, Vitale M, Cantoni C, Petretto A, et al. Melanoma cells inhibit natural killer cell function by modulating the expression of activating receptors and cytolytic activity. *Cancer Res* (2012) **72**:1407–15. doi:10.1158/0008-5472.CAN-11-2544
167. Spaggiari GM, Capobianco A, Abdelrazik H, Becchetti F, Mingari MC, Moretta L. Mesenchymal stem cells inhibit natural killer-cell proliferation, cytotoxicity, and cytokine production: role of indoleamine 2,3-dioxygenase and prostaglandin E2. *Blood* (2008) **111**:1327–33. doi:10.1182/blood-2007-02-074997
168. Sivori S, Pende D, Bottino C, Marcenaro E, Pessino A, Biassoni R, et al. NKp46 is the major triggering receptor involved in the natural cytotoxicity of fresh or cultured human NK cells. Correlation between surface density of NKp46 and natural cytotoxicity against autologous, allogeneic or xenogeneic target cells. *Eur J Immunol* (1999) **29**:1656–66. doi:10.1002/(SICI)1521-4141(199905)29:05<1656::AID-IMMU1656>3.0.CO;2-1
169. Bournazou E, Bromberg J. Targeting the tumor microenvironment: JAK-STAT3 signaling. *JAKSTAT* (2013) **2**:e23828. doi:10.4161/jkst.23828
170. Fang H, Declerck YA. Targeting the tumor microenvironment: from understanding pathways to effective clinical trials. *Cancer Res* (2013) **73**:4965–77. doi:10.1158/0008-5472.CAN-13-0661
171. Samples J, Willis M, Klauber-Demore N. Targeting angiogenesis and the tumor microenvironment. *Surg Oncol Clin N Am* (2013) **22**:629–39. doi:10.1016/j.soc.2013.06.002
172. Giordano G, Febbraro A, Venditti M, Campidoglio S, Olivieri N, Raieta K, et al. Targeting angiogenesis and tumor microenvironment in metastatic colorectal cancer: role of aflibercept. *Gastroenterol Res Pract* (2014) **2014**:526178. doi:10.1155/2014/526178
173. Nelson D, Fisher S, Robinson B. The "Trojan Horse" approach to tumor immunotherapy: targeting the tumor microenvironment. *J Immunol Res* (2014) **2014**:789069. doi:10.1155/2014/789069
174. Omidi Y, Barar J. Targeting tumor microenvironment: crossing tumor interstitial fluid by multifunctional nanomedicines. *Bioimpacts* (2014) **4**:55–67. doi:10.5681/bi.2014.021
175. Lanier LL. Evolutionary struggles between NK cells and viruses. *Nat Rev Immunol* (2008) **8**:259–68. doi:10.1038/nri2276
176. Alici E, Sutlu T, Sirac Dilber M. Retroviral gene transfer into primary human natural killer cells. *Methods Mol Biol* (2009) **506**:127–37. doi:10.1007/978-1-59745-409-4_10
177. Brandstadter JD, Yang Y. Natural killer cell responses to viral infection. *J Innate Immun* (2011) **3**:274–9. doi:10.1159/000324176

178. Zhang J, Sun R, Wei H, Zhang J, Tian Z. Characterization of interleukin-15 gene-modified human natural killer cells: implications for adoptive cellular immunotherapy. *Haematologica* (2004) **89**:338–47.
179. Jiang W, Zhang J, Tian Z. Functional characterization of interleukin-15 gene transduction into the human natural killer cell line NKL. *Cytotherapy* (2008) **10**:265–74. doi:10.1080/14653240801965156
180. Clemenceau B, Vivien R, Pellat C, Foss M, Thibault G, Vie H. The human natural killer cytotoxic cell line NK-92, once armed with a murine CD16 receptor, represents a convenient cellular tool for the screening of mouse mAbs according to their ADCC potential. *MAbs* (2013) **5**:587–94. doi:10.4161/mabs.25077
181. Mentlik James A, Cohen AD, Campbell KS. Combination immune therapies to enhance anti-tumor responses by NK cells. *Front Immunol* (2013) **4**:481. doi:10.3389/fimmu.2013.00481
182. Lokhorst HM, Plesner T, Laubach JP, Nahi H, Gimsing P, Hansson M, et al. Targeting CD38 with daratumumab monotherapy in multiple myeloma. *N Engl J Med* (2015) **373**(13):1207–19. doi:10.1056/NEJMoa1506348
183. Sliwkowski MX, Mellman I. Antibody therapeutics in cancer. *Science* (2013) **341**:1192–8. doi:10.1126/science.1241145
184. Morschhauser F, Leonard JP, Fayad L, Coiffier B, Petillon MO, Coleman M, et al. Humanized anti-CD20 antibody, veltuzumab, in refractory/recurrent non-Hodgkin's lymphoma: phase I/II results. *J Clin Oncol* (2009) **27**:3346–53. doi:10.1200/JCO.2008.19.9117
185. Kalaycio ME, Negrea OG, Allen SL, Rai KR, Abbasi RM, Horne H, et al. Subcutaneous injections of low doses of humanized anti-CD20 veltuzumab: a phase I study in chronic lymphocytic leukemia. *Leuk Lymphoma* (2015) **19**:1–9. doi:10.3109/10428194.2015.1085531
186. Wierda WG, Padmanabhan S, Chan GW, Gupta IV, Lisby S, Osterborg A, et al. Ofatumumab is active in patients with fludarabine-refractory CLL irrespective of prior rituximab: results from the phase 2 international study. *Blood* (2011) **118**:5126–9. doi:10.1182/blood-2011-04-348656
187. Czuczman MS, Fayad L, Delwail V, Cartron G, Jacobsen E, Kuliczkowski K, et al. Ofatumumab monotherapy in rituximab-refractory follicular lymphoma: results from a multicenter study. *Blood* (2012) **119**:3698–704. doi:10.1182/blood-2011-09-378323
188. Coiffier B, Radford J, Bosly A, Martinelli G, Verhoef G, Barca G, et al. A multicentre, phase II trial of ofatumumab monotherapy in relapsed/progressive diffuse large B-cell lymphoma. *Br J Haematol* (2013) **163**:334–42. doi:10.1111/bjh.12537
189. Ogawa Y, Ogura M, Suzuki T, Ando K, Uchida T, Shirasugi Y, et al. A phase I/II study of ofatumumab (GSK1841157) in Japanese and Korean patients with relapsed or refractory B-cell chronic lymphocytic leukemia. *Int J Hematol* (2013) **98**:164–70. doi:10.1007/s12185-013-1393-x
190. Ogura M, Hatake K, Tobinai K, Uchida T, Suzuki T, Terui Y, et al. Phase I study of ofatumumab, a human anti-CD20 antibody, in Japanese patients with relapsed or refractory chronic lymphocytic leukemia and small lymphocytic lymphoma. *Jpn J Clin Oncol* (2013) **43**:466–75. doi:10.1093/jjco/hyt022
191. Furtado M, Dyer MJ, Johnson R, Berrow M, Rule S. Ofatumumab monotherapy in relapsed/refractory mantle cell lymphoma – a phase II trial. *Br J Haematol* (2014) **165**:575–8. doi:10.1111/bjh.12769
192. Moreno C, Montillo M, Panayiotidis P, Dimou M, Bloor A, Dupuis J, et al. Ofatumumab in poor-prognosis chronic lymphocytic leukemia: a phase IV, non-interventional, observational study from the European research initiative on chronic lymphocytic leukemia. *Haematologica* (2015) **100**:511–6. doi:10.3324/haematol.2014.118158
193. Osterborg A, Wierda WG, Mayer J, Hess G, Hillmen P, Schetelig J, et al. Ofatumumab retreatment and maintenance in fludarabine-refractory chronic lymphocytic leukaemia patients. *Br J Haematol* (2015) **170**:40–9. doi:10.1111/bjh.13380
194. Forero-Torres A, De Vos S, Pohlman BL, Pashkevich M, Cronier DM, Dang NH, et al. Results of a phase 1 study of AME-133v (LY2469298), an Fc-engineered humanized monoclonal anti-CD20 antibody, in FcgammaRIIIa-genotyped patients with previously treated follicular lymphoma. *Clin Cancer Res* (2012) **18**:1395–403. doi:10.1158/1078-0432.CCR-11-0850
195. Ganjoo KN, De Vos S, Pohlman BL, Flinn IW, Forero-Torres A, Enas NH, et al. Phase 1/2 study of ocaratuzumab, an Fc-engineered humanized anti-CD20 monoclonal antibody, in low-affinity FcgammaRIIIa patients with previously treated follicular lymphoma. *Leuk Lymphoma* (2015) **56**:42–8. doi:10.3109/10428194.2014.911859
196. Salles G, Morschhauser F, Lamy T, Milpied N, Thieblemont C, Tilly H, et al. Phase 1 study results of the type II glycoengineered humanized anti-CD20 monoclonal antibody obinutuzumab (GA101) in B-cell lymphoma patients. *Blood* (2012) **119**:5126–32. doi:10.1182/blood-2012-01-404368
197. Morschhauser FA, Cartron G, Thieblemont C, Solal-Celigny P, Haioun C, Bouabdallah R, et al. Obinutuzumab (GA101) monotherapy in relapsed/refractory diffuse large b-cell lymphoma or mantle-cell lymphoma: results from the phase II GAUGUIN study. *J Clin Oncol* (2013) **31**:2912–9. doi:10.1200/JCO.2012.46.9585
198. Ogura M, Tobinai K, Hatake K, Uchida T, Suzuki T, Kobayashi Y, et al. Phase I study of obinutuzumab (GA101) in Japanese patients with relapsed or refractory B-cell non-Hodgkin lymphoma. *Cancer Sci* (2013) **104**:105–10. doi:10.1111/cas.12040
199. Cartron G, De Guibert S, Dilhuydy MS, Morschhauser F, Leblond V, Dupuis J, et al. Obinutuzumab (GA101) in relapsed/refractory chronic lymphocytic leukemia: final data from the phase 1/2 GAUGUIN study. *Blood* (2014) **124**:2196–202. doi:10.1182/blood-2014-07-586610
200. Sehn LH, Goy A, Offner FC, Martinelli G, Caballero MD, Gadeberg O, et al. Randomized phase II trial comparing obinutuzumab (GA101) with rituximab in patients with relapsed CD20+ indolent B-cell non-Hodgkin lymphoma: final analysis of the GAUSS study. *J Clin Oncol* (2015) **33**(30):3467–74. doi:10.1200/JCO.2014.59.2139
201. Le Garff-Tavernier M, Decocq J, De Romeuf C, Parizot C, Dutertre CA, Chapiro E, et al. Analysis of CD16+CD56dim NK cells from CLL patients: evidence supporting a therapeutic strategy with optimized anti-CD20 monoclonal antibodies. *Leukemia* (2011) **25**:101–9. doi:10.1038/leu.2010.240
202. Kern DJ, James BR, Blackwell S, Gassner C, Klein C, Weiner GJ. GA101 induces NK-cell activation and antibody-dependent cellular cytotoxicity more effectively than rituximab when complement is present. *Leuk Lymphoma* (2013) **54**:2500–5. doi:10.3109/10428194.2013.781169
203. Cheney CM, Stephens DM, Mo X, Rafiq S, Butchar J, Flynn JM, et al. Ocaratuzumab, an Fc-engineered antibody demonstrates enhanced antibody-dependent cell-mediated cytotoxicity in chronic lymphocytic leukemia. *MAbs* (2014) **6**:749–55. doi:10.4161/mabs.28282
204. Alici E. IPH-2101, a fully human anti-NK-cell inhibitory receptor mAb for the potential treatment of hematological cancers. *Curr Opin Mol Ther* (2010) **12**:724–33.
205. Benson DM Jr, Cohen AD, Jagannath S, Munshi NC, Spitzer G, Hofmeister CC, et al. A phase I trial of the anti-KIR antibody IPH2101 and lenalidomide in patients with relapsed/refractory multiple myeloma. *Clin Cancer Res* (2015) **21**(18):4055–61. doi:10.1158/1078-0432.CCR-15-0304
206. Gleason MK, Verneris MR, Todhunter DA, Zhang B, Mccullar V, Zhou SX, et al. Bispecific and trispecific killer cell engagers directly activate human NK cells through CD16 signaling and induce cytotoxicity and cytokine production. *Mol Cancer Ther* (2012) **11**:2674–84. doi:10.1158/1535-7163.MCT-12-0692
207. Wiernik A, Foley B, Zhang B, Verneris MR, Warlick E, Gleason MK, et al. Targeting natural killer cells to acute myeloid leukemia in vitro with a CD16 x 33 bispecific killer cell engager and ADAM17 inhibition. *Clin Cancer Res* (2013) **19**:3844–55. doi:10.1158/1078-0432.CCR-13-0505
208. Jensen MC, Popplewell L, Cooper LJ, Digiusto D, Kalos M, Ostberg JR, et al. Antitransgene rejection responses contribute to attenuated persistence of adoptively transferred CD20/CD19-specific chimeric antigen receptor redirected T cells in humans. *Biol Blood Marrow Transplant* (2010) **16**:1245–56. doi:10.1016/j.bbmt.2010.03.014
209. Morgan RA, Yang JC, Kitano M, Dudley ME, Laurencot CM, Rosenberg SA. Case report of a serious adverse event following the administration of T cells transduced with a chimeric antigen receptor recognizing ERBB2. *Mol Ther* (2010) **18**:843–51. doi:10.1038/mt.2010.24
210. Savoldo B, Ramos CA, Liu E, Mims MP, Keating MJ, Carrum G, et al. CD28 costimulation improves expansion and persistence of chimeric antigen receptor-modified T cells in lymphoma patients. *J Clin Invest* (2011) **121**:1822–6. doi:10.1172/JCI46110
211. Kochenderfer JN, Dudley ME, Carpenter RO, Kassim SH, Rose JJ, Telford WG, et al. Donor-derived CD19-targeted T cells cause regression of malignancy

persisting after allogeneic hematopoietic stem cell transplantation. *Blood* (2013) **122**:4129–39. doi:10.1182/blood-2013-08-519413

212. Shortt J, Hsu AK, Johnstone RW. Thalidomide-analogue biology: immunological, molecular and epigenetic targets in cancer therapy. *Oncogene* (2013) **32**:4191–202. doi:10.1038/onc.2012.599
213. Romee R, Leong JW, Fehniger TA. Utilizing cytokines to function-enable human NK cells for the immunotherapy of cancer. *Scientifica (Cairo)* (2014) **2014**:205796. doi:10.1155/2014/205796
214. Miller JS. Therapeutic applications: natural killer cells in the clinic. *Hematology Am Soc Hematol Educ Program* (2013) **2013**:247–53. doi:10.1182/asheducation-2013.1.247

Permissions

All chapters in this book were first published in TNKCRLIAAE, by Frontiers; hereby published with permission under the Creative Commons Attribution License or equivalent. Every chapter published in this book has been scrutinized by our experts. Their significance has been extensively debated. The topics covered herein carry significant findings which will fuel the growth of the discipline. They may even be implemented as practical applications or may be referred to as a beginning point for another development.

The contributors of this book come from diverse backgrounds, making this book a truly international effort. This book will bring forth new frontiers with its revolutionizing research information and detailed analysis of the nascent developments around the world.

We would like to thank all the contributing authors for lending their expertise to make the book truly unique. They have played a crucial role in the development of this book. Without their invaluable contributions this book wouldn't have been possible. They have made vital efforts to compile up to date information on the varied aspects of this subject to make this book a valuable addition to the collection of many professionals and students.

This book was conceptualized with the vision of imparting up-to-date information and advanced data in this field. To ensure the same, a matchless editorial board was set up. Every individual on the board went through rigorous rounds of assessment to prove their worth. After which they invested a large part of their time researching and compiling the most relevant data for our readers.

The editorial board has been involved in producing this book since its inception. They have spent rigorous hours researching and exploring the diverse topics which have resulted in the successful publishing of this book. They have passed on their knowledge of decades through this book. To expedite this challenging task, the publisher supported the team at every step. A small team of assistant editors was also appointed to further simplify the editing procedure and attain best results for the readers.

Apart from the editorial board, the designing team has also invested a significant amount of their time in understanding the subject and creating the most relevant covers. They scrutinized every image to scout for the most suitable representation of the subject and create an appropriate cover for the book.

The publishing team has been an ardent support to the editorial, designing and production team. Their endless efforts to recruit the best for this project, has resulted in the accomplishment of this book. They are a veteran in the field of academics and their pool of knowledge is as vast as their experience in printing. Their expertise and guidance has proved useful at every step. Their uncompromising quality standards have made this book an exceptional effort. Their encouragement from time to time has been an inspiration for everyone.

The publisher and the editorial board hope that this book will prove to be a valuable piece of knowledge for researchers, students, practitioners and scholars across the globe.

List of Contributors

Arianexys Aquino-López
Department of Pediatric Research, The University of Texas M.D. Anderson Cancer Center, Graduate School of Biomedical Sciences, Houston, TX, USA
School of Medicine, University of Puerto Rico, San Juan, Puerto Rico

Vladimir V. Senyukov, Zlatko Vlasic and Eugenie S. Kleinerman
Department of Pediatric Research, The University of Texas M.D. Anderson Cancer Center, Graduate School of Biomedical Sciences, Houston, TX, USA

Dean A. Lee
Center for Childhood Cancer and Blood Disorders, The Research Institute, Nationwide Children's Hospital, Columbus, OH, USA

Kyra Fischer, Sara Tognarelli, Evelyn Ullrich and Peter Bader
University Hospital Frankfurt, Department for Children and Adolescents Medicine, Division of Stem Cell Transplantation and Immunology, Goethe University, Frankfurt, Germany
LOEWE Center for Cell and Gene Therapy, Goethe University, Frankfurt, Germany

Stefanie Roesler, Cathinka Boedicker and Simone Fulda
Institute for Experimental Cancer Research in Pediatrics, Goethe University, Frankfurt, Germany
German Cancer Consortium (DKTK), Heidelberg, Germany
German Cancer Research Center (DKFZ), Heidelberg, Germany

Ralf Schubert
University Hospital Frankfurt, Department for Children and Adolescents Medicine, Goethe University, Frankfurt, Germany
University Hospital Frankfurt/Main, Department for Children and Adolescents Medicine, Division of Pulmonology, Allergy and Cystic Fibrosis, Goethe University, Frankfurt, Germany

Alexander Steinle
University Hospital Frankfurt, Department for Molecular Medicine, Goethe University, Frankfurt, Germany

Thomas Klingebiel
LOEWE Center for Cell and Gene Therapy, Goethe University, Frankfurt, Germany
University Hospital Frankfurt, Department for Children and Adolescents Medicine, Goethe University, Frankfurt, Germany

John P. Veluchamy
Department of Medical Oncology, VU University Medical Center, Cancer Center Amsterdam, Amsterdam, Netherlands
Glycostem Therapeutics, Oss, Netherlands

Nina Kok and Jan Spanholtz
Glycostem Therapeutics, Oss, Netherlands

Hans J. van der Vliet, Henk M. W. Verheul and Tanja D. de Gruijl
Department of Medical Oncology, VU University Medical Center, Cancer Center Amsterdam, Amsterdam, Netherlands

Anushruti Sarvaria, J. Alejandro Madrigal and Aurore Saudemont Dunia Jawdat
Anthony Nolan Research Institute, London, UK
Cancer Institute, University College London, London, UK
King Abdullah International Medical Research Center, Riyadh, Saudi Arabia

Congcong Zhang and Winfried S. Wels
Georg-Speyer-Haus, Institute for Tumor Biology and Experimental Therapy, Frankfurt am Main, Germany
German Cancer Consortium (DKTK), Partner Site Frankfurt/Mainz, Frankfurt am Main, Germany
German Cancer Research Center (DKFZ), Heidelberg, Germany

Pranav Oberoi, Sarah Oelsner, Anja Waldmann and Aline Lindner
Georg-Speyer-Haus, Institute for Tumor Biology and Experimental Therapy, Frankfurt am Main, Germany

Torsten Tonn
German Cancer Research Center (DKFZ), Heidelberg, Germany
German Red Cross Blood Donation Service North-East, Institute for Transfusion Medicine, Dresden, Germany
Medical Faculty Carl Gustav Carus, TU Dresden, Dresden, Germany

German Cancer Consortium (DKTK), Partner Site Dresden, Dresden, Germany

Olaf Oberschmidt, Stephan Kloess and Ulrike Koehl
Institute of Cellular Therapeutics, Hannover Medical School, Hannover, Germany

John P. Veluchamy
Department of Medical Oncology, VU University Medical Center, Cancer Center Amsterdam, Amsterdam, Netherlands
Glycostem Therapeutics, Oss, Netherlands

Silvia Lopez-Lastra
Innate Immunity Unit, Institut Pasteur, Paris, France
Institut National de la Santé et de la Recherche Médicale (INSERM) U1223, Paris, France
Université Paris-Sud (Paris-Saclay), Paris, France

Jan Spanholtz, Fenna Bohme and Nina Kok
Glycostem Therapeutics, Oss, Netherlands

Daniëlle A. M. Heideman
Department of Pathology, VU University Medical Center, Amsterdam, Netherlands

Henk M. W. Verheul, Tanja D. de Gruijl and Hans J. van der Vliet
Department of Medical Oncology, VU University Medical Center, Cancer Center Amsterdam, Amsterdam, Netherlands

James P. Di Santo
Innate Immunity Unit, Institut Pasteur, Paris, France
Institut National de la Santé et de la Recherche Médicale (INSERM) U1223, Paris, France

Stephan Kloess, Alessa Ede Valverde da Silva, Olaf Oberschmidt, Tanja Gardlowski, Nadine Matthies, Lubomir Arseniev and Ulrike Köhl
Institute for Cellular Therapeutics, IFB-Tx, Hannover Medical School (MHH), Hannover, Germany

Maulik Vyas
Department I of Internal Medicine, University Hospital of Cologne, Cologne, Germany

Elke Pogge von Strandmann
Experimental Tumor Research, Center for Tumor Biology and Immunology, Philipps University Marburg, Marburg, Germany

Michael Heuser
Department of Hematology, Hemostasis, Oncology, and Ste Cell Transplantation, Hannover Medical School (MHH), Hannover, Germany

Maulik Vyas and Elke Pogge von Strandmann
Experimental Tumor Research, Center for Tumor Biology and Immunology, Clinic for Hematology, Oncology and Immunology, Philipps University, Marburg, Germany

Rolf Müller
Institute of Molecular Biology and Tumor Research, Center for Tumor Biology and Immunology, Philipps University, Marburg, Germany

Gianfranco Pittari and Deena Mudawi
Department of Medical Oncology, National Center for Cancer Care and Research, HMC, Doha, Qatar

Luca Vago
Unit of Immunogenetics, Leukemia Genomics and Immunobiology, IRCCS San Raffaele Scientific Institute, Milano, Italy
Hematology and Bone Marrow Transplantation Unit, IRCCS San Raffaele Scientific Institute, Milano, Italy

Moreno Festuccia, Luisa Giaccone and Benedetto Bruno
Department of Oncology/Hematology, A.O.U. Città della Salute e della Scienza di Torino, Presidio Molinette, Torino, Italy
Department of Molecular Biotechnology and Health Sciences, University of Torino, Torino, Italy

Chiara Bonini
Experimental Hematology Unit, Division of Immunology, Transplantation and Infectious Diseases, IRCCS San Raffaele Scientific Institute, Milano, Italy
Vita-Salute San Raffaele University, Milano, Italy

Meriem Messaoudene
INSERM U1015, Gustave Roussy Cancer Campus, Villejuif, France
Gustave Roussy Cancer Campus, Villejuif, France

Alexandra Frazao, Pierre Jean Gavlovsky, Antoine Toubert, Nicolas Dulphy and Anne Caignard
INSERM U1160, Institut Universitaire d'Hématologie, Hôpital Saint Louis, Paris, France

Silvia Lopez-Lastra
Innate Immunity Unit, Institut Pasteur, Paris, France
Inserm U1223, Paris, France
Université Paris-Sud (Paris-Saclay), Paris, France

James P. Di Santo
Innate Immunity Unit, Institut Pasteur, Paris, France
Inserm U1223, Paris, France

Erhard Hofer
Department of Vascular Biology, Medical University of Vienna, Vienna, Austria

Ulrike Koehl
Institute of Cellular Therapeutics, IFB-Tx, Hannover Medical School, Hannover, Germany

Stephan Klöss, Ruth Esser, Tanja Gardlowski, Lubomir Arseniev and Ulrike Köhl
Institute of Cellular Therapeutics, Integrated Research and Treatment Center Transplantation (IFB-Tx), Hannover Medical School, Hannover, Germany

Nicole Chambron and Oliver Seitz
Department of Oral, Cranio-Maxillofacial and Facial Plastic Surgery, Klinikum Hanau GmbH, Hanau, Germany

Sandra Weil and Joachim Koch
Georg-Speyer-Haus Institute for Tumor Biology and Experimental Therapy, Frankfurt, Germany

Elke Pogge von Strandmann
Klinik Ifür Innere Medizin, Uniklinik Köln, Cologne, Germany

Michael A. Morgan
Institute of Experimental Haematology, Hannover Medical School, Hannover, Germany

Katayoun Rezvani
Department of Stem Cell Transplantation, Division of Cancer Medicine, The University of Texas MD Anderson Cancer Center, Houston, TX, USA

Rayne H. Rouce
Department of Pediatrics, Texas Children's Cancer and Hematology Centers, Baylor College of Medicine, Houston, TX, USA,
Center for Cell and Gene Therapy, Baylor College of Medicine Houston Methodist Hospital and Texas Children's Hospital, Houston, TX, USA

Wei Wang, Amy K. Erbe, Jacquelyn A. Hank and Zachary S. Morris
Department of Human Oncology, University of Wisconsin-Madison, Madison, WI, USA,

Paul M. Sondel
Department of Human Oncology, University of Wisconsin-Madison, Madison, WI, USA,
Department of Pediatrics, University of Wisconsin-Madison, Madison, WI, USA

Benedikt Jacobs
Department of Cancer Immunology, Institute for Cancer Research, Oslo University Hospital, Radiumhospital, Oslo, Norway
The KG Jebsen Center for Cancer Immunotherapy, Institute of Clinical Medicine, University of Oslo, Oslo, Norway
Department of Haematology and Oncology, University Hospital Erlangen, Erlangen, Germany

Sara Tognarelli and Peter Bader
Department of Pediatric Stem Cell Transplantation and Immunology, Children's Hospital, Johann Wolfgang Goethe-University, Frankfurt, Germany
LOEWE Center for Cell and Gene Therapy, Johann Wolfgang Goethe-University, Frankfurt, Germany

Kerstin Poller and Andreas Mackensen
Department of Haematology and Oncology, University Hospital Erlangen, Erlangen, Germany

Evelyn Ullrich
Department of Pediatric Stem Cell Transplantation and Immunology, Children's Hospital, Johann Wolfgang Goethe-University, Frankfurt, Germany
LOEWE Center for Cell and Gene Therapy, Johann Wolfgang Goethe-University, Frankfurt, Germany
Department of Haematology and Oncology, University Hospital Erlangen, Erlangen, Germany

Carin I. M. Dahlberg, Michael Chrobok and Adil D. Duru
Cell Therapies Institute, Nova Southeastern University, Fort Lauderdale, FL, USA,
Cell and Gene Therapy Group, Center for Hematology and Regenerative Medicine (HERM), Karolinska University Hospital Huddinge, NOVUM, Stockholm, Sweden,

Dhifaf Sarhan
Oncology-Pathology, Cancer Center Karolinska, Karolinska Institutet, Stockholm, Sweden,
Division of Hematology, Oncology and Transplantation, Masonic Cancer Research Center, University of Minnesota, Minnesota, MN, USA,

Evren Alici
Cell Therapies Institute, Nova Southeastern University, Fort Lauderdale, FL, USA,
Cell and Gene Therapy Group, Center for Hematology and Regenerative Medicine (HERM), Karolinska University Hospital Huddinge, NOVUM, Stockholm, Sweden,
Hematology Center, Karolinska University Hospital Huddinge, Stockholm, Sweden

Index

www.ingramcontent.com/pod-product-compliance
Lightning Source LLC
Chambersburg PA
CBHW081216180926
51613CB00003B/251
9781639893850